Probability and Statistics for Economics and Business

Probability and Statistics for Economics and Business

An Introduction Using R

Jason Abrevaya

The MIT Press
Cambridge, Massachusetts
London, England

The MIT Press
Massachusetts Institute of Technology
77 Massachusetts Avenue, Cambridge, MA 02139
mitpress.mit.edu

This book was set in Times New Roman by Westchester Publishing Services. Printed and bound in the United States of America.

Library of Congress Cataloging-in-Publication Data

Names: Abrevaya, Jason author
Title: Probability and statistics for economics and business : an introduction using R / Jason Abrevaya.
Description: Cambridge, Massachusetts : The MIT Press, [2025] | Includes bibliographical references and index.
Identifiers: LCCN 2024060243 (print) | LCCN 2024060244 (ebook) | ISBN 9780262553360 paperback | ISBN 9780262384742 epub | ISBN 9780262384759 pdf
Subjects: LCSH: Managerial economics–Study and teaching (Higher) | Probabilities–Study and teaching (Higher) | R (Computer program language)
Classification: LCC HD30.22 .A25 2025 (print) | LCC HD30.22 (ebook) | DDC 338.5/024658–dc23/eng/20250804
LC record available at https://lccn.loc.gov/2024060243
LC ebook record available at https://lccn.loc.gov/2024060244

10 9 8 7 6 5 4 3 2 1

EU Authorised Representative: Easy Access System Europe, Mustamäe tee 50,
10621 Tallinn, Estonia | Email: gpsr.requests@easproject.com

To Allison, Logan, and Hayden

Contents

Preface

This book is intended primarily for an introductory course in probability and statistics for economics and business undergraduates. It introduces students to the *R* statistical programming language and uses *R* to illustrate the concepts covered. While the book explores many of the standard topics traditionally seen in an introductory probability and statistics course, it takes a modern approach that includes coverage of simulation methods and the bootstrap, facilitated by *R*, and it has a greater focus on concepts useful to economists and data analysts, including general statistical-inference results that apply well beyond averages and variances. The book covers finite-sample theory, as it is instructive for pedagogical reasons, but it places a much greater emphasis on large-sample or asymptotic theory, as the latter is used almost exclusively in modern statistical analysis. For example, the treatment of linear regression models in chapters 17 and 18 does not consider finite-sample theory at all. While its primary audience is expected to be undergraduates, the book would also be a good fit for an introductory statistics course in master's programs (e.g., economics, applied economics, business disciplines, and public policy) or as preparation for doctoral programs.

The book assumes a moderate level of mathematical maturity (e.g., comfort with set theory, mathematical notation, and algebra) and uses calculus for the treatment of continuous random variables (chapters 10 and 11). (For a course without a calculus prerequisite, an instructor could cover this material in a way that does not use calculus.) The mathematical rigor provided is at a higher level than traditional business statistics textbooks, with the intent of developing deeper understanding of concepts and not just offering "plug-and-chug" methods for statistical inference. Proofs of important results are provided where possible and appropriate. But some results are stated without proof since it would involve topics and methods beyond the scope of the book. The Central Limit Theorem (CLT, covered in chapter 13) is one example, but there are many others, including the various asymptotic-theory results for descriptive statistics (chapter 14). To the extent possible, intuition is provided for any results that are provided without proof. The goal is to provide students with more advanced tools than they would ordinarily encounter in an introductory course, with the trade-off that some of the results will have to be taken on faith. For students interested in proofs of such results, I recommend the book *Introduction to Mathematical Statistics* (Hogg, McKean, & Craig, 2019), which covers many of the topics in this book with a more theoretical treatment and proves many of the stated propositions, including the Law of Large

Numbers (LLN), the CLT, the delta method, and many of the asymptotic properties of various descriptive statistics and estimators. In some cases, to provide a bit of a historical perspective, the end notes also provide references to the original sources for some of the results.

The book bridges what I perceive to be a disconnect between how an introductory statistics course is taught to undergraduate students and how more advanced econometrics courses are taught to both undergraduate and graduate students. There is much agreement that an introductory statistics course must develop in students an understanding of random variables and sampling-distribution theory. While most introductory courses discuss various types of descriptive statistics when it comes to actual data, the bulk of the focus for sampling distribution theory and statistical inference is for the sample average, and perhaps the sample variance. In part, this approach is related to the availability of finite-sample results for the sample average, but strangely, the focus on the sample average remains even when the course or book moves to the CLT and large-sample theory. As an example, we teach every student about sample medians and sample correlations and how to calculate them, but how many students learn that sample medians and sample correlations also have standard errors that can be calculated? The same is true of the sample interquartile range (IQR), the sample standard deviation, and basically any descriptive statistic that a student would use in practice. As such, this book stresses the generality of the normal-distribution approximation that can be used for a wide range of descriptive statistics and estimators in the presence of large samples. After all, nearly all the samples that economists and data scientists work with in practice are large in the sense that is required for the large-sample results to hold. This generality is more in line with how we teach regression-type estimators and more advanced estimators in later courses. The emphasis on the large-sample or asymptotic approach is evident in the book's treatment of estimation and confidence intervals (chapter 14) and hypothesis testing (chapter 16), where the finite-sample (sample average) case is considered briefly before moving to general large-sample results. Also, the treatment of linear regression (chapters 17 and 18) is completely within a large-sample framework, so the reader will not find t or F distributions in these chapters.

Why *R*? This book could have been written without an emphasis on *R*, and I imagine that a course could even be taught using this book without *R*. However, the use of *R* allows students to fully engage with the probability and statistics concepts that are covered. The ability to simulate randomness, such as the toss of a coin, the roll of a die, or a draw of a random number between 0 and 1, is an extremely powerful tool. With *R* and its built-in mathematical and statistical functions, students can conduct simulations with 100,000 replications or more with just one or two lines of code. Sample code is provided throughout the book and on the companion website, providing a road map for students to do related analysis. This book provides more emphasis on simulation methods than do other introductory statistics textbooks. The simulations in *R* help to develop a greater understanding of sampling distribution theory, which is often a difficult topic for students due to its theoretical nature. For instance, in many of the exercises in these chapters, students are asked to use simulations to approximate population quantities, like expected values or population standard deviations, that they have derived analytically. Of course, *R* also provides tools for standard statistical inference, with built-in functions for the calculation of descriptive statistics, determination of critical values, implementation of linear regression, and other tasks. Beyond its pedagogical benefits, *R* is also free and used widely both inside and outside academia.

Companion Website

The companion website, https://www.probstats4econ.com, has the following resources available:

- Data sets used in the chapters and/or end-of-chapter exercises, including complete documentation of each one.
- The *R* code used in the book, as well as additional code (in some cases used to create figures) that is not shown explicitly in the text. The code is available to be downloaded as a batch file and can also be viewed on a section-by-section basis.
- Description of the *R* package `probstats4econ`, which is also a companion for the book. The *R* package includes the data sets from the website, as well as several helpful *R* functions that are introduced in the book. The package is available to be installed from the Comprehensive *R* Archive Network (CRAN), and instructions to do so are provided in chapter 1.

Outline of the Book

The book has an ordering of topics that is fairly standard for a probability and statistics book. Probability theory and combinatorics are covered in chapters 2–4, sampling and descriptive statistics in chapters 5–7, random variables in chapters 8–11, sampling distributions in chapters 12–13, statistical inference in chapters 14–16, and linear regression in chapters 17–18. We provide additional detail on the material covered in these chapters before discussing how the chapters might be used (or not) in different courses. End-of-chapter exercises are provided for each chapter, with asterisks denoting questions or parts of questions that are more challenging.

Chapter 1 provides an introduction to the *R* programming language, a powerful tool for statistical analysis. The goal here is not to provide an exhaustive treatment of *R*, but rather to give students the basic skills needed to navigate and use *R* for the topics covered in the subsequent chapters. This chapter is stand-alone in the sense that its material is not crucially needed for the next few chapters, allowing students to learn *R* at the same time that probability theory is covered in a course.

Chapters 2 and 3 cover probability theory. Chapter 2 discusses the meaning of probability and the fundamental properties of probabilities. A frequentist intepretation of probabilities is provided and serves as a preview of the LLN, covered later in chapter 13. The properties of probabilities follow from the three Axioms of Probability. Chapter 3 introduces concepts related to probabilities of multiple events, including conditional probabilities, joint probabilities, and the notion of independence. These two chapters rely on the use of set theory, as possible outcomes are modeled as a set. While most students will already be familiar with set theory, the treatment of the material starts from first principles, so it is not necessary to have such familiarity.

Chapter 4 covers combinatorics, or counting methods. The focus is on counting the number of ways that a group of options/objects can be selected from a larger group, where that number may depend on (1) whether the ordering of the options matters and (2) whether an option can be chosen more than once. The final section of the chapter considers probability problems involving equally likely choices among the set of options enumerated by combinatorics. This section provides the first illustration of the power of simulation, where a difficult-to-derive probability is approximated through the use of computer simulations.

Chapters 5–7 shift the focus to real-world data. Chapter 5 offers a brief introduction to the types of data sets commonly used in economics and business, including cross-sectional data, time-series data, and panel data. This chapter also introduces the concept of the population and random sampling. The discussion of random sampling considers simple random samples, stratified random samples, and sample-selection bias. Chapters 6 and 7 introduce descriptive statistics and visuals for data, with chapter 6 focused on univariate data (a single variable) and chapter 7 focused on bivariate data (two variables). The descriptive statistics include measures of location (sample mean, sample median, and sample quantiles), measures of dispersion (sample variance, sample standard deviation, and sample IQR), and measures of association (sample covariance and sample correlation). The visuals include bar charts, histograms, density plots, time-series plots, and scatter plots. The descriptive statistics considered in these chapters are the ones for which statistical inference is considered in later chapters.

Chapters 8–11 cover random variables in detail, splitting the discussion into discrete (chapters 8 and 9) and continuous (chapters 10 and 11) random variables. While the approaches for the two types of random variables have similarities, the mathematics are sufficiently different that it is useful to consider them separately. For instance, discrete summations are sufficient for the analysis of discrete random variables, but integration is necessary for some of the analysis of continuous random variables. Chapters 10 and 11 are the two chapters where students are assumed to have a knowledge of calculus so they can understand integrals and, to a lesser extent, derivatives. For the pairs of chapters on discrete and continuous random variables, the first pair (chapters 8 and 10) covers the general theory, while the second pair (chapters 9 and 11) examines specific models that are widely used in practice. The general theory in chapters 8 and 10 covers: (1) distribution functions; (2) population quantities, including population means, population quantiles, population variances, and population standard deviations; (3) bivariate relationships, including population covariances and population correlations; (4) linear transformations of univariate random variables; and (5) linear combinations of multiple random variables. For discrete random variables, chapter 9 considers specific models associated with a binary (Bernoulli) random variable and several other cases (binomial, geometric, negative binomial, and Poisson). For continuous random variables, chapter 11 focuses primarily on the important case of the normal distribution, but it also discusses other useful distributions (log-normal, chi-square, exponential, and mixture of normals).

Chapters 12 and 13 address sampling distribution theory, which provides the foundation for the statistical inference of later chapters. Both chapters consider a framework with underlying random variables that are independent and identically distributed (i.i.d.). Chapter 12 focuses on exact sampling distributions, also known as "finite-sample distributions." It considers some cases where the exact sampling distribution can be described analytically (the sample average of i.i.d. binary random variables and the sample average and the sample variance of i.i.d. normal random variables), before moving to cases where analytical descriptions are not easily available. For the latter, simulation methods are used to approximate and visualize the exact sampling distributions for various examples. Chapter 13 focuses on asymptotic sampling distributions, also known as large-sample distributions. It starts by introducing two classic results in statistics, the LLN and the CLT, results that directly allow a normal-distribution approximation of the sampling distribution of the sample average in large samples (regardless of the nature of the underlying i.i.d. random variables). The remainder of the chapter stresses the generality of the idea of asymptotically normal sampling distributions, providing the distributions and examples for other descriptive statistics like the sample standard deviation, sample quantiles, and sample correlation.

Chapter 14 builds on the sampling-distribution theory developed in the preceding chapters to discuss estimation and confidence intervals, two of the most important topics in statistical inference. It introduces properties of statistical estimators, stressing the importance of consistency and asymptotic normality. Methods for constructing two-sided and one-sided confidence intervals are discussed. Finite-sample confidence intervals are considered for the case of the population mean of i.i.d. normal random variables. The remainder of the chapter focuses on parameter estimation and the calculation of standard errors and confidence intervals for large samples, which includes many important cases of practical interest. Asymptotic (large-sample) confidence intervals and standard errors are provided for population quantities, including the population mean, population standard deviation, population quantiles, population correlation, the difference in population means for two variables from the same sample, and the difference in population parameters (e.g., means or standard deviations) for variables from two different samples.

Chapter 15 introduces the bootstrap, a computer simulation method for statistical inference that offers an alternative to either finite-sample inference or asymptotic inference. The bootstrap is a very general approach, requiring no mathematical analysis, and it is particularly appealing when asymptotic inference involves a standard error for which the formula is either unknown or intractable. The chapter shows how to create bootstrap samples, calculate bootstrap standard errors, and construct confidence intervals (normal-based confidence intervals and bootstrap percentile intervals). The methods are illustrated with inference related to population quantiles, population correlations, and the difference in population correlations for variables within the same sample.

Chapter 16 covers hypothesis testing. It starts with finite-sample theory for testing hypotheses about the population mean of i.i.d. normal random variables. Then, asymptotic theory is used to consider testing a hypothesis about *any* parameter that has an asymptotically normal estimator. For these single-parameter tests, the testing process is described in detail, starting with the statement of a null hypothesis to be tested. The discussion describes rejection rules (based on test statistics, critical values, and p-values) and explains both the level and the power of a hypothesis test. The single-parameter asymptotic tests are quite general, similar to chapter 14, and include cases like the difference in population parameters for both one-sample and two-sample situations. The last section of the chapter extends the hypothesis testing framework to the case of multiple hypotheses rather than a single hypothesis. The Wald test is introduced, with the more technical mathematical material relegated to the chapter's appendix.

Finally, chapters 17 and 18 provide an introduction to linear regression models, covering simple linear regression (SLR; one explanatory variable) in chapter 17 and multiple linear regression (MLR; multiple explanatory variables) in chapter 18. Linear regression is one of the most important and powerful tools for a data analyst. While an exhaustive treatment of regression models is not possible in just two chapters, they do introduce some of the most important topics in regression analysis, particularly for cross-sectional i.i.d. data. The introduction of the regression models shares the theoretical formality of previous chapters, viewing the least-squares regression estimators as random variables and describing their sampling distributions. In these two chapters, statistical inference is considered entirely from an asymptotic perspective; finite-sample theory is not covered. The treatment is consistent with modern econometric practice, focusing on inference with heteroskedastic errors and not assuming either homoskedasticity or the normality of regression-model residuals. Homoskedasticity is considered only for pedagogical reasons, to provide better intuition for the factors affecting standard errors, and normality is considered only in the context of forecasting. These chapters also provide a fairly rigorous approach to thinking about the regression model's exogeneity assumption and whether the estimates of the model parameters have a causal interpretation.

Using the Book for a Course

The entire book is probably too much material for a one-semester course. All the chapters could serve as the basis for an in-depth, two-semester sequence, and it's possible that advanced students (e.g., an honors undergraduate course or a master's course) could handle the material in a single semester.

The two chapters that are easiest to omit are chapter 4 (combinatorics) and chapter 15 (bootstrap). And, for a course that does not cover linear regression, chapters 17 and 18 would be omitted.

I expect that the book will be most useful for an introductory statistics course for economics or business undergraduates. For example, my own introductory statistics course, required for economics majors at the University of Texas, covers most of chapters 1–14 and 16, but several topics are omitted in the interest of time, including some of the random variable models in chapters 9 and 11, some of the sampling distributions in chapter 13, and the Wald test in chapter 16.

To spend less time on probability theory and more time on statistical inference, a course could omit chapters 2–4 and jump straight to sampling and descriptive statistics, as the discrete random variable material (chapter 8) covers many of the probability concepts in chapters 2 and 3. Such a course could cover most of the topics through chapter 16, including the bootstrap (chapter 15), and perhaps even some topics in linear regression.

A course using this book could also focus entirely on asymptotic theory and omit finite-sample theory altogeher. While it may still be instructive for students to read chapter 12, such a course could omit the finite-sample sections about estimation (section 14.2) and hypothesis testing (section 16.1). Students are sometimes confused by jumping back and forth between finite-sample (e.g., t-distribution) results and asymptotic results. Also, omission of finite-sample theory topics would allow a greater focus on other inference topics.

For a course that does not have calculus as a prerequisite, the main issue is the book's treatment of continuous random variables in chapters 10 and 11. An instructor would need to teach the cumulative distribution function (cdf) material differently, which is possible by using visual methods and relying on R to carry out the necessary calculations.

For a course covering more advanced econometrics topics, the regression chapters (chapters 17 and 18) likely do not cover enough material. For instance, they do not consider instrumental variables models, panel-data models, or non linear models. I highly recommend either *Introductory Econometrics: A Modern Approach* (Wooldridge 2025) or *Introduction to Econometrics* (Stock & Watson 2018) as a logical follow-up to this book. This book, with its use of *R*, may still serve as a useful companion to either of these textbooks since the regression chapters show how to do many of the most important regression analyses in *R*.

Acknowledgments

I am grateful to the University of Texas for the research sabbatical in fall 2022 and spring 2023 that made this book possible. There is no way that the project could have been started or completed without the freedom to focus on it. During my sabbatical, I was fortunate enough to be in many amazing places, which made the process of writing more enjoyable. I owe special thanks to Karen Weiss for her generosity in allowing my wife and me to spend time in her Ventura home, as the first chapter drafts took shape with the sound of the crashing waves of the Pacific Ocean in the background.

It's impossible to mention everyone who played a role in inspiring my interest in econometrics and statistics, but I will mention a few: Herman Chernoff, from whom I was fortunate to learn about stochastic processes; Gary Chamberlain and Josh Angrist, from whom I took my first undergraduate econometrics course; and Jerry Hausman, who was my instructor, my mentor, and my co-author during my PhD studies at the Massachusetts Institute of Technology (MIT).

I've taught econometrics and statistics at all levels (undergraduate, master's, PhD) for nearly 30 years at three universities. My many students and colleagues over the years have made me a better teacher and have shaped the way that I explain concepts and connect those concepts to real-world applications. This book grew out of notes that I prepared for the Economic Statistics course at the University of Texas, a course that I first taught in fall 2021. I'm thankful to Valerie Bencivenga for providing me with the many materials, including notes and problems, that she prepared for previous iterations of the course.

I'd also like to thank several people who have assisted during the publication process. I thank Jeff Wooldridge for his encouragement to turn the manuscript into a full-blown textbook and to speak with potential publishers. Philip Good, from Cambridge University Press, was generous with his time and insights, and he urged me to make *R* an integral part of the book. Catherine Woods has been my editor at MIT Press, and she provided useful advice along the way and was able to solicit reviews in a very timely fashion. I am very grateful to the anonymous reviewers who took the time to provide their valuable feedback and also to Susan McClung for her careful copy editing of the manuscript.

Last but not least, I'd like to express gratitude to my wonderful family. I'm incredibly proud of my two children, Logan and Hayden, whose kindness, intelligence, and talents inspire me. And, my wife, Allison, is my everything, and she has provided support and love during the (sometimes arduous) process of writing this book. Her belief in me has not wavered, and she made this journey possible and rewarding.

1

The Basics of *R*

1.1 Installing *R*

R is a statistical programming language for data analysis and visualization widely used by economists and data scientists. This book uses *R* to illustrate statistical concepts and implement analytical methods. For the best experience of working with *R*, readers should also install the software package *RStudio*. While *RStudio*, strictly speaking, is not required to run *R*, it provides a user-friendly graphical interface that makes running *R* much easier. *RStudio* provides an advanced editor with features that include syntax highlighting, code completion, and debugging. It also has several tools to streamline data analysis, including a workspace viewer and a plotting window.

Both *R* and *RStudio* are available for standard operating systems (Microsoft Windows, macOS, Linux) and can be downloaded for free at https://cran.r-project.org and https://posit.co/downloads, respectively. Be sure to download and install *R* prior to downloading and installing *RStudio*.

RStudio has four main panes as part of its interface, as seen in figure 1.1 and described here:

1. Source (top left): This pane allows the user to create, edit, and save *R* scripts. Data sets can also be browsed in this pane. It appears only when a script or data set has been opened.
2. Console (bottom left): This pane allows the user to enter commands directly in *R* and see the output.
3. Environment/History (top right): This pane displays the current workspace, including data sets, variables, and functions that have been introduced. It also provides access to previous commands and output.
4. Plots/Help/Packages/Files (bottom right): This pane displays plots, provides access to *R* documentation, allows the installation and management of *R* packages, and allows direct access to file directories.

Figure 1.1
The four panes in *RStudio*

Within *RStudio*, there are two main options for writing *R* commands:

- The console: This command-line interface appears when *RStudio* is opened. Commands can be typed directly into the console, and *R* executes them immediately. Using the console is a good option for simple calculations or experimentation with *R* functions.
- *R* scripts: An *R* script is a text file containing a series of *R* commands. A script file can be edited and saved, making it useful for more complex data analysis and programming tasks. With *RStudio*, it is straightforward to run an entire script or a selected section of a script.

The two options have advantages and disadvantages. The console is quick and easy to use, but it can be difficult to keep track of all the commands that have been issued. An *R* script, on the other hand, makes it easier to organize code, make it more readable, and save it. But, since many commands from an *R* script are executed at once, it may take more time to figure out why code is not performing as expected.

Throughout this book, we use the console option for *R* commands and output to demonstrate how commands work. However, all *R* code is accessible as script files on the book's companion website, https://www.probstats4econ.com, organized by chapter and section. These files enable readers to run large chunks of code conveniently. Moreover, readers can modify scripts to see the impact of code edits on output, add analysis, or cut and paste from script files to create new scripts.

1.2 Arithmetic Operations and Mathematical Functions

R performs a wide range of mathematical calculations. It can be used like a calculator, performing arithmetic operations using standard mathematical operators: addition (+), subtraction (−), multiplication (`*`), division (/), and exponentiation (^). For example, typing `5+3` and pressing Enter in the *R* console returns the result `8`. Some other simple examples are illustrated in the *R* code that follows.

```
# simple arithmetic operations
5+3

## [1] 8

5-3

## [1] 2

5*3

## [1] 15

5/3

## [1] 1.666667

5^3

## [1] 125
```

In this block of code, the first line is a "comment," which is not executed by *R*. When writing *R* code, we can start any line with the hashtag character (#) to indicate that the line is a descriptive comment rather than a command to be executed. The first command that gets evaluated is `5+3`, and *R* returns the answer `8`. Any printed *R* output in this book is on a line starting with two hashtag characters (##).[1] How about the `[1]` that appears right after ##? In cases where a single value is returned, `[1]` appears before the ouput. Later in this chapter, we consider cases in which multiple values are returned simultaneously (as part of a "vector"), and when there are multiple lines of output, the number within the `[]` brackets for a particular line indicates which element is being shown first.

R follows the standard mathematical order of operations, which is (listed from highest to lowest priority):

1. Parentheses
2. Exponentiation
3. Multiplication and division (performed from left to right)
4. Addition and subtraction (performed from left to right)

```
(3+4)^2

## [1] 49

3+4^2

## [1] 19

3+2*4^2

## [1] 35

3+2*4+2
```

```
## [1] 13
(3+2)*(4+2)
## [1] 30
```

For the first expression, the addition within the parentheses occurs before the exponentiation. For the second expression, the exponentiation occurs before the addition. For the third expression, the exponentiation occurs first, followed by the multiplication and then the addition. For the fourth expression, the expressions within both parentheses are evaluated before being multiplied together.

In addition to arithmetic operators, *R* has many mathematical functions that facilitate calculations and data analysis. Here are some examples of commonly used mathematical functions:

- `abs(x)`: Calculates the absolute value $|x|$.
- `sqrt(x)`: Calculates the square root $\sqrt{x}$.
- `exp(x)`: Calculates the exponential value e^x. `exp(1)` returns Euler's constant $e \approx 2.718282$ since $e^1 = e$.
- `log(x)`: Calculates the natural logarithm $\ln(x)$.
- `factorial(x)`: Calculates the factorial $x! = x(x-1)(x-2)\cdots(3)(2)(1)$.

The following output shows examples using `abs(x)` and `sqrt(x)`:

```
5*abs(-3)
## [1] 15
sqrt(17)/2
## [1] 2.061553
```

R functions, like the mathematical functions given here, can have a required argument or multiple required arguments, along with optional arguments that can be specified during function calls. To determine these arguments for any function, *R* documentation or "help" can be requested by typing a question mark (`?`) followed by the function name. For example, the command `?sqrt` requests the documentation for the `sqrt` function, which then appears in the bottom-right window of the *RStudio* interface. The documentation indicates that the usage for the function is `sqrt(x)`, meaning that it has a single required argument `x` and no optional arguments. In the example here, the argument is a single number (`17`), although the documentation specifies that the argument `x` can be a "numeric or complex vector or array." The concept of vectors is described in section 1.4.

The `log` function has an optional argument, which can be confirmed by requesting documentation with `?log`. The documentation indicates that the usage for the function is `log(x, base = exp(1))`, meaning that it has a required argument `x` and an optional argument `base`. If the optional argument is omitted, then its default value is `exp(1)`, resulting in the natural logarithm (base *e*). To calculate a base 10 logarithm, there are two ways to call the `log` function with the optional argument:

```
log(100,base=10)

## [1] 2

log(100,10)

## [1] 2

log(100)

## [1] 4.60517
```

The first and second commands are equivalent, with the first explicitly using the name of the `base` argument and the second relying on the usage having the base as the second argument. The former approach is generally preferred, as a function may have multiple optional arguments and specifying the name of the optional argument avoids confusion and mistakes. The third command uses no optional argument, so the natural logarithm (base *e*) of `100` is calculated.

Two additional examples of mathematical functions with optional arguments are `round` and `signif`:

- `round(x, digits = 0)`: Rounds a number `x` to a specified number of `digits`. If the argument `digits` is not specified, the default value is `0`, in which case the function rounds to the nearest integer.
- `signif(x, digits = 6)`: Returns a number `x` with a specified number of significant digits that is equal to `digits`. If the argument `digits` is not specified, the default value is `6`.

The following output illustrates the use of `round` and `signif`:

```
round(50/3)

## [1] 17

round(exp(1),digits=4)

## [1] 2.7183

signif(50/3,digits=5)

## [1] 16.667
```

1.3 Variables and Data Types

1.3.1 Variables

In *R*, variables are used to store data and facilitate data manipulation. A variable is basically a container that holds a value, like a number or a text string. When we create a variable in *R*, we give the variable a name and assign a value to it, using the assignment operator <-:

```
x <- 8
x

## [1] 8
```

```
x+5

## [1] 13

x <- 2*x
x

## [1] 16

frog

## Error in eval(expr, envir, enclos): object 'frog' not found
```

The first command assigns the value `8` to variable `x`. This command has no output associated with it. The next command, simply `x`, does provide output, corresponding to the value `8` stored in variable `x`. The variable can then be used in other expressions. The third command outputs the value of `x+5`, which is `13`. Importantly, this command does not change the value of variable `x`, which is still `8`. The fourth command *does* change the value of the variable `x` with another assignment operator `<-`. Specifically, the new value assigned to `x` is 2 times the old value of `x`, or `16`. The last command shows that an error message is returned when we refer to a variable name (in this case, `frog`) that has not been assigned a value.

Variables can store different types of data. The variable `x` has a numeric value, but variables can also contain text strings, logical values (indicating true or false), and other types of data. The data type of a variable can be determined by the `class()` function:

```
y <- 3.4
class(y)

## [1] "numeric"

str <- "Economic statistics"
class(str)

## [1] "character"
```

The most common data types in *R* are the following:

- `numeric`: Numeric values, including real numbers (e.g., `3.4`) and integers (e.g., `8`).[2]
- `logical`: Logical values, which can be either `TRUE` or `FALSE`.
- `character`: Text strings (e.g., `"intro to statistics"`).
- `date`: Date values (e.g., `2022-04-30`).[3]
- `factor`: A data type for a *categorical variable* with a fixed set of possible values (e.g., a variable for eye color, with the three possible values `Blue`, `Brown`, and `Other`; or a variable for restaurant ratings, with the four possible values `Excellent`, `Good`, `Fair`, and `Poor`).

If a variable is no longer needed, we can delete it with the `rm` function. For example, if variable `x` has been assigned a value, the command `rm(x)` removes the variable from the *R* working environment. The command `rm(list = ls())` removes *all* variables from the *R* working environment, although it should be used with caution since *R* does not ask for confirmation when the `rm` function is called.

1.3.2 Logical Values (`logical` Data Type)

The `logical` data type is used for the logical values `TRUE` and `FALSE`. Logical data types are commonly used to compare two values to each other. For example, the == operator tests whether two values are equal to each other, in which case `TRUE` is returned; or not equal to each other, in which case `FALSE` is returned. In contrast, the operator != tests for the inequality of two values, returning `TRUE` if they are not equal and `FALSE` if they are equal. Similarly, the operators <, >, <=, and >= are used to test for strict inequality or weak inequality of two values:

```
8 == 3+5

## [1] TRUE

x <- 16
x > 12

## [1] TRUE

xsmall <- (x<=9)
xsmall

## [1] FALSE
```

The first command returns `TRUE` since `8` is exactly equal to `3+5`. The second command assigns the value `16` to variable `x`, and the third command returns `TRUE` since `x` is strictly greater than `12`. The fourth command assigns the value of `x<=9`, which is `FALSE` since `x` is not less than or equal to `9`, to the variable `xsmall`. As a result, the variable `xsmall` has a logical data type, and the last command outputs the value of `xsmall`.

The logical operators *and* and *or* are represented by the symbols `&` and `|`, respectively, and can be used to combine multiple logical values and return a new logical value based on the combination. The *and* operator (`&`) returns `TRUE` if both of the logical values being considered are `TRUE` and `FALSE` otherwise. The *or* operator (`|`) returns `TRUE` if at least one of the logical values being considered is `TRUE`, and `FALSE` otherwise:

```
x <- 16
(x>12) & (x<=9)

## [1] FALSE

(x>12) | (x<=9)

## [1] TRUE
```

The *not* operator (`!`) can be applied to a logical value to return the opposite value. If `v` is a logical variable, the expression `!v` is `FALSE` if `v` is `TRUE` and `TRUE` if `v` is `FALSE`:

```
x <- 16
(x>12)

## [1] TRUE

!(x>12)

## [1] FALSE
```

Parentheses can be used for more complex logical expressions and to control the order of operations:

```
x <- 16
y <- 30
((y>2*x)|(y<3*x)) & (abs(x-y)!=10)

## [1] TRUE
```

In this example, `((y>2*x)|(y<3*x))` is `TRUE` since `(y<3*x)` is `TRUE`. The absolute value of the difference between `x` and `y` is not equal to `10`, so the expression `(abs(x-y)!=10)` is also `TRUE`, meaning that the overall expression obtained by applying the *and* (`&`) operator is also `TRUE`.

A useful feature of logical values is that we can perform mathematical operations on them. When a logical value is included in a mathematical expression in *R*, a `TRUE` value is treated as a 1 and a `FALSE` value is treated as a 0. The following *R* code provides a few simple examples:

```
TRUE+FALSE

## [1] 1

x <- 4.3
y <- 8.7
1*(x<4)

## [1] 0

(x>4)*(x<7)

## [1] 1

(y>4)*(y<7)

## [1] 0
```

The first expression, `TRUE+FALSE`, evaluates as `1+0`. After setting the values of the `x` and `y` variables, the `1*(x<4)` expression returns `0` since `(x<4)` is false, `(x>4)*(x<7)` returns `1` since `(x>4)` and `(x<7)` are both true, and `(y>4)*(y<7)` returns `0` since `(y>4)` is true and `(y<7)` is false.

As will be seen in section 1.6, logical data types also provide a convenient way to control the flow of a program or to make decisions based on whether certain conditions hold.

1.3.3 String Values (`character` Data Type)

In *R*, a *string* is a sequence of characters, such as letters, numbers, spaces, and punctuation marks. To create a string, we enclose the text in quotation marks:

```
str <- "intro to statistics"
str

## [1] "intro to statistics"
```

The first command assigns the string value `"intro to statistics"` to a variable named `str`, and the second command outputs the value of the variable `str`.

There are several useful functions for manipulating strings, including the following:

- `nchar(x)`: Returns the number of characters in string `x`.
- `toupper(x)` and `tolower(x)`: Converts all the characters in string `x` to uppercase or lowercase, respectively.
- `substr(x, start, stop)`: Extracts a portion of string `x`, called a "substring," that starts at the character indicated by the `start` parameter and ends at the character indicated by the `stop` parameter.
- `paste(..., sep = " ")`: Takes one or more values (strings, numbers, etc.) as arguments and pastes them together as a single string. The optional parameter `sep`, whose default value is a space (`" "`), is inserted between each of the strings being pasted together.

The following output illustrates the use of these string functions:

```
str <- "intro to statistics"
nchar(str)

## [1] 19

toupper(str)

## [1] "INTRO TO STATISTICS"

substr(str,3,12)

## [1] "tro to sta"

x <- 6
y <- exp(1)
paste("The value of x is", x, "and the value of y is", round(y,4))

## [1] "The value of x is 6 and the value of y is 2.7183"

paste("The value of x is", x, "and the value of y is", round(y,4), sep="")

## [1] "The value of x is6and the value of y is2.7183"

str2 <- paste(str,str)
str2

## [1] "intro to statistics intro to statistics"
```

This example creates the `str` variable as a string, with a `char` data type as before. The outputs of the `nchar` and `toupper` functions are self-explanatory. For the `substr(str,3,11)` command, the output is the substring that starts at the third character of `str`, which is the first `t`, and ends at the 12th character of `str`, which is the `a`. The remaining commands illustrate how the `paste` function can be used. After the `x` and `y` values are assigned values, the first `paste` function results in a string that is output with the default space (`" "`) separator, and the second `paste` function results in a string that is output with no separator between the arguments. The command `str2 <- paste(str,str)` pastes two copies of the `str` together, with the default space separator.

Sometimes it is useful to know whether a particular substring is contained within a string. The `grepl` function provides one way to do this:

- `grepl(pattern, x)`: Returns `TRUE` if the string `pattern` is contained within string `x` and `FALSE` otherwise. The function `grepl` has several optional arguments, including, for instance, `ignore.case`, which indicates whether the case of the letters (uppercase versus lowercase) should be ignored and whose default value is `FALSE`. The interested reader can use the *R* documentation, by typing `?grepl`, to get more information about `grepl` and related functions:

```
answers <- "ACBBAD"
grepl("AA", answers)

## [1] FALSE

grepl("BB", answers)

## [1] TRUE

grepl("AAA", answers)

## [1] FALSE

grepl("AAA", answers) | grepl("BBB", answers) |
   grepl("CCC", answers) | grepl ("DDD", answers)

## [1] FALSE
```

In this example, `answers` is assigned the value `"ACBBAD"`, corresponding to a student's answers to six multiple-choice exam questions, each of which has the possible answers `A`, `B`, `C`, and `D`. The first `grepl` command asks whether the substring `"AA"` (two straight `A` responses) is within `answers`. The second `grepl` command asks whether two straight `B` responses are within `answers`. The third `grepl` command asks whether three straight `A` responses are within `answers`. And the fourth `grepl` command asks whether there is a sequence of three straight responses of *any* of `A`, `B`, `C`, or `D`.

1.4 Vectors

A *vector* is a collection of elements of the same data type. For example, a vector can be a collection of numerical values, a collection of logical values, or a collection of strings. Vectors can be created and manipulated in various ways.

1.4.1 Creating Vectors

The simplest method to create a vector is with the `c` function, and we can determine the length of a vector with the `length` function as follows:

- `c(...)`: Returns a vector that is constructed from one or more arguments, with the order of the vector elements corresponding to the order of the arguments.
- `length(x)`: Returns the length (i.e., the number of elements) of vector `x`.

```
numvec <- c(8,12,5,10,3)
numvec

## [1]  8 12  5 10  3
```

```
numvec <- c(numvec,18)
numvec

## [1]  8 12  5 10  3 18

length(numvec)

## [1] 6

answers <- c("A","C","B","B","A","D")
answers

## [1] "A" "C" "B" "B" "A" "D"

tfvec <- c(TRUE,FALSE,FALSE,TRUE)
tfvec

## [1]  TRUE FALSE FALSE  TRUE

length(tfvec)

## [1] 4

c(8,12,"A")

## [1] "8"  "12" "A"
```

The first command creates a vector with five numeric elements and assigns it to the variable `numvec`. In the third command, `c(numvec,18)` constructs a vector consisting of the original `numvec` vector with an additional numeric value (`18`) added to the end of it. The assignment of the `answers` variable shows that a vector can consist of strings, each of which happen to be one-character strings here. The assignment of the `tfvec` variable shows that a vector can consist of logical values (`TRUE` or `FALSE`). The last command, `c(8,12,"A")`, illustrates that the elements of a vector must all be of the same type. Notice that, unlike `numvec`, the `8` and `12` values are contained within quotation marks, meaning that `"8"` and `"12"` are strings. When *R* sees that string `"A"` is part of the vector being created, it forces the other values (which were numeric in this case) to be strings.

There are other useful functions for creating vectors. For instance, we can create a numeric vector containing a sequence of numbers using either the `seq` function or the `:` operator as follows:

- `seq(from, to, by = 1)`: Returns a numeric vector consisting of a sequence of numbers that starts at the value of the argument `from`, with each successive element of the sequence obtained by adding the increment `by` (an optional argument whose default value is `1`) and the argument `to`, indicating when the sequence should end. For an increasing sequence of numbers (a positive `by` value), the last number in the sequence will be less than or equal to the argument `to`. For a decreasing sequence of numbers (a negative `by` value), the last number in the sequence will be greater than or equal to the argument `to`. (When `from` is larger than `to`, the default value of `by` is `-1`.)
- The `:` operator: Returns a numeric vector consisting of a sequence of numbers between the two operands, where the increment is equal to 1. Specifically, the command `from:to` is shorthand for `seq(from, to, by = 1)` if `from` is less than `to` and shorthand for `seq(from, to, by = -1)` if `from` is less than `to`.

The following examples illustrate the use of the `seq` function and the `:` operator:

```
seq(1,10)

##  [1]  1  2  3  4  5  6  7  8  9 10

seq(1,10,2)

## [1] 1 3 5 7 9

seq(5,4,-0.1)

##  [1] 5.0 4.9 4.8 4.7 4.6 4.5 4.4 4.3 4.2 4.1 4.0

1:6

## [1] 1 2 3 4 5 6

1.2:6

## [1] 1.2 2.2 3.2 4.2 5.2

6:1

## [1] 6 5 4 3 2 1

vec <- c(1:6,seq(10,20,2))
vec

##  [1]  1  2  3  4  5  6 10 12 14 16 18 20
```

For the `seq(1,10,2)` command, the increment for the sequence is `2`, and the last element of the sequence is `9` since the next element of the sequence (`11`) would be larger than the end of the sequence (`10`) that was specified. Similarly, for the `1.2:6` command, where the increment is automatically equal to 1, the last element of the sequence is `5.2` since the next element would be larger than the end of the sequence (`6`) specified. The creation of the `vec` variable illustrates that sequences can be stored in a vector variable and can also be combined through the use of the `c` function. In this case, the `vec` variable has been assigned to a vector that consists of the integers between `1` and `6` (inclusive) and then the even integers between `10` and `20` (inclusive).

The `rep` function is a convenient way to create or *initialize* a vector that, unlike the `seq` function, can be used for any type of vector, including numeric, logical, and string.

- `rep(x, times = 1)`: Returns a vector created by repeating the value `x` a certain number of times, specified by the optional argument `times` (whose default value is `1`):

```
rep(0,5)

## [1] 0 0 0 0 0

rep(1,5)

## [1] 1 1 1 1 1

rep(FALSE,8)

## [1] FALSE FALSE FALSE FALSE FALSE FALSE FALSE FALSE

rep("abcd",6)

## [1] "abcd" "abcd" "abcd" "abcd" "abcd" "abcd"
```

1.4.2 Accessing Elements of Vectors

To access specific elements of a vector or even a single element, square brackets `[]` are used. The convention is that the *index* of the first element of a vector is `1`, the index of the second element is `2`, and so on. To access a single element of the vector, we specify a single number within the square brackets (e.g., `x[2]` to access the second element of vector `x`). To access multiple elements of the vector, we specify a vector of indices within the square brackets, as illustrated next:

```
numvec <- c(8,12,5,10,3,18,7,10,2,8)
length(numvec)

## [1] 10

numvec[3]

## [1] 5

numvec[2:5]

## [1] 12  5 10  3

numvec[(length(numvec)-2):length(numvec)]

## [1] 10  2  8

numvec[c(4,1,9)]

## [1] 10  8  2
```

In this example, the variable `numvec` is assigned to be a vector containing 10 numeric values. The `numvec[3]` command returns the third element of `numvec`. The `numvec[2:5]` command returns a vector consisting of the the second through fifth elements (inclusive) of `numvec`. The next command uses the `length` function to assist in returning a vector with the last three elements of `numvec`. The `numvec[c(4,1,9)]` command illustrates that the chosen elements need not be consecutive, as this command returns a vector with the fourth element, first element, and ninth element of `numvec`.

1.4.3 Commonly Used Vector-Related Functions

Here are some additional vector-related functions that are frequently used in *R*:

- `min(x)` and `max(x)`: Returns the minimum and maximum values of vector `x`, respectively.
- `sort(x, decreasing = FALSE)`: Returns a vector with the elements of vector `x` sorted in ascending order (if the optional argument `decreasing` is omitted) or in descending order (if the optional argument is specified as `decreasing = TRUE`).
- `unique(x)`: Returns a vector with only the unique values of vector `x`.
- `sum(x)`: Returns the sum of the elements of vector `x`. (If `x` is a logical vector, `TRUE` and `FALSE` are treated as the values `1` and `0`, respectively.)
- `mean(x)`: Returns the arithmetic mean of the elements of vector `x`, which is the sum of the elements divided by the number of elements. (If `x` is a logical vector, `TRUE` and `FALSE` are treated as the values `1` and `0`, respectively.)

- `cumsum(x)`: Returns a vector containing the cumulative or running sum of the elements of vector `x`. The first element is the first element of `x`, the second element is the sum of the first two elements of `x`, the third element is the sum of the first three elements of `x`, and so on. (If `x` is a logical vector, `TRUE` and `FALSE` are treated as the values `1` and `0`, respectively.)

The functions `min`, `max`, `sort`, and `unique` can be used for numeric, logical, and string vectors, whereas the functions `sum`, `mean`, and `cumsum` are not meant for string vectors:

```
numvec <- c(8,12,5,10,3,18,7,10,2,8)
min(numvec)

## [1] 2

max(numvec)

## [1] 18

sort(numvec)

##  [1]  2  3  5  7  8  8 10 10 12 18

sort(numvec, decreasing = TRUE)

##  [1] 18 12 10 10  8  8  7  5  3  2

unique(numvec)

## [1]  8 12  5 10  3 18  7  2

sort(unique(numvec))

## [1]  2  3  5  7  8 10 12 18

sum(numvec)

## [1] 83

mean(numvec)

## [1] 8.3

cumsum(numvec)

##  [1]  8 20 25 35 38 56 63 73 75 83
```

1.4.4 Element-by-Element Operations for Vectors

A convenient feature of *R* is that it automatically applies mathematical/logical operators and other functions to vectors on an element-by-element basis. We first illustrate this idea with numerical vectors:

```
numvec <- c(8,12,5,10,3,18,7,10,2,8)
numvec + 1

##  [1]  9 13  6 11  4 19  8 11  3  9

numvec + 1:10

##  [1]  9 14  8 14  8 24 14 18 11 18

sqrt(numvec)
```

```
##  [1] 2.828427 3.464102 2.236068 3.162278 1.732051 4.242641 2.645751 3.162278
##  [9] 1.414214 2.828427

sqrt(numvec)[2:4]

## [1] 3.464102 2.236068 3.162278

numvec >= 9

##  [1] FALSE  TRUE FALSE  TRUE FALSE  TRUE FALSE  TRUE FALSE FALSE

prices <- c(1.24,3.12,0.78,2.22,4.57,2.89,4.08,1.83,3.78,2.66)
numvec*prices

##  [1]  9.92 37.44  3.90 22.20 13.71 52.02 28.56 18.30  7.56 21.28

revenue <- sum(numvec*prices)
revenue

## [1] 214.89
```

The `numvec` variable is a vector with 10 numerical values, and the result of the command `numvec + 1` is a vector that takes each element of `numvec` and adds the value `1` to it. In contrast, for the command `numvec + 1:10`, where both operands are 10-element vectors, the vector that is returned is an element-by-element sum of the two operands. The first element is the sum of the first element of `numvec` (`8`) and the first element of `1:10` (`1`), the second element is the sum of the second element of `numvec` (`12`) and the second element of `1:10` (`2`), and so on. A mathematical function like `sqrt` returns a vector with the function applied to each element of the vector argument. In this case, we can confirm that the resulting vector has the square root of `8` as its first element, the square root of `12` as its second element, and so on. The command `sqrt(numvec)[2:4]` demonstrates the generality of indexing vectors; since `sqrt(numvec)` is itself a vector, with the same length as the original `numvec`, the `[2:4]` indexing returns the second through fourth elements of the square root applied to the `numvec` vector. The `numvec >= 9` command illustrates how logical operators are also applied on an element-by-element basis. The resulting vector is a logical vector, where the first element indicates whether the first element of `numvec` (`8`) is greater than `9`, the second element indicates whether the second element of `numvec` (`12`) is greater than `9`, and so on. The last set of commands, involving the `prices` and `revenue` variables, illustrate how vector operations can be used in a simple economic example. If the variable `numvec` represents the quantities of 10 different goods that are purchased at a certain store and the variable `prices` represents the corresponding prices of these 10 goods, the command `numvec*prices` returns a vector with 10 elements, each corresponding to the revenue associated with a given good. Then, the variable `revenue`, obtained with the `sum` function, provides the total revenue for all the goods.

Here are some examples of element-by-element operations and functions for string and logical vectors:

```
strvec <- c("ab","cd","ef")
toupper(strvec)

## [1] "AB" "CD" "EF"
```

```
paste(strvec,".",sep="")

## [1] "ab." "cd." "ef."

strvec=="ab"

## [1]  TRUE FALSE FALSE

logicvec1 <- c(TRUE,FALSE,FALSE,TRUE)
logicvec2 <- c(TRUE,TRUE,FALSE,TRUE)
logicvec1|logicvec2

## [1]  TRUE  TRUE FALSE  TRUE

logicvec1&logicvec2

## [1]  TRUE FALSE FALSE  TRUE
```

For the string vector `strvec`, `toupper(strvec)` capitalizes each element, `paste(strvec,".",sep="")` appends a period (`.`) to the end of each element, and `strvec=="ab"` does a logical comparison of each element with the string `"ab"`. For the logical vectors, `logicvec1` and `logicvec2` are created as logical vectors with four elements. The commands `logicvec1|logicvec2` and `logicvec1&logicvec2` apply the *or* (`|`) and the *and* (`&`) operators, respectively, to the two vectors on an element-by-element basis.

1.4.5 Logical Values and Vectors

We have seen the power of logical operators when applied to vectors. For example, for the numerical vector `numvec` considered previously, the expression `numvec >= 9` returns a logical vector of the same length as `numvec`, with each element indicating whether the corresponding element of `numvec` is greater than or equal to `9`. There are some additional functions that facilitate the use of logical vectors and the use of logical operators for other vectors, including the following:

- `all(x)`: Returns `TRUE` if all the values in logical vector `x` are `TRUE`, and `FALSE` otherwise.
- `any(x)`: Returns `TRUE` if any of the values in logical vector `x` is `TRUE`, and `FALSE` otherwise.
- `which(x)`: Returns a vector of indices corresponding to `TRUE` elements of logical vector `x`.
- `ifelse(test, yes, no)`: Returns a vector based upon a logical condition, given by the argument `test`, where the value given by the `yes` argument is used if the condition is `TRUE` and the value given by the `no` argument is used if the condition is `FALSE`.

```
numvec <- c(8,12,5,10,3,18,7,10,2,8)
prices <- c(1.24,3.12,0.78,2.22,4.57,2.89,4.08,1.83,3.78,2.66)
all(numvec >= 9)

## [1] FALSE

any(numvec >= 9)
```

```
## [1] TRUE

ifelse(prices > 3.50, "high price", "low price")

##  [1] "low price"  "low price"  "low price"  "low price"  "high price"
##  [6] "low price"  "high price" "low price"  "high price" "low price"

ifelse(prices > 3.50, 3.50, prices)

##  [1] 1.24 3.12 0.78 2.22 3.50 2.89 3.50 1.83 3.50 2.66

which(prices > 3.50)

## [1] 5 7 9

high_prices <- prices[which(prices > 3.50)]
high_prices

## [1] 4.57 4.08 3.78
```

Assume again that `numvec` represents the quantities of 10 different goods purchased at a store and the variable `prices` represents the corresponding prices of these 10 goods. The `all(numvec >= 9)` command returns `FALSE` since the quantities are not all greater than or equal to `9`, and the `any(numvec >= 9)` command returns `TRUE` since at least one of the quantities is greater than or equal to `9`. The first `ifelse` command returns a string vector, with the value `"high price"` corresponding to any element of `prices` greater than `3.50` and the value `"low price"` corresponding to any element of `prices` not greater than `3.50`. The second `ifelse` does something slightly different, creating a numeric vector where each element is either exactly `3.50` (when the corresponding element of `prices` is greater than `3.50`) or equal to the corresponding element of `prices` (when this element is not greater than `3.50`). The `which(prices > 3.50)` command returns a vector of the indices corresponding to the elements of `prices` that are greater than `3.50`. There are three such elements, corresponding to the indices `5`, `7`, and `9` of the `prices` vector. The usefulness of the `which` function command is illustrated in the subsequent command, where `which(prices > 3.50)` it itself used within the square brackets `[]` to select elements of the `prices` vector. Specifically, the new variable `high_prices` as a vector that consists of only the elements of `prices` for which the condition `prices > 3.50` is true.

A particularly useful function to use for a logical vector is the `sum` function. When `x` is a logical vector, `sum(x)` treats `TRUE` values as 1s and `FALSE` values as 0s, which means that `sum(x)` returns the total number of `TRUE` values in vector `x`. As an example, using the `numvec` and `prices` vectors defined in this discussion, the following code uses the `sum` function to count how many `numvec` elements are greater than or equal to `9` and how many `prices` elements are greater than `3.50`:

```
sum(numvec >= 9)

## [1] 4

sum(prices > 3.50)

## [1] 3
```

For a logical vector x, the mean(x) returns the number of TRUE elements in x divided by the total number of elements of x, which is the proportion or fraction of elements of x that are TRUE:

```
mean(numvec >= 9)

## [1] 0.4

mean(prices > 3.50)

## [1] 0.3
```

Throughout this book, we will apply the sum or mean functions to logical vectors as part of computer simulations involving random numbers. To preview this type of calculation, we briefly introduce the function runif. When called with a single argument n, the function runif(n) returns a vector of n random (real) numbers between 0 and 1. This "uniform" random variable will be discussed in chapter 10, but for now, we can think of any real number between 0 and 1 as being equally likely to be chosen:

```
set.seed(1234)
x <- runif(1000)
x[1:10]

##  [1] 0.113703411 0.622299405 0.609274733 0.623379442 0.860915384 0.640310605
##  [7] 0.009495756 0.232550506 0.666083758 0.514251141

sum(x<0.3)

## [1] 291

mean(x<0.3)

## [1] 0.291

mean((x>0.6)*(x<0.8))

## [1] 0.198
```

Ignore the first command for now, as the set.seed function will be discussed in chapter 2. The second command creates vector x containing 1,000 random numbers between 0 and 1, and the third command outputs the first 10 values of x. The sum(x<0.3) command outputs the number of random numbers, out of 1,000, that are less than 0.3, and the mean(x<0.3) command outputs the sum divided by the length of x (1,000). The final command, mean((x>0.6)*(x<0.8)), returns the fraction of the 1,000 random numbers that are between 0.6 and 0.8 since (x>0.6)*(x<0.8) evaluates to 1 when both x>0.6 and x<0.8 are TRUE and 0 otherwise. For this simulation, 29.1% of the 1,000 random numbers are less than 0.3, and 19.8% are between 0.6 and 0.8.

1.4.6 Factor Variables and Vectors

The factor data type is a data type for a categorical variable with a fixed set of possible values. In *R*, such a categorical variable is called a *factor variable*. As an example, suppose that a vector contains a student's responses to five questions, with possible values of True and False:

```
answers <- c("True","False","True","True","False")
answers

## [1] "True"  "False" "True"  "True"  "False"

tfvec <- factor(answers)
tfvec

## [1] True  False True  True  False
## Levels: False True

levels(tfvec)

## [1] "False" "True"
```

The first command assigns `answers` to a string vector with the student's answers, and the second command outputs this vector. The third command creates `tfvec` by applying the `factor` function to the original vector `answers`. The `factor` function converts the original string vector into a vector that has the factor data type. As the output shows, `tfvec` has the same values as `answers`, but *R* has automatically detected that the `Levels` of the factor variable are `False` and `True`. The `levels` function can be used to return the levels associated with a factor variable. By default, *R* orders the levels of the factor variable alphabetically unless specified otherwise.

If a categorical variable has a natural ordering, we can also explicitly specify the factor levels in the correct order. For example, suppose that we have customer ratings for two different restaurants, contained in vectors `vec1` and `vec2`. There are four possible ratings, which are (in ascending order) `Poor`, `Fair`, `Good`, and `Excellent`:

```
vec1 <- c("Good","Excellent","Good","Poor","Good")
vec2 <- c("Fair","Excellent","Excellent","Poor","Good","Excellent")

rating_levels <- c("Poor","Fair","Good","Excellent")
ratings1 <- factor(vec1, levels=rating_levels)
ratings2 <- factor(vec2, levels=rating_levels)

ratings1

## [1] Good      Excellent Good      Poor      Good
## Levels: Poor Fair Good Excellent

ratings2

## [1] Fair      Excellent Excellent Poor      Good      Excellent
## Levels: Poor Fair Good Excellent

table(ratings1)

## ratings1
##      Poor      Fair      Good Excellent
##         1         0         3         1

table(ratings2)

## ratings2
##      Poor      Fair      Good Excellent
##         1         1         1         3
```

The assignment of `rating_levels` to the four ratings categories tells *R* the order of the categories. Then, `ratings1` and `ratings2` are created as vectors of factor variables, based upon the original `vec1` and `vec2` string vectors and using the levels specified by `rating_levels`. When `ratings1` is output, the levels are shown in the order that we specified, and moreover the output shows the level `Fair` even though there are no `Fair` ratings in the `ratings1` vector. Another way to output and view the data within the factor-variable vector is with the `table` command, which provides a tabulation (count) of the number of vector elements that are within each category. For the table of `ratings1`, all four categories are shown, with a `0` indicating no elements with a `Fair` rating.

1.5 Output

In the examples considered thus far, *R* supplies output automatically for most expressions, with the exception of variable assignments:

```
x <- 5
x

## [1] 5

x/3

## [1] 1.666667
```

In this example, the variable assignment `x <- 5` does not result in output, but the commands `x` and `x/3` both result in output, corresponding to the values of the two expressions.

The `print` function provides an alternative method for providing output. In many cases, the command `print(x)` leads to the same output as the command `x`, but the `print` function is sometimes preferred by *R* users since (1) it has optional arguments that can be useful and (2) it can provide output that is better formatted for tables, regressions, and other elements.

1.6 Programming

1.6.1 Conditional (`if-else`) Execution

The simplest version of conditional execution involves a single `if` statement, where a command or series of commands gets executed if a certain logical condition holds:

```
#  if (logical condition) {
#     ... commands ...
#  }
```

The syntax for the `if` statement has a logical condition, which evaluates to `TRUE` or `FALSE`, within parentheses after the `if` keyword. If the logical condition evaluates to `TRUE`, the sequence of commands within the curly braces { and } are executed:

```
str <- "intro to statistics"

if (nchar(str) > 10) {
  print(paste(str,"is a long string."))
  print(paste(str,"has",nchar(str),"characters."))
```

```
}

## [1] "intro to statistics is a long string."
## [1] "intro to statistics has 19 characters."
```

The `if` statement checks whether the string `str` has more than 10 characters. Since it does, the commands within { and } are executed, resulting in two lines of output. Had `str` been a string with 10 or fewer characters, the commands within { and } would have been skipped, and no output would have resulted.

What if we want to execute different commands when the logical condition within the `if` statement does *not* hold? In this case, we use an `if-else` statement that executes one set of commands if the logical condition holds and a different set of commands if it does not:

```
#  if (logical condition) {
#      ... commands executed if condition is TRUE...
#  } else {
#      ... commands executed if condition is FALSE...
#  }
```

This syntax has the `else` keyword appearing after the right curly brace } of the original `if` statement and additional commands contained within a second set of curly braces { and } after the `else` keyword:

```
price <- 25.78

if (price > 20) {
  print("The price is too high.")
  price <- 0.9*price
} else {
  print("The price is not too high.")
}

## [1] "The price is too high."
```

The `if` statement checks if `price` is greater than `20`, which is the case here. As a result, the output `"The price is too high."` is provided, and the variable `price` is set to 90% of its original value. Had price been less than or equal to `20`, the output `"The price is not too high."` would have been provided, and the variable `price` would have been left unchanged.

We can check additional logical conditions within an `if-else` statement by using the keywords `else if` rather than `else`, as illustrated in the following example:

```
price <- 25.78

if (price > 20) {
  print("The price is too high.")
  price <- 0.9*price
} else if (price < 10) {
  print("The price is too low.")
  price <- 1.1*price
} else {
  print("The price is not too high or too low.")
```

```
}
## [1] "The price is too high."
```

Like the previous example, the first logical condition (`price > 20`) is checked by the initial `if` statement, with the two commands below it executed if this condition is `TRUE`. In this example, however, if the (`price > 20`) condition is `FALSE`, the subsequent `else if` checks whether the logical condition (`price < 10`). If this condition is `TRUE`, the output `"The price is too low."` is provided, and the variable `price` is set to 110% of its original value. If this condition is `FALSE`, in which case `price` is between `10` and `20`, the output `"The price is not too high or too low."` is provided.

1.6.2 `for` Loops

A `for` loop is a programming structure used to repeat a set of commands a specified number of times or for each element in a vector. This type of loop will be especially useful in this book when we want to conduct a specific number of simulations of a certain process:

```
#  for (var in sequence) {
#       ... commands ...
#  }
```

In this syntax, `var` is a variable used to store values during the loop, and `sequence` is a vector specifying the sequence of values over which the `for` loop will iterate. For each value in `sequence`, the `for` loop executes the commands specified within the curly braces `{` and `}`. For instance, to conduct 10,000 simulations of a certain process, a `for` loop can loop over the simulation number (from 1 to 10,000), and the commands within the loop will be executed repeatedly for each simulation.

As an example, we consider how a `for` loop can be used to calculate the *Fibonacci sequence*, which is a sequence of numbers in which each number is the sum of the two preceding numbers. The sequence starts with the numbers 1 and 1, so that the Fibonacci sequence is $1, 1, 2, 3, 5, 8, 13, 21, \ldots$:

```
sequence_length <- 10
fib_sequence <- rep(0,sequence_length)
fib_sequence[1] <- 1
fib_sequence[2] <- 1

for (i in 3:sequence_length) {
  fib_sequence[i] <- fib_sequence[i-2]+fib_sequence[i-1]
}

print(fib_sequence)
##  [1]  1  1  2  3  5  8 13 21 34 55
```

The variable `sequence_length` is the desired length of the Fibonacci sequence, set to `10` here. Then, the variable `fib_sequence` is initialized as a numeric vector having all 0s and with a length equal to the specified `sequence_length`. The next two commands assign the value 1 to both the first and second elements of fib_sequence, which corresponds to 1 and 1 being the first two numbers in the Fibonacci sequence. The for loop does the rest of the work in determining the Fibonacci sequence. The variable i loops from 3 to sequence_length. With sequence_length being 10, variable i has the value 3 on the first iteration of the loop, 4 on the second iteration of the loop, and so on, through 10 on the eighth iteration of the loop. On the first iteration of the loop, fib_sequence[3] is assigned to the sum of fib_sequence[1] and fib_sequence[2], which is 2. On the second iteration of the loop, fib_sequence[4] is assigned to the sum of fib_sequence[2] and fib_sequence[3], which is 3. This process continues until the final iteration of the loop, where fib_sequence[10] is assigned to the sum of fib_sequence[8] and fib_sequence[9]. The last print command provides the output with the Fibonacci sequence of length sequence_length.

The following example involves the use of strings and illustrates how the sequence vector can be something other than a simple sequence of numeric values:

```
strvec <- c("cow","horse","pig","chicken","goat")

first_letters <- ""

for (str in strvec) {
  first_letters <- paste(first_letters,substr(str,1,1),sep="")
}

print(first_letters)

## [1] "chpcg"
```

The for loop builds up a string first_letters, which consists of the first letters from each string in the string vector strvec. The variable first_letters is initialized as an empty string, "". In the first iteration of the for loop, variable str is equal to "cow", the first element of strvec. The expression substr(str,1,1) yields the first character of str, which is "c", and the paste function pastes it at the end of first_letters, such that the value of first_letters after the first iteration is "c". In the second iteration, variable str is equal to "horse", and the character "h" gets added to the end of first_letters, such that its value is "ch" after the second iteration. This process continues, with a total of five iterations, and the print command outputs the final value of first_letters.

1.6.3 while Loops

A while loop is used to repeatedly execute a set of commands so long as a logical condition is true. Whereas a for loop is most often used when the number of iterations is known (e.g., a loop of 10,000 simulations), a while loop is used in situations where the total number of iterations may be unknown at the start. Instead, the while loop continues until the specified logical condition doesn't hold:

```
#  while (logical condition) {
#      ... commands ...
#  }
```

The syntax for the `while` loop has a logical condition, which evaluates to `TRUE` or `FALSE`, within parentheses after the `while` keyword. If the logical condition is `TRUE`, the commands within the curly braces { and } are executed. After the commands are executed, the computer goes back to the start of the `while` loop to check the logical condition again. If the logical condition is still `TRUE`, the commands within the curly braces { and } are executed again. This process continues until the `while` logical condition is `FALSE`, at which point the computer stops the loop and, if there are commands after the loop (i.e., after the right curly brace }), jumps to those commands:

```
sales <- c(38,52,24,61,47,18,29,44,41)
total_sales <- 0
idx <- 0

while (total_sales <= 200) {
  idx <- idx + 1
  total_sales <- total_sales + sales[idx]
}

print(paste("It took",idx,"days to exceed 200 sales. There were",total_sales,
      "sales after",idx,"days."))

## [1] "It took 5 days to exceed 200 sales. There were 222 sales after 5 days."
```

The numeric vector `sales` contains data on daily sales for a certain company, and a `while` loop determines how many days it takes for the company's total (cumulative) sales to be at least 200 units. The `total_sales` variable, used to keep track of cumulative daily sales, is initialized to `0`. The `idx` variable, used to keep track of the number of the day (i.e., the index of the `sales` vector), is also initialized to `0`. When the `while` loop is first reached, the logical condition is `TRUE` since `total_sales` is equal to `0`. The commands within the loop increment `idx`, so it now has a value of `1`, and add `sales[1]` to `total_sales`, which now has a value of `38`. The logical condition remains true, so the commands within the loop are executed again, leading to an `idx` value of `2` and a `total_sales` of `90`. This process continues until `idx` is equal to `5`, and `total_sales` is equal to `222`, at which point the logical condition for the `while` loop is `FALSE`, so the loop is ended and the `print` command outputs the information.

1.7 Writing Functions

A *function* is a block of code that performs a specific task. There are many built-in *R* functions that facilitate calculations and data manipulation. But new functions can also be created to perform tasks not already built into *R*. Here are the steps for creating a new function:

- **Specify the function name:** Choose a name that describes what the function does. For example, if the function calculates the distance between two points (x_1, y_1) and (x_2, y_2), an appropriate name might be `distance` or something similar. (Do not pick a function name already used by *R*, as defining a function with an existing name will replace the existing *R* function.)
- **Specify the arguments:** Arguments are the values that the function takes as input. When the function is called, these arguments are enclosed in parentheses after the

function name. Some arguments may be required and some may be optional, as we have seen for built-in functions like `round` and `log`.

- **Write the code:** The code for the function, which is contained within curly braces { and }, performs the desired task using the provided arguments.
- **Return the output:** The `return` function returns the output that results from execution of the function's code. The output could be any type of object, like a number, a string, a logical value, or a vector.[4]

The following example shows how to define a function that calculates the area of a triangle based upon the values of its base and its height:

```
triangle_area <- function(base, height) {
  area <- 0.5*base*height
  return(area)
}

triangle_area(10,5)

## [1] 25

area

## Error in eval(expr, envir, enclos): object 'area' not found
```

The name of the function is `triangle_area`, and it is assigned to be a function with two required arguments, `base` and `height`. Within the function, the variable `area` is set to be equal to the area of the triangle, which is 1/2 times base times height. Then the `return` function returns the function's output, which is the value of `area`. The `triangle_area(10,5)` command makes a call to the function `triangle_area` with argument values `10` (for `base`) and `5` (for `height`), and the resulting area of `25` is printed. The final command, `area`, results in an error message. Even though the variable `area` is defined within the function `triangle_area`, it is not recognized by *R* once the function evaluation is complete; we say that the variable `area` is "local" to the function. The same is true for any variables created within a function. It is also true for the arguments of the function; the variables `base` and `height` would not be recognized after the function evaluation is complete. When a variable or argument is defined inside a function, it becomes a local variable by default. As such, the variable or argument name can be used freely within the function without affecting any variables outside the function. For example, if `height` were a variable that already existed before the call to `triangle_area`, the use of the argument name `height` would not affect the value of the variable `height`. This idea is illustrated here:

```
height <- 8
triangle_area(10,5)

## [1] 25

height

## [1] 8
```

The variable `height` is assigned the value `8`. Then, even though the `triangle_area` function uses an argument with the same name (`height`) and assigns the value `10` to the argument `height`, the value of the original variable remains `8` after the function evaluation is complete.

Without any changes, the `triangle_area` function can take vectors as its two arguments, with `base` being a vector of triangle bases and `height` being a vector of triangle heights. The first elements of `base` and `height` correspond to the first triangle, the second elements correspond to the second triangle, and so on. Then, since *R* automatically performs arithmetic operations on an element-by-element basis, the function `triangle_area` returns a vector of the areas for each of the triangles:

```
base_vec <- c(10,15,20)
height_vec <- c(5,8,10)
triangle_area(base_vec,height_vec)

## [1]  25  60 100
```

The following example shows how to define optional arguments in a function. We define the function `distance` to calculate the distance between any two points (x_1, y_1) and (x_2, y_2), given by $\sqrt{(x_1 - x_2)^2 + (y_1 - y_2)^2}$:

```
distance <- function(x1, y1, x2 = 0, y2 = 0) {
  return(sqrt((x1-x2)^2+(y1-y2)^2))
}

distance(3,4)

## [1] 5

distance(3,4,5,-1)

## [1] 5.385165
```

Here, `x1` and `y1` are required arguments for `distance`, while `x2` and `y2` are optional arguments. For the optional arguments, an equal sign (=) appears after the argument and is followed by the default value. The default values for `x2` and `y2` both equal `0`, corresponding to $(x_2, y_2) = (0, 0)$ being the origin if the function is called without the optional arguments. Thus, the `distance(3,4)` command returns the distance from $(3, 4)$ to $(0, 0)$, which is equal to `5`, and the `distance(3,4,5,-1)` command returns the distance from $(3, 4)$ to $(5, -1)$, which is $\sqrt{29}$, or approximately `5.385165`.

1.8 Data Frames and File Input

A *data frame* is a two-dimensional tablelike data structure, similar to a spreadsheet, where each column can be a different data type (e.g., numeric, logical, character, or factor). A data frame can be created using the `data.frame` function, as in the following example:

```
df <- data.frame(name = c("Amy", "Blake", "Chloe"), age = c(28, 41, 32),
        employed = c("Yes", "No", "Yes"))
```

```
print(df)

##    name age employed
## 1   Amy  28      Yes
## 2 Blake  41       No
## 3 Chloe  32      Yes

mean(df$age)

## [1] 33.66667

sapply(df,class)

##        name         age    employed
## "character"   "numeric" "character"

df$employed <- factor(df$employed)
sapply(df,class)

##        name         age    employed
## "character"   "numeric"    "factor"
```

The first command, using the `data.frame` function, assigns the data frame to variable `df`. This data frame is specified to have three variables (columns): `name`, `age`, and `employed`. The data frame has three rows, or three observations for each variable, as can be seen most clearly from the output of the `print(df)` command. The `$` syntax refers to an individual variable within the data frame, so `df$age` refers to the `age` variable (column). The command `mean(df$age)` outputs the arithmetic mean of the three values of `age`. The next command, `sapply(df,class)`, indicates the data type associated with each of the three variables in `df`. The function `sapply` applies the function specified by its second argument (in this case, `class`) to each variable in the data frame specified in its first argument (in this case, `df`). The resulting output indicates that `name` and `employed` are strings and `age` is numeric. Since it is more appropriate to treat `employed` as a factor variable, with possible values `"Yes"` and `"No"`, the `factor` function in the next command reassigns the `employed` variable so its values are factors rather than strings. The output of the second `sapply` command confirms that the data type of `employed` has changed.

R provides various ways to access data frame elements, including rows, columns, and other selected data cells. For the data frame `df` created here, the following examples show some different possibilities. Just as square bracket syntax (`[` and `]`) references elements of vectors, square brackets can also be used to reference elements of data frames. The difference is that data frames are two-dimensional objects, whereas vectors are one-dimensional:

```
df$name

## [1] "Amy"   "Blake" "Chloe"

df$age[1]

## [1] 28

df[1,2]
```

```
## [1] 28

df[,2]

## [1] 28 41 32

df[,c("name","employed")]

##    name employed
## 1   Amy      Yes
## 2 Blake       No
## 3 Chloe      Yes

df[1:2,]

##    name age employed
## 1   Amy  28      Yes
## 2 Blake  41       No

df[df$age>30,]

##    name age employed
## 2 Blake  41       No
## 3 Chloe  32      Yes
```

The expression `df$name` refers to the `name` variable and outputs a vector with all the values for that variable in the data frame. The expression `df$age[1]` refers to the `age` variable, with `df$age` being a vector of all the `age` values and `df$age[1]` being the first element of that vector, which is `28`. An alternative method to access the same `age` value from the data frame is to directly reference the appropriate element of the data frame, which is what the `df[1,2]` expression does. Note that `df[1,2]` has two indices specified within the square brackets and separated by a comma. For this syntax, the first index (or indices) refers to the rows of the data frame, and the second index (or indices) refers to the columns or variables of the data frame. Therefore, `df[1,2]` is the first element (row `1`) of the second variable (column `2`, which is the `age` variable), which has the same value (`28`) as the `df$age[1]` expression. The expression `df[,2]` omits the first index within the square brackets, which tells *R* to provide *all* the rows for the specified columns. Therefore, the value of `df[,2]` is the vector of all the values for the second (`age`) variable. In the following expression, `df[,c("name","employed")]`, the variables are specified by name rather than index number, and all rows for `name` and `employed` are returned since the first index is again omitted. The result is itself a data frame with two columns, in contrast to the vector that was returned by `df[,2]` when only a single column was specified. The expression `df[1:2,]` omits the second index within the square brackets, which tells *R* to provide *all* the columns for the specified rows. As such, the result of `df[1:2,]` is itself a data frame that contains the first and second rows, as specified by `1:2`, of the original data frame `df`. This notation provides a convenient way to select certain rows from the data frame, which can be extremely useful for data analysis. The final expression, `df[df$age>30,]`, provides a simple example of selecting a subset of data based upon a logical condition. This expression returns a data frame with all rows for which the `age` variable is greater than `30`.

While the `data.frame` function can be used to directly create a data frame, it is much more common to create a data frame based on an existing file that contains data. *R* has the ability to read many types of files, including text files and spreadsheet (e.g., Microsoft

Excel) files. In the interest of space, we focus on two specific file formats in this section: the `csv` format and the `RData` format.

1.8.1 Reading Data from a `csv` File

A `csv` (which means "comma-separated values") file is a text file where each line of the file corresponds to a row of data, with values separated by a `,` character. An advantage of a `csv` file is that it is "universal," in the sense that all statistical packages have the ability to read `csv` data files. This book's data sets are available on the companion website as `csv` files. For each data set, the first line of the `csv` file contains the names of the variables to be used for the data frame, again separated by commas. As an example, the following box shows the first few lines from the file `exams.csv`, which contains exam scores for 77 students. The file has 78 lines in it, with the first line having the variable names (`exam1` and `exam2`) and the remaining lines having the values for the 77 students:

```
exam1,exam2
63,71
86,84
68,63
94,82
...
...
```

To load a `csv` file as a data frame, here are the necessary steps:[5]

1. **Set your "working directory":** *R* uses the working directory when trying to load a file. The current working directory can be checked by typing `getwd()`. The working directory can be set by either (1) changing it in the "Preferences..." settings in *RStudio* or (2) using the `setwd` function to specify the directory (e.g., `setwd("C:/stat-course/data/")`).
2. **Load the `csv` file:** With the working directory set, the `read.csv` function loads the `csv` file. The first argument of the `read.csv` function is the file name, provided within double quotes. By default, `read.csv` assumes that the first line of the file contains the variable names. (If the first line of the file contains the first row of data rather than variable names, the optional argument `header = FALSE` should be specified. Then, *R* reads in the data and create the variable names as `V1`, `V2`, `V3`, etc.) There are several optional arguments that can be specified, with details available from the `?read.csv` documentation. Two commonly used options can help to make sure that the data types of the variables are appropriate:
    - `stringsAsFactors`: Specifying `stringsAsFactors = TRUE` ensures that string data are read in as factor variables. For example, if a variable has only the values `"Yes"` and `"No"`, the default is for `read.csv` to read the variable as a string variable, but `stringsAsFactors = TRUE` instead reads it as a factor variable with two possible values.
    - `colClasses`: Specifying the `colClasses` argument directly indicates the desired data types for each of the variables. For example, if the file `zipcode.csv` contains just a single variable with five-digit US ZIP

codes in it, those ZIP codes can be read in as strings rather than numbers with the expression `read.csv("zipcode.csv", colClasses = c("character"))`. With multiple variables, the argument `colClasses` is a vector with length equal to the number of variables and data types specified, in order, corresponding to the variables. For example, `colClasses = c("character","numeric","factor")` would read as three variables—the first as a string variable, the second as a numeric variable, and the third as a factor variable.

The following example loads the file `exams.csv` into *R*. As `exam1` and `exam2` will be treated as numeric variables, no additional arguments are specified for the `read.csv` function since *R* automatically reads them as numeric variables:

```
exams <- read.csv("exams.csv")
exams[5:9,]

##   exam1 exam2
## 5    60    59
## 6    75    80
## 7    79    68
## 8    72    67
## 9    90    79

print(paste("Top score on exam #1:",max(exams$exam1)))

## [1] "Top score on exam #1: 99"

print(paste("Top score on exam #2:",max(exams$exam2)))

## [1] "Top score on exam #2: 96"
```

The variable `exams` is assigned to be a data frame associated with the data in `exams.csv`. Then, any of the usual commands can be used to access or manipulate the data in `exams`. The `exams[5:9,]` command returns a data frame consisting of the data for the fifth through ninth students. The two `print` commands output the maximum scores for the `exam1` and `exam2` variables.

Once a data frame has been created from a file, there are several useful *R* functions that can be used to examine and summarize the data frame, including the following:

- `View(df)`: Displays `df` as a spreadsheet in the *RStudio* script window.
- `head(df, n = 6)`: Returns the first six rows of `df`. The optional argument `n`, whose default value is `6`, can be changed to return a different number of rows.
- `tail(df, n = 6)`: Returns the last six rows of `df`. The optional argument `n`, whose default value is `6`, can be changed to return a different number of rows.
- `str(df)`: Shows the structure of `df`, including the number of rows, the number of variables, the variable names, the variable data types, and some sample values for each variable.
- `summary(df)`: Provides a summary of the variables in `df`. The information provided for a variable depends on its data type. For numeric variables, descriptive measures (including minimum value, maximum value, and arithmetic mean) are

provided. For factor variables, observation counts for some of or all the categories are provided.

- `nrow(df)` and `ncol(df)`: Return the number of rows and columns in `df`, respectively.
- `names(df)`: Returns a string vector containing the names of the variables in `df`.

```
head(exams)

##   exam1 exam2
## 1    63    71
## 2    86    84
## 3    68    63
## 4    94    82
## 5    60    59
## 6    75    80

str(exams)

## 'data.frame': 77 obs. of  2 variables:
##  $ exam1: int  63 86 68 94 60 75 79 72 90 79 ...
##  $ exam2: int  71 84 63 82 59 80 68 67 79 85 ...

summary(exams)

##      exam1           exam2
##  Min.   :33.00   Min.   :20.0
##  1st Qu.:68.00   1st Qu.:64.0
##  Median :79.00   Median :73.0
##  Mean   :77.31   Mean   :70.9
##  3rd Qu.:89.00   3rd Qu.:81.0
##  Max.   :99.00   Max.   :96.0

names(exams)

## [1] "exam1" "exam2"
```

We can add variables to an existing data frame. For example, to add the variable `avg` (the average of the two exam scores) to the `exams` data frame, we do an assignment command for `exams$avg`:

```
exams$avg <- (exams$exam1+exams$exam2)/2
str(exams)

## 'data.frame': 77 obs. of  3 variables:
##  $ exam1: int  63 86 68 94 60 75 79 72 90 79 ...
##  $ exam2: int  71 84 63 82 59 80 68 67 79 85 ...
##  $ avg  : num  67 85 65.5 88 59.5 77.5 73.5 69.5 84.5 82 ...

write.csv(exams,file="exams-edited.csv")
```

The output from `str(exams)` confirms that a third variable, named `avg`, has been added to the `exams` data frame. Had we instead done the assignment as `avg <-(exams$exam1+exams$exam2)/2`, without the `exams$` prefix, the vector `avg` would have been created correctly, but it would simply be another variable in *R* and not part of the `exams` data frame. Since the data set has been changed, we might want to save the new data frame to a file. The `write.csv` command saves the data frame `exams` to a new `csv` file called `exams-edited.csv`; alternatively, we could over write the original file with the command `write.csv(exams,file="exams.csv")`.

As with other types of variables, a data frame can be removed from the *R* working environment using the `rm` function. For example, if we are finished using the `exams` data frame, `rm(exams)` removes it from the *R* working environment.

1.8.2 Reading Data from a RData File

The `RData` file format is designed to be used specifically for *R*. While we will focus on reading and writing a single data frame, the `RData` format is much more general and allows *R* users to save multiple data frames and other objects (variables, graphs, etc.) to a single `RData` file, specifed with the `.RData` file extension.

The `load` function is used to load the contents of an `RData` file into the *R* environment. The following code shows the simplest use of the `load` function, where the single argument is the `RData` file being loaded:

```
load("exams.RData")
str(exams)

## 'data.frame': 77 obs. of  2 variables:
##  $ exam1: int  63 86 68 94 60 75 79 72 90 79 ...
##  $ exam2: int  71 84 63 82 59 80 68 67 79 85 ...
```

The `exams.RData` file contains a data frame that is *already stored within the* `exams` *variable*. As such, there is no assignment necessary in the first command, in contrast to the assignment that was necessary when using `read.csv` for a `csv` file. The command `str(exams)` confirms that the data frame is loaded correctly. The `save` function is used to save a data frame (and/or other *R* objects) to an `RData` file. In the code here, we add the variable `avg` to the `exams` data frame and then save the `exams` data frame to a different `RData` file:

```
exams$avg <- (exams$exam1+exams$exam2)/2
str(exams)

## 'data.frame': 77 obs. of  3 variables:
##  $ exam1: int  63 86 68 94 60 75 79 72 90 79 ...
##  $ exam2: int  71 84 63 82 59 80 68 67 79 85 ...
##  $ avg  : num  67 85 65.5 88 59.5 77.5 73.5 69.5 84.5 82 ...

save(exams, file="exams-edited.RData")
```

Even though the file name is different, the name of the data frame (`exams`) remains the same, and the new file `exams-edited.RData` contains the `exams` data frame.

The following code shows a simple example where we save more than one object to an `RData` file:

```
instructor <- "Steph Stats"
save(exams, instructor, file="exams-edited.RData")
```

We first create a string variable `instructor`. The `save` command now has two arguments before the `file` argument is specified, causing both the `exams` data frame and the `instructor` string variable to be saved in `exams-edited.RData`. Now, when this file gets loaded by the `load` command, both objects are created in the *R* environment. To save more objects in an RData file, this basic idea can be extended to include many objects as arguments before the file argument is specified for the save function.

1.9 Missing Values

Missing data are commonly encountered in real-world data sets. For instance, the values of a particular variable may be missing (i.e., unavailable) if a survey question is unanswered, a statistical agency has not released the value, or for a variety of other reasons. We can also have a missing value when the variable is not applicable for a certain observational unit. An example is a variable measuring the hourly wage of workers, which has a value only for workers paid on an hourly basis and is missing for workers who are not paid on an hourly basis.

A missing value is indicated by the special syntax NA, short for "not available." Let's consider an example with data on eight workers, with their hourly wages in dollars (which may be missing for those who are not paid on an hourly basis) and their ages. For this example, the following two functions for dealing with missing values are also illustrated:

- is.na(x): If x is a vector, returns a logical vector indicating which elements of x are missing.
- na.omit(x): If x is a data frame, returns a data frame with only those rows having nonmissing values for *all* of the variables.

```
hourly_wage <- c(12.50, NA, 10.75, 11.00, NA, NA, 14.80, 13.25)
age <- c(24, 42, 31, 61, 55, 26, 34, 59)
is.na(hourly_wage)

## [1] FALSE  TRUE FALSE FALSE  TRUE  TRUE FALSE FALSE

hourly_wage[!is.na(hourly_wage)]

## [1] 12.50 10.75 11.00 14.80 13.25

sum(is.na(hourly_wage))

## [1] 3

worker_df <- data.frame(wagehr = hourly_wage, age = age)
na.omit(worker_df)

##   wagehr age
## 1  12.50  24
## 3  10.75  31
## 4  11.00  61
## 7  14.80  34
## 8  13.25  59
```

The hourly_wage vector has three missing values and five nonmissing values. The expression is.na(hourly_wage) returns a logical vector having TRUE values corresponding to the missing elements of hourly_wage and FALSE values for the other elements. The expression sum(is.na(hourly_wage)) returns the total number of missing values in hourly_wage since TRUE and FALSE are treated as 1 and 0, respectively, by the sum function. The hourly_wage[!is.na(hourly_wage)] expression returns a vector with all nonmissing elements of `hourly_wage`. The `data.frame` function creates the data frame `worker_df`, consisting of the hourly wage and age variables. Then the `na.omit(worker_df)` command returns a data frame consisting of only those rows with nonmissing hourly wage values.

Several *R* functions that take vectors as arguments do not work correctly in the presence of missing (`NA`) values. For instance, to calculate the average hourly wage for workers, the expression `mean(hourly_wage)` would seem appropriate. Unfortunately, that expression doesn't perform as desired, and the optional argument `na.rm = TRUE` needs to be specified:

```
mean(hourly_wage)

## [1] NA

mean(hourly_wage, na.rm = TRUE)

## [1] 12.46

max(hourly_wage, na.rm = TRUE)

## [1] 14.8
```

The expression `mean(hourly_wage)` returns `NA`, indicating that *R* is unable to calculate the mathematical average when the vector has missing values. Adding the optional argument `na.rm = TRUE` fixes the problem, with `12.46` reported as the average of the five nonmissing values. Similarly, using `na.rm = TRUE` for the `max` function allows *R* to calculate the largest value (`14.8`) of the nonmissing values. If the expression `max(hourly_wage)` had been used without `na.rm = TRUE`, the expression would also have a result of `NA`.

1.10 *R* Packages

An *R* package is a collection of functions, data, and/or documentation that are bundled so they can be easily installed and used. Such packages provide add-on capabilities to the standard *R* statistical software, and they are a convenient way to share code with others and to organize and reuse your own code. Before an *R* package can be used, it first needs to be installed using the `install.packages` function. Once the package has been installed, it can be used in an *R* session with the `library` function, and the description for the package can be accessed using the optional `help` argument for the `library` function.

As an example, we consider the installation and use of `stringr`, an *R* package that provides several useful functions for string manipulation. First, the `stringr` package must be installed:

```
install.packages("stringr")
```

The package name `stringr` is enclosed in double quotes when used as an argument of the `install.packages` function. With the `stringr` package installed, its contents can be used in *R* after the command `library(stringr)` is executed:

```
library(stringr)
library(help = stringr)
str_trim("  testing this out ")

## [1] "testing this out"
```

The double quotes are unnecessary for `stringr` when it is used as an argument for the `library` function. The second command, `library(help = stringr)`, opens a window in the *RStudio* script pane with information about the `stringr` library and its included functions. For example, the function `str_trim` removes any extra spaces from the start and end of a string; in the example given here, there are two spaces at the start of the string and one at the end of the string that are removed. In most cases, we can access detailed documentation for new functions that are added through packages. The command `?str_trim`, for instance, provides documentation for the `str_trim` function.

1.10.1 The `probstats4econ` Package

This book has a companion *R* package, `probstats4econ`, that includes all the data sets (data frames) in the book and several user-defined functions introduced throughout the book. We install the package as follows:

```
install.packages("probstats4econ")
library(probstats4econ)
```

The data frames in the `probstats4econ` package are "lazily loaded" into *R*, which means that they are available to the user even though they will not initially appear in the upper-right Environment/History pane. As an example, after loading the package with the `install.packages` and `library` functions, we can immediately use the `exams` data frame within the console. Even after using the data set, it still will not appear in the Environment/History pane unless a change (e.g., adding a new variable) is made to the data frame. *R* itself has several data sets lazily loaded when *RStudio* is launched. A complete list of data sets can be seen by using the command `library(help=datasets)`.

Aside from the `probstats4econ` package, this book primarily considers the use of *R* without the installation of additional packages, which is known as "base *R*." That said, there are several *R* packages that are very useful for data analysis. Examples include `ggplot`, a package for creating high-quality graphics and data visualizations, and `dplyr`, a package for data manipulation. Both `ggplot` and `dplyr` are part of a larger package, `tidyverse`, that is popular among data scientists.

Notes

1. The *RStudio* console window does *not* display the `##` at the beginning of output lines. This book puts the `##` at the beginning of output so the full block can be copied and, when executed, only the commands and not the output lines (since they are now "comments") are executed.
2. *R* has an `integer` data type, which provides a more efficient way to store integer-valued variables in memory.
3. *R* has a `time` data type for storing time values (e.g., `11:49:20`).
4. It is also possible for a function to have multiple return values, which can be done by returning a *list* object. An *R* list is a collection of different objects, which need not be of the same data type. The code for `wald_test` in chapter 14 shows an example having a list of two objects returned.

5. There are point-and-click alternatives in *RStudio* to read data files as well, including (1) click on "Import Dataset" in the upper-right Environment/History pane; (2) click on the "Files" tab in the lower-right Plots/Help/Packages/Files pane and select the desired directory and file; and (3) click on the "Import Dataset" submenu from the main "File" menu.

Exercises

1. For each of the following sections of *R* code, indicate what output, if any, is provided by *R* and the final value of variable `x`. Answer without using *R*.
   a. Code section 1:

```
x <- 16
sqrt(x)
```

   b. Code section 2:

```
x <- 16
x <- sqrt(x)
x <- c(x,x)
```

   c. Code section 3:

```
x <- c(79,16,53,44)
x <- sort(x, decreasing = TRUE)
length(x)
```

   d. Code section 4:

```
y <- c(6.7,-3.3,4.2)
x <- (1:3)*y
min(x)
x <- (x > 0)
sum(x)
```

2. A company has the following daily sales over the course of nine days: 38, 52, 24, 61, 47, 18, 29, 44, and 41.
   a. Create a numerical vector `sales` that contains the daily sales.
   b. Provide a single command to calculate the average daily sales.
   c. Provide a single command to calculate the number of days for which sales are strictly between 40 units and 60 units.
   d. Provide a single command to calculate the proportion of days for which sales are strictly between 40 units and 60 units.
3. Provide a single command, involving the `sqrt` function, to output a vector of all perfect squares that are less than 1,000, given in ascending order. The perfect squares, in ascending order, are $1^2, 2^2, 3^2, 4^2, 5^2, \ldots$ or $1, 4, 9, 16, 25, \ldots$
4. Answer the following questions that involve the creation and use of numerical vectors.
   a. Create a vector `mult_two` consisting of all multiples of 2 (i.e., even numbers) between 1 and 200 (inclusive).

b. Create a vector `mult_three` consisting of all multiples of 3 between 1 and 200 (inclusive).

c. Provide a single command to calculate how much longer `mult_two` is than `mult_three`.

d. Using `mult_two` and `mult_three`, create a new vector `mult_vec` consisting of all the numbers between 1 and 200 (inclusive) that are a multiple of 2, a multiple of 3, or both. The vector `mult_vec` should have the vector elements in increasing order and no duplicate elements (e.g., the number 6 should appear only once even though it is in both `mult_two` and `mult_three`). How many elements does `mult_vec` have?

5. Write a `rectangle_area` function that calculates the area of a rectangle based upon two arguments. The first argument is `base`, which is the length of the rectangle base. The second argument is `height`, which is the rectangle height. Make the second argument optional, with the default value specified by `height = base`, corresponding to a square. Confirm that `rectangle_area(3)` returns the area of a 3×3 square and `rectangle_area(4,5)` returns the area of a 4×5 rectangle.

6. Write an `even_product` function that takes a single numerical (integer) argument `x` and returns the product of the first `x` even integers. For example, `even_product(4)` should return the product of 2, 4, 6, and 8, which is `384`.

7. Refer to the *R* code (section 1.6.2) that calculates the first 10 numbers in the Fibonacci sequence.

   a. Modify the code to create a `fibonacci` function that takes `sequence_length` as an argument and returns the Fibonnaci sequence as a vector. Confirm that `fibonacci(10)` returns the correct sequence of 10 numbers.

   b. Does your `fibonacci` function work when called with `1` or `2` as its argument? If not, modify the code to handle these two cases.

   c. The *Tribonacci sequence* is similar to the Fibonacci sequence except that each element of the Tribonacci sequence is the sum of the previous *three* numbers in the sequence. The beginning of the Tribonacci sequence is $1, 1, 2, 4, 7, 13, 24, 44, \ldots$ Write a `tribonacci` function that takes a single positive-integer argument and returns the Tribonacci sequence of that length.

8. Use a `while` loop to determine the smallest positive integer n for which $\ln(n) + \ln(2n)$ is greater than 7.

9. Write a function called `longest_trend`, with a numerical vector `x` as its single argument, that returns the number associated with the longest subvector of strictly increasing values in `x`. For instance, for the vector `c(3,-1,1,0,-2)`, the function would return `2` since the sequence `(-1,1)` is the longest subvector of strictly increasing values. Similarly, for the vector `c(78,43,21,-5)`, the function would return `1` since the elements are decreasing; for the vector `c(-10,5,0,3,21,56,8)`, the function would return `4` since the sequence `(0,3,21,56)` is the longest subvector of strictly increasing values.

a. Write the `longest_trend` function using either a `for` loop or a `while` loop. Confirm that the function returns the correct values for the three example vectors specified in the question.

b. Modify the `longest_trend` function to include a second optional argument `decrease` with a default value of `FALSE`. When `decrease` is `FALSE`, the function should work as described previously. When `decrease` is `TRUE`, the function should return the number associated with the longest subvector of strictly *decreasing* values in `x`. In the three example vectors given in this question, the function should return `3`, `4`, and `2`, respectively, when `decrease=TRUE`.

10. Using the `sales` vector created in question 2, answer the following questions.

a. Create a vector called `sales_yesterday` that contains the daily sales that occurred yesterday, which should be `c(NA,38,52,24,61,47,18,29,44)`, where the first element is missing (`NA`) since there is no observation before the first day. Create this new vector by using the original `sales` vector (rather than the brute-force method of assigning the new vector to the list of values specified).

b. Use a single command that, ignoring the first day, calculates the number of days for which sales are strictly greater than the previous day's sales.

c. Using the `sales` and `sales_yesterday` vectors, calculate the average daily sales on days for which the previous day's sales were stricly less than 30 units.

11. Create a vector of 5,000 random numbers between 0 and 1 with the following two commands: `set.seed(1234)`, followed by `x <- runif(5000)`.

a. Use the `sum` function to calculate the sum of the 5,000 numbers.

b. Use the `mean` function to calculate the proportion of random numbers between 0.15 and 0.40.

c. Thinking about the vector as an ordered sequence of random numbers, determine the proportion of times that an element of the sequence is within 0.1 of the previous element of the sequence. The first element has no previous element, so the proportion should be calculated for the remaining 4,999 elements of the sequence.

12. Suppose that the 20-character string `returns`, defined next, indicates whether the stock price of a certain company goes up, indicated by `U`, or down, indicated by `D`, over the course of 20 days of trading on the stock market.

```
returns <- "UDUUUDDUDUDUUUUDUDUU"
```

a. Using a `for` loop to loop over the characters of `returns`, determine how many days the stock goes up.

b. Using another `for` loop, determine how many times a `D` (stock-price drop) is followed immediately by a `U` (stock-price increase).

c. Use the `grepl` command to determine whether there is a streak of four consecutive days of stock-price increases. Use the `grepl` command to determine whether there is a streak of four consecutive days of stock-price drops.

d. Write a `strtovec` function that takes a string called `string` as a single argument and returns a vector consisting of the single characters that comprise `string`. For example, `strtovec("abc")` should return the vector consisting of the elements `"a"`, `"b"`, and `"c"`.

e. Using the `strtovec` function from part (d), provide a single command to determine the number of days that the stock goes up (based upon `returns`).

13. Use the **exams** data set, read into the data frame `exams`, for this question.

a. Create a vector `scorediff` that contains the difference between the `exam2` score and the `exam1` score for each student. What is the average of `scorediff`? What is the maximum value of `scorediff`, and which student (i.e., which row number) has this maximum value? What is the minimum value of `scorediff`, and which student has this minimum value?

b. The professor would like to reward students who show improvement on the second exam. Specifically, she will place 70% weight on the second exam if a student's second exam score is at least 5% higher than the first exam score. Otherwise, she will just place 50% weight on both exams. Create a vector `composite_score` that calculates a composite score based on these grading guidelines. For a student who gets 50% weight on both exams, the composite score should be the sum of the two exam scores; for a student who gets 70% weight on the second exam, the composite score should be 0.6 times the first exam score plus 1.4 times the second exam score. What is the average of `composite_score`? Which student benefits the most from the 70% weighting rule?

14. Use the **cigdata** data set, read into the data frame `cigdata`, for this question. This data set contains information on cigarette taxes, prices, and sales in 2019 for each state (plus the District of Columbia) in the US.

a. Use the `nrow` function to confirm the number of observations.

b. The variables `cigprice` and `cigtax` are the average price of a pack of cigarettes and the tax per pack of cigarettes, respectively, for each state. Which states have the highest and lowest values for these two variables? (The states can be identified by either of the string variables `state` or `statename`.)

c. What is the average tax per pack according to the data?

d. Write a command that gives a vector of the five highest values of the tax per pack.

e. Create a variable `cigdata$statetax_pct` that is equal to the state-tax percentage, defined as the tax per pack divided by the average pack price for each state. What are the minimum, maximum, and average of this variable?

2

Introduction to Probability Theory

Probability theory underlies all of statistics, providing a mathematical framework for quantifying uncertainty and randomness in real-world phenomena. It allows practitioners to model and analyze the likelihood of different outcomes in a given situation. Key statistical concepts, like estimation, confidence intervals, and hypothesis tests, are all based on probability theory. Therefore, to understand statistics and apply statistical inference properly, it is important to have a solid foundation in the basic concepts of probability theory. This chapter provides an introduction to probability theory by discussing the meaning of probability and introducing the fundamental properties of probabilities. Chapter 3 builds on the material of this chapter and introduces the concepts of conditional probability and independence.

Before jumping into terminology and definitions, we first consider a motivating example.

Example 2.1 (Widget website) *The website Widgets.com has* 3,000 *total registered users and wants to test the effectiveness of two possible e-mail campaigns. E-mail A is sent to* 300 *users at random, e-mail B is sent to* 300 *users at random, and the other* 2,400 *users receive no e-mail. For each registered user, Widgets.com has the following information at some point (say, one week) after the e-mail campaigns:*

$$campaign = \begin{cases} A & \textit{if user receives e-mail A} \\ B & \textit{if user receives e-mail B} \\ None & \textit{if user receives no e-mail} \end{cases}$$

$$purchase = \begin{cases} Y & \textit{if user has made a purchase in the last week} \\ N & \textit{if user has not made a purchase in the last week} \end{cases}$$

Despite being a relatively simple example, there are many probability concepts associated with this experiment. First, what do we mean when we say that an e-mail is send to users "at random"? Intuitively, we mean that any user has the same chance of receiving the e-mail as any other user. One way to choose the users at random for the two e-mail campaigns is as follows:

- *E-mail A: The first recipient is chosen randomly from the* 3,000 *users, such that everyone has a* $1/3000$ *chance of being chosen. The second recipient is chosen randomly from the 2,999 remaining users, so that each remaining user has a 1/2999 chance of being chosen. This process continues through the 300th recipient, with each of the 2,701 remaining users having a 1/2701 chance of being chosen.*

- *E-mail B: The first recipient is chosen randomly from the 2,700 remaining users, such that each remaining user has a 1/2700 chance of being chosen. The second recipient is chosen randomly from the 2,699 remaining users, such that each remaining user has a 1/2699 chance of being chosen. This process continues through the 300th recipient, with each of the 2,401 remaining users having a 1/2401 chance of being chosen.*
- *The 2,400 remaining users who did not receive e-mail A or e-mail B are the users for which the value of campaign is None. We can think of this group of users as a "control group" to which we can compare the e-mail A recipients and the e-mail B recipients.*

This method of selecting the e-mail recipients is called ***sampling without replacement****. Starting with the full sample of 3,000 users, the 300 e-mail A recipients are chosen one by one. Once a user is randomly selected to receive e-mail A, that user cannot be randomly selected again. The user is not replaced back into the sample, from which the terminology "without replacement" comes. Instead, the user is removed from the sample that is used to randomly select the subsequent e-mail recipients.*

The `sample` *function in R can be used to implement the random selection of e-mail A and e-mail B recipients. Here is a description of the usage of* `sample`*:*

- `sample(x, size, replace = FALSE, prob = NULL)`: Returns a random sample of size `size` from the elements in vector `x`. The optional argument `replace` indicates whether sampling should be done without replacement (which is the default, `replace = FALSE`) or with replacement (which is `replace = TRUE`). The optional argument `prob` specifies how likely each element of `x` is to be sampled, with the default that each element is equally likely to be chosen.

If the 3,000 registered users are uniquely identified by a user number, ranging from `1` *to* `3000`*, the following code randomly selects the e-mail recipients:*

```
temp <- sample(1:3000,600)
emailA_recipient <- temp[1:300]
emailB_recipient <- temp[301:600]
```

The vector `1:3000` *represents the full set of users. The vector* `temp` *is created with the* `sample` *function by randomly picking* `600` *elements from* `1:3000`*, where each of the elements is equally likely to be chosen. Then the* `emailA_recipient` *vector is assigned to be the first 300 elements of* `temp`*, and the* `emailB_recipient` *vector is assigned to be the last 300 elements of* `temp`*. The nonrecipients are the elements of* `1:3000` *that are not in either* `emailA_recipient` *or* `emailB_recipient`*.*

Before implementation of this e-mail experiment, there are some known probabilities and some unknown probabilities:

- *For any given user, the chance or probability that she receives e-mail A is known. Since each user has the same chance of being chosen to receive e-mail A, this* known probability *is* $300/3000 = 1/10$ *or 10%. Similarly, the known probability of receiving e-mail*

B is 1/10 *or* 10%, *and the known probability of receiving no e-mail is* 2400/3000 = 4/5 *or* 80%.

- *For any user who receives e-mail A, the chance or probability that she makes a purchase is an* unknown probability. *The same is true of the probability that an e-mail B recipient makes a purchase and the probability that a user who received neither e-mail makes a purchase.*

Ultimately, the goal of the e-mail experiment by Widgets.com is to determine how effective the e-mail campaigns are. Specifically, the following questions are of interest:

- *Is campaign A more (or less) effective than campaign B? In terms of probability, is the probability of purchase by an e-mail A recipient higher (or lower) than the probability of purchase by an e-mail B recipient?*
- *Is campaign A more (or less) effective than no campaign?*
- *Is campaign B more (or less) effective than no campaign?*

Suppose that Widgets.com finds that 60 *e-mail A recipients (*20% *of the* 300*) made a purchase,* 66 *e-mail B recipients (*22% *of the* 300*) made a purchase, and* 360 *of the nonrecipients (*15% *of the* 2,400*) made a purchase. For these specific users, the results indicate that the e-mail B campaign is slightly more successful than the e-mail A campaign (*22% *versus* 20%*) in leading to purchases, and even more so when compared to the nonrecipients (*22% *versus* 15%*). But is there enough evidence to conclude that the e-mail B campaign is truly better than the e-mail A campaign (or no e-mail at all)? Should Widgets.com use e-mail B for its future campaigns instead of e-mail A? Perhaps the higher purchase rate (*22%*) for e-mail B recipients happened by chance, and it might not be the case that we would see the same result for a new batch of users receiving the two e-mail campaigns. As seen later in this book, the power of statistical inference is the ability to take outcomes such as these (i.e., the purchasing outcomes for the three groups) and provide an analysis of what might happen for new users who are presented with e-mail A, e-mail B, or neither.*

Before leaving this example, it is worth noting that Widgets.com might not be simply interested in whether a purchase is made, but also the revenue from such purchases. For example, the following information, in addition to the campaign and purchase described here, could be collected:

amount = *dollar amount of purchases by the user*
(equal to zero if no purchases) in the last week

Whereas purchase is a binary outcome, with the two possibilities being purchase (Y) or no purchase (N), amount can potentially take on many different values, depending upon the prices and quantities of the widgets available for purchase. Based upon the results of their e-mail campaign experiment, Widgets.com might like to determine which campaign is likely to be more successful with respect to revenues or profits for new users.

2.1 Experiments and Sample Spaces

The formal introduction to probability theory begins by defining an **experiment**.

Definition 2.1 *An **experiment** is any process whose outcome is subject to uncertainty.*

Some examples of experiments include the following:

- Tossing a coin once or several times
- Obtaining income information from one or more individuals
- Measuring the return on one or more financial assets during a given year

Example 2.1 consists of multiple experiments. The random assignment of 3,000 users to three groups (e-mail A, e-mail B, or neither) creates uncertainty for each user's group placement, and there is also uncertainty about each user's purchase behavior.

To specify the possible outcomes associated with any given experiment, we define the **sample space**.

Definition 2.2 *The **sample space**, denoted as $\mathcal{S}$, is the set of all possible outcomes for an experiment.*

The mathematical concept of a **set** is used in definition 2.2. To review, a set is a collection of distinct objects that is specified by listing its elements within curly braces { and }. The ordering of the elements within a set does not matter, and a set does not have duplicate elements.

The simplest sample space has two possible outcomes. If there is only one outcome, there is no uncertainty, and therefore it does not constitute an experiment. A classic example of a two-outcome sample space is a coin toss, where the possible outcomes are heads (denoted as H) and tails (denoted as T). The sample space for a coin toss is $\mathcal{S}=\{H,T\}$. Since the order of outcomes doesn't matter, $\mathcal{S}=\{T,H\}$ is equivalent to $\mathcal{S}=\{H,T\}$.

Here are some additional examples of two-outcome sample spaces:

- Asset return for the year: $\mathcal{S}=\{U,D\}$, where U denotes a positive return ("up") and D denotes a negative return ("down")
- Website visitor purchase behavior: $\mathcal{S}=\{Y,N\}$, where Y indicates purchase and N indicates no purchase
- Student exam result: $\mathcal{S}=\{P,F\}$, where P indicates pass and F indicates fail
- Worker's union status: $\mathcal{S}=\{U,NU\}$, where U indicates a union worker and NU indicates a nonunion worker

Of course, sample spaces can have more than two outcomes. Some examples include the following:

- Roll of a six-sided die: $\mathcal{S}=\{1,2,3,4,5,6\}$
- Day of the week of a baby's birth: $\mathcal{S}=\{Mon, Tue, Wed, Thu, Fri, Sat, Sun\}$
- Number of days during January for which a city's temperature is above freezing: $\mathcal{S}=\{0,1,2,\ldots,31\}$, where the "..." shorthand indicates that the sample space contains 32 outcomes, consisting of all integers between 0 and 31 (inclusive)

Often, an experiment involves multiple "trials" of the same underlying process:

- Tossing two coins: $\mathcal{S} = \{HH, HT, TH, TT\}$
- Purchase behavior of the first three visitors to a website on a given day:

$$\mathcal{S} = \{YYY, YYN, YNY, YNN, NYY, NYN, NNY, NNN\}$$

In these last two examples, we've implicitly assumed that the order of the outcomes matters, with the outcome *HT* (heads, then tails) different from the outcome *TH* (tails, then heads). The number of possible outcomes for the two coin tosses is four, which is the number of possible outcomes for the first toss (two) times the number of possible outcomes for the second toss (two). The number of possible outcomes for the purchase behavior of the three website visitors is eight, which is the number of possible outcomes for the first visitor (two), times the number of possible outcomes for the second visitor (two), times the number of possible outcomes for the third visitor (two). Using this same logic, the size of the sample space for an experiment involving the tossing of five coins is $2 \cdot 2 \cdot 2 \cdot 2 \cdot 2 = 2^5 = 32$, and the size of the sample space for an experiment involving the tossing of 10 coins is $2^{10} = 1{,}024$.

For the coin toss and website purchase examples, if we care only about the total number of heads or purchases, the sample spaces would be:

- Number of heads from two coin tosses: $\mathcal{S} = \{0, 1, 2\}$
- Number of purchases from the first three visitors to a website on a given day: $\mathcal{S} = \{0, 1, 2, 3\}$

Compared with the sample spaces for the sequences of coin tosses or website purchases, these sample spaces have fewer possible outcomes. Why? Here, the order of the outcomes doesn't matter since only the total number of heads or purchases matters. For the coin tosses, the outcome of 1 total head arises when the sequence *HT* or the sequence *TH* occurs. For the website purchases, the outcome of 1 total purchase arises when *YNN*, *NYN*, or *NNY* occurs.

In *R*, we can use a vector to represent a finite sample space. For example, the following code creates vectors for the sample spaces for a coin toss, a roll of a die, and a sequence of two coin tosses, in that order:

```
coin <- c("H","T")
coin

## [1] "H" "T"

die <- 1:6
die

## [1] 1 2 3 4 5 6

twocoins <- c("HH","HT","TH","TT")
twocoins

## [1] "HH" "HT" "TH" "TT"
```

If there is any concern that vector `vec` contains duplicate elements, `unique(vec)` returns a vector (or set) of the distinct elements. The `sort` function can be used to order the elements if desired.

Example 2.2 (Two car dealerships) *Suppose that you own two car dealerships: dealership A and dealership B, where dealership A has four salespeople and dealership B has three salespeople. Consider the experiment associated with the number of salespeople at dealership A and the number of salespeople at dealership B who sell a car on a given day. The following table enumerates all the possible outcomes associated with the sample space for this experiment:*

		Dealership B			
		0	1	2	3
	0	$(0,0)$	$(0,1)$	$(0,2)$	$(0,3)$
	1	$(1,0)$	$(1,1)$	$(1,2)$	$(1,3)$
Dealership A	2	$(2,0)$	$(2,1)$	$(2,2)$	$(2,3)$
	3	$(3,0)$	$(3,1)$	$(3,2)$	$(3,3)$
	4	$(4,0)$	$(4,1)$	$(4,2)$	$(4,3)$

For instance, the outcome $(2, 1)$ *occurs when two salespeople at dealership A and one salesperson at dealership B sell a car on a given day, and the outcome* $(0, 0)$ *occurs when no one sells a car on a given day. The sample space* $\mathcal{S}$ *is the set of the* 20 *possible outcomes in the table. The number of possible outcomes is equal to the five possible outcomes for dealership A times the four possible outcomes for dealership B.*

Although the examples considered so far have had a finite number of outcomes, definition 2.2 does not restrict the sample space to being a finite set. For example, the sample space corresponding to the number of patents that a company receives in a given year can be written as $\mathcal{S} = \{0, 1, 2, 3, \ldots\}$, the set of all nonnegative integers. It may seem strange to specify this sample space to be infinite, as it is seemingly impossible for a company to receive 1 million patents in a given year. But there's no obvious way to specify what the maximum number of patents should be (1,000? 10,000?), so it makes more sense to allow arbitrarily large values, even though the larger values are extremely unlikely. The sample space $\mathcal{S} = \{0, 1, 2, 3, \ldots\}$ is useful for many other situations where it is not obvious how we would arbitrarily set a maximum possible value, such as the number of children in a given family, the number of times that an individual is arrested, or the number of initial public offerings (IPOs) on the New York Stock Exchange in a given year.

Infinite sample spaces can arise in other ways. Thinking about website purchases again, if we are interested in the sequence of website visitors on a given day *until a purchase is made*, the sample space is

$$\mathcal{S} = \{Y, NY, NNY, NNNY, NNNNY, \ldots\}.$$

An equivalent approach is to consider the sample space corresponding to the number of visitors observed before a purchase is made, which is

$$\mathcal{S} = \{1, 2, 3, 4, 5, \ldots\}.$$

These two approaches are equivalent since there is a direct one-to-one relationship between the outcomes in the two sample spaces: Y corresponds to 1, NY corresponds to 2, NNY corresponds to 3, and so on.

An infinite sample space also naturally arises when the underlying experiment involves a quantity that can be a real number. Some examples include the following:

- The income of a given individual: $\mathcal{S} = [0, \infty)$, where $[0, \infty)$ denotes all nonnegative real numbers
  - The outcome 0 corresponds to an individual with no income. The specification of $\mathcal{S}$ here allows an arbitrarily large income value, as there's no way to set a reasonable maximum, as mentioned before.
- The fraction of income that a given employed individual saves in a given year: $\mathcal{S} = [0, 1]$, where $[0, 1]$ denotes all real numbers between 0 and 1 (inclusive)
  - The outcomes 0 and 1 correspond to the individual saving none of their income or all of their income, respectively. An outcome of 0.2 corresponds to the individual saving 20% of their income. Allowing for all real numbers between 0 and 1 is convenient here since it places no restrictions on what the savings (numerator) or the income (denominator) can be. Even in cases where the observed data are rounded (e.g., to two decimal places), it is still useful to specify the sample space in terms of real numbers for conceptual purposes.
- The return of an asset in a given year: $\mathcal{S} = [-1, \infty)$, where $[-1, \infty)$ denotes all real numbers greater than or equal to -1
  - If the asset price is p_0 at the beginning of the year and p_1 at the end of tthe year, the asset return is $\frac{p_1 - p_0}{p_0}$. Assuming that the asset price cannot be negative, the lowest possible asset return is -1, which occurs when $p_1 = 0$. There is no upper bound on the sample space $\mathcal{S}$ since p_1 can be arbitrarily large.

2.2 Events

As defined previously, the sample space is the set of the possible outcomes for a given experiment. To be able to consider the possible combinations of an experiment's outcomes, we define an **event** and some specific types of events.

Definition 2.3 *An **event** is any subset of outcomes within sample space $\mathcal{S}$. A **simple event** is exactly one outcome from $\mathcal{S}$, whereas a **composite event** consists of more than one outcome from $\mathcal{S}$.*

If the sample space $\mathcal{S}$ is finite, the number of simple events is the size of the sample space $\mathcal{S}$. The sample space $\mathcal{S}$ is itself an event since $\mathcal{S}$ is a subset of itself and, moreover, is a composite event since $\mathcal{S}$ has more than one outcome.

Definition 2.4 *For any event E with a finite number of outcomes, let $|E|$ denote the number of outcomes in event E.*

Using this notation, a simple event E has $|E| = 1$, and a composite event E has $|E| > 1$.

Example 2.3 (Three website visitors) *For the first three visitors to a website on a given day, the sample space for their purchase behavior is*

$$\mathcal{S} = \{YYY, YYN, YNY, YNN, NYY, NYN, NNY, NNN\}.$$

Each element of $\mathcal{S}$ is a simple event. For example, event $E = \{YYY\}$ is a subset of $\mathcal{S}$ corresponding to three purchases in a row. There are eight different simple events since $|\mathcal{S}| = 8$.

How about event A, where the first two visitors make a purchase? $A = \{YYY, YYN\}$ *is a composite event, with two possible outcomes from* $\mathcal{S}$. *How about event B, where two total purchases are made?* $B = \{YYN, YNY, NYY\}$ *is also a composite event, with three possible outcomes from* $\mathcal{S}$.

Example 2.4 (Two car dealerships) *In example 2.2, the sample space* $\mathcal{S}$ *has* 20 *outcomes, so* $|\mathcal{S}| = 20$. *There are* 20 *simple events associated with* $\mathcal{S}$. *One example is event* $E = \{(2,2)\}$, *where two salespeople at each dealership sell a car on a given day. How about the event, denoted as* E', *where the* total *number of salespeople who sell a car is equal to two? This composite event is* $E' = \{(0,2), (1,1), (2,0)\}$. *How about the event, denoted as* E'', *where the number of salespeople at dealership B who sell a car is greater than the number of salespeople at dealership A who sell a car? This composite event is* $E'' = \{(0,1), (0,2), (0,3), (1,2), (1,3), (2,3)\}$.

For any event, the result of the experiment is that either the event occurs or it does not occur. If the event does not occur, we say that the **complement** of the event occurs, as defined next.

Definition 2.5 *The* ***complement*** *of an event A, denoted as* A^c, *is the set of all outcomes in sample space* $\mathcal{S}$ *that are not in A.*

Example 2.5 (Three website visitors) *From example 2.3, event* $B = \{YYN, YNY, NYY\}$ *corresponds to two total purchases. Its complement is* $B^c = \{YYY, YNN, NYN, NNY, NNN\}$, *which contains all the outcomes for which there were* not *two total purchases made, or equivalently* B^c *contains all the outcomes for which either zero, one, or three total purchases were made. The size of the event* B^c *(5 outcomes) is the size of the sample space* $\mathcal{S}$ *(9 outcomes) minus the size of the event B (4 outcomes).*

Since any outcome in sample space $\mathcal{S}$ must be either in A or its complement A^c, we have proposition 2.1:

Proposition 2.1 *For a sample space* $\mathcal{S}$ *with a finite number of outcomes,* $|A| + |A^c| = |\mathcal{S}|$ *for any event A.*

Since many possible events may be associated with sample space $\mathcal{S}$, it is useful to have ways to think about multiple events at once. The following definitions introduce the concepts of the *union* of two events and the *intersection* of two events, which correspond to either of two events occurring (the union) and both events occurring (the intersection).

Definition 2.6 *The* ***union*** *of events A and B, denoted as* $A \cup B$, *is the set of all outcomes in event A, event B, or both. The union is sometimes read as "A or B."*

Example 2.6 (Three website visitors) *Recall that event* $A = \{YYY, YYN\}$ *corresponds to the first two visitors making a purchase and event* $B = \{YYN, YNY, NYY\}$ *to a total of two purchases being made. The event* $A \cup B = \{YYY, YYN, YNY, NYY\}$ *corresponds to the first two visitors making a purchase* or *a total of two purchases being made. The outcome YYN is in both event A and event B, but it appears only once in event* $A \cup B$ *since* $A \cup B$ *is a set.*

If A and B are events with a finite number of outcomes, $|A \cup B|$ must always be less than or equal to the sum of $|A|$ and $|B|$. In example 2.6, $|A| = 2$, $|B| = 3$, and $|A \cup B| = 4$. The only case in which $|A \cup B| = |A| + |B|$ is when A and B have no outcomes in common; in such a case, A and B are said to be **disjoint events**.

Definition 2.7 *Events A and B are **disjoint events** (or **mutually exclusive events**) if they have no outcomes in common; that is, there is no outcome in sample space $\mathcal{S}$ that is an element of both event A and event B.*

In example 2.6, A and B are not disjoint events since they share the outcome YYN. Since disjoint events have no outcomes in common, they can be thought of as events that cannot possibly occur at the same time. In the example of tossing two coins, where $\mathcal{S} = \{HH, HT, TH, TT\}$, event A, where an equal number of heads and tails appear ($A = \{HT, TH\}$), and event B, where two heads appear ($B = \{HH\}$), are disjoint events since they have no outcomes in common and cannot possibly occur at the same time.

Proposition 2.2 *If A and B are events with a finite number of outcomes, $|A \cup B| \le |A| + |B|$. Moreover, $|A \cup B| = |A| + |B|$ if A and B are disjoint events, and $|A \cup B| < |A| + |B|$ if A and B are not disjoint events.*

We now move to the **intersection** of two events.

Definition 2.8 *The **intersection** of events A and B, denoted as $A \cap B$, is the set of all outcomes in both event A and event B. The intersection is sometimes read as "A and B."*

Example 2.7 (Three website visitors) *Recall that event $A = \{YYY, YYN\}$ corresponds to the first two visitors making a purchase and event $B = \{YYN, YNY, NYY\}$ to a total of two purchases being made. The event $A \cap B$ corresponds to the first two visitors making a purchase* and *a total of two purchases being made. In this case, $A \cap B = \{YYN\}$ since YYN is the only outcome in both events.*

If A and B are events with a finite number of outcomes, $|A \cap B|$ cannot possibly be larger than either $|A|$ or $|B|$.

Proposition 2.3 *If A and B are events with a finite number of outcomes, $|A \cap B| \le |A|$ and $|A \cap B| \le |B|$. The only case in which $|A \cap B| = |A|$ is when event A is a subset of event B (i.e., all outcomes in A are also outcomes in B). Similarly, the only case in which $|A \cap B| = |B|$ is when event B is a subset of event A.*

Again focusing on events with a finite number of outcomes, there is an interesting relationship between the size of union $A \cup B$ and the size of intersection $A \cap B$. We know that all the outcomes in event A are in $A \cup B$, and all the outcomes in event B are in $A \cup B$. However, the size of $A \cup B$ is not necessarily equal to $|A| + |B|$ since proposition 2.2 states that $|A \cup B| < |A| + |B|$ when A and B are not disjoint (i.e., they have outcomes in common). Note that $A \cup B$ contains three types of outcomes: (1) outcomes in both A and B, of which there are $|A \cap B|$; (2) outcomes in A but not B, of which there are $|A| - |A \cap B|$; and (3) outcomes in B but not A, of which there are $|B| - |A \cap B|$. Therefore, the number of outcomes in $A \cup B$ is $|A \cap B| + |A| - |A \cap B| + |B| - |A \cap B|$, or $|A| + |B| - |A \cap B|$. In this last expression, the subtraction of $|A \cap B|$ effectively eliminates the double-counting in $|A|+|B|$ for the outcomes that appear in both A and B. This result is formally stated in proposition 2.4.

Proposition 2.4 *If A and B are events with a finite number of outcomes,*

$$|A \cup B| = |A| + |B| - |A \cap B|,$$

or, equivalently,

$$|A \cap B| = |A| + |B| - |A \cup B|.$$

Previously, proposition 2.2 stated that $|A \cup B| = |A| + |B|$ only in the case that A and B are disjoint. When A and B are disjoint, there are no outcomes in $A \cap B$, so $|A \cap B| = 0$, which agrees with the result in proposition 2.4.

An event that has no outcomes is called the **null event** and is formally defined as follows.

Definition 2.9 *The **null event** consists of no outcomes and is denoted as $\varnothing$. The size of the null event is $|\varnothing| = 0$.*

Disjoint events A and B have $A \cap B = \varnothing$.

Several *R* functions facilitate the use and manipulation of (finite) events or sets, including the following:

- `elt %in% x`: Returns `TRUE` if the element `elt` is in set `x` and `FALSE` otherwise.
- `union(x, y)`: Returns the union of sets `x` and `y`.
- `intersect(x, y)`: Returns the intersection of sets `x` and `y`.
- `setdiff(x, y)`: Returns the elements of set `x` that are not in set `y`. If `x` is the sample space, then the function returns the complement of `y`.
- `setequal(x, y)`: Returns `TRUE` if sets `x` and `y` are the same.

Also, since a vector represents an event or set in *R*, the number of outcomes in the event is the size of the vector and can be determined with the `length` function. The following code illustrates the use of these functions for the example 2.7, where S is the sample space for the purchase behavior of three customers, A is the event that the first two customers makes a purchase, and B is the event that a total of two customers make a purchase:

```
S <- c("YYY","YYN","YNY","YNN","NYY","NYN","NNY","NNN")
A <- c("YYY","YYN")
B <- c("YYN","YNY","NYY")
length(S)

## [1] 8

"YYY" %in% S

## [1] TRUE

union(A,B)

## [1] "YYY" "YYN" "YNY" "NYY"

intersect(A,B)

## [1] "YYN"
```

```
setdiff(S,B)

## [1] "YYY" "YNN" "NYN" "NNY" "NNN"

no_purchases <- c("NNN")
intersect(no_purchases,A)

## character(0)

length(intersect(no_purchases,A))

## [1] 0
```

The total number of outcomes, returned by `length(S)`, is `8`. The `union` and `intersect` functions return $A \cup B$ and $A \cap B$, respectively. The `setdiff(S,B)` expression returns the elements of `S` not in `B`, which is the complement B^c. The last two expressions consider the intersection of the event of no purchases, $\{NNN\}$, with event A. The intersection $\{NNN\} \cap A = \varnothing$ is indicated by the output of `character(0)` from the `intersect(no_purchases,A)` expression. The size of $\{NNN\} \cap A = \varnothing$ is zero, as indicated by the `length` function.

For another example of the union and the intersection of events, we return to the car dealership example.

Example 2.8 (Two car dealerships) *Recall from example 2.4 that* $E' = \{(0,2),(1,1),(2,0)\}$ *is the event that the* total *number of salespeople selling a car is* 2, *and* $E'' = \{(0,1),(0,2),(0,3),(1,2),(1,3),(2,3)\}$ *is the event that the number of salespeople at dealership B selling a car is greater than the number of salespeople at dealership A selling a car. The events* E' *and* E'' *are not disjoint since they share the outcome* $(0,2)$. *The intersection* $E' \cap E''$ *is the event containing this one outcome,* $E' \cap E'' = \{(0,2)\}$. *The union* $E' \cup E''$, *which contains all outcomes for which the total number of salespeople selling a car is* 2 or *for which more dealership B salespeople sell a car than dealership A salespeople, is*

$$E' \cup E'' = \{(0,1),(0,2),(0,3),(1,1),(1,2),(1,3),(2,0),(2,3)\}.$$

We have $|E'| = 3$, $|E''| = 6$, $|E' \cap E''| = 1$, *and* $|E' \cup E''| = 3 + 6 - 1 = 8$, *where the* -1 *eliminates the double-counting of the* $(0,2)$ *outcome in the union.*

The notion of unions and intersections can be extended to more than two events, as follows.

Definition 2.10 *Suppose there are k events, denoted as* $A_1, A_2, \ldots, A_k$. *The union of these k events, denoted as* $A_1 \cup A_2 \cup \cdots \cup A_k$, *is the set of all outcomes that are in* any *of the events* $A_1, A_2, \ldots, A_k$.

Definition 2.11 *Suppose there are k events, denoted as* $A_1, A_2, \ldots, A_k$. *The intersection of these k events, denoted as* $A_1 \cap A_2 \cap \cdots \cap A_k$, *is the set of all outcomes that are in* all *of the events* $A_1, A_2, \ldots, A_k$.

2.3 What Is a Probability?

Having formalized sample spaces and events associated with those sample spaces, we now introduce the concept of a probability. First, we introduce notation for the *probability of an event*, letting $P(A)$ denote the probability of event A occurring. Although we haven't formally defined probability $P(A)$, we can think of it as the chance or likelihood of event A occurring.

Example 2.9 (Coin toss) *The sample space is* $S = \{H, T\}$, *where* H *is heads and* T *is tails. When a fair coin is tossed, everyone can agree that the likelihood that the coin lands on heads is the same as the likelihood that the coin lands on tails. In fact, saying that a coin is "fair" means by definition that there is an equal chance of heads and tails (and no chance of landing on its edge!). If* $A = \{H\}$ *is the event associated with heads,* $P(A) = 50\%$ *or* $P(A) = 0.5$. *Similarly, if* $B = \{T\}$ *is the event associated with tails,* $P(B) = 50\%$ *or* $P(B) = 0.5$.

When we say that the probability of tossing a head is 50% or 0.5, what does that mean? If heads occurs 50% of the time, a series of repeated coin tosses should have roughly half the tosses resulting in heads and half resulting in tails. But how many tosses are necessary to get close to 50%? With a small number of tosses, the inherent randomness makes it so that we are not necessarily going to see very close to 50% heads. We might, depending on chance, but we also might not. Consider just two tosses. You can convince yourself that the chances of seeing *HH*, *HT*, *TH*, and *TT* are equally likely, each with a 25% chance, meaning that there is a 50% chance of having exactly half heads (one head, one tail) but also a 50% of being "far away" from exactly half heads (*HH* is 100% heads, *TT* is 0% heads). The situation, however, changes when the number of coin tosses increases. Thinking about 500 coin tosses, it's not necessarily the case that there will be an exactly even split of 250 heads and 250 tails, but it becomes more likely that there will be *approximately* 50% heads as compared to a situation with far fewer coin tosses.

To illustrate what happens as the number of coin tosses increases, we use *R* to simulate the coin tosses. The following code uses the `sample` function to simulate 10 coin tosses:

```
sample(c("H","T"), 10, replace = TRUE)

##  [1] "H" "H" "H" "T" "T" "H" "H" "H" "T" "H"
```

The sample space $\{H, T\}$, represented by the vector `c("H","T")`, is the first argument for the `sample` function. The second argument (`10`) indicates the number of coin tosses to be simulated. The argument `replace = TRUE` specifies **sampling with replacement**; that is, for each coin toss, it is always possible to get heads or tails regardless of what has occurred previously. Finally, there is no need to specify the optional `prob` argument since the two outcomes are equally likely. If we had an "unfair" coin that was more likely to come up heads (say with a 60% chance), then the command `sample(c("H","T"), 10, replace = TRUE, prob = c(0.6,0.4))` would be used, with the probabilities `0.6` and `0.4` associated with `"H"` and `"T"`, respectively.

Using the `sample` function, we can conduct many random simulations and summarize the results with very few lines of code, sometimes even just one line of code. For example, to simulate 10,000 coin tosses, we change the second argument of the `sample` function. And to check how often heads occurs, we use the `mean` function to calculate the proportion of the vector elements equal to `H`:

```
set.seed(1234)
cointosses <- sample(c("H","T"), 10000, replace = TRUE)
mean(cointosses=="H")

## [1] 0.4963
```

The `set.seed(1234)` command sets the **random seed** in the computer. The argument `1234` has been arbitrarily chosen, but the purpose of the `set.seed` function is to ensure that we get the same simulation results if we run it at a different time or on a different computer. Setting a random seed is a good practice, as it makes it easier for you and others to replicate results from simulations involving randomness. Throughout this book, the `set.seed(1234)` command is used before any simulations involving randomness. For each of the 10,000 tosses, the computer randomly chooses between heads (`H`) and tails (`T`). The vector `cointosses` is assigned to the full vector of 10,000 coin tosses, and the `mean(cointosses=="H")` command calculates the proportion of heads over the 10,000 tosses. The result is a proportion of 49.63%, or 4,963 heads out of the 10,000 tosses.

To see how the calculated frequency of heads changes as the number of simulations increases, figure 2.1 plots calculated frequencies against the simulation number. After each of the 10,000 tosses, we calculate the fraction of heads that have occurred so far. The x-axis has the number of the coin toss, and the y-axis has the fraction of heads after each toss. A dotted line is drawn at the value of 0.5. When very few tosses have occurred, which corresponds to the leftmost part of the x-axis, the fraction of heads can be quite far from 0.5, with the points appearing to be more "noisy." As the number of tosses increases, moving to the right along the x-axis, the points get less "noisy," with the fraction of heads stabilizing around 0.5. The complete *R* code is as follows:

```
set.seed(1234)

# toss a coin 10,000 times
cointosses <- sample(c("H","T"), 10000, replace = TRUE)

# show the results of the first 10 tosses
cointosses[1:10]

##  [1] "T" "T" "T" "T" "H" "T" "H" "H" "H" "T"

# calculate the cumulative sum of heads
# (e.g. cumul_heads[100] is number of heads through first 100 tosses)
cumul_heads <- cumsum(cointosses=="H")

# calculate the cumulative frequency of heads
# (e.g. freqheads[100] is the frequency of heads through first 100 tosses)
freq_heads <- cumul_heads/(1:10000)

# plot the freqheads vector (when plot has only a single variable (vector) argument,
# it plots the variable versus 1 through the length of the vector)
plot(freq_heads,xlab="Simulation number",
     ylab="Cumulative frequency of heads",cex=0.5)
abline(h=0.5,lty=3)
```

The function `cumsum` calculates the cumulative number of heads through any number of tosses, which is stored in the vector `cumul_heads`. Figure 2.1 is created by the `plot`

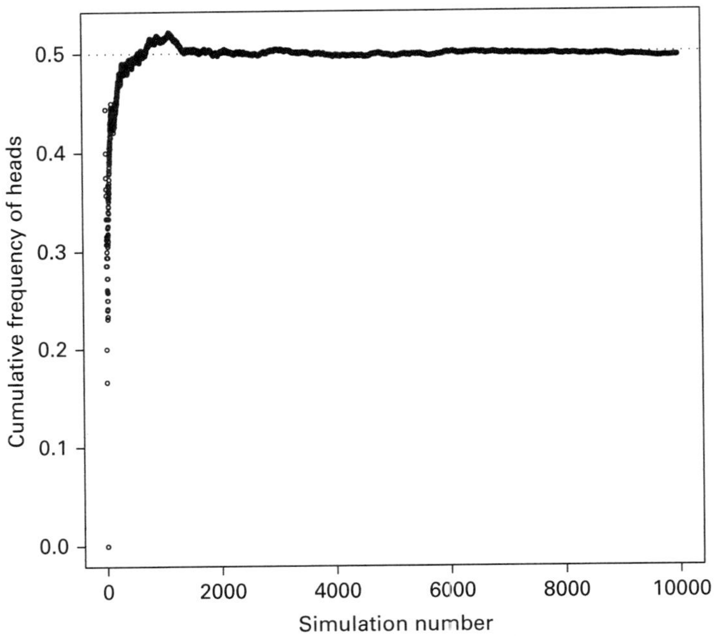

Figure 2.1
Head frequency for 10,000 simulated coin tosses

and `abline` functions, with the `plot` command plotting the vector `freq_heads` against the number of the simulation and the `abline` command drawing a horizontal line at the value 0.5 (`h=0.5` argument) that is dotted (`lty=3` argument). To highlight the dotted line, we could also change its color; for example, adding `col="blue"` as an argument to the `abline` function would make it blue.

How many coin tosses are required for the fraction of heads to be close to 0.5 and stabilize near that value? While we don't have the tools to answer that question yet, the takeaway from figure 2.1 is that a higher number of tosses has a realized fraction of heads that is more likely to be closer to the probability $P(A) = 0.5$. Additional examples of computer-simulated experiments are provided here, but at this point, we can provide a more formal description of what is meant by $P(A)$, the probability of an event A associated with an experiment and sample space.

Imagine being able to repeat an experiment a *large* number of times, such as n times. Think really large—a million, a billion, a trillion, or even more! For each experiment, record whether event A has occurred. Let n_A be the total number of times that event A occurs over the n experiments. The fraction, or frequency, of experiments in which event A occurs is equal to $\frac{n_A}{n}$. Then $P(A)$ is the number that $\frac{n_A}{n}$ approaches as n gets arbitrarily large. [We are implicitly assuming that there is a number (a "limit") that $\frac{n_A}{n}$ approaches as n gets arbitrarily large. This idea, known as the Law of Large Numbers (LLN), is discussed in chapter 13.]

Thinking about probability in terms of a large number of repeated experiments is known as the **frequentist interpretation** of probability. The word "frequentist" is used since the probability of event A is viewed as the long-run frequency of A occurring in a large number of repeated experiments.

Before discussing the properties of probabilities, a few more examples are considered to illustrate the frequentist interpretation of probability.

Example 2.10 (Tossing two coins) *Consider the experiment of tossing two fair coins. The sample space $S = \{HH, HT, TH, TT\}$ has four possible outcomes. If $A = \{HH\}$ is the event corresponding to two heads, what is $P(A)$? For fair coins, it's perhaps not surprising that each of the four outcomes in S is going to be equally likely, so we would expect $P(A) = 0.25$. (In chapter 3, we formalize why this probability is equal to 0.25.) Figure 2.2 shows the results from 10,000 simulations, with the computer randomly tossing two coins for each simulation. The* x*-axis is the number of the simulation, and the* y*-axis is the calculated frequency of HH occurring through that number of simulations. Similar to the simple coin toss example, the frequencies appear to stabilize when the number of experiments gets larger, but here, the stabilization occurs at a level close to the $P(A) = 0.25$ value, indicated by the horizontal dotted line.*

```
set.seed(1234)

# toss two coins 10,000 times
coin1tosses <- sample(c("H","T"), 10000, replace=TRUE)
coin2tosses <- sample(c("H","T"), 10000, replace=TRUE)

# calculate the cumulative number of "two heads" occurrences
cumul_twoheads <- cumsum(coin1tosses=="H" & coin2tosses=="H")

# calculate the cumulative frequency of "two heads" occurrences
freq_twoheads <- cumul_twoheads/(1:10000)

# plot the freq_twoheads vector
plot(freq_twoheads,xlab="Simulation number",
     ylab="Cumulative frequency of two-heads",cex=0.5)
abline(h=0.25,lty=3)
```

Each of the vectors `coin1tosses` *and* `coin2tosses` *contains the outcomes of 10,000 coin tosses. The occurrence of two heads happens when the corresponding elements of these two vectors are both* `"H"`*. For example, the fifth simulation of two coin tosses results in two heads when both* `coin1tosses[5]` *and* `coin2tosses[5]` *are* `"H"`*. The vector* `cumul_twoheads` *contains the cumulative number of times that the two-heads event has occurred.*

If we are only interested in approximating the probability of getting two heads in two tosses, rather than tracking the cumulative frequencies, the following code suffices:

```
set.seed(1234)
coin1tosses <- sample(c("H","T"), 10000, replace=TRUE)
coin2tosses <- sample(c("H","T"), 10000, replace=TRUE)
mean(coin1tosses=="H" & coin2tosses=="H")

## [1] 0.2513
```

Example 2.11 (Six-sided die) *Consider the experiment of rolling a fair, six-sided die. The sample space $S = \{1, 2, 3, 4, 5, 6\}$ has six possible outcomes. Let $A = \{6\}$ be the simple event that a 6 is rolled. The die being fair means that each of the six outcomes is equally likely, so $P(A) = 1/6$. Figure 2.3 shows the results from 10,000 simulations of this experiment, with the*

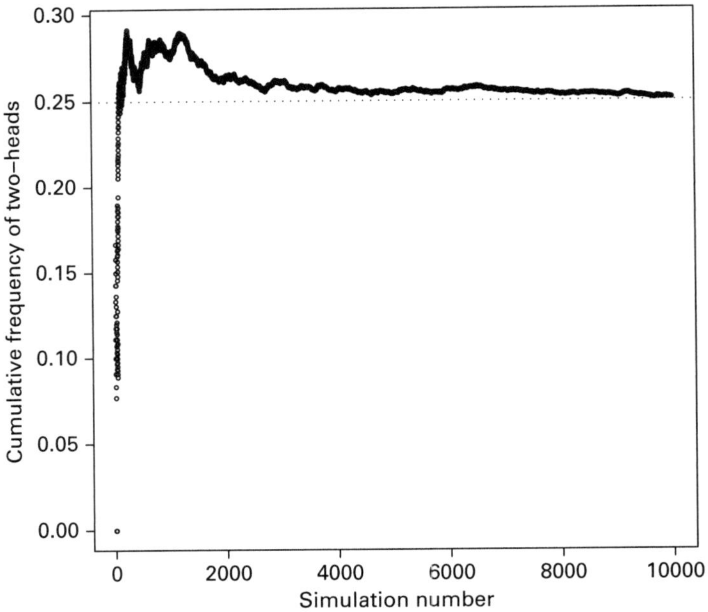

Figure 2.2
Double-head frequency for 10,000 simulations of two coin tosses

computer randomly rolling a six-sided die for each of the 10,000 *simulations. The frequency of a* 6 *being rolled stabilizes as the number of experiments gets larger, and it appears to stabilize at around* 1/6, *which is the level of the horizontal dotted line.*

```
set.seed(1234)

# roll six-sided die 10,000 times
dierolls <- sample(1:6, 10000, replace = TRUE)

# show the results of the first 10 rolls
dierolls[1:10]

##  [1] 4 2 6 5 4 1 5 6 4 2

# calculate the cumulative number of "six" rolls
cumul_six <- cumsum(dierolls==6)

# calculate the cumulative frequency of "six" rolls
freq_six <- cumul_six/(1:10000)

# plot the freq_six vector
plot(freq_six,xlab="Simulation number",ylab="Cumulative frequency of sixes",cex=0.5)
abline(h=1/6,lty=3)
```

The sample space is represented by the vector `1:6`, *consisting of the integers between* 1 *and* 6 *(inclusive), and this argument is the only difference from the coin-toss example with sample space* `c("H","T")`. *The optional* `prob` *argument is not specified since each outcome is equally likely.*

If we are only interested in approximating the probability of rolling a 6, *the following code suffices:*

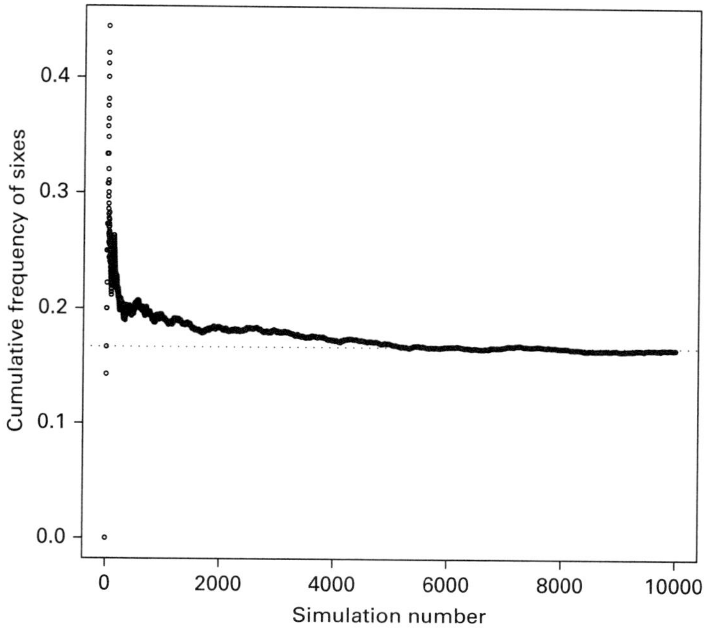

Figure 2.3
The frequency of sixes for 10,000 simulations of a die roll

```
set.seed(1234)
dierolls <- sample(1:6, 10000, replace = TRUE)
mean(dierolls==6)

## [1] 0.1652
```

2.4 Properties of Probabilities

To discuss the properties of probabilities, we first state the three **Axioms of Probability**:

Axiom 1 For any event A, $0 \leq P(A) \leq 1$.

Axiom 2 $P(\mathcal{S}) = 1$.

Axiom 3 If $A_1, A_2, \ldots, A_k$ is a finite collection of disjoint events,

$$P(A_1 \cup A_2 \cup \cdots \cup A_k) = \sum_{i=1}^{k} P(A_i).$$

Axiom 1 says that an event cannot have a negative probability or a probability greater than 1. The extremes of $P(A) = 0$ (no chance that A occurs) and $P(A) = 1$ (A occurs with 100% certainty) are permitted by this axiom.

Axiom 2 says that the probability associated with sample space $\mathcal{S}$ equals 1. Since $\mathcal{S}$ contains all possible outcomes of the experiment, there is a 100% chance that one of them will occur.

Axiom 3 is the "additivity" axiom.[1] In this axiom, the events $A_1, A_2, \ldots, A_k$ are specified to be a "finite collection of disjoint events." Recall that definition 2.7 introduced the notion of disjoint events in the context of two events, which can be generalized as in definition 2.12.

Definition 2.12 *$A_1, A_2, \ldots, A_k$ are a* ***collection of disjoint events*** *if there is no pair of events within the collection for which there is a shared outcome. Mathematically, $A_1, A_2, \ldots, A_k$ are a collection of disjoint events if $A_i \cap A_j = \varnothing$ for any $i, j \in \{1, \ldots, k\}$ with $i \neq j$.*

Intuitively, $A_1, A_2, \ldots, A_k$ being disjoint means that any given event, such as A_j, cannot possibly happen if any of the other events happens. Axiom 3 states that the probability that *any* of the disjoint events occurs is equal to the sum of the probabilities of each of the individual events. A simple case of disjoint $A_1, A_2, \ldots, A_k$ arises when each A_j contains a single outcome (i.e., A_j is a simple event) that is different from the other (simple) events.

Definition 2.13 *$A_1, A_2, \ldots, A_k$ are a* ***collection of exhaustive events*** *if $A_1 \cup A_2 \cup \cdots \cup A_k = \mathcal{S}$ or, equivalently, every outcome in the sample space $\mathcal{S}$ is in at least one of the events $A_1, A_2, \ldots, A_k$.*

For exhaustive events $A_1, A_2, \ldots, A_k$, the union of the events "exhaust" the outcomes in the sample space, and applying axiom 2, it follows that $P(A_1 \cup A_2 \cup \cdots \cup A_k) = 1$.

Example 2.12 (Three website visitors) *Consider the following events that correspond to the total number of purchases made by the first three website visitors on a given day:*

Event that no purchases are made: $A_0 = \{NNN\}$
Event that one purchase is made: $A_1 = \{YNN, NYN, NNY\}$
Event that two purchases are made: $A_2 = \{YYN, YNY, NYY\}$
Event that three purchases are made: $A_3 = \{YYY\}$

These events are disjoint since there is no outcome that is an element of two different events. The events are also collectively exhaustive, with $A_0 \cup A_1 \cup A_2 \cup A_3 = \mathcal{S}$ and $P(A_0 \cup A_1 \cup A_2 \cup A_3) = 1$. How about event B, where at least *one purchase is made? This event is $B = A_1 \cup A_2 \cup A_3$. Applying axiom 3, the probability of* at least *one purchase being made is $P(B) = P(A_1 \cup A_2 \cup A_3) = P(A_1) + P(A_2) + P(A_3)$, which is the probability of exactly one purchase being made plus the probability of exactly two purchases being made plus the probability of exactly three purchases being made. At this point, we can't say anything more about the* value *of this probability since the purchase probability is unknown.*

The probability axioms lead to several other interesting properties of probabilities, some of which are stated in proposition 2.5.

Proposition 2.5 (Properties of probabilities) *Let A and B be any two events. The following properties are implied by the Axioms of Probability:*

i. *Probability of a complement: $P(A^c) = 1 - P(A)$.*
ii. *Probability of the null event: $P(\varnothing) = 0$.*
iii. *Partitioning an event: $P(A) = P(A \cap B) + P(A \cap B^c)$.*
iv. *Probability of the union of events: $P(A \cup B) = P(A) + P(B) - P(A \cap B)$.*
v. *If A and B are disjoint events, $P(A \cap B) = 0$ and $P(A \cup B) = P(A) + P(B)$.*

vi. *De Morgan's Laws for probabilities:* $P((A \cup B)^c) = P(A^c \cap B^c)$ *and* $P((A \cap B)^c) = P(A^c \cup B^c)$.

vii. *Equally likely outcomes: Suppose that* $\mathcal{S}$ *is a finite sample space with k possible outcomes. If every outcome in* $\mathcal{S}$ *is equally likely to occur,*

$$P(A_1) = P(A_2) = \cdots = P(A_k) = 1/k$$

for each of the simple events $A_1, A_2, \ldots, A_k$ *in* $\mathcal{S}$.

Property (i) follows from the facts that A^c and A are disjoint, $A^c \cup A = \mathcal{S}$, and $P(\mathcal{S}) = 1$, which, taken together, imply $P(A^c) + P(A) = 1$ or, equivalently, $P(A^c) = 1 - P(A)$.

Example 2.13 (Three website visitors) *In example 2.12, we considered the probability for event B that* at least *one purchase is made. An alternative approach to finding that probability is to consider* B^c*, which is the event* A_0 *such that no purchases are made,* $B^c = A_0 = \{NNN\}$. *Applying property (i), we have* $P(B) = 1 - P(A_0)$. *Since we know that* $P(A_0 \cup A_1 \cup A_2 \cup A_3) = P(A_0) + P(A_1) + P(A_2) + P(A_3) = 1$, *note that* $P(B) = 1 - P(A_0)$ *is equivalent to* $P(B) = P(A_1) + P(A_2) + P(A_3)$.

Property (ii) says that the probability of the null event is equal to zero, which intuitively makes sense since it is impossible for the experiment to have *no* outcome. This property follows from $\varnothing = \mathcal{S}^c$, to which we apply property (i): $P(\varnothing) = 1 - P(\mathcal{S}) = 1 - 1 = 0$.

Property (iii) involves partitioning event A into two disjoint events: one event with outcomes also in event B, which is $A \cap B$, and one event with outcomes not in event B, which is $A \cap B^c$. The two events $A \cap B$ and $A \cap B^c$ are disjoint since no outcome can be in both B and B^c. The union of $A \cap B$ and $A \cap B^c$ is event A since every outcome in A is either in $A \cap B$ or $A \cap B^c$. Applying axiom 3, then, implies that $P(A) = P(A \cap B) + P(A \cap B^c)$.

Example 2.14 (Six-sided die) *Let* $A = \{3, 4, 5, 6\}$ *be the event of rolling at least a* 3. *Let* $B = \{2, 4, 6\}$ *be the event of rolling an even number. A can be partitioned into its even numbers,* $A \cap B = \{4, 6\}$, *and its odd numbers,* $A \cap B^c = \{3, 5\}$.

Property (iv), which states $P(A \cup B) = P(A) + P(B) - P(A \cap B)$ for events A and B, provides the relationship between the probability of a union of events and the probability of an intersection of events. We already know from axiom 3 that for *disjoint* events A and B, we have $P(A \cup B) = P(A) + P(B)$. But property (iv) is more general, as it covers events that are disjoint, when $P(A \cap B) = \varnothing$, and events that are not disjoint, when $P(A \cap B) \neq \varnothing$.[2] As we saw before in the context of counting the number of outcomes in $A \cup B$ (recall proposition 2.4, where $|A \cup B| = |A| + |B| - |A \cap B|$), subtracting probability $P(A \cap B)$ avoids the double-counting that occurs from the $P(A)$ and $P(B)$ terms for outcomes that are in both A and B.

One way to visualize property (iv) is through the use of a Venn diagram, as shown in figure 2.4. This figure depicts two scenarios: one in which intersection $A \cap B$ contains one or more outcomes (the Venn diagram on the left) and one in which intersection $A \cap B$ contains no outcomes (the Venn diagram on the right, having $A \cap B = \varnothing$). For both Venn diagrams, the light-gray circle corresponds to event A and the dark-gray circle corresponds to event B. In the left Venn diagram, the intersection $A \cap B$ is the region where the two circles overlap. If the area of each circle is equal to the probability of the corresponding event, with the area

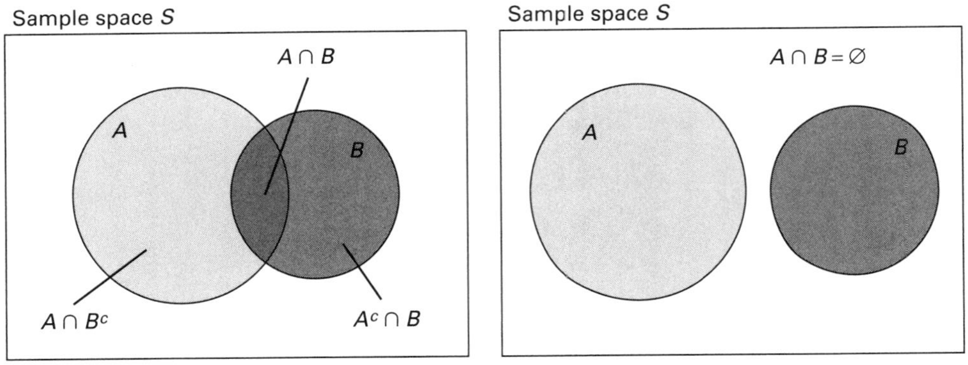

Figure 2.4
Venn diagrams and $P(A \cap B)$

of the overlap $A \cap B$ being equal to $P(A \cap B)$, property (iv) follows intuitively. The quantity $P(A) + P(B)$ is the sum of the areas of the two circles, but this sum double-counts region $A \cap B$. Therefore, to get $P(A \cup B)$, which is the area of the the full shaded region that counts the region $A \cap B$ only once, we must subtract off $P(A \cap B)$. This leads to $P(A \cup B) = P(A) + P(B) - P(A \cap B)$. In the right Venn diagram, events A and B have no outcomes in common, so $A \cap B = \varnothing$ and, by property (v), $P(A \cap B) = 0$. In this case, property (iv) simplifies to $P(A \cup B) = P(A) + P(B)$ since there is no issue with double-counting.

For the Venn diagram on the left, event A is partitioned into two parts, $A \cap B$ (events in A that are also in B) and $A \cap B^c$ (events in A that are not in B). This partitioning corresponds to property (iii), $P(A) = P(A \cap B) + P(A \cap B^c)$. Similarly, the event B is partitioned into two parts, $A \cap B$ (events in B that are also in A) and $A^c \cap B$ (events in B that are not in A), so that $P(B) = P(A \cap B) + P(A^c \cap B)$.

Example 2.15 (Six-sided die) *Let* $A = \{3, 4, 5, 6\}$ *be the event of rolling at least a* 3. *Let* $B = \{2, 4, 6\}$ *be the event of rolling an even number. Then,* $A \cup B = \{2, 3, 4, 5, 6\}$ *and* $A \cap B = \{4, 6\}$. *Note that* $P(A) + P(B)$ *double-counts the probability of rolling a* 4 *or rolling a* 6, *and the subtraction of* $P(A \cap B)$ *corrects this double-counting in the formula* $P(A \cup B) = P(A) + P(B) - P(A \cap B)$. *With respect to the left Venn diagram in figure 2.4, the light-gray circle represents* $\{3, 4, 5, 6\}$, *the dark-gray circle represents* $\{2, 4, 6\}$, *and the overlap represents* $\{4, 6\}$.

Property (v) states that it is impossible for two disjoint events to occur at the same time, which follows directly from property (ii) since $A \cap B = \varnothing$ for disjoint A and B.

Property (vi) states the probability properties associated with famous set-theory properties known as *De Morgan's Laws*. The two properties for sets are that

$$(A \cup B)^c = A^c \cap B^c \text{ and } (A \cap B)^c = A^c \cup B^c.$$

The first property is that the complement of a union of two events is the intersection of the complement of the two events, which holds since any outcome that is not in either A or B must be in both A^c and B^c (and vice versa). Both $(A \cup B)^c$ and $A^c \cap B^c$ consist of outcomes that belong to neither A nor B. Similarly, the second property is that the complement of the intersection of two events is the union of the complement of the two events, which holds since any outcome that is not in both A and B must be in either A^c or B^c (and vice versa). Both $(A \cap B)^c$ and $A^c \cup B^c$ consist of outcomes that are not in both A and B. Property (vi) follows immediately once we know $(A \cup B)^c = A^c \cap B^c$ and $(A \cap B)^c = A^c \cup B^c$.

Property (vii) provides the basis for calculating event probabilities when the outcomes of an experiment are equally likely, as with the toss of a fair coin or the roll of a fair six-sided die. The result itself is a direct implication of axioms 2 and 3. The simple events $A_1, \ldots, A_k$ are disjoint, so $A_1 \cup \cdots \cup A_k = \mathcal{S}$ and, applying axioms 2 and 3, we have $P(A_1) + P(A_2) + \cdots + P(A_k) = P(\mathcal{S}) = 1$. Since each of the k events is equally likely, each event probability is $1/k$.

Example 2.16 (Six-sided die) *Let $A = \{3, 4, 5, 6\}$ be the event of rolling at least a 3. If all outcomes of $\mathcal{S} = \{1, 2, 3, 4, 5, 6\}$ are equally likely, as for a fair die, the probability of any outcome is $1/6$. Event A can be thought of as the union of the four disjoint simple events $\{3\}$, $\{4\}$, $\{5\}$, and $\{6\}$, so that $P(A)$ is the sum of the probability of these four simple events, which is $4/6$ or $2/3$. Similarly, for event $B = \{2, 4, 6\}$ that an even number is rolled, $P(B) = 3/6 = 1/2$.*

Property (iv) from proposition 2.5, which is the union-of-events property, can be extended to more than two events, as stated in proposition 2.6.

Proposition 2.6 (Generalization of the union-of-events property)
For two events A and B,

$$P(A \cup B) = P(A) + P(B) - P(A \cap B).$$

For three events A, B, and C,

$$P(A \cup B \cup C) = P(A) + P(B) + P(C) - P(A \cap B) - P(A \cap C) - P(B \cap C) + P(A \cap B \cap C).$$

For four events A, B, C, and D,

$$P(A \cup B \cup C \cup D) = \begin{array}{l} P(A) + P(B) + P(C) + P(D) \\ -P(A \cap B) - P(A \cap C) - P(A \cap D) - P(B \cap C) - P(B \cap D) \\ -P(C \cap D) + P(A \cap B \cap C) + P(A \cap B \cap D) + P(A \cap C \cap D) \\ +P(B \cap C \cap D) - P(A \cap B \cap C \cap D), \end{array}$$

and so on, for larger numbers of events.

For the case of three events in proposition 2.6, where

$$P(A \cup B \cup C) = P(A) + P(B) + P(C) - P(A \cap B) - P(A \cap C) - P(B \cap C) + P(A \cap B \cap C),$$

Figure 2.5 provides a Venn diagram that illustrates how this property works. This figure is similar to the left diagram in figure 2.4 except that the third event C has been added. Again, the area of each circle should be thought of as the probability of the associated event, as should any of the areas of the overlaps of the circles. The intersection events $A \cap B$, $A \cap C$, $B \cap C$, and $A \cap B \cap C$ are indicated in the figure. To be clear, the region for intersection $A \cap B$ includes the region for $A \cap B \cap C$; similarly, the regions for $A \cap C$ and $B \cap C$ also include the region for $A \cap B \cap C$. When the areas of the three circles are added up, giving $P(A) + P(B) + P(C)$, the sum double-counts the areas associated with $A \cap B$, $A \cap C$, and $B \cap C$. As before, subtracting the quantities $P(A \cap B)$, $P(A \cap C)$, and $P(B \cap C)$ accounts for

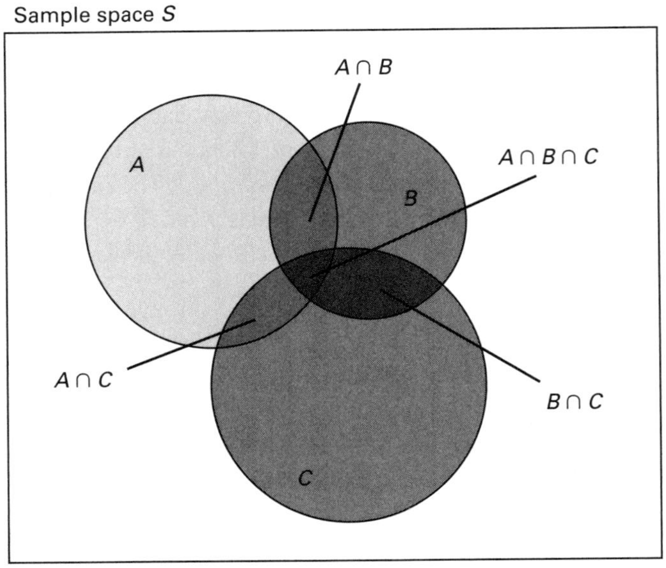

Figure 2.5
Venn diagram and $P(A \cap B \cap C)$

the double-counting. Unfortunately, when these three probabilities are subtracted, the area associated with $A \cap B \cap C$, which is $P(A \cap B \cap C)$, is subtracted one too many times. The region corresponding to $A \cap B \cap C$ gets counted three times in $P(A) + P(B) + P(C)$ and then subtracted three times when $P(A \cap B)$, $P(A \cap C)$, and $P(B \cap C)$ are subtracted, so the quantity $P(A \cap B \cap C)$ is added back to get the correct $P(A \cup B \cup C)$, as in the formula above.

Notes

1. A more general version of the third axiom adds the following to allow for an infinite number of disjoint events: If $A_1, \ldots, A_k, \ldots$ is a *countably infinite* collection of disjoint events,

$$P(A_1 \cup A_2 \cup A_3 \cup \cdots) = \sum_{i=1}^{\infty} P(A_i).$$

"Countably infinite" means that there is a natural way to count the events or, put another way, to label the events 1, 2, 3, and so on.

2. To show that property (iv) holds, consider the two events B and $A \cap B^c$. These events are disjoint since any outcome in B can't possibly be in B^c or, therefore, $A \cap B^c$. We also know that $(A \cap B^c) \cup B = A \cup B$, so

$$P(A \cup B) = P(A \cap B^c) + P(B) = P(A) - P(A \cap B) + P(B) = P(A) + P(B) - P(A \cap B)$$

from applying property (iii), plugging in $P(A) - P(A \cap B)$ for $P(A \cap B^c)$.

Exercises

1. You have five songs on your playlist, with songs 1 and 2 by Beyoncé and songs 3, 4, and 5 by Pink. You listen to the playlist in random order, but without repeats (e.g., once song 1 is played, it doesn't get played again). You continue to listen until a song by Pink is played. For example, 214 is one possible song sequence (outcome).

a. What is the sample space $\mathcal{S}$?

b. What is the event A in which song 5 is played?

c. What is the event B in which song 2 is not played?

2. For two events A and B, suppose that every outcome in A is also in B. State whether each of the following statements is true or false, and explain why.

   a. The size of B, denoted as $|B|$, is strictly less than the size of A, denoted as $|A|$.

   b. The union of A and B is B.

   c. Every outcome in B^c is in A^c.

3. Consider randomly selecting a college student. Let V be the event that the student has a paid video-streaming account, and let M be the event that the student has a paid music-streaming account. Suppose that $P(V) = 0.7$ and $P(M) = 0.5$.

   a. Is it possible that $P(V \cap M) = 0.6$? Why or why not?

   b. Suppose that the probability of the student having both a video-streaming account and a music-streaming account is 0.35.

      i. What is the probability that the student has at least one of the two types of accounts?

      ii. What is the probability that the student has neither type of account?

      iii. In terms of V and M, what is the event in which the student has a video-streaming account but no music-streaming account? What is the probability of this event?

4. Consider the experiment of picking a number randomly from the sample space $\mathcal{S} = \{1, 2, 3, \ldots, 999, 1000\}$. Let A be the event that the number is a multiple of three, let $B = \{500, 501, \ldots, 699, 700\}$ be the event that the number is between 500 and 700 (inclusive), and let C be the event that the number is a perfect square (i.e., $C = \{1, 4, 9, 16, \ldots, 900, 961\}$). For this question, use R to create sets (vectors) and perform the necessary set operations.

   a. Create three vectors `eventA`, `eventB`, and `eventC` for events A, B, and C and a vector `samplespace` for $\mathcal{S}$.

   b. Create a vector containing $A \cup B$ and calculate $|A \cup B|$.

   c. Create a vector containing $A \cap C$ and calculate $|A \cap C|$.

   d. Create a vector containing C^c and calculate $|C^c|$.

   e. Create a vector containing $A \cap B \cap C$ and calculate $|A \cap B \cap C|$.

   f. Create a vector containing $(A \cap B^c) \cup C$ and calculate $|(A \cap B^c) \cup C|$.

5. Suppose that on any given weekday, 70% of college students eat breakfast, 60% do homework, and 85% do at least one of these two things.

   a. What is the probability that a randomly selected student eats breakfast and does homework?

   b. What is the probability that a randomly selected student does neither activity?

6. The probability of the union, $P(A \cup B) = P(A) + P(B) - P(A \cap B)$, includes outcomes that are in both A and B. Provide a formula for the probability that *exactly one* of the events A and B occurs (but not both) in terms of $P(A)$, $P(B)$, and $P(A \cup B)$.
7. A company has two research projects R_1 and R_2, each of which either results in a patent or does not. The probability that project R_1 results in a patent is 0.45, the probability that project R_2 results in a patent is 0.15, and the probability that both projects result in a patent is 0.05.
   a. What is the probability that at least one of the two projects results in a patent?
   b. What is the probability that neither of the two projects results in a patent?
   c. What is the probability that exactly one of the two projects results in a patent?
8. The sample space for the price of a company's stock on a given day is $\mathcal{S} = [0, \infty)$, which consists of all nonnegative real numbers (including zero). In this case, $\mathcal{S}$ has an infinite number of possible outcomes. Consider the following events: $A = [80, 100]$, $B = (60, 90]$, and $C = (95, \infty)$. A square bracket [or] indicates that an interval is inclusive for that end point, and a parenthesis (or) indicates that an interval is exclusive for that end point. Therefore, A is the set of prices p such that $80 \leq p \leq 100$, B is the set of prices p such that $60 < p \leq 90$, and C is the set of prices p such that $p > 95$. Using the bracket and parenthesis notation and (when necessary) the union operator, what are the following events?
   a. C^c
   b. B^c
   c. $A \cap B$
   d. $A \cup C$
   e. $A^c \cup B$
   f. $A \cap B \cap C^c$
9. At a given company, worker salaries for the following year have just been decided. A worker's salary can either increase (I), decrease (D), or remain the same (R). Consider observing whether it increases, decreases, or stays the same for three employees.
   a. What is the sample space $\mathcal{S}$?
   b. What is the event A in which all three workers have different outcomes?
   c. What is the event B in which exactly one of the three workers has a salary increase?
   d. What is the event C in which exactly two of the three workers have the same outcome?
   e. Determine the following events: C^c and $B \cap C$.
   f. Are A and C disjoint events?
   g. Are A and C collectively exhaustive?
10. A small information systems firm has the resources to respond to two invitations to submit a proposal for a contract. When a proposal is submitted, it may be accepted outright, it may be rejected outright, or a revision of the proposal may be requested. If

a revision is requested, submission of the revision leads to acceptance or rejection. You may assume that the firm always submits a revision if one is requested.

a. What is the sample space? Develop your own notation for this part.

b. Define event A to be "both proposals are eventually accepted," event B to be "both proposals are eventually rejected," and event C to be "a revision is submitted." Which outcomes belong to each of these events?

c. Are A and B disjoint events? Are A and B collectively exhaustive?

d. Are A and C disjoint events? Are A and C collectively exhaustive?

11. Burger Barn and Patty Palace will both open one new restaurant in Texas next year. Burger Barn is choosing among four cities: A(ustin), D(allas), H(ouston), and S(an Antonio). Patty Palace is choosing among only three cities (A, D, H) since it already has too many locations in San Antonio. Here is the probability table associated with their decisions:

		Burger Barn			
		A	D	H	S
	A	0.07	0.12	0.16	0.12
Patty Palace	D	0.03	0.02	0.15	0.15
	H	0.06	0.04	0.03	0.05

a. What is the probability that Burger Barn locates in Dallas and Patty Palace locates in Houston?

b. What is the probability that Burger Barn locates in Dallas?

c. What is the probability that Patty Palace locates in Houston?

d. What is the probability that Burger Barn and Patty Palace locate in the same city?

e. What is the probability that Burger Barn and Patty Palace locate in different cities?

12. An investor has three investment opportunities (A, B, and C) and is asked to rank them in order of preference, with the most preferred listed first. Since she is indifferent among the three investments, she ranks them randomly.

a. What are the outcomes in S, and what are the probabilities for each of the outcomes?

b. What is the probability that investment C is ranked first?

c. What is the probability that investment C is ranked first and investment A is ranked last?

13. On a particular day, Steve's Sneaker Shop is beginning to stock pairs of a highly anticipated new sneaker. Hundreds of customers are expected at the store. The owner (Steve) decides that one of the first 50 customers will receive a free pair of the sneakers, and he randomly picks a number between 1 and 50 (inclusive), each being equally likely, before the store opens.

a. What is the probability that the fifth customer gets the free pair of sneakers?

b. What is the probability that the free pair of sneakers is given away before the 20th customer?

c. Conduct 10,000 simulations in *R* to approximate the probabilities in (a) and (b), where each simulation involves a random number being chosen from the vector `1:50`.

14. a. Modify the *R* code for 10,000 coin-toss simulations, used to create figure 2.1, to replace the fair coin with an unfair coin. Specifically, consider an unfair coin that comes up heads with 60% probability. In addition to changing the random tosses, change where the dotted line appears.

    b. Repeat (a), but with only 100 simulations. How does the figure compare to the one in (a)?

15. At a large technology company, 80% of employees work full time (at least 40 hours per week), 70% have a flexible work arrangement (that allows them to work partially from home), and 50% have a company-issued laptop computer. Refer to these three employee characteristics as A, B, and C, respectively, so $P(A)=0.8$, $P(B)=0.7$, and $P(C)=0.5$. The following things are also true: 60% of employees work full time *and* have a flexible work arrangement, 38% of employees work full time *and* have a company-issued laptop computer, 42% of employees have a flexible work arrangement *and* have a company-issued laptop computer, and 35% work full time *and* have a flexible work arrangement *and* have a company-issued laptop computer.

    a. What is the probability that an employee has *at least one* of the three characteristics?

    b. What is the probability that an employee has *exactly one* of the three characteristics? (For this part, you might find it easiest to use a Venn diagram.)

    c. What is the probability that an employee has *at least two* of the three characteristics?

3

Conditional Probabilities and Independence

This chapter builds on the basic probability properties from chapter 2 to say more about what happens when there are multiple events. What can be said about the probability that two events both occur? What can be said about the probability of one event occurring if it is known that the other event occurs? Various types of probabilities are discussed, including *joint probabilities*, *conditional probabilities*, and *marginal probabilities*, and the important concept of *independence* is introduced.

3.1 Definition and Properties of Conditional Probabilities

Before introducing conditional probabilities, we introduce some additional terminology associated with probabilities.

Definition 3.1 *$P(A)$, the probability of an event A, is also called the* ***unconditional probability*** *or* ***marginal probability*** *of A.*

We say that $P(A)$ is an "unconditional probability" since we do not condition on any other event occurring, in contrast to the conditional probability introduced next. And, as seen later in this chapter, the terminology "marginal probability" relates to the idea that this unconditional probability will sometimes appear in the margin of a probability table.

Each outcome in the sample space has an unconditional or marginal probability, and the collection of these probabilities is the **probability distribution**:

Definition 3.2 *A* ***probability distribution*** *is a complete description of the probabilities associated with every outcome in sample space $\mathcal{S}$.*

Example 3.1 (Six-sided die) *For a fair die and $\mathcal{S} = \{1, 2, 3, 4, 5, 6\}$, the probability distribution is*

$$P(1) = P(2) = P(3) = P(4) = P(5) = P(6) = 1/6.$$

Definition 3.3 *The* ***joint probability*** *of events A and B is the probability $P(A \cap B)$ that both events occur.*

Examples of joint probabilities were seen in chapter 2. For instance, example 2.15 considered the joint probability of rolling at least a 3 (event A) and rolling an even number (event B).

We can now formally define the **conditional probability**, as in definition 3.4.

Definition 3.4 *The* ***conditional probability*** $P(A|B)$, *which is the probability of event A given that event B has occurred, is*

$$P(A|B) = \frac{P(A \cap B)}{P(B)}$$

if $P(B) > 0$. $P(A|B)$ *is often read as "the probability of A given B" or "the probability of A conditional on B."*

The $P(B) > 0$ condition ensures that division by zero will not occur. Having $P(B) > 0$ is *not* restrictive since it just means that the event being conditioned upon can actually happen. When we condition on event B, we focus only on outcomes in B and ignore any outcomes in the rest of sample space $\mathcal{S}$ (given by B^c).

The conditional probability is quite different from the joint probability. For the six-sided die example, $P(A|B)$ is the probability that the die roll is at least 3 if we know that it is an even number. Knowing that B has occurred means that the possible outcomes are $\{2, 4, 6\}$, so now what is the probability that the die roll is at least 3? Flipping things around, $P(B|A)$ is the probability the die roll is an even number if we know that it is at least 3. Knowing that A has occurred tells us that the possible outcomes are $\{3, 4, 5, 6\}$, so now what is the probability that the die roll is even? We can calculate both of these conditional probabilities using definition 3.4.

Example 3.2 (Six-sided die) *The probabilities of* $A = \{3, 4, 5, 6\}$, *a roll of at least* 3, *and* $B = \{2, 4, 6\}$, *an even roll, are* $P(A) = 2/3$ *and* $P(B) = 1/2$. *The joint probability is* $P(A \cap B) = 1/3$ *since* $A \cap B = \{4, 6\}$. *The conditional probability that the die roll is at least* 3, *given that the die roll is even, is*

$$P(A|B) = \frac{P(A \cap B)}{P(B)} = \frac{1/3}{1/2} = 2/3,$$

which makes sense since $A \cap B = \{4, 6\}$ *has two of the three outcomes in* $B = \{2, 4, 6\}$. *Reversing the roles of A and B, the conditional probability that the die roll is even, given that the die roll is at least* 3, *is*

$$P(B|A) = \frac{P(B \cap A)}{P(A)} = \frac{1/3}{2/3} = 1/2.$$

We have $P(B \cup A) = P(A \cup B)$ *since* $B \cup A = A \cup B$. *Again, the answer is intuitive since* $A \cap B = \{4, 6\}$ *has two of the four outcomes in* $A = \{3, 4, 5, 6\}$.

The conditional probability is itself a probability by definition, and its main properties follow directly from the Axioms of Probability introduced in chapter 2:

Proposition 3.1 (Properties of conditional probabilities) *If B is an event with* $P(B) > 0$, *the following properties hold:*

i. $0 \le P(A|B) \le 1$ *for any event A.*

ii. *If* $A_1, A_2, \ldots, A_k$ *is a collection of disjoint events,*

$$P(A_1 \cup A_2 \cup \cdots \cup A_k|B) = \sum_{i=1}^{k} P(A_i|B).$$

iii. *If* $A_1, A_2, \ldots, A_k$ *is a collection of disjoint and exhaustive events,*

$$P(A_1 \cup A_2 \cup \cdots \cup A_k|B) = 1.$$

iv. $P(A^c|B) = 1 - P(A|B)$ *for any event A.*

v. $P(A_1|B) = P(A_1 \cap A_2|B) + P(A_1 \cap A_2^c|B)$ *for any events* A_1 *and* A_2.

vi. $P(A_1 \cup A_2|B) = P(A_1|B) + P(A_2|B) - P(A_1 \cap A_2|B)$ *for any events* A_1 *and* A_2.

vii. *If* A_1 *and* A_2 *are disjoint events,* $P(A_1 \cap A_2|B) = 0$.

The properties look very similar to the first and third probability axioms and the properties from proposition 2.5, with the difference being that all the probabilities in proposition 3.1 are conditional probabilities, conditioned on event B.

3.2 Multiplication Rule and Bayes' Theorem

Definition 3.4 shows how a conditional probability ("A occurs given B occurs") is related to a joint probability ("A and B occur") and an unconditional probability ("B occurs"). Rewriting this definition yields a result known as the **multiplication rule**, which is useful for determining a joint probability of two events when information is available on a conditional probability.

Proposition 3.2 (Multiplication rule) *For any two events A and B,*

$$P(A \cap B) = P(A|B)P(B) \text{ if } P(B) > 0$$

and

$$P(A \cap B) = P(B|A)P(A) \text{ if } P(A) > 0.$$

Proposition 3.2 follows directly from definition 3.4 since $P(A|B) = \frac{P(A \cap B)}{P(B)}$ and $P(B|A) = \frac{P(A \cap B)}{P(A)}$, respectively. It can be useful to have the two alternative equations since in some cases, you might know $P(B)$ and $P(A|B)$ but not $P(A)$ and $P(B|A)$, or vice versa.

Example 3.3 (Product returns) *Suppose Widgets.com sells three types of widgets, each with different purchase probabilities and return rates, as follows:*

- *Widget 1: The probability that a widget purchase is widget 1 is* 60%. *The probability that a widget 1 purchase is returned is* 15%.
- *Widget 2: The probability that a widget purchase is widget 2 is* 30%. *The probability that a widget 2 purchase is returned is* 25%.
- *Widget 3: The probability that a widget purchase is widget 3 is* 10%. *The probability that a widget 3 purchase is returned is* 35%.

Let events A_1, A_2, and A_3 correspond to a purchase being widget 1, widget 2, and widget 3, respectively. We know that $P(A_1) = 0.6$, $P(A_2) = 0.3$, and $P(A_3) = 0.1$. Let R be the event that a widget purchase is returned. We know the conditional probabilities $P(R|A_1) = 0.15$, $P(R|A_2) = 0.25$, and $P(R|A_3) = 0.35$. Given this information, what is the joint probability that a widget purchase is widget 1 and the purchase is returned? We have

$$P(A_1 \cap R) = P(R|A_1)P(A_1) = (0.15)(0.6) = 0.09.$$

(Why not use the alternative formula $P(A_1 \cap R) = P(A_1|R)P(R)$ here?)

How about the joint probability that a widget purchase is widget 2 and the purchase is not returned? The event of the purchase not being returned is R^c, so

$$P(A_2 \cap R^c) = P(R^c|A_2)P(A_2) = (1 - 0.25)(0.3) = 0.225.$$

We can do even more with conditional probabilities after introducing two important results, the Law of Total Probability and Bayes' theorem. We start with the **Law of Total Probability**, which is based on the concept of partitioning introduced in chapter 2.

Proposition 3.3 (Law of Total Probability) *If $A_1, A_2, \ldots, A_k$ are disjoint events, each with positive probability, and also exhaustive events (i.e., $P(A_1) + P(A_2) + \cdots + P(A_k) = 1$), then for any event B,*

$$P(B) = P(B|A_1)P(A_1) + P(B|A_2)P(A_2) + \cdots + P(B|A_k)P(A_k) = \sum_{j=1}^{k} P(B|A_j)P(A_j).$$

Using the disjoint and exhaustive events $A_1, \ldots, A_k$, event B can itself be partitioned into k events—namely, $B \cap A_1, B \cap A_2, \ldots, B \cap A_k$. Since $A_1, \ldots, A_k$ are exhaustive events, the union of these k events is equal to B since any outcome in B must be in one of the k partitions. Applying axiom 3, then, yields

$$P(B) = P(B \cap A_1) + P(B \cap A_2) + \cdots + P(B \cap A_k).$$

The Venn diagram in figure 3.1 illustrates the idea of partitioning. The Venn diagram depicts a partition of the sample space $\mathcal{S}$ into five disjoint and exhaustive events A_1, A_2, A_3, A_4, and A_5. Event B is represented by the gray circle, and B is itself partitioned into five events: $B \cap A_1, B \cap A_2, B \cap A_3, B \cap A_4$, and $B \cap A_5$. (In figure 3.1, each of the $B \cap A_j$ events is nonempty, although it is certainly possible in other examples to have $B \cap A_j = \varnothing$.) Then the overall probability of event B is the sum of the five probabilities $P(B \cap A_1)$, $P(B \cap A_2)$, $P(B \cap A_3)$, $P(B \cap A_4)$, and $P(B \cap A_5)$.

Having $P(B) = P(B \cap A_1) + P(B \cap A_2) + \cdots + P(B \cap A_k)$ for disjoint and exhaustive events $A_1, \ldots, A_k$ leads directly to the Law of Total Probability result. Specifically, from the multiplication rule (proposition 3.2), $P(B \cap A_1) = P(B|A_1)P(A_1)$ and simililarly for the other partitions, so

$$P(B) = P(B|A_1)P(A_1) + P(B|A_2)P(A_2) + \cdots + P(B|A_k)P(A_k).$$

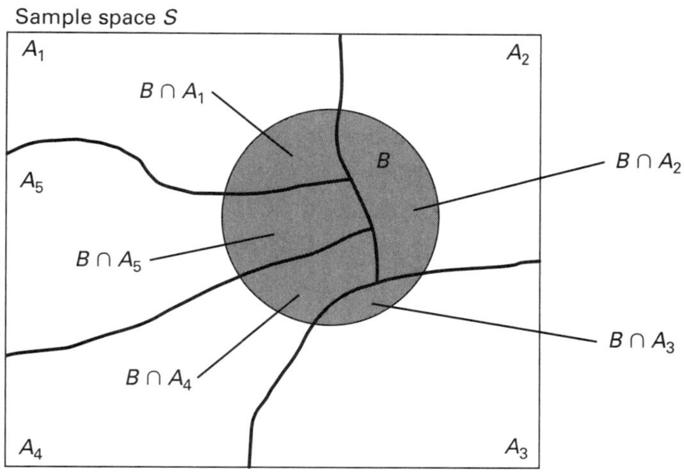

Figure 3.1
Venn diagram for event partitions

Example 3.4 (Product returns) *Continuing example 3.3, the following question can now be answered: For a widget purchase, what is the probability that the widget is returned? In other words, what is the unconditional probability $P(R)$? A direct application of the Law of Total Probability gives*

$$P(R) = P(R|A_1)P(A_1) + P(R|A_2)P(A_2) + P(R|A_3)P(A_3)$$

$$= (0.15)(0.6) + (0.25)(0.3) + (0.35)(0.1) = 0.2.$$

The unconditional probability of a return is 0.2, or 20%. From this equation, this unconditional probability is a weighted average of the three conditional return probabilities for the different types of widgets, where the weights are the probabilities of the partitions (here, the probabilities of the three types of widgets). With the unconditional probability $P(R)$, we can also determine the conditional probability of each widget type given that a widget is returned. If a Widgets.com employee receives a return in a sealed package, what is the probability that this return is widget 1, or widget 2, or widget 3? These conditional probabilities are determined as follows:

$$P(A_1|R) = \frac{P(A_1 \cap R)}{P(R)} = \frac{P(R|A_1)P(A_1)}{P(R)} = \frac{(0.15)(0.6)}{0.2} = 0.45$$

$$P(A_2|R) = \frac{P(A_2 \cap R)}{P(R)} = \frac{P(R|A_2)P(A_2)}{P(R)} = \frac{(0.25)(0.3)}{0.2} = 0.375$$

$$P(A_3|R) = \frac{P(A_3 \cap R)}{P(R)} = \frac{P(R|A_3)P(A_3)}{P(R)} = \frac{(0.35)(0.1)}{0.2} = 0.175.$$

The conditioning information changes the probabilities. Whereas the unconditional probability of A_1 is 0.6, the probability of A_1, given that it is a returned widget, is 0.45. For A_2, the unconditional probability is 0.3, and the conditional probability given R is 0.375. For A_3, the unconditional probability is 0.1, and the conditional probability given R is 0.175.

As seen in example 3.4, probabilities can be updated once new information is incorporated. We move from an unconditional probability, with no information given, to a conditional probability, where we condition upon new information. This approach to updating probabilities is embodied in a famous result known as **Bayes' theorem**.

Proposition 3.4 (Bayes' theorem) *If $A_1, A_2, \ldots, A_k$ are disjoint and exhaustive events, each with positive probability, then for any event B with $P(B) > 0$,*

$$P(A_i|B) = \frac{P(B|A_i)P(A_i)}{\sum_{j=1}^{k} P(B|A_j)P(A_j)}.$$

To see why Bayes' theorem holds, recall that $P(A_i|B) = \frac{P(A_i \cap B)}{P(B)}$ by the definition of a conditional probability. For the numerator, $P(A_i \cap B) = P(B|A_i)P(A_i)$ by the multiplication rule. For the denominator, $P(B) = \sum_{j=1}^{k} P(B|A_j)P(A_j)$ by the Law of Total Probability.

In the context of Bayes' theorem, the unconditional probability $P(A_i)$ is sometimes referred to as the **prior probability** of event A_i since it is the probability prior to considering any additional information. The conditional probability $P(A_i|B)$ is sometimes referred to as the **posterior probability** of event A_i since the probability has been updated with the information that event B has occurred. Incorporation of conditioning information and moving from a prior probability to a posterior probability is called **Bayesian updating**.

Example 3.5 (Vaccination and infection) *Suppose that the probability that an individual is vaccinated against a certain disease is 70%. Let V be the event that an individual is vaccinated, with $P(V) = 0.7$, and $NV = V^c$ that an individual is unvaccinated, with $P(NV) = 0.3$. Let D be the event that an individual is infected with the underlying disease. Suppose that the probability of infection is 5% for vaccinated individuals ($P(D|V) = 0.05$) and 50% for unvaccinated individuals ($P(D|NV) = 0.5$). Then, what is the unconditional probability of infection P(D)? Since V and NV are disjoint and exhaustive, the Law of Total Probability yields*

$$P(D) = P(D|V)P(V) + P(D|NV)P(NV) = (0.05)(0.7) + (0.5)(0.3) = 0.185.$$

The probability of infection is a weighted average of the conditional probabilities of infection. With P(D) determined, we use Bayes' theorem to determine the conditional probability that an individual is vaccinated, given that the individual is infected:

$$P(V|D) = \frac{P(D|V)P(V)}{P(D|V)P(V) + P(D|NV)P(NV)} = \frac{(0.05)(0.7)}{(0.05)(0.7) + (0.5)(0.3)} \approx 0.189.$$

The conditional probability of vaccination given infection is approximately 18.9%. We could also determine $P(NV|D)$ using Bayes' theorem, but once we know $P(V|D) \approx 0.189$, we immediately have $P(NV|D) = 1 - P(V|D) \approx 0.811$ or 81.1%.

What would happen if the vaccination rate were much higher, say 90%, so $P(V) = 0.9$ and $P(NV) = 0.1$? With this higher vaccination rate, the conditional probability of vaccination given infection is

$$P(V|D) = \frac{P(D|V)P(V)}{P(D|V)P(V) + P(D|NV)P(NV)} = \frac{(0.05)(0.9)}{(0.05)(0.9) + (0.5)(0.1)} \approx 0.474.$$

We get a much *higher conditional probability of vaccination given infection in this case. Even though the infection rate for vaccinated individuals is quite low* (5%), *the very high unconditional probability of vaccination leads to the vaccinated making up a very large proportion of the infected individuals, much larger than with a* 70% *vaccination rate. In the partition terminology, the vaccinated partition of the infected event gets larger when the vaccinated probability goes up. The interested reader can try this exercise with even higher vaccination rates, like* 95% *and* 99%, *and determine* $P(V|D)$.

3.3 Probability Tables

Probability tables provide a convenient way to present the joint probabilities of events. Specifically, they are used to cross-tabulate events where the rows represent a finite collection of disjoint and exhaustive events and the columns represent a finite collection of disjoint and exhaustive events. To illustrate, we return to the widget returns example and consider a probability table with the widget type as rows and the return status as columns.

Example 3.6 (Product returns) *From example 3.3, recall that the three types of widgets made up* 60%, 30%, *and* 10% *of purchases with corresponding return rates of* 15%, 25%, *and* 35%. *Here is a probability table that corresponds to these probabilities:*

		Returned		
		Yes (R)	*No (R^c)*	*Total*
	Widget 1 (A_1)	***0.090***	***0.510***	*0.600*
Purchase	*Widget 2 (A_2)*	***0.075***	***0.225***	*0.300*
	Widget 3 (A_3)	***0.035***	***0.065***	*0.100*
	Total	*0.200*	*0.800*	

Events A_1, A_2, *and* A_3 *are disjoint and exhaustive, and events* R *and* R^c *are disjoint and exhaustive. Each of the bold numbers in the table corresponds to a joint probability. For instance, the* 0.090 *value is the joint probability of a widget 1 purchase and a return:* $P(A_1 \cap R) = 0.090$. *The* 0.225 *value is the joint probability of a widget 2 purchase and no return:* $P(A_2 \cap R^c) = 0.225$. *Since both the rows and columns have collections of joint and exhaustive events, the joint probabilities in the table must sum to* 1. *In this case,* $0.090 + 0.510 + 0.075 + 0.225 + 0.035 + 0.065 = 1$.

Both a row and a column labeled "Total" have been included in this probability table. The "Total" column provides the totals for the joint probabilities in each row, and the "Total" row provides the totals for the joint probabilities in each column. These "marginal" probabilities are the unconditional probabilities corresponding to the given row or column that is being summed. For the "Widget 1 (A_1)" row, the 0.600 *value is the sum of* $P(A_1 \cap R) = 0.090$ *and* $P(A_1 \cap R^c) = 0.510$, *so that* $P(A_1) = 0.6$. *For the "Yes (R)" column, the* 0.200 *value is the sum of* $P(A_1 \cap R) = 0.090$, $P(A_2 \cap R) = 0.075$, *and* $P(A_3 \cap R) = 0.035$, *so that* $P(R) = 0.200$.

When the joint probabilities are completely specified, as in this case, conditional probabilities can be calculated quite easily. For instance, the conditional probability of a widget

1 purchase (A_1), given that the purchase is returned (R), is

$$P(A_1|R) = \frac{P(A_1 \cap R)}{P(R)} = \frac{0.090}{0.200} = 0.45,$$

where the numerator is the joint probability 0.090 *from the table and the denominator is the marginal probability* $P(R) = 0.200$ *in the "Total" row. Similarly, the conditional probability of a return (R) given a widget 1 purchase (A_1) is*

$$P(R|A_1) = \frac{P(R \cap A_1)}{P(A_1)} = \frac{0.090}{0.600} = 0.15,$$

where the numerator is the joint probability 0.090 *from the table and the denominator is the marginal probability* $P(A_1) = 0.600$ *in the "Total" column.*

Example 3.7 (Vaccination and infection) *From example 3.5, recall that the vaccination probability was* $P(V) = 0.7$, *the probability of infection given vaccination was* $P(D|V) = 0.05$, *and the probability of infection given nonvaccination was* $P(D|NV) = 0.5$. *The following probability table corresponds to these vaccination/infection probabilities:*

		Infected		
		Yes (D)	*No (D^c)*	*Total*
Vaccinated	*Yes (V)*	***0.035***	***0.665***	*0.700*
	No (NV)	***0.150***	***0.150***	*0.300*
	Total	*0.185*	*0.815*	

Each of the four joint probabilities comes directly from the information on the vaccination probability and the conditional infection probabilities. For instance, the 0.035 *value is* $P(V \cap D) = P(D|V)P(V) = (0.05)(0.7) = 0.035$.

From the probability table, the conditional probability of vaccination given infection is

$$P(V|D) = \frac{P(V \cap D)}{P(D)} = \frac{0.035}{0.185} \approx 0.189,$$

whereas the probability of nonvaccination given infection is

$$P(NV|D) = \frac{P(NV \cap D)}{P(D)} = \frac{0.150}{0.185} \approx 0.811.$$

With a higher vaccination probability of $P(V) = 0.9$ *and the same conditional infection probabilities (5% for vaccinated, 50% for unvaccinated), the probability table would be:*

		Infected		
		Yes (D)	*No (D^c)*	*Total*
Vaccinated	*Yes (V)*	***0.045***	***0.855***	*0.900*
	No (NV)	***0.050***	***0.050***	*0.100*
	Total	*0.095*	*0.905*	

Example 3.8 (Two car dealerships) *Example 2.2 considered an experiment associated with the number of salespeople at dealership A and the number of salespeople at dealership B*

who sell cars on a given day. Suppose that the joint probabilities are given by the values in the following probability table, without marginal probabilities specified:

		Dealership B			
		0	1	2	3
Dealership A	0	***0.02***	***0.04***	***0.03***	***0.01***
	1	***0.03***	***0.08***	***0.05***	***0.02***
	2	***0.07***	***0.15***	***0.11***	***0.06***
	3	***0.04***	***0.09***	***0.06***	***0.02***
	4	***0.02***	***0.05***	***0.04***	***0.01***

Let A_0, A_1, A_2, A_3, A_4 denote the events associated with the number of dealership A salespeople selling a car on a given day. Let B_0, B_1, B_2, B_3 denote the events associated with the number of dealership B salespeople selling a car on a given day. The most likely joint event is $A_2 \cap B_1$ (sales by two salespeople at dealership A and one salesperson at dealership B), with probability 15%. *The least likely joint events are $A_0 \cap B_3$ and $A_4 \cap B_3$, each with probability* 1%.

The marginal probabilities of dealership A events and dealership B events can be calculated from the joint probabilities. For instance, the unconditional or marginal probability that dealership A has two salespeople sell a car is obtained by adding up the joint probabilities in the row labeled "2":

$$P(A_2) = P(A_2 \cap B_0) + P(A_2 \cap B_1) + P(A_2 \cap B_2) + P(A_2 \cap B_3)$$
$$= 0.07 + 0.15 + 0.11 + 0.06 = 0.39.$$

The unconditional or marginal probability that dealership B has one salesperson sell a car is obtained by adding up the joint probabilities in the column labeled "1":

$$P(B_1) = P(B_1 \cap A_0) + P(B_1 \cap A_1) + P(B_1 \cap A_2) + P(B_1 \cap A_3) + P(B_1 \cap A_4)$$
$$= 0.04 + 0.08 + 0.15 + 0.09 + 0.05 = 0.41.$$

Conditional probabilities are calculated in the usual way. For instance, the conditional probability of two dealership A salespeople selling a car (A_2), given that three dealership B salespeople sell a car (B_3), is:

$$P(A_2|B_3) = \frac{P(A_2 \cap B_3)}{P(B_3)} = \frac{0.06}{0.01 + 0.02 + 0.06 + 0.02 + 0.01} = 0.5.$$

We can also calculate conditional probabilities that are not based solely upon simple events. An example is the conditional probability of at least *two dealership A salespeople selling a car ($A_2 \cup A_3 \cup A_4$), given that three dealership B salespeople sell a car (B_3), which is*

$$P(A_2 \cup A_3 \cup A_4|B_3) = P(A_2|B_3) + P(A_3|B_3) + P(A_4|B_3)$$
$$= \frac{P(A_2 \cap B_3)}{P(B_3)} + \frac{P(A_3 \cap B_3)}{P(B_3)} + \frac{P(A_4 \cap B_3)}{P(B_3)}$$
$$= \frac{0.06 + 0.02 + 0.01}{0.01 + 0.02 + 0.06 + 0.02 + 0.01} = 0.75.$$

Another example is the conditional probability of two dealership A salespeople selling a car (A_2), given that the number of salespeople selling a car at the two dealerships are equal to each other. Let E be the event that the number of salespeople selling a car at the two dealerships are equal to each other, which is

$$E = (A_0 \cap B_0) \cup (A_1 \cap B_1) \cup (A_2 \cap B_2) \cup (A_3 \cap B_3),$$

the union of "both 0" and "both 1" and "both 2" and "both 3." The conditional probability is

$$P(A_2|E) = \frac{P(A_2 \cap E)}{P(E)} = \frac{0.11}{0.02 + 0.08 + 0.11 + 0.02} = \frac{11}{23}.$$

3.4 Independence

In this section, we consider the concept of **independence** of events, which essentially means that knowing that one event occurs does not provide any additional information about the other event. In situations where events are independent of each other, the calculation of joint probabilities is greatly simplified. We start with the definition of independent events.

Definition 3.5 *Events A and B are **independent** if $P(A|B) = P(A)$. Events A and B are **dependent** if $P(A|B) \neq P(A)$.*

The $P(A|B) = P(A)$ condition says that knowing that B occurs does not affect the probability of A occurring. The conditional probability of A given B is the same as the unconditional probability of A. Even though definition 3.5 specifies that only $P(A|B) = P(A)$ is required for independence, $P(A|B) = P(A)$ immediately implies that $P(B|A) = P(B)$ (i.e., knowing A does not affect the probability of B):

$$P(B|A) = \frac{P(B \cap A)}{P(A)} = \frac{P(A|B)P(B)}{P(A)} = P(B).$$

Therefore, the independence of A and B can be established by checking *either* $P(A|B) = P(A)$ or $P(B|A) = P(B)$. It is not necessary to check both. Likewise, to show the dependence of A and B, either $P(A|B) \neq P(A)$ or $P(B|A) \neq P(B)$ can be checked.

If events A and B are independent, knowing that B occurs doesn't affect the probability of A. It is also the case that knowing that B does *not* occur doesn't affect the probability of A.

Proposition 3.5 *If events A and B are independent, $P(A|B^c) = P(A)$.*

Example 3.9 (Product returns) *Based upon the probability table in example 3.6, are events A_1 (widget 1 purchase) and R (return) independent? Note that $P(A_1|R) = 0.09/0.20 = 0.45$, which is not equal to the unconditional probability $P(A_1) = 0.6$. Alternatively, note that $P(R|A_1) = 0.15$, which is not equal to the unconditional probability $P(R) = 0.2$. Intuitively, it makes sense that A_1 and R are dependent since the return probability depends on the type of widget purchased; that is, knowing that widget 1 is purchased provides additional information about the return probability.*

Example 3.10 (Vaccination and infection) *From the probability table in example 3.7, the vaccination event (V) and infection event (D) are dependent since $P(D|V) = 0.05$ and*

$P(D) = 0.095$. The dependence of these events arises since the infection probability changes depending on whether an individual is vaccinated or nonvaccinated.

Example 3.11 (Two coin tosses) *Recall experiment 2.10 where two coins are tossed, with sample space $\mathcal{S} = \{HH, HT, TH, TT\}$. Let event H_1 correspond to the first toss being heads and event H_2 correspond to the second toss being heads. If the outcome of one toss has no effect on the outcome of the other toss, then H_1 and H_2 are independent events, with $P(H_1|H_2) = P(H_1) = 0.5$ and $P(H_2|H_1) = P(H_2) = 0.5$. Due to independence, each of the probabilities for the outcomes in $\mathcal{S}$ is equal to 0.25:*

$$P(HH) = P(H_1|H_2)P(H_2) = P(H_1)P(H_2) = (0.5)(0.5) = 0.25$$

$$P(HT) = P(H_1|H_2^c)P(H_2^c) = P(H_1)P(H_2^c) = (0.5)(0.5) = 0.25$$

$$P(TH) = P(H_1^c|H_2)P(H_2) = P(H_1^c)P(H_2) = (0.5)(0.5) = 0.25$$

$$P(TT) = P(H_1^c|H_2^c)P(H_2^c) = P(H_1^c)P(H_2^c) = (0.5)(0.5) = 0.25.$$

In example 3.11, the joint probabilities for the two coin tosses simplify to the product of two unconditional probabilities. This simplification is a general property of independent events, as in proposition 3.6.

Proposition 3.6 *Events A and B are independent if and only if $P(A \cap B) = P(A)P(B)$.*

This proposition provides an alternative method to check independence without conditional probabilities.

Example 3.12 (Product returns) *A_1 and R are dependent since $P(A_1 \cap R) = 0.09$ and $P(A_1)P(R) = (0.6)(0.2) = 0.12$.*

Example 3.13 (Vaccination and infection) *V and D are dependent since $P(V \cap D) = 0.035$ and $P(V)P(D) = (0.7)(0.185) = 0.1295$.*

The concept of independence can be extended beyond two events, as in definition 3.6.

Definition 3.6 (Independence of more than two events) *Events $A_1, A_2, \ldots, A_k$ are* ***mutually independent*** *if, for* any *subset of the events, the joint probability is equal to the product of the individual probabilities. Equivalently, the conditional probability of any event A_i, given* any *subset of other events, is equal to the unconditional probability of A_i.*

For the case of two events, this definition corresponds exactly to the notion of independence in proposition 3.6. For three events (A_1, A_2, A_3), the events are mutually independent if

$$P(A_1 \cap A_2) = P(A_1)P(A_2),\ P(A_1 \cap A_3) = P(A_1)P(A_3),\ P(A_2 \cap A_3) = P(A_2)P(A_3),$$

and

$$P(A_1 \cap A_2 \cap A_3) = P(A_1)P(A_2)P(A_3).$$

Or, in terms of conditional probabilities, the three events are mutually independent if

$$P(A_1|A_2) = P(A_1|A_3) = P(A_1|A_2 \cap A_3) = P(A_1),$$

$$P(A_2|A_1) = P(A_2|A_3) = P(A_2|A_1 \cap A_3) = P(A_2),$$

and

$$P(A_3|A_1) = P(A_3|A_2) = P(A_3|A_1 \cap A_2) = P(A_3).$$

At this point, it is useful to introduce more concise notation for the intersection ($\cap$) of events. Specifically, we put a comma between events to indicate intersection. For example, we write

$$P(A_1, A_2) \text{ for } P(A_1 \cap A_2),$$

$$P(A_1, A_2, A_3) \text{ for } P(A_1 \cap A_2 \cap A_3),$$

$$P(A_1|A_2, A_3) \text{ for } P(A_1|A_2 \cap A_3),$$

$$P(A_1, A_2|A_3) \text{ for } P(A_1 \cap A_2|A_3),$$

and so on.

When there are many events, the number of possible products involved in the definition of mutual independence can be quite large. Consider the case of 50 events $(A_1, A_2, \ldots, A_{50})$, for which mutual independence is equivalent to the following:

$$P(A_1, A_2) = P(A_1)P(A_2) \text{ and similar for any two events in } \{A_1, \ldots, A_{50}\},$$

$$P(A_1, A_2, A_3) = P(A_1)P(A_2)P(A_3) \text{ and similar for any three events in } \{A_1, \ldots, A_{50}\},$$

$$\vdots \; \vdots \; \vdots$$

$$\vdots \; \vdots \; \vdots$$

$$P(A_1, A_2, \ldots, A_{49}) = P(A_1)P(A_2)\cdots P(A_{49}) \text{ and similar for any 49 events in } \{A_1, \ldots, A_{50}\},$$

$$\text{and } P(A_1, A_2, \ldots, A_{50}) = P(A_1)P(A_2)\cdots P(A_{50}).$$

Thankfully, in practice, we rarely check all the probability-product possibilities for such a large number of events. Instead, the more common situation is that we *assume* mutual independence of a collection of events and then use the implied probability-product equalities. For instance, for 50 tosses of a coin, we might assume that the events corresponding to heads on each toss (denoted as $H_1, H_2, \ldots, H_{50}$) are mutually independent since the result of one coin toss should not depend on the result of any other coin tosses. With the independence of the 50 coin tosses, it is easy to calculate joint probabilities involving the outcomes of any of the individual tosses. As an example, the probability of heads on tosses 10, 20, and 30 is $P(H_{10}, H_{20}, H_{30}) = P(H_{10})P(H_{20})P(H_{30}) = (0.5)(0.5)(0.5) = 0.125$. In fact, since heads and tails have the same probability 0.5 for every toss, the probability of *any* joint outcome of three tosses is equal to 0.125.

Example 3.14 (Three website visitors) *Recall example 2.3, which considered the purchase behavior of the first three visitors to a website on a given day, with sample space*

$$\mathcal{S} = \{YYY, YYN, YNY, YNN, NYY, NYN, NNY, NNN\}.$$

Let A_1, A_2, and A_3 denote the events that the first, second, and third visitor makes a purchase, respectively:

$$A_1 = \{YYY, YYN, YNY, YNN\}$$

$$A_2 = \{YYY, YYN, NYY, NYN\}$$

$$A_3 = \{YYY, YNY, NYY, NNY\}.$$

Assume that A_1, A_2, and A_3 are mutually independent, which is sensible if the purchase behavior of one website visitor is not affected by the purchase behavior of other website visitors. Moreover, assume that the unconditional purchase probability of any given website visitor is 20%, so $P(A_1) = P(A_2) = P(A_3) = 0.2$. The unconditional probability of nonpurchase by any website visitor is 80%, so $P(A_1^c) = P(A_2^c) = P(A_3^c) = 0.8$. With the mutual independence assumption, we can determine the probability for any outcome in $\mathcal{S}$. For instance, the probability of YNY (the first and third visitors make a purchase, and the second visitor does not) is $P(A_1)P(A_2^c)P(A_3) = (0.2)(0.8)(0.2) = 0.032$.

With the ability to calculate these joint probabilities, the probabilities associated with the total number of purchases *can also be determined. Let B_0, B_1, B_2, and B_3 denote the events that zero, one, two, and three total purchases are made, respectively:*

$$B_0 = \{NNN\},\ B_1 = \{YNN, NYN, NNY\},\ B_2 = \{YYN, YNY, NYY\},\ B_3 = \{YYY\}.$$

Their probabilities are

$$P(B_0) = \overbrace{(0.8)(0.8)(0.8)}^{NNN} = 0.512$$

$$P(B_1) = \overbrace{(0.2)(0.8)(0.8)}^{YNN} + \overbrace{(0.8)(0.2)(0.8)}^{NYN} + \overbrace{(0.8)(0.8)(0.2)}^{NNY} = 0.384$$

$$P(B_2) = \overbrace{(0.2)(0.2)(0.8)}^{YYN} + \overbrace{(0.2)(0.8)(0.2)}^{YNY} + \overbrace{(0.8)(0.2)(0.2)}^{NYY} = 0.096$$

$$P(B_3) = \overbrace{(0.2)(0.2)(0.2)}^{YYY} = 0.008.$$

These four probabilities sum to 1 since B_0, B_1, B_2, and B_3 are disjoint and exhaustive events.

Does the total number of purchases convey any useful information about whether a given individual makes a purchase? Let's focus on the first website visitor and consider the conditional probabilities $P(A_1|B_0)$, $P(A_1|B_1)$, $P(A_1|B_2)$, and $P(A_1|B_3)$:

$$P(A_1|B_0) = \frac{P(A_1, B_0)}{P(B_0)} = \frac{0}{0.512} = 0$$

$$P(A_1|B_1) = \frac{P(A_1, B_1)}{P(B_1)} = \frac{(0.2)(0.8)(0.8)}{0.512} = \frac{1}{3}$$

$$P(A_1|B_2) = \frac{P(A_1, B_2)}{P(B_2)} = \frac{(0.2)(0.2)(0.8) + (0.2)(0.8)(0.2)}{0.384} = \frac{2}{3}$$

$$P(A_1|B_3) = \frac{P(A_1, B_3)}{P(B_3)} = \frac{(0.2)(0.2)(0.2)}{0.008} = 1.$$

These four conditional probabilities make intuitive sense. If no purchases occur (B_0), the first visitor can't possibly make a purchase, so $P(A_1|B_0) = 0$. If three purchases occur (B_3), the first visitor must make a purchase, so $P(A_1|B_3) = 1$. When one total purchase is made (B_1), *the probability that the purchaser is the first visitor is* 1/3 *since there's nothing inherently different about the three visitors, meaning that their chance of being the purchaser should be the same (*1/3 *each). When two total purchases are made (B_2), the probability that the purchaser is the first visitor is* 2/3 *since the chance of any of the three visitors being the nonpurchaser should be the same (*1/3 *each). So it's certainly the case that information about the total number of purchases is useful for updating the probability that the first visitor makes a purchase (A_1). One can check that A_1 and B_0 are dependent, as are A_1 and B_1, A_1 and B_2, and A_1 and B_3.*

Likewise, whether a given individual makes a purchase is also informative about the number of total purchases that are made. For instance, if the first visitor makes a purchase (A_1), then

$$P(B_0|A_1) = \frac{P(A_1, B_0)}{P(A_1)} = 0$$

$$P(B_1|A_1) = \frac{P(A_1, B_1)}{P(A_1)} = \frac{(0.2)(0.8)(0.8)}{0.2} = 0.64$$

$$P(B_2|A_1) = \frac{P(A_1, B_2)}{P(A_1)} = \frac{(0.2)(0.2)(0.8) + (0.2)(0.8)(0.2)}{0.2} = 0.32$$

$$P(B_3|A_1) = \frac{P(A_1, B_3)}{P(A_1)} = \frac{(0.2)(0.2)(0.2)}{0.2} = 0.04.$$

These conditional probabilities differ from the unconditional probabilities of B_0, B_1, B_2, and B_3.

When events are not mutually independent, it can be more complicated to determine joint probabilities. The usual approach is to consider successive conditioning. This idea has been seen in the context of two events, where

$$P(A_1, A_2) = P(A_2|A_1)P(A_1) \text{ or } P(A_1, A_2) = P(A_1|A_2)P(A_2).$$

For three events, this idea can be extended to

$$P(A_1, A_2, A_3) = P(A_3|A_1, A_2)P(A_2|A_1)P(A_1).$$

The joint probability of $A_1 \cap A_2 \cap A_3$, which is $P(A_1, A_2, A_3)$, is the probability that A_1 occurs times the probability A_2 occurs given A_1 occurring, times the probability A_3 occurs given both A_1 and A_2 occurring. The order of events can be switched around however we would like. Depending upon the problem, it might be easier to find the joint probability with one ordering rather than another. For three events, the following successive conditioning equations are also true:

$$P(A_1, A_2, A_3) = P(A_2|A_1, A_3)P(A_3|A_1)P(A_1),$$

$$P(A_1, A_2, A_3) = P(A_3|A_1, A_2)P(A_1|A_2)P(A_2),$$

$$P(A_1, A_2, A_3) = P(A_1|A_2, A_3)P(A_3|A_2)P(A_2),$$

$$P(A_1, A_2, A_3) = P(A_2|A_1, A_3)P(A_1|A_3)P(A_3),$$

$$P(A_1, A_2, A_3) = P(A_1|A_2, A_3)P(A_2|A_3)P(A_3).$$

With four events,

$$P(A_1,A_2,A_3,A_4) = P(A_4|A_1,A_2,A_3)P(A_3|A_1,A_2)P(A_2|A_1)P(A_1),$$

and it is similar for any ordering of the conditioning. This idea generalizes to any number of events.

Example 3.15 (Drawing aces) *A standard deck of playing cards has 52 cards, including four aces. If a single card is randomly chosen from the deck, the chance of drawing an ace is 4/52, or 1/13. If two cards are randomly chosen from the deck (without replacement), what is the probability of drawing two aces? If A_1 denotes the event of drawing an ace on the first card and A_2 denotes the event of drawing an ace on the second card,*

$$P(A_1,A_2) = P(A_1)P(A_2|A_1) = \frac{4}{52}\cdot\frac{3}{51} = \frac{12}{2652} \approx 0.0045, \text{ or } 0.45\%.$$

The conditional probability $P(A_2|A_1)$ is 3/51 since there are only three aces left if A_1 has occurred. This idea can be extended to determine the probability of drawing three aces when three cards are randomly chosen from the deck (without replacement):

$$P(A_1,A_2,A_3) = P(A_1)P(A_2|A_1)P(A_3|A_1,A_2) = \frac{4}{52}\cdot\frac{3}{51}\cdot\frac{2}{50} \approx 0.00018, \text{ or } 0.018\%.$$

The conditional probability $P(A_3|A_1,A_2)$ is 2/50 since there are only two aces left if both A_1 and A_2 have occurred.

3.5 Examples with an Infinite Number of Outcomes

In some circumstances, we need to consider the probabilities associated with an infinite number of possible outcomes. Some infinite sample spaces were considered in section 2.1. As examples, we might be interested in the number of coin tosses until a heads is observed or the number of website visitors until a purchase is observed, both of which have $\mathcal{S} = \{1,2,3,\ldots\}$. Or we might be interested in the number of patents that a firm receives in a given year, the number of children in a family, the number of arrests of an individual, or the number of initial public offerings on the New York Stock Exchange in a given year. In this section, we discuss the examples of coin tosses and website purchases in more detail to develop an understanding of the issues involved with the number of possible outcomes being infinite.

Example 3.16 (Coin tosses until the first heads appears) *Suppose that a coin is repeatedly tossed. We assume that the coin tosses are mutually independent, specifically that the probability of a heads on any given toss is not affected by the outcome of any other toss. For these repeated coin tosses, how many tosses occur before a heads is observed? The sample space is $\mathcal{S} = \{1,2,3,\ldots\}$, with 1 corresponding to the first toss being heads, 2 corresponding to the first toss being tails and the second toss being heads, 3 corresponding to the first and second tosses being tails and the third toss being heads, and so on. Let A_k denote that event that the first heads occurs on the kth coin toss. The probability distribution is*

$$P(A_1) = \overbrace{\frac{1}{2}}^{H} = \frac{1}{2}$$

$$P(A_2) = \overbrace{\frac{1}{2} \cdot \frac{1}{2}}^{TH} = \frac{1}{4}$$

$$P(A_3) = \overbrace{\frac{1}{2} \cdot \frac{1}{2} \cdot \frac{1}{2}}^{TTH} = \frac{1}{8}$$

$$\vdots$$

$$P(A_k) = \frac{1}{2^k}$$

$$\vdots$$

The probability that the first heads appears on the 10*th toss is* $P(A_{10}) = \frac{1}{2^{10}} = \frac{1}{1024} \approx 0.00098$ *or approximately* 0.098%. *While this probability of* A_{10} *is quite low, it's not equal to zero, so it is possible that* A_{10} *occurs. This is true even for much larger k. While the probability of* A_k *gets smaller and smaller, it never reaches zero.*

In the repeated coin toss example, there are an infinite number of possible events $A_1, A_2, A_3, \ldots$ corresponding to when the first heads toss is observed. These events are disjoint. For their probabilities to constitute a probability distribution, it must be the case that they sum to 1. To show that $P(A_1) + P(A_2) + P(A_3) + \cdots = \sum_{j=1}^{\infty} P(A_j) = \sum_{j=1}^{\infty} \frac{1}{2^j}$ is equal to 1, we introduce a general fact about the sum of an **infinite geometric series**.

Proposition 3.7 *For real numbers a and r, with* $|r| < 1$, *the sum of an infinite geometric series is*

$$\sum_{j=0}^{\infty} ar^j = a + ar + ar^2 + ar^3 + \cdots = \frac{a}{1-r}.$$

The series is "geometric" since each successive term in the series is multiplied by the same constant r, and "infinite" since the terms in the series continue forever. The condition $|r| < 1$ ensures that the sum of the series has a well-defined value rather than diverging to infinity or negative infinity.

Example 3.17 (Coin tosses until the first heads appears) *Proposition 3.7 can be applied with* $a = 1/2$ *and* $r = 1/2$ *to yield*

$$P(A_1) + P(A_2) + P(A_3) + \cdots = \sum_{j=1}^{\infty} P(A_j) = \frac{1/2}{1 - 1/2} = 1,$$

confirming that the probabilities constitute a probability distribution over the sample space $S = \{1, 2, 3, \ldots\}$. *We can use proposition 3.7 to determine other probabilities for this experiment. For example, the probability that it takes at least four tosses to see heads is*

$$P(A_4) + P(A_5) + P(A_6) + \cdots = \frac{1}{16} + \frac{1}{32} + \frac{1}{64} + \cdots = \frac{1/16}{1 - 1/2} = \frac{1}{8}.$$

Alternatively, to determine this probability, we can think about the complement of the event that the heads occurs during the first three tosses, whose probability is

$$1-(P(A_1)+P(A_2)+P(A_3))=1-\left(\frac{1}{2}+\frac{1}{4}+\frac{1}{8}\right)=\frac{1}{8}.$$

Example 3.18 (Website visitors until a purchase) *The example of purchase behavior for three website visitors can be extended to allow an infinite stream of website visitors. Assume the independence of each visitor's purchase behavior; that is, for any subset of visitors, the individual purchase events are mutually independent. Assume that the probability of a purchase by any given visitor is 20%. Similar to the coin toss example, let A_k denote the event corresponding to the first purchase being made by the kth website visitor. The probabilities of $A_1, A_2, \ldots$ are*

$$P(A_1)=\overbrace{0.2}^{Y}=0.2$$

$$P(A_2)=\overbrace{(0.8)(0.2)}^{NY}=0.16$$

$$P(A_3)=\overbrace{(0.8)(0.8)(0.2)}^{NNY}=0.128$$

$$\vdots$$

$$P(A_k)=(0.8)^{k-1}(0.2)$$

$$\vdots$$

For A_k to occur, the first $k-1$ visitors do not make a purchase and then the kth visitor makes a purchase, which corresponds to $P(A_k)=(0.8)^{k-1}(0.2)$. These probabilities sum to 1 since proposition 3.7 can be applied with $a=0.2$ and $r=0.8$. What is the probability that it takes at least 10 website visitors before a purchase is observed? Applying proposition 3.7 yields

$$P(A_{10})+P(A_{11})+\cdots=(0.8)^9(0.2)+(0.8)^{10}(0.2)+\cdots=\frac{(0.8)^9(0.2)}{1-0.8}=(0.8)^9\approx 0.134.$$

We get $(0.8)^9$ as the probability here, which is the equal to the probability that the first nine visitors are nonpurchasers, an event that is equivalent to seeing at least 10 visitors before a purchase.

Example 3.19 (Number of patents) *Let the sample space $\mathcal{S}=\{0,1,2,3,\ldots\}$ contain the possible outcomes for the number of patents that a firm is awarded in a given year. Let A_k denote the event that the firm is awarded k patents in a given year. Assume that the probability of being awarded no patents is $P(A_0)=0.5$ and the probability of being awarded one patent is $P(A_1)=0.3$. If the probabilities for A_k for $k\geq 2$ (events associated with two or more patents) are $P(A_k)=0.3c^{k-1}$ for some constant c, what must the value of c be? Note that $P(A_2)=0.3c$, $P(A_3)=0.3c^2$, $P(A_4)=0.3c^3$, and so on. Since the events $A_0, A_1, A_2, \ldots$ are disjoint and exhaustive, their probabilities must sum to 1:*

$$P(A_0)+P(A_1)+P(A_2)+P(A_3)+\cdots=0.5+0.3+0.3c+0.3c^2+\cdots=0.5+\frac{0.3}{1-c},$$

so that $c = 0.4$. *For this value of c,* $P(A_2) = 0.12$, $P(A_3) = 0.048$, $P(A_4) = 0.0192$, *and so on.*

Exercises

1. A survey of college students with cell phones finds that 50% have an iPhone (I), 30% have a Samsung phone (S), and 20% have another brand of phone (A). Moreover, 70% of iPhone owners have unlimited data plans, 60% of Samsung owners have unlimited data plans, and 45% of other phone owners have unlimited data plans. Let U denote the event for having an unlimited data plan.
   a. What is the probability that a surveyed student has a Samsung phone and an unlimited data plan?
   b. What is the probability that a surveyed student has an unlimited data plan?
   c. For a surveyed student with an unlimited data plan, what is the probability that the student owns an iPhone?
2. The percentages of blood types in the US (ignoring positive and negative subtypes) are as follows: A (40%), B (11%), AB (4%), and O (45%). Suppose that two individuals are drawn randomly from the population and that their blood types are independent of each other.
   a. What is the probability that both individuals have blood type A?
   b. What is the probability that the blood types of the two individuals are the same?
   c. What is the probability that the blood types of the two individuals are different?
3. You have three different e-mail accounts, with 70% of e-mails coming into account A, 20% coming into account B, and 10% coming into account C. The likelihood that an e-mail is spam is 1%, 2%, and 5% for the three accounts, respectively.
   a. What is the probability that a randomly selected e-mail is spam?
   b. If you know that a received e-mail is spam, what is the probability that it is in account A?
4. An inventor submits two patent applications. Let P_1 be the event that the first patent application is successful, and let P_2 be the event that the second patent application is successful. Assume that P_1 and P_2 are independent, with $P(P_1) = 0.6$ and $P(P_2) = 0.3$.
   a. If the first patent application is unsuccessful, what is the probability that the second application is successful?
   b. What is the probability that at least one of the two patent applications is successful?
   c. If you know that at least one of the patent applications is successful, what is the probability that only the second application is successful?
5. One of two possible suspects (Bert or Ernie) commits a crime. Bert has more of a checkered past than Ernie, so suppose that the prior probability of Bert committing the crime is 70% and the prior probability of Ernie committing the crime is 30%. There was

an eyewitness to the crime who says that she saw Ernie do it. The probability that the eyewitness is correct equals 90%. That is, the probability of the eyewitness saying that she saw Ernie, given that Ernie committed the crime, is 0.9. Similarly, the probability of the eyewitness saying she saw Bert, given that Bert committed the crime, is 0.9. What is the posterior probability that Ernie committed the crime, given the eyewitness testimony?

6. The probability that an individual in the US has a certain disease is 5%. There is a diagnostic test that correctly detects the disease 97% of the time (3% false negative) and correctly detects absence of the disease 99% of the time (1% false positive). A randomly selected individual has a diagnostic test, and the result is positive (i.e., it detects the disease). What is the posterior probability that the individual has the disease?
7. At a certain large company, 70% of the economists and 50% of the data scientists have a graduate degree, respectively. Furthermore, 80% of the employees at the company are economists and 20% are data scientists. If a randomly chosen employee has a graduate degree, what is the probability that the employee is an economist?
8. For a certain election, the probability of a registered voter being a "young voter" (i.e., under 40) is 35%, and the probability of a registered voter being an "older voter" (i.e., at least 40) is 65%. Registered young voters are less likely to actually vote (55% probability) in the election than registered older voters (75% probability). If a randomly selected registered voter did not vote in the election, what is the probability that it is a young voter?
9. The use of generative artificial intelligence (AI) tools to assist with writing has become prevalent. It is sometimes possible to detect the use of an AI tool based upon words that are more frequently used by an AI tool as compared to a human. This question considers a simple example. Suppose that, prior to the introduction of AI tools, the probability that a student uses the word "delve" in a submitted paper is 3%, whereas after the introduction of AI tools, the probability jumps to 15%. Let g denote the probability that a student uses an AI tool (after its introduction). Assume that the probability that a student uses the word "delve" remains at 3% conditional on the student not using the AI tool (after its introduction).
   a. If $g = 1$ (all students use the AI tool), what is the probability that a student uses "delve" in a submitted paper after the introduction of AI tools?
   b. For this part, assume that the probability that a student uses the AI tool is 40% (i.e., $g = 0.4$).
      i. What is the probability that a student uses "delve" in a submitted paper if the student uses an AI tool?
      ii. If "delve" is used in a submitted paper, what is the probability that the student used an AI tool?
   c. *For this part, assume that g is unknown. What is the lowest possible value of g?
10. Two companies, Acme Manufacturing and Bolts Emporium, have shares of stock that trade on an exchange. On any given day, the following probability table describes the performance of the two stocks, where each stock can go up in price, go down in price, or stay the same:

		Bolts Emporium		
		Down	No Change	Up
	Down	0.18	0.02	0.22
Acme Manufacturing	No Change	0.03	0.02	0.04
	Up	0.14	0.05	0.30

a. What is the probability that Acme Manufacturing's stock goes up on a given day?

b. What is the probability that Bolts Emporium's stock has some change ("Down" or "Up") on a given day?

c. What is the probability that at least one of the two stocks has no change on a given day?

d. Are the events "Acme Manufacturing's stock goes up" and "Bolts Emporium's stock goes up" independent?

e. What is the probability that Acme Manufacturing's stock goes up on a day that Bolts Emporium's stock goes up?

f. What is the probability that Acme Manufacturing's stock goes up on a day that Bolts Emporium's stock doesn't go down?

g. If exactly one of the two stocks goes up on a given day, what is the probability that it is Acme Manufacturing?

11. The following table gives the joint counts of age and rank of the faculty of a major university in a recent year. There are a total of 1,164 faculty members in the table:

		Rank		
		Assistant	Associate	Full
	Under 30	57	3	2
	30–39	163	170	54
Age	40–49	61	125	160
	50–59	36	68	155
	60 and over	3	15	92

Interpret the relative frequencies as true probabilities for the purposes of this question. For example, the joint probability that a randomly chosen professor is under 30 and an assistant professor is $\frac{57}{1164}$.

a. What is the probability that a randomly chosen faculty member is a full professor?

b. What is the probability that a randomly chosen faculty member is 50+ years old?

c. What is the probability that a randomly chosen faculty member is 50+ years old and a full professor?

d. What is the probability that a randomly chosen faculty member is 50+ years old or a full professor?

e. What is the probability that a professor who is 50+ years old is a full professor?

f. What is the probability that a full professor is 50+ years old?

12. Suppose that on any given weekday, 70% of college students eat breakfast, 60% do homework, and 85% do at least one of these two things.
    a. Are the events "Student eats breakfast" and "Student does homework" independent?
    b. If a randomly selected student eats breakfast, what is the probability that she or he does homework?
    c. Suppose that two students are selected at random and their behaviors are independent of each other. What is the probability that exactly one of the two students does homework?
    d. Suppose that 80% of students who eat breakfast on a given day also eat lunch that day, while 90% of students who don't eat breakfast on a given day eat lunch that day. For a student who eats lunch on a given day, what is the probability that she or he eats breakfast on that day?
13. Suppose that A and B are independent events.
    a. Show that A and B^c are independent events.
    b. Show that A^c and B^c are independent events.
14. Consider a group of 10 stocks, with 20% (2 stocks) corresponding to "tech" companies and 80% (8 stocks) corresponding to "nontech" companies. Suppose that you randomly pick 2 different stocks, with T_1 denoting that the first stock is tech and T_2 denoting that the second stock is tech.
    a. Determine $P(T_1)$, $P(T_2)$, and $P(T_1 \cup T_2)$. Are T_1 and T_2 independent?
    b. Repeat (a), but assume that there are 100 stocks, with 20% (20 stocks) corresponding to tech companies.
    c. Repeat (a), but assume that there are 1,000 stocks, with 20% (200 stocks) corresponding to tech companies.
    d. In considering the independence checks in (a)–(c), what can you say about independence becoming approximately correct as the number of stocks increases?
15. Four employees at a given company individually go out for lunch. Suppose that they each randomly and independently choose a restaurant from among seven choices.
    a. What is the probability that each employee goes to a different restaurant?
    b. What is the probability that all four employees go to the same restaurant? (The restaurant can be any of the seven possibilities.)
    c. Conduct 10,000 computer simulations in *R* to confirm your answers to (a) and (b).
16. A space technology company is attempting to launch four satellites into orbit, using four separate rockets. Assume that the probability of success for each launch is 90% and the launches are independent of each other.
    a. What is the probability distribution of the total number of successful launches?
    b. Conditional on knowing that at least three launches were successful, what is the probability that exactly three launches were successful?

17. You have three quarters, two dimes, three nickels, and one penny in your pocket. When removing a coin from your pocket, you may assume that there is an equal chance of picking any of the coins that are in your pocket.
    a. If you randomly remove one coin, what is the probability that a quarter is not chosen?
    b. If you randomly remove one coin at a time, without putting coins back into your pocket, what is the probability that the first quarter is not chosen until the third coin or later?
    c. Rather than being interested in the sequence of coins, suppose that you are interested in the total amount of money that you take out of your pocket before the first quarter is taken out. What are the outcomes in the sample space S for this experiment?
    d. Conduct 10,000 computer simulations in R to approximate the probability that exactly \$0.15 is taken out of your pocket before the first quarter is taken out.
18. Suppose that you repeatedly and independently roll a fair, six-sided die. Let S_1 denote the event that the first six appears on the first roll, S_2 denote the event that the first six appears on the second roll, and so on.
    a. What is $P(S_1)$?
    b. What is $P(S_k)$ for an arbitrary $k \in \{1, 2, 3, \ldots\}$?
    c. Show that the $P(S_k)$ probabilities sum to 1.
    d. What is the probability that it takes at least 10 rolls for a 6 to appear?
    e. What is the probability that the first 6 appears on an even roll (i.e., for $k \in \{2, 4, 6, 8, \ldots\}$)?
    f. What is the probability that the first 6 appears on the second roll, given that the first 6 appears on an even roll?
    g. Conduct 10,000 computer simulations in R to confirm your answer in (d). For each simulation, use a `while` loop to repeatedly simulate a die roll until a 6 appears.
    h. Modifying your code from (g) as necessary, what is the average number of rolls that it takes for a 6 to appear over the 10,000 simulations?
19. Two individuals A and B play the following dice game: the players alternate rolling a fair, six-sided die, and a player wins if she rolls the same number as the previous player did. The players continue alternating rolls until someone wins. Assume that player A starts and can't win on the first roll (since there was no previous roll). What is the probability that player A wins?
20. An electronics store (Electric City) sells laptop computers. On any given day, the probability that Electric City doesn't sell a laptop computer is 0.35, and the probability that it sells k laptop computers (for $k > 0$) is $0.25c^{k-1}$.
    a. What is the value of c?
    b. What is the probability that the store sells at least three laptop computers?
    c. Another store in the same city (Computer King) also sells laptop computers. On any given day, this store has probabilities 0.20, 0.50, and 0.30 of selling zero, one, and two laptop computers, respectively. If sales at the two stores are independent

of each other, what is the probability that Electric City sells more laptop computers than Computer King on any given day?

d. Using the same information provided in (c), on any given day, what is the probability that Electric City sells exactly two laptop computers, given that Electric City sells more laptop computers than Computer King?

21. *A gambler is playing a casino game where they win \$100 with probability p and lose \$100 with probability $1-p$, with $p<0.5$ (so that the casino has an advantage).

a. Suppose that the gambler starts with \$200. If the gambler repeatedly plays the game until she loses all her money or until she has \$400, what is the probability (in terms of p) that she loses all her money?

b. Conduct 10,000 simulations in *R* to confirm your answer in (a) for the values $p=0.48$ (small casino advantage) and $p=0.45$ (large casino advantage).

22. *In college volleyball, the first team to score 25 points wins a set. But, if the teams are tied at 24–24, the set continues until one team has 2 points more than the other team (i.e., 26–24, 27–25, or 28–26, and so on). Suppose that two teams, call them team *A* and team *B*, have reached a 24–24 score. Team *A* is the better team, and the probability that team *A* wins any future point is 60%. Assume that all future points are mutually independent.

a. What is the probability that team A wins the set by a 26–24 score?

b. What is the probability that the set becomes tied at 25-25?

c. What is the probability that team A wins the set?

d. Conditional on team A winning the set, what is the probability that team A wins the set by a 26–24 score?

e. Conduct 10,000 simulations in *R* to confirm your answer in (c). For each simulation, use a `while` loop to repeatedly simulate points until one of the two teams wins.

f. Modify your code from (e) to record, for each simulated game, the number of points scored by the winning team. What is the average winning score over the 10,000 simulations?

4

Combinatorics (Counting Methods)

Combinatorics is a topic in mathematics that involves counting methods. In many interesting situations, our ability to determine probabilities requires that we count and/or enumerate the possible outcomes. Chapters 2 and 3 considered some simple examples (e.g., sequences of coin-toss outcomes and sequences of customer-purchase outcomes) of counting different types of outcomes. As another example, consider a lottery where the winning numbers are 4 distinct numbers that the lottery authority chooses from the set $\{1, 2, \ldots, 30\}$. How many ways can the 4 numbers be drawn? If you were to play the lottery at random (i.e., randomly picking 4 numbers from the 30 possible numbers), what is your chance of winning the lottery?

4.1 Product Rule and Sum Rule

The first counting tools introduced here are the **product rule** and the **sum rule**. In the simplest case, we consider a situation in which two actions are possible.

Proposition 4.1 (Product rule) *If there are m choices for the first action and n choices for the second action, the number of possible choices for both actions is mn.*

Example 4.1 (Retirement savings) *Your company's retirement plan has 8 different bond funds and 12 different stock funds. For your portfolio, you are required to pick one bond fund and one stock fund. How many possible ways are there to pick a portfolio? Since there are 8 ways to take the first action (choice of bond fund) and 12 ways to take the second action (choice of stock fund), the product rule implies that the number of ways to pick a portfolio is* $(8)(12) = 96$.

Proposition 4.2 (Sum rule) *If there are m choices for the first action, n choices for the second action, and* only one *of the actions can be taken, the number of possible choices is $m + n$.*

Example 4.2 (Retirement savings) *If instead of choosing both a bond fund and a stock fund, suppose that you are restricted to choose only one fund overall. With 8 options for the bond fund and 12 options for the stock fund, the number of possible choices for your one fund is* $8 + 12 = 20$.

Both the product rule and the sum rule generalize to more actions, as in proposition 4.3 and proposition 4.4., respectively.

Proposition 4.3 (Generalized product rule) *Suppose that there are k actions. If there are n_1 choices for the first action, n_2 choices for the second action, and so on through n_k choices for the kth action, the number of possible choices for the k actions is the product of the numbers of choices, $n_1 n_2 \cdots n_k$.*

Proposition 4.4 (Generalized sum rule) *Suppose that there are k actions. If there are n_1 choices for the first action, n_2 choices for the second action, and so on through n_k choices for the kth action, the number of possible choices when only one action can be chosen is the sum of the numbers of choices, $n_1 + n_2 + \cdots + n_k$.*

Example 4.3 (Retirement savings) *The retirement plan adds money-market fund options, such that there are* 8 *bond funds,* 12 *stock funds, and* 4 *money-market funds. How many ways are there to pick a portfolio with one bond fund, one stock fund, and one money-market fund? The product rule implies that the total number of choices for the portfolio is* $(8)(12)(4) = 384$. *If you're restricted to picking just one of the funds, the number of choices from the sum rule is* $8 + 12 + 4 = 24$.

4.2 Permutations and Combinations

Consider a situation in which there is a group with n distinct objects. We are interested in knowing how many ways there are to choose a group of k elements from the group, where $k \leq n$. To count the number of ways, we need to be more precise about (1) whether repeats among the chosen elements are allowed and (2) whether the order of the chosen elements matters.

Here is an example where repeats are allowed and the order of the elements matters.

Example 4.4 (Four-letter password) *You are asked to provide a four-letter password, where each letter must be uppercase. The order of the letters matters here (e.g., ASJV is different from SVJA). If repeated letters are allowed, the number of possible four-letter passwords is* $(26)(26)(26)(26) = 26^4 = 456{,}976$ *by the product rule.*

Here is an example where repeats are allowed and the order of the elements does not matter.

Example 4.5 (Choice of bonus stocks) *An online trading app is offering a sign-up bonus, where you receive three shares from a group of* 10 *possible stocks* $\{S_1, S_2, \ldots, S_{10}\}$. *You choose any three stocks and are allowed to have repeats. That is, you can choose to receive three shares of one company, such as* S_6. *The order in which you pick the three shares is unimportant here. The product rule cannot be directly applied here. The number of possible three-share portfolios is* not $(10)(10)(10) = 1{,}000$ *since the order of the shares doesn't matter. For instance, the choice* $S_1S_7S_{10}$ *is the same as any of the following choices:* $S_1S_{10}S_7$, $S_7S_1S_{10}$, $S_7S_{10}S_1$, $S_{10}S_1S_7$, *or* $S_{10}S_7S_1$.

Here is an example where repeats are not allowed and the order of the elements matters.

Example 4.6 (Board of directors with titled positions) *A company needs to select a three-member board of directors and must first designate one member as Chair, then one member as Vice Chair, and then one member as Secretary. If there are* 20 *candidates available for the board positions, how many ways can the board be selected? There are* 20 *ways to fill the President position, after which there are* 19 *possible ways to fill the Vice Chair position, after which there are* 18 *possible ways to fill the Secretary position. The product rule then implies that the total number of ways to select the board is* $(20)(19)(18) = 6{,}840$*. It doesn't matter how we think about filling the positions. We could start with Vice Chair, then Secretary, and then President, and it would still be* 20 *choices, then* 19 *choices, and then* 18 *choices. An important feature here is that you can't have repeats in the three positions since three distinct board members must be selected.*

Here is an example where repeats are not allowed and the order of the elements does not matter.

Example 4.7 (Board of directors with no titled positions) *Consider the same setting as example 4.6, but without titled positions. The company needs to select a three-member board of directors from a group of* 20 *candidates, but the order in which the candidates is chosen is irrelevant. Let's say that the possible candidates are* $\{C_1, C_2, \ldots, C_{20}\}$*. Since the order doesn't matter, the number of possible three-member boards is going to be considerably less than* $(20)(19)(18) = 6{,}840$*. Any ordered choice of three members is equivalent to five other ordered choices of three members. For example, the choice* $C_3C_{11}C_{19}$ *is equivalent to the following five choices:* $C_3C_{19}C_{11}$, $C_{11}C_3C_{19}$, $C_{11}C_{19}C_3$, $C_{19}C_3C_{11}$, *or* $C_{19}C_{11}C_3$*. As a result, the total number of possible unordered three-person boards is equal to* $6{,}840/6 = 1{,}140$.

Using the last two examples as motivation, we formally define the terms **permutation** and **combination**, which refer to an ordered subset of distinct choices and an unordered subset of distinct choices, respectively.

Definition 4.1 *An ordered subset of distinct choices is called a* ***permutation****, and* $P_{n,k}$ *denotes the number of permutations of size* k *that can be formed from* n *objects (for* $k \leq n$*).*

Proposition 4.5 (Number of permutations) *The number of permutations of size* k *that can be formed from* n *objects (for* $k \leq n$*), denoted as* $P_{n,k}$*, is*

$$P_{n,k} = \frac{n!}{(n-k)!},$$

where $j! = (1)(2)\cdots(j)$*, read as "*j *factorial," is the product of all positive integers through* j*. An equivalent formula is*

$$P_{n,k} = (n)(n-1)\cdots(n-k+1).$$

Example 4.8 (Board of directors with titled positions) *The number of permutations for the three titled board positions in example 4.6 is* $P_{20,3} = \frac{20!}{17!} = (20)(19)(18) = 6{,}840$.

The number of permutations can be calculated in *R* using either formula in proposition 4.5, with the function `factorial` used for the first formula.

```
factorial(20)/factorial(17)

## [1] 6840

20*19*18

## [1] 6840
```

Example 4.9 (Stock portfolio with different weights) *Suppose that there are* 100 *possible stocks, and you want to form a* 30%/25%/20%/15%/10% *weighted portolio of five stocks. How many possible portfolios are there? The order of the five stocks matters here since a different weight is being placed on each of the five choices. Then the number of portfolio choices is* $P_{100,5} = \frac{100!}{95!} = (100)(99)(98)(97)(96) = 9{,}034{,}502{,}400$ *(over 9 billion!).*

Moving to the case of unordered subsets of choices, a combination is defined as in definition 4.2.

Definition 4.2 *An unordered subset of choices is called a* ***combination****, and* $C_{n,k}$ *denotes the number of combinations of size k that can be formed from n objects (for* $k \leq n$*).*

Proposition 4.6 (Number of combinations) *The number of combinations of size k that can be formed from n objects (for* $k \leq n$*), denoted as* $C_{n,k}$*, is*

$$C_{n,k} = \frac{P_{n,k}}{k!} = \frac{n!}{k!(n-k)!}.$$

Sometimes the notation $\binom{n}{k}$*, read as "n choose k," is used as an alternative to the* $C_{n,k}$ *notation, so*

$$\binom{n}{k} = C_{n,k} = \frac{n!}{k!(n-k)!}.$$

Example 4.10 (Board of directors with no titled positions) *The number of combinations for the three untitled board positions in example 4.7 is* $C_{20,3} = \binom{20}{3} = \frac{20!}{3!17!} = \frac{(20)(19)(18)}{(3)(2)(1)} = 1{,}140$*. Proposition 4.6 provides an exact relationship between* $C_{20,3}$ *and* $P_{20,3}$*, specifically* $C_{20,3} = \frac{P_{20,3}}{3!} = \frac{P_{20,3}}{6}$*. As discussed previously, any ordered choice of three members is equivalent to five other ordered choices of those three members. Why is that the case? For a group of three members, there are six different orderings: three choices for the first member, then two choices for the second member, and then one choice for the third member, which is a total of* $(3)(2)(1) = 3! = 6$*. The division of* $P_{20,3}$ *by* $3!$ *in the* $C_{20,3}$ *formula accounts for the fact that the* 6 *possible orderings of any three members is equivalent to just one unordered group of the three members. The R function* `choose(n, k)` *calculates* $\binom{n}{k} = C_{n,k}$*.*

```
choose(20,17)

## [1] 1140
```

Example 4.11 (Stock portfolio with equal weights) *Suppose that there are* 100 *possible stocks. You want to invest equally (*20% *each) in five different stocks. How many possible portfolios are there? The order of the five stocks does not matter since it's an* *equally weighted portfolio. Then the number of portfolio choices is* $C_{100,5} = \binom{100}{5} = \frac{100!}{5!95!} = \frac{(100)(99)(98)(97)(96)}{(5)(4)(3)(2)(1)} = 75{,}287{,}520$*, still a large number but smaller than* $P_{100,5}$ *by a factor of* $5! = 120$*).*

Example 4.12 (Choice of bonus stocks) *We return to example 4.5, where we considered an online trading app bonus in which the user is allowed to choose three shares from the set of* 10 *possible stocks* $\{S_1, S_2, \ldots, S_{10}\}$*. The important difference from example 4.11 is that the user is not restricted to picking three distinct stocks. There are three possibilities: (1) the user's choice of shares involves three distinct stocks; (2) the user's choice of shares involves two distinct stocks, with two shares of one stock and one share of the other stock; and (3) the user's choice of shares involves one stock, with all three shares being that stock. If we can determine the number of possible choices associated with each of these three possibilities, the sum rule can be used to determine the total number of possible choices. For case (1), where the choice of shares involves three distinct stocks, the number of possible choices is* $C_{10,3} = \binom{10}{3} = 120$ *since the order of the three stocks doesn't matter (each has one share). For case (2), where the choice of shares involves two distinct stocks, the number of possible choices is* $P_{10,2} = (10)(9) = 90$ *since the order of the two stocks matters (one has two shares, the other has one share). For case (3), where the choice of shares involves one stock, the number of possible choices is just* 10*. Thus, by the sum rule, the total number of possible choices is* $120 + 90 + 10 = 220$*.*

4.2.1 Generalizing the Combination Formula

The combination formula $\binom{n}{k} = C_{n,k} = \frac{n!}{k!(n-k)!}$, from proposition 4.6, gives the number of combinations of k objects that can be formed from n objects ($k \leq n$), where the ordering of k objects does not matter. Alternatively, one can think of $\binom{n}{k}$ as the number of ways that n objects can be split into two subsets of k and $n-k$ objects. Viewed in this way, we can generalize the idea and formula to situations where n objects are to be split into more than two subsets. Proposition 4.7 provides this generalization.

Proposition 4.7 *The number of ways that n objects can be split into* $m \geq 2$ *different subsets of size* $k_1, k_2, \ldots, k_m$*, where* $k_1 + k_2 + \cdots + k_m = n$*, is*

$$\binom{n}{k_1, k_2, \ldots, k_m} = \frac{n!}{k_1!\, k_2! \cdots k_m!}.$$

Example 4.13 (Academic committees) *An economics department has* 26 *faculty members, and it needs to create three different committees: an admissions committee with* 12 *members, a curriculum committee with* 8 *members, and an alumni relations committee with* 4 *members. Suppose that no faculty member can serve on more than one committee, so there are* 2 *faculty members who are not on a committee. Then, the total number of ways that the committees can be formed is*

$$\binom{26}{12, 8, 4, 2} = \frac{26!}{12!8!4!2!} = 435{,}031{,}096{,}500.$$

In this example, there are $m = 4$ *subsets, three for the three committees and the fourth for noncommittee members. While the generalized combination formula provides a direct way to calculate the number of combinations, it would also be possible to calculate it using the original* $\binom{n}{k}$ *definition. Specifically, the number of ways to pick the admissions committee is* $\binom{26}{12}$, *which allows* $\binom{14}{8}$ *ways to choose the curriculum committee, which allows* $\binom{6}{4}$ *ways to choose the alumni relations committee. The reader can verify that* $\binom{26}{12}\binom{14}{8}\binom{6}{4}$, *obtained by the multiplication rule, is also equal to* $\frac{26!}{12!8!4!2!}$.

What if the department also needs to specify a chair for each of the three committees? We can think of there being $m = 7$ *subsets now, rather than* $m = 4$ *subsets, as each of the three committees is effectively split into two subsets, one for the chair member (*1 *faculty member) and one for the nonchair members (*11 *for admissions,* 7 *for curriculum,* 3 *for alumni relations). The number of possible ways to form the committees, with the chairs specified, is considerably larger than the answer above and is equal to*

$$\binom{26}{11,1,7,1,3,1,2} = \frac{26!}{11!1!7!1!3!1!2!} = 167{,}051{,}941{,}056{,}000.$$

4.3 Probabilities for Equally Likely Choices

In some situations, we are interested in the probability of seeing certain permutations or combinations. In the case where any possible permutation or combination is equally likely, it is often possible to figure out the probability either directly, *using math*, or indirectly, *using computer simulation*. For now, let's focus on what's possible with mathematical reasoning.

Example 4.14 (Website names) *It's well known that shorter website names are more valuable. In fact, it's highly unlikely that a three-character ,com website would be available for registration. How many three-character ,com websites are there? By "character," we mean any letter a through z or number* 0 *through* 9. *Website names are not case-sensitive, so we'll just assume that the letters are lowercase. Since repeated characters are allowed, the total number of three-character websites is* $(36)(36)(36) = 46{,}656$ *by the product rule.*

If one of the three-character websites is chosen at random, each with the same probability $\frac{1}{46{,}656}$, *what is the probability that the chosen website is composed entirely of alphabetic characters and no numbers? To answer this question, the simplest approach is to determine how many of the* 46,656 *three-character websites are composed entirely of alphabetic characters. By the product rule, the total number of websites composed entirely of alphabetic characters is* $(26)(26)(26) = 17{,}576$. *Therefore, the probability that a randomly chosen three-character website consists of only alphabetic characters is* $\frac{17{,}576}{46{,}656} \approx 0.377$ *or* 37.7%.

*If one of the three-character websites is chosen at random, what is the probability that the chosen website has distinct alphabetic characters? Again, we can count the number of three-character websites with three distinct alphabetic characters and then divide by the total number of websites (*46,656*). Since the alphabetic characters must be distinct and order matters, the number of three-character websites with three distinct alphabetic characters is equal to* $C_{26,3} = (26)(25)(24) = 15{,}600$. *Therefore, the probability that a randomly chosen three-character website consists of three distinct alphabetic characters is* $\frac{15{,}600}{46{,}656} \approx 0.334$ *or* 33.4%.

Example 4.15 (Stock portfolio with sectors) *Consider again the problem of forming a five-stock, equally weighted portfolio, with* 20% *weight on each stock, but now suppose that* 20 *of the* 100 *stocks are classified as "tech" stocks and* 80 *of the* 100 *stocks are classified as "nontech" stocks. How many equally weighted, five-stock portfolios have exactly two tech stocks and three nontech stocks? There are* $C_{20,2} = \binom{20}{2}$ *ways of choosing the tech stocks (the ordering of the two stocks doesn't matter) and* $C_{80,3} = \binom{80}{3}$ *ways of choosing the nontech stocks, so by the product rule, the total number of five-stock portfolios with two tech stocks and three nontech stocks is* $\binom{20}{2}\binom{80}{3}$.

If we form a five-stock equally weighted portfolio by picking five stocks from the 100 *total at random, what is the probability the portfolio has two tech stocks and three nontech stocks? We've already found the relevant numerator, which is* $\binom{20}{2}\binom{80}{3}$. *The relevant denominator is the total number of possible five-stock portfolios, which is* $C_{100,5} = \binom{100}{5}$. *Thus, the probability is*

$$\frac{\binom{20}{2}\binom{80}{3}}{\binom{100}{5}} = \frac{\frac{(20)(19)}{(2)(1)} \cdot \frac{(80)(79)(78)}{(3)(2)(1)}}{\frac{(100)(99)(98)(97)(96)}{(5)(4)(3)(2)(1)}} \approx 0.207 \text{ or } 20.7\%.$$

How about the probability that a randomly chosen five-stock portfolio has at most *two tech stocks? Using similar reasoning, the number of five-stock portfolios with zero tech stocks is* $\binom{20}{0}\binom{80}{5}$ *and the number with one tech stock is* $\binom{20}{1}\binom{80}{4}$. *Thus, the probability of having* at most *two tech stocks in a randomly chosen, five-stock portfolio is*

$$\frac{\binom{20}{0}\binom{80}{5} + \binom{20}{1}\binom{80}{4} + \binom{20}{2}\binom{80}{3}}{\binom{100}{5}} \approx 0.947, \text{ or } 94.7\%.$$

Although we have used combinatorics to answer interesting probability questions in the examples given in this chapter, some problems are too complicated to solve analytically in this way (or at least they are so complicated that you would rather not try!). Such an example is considered next and illustrates how computer simulation offers an alternative method for calculating probabilities.

Example 4.16 (The likelihood of streaks) *If a coin is tossed* 100 *times, what is the probability of observing a streak of at least five consecutive heads during the* 100 *tosses? (Try guessing before reading any further.) With two possibilities for each toss, there are* 2^{100} *possible* 100*-coin sequences by the product rule. Moreover, if the coin tosses are independent of each other, each of these* 2^{100} *sequences must be equally likely. Therefore, the probability of observing a streak of at least five consecutive heads during the* 100 *tosses is equal to* $\frac{S}{2^{100}}$, *where S is the number of* 100*-coin sequences with a streak of at least five consecutive heads. Unfortunately, it is extremely difficult to analytically determine the value of S. An alternative approach is to use computer simulation, as follows:*

1. *Step 1: Simulate the experiment of flipping* 100 *coins, with probability* 1/2 *of heads and probability* 1/2 *of tails for each toss.*
2. *Step 2: For the simulated sequence of* 100 *coin tosses, check and record whether there is a streak of at least five heads.*
3. *Repeat Steps 1 and 2* many (100,000) *times, and then determine the frequency or proportion of simulated* 100*-coin sequences that include a streak of at least five heads.*

Figure 4.1 shows results based upon 100,000 *computer simulations, repeating steps 1 and 2 for each simulation. It shows how the calculated frequency of five-head streaks changes*

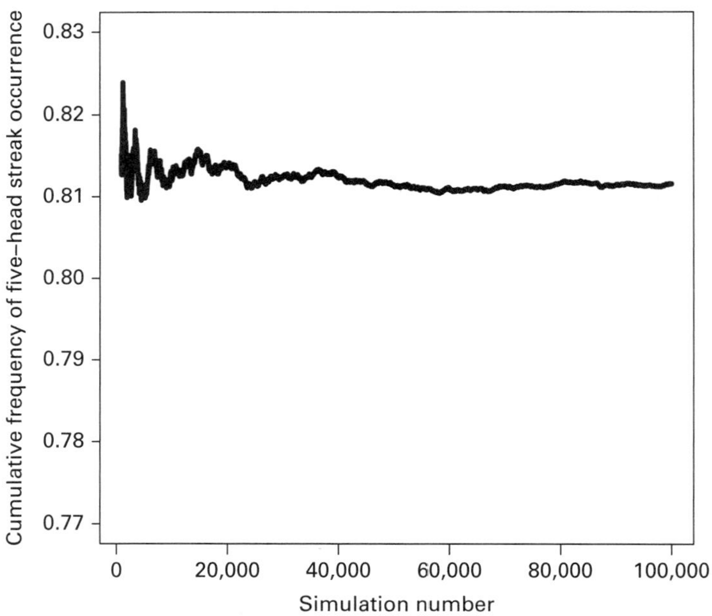

Figure 4.1
Simulated frequency of a five-head streak in 100 coin tosses

over the course of the 100,000 *simulations, with the x-axis indicating the number of the simulation. (The figure starts the x-axis at the 1,000th simulation since the frequencies are inherently noisier after a small number of simulations. Dropping these early simulations allows us to focus more clearly on the relevant range where the simulated frequencies end up.)*

Figure 4.1 is generated by the following R code:

```
set.seed(1234)

# initialize the number of simulations and number of tosses
num_simulations <- 100000
num_tosses <- 100

# initialize a vector to store the cumulative count of streaks
cumul_streaks <- rep(0, num_simulations)

# initialize a counter for the number of streaks observed
streak_counter <- 0

# loop with simulated coin tosses and check for streak
for (i in 1:num_simulations) {
        # simulate the coin tosses
        tosses <- sample(c("H","T"), num_tosses, replace = TRUE)

        # paste together the toss outcomes into a single string
        toss_string <- paste(tosses, collapse = "")

        # add one to streak counter if streak of five heads occurs
        streak_counter <- streak_counter + grepl("HHHHH", toss_string)
```

```
        cumul_streaks[i] <- streak_counter
}

# create a vector of the cumulative frequencies
freq_streaks <- cumul_streaks/(1:num_simulations)

# output the final frequency (approximated probability)
freq_streaks[num_simulations]

## [1] 0.81156

# plot the cumulative frequencies against the simulation number
plot(1000:num_simulations,freq_streaks[1000:num_simulations],
     xlab="Simulation number",ylab="Cumulative frequency of five-head
     streak occurrence", cex=0.5, ylim=c(0.77,0.83))
```

We use a `for` *loop to repeatedly conduct the* 100*-toss experiment. The loop is executed* 100,000 *times. During each iteration of the loop, (1)* 100 *coin tosses are simulated, (2) a string of the results is created and stored in* `toss_string`*, using the* `collapse = ""` *option for the* `paste` *function, (3) the variable* `streak_counter` *is incremented by one if the string* `"HHHHH"` *occurs within* `toss_string`*, and (4) the cumulative streak count is set to the value of* `streak_counter`*. After the loop, the vector of cumulative frequencies* `freq_streaks` *is created and plotted using the* `plot` *function.*

The simulated frequencies level off at around 0.81*, or* 81%*. The actual calculated frequency after the* 100,000 *simulations, resulting from the command* `freq_streaks [num_simulations]`*, is* 0.81156*. How can we determine the accuracy of the computer-simulated frequency? That is, how close is the computer's answer is to the true probability of the event, which is the probability of seeing a five-head streak in a* 100*-coin sequence? While a meaningful answer can't be provided yet, the statistical tools developed in this book will allow simulation accuracy to be quantified. There is a* ***simulation error*** *whenever a computer simulates probabilities, and we will quantify how large this simulation error is likely to be.*

Exercises

1. Your phone has 30 songs available: 10 by artist A, 8 by artist B, and 12 by artist C.
    a. How many different ways can you play first a song by artist A and then a song by artist B?
    b. How many different ways can you play first a song by artist A, then a song by artist B, and then a song by artist C?
    c. If you no longer restrict the order of the artists, how many different ways can you play three songs that each have a different artist?
2. A 12-person jury is being selected from a pool of 50 potential jurors. Among the potential jurors, there are 20 men and 30 women.
    a. How many possible ways are there to select the jury?
    b. If the judge insists on there being an equal number of men and women on the jury, how many possible ways are there to select the jury?

c. The jury needs a foreperson, who is one of the 12 jurors. Ignoring the gender restriction for now, how many possible ways are there to select the jury and foreperson?

d. Now assume that the judge insists that there are an equal number of men and women on the jury *and* the foreperson is woman. How many possible ways are there to select the jury and foreperson?

3. A company visits a college campus to interview students. The company has seven economics majors, six finance majors, and five accounting majors from which to choose. Unfortunately, the company has lost everyone's résumés, so they randomly pick three students to interview.

   a. What is the probability that all three interviewees are economics students?

   b. What is the probability that all three interviewees are from the same major?

   c. What is the probability that the set of three interviewees has either no economics students *or* no finance students?

   d. What is the probability that at least one of the majors has no students interviewed?

4. At the Pick One restaurant, you have to choose one dish option from each of the following dinner courses: Appetizer (Vegetarian, Chicken, Pork), Salad (Vegetarian, Chicken), Small Plate (Vegetarian, Beef, Seafood), and Large Plate (Vegetarian, Beef, Chicken, Pork, Seafood). So, in order, $CVBS$ is one possible choice.

   a. How many possible ways are there to order?

   b. How many possible ways are there to order if the Appetizer and Salad are both Chicken?

   c. How many possible ways are there to order if exactly two of the dishes are Vegetarian?

   d. If you order completely at random, what is the chance of having at least two Chicken dishes?

5. An office manager needs to assign offices to nine employees. There are three offices available: office A, which holds two workers; office B, which holds three workers; and office C, which holds four workers.

   a. How many possible office assignments of the employees are there?

   b. McKenna is one of the nine workers. If assignments are made at random, what is the probability that McKenna gets assigned to office C?

   c. If assignments are made at random, what is the probability that McKenna is in the same office as her coworker Daniel?

   d. If assignments are made at random, what is the probability that McKenna is in the same office as her coworkers Daniel and Alma?

   e. Conduct 100,000 simulations in R to approximate the three probabilities in (b)–(d).

6. A lottery game awards a jackpot to any player who correctly guesses 4 distinct numbers from 30 possible numbers. The lottery employee picks 4 numbers at random, without replacement, from $\{1, 2, \ldots, 30\}$.

a. If a player randomly chooses 4 numbers, what is the probability that she wins the jackpot?

b. If a player likes small numbers and randomly chooses 4 numbers only from $\{1, 2, \ldots, 10\}$, what is the probability that she wins the jackpot?

c. Suppose that the lottery adds a Bonus Number to its game. For this new game, the lottery employee picks 5 numbers at random, without replacement, from $\{1, 2, \ldots, 30\}$. The 5th number chosen is set aside as the Bonus Number. The player again guesses 4 numbers, but now also specifies a 5th number (different from the first 4) as their guess for the Bonus Number. If they get the first 4 numbers correct, they win the jackpot; but now, if they also get the Bonus Number correct, they win double the jackpot. (If they get the "Bonus Number" correct but not the first 4 numbers, they get nothing.) What is the probability that a player wins the double jackpot? What is the probability that a player wins the jackpot, but not the double jackpot?

7. *An online banking app requires its users to enter a six-digit personal identification number (PIN). Each digit can be $0, 1, \ldots, 9$.

   a. If there are no restrictions on picking a PIN, how many possible PINs are there?

   b. If the app does not allow any PIN with a digit appearing in three consecutive places, how many possible PIN's are there? (To be clear, we are ruling out a PIN like 124449 since 4 appears three times in a row.)

   c. Suppose that you chose your PIN randomly (i.e., you randomly picked six-digit numbers until you had a PIN satisfying the restriction from (b)). Someone is looking over your shoulder when you type your PIN into the app, and they notice that the first two digits are 42 and the last two digits are 35. The onlooker tries to guess your PIN by randomly picking two digits for the middle two digits, but he is unaware of the app's restriction on PINs. What is the probability that he correctly guesses your PIN?

   d. Same as (c), except now assume that the onlooker is aware of the app's restriction on PINs. As such, he randomly picks the middle two digits, but he continues to do so until the PIN does not have a digit appearing in three consecutive places. What is the probability that he correctly guesses your PIN?

8. (*Birthday problem*) Consider a class of students. Assume that the birthdays of students are distributed randomly and independently of each other. Ignore leap years, so the chance of a student having any given birthday is $1/365$.

   a. For two specific students, what is the probability that they have the same birthday?

   b. For three specific students, what is the probability that at least two of them share a birthday? (Hint: Think first about the probability that the three students each have a different birthday. How does the probability in the question relate to this probability?)

   c. Following the logic from part (b), if the size of the class is 20 students, what is the probability that at least two of them share a birthday? Try to simplify the formula using factorial functions, but don't worry about calculating the exact answer.

d. Using this approach for any class size N, one can show that the probability that at least two of the students share a birthday is

$$1 - \frac{N!\binom{365}{N}}{365^N}.$$

Apply this formula in R with $N = 20$ to calculate the probability for (c).

e. Write a function `birthday` that takes an integer `n` ≥ 2 as its single argument and returns the probability from the formula in (d).

f. Using the `birthday` function from (e), how many students would need to be in a class for the probability of two students sharing a birthday to be greater than 90%?

g. Using the `birthday` function from (e), plot the probabilities of at least two students sharing a birthday against class size N for all values of N between 20 and 60 (inclusive).

9. (*Matching problem*) At the beginning of an exam, every student is required to leave her or his phone on the professor's desk. After the exam is over, the professor randomly gives back the phones to students when they turn in their exams.

 a. If there are only three students in the class, what is the probability that at least one student gets his or her own phone back after the exam?

 b. If there are only four students in the class, what is the probability that at least one student gets her or his own phone back after the exam? (If you're having trouble with this part, list all $P_{4,4} = 4! = 24$ possible permutations for the order of the phones being returned.)

 c. While it is possible to mathematically generalize the results from (a) and (b) to an arbitrary class size N, an alternative approach is to use computer simulation to approximate the probability. Conduct 10,000 computer simulations in R to approximate the probability that at least one student gets his or her own phone back in a class of 20 students.

 d. Generalize your code from (c) by writing a function `matching` that takes an integer `n` ≥ 2 as its single argument and returns the approximate probability (based on 10,000 simulations) that at least one student gets her or his own phone back in a class of `n` students.

10. A business magazine rates mutual fund managers based upon how often their funds beat the return on the Standard & Poor's (S&P) 500 index. Any given manager has a 50% of beating the S&P index in a given year (and 50% of not beating it), and each year's performance is mutually independent.

 a. The magazine gives its "Gold Star" rating to any manager who has beaten the S&P index in at least five of the last six years. What is the probability of any given manager getting the "Gold Star" rating?

 b. The magazine gives its "Silver Star" rating to any manager who has beaten the S&P index in exactly four of the last six years. What is the probability of any given manager getting the "Silver Star" rating?

c. If you know that a manager has beaten the S&P index in at least four of the last six years, what is the probability that the manager gets the "Gold Star" rating? The "Silver Star" rating?

11. A publisher has six economics textbooks in its catalog. The publisher is deciding on its advertising strategy, and each textbook can be advertised or not.

    a. How many distinct outcomes are possible? (Order does not matter.)
    b. How many outcomes have four or more of the economics textbooks advertised?
    c. A marketing specialist is hired to rank the six textbooks according to projected sales over the next two years. How many distinct rankings are possible?
    d. The publisher decides to commission new editions of three of the six textbooks, one in each of the next three years. The marketing specialist is asked to choose three textbooks for new editions, specifying the textbook to be updated in the first year, the textbook to be updated in the second year, and the textbook to be updated in the third year. How many distinct choices of the three textbooks are possible? (Order matters.)
    e. The publisher decides to start an annual mailing to economics professors of an advertising pamphlet that focuses on three of its six textbooks. The company does not want to focus on the same three textbooks more than once. It does not mind repeating one or two of them, but it does not want all three to be the same as in a previous mailing. How many years can the publisher stick to this policy before it is forced to change it?
    f. How would your answer to (e) change if one of the authors just won the Nobel Memorial Prize in Economic Sciences, and it is decided that this author's textbook must be included every year?
    g. Suppose that, of the five remaining textbooks (other than the Nobel Prize–winning author's book), three are advanced and two are introductory. How does your answer to (e) change if, in addition to the text by the Nobel Prize winner, one text must be advanced and the other must be introductory?
    h. If the three textbooks are chosen randomly, what is the probability that the textbook by the Nobel Prize winner is included?
    i. If the three textbooks are chosen randomly, what is the probability that at least one introductory textbook is included?
    j. If three textbooks are chosen randomly for four years in a row, what is the probability that the textbook by the Nobel Prize winner is included in all four years?

12. Modify the *R* code from example 4.16 to approximate the probability of having a streak of at least *six* consecutive heads in 100 tosses. Change the limits of the *y*-axis, using the `ylim=c(0,1)` option. What is the simulated frequency of streaks? Now do the same exercise for a streak of at least *seven* consecutive heads in 100 tosses.

13. Refer to example 4.14 for this question.

    a. Conduct 100,000 simulations in *R* to approximate the two probabilities found analytically in example 4.14: (i) the probability that a randomly chosen three-letter website name has only alphabetic characters, and (ii) the probability that a randomly chosen three-letter website name has three *distinct* alphabetic characters. (Hint: Sample from the vector `0:35`, where `0` through `9` correspond to the numerical characters, and `10` through `35` correspond to the alphabetic characters.)

b. Conduct 100,000 simulations in *R* to approximate the probability that the sum of any numerical characters in a randomly chosen three-letter website name is greater than 10. (If there are no numerical characters in a website name, the sum should be treated as zero.)

5

Economic Data and Sampling

This chapter discusses several types of economic data that are typically encountered in practice. It also formalizes the concept of *sampling*, providing a framework to explain how data are generated and observed.

5.1 Types of Data

There are three main types of economic data that make up the vast majority of the data sets used in practice: cross-sectional data, time-series data, and panel (or longitudinal) data. We begin with some definitions and examples for each of the three types.

5.1.1 Cross-sectional Data

Definition 5.1 ***Cross-sectional data*** *consist of observations on different units that are measured at the same point in time or within the same time interval.*

The cross-sectional units can be individuals, households, firms, countries, states, or any other units for which data are observed. The key defining feature of cross-sectional data is that the data represent a "snapshot," at a given point in time or within some time interval, of what is known about the observed units. Here are some examples.

Example 5.1 (Labor-force data) *To study the determinants of individual earnings, we can use data for individuals that contain information on labor-force status (working? unemployed? not in the labor force?), earnings (if any), hours worked (if any), and other characteristics. An example is the Current Population Survey (CPS), which is a survey conducted in the US by the Bureau of Labor Statistics.*

Example 5.2 (Firm-level data) *Cross-sectional data on firms might include revenues, profits, research and development expenditures, patents awarded, and other firm characteristics during a given year.*

Example 5.3 (State-level cigarette data) *In the US, there is huge variation in the tax on cigarettes assessed by individual states. In* 2022, *the tax on a pack of cigarettes was* \$4.35 *in New York, but only* \$0.45 *in North Carolina. A study of cigarette taxes/prices and smoking rates might involve collecting data for each of the* 50 *states at a given point in time (such* *as* 2022*) on cigarette taxes (tax per pack), cigarette prices (average price per pack), and smoking rates among certain age groups.*

The statistical theory in this book is most applicable to cross-sectional data. The reason is that the cross-sectional units can generally be treated as if they are not related to each other, so we will think of these units as "independent" in a sense that is related to the concept of independence introduced in section 3.4. For instance, for the labor-force data in example 5.1, we might assume that knowing one individual's wage doesn't provide any useful information about another individual's wage.

5.1.2 Time-Series Data

In contrast to cross-sectional data, **time-series data** consist of quantities that are observed over time rather than as a snapshot.

Definition 5.2 ***Time-series data*** *consist of observations on the same unit that are measured at different points in time. When the time series consists of a single variable, the data are* ***univariate time-series data****. When the time series consists of more than one variable, the data are* ***multivariate time-series data****.*

Again, the idea of a "unit" is general and can be many things in practice (an individual, a firm, a country, etc.). Some examples of time-series data sets include the following.

Example 5.4 (Macroeconomic indicators for the US) *To track and analyze the overall macroeconomy of the US, annual data for the unemployment rate, inflation, gross domestic product (GDP) growth, and the budget deficit can be collected.*

Example 5.5 (Asset returns) *There is a wealth of time-series data available for financial markets, which can be collected at almost any time frequency—annually, monthly, daily, by the hour, and even by the minute. An example of a time-series data set would be the daily returns for a given stock, like Apple, or for some other financial asset.*

Whereas the units, like individuals or firms, in cross-sectional data can generally be treated as unrelated to each other, it is often the case that the observations in a time-series data set are related to each other. For example, the price of the stock on a given day is going to be closely related to the price of the stock on the previous day. Likewise, the inflation rate in the US for a given year is likely to be related to the inflation rate for the previous year.

5.1.3 Panel Data

Panel data, also called **longitudinal data**, combines features from both cross-sectional data and time-series data. The term "panel data" is the one most frequently used within the field of economics, though other fields of study have traditionally used "longitudinal data." A key feature of panel data is that the every cross-sectional unit is observed repeatedly in multiple time periods.

Definition 5.3 ***Panel (or longitudinal) data*** *have both a cross-sectional dimension and a time-series dimension, with information about the same cross-sectional units being observed at different points in time. The number of times that each cross-sectional unit is observed is at least two.*

Here are some examples.

Example 5.6 (State-level panel data on cigarettes) *Building on example 5.3, annual data can be collected for all 50 states on cigarette taxes (tax per pack), cigarette prices (average price per pack), and smoking rates among certain age groups. Whereas the cross-sectional data set allows analysis of data collected at only one point in time, the panel data set allows analysis of how cigarette taxes and smoking rates have changed overtime, and perhaps whether there is a relationship between tax changes and smoking-rate changes.*

Example 5.7 (Influenza data) *To analyze trends in influenza infections and vaccinations in the US, monthly data over the course of several years can be collected for all 50 states on influenza vaccination rates, deaths caused by influenza, and hospitalizations caused by influenza.*

Example 5.8 (Global macroeconomic panel data) *Example 5.4 considered time-series data of macroeconomic variables for the US. To build a panel data set for the global economy, time-series data with the same variables, at the same annual frequency, could be collected for other countries. The resulting panel data set would have annual data on the unemployment rate, inflation rate, and GDP growth rate for many countries over a period of several years.*

The three types of data discussed here (cross-sectional, time-series, and panel) do not cover all the possibilities of interest. For example, in cases where we have *different* cross sections that are observed at different points in time, the data create a *repeated cross section*. Whereas each cross-sectional unit in a panel data set is observed more than once, each unit is observed only once in the repeated cross section (i.e., in one of the time periods but not the others).

5.2 Types of Variables

Any given data set contains observations on one or more **variables**, where a variable is defined as follows.

Definition 5.4 *A **variable** is any measurable characteristic of a given unit of observation.*

The two main types of variables are **numerical variables** and **categorical variables**, where the former is described by a number and the latter is not. We introduce numerical variables in section 5.2.1 and categorical variables in section 5.2.2.

5.2.1 Numerical Variables

Definition 5.5 *A **numerical variable** is a variable that has a number associated with the characteristic.*

Some examples of numerical variables include the following:

- The weekly earnings of an individual
- The annual GDP growth of a country
- The number of website visitors on a given day
- The number of children in a household

Even among numerical variables, there are different types. In the examples given previously, the number of children in a household can only be a nonnegative integer, whereas the annual GDP growth of a country can be a positive or negative real number. It is useful to further break down numerical variables into three types: discrete numerical variables, continuous numerical variables, and approximately continuous numerical variables.

Definition 5.6 *A* ***discrete variable*** *(or* ***discrete numerical variable****) is a variable where the number of possible values can be counted, even if the number of possible values is infinite.*

The number of children in a given household and the number of patents awarded to a given firm in a given year are both examples of discrete variables. Even though these variables will usually have pretty low values in any observed data, we can still think of these variables as having an infinite number of possible values, with any value from the set $\{0, 1, 2, \ldots\}$ possible. In other cases, a discrete variable may be inherently finite. Some examples of finite discrete variables include a student's score on an Advanced Placement (AP) exam, which is in $\{1, 2, 3, 4, 5\}$; the number of states that have a Republican governor in a given year, which is in $\{0, 1, 2, \ldots, 50\}$; or the number of months for which a given stock has a positive return in a given year, which is in $\{0, 1, 2, \ldots, 12\}$. Although these examples all involve integer values, nothing in the definition of discrete variables restricts values to be integers. For example, shoe size is a discrete variable that can take noninteger values for half sizes.

In contrast to discrete variables, a continuous variable has values along some portion, or all, of the real line, so the number of possible values is not countable.

Definition 5.7 *A* ***continuous variable*** *(or* ***continuous numerical variable****) is a variable that can take on any value on some interval or intervals of the real line, including perhaps the entire real line.*

Examples of continuous variables include the monthly rainfall in a given city (measured in inches, but not rounded), the daily stock return for a given stock, the fraction of monthly income that a given employed individual saves in a given month, and the annual GDP of a given country. The possible values are different in these examples, with monthly rainfall and annual GDP being nonnegative real numbers in $[0, \infty)$, the daily stock return being any real number in $[-1, \infty]$, and the savings fraction being any real number in $[0, 1]$.

As will be seen in future chapters, the probability models used to model discrete variables and continuous variables are fundamentally different. Think about the comparison between one of our discrete variable examples (the number of children in a household) and one of our continuous variable examples (the amount of monthly rainfall in a city). Suppose that we are interested in knowing the probability that the value of the variable is between 1 and 3 (inclusive). For the "number of children" variable, we add together three probabilities: the probability of one child, the probability of two children, and the probability of three children. For the "monthly rainfall" variable, that approach is not appropriate since the variable can take any noninteger value between 1 and 3. Instead, we use calculus to "add up" (integrate) the probabilities over all the possible values between 1 and 3.

Since probability models for continuous variables are often easier to work with than those for discrete variables, it is often the case that a discrete variable will be modeled as continuous when it is reasonable to do so. The concept of an "approximately continuous" variable, as given in definition 5.8, covers this case.

Definition 5.8 *An* ***approximately continuous variable*** *(or* ***approximately continuous numerical variable****) is a discrete variable that can be treated and modeled as continuous, which is the case when two conditions are met: (1) the unit of measurement is small compared to typical values of the variable, and (2) the number of values in the data set is large, with relatively few repeats.*

Some examples of approximately continuous variables are the weekly earnings of a given individual, the number of employees in a given firm, and the credit score for a given individual. Thinking about the weekly earnings variable, the values would generally be reported in dollars (perhaps as an integer, or perhaps reported to two decimal places), so formally speaking, the variable is discrete. But the number of possible values is extremely large, and the unit of measurement (dollars) is very small relative to the typical weekly wage, which can be in the hundreds or thousands of dollars. Likewise, the number of employees in a given firm can take on many possible values and be in the hundreds or even thousands, whereas the unit of measurement (one worker) is relatively small.

5.2.2 Categorical Variables

Definition 5.9 *A* ***categorical variable*** *is a variable that has two or more categories, but no obvious numerical measure, associated with the characteristic. A* ***binary variable*** *is a special case of a categorical variable with only two categories.*

Some examples of categorical variables include the following:

- Labor force status of an individual (employed, unemployed, not in labor force)
- Home ownership status of an individual (yes, no)
- Vaccination status of an individual (yes, no)
- Health rating of an individual (excellent health, good health, fair health, poor health)
- State identifier (AL, AK, AZ, …, WI, WV, WY)

The "home ownership status" and "vaccination status" variables are both binary variables since they take on only two possible values. The other examples have more categories, with the "labor force status" variable having 3, the "health rating" variable having 4, and the "state identifier" variable having 50. In looking at the nonbinary examples, the "health rating" variable has a natural ordering, with excellent health being better than good health, good health better than fair health, and fair health better than poor health, whereas the "labor force status" variable and the "state identifier" variable have no obvious ordering to their values. We formally introduce terminology to distinguish these two cases as follows.

Definition 5.10 *A categorical variable is* ***ordered*** *if there is a natural ordering to the choices. A categorical variable is* ***unordered*** *if there is no natural ordering to the choices.*

In practice, it is often useful to use numerical values to indicate that a categorical variable has a certain value. Specifically, a discrete 0–1 variable can be used to indicate if a categorical variable has a given value. A 1 indicates that it has that value, and a zero indicates that it does not have that value. This 0–1 variable can be called an **indicator variable**, a **dummy variable**, or a binary variable. Here, the binary variable is numerical, in contrast to the binary categorical variable in definition 5.9.

To see how this works, first consider one of the binary categorical variable examples. For the "home ownership" variable, which has the possible values "yes" and "no," we can define the variable *owner* to be

$$owner = \begin{cases} 1 & \text{if individual owns a home ("yes")} \\ 0 & \text{if not ("no")} \end{cases}$$

While we could also define the variable *nonowner* to be 1 if the individual does not own a home and 0 if the individual does, there's no need to do so since the value of *nonowner* is known if the value of *owner* is known.

Now consider the "labor force status" variable, which is a categorical variable with more than two categories. In this case, three indicator variables, *employed*, *unemployed*, and *notinlf*, can be defined as follows:

$$employed = \begin{cases} 1 & \text{if individual is employed} \\ 0 & \text{if not (unemployed or not in labor force)} \end{cases}$$

$$unemployed = \begin{cases} 1 & \text{if individual is unemployed} \\ 0 & \text{if not (employed or not in labor force)} \end{cases}$$

$$notinlf = \begin{cases} 1 & \text{if individual is not in labor force} \\ 0 & \text{if not (employed or unemployed)} \end{cases}$$

Using three indicator variables here is overkill, as once the values of two indicator variables are known, the value of the third variable is known. Since an individual is in one and only one of the three categories, we have

$$employed + unemployed + notinlf = 1.$$

Therefore, if *employed* and *unemployed* are already defined, it follows that $notinlf = 1 - employed - unemployed$, so it's unnecessary to specify *notinlf* as a third variable. For three categories, only two indicator variables (for two of the categories) are needed to completely characterize the categorical variable. It doesn't matter which of the three categories is the "omitted" category.

This basic idea generalizes to categorical variables with more categories. If a categorical variable has C different categories that are disjoint and exhaustive, using the same terminology introduced for events in section 2.2, $C - 1$ indicator variables completely describe the categorical variable, and it doesn't really matter which category is omitted. Of course, it is important to have disjoint and exhaustive categories for this to be true, as that implies that one and only one category may be the actual value for the categorical variable.

5.3 The Population and Sampling

Having introduced different types of variables that might be part of the observed data, we now formalize how these data might be collected and observed. As a starting point, we need to decide the group for which we want to conduct statistical analysis and from which we are going to pick representative observations. In statistics, this group is called the **population**, and the representative observations constitute the **sample**.

Definition 5.11 *The **population** is the entire group for which conclusions are to be made from statistical analysis.*

Definition 5.12 *The **sample** consists of the specific units from the population for which data are collected and observed.*

Definition 5.13 *The **sample size**, usually denoted as n, is the number of units collected and observed.*

The collected and observed data in the sample, consisting of n units, are "drawn" or taken from the underlying population of interest. In many cases, the population is very large, and it is impossible to collect data on the entire population due to cost, logistics, and/or other factors.

Example 5.9 (Election polling) *We are interested in predicting the outcome of a political election with two candidates, candidate A and candidate B. We conduct a preelection poll, where the population of interest consists of all likely voters. The collected sample is a subset of the population of likely voters who are surveyed to ask if they intend to vote for candidate A or candidate B.*

Example 5.10 (Salary expectations) *We are interested in determining what students at Capita University think about their postgraduation jobs and salaries. The population consists of all Capita University students. The sample is a subset of Capita University students who are surveyed about their predicted postgraduation salary and job, and possibly additional items.*

In both of these examples, it's impractical to survey the entire population of interest, so drawing a representative sample makes sense. In other cases, we can collect data on all the units available, as would be the case if we want to collect data for all 50 US states.

Example 5.11 (State-level cigarette data) *We are interested in collecting data on cigarette taxes/prices and smoking rates for individual states in a given year. The population of interest consists of the states in the US. Since it's feasible to collect data on all 50 states, the sample consists of the cigarette taxes/prices and smoking rates for all 50 states, with sample size $n = 50$.*

5.3.1 Simple Random Sampling and Selection Bias

Intuitively, we want the observed sample to be representative of the units in the population of interest. The ideal version of a representative sample is known as a **simple random sample**.

Definition 5.14 *A sample of size n is called a* ***simple random sample*** *if each element of the population is equally likely to be sampled or, equivalently, that any possible sample of size n is equally likely to be chosen from the population.*

While a thorough discussion of nonrandom sampling procedures is beyond the scope of this book, it is important to understand why a sample might not be a simple random sample and how a nonrandom sample may affect the analysis being conducted. When the sample is not a simple random sample, it may introduce **sampling bias** (or **sample selection bias**) if the sampling method is related to the variable(s) of interest.

Example 5.12 (Election polling) *Returning to example 5.9, would a simple random sample be drawn from the population of likely voters if a land-line phone survey is used? The answer depends on whether someone having a land line (i.e., a noncellular phone in one's residence) is related to whether they would favor candidate A or candidate B. If it's known that older voters are much more likely than younger voters to both support candidate A and have a land line, the land-line survey is not going to yield a simple random sample. Actually, surveying through land lines by itself would violate the definition of a sample random sample given in definition 5.14 since older voters are more likely to have land lines, and therefore more likely than younger voters to be sampled in the first place. However, that fact would not mean that the sampling approach is a poor one by itself if it weren't also the case that the use of a land line is related to candidate preference. In this case, since the goal is to infer the proportion of likely voters who favor each candidate, the land-line survey is problematic because land-line use is related to candidate preference. Sampling bias or sample selection bias is introduced since the sample is more likely to have candidate A supporters than a simple random sample would have.*

Example 5.13 (Salary expectations) *Returning to example 5.10, let's assume that every Capita University student is assigned a random student ID number when they first arrive on campus. Consider two alternative sampling approaches for gathering data about job/salary expectations: method 1, which involves surveying all Capita University students whose student ID number ends in a "5"; and method 2, which involves surveying all Capita University students who are in an advanced economics course. Method 1 does not lead to sampling bias since the last number of the student ID number is chosen at random and, thus, should not be systematically related to any of the variables (salary expectation, job expectation, etc.) in which we might be interested.*[1] *Method 2 is likely to lead to sampling bias since students in the advanced economics course are not representative of the population of all Capita University students, except in the unlikely case that the advanced economics course is required for all students. For instance, if advanced economics students are more likely to end up in high-paying jobs than other students, surveying only those students will most likely lead to higher salary expectations than we would expect from representative students in the population.*

5.3.2 Stratified Sampling

With simple random sampling, the sample is drawn randomly from the entire population of interest. In some cases, however, there are reasons that practitioners might want to first split the population into a set of **subpopulations** or **strata** before collecting a sample.

Definition 5.15 *A* ***stratified random sample*** *is created by splitting the population into defined subpopulations or strata and drawing a simple random sample from each subpopulation or stratum. Such a sample is called a* ***proportionate stratified random sample*** *if the sizes of the strata samples are proportional to the true probabilities of the strata in the population, and a* ***disproportionate stratified random sample*** *if they are not.*

The use of a proportionate stratified random sample ensures that the obtained sample is representative of the population with respect to the defined strata.

Example 5.14 (Political polling) *Suppose that a political poll of* 100 *individuals is being conducted in a particular area, where all the individuals are asked whether they prefer the candidate from political party A or the candidate from political party B. Within the area,* 60% *of the population are registered with political party A and* 40% *are registered with political party B. A proportionate stratified random sample results from drawing a simple random sample of* 60 *party A individuals (from the subpopulation of party A individuals) and a simple random sample of* 40 *party B individuals (from the subpopulation of party B individuals). Without imposing the proportionate strata, a simple random sample of* 100 *individuals might overrepresent party A individuals or party B individuals. For instance, if there were* 65 *party A individuals in the simple random sample, we'd intuitively get an overestimate of the preference for the party A candidate from the poll (assuming that party A individuals are more likely than party B individuals to support the party A candidate).*

While the use of proportionate stratified random sampling is appealing in certain circumstances, there are also sometimes reasons for using disproportionate sampling instead. With disproportionate sampling, one or more strata will be **oversampled** relative to their true proportions in the population, and one or more strata will be **undersampled** relative to their true proportions in the population.

Example 5.15 (Rare disease) *In a certain population, suppose the probability that an individual has a particular genetic disease is* 0.05% *(*1*-in-*2,000 *chance). If a researcher is interested in studying the characteristics of individuals with the disease or making comparisons of those characteristics to individuals without the disease, she needs to have a considerable number of individuals with the disease in her sample. Thinking about a simple random sample, even a sample of size* 10,000 *would have very few individuals (*5*, on average) with the disease. Instead, the researcher would want to oversample the subpopulation or stratum of individuals with the disease. For example, she might construct a disproportionate stratified random sample with* 20% *of the individuals having the disease and* 80% *not having the disease. Since the subpopulation of individuals with the disease is oversampled, the subpopulation of individuals without the disease is undersampled.*

Example 5.16 (Union and nonunion wages) *In the population of employed individuals in the US, the proportion that are union members is* 10%*, and it is well-known that union workers earn more, on average, than nonunion workers. If a researcher is interested in estimating the average earnings in the overall population of workers, it may be sensible to construct a proportionate stratified random sample based upon the true proportions of* *union workers (10%) and nonunion workers (90%) in the population. On the other hand, if a researcher wants to compare the average earnings between union workers and nonunion workers, it might make more sense to oversample union workers such that the size of the union-worker and nonunion-worker samples are comparable.*

Note

1. Whether method 1 yields a simple random sample depends upon how you think about the population. If the population consists of students before they arrive on campus, method 1 yields a simple random sample since the assignment of the student ID hasn't occurred yet, and any student is equally likely to have a 5 as the last digit. If the population consists of students after they arrive on campus, method 1 won't pick any students who don't have a 5 as the last digit, so in that sense, the units of the population are not equally likely to be chosen. This distinction is unimportant here since, as the discussion has highlighted, there is no sampling bias introduced by method 1.

Exercises

1. For each of the following examples, indicate whether the data are cross-sectional, time-series, or panel data.
   a. An avid runner records the number of miles that she runs every day for 100 straight days.
   b. A random sample of 100 older adults, aged 65 and older, were surveyed in January 2023 about their health status and medical expenditures.
   c. A financial analyst randomly picks 20 companies that are listed on the New York Stock Exchange and gathers data on their profits and sales for the year 2022.
   d. A Canadian economist gathers annual data on each of Canada's 13 administrative divisions, including the population and the unemployment rate for each division, for each year between 2000 and 2020.
   e. A random sample of 100 college students is asked whether they have received an influenza vaccine in the last year.
   f. A random sample of 100 college seniors is asked for their semester grade point average for each of their first six semesters at the university.
2. A survey is taken of economics majors at a particular university. For each of the following variables, indicate whether the variable is (i) categorical, (ii) discrete but not approximately continuous, (iii) discrete and approximately continuous, or (iv) continuous.
   a. Favorite economics professor
   b. Number of economics courses taken prior to taking econometrics (a required course for majors)
   c. Born in the US or not
   d. Cumulative GPA prior to the current semester (not rounded)
3. For each of the following variables, indicate whether the variable is (i) categorical, (ii) discrete but not approximately continuous, (iii) discrete and approximately continuous, or (iv) continuous.

a. The daily number of deliveries that a major online retailer makes to US residential addresses
b. Whether a person has a flu vaccination in a given year
c. The time that it takes a personal shopper at a grocery store to complete a customer's order
d. The number of cellphones that an individual has owned in their lifetime

4. Consider drawing a simple random sample of $n = 10$ observations from a population consisting of 100 units.
   a. How many possible ways are there to draw the simple random sample?
   b. What is the probability that a given observation from the population is in the simple random sample that is drawn?
   c. What is the probability that any two given observations from the population are in the simple random sample that is drawn?
5. A major city has taken a complete census of all restaurants within its city limits and is interested in the percentage that have public health violations. It doesn't have the resources to conduct inspections at all restaurants, so it must draw a sample from the full census (population). For each of the following sampling possibilities, explain whether there would be concern about sample selection bias, and explain your answer.
   a. The city has six different ZIP codes and conducts inspections at all the restaurants in one of these ZIP codes.
   b. The city conducts inspections at 15% of the restaurants, chosen completely at random.
   c. The city conducts inspections at all restaurants having a street address number ending in 3.
   d. The city conducts inspections at the 30 largest restaurants in the census.
6. For each of the following examples, discuss whether there is a potential for sample selection bias and if so, explain why. For a given example, there may be multiple reasons for sample selection bias.
   a. A pharmaceutical company wants to test the efficacy of a new medication for treating a disease. The company is interested in the population of all individuals having the disease, and they recruit participants through advertisements posted in medical facilities and in online forums.
   b. A firm wants to determine the effectiveness of an employee training program (how effective the training is for a representative employee at the firm). The program is voluntary, so the firm measures the effectiveness (the difference between preprogram productivity and postprogram productivity) for employees who enroll in the program.
   c. The Social Security Administration (SSA) has comprehensive earnings-history data for every US citizen. A researcher with access to SSA data is interested in the average earnings of 30-year-old citizens in 2020. She randomly draws 5,000 individuals from the SSA data who were 30 years old in 2020 and gathers their earnings data for that year.

d. A magazine reports the average salaries of graduates from law schools. To gather their data, it conducts an online survey of law school graduates, who are asked to voluntarily share their own salary information.

e. An economist would like to analyze the economic impact of recent tax cuts on small businesses. She collects data from businesses that voluntarily respond to a survey.

7. The following table summarizes the political party affiliation (*A* or *B*) and age group ("Under 30," "30–50," and "Over 50") for the population of registered voters in a particular voting district:

		Party	
		A	*B*
Age	Under 30	15%	5%
	30–50	25%	15%
	Over 50	20%	20%

For example, 15% of the registered voters are affiliated with party *A* and under 30 years old.

a. If a sample of 1,000 voters is stratified on the basis of party affiliation alone, how many voters in the sample are affiliated with party *A* versus party *B*?

b. If a sample of 1,000 voters is stratified on the basis of *both* party affiliation and age group, how many strata are there, and how many voters are in each stratum?

c. A sample of 1,000 voters has exactly 380 voters in the "Over 50" age group. Which one (or more) of the following are possible:

i. The sample is a simple random sample.

ii. The sample is a proportionate stratified random sample, stratified by age group.

iii. The sample is a proportionate stratified random sample, stratified by party affiliation.

6

Descriptive Statistics and Visuals: Univariate Data

Chapter 5 introduced different types of variables that might be observed in a data set. This chapter considers actual data set examples and introduces common descriptive statistics and visual devices that can be used to summarize variables. It focuses on descriptive statistics and visuals for univariate data, considering summary measures of a single variable, and chapter 7 focuses on descriptive statistics and visuals that summarize the relationship *between* two or more variables.

Recall that n denotes the sample size. For a single variable, generically denoted as x, the following notation denotes a sample of observations:

$$\{x_1, x_2, \ldots, x_n\},$$

or, more concisely, $\{x_i\}_{i=1}^n$. The term **statistic** is defined to be any numerical measure that is based upon the sample.

Definition 6.1 *A* ***statistic*** *is a function of the observed sample data. For univariate data, a statistic takes the form* $s(x_1, x_2, \ldots, x_n)$ *for some function* $s(\cdot)$*.*

A **descriptive statistic** is one whose purpose is to summarize data in some way. Descriptive statistics can be used to describe how likely certain values are, where the "center" of the sample observations is, how "noisy" the sample observations are, and other characteristics. This chapter considers descriptive statistics for categorical data and numerical data separately. While it's not possible to cover the full universe of descriptive statistics used in real-world applications here, several of the most prevalent ones are considered.

6.1 Data Set Examples

Throughout this chapter and chapter 7, two data sets will be used to illustrate how descriptive statistics can summarize different types of variables. The first, introduced in example 6.1, is a cross-sectional labor force data set taken from the 2019 Current Population Survey (CPS), and the second, introduced in example 6.2, is a time-series data set on monthly stock returns for a set of stocks between 1991 and 2021.

Example 6.1 (Labor force data) *The **cps** data set is a cross-sectional sample from the* 2019 *CPS, consisting of* 4,013 *observations on adults aged 30 to 59 (inclusive). The sample size is* $n = 4013$*, and there are 17 variables for each observation. Some variables are numerical, like age and education, and some are categorical, like labor-force status and union status. For numerical variables, it's important to clearly specify the units of the underlying variable. In certain cases, a variable may have a missing value rather than a numerical value (for numerical variables) or a category (for categorical variables). For example, if an individual is unemployed or not in the labor force, the weekly earnings variable and the union status variable are recorded as "missing" since the individual has no job. The complete list of variables, split into categorical variables and numerical variables, is as follows:*

- *Categorical variables:*
  - *statefips: 51 possible two-character codes ("AL", "AK", "AZ", ..., "WI", "WV", "WY") for the 50 US states plus Washington, DC*
  - *gender: 2 values ("Female", "Male")*
  - *metro: 2 values ("Metro", "Nonmetro"), indicating whether individual lives in a metropolitan area or not*
  - *race: 3 values ("Black", "White", "Other")*
  - *hispanic: 2 values ("Hispanic", "Nonhispanic")*
  - *marstatus: 4 values ("Married", "Divorced", "Widowed", "Never married")*
  - *lfstatus: 3 values ("Employed", "Unemployed", "Not in LF")*
  - *ottipcomm: 2 values ("Yes", "No"), indicating whether earnings include overtime, tips, and/or commissions; missing if lfstatus is "Unemployed" or "Not in LF"*
  - *hourly: 2 values ("Hourly", "Nonhourly"); missing if lfstatus is "Unemployed" or "Not in LF"*
  - *unionstatus: 2 values ("Union", "Nonunion"); missing if lfstatus is "Unemployed" or "Not in LF"*
- *Numerical variables:*
  - *age (in years): Age of the individual; values provided as integers*
  - *hrslastwk (in hours): Hours worked by the individual last week; values provided as integers; missing if lfstatus is "Unemployed" or "Not in LF"*
  - *unempwks (in weeks): Number of weeks that an individual has been unemployed; values provided as integers; missing if lfstatus is "Employed" or "Not in LF"*
  - *wagehr (in dollars): Hourly wage for the individual; values provided to two decimal places; missing if hourly is "Nonhourly" or if lfstatus is "Employed" or "Not in LF"*
  - *earnwk (in dollars): Earnings for the individual last week; values provided as integers; missing if lfstatus is "Unemployed" or "Not in LF"*
  - *ownchild: Number of children in the individual's household*
  - *educ (in years): Highest level of education attained by the individual*

Having loaded the **cps** *data set into a data frame called* `cps`*, the following R code uses the functions* `str` *and* `summary` *to show the structure of the data frame and to summarize its variables, respectively:*

```
str(cps)

## 'data.frame': 4013 obs. of  17 variables:
##  $ statefips  : Factor w/ 51 levels "AK","AL","AR",..:
##  5 1 27 35 41 43 2 6 21 44 ...
##  $ age        : int  50 34 50 30 40 35 56 42 55 58 ...
##  $ hrslastwk  : int  40 40 NA 44 NA 30 40 25 40 46 ...
##  $ unempwks   : int  NA NA NA NA NA NA NA NA NA NA ...
##  $ wagehr     : num  12 NA NA NA NA NA 25 8 NA NA ...
##  $ earnwk     : num  577 3049 NA 2500 NA ...
##  $ ownchild   : int  0 1 4 0 0 0 0 0 0 0 ...
##  $ educ       : num  14 18 16 18 12 12 12 7.5 12 13 ...
##  $ gender     : Factor w/ 2 levels "Female","Male": 2 1 2 1 1 2 2 1 1 1 ...
##  $ metro      : Factor w/ 2 levels "Metro","Non-metro": 1 1 1 1 1 1 1 1 1 1 ...
##  $ race       : Factor w/ 3 levels "Black","Other",..: 1 3 3 3 1 3 3 1 2 3 ...
##  $ hispanic   : Factor w/ 2 levels "Hispanic","Non-hispanic":
##  2 2 1 2 2 2 2 2 2 1 ...
##  $ marstatus  : Factor w/ 4 levels "Divorced","Married",..:
##  3 2 2 3 3 1 2 4 3 2 ...
##  $ lfstatus   : Factor w/ 3 levels "Employed","Not in LF",..:
##  1 1 2 1 2 1 1 1 1 1 ...
##  $ ottipcomm  : Factor w/ 2 levels "No","Yes": 1 1 NA 1 NA 1 1 1 1 1 ...
##  $ hourly     : Factor w/ 2 levels "Hourly","Non-hourly":
##  1 2 NA 2 NA 2 1 1 2 2 ...
##  $ unionstatus: Factor w/ 2 levels "Non-union","Union": 1 1 NA 1 NA 1 1 1 1 1 ...

summary(cps)

##    statefips        age          hrslastwk        unempwks
##  CA     : 341   Min.   :30.00   Min.   : 1.00   Min.   :  1.00
##  TX     : 250   1st Qu.:37.00   1st Qu.:40.00   1st Qu.:  4.00
##  FL     : 196   Median :45.00   Median :40.00   Median :  8.50
##  NY     : 162   Mean   :45.02   Mean   :40.34   Mean   : 17.75
##  OH     : 109   3rd Qu.:53.00   3rd Qu.:42.00   3rd Qu.: 20.00
##  GA     : 108   Max.   :59.00   Max.   :99.00   Max.   :119.00
##  (Other):2847                   NA's   :1204    NA's   :3907
##      wagehr          earnwk          ownchild           educ
##  Min.   : 1.01   Min.   :  12.0   Min.   :0.0000   Min.   : 0.00
##  1st Qu.:12.78   1st Qu.: 520.0   1st Qu.:0.0000   1st Qu.:12.00
##  Median :16.41   Median : 770.0   Median :0.0000   Median :12.00
##  Mean   :18.60   Mean   : 971.2   Mean   :0.7478   Mean   :12.57
##  3rd Qu.:22.00   3rd Qu.:1193.6   3rd Qu.:1.0000   3rd Qu.:14.00
##  Max.   :90.00   Max.   :8779.7   Max.   :7.0000   Max.   :18.00
##  NA's   :2174    NA's   :1204
##     gender           metro          race              hispanic
##  Female:2093   Metro    :3175   Black: 476   Hispanic    : 745
##  Male  :1920   Non-metro: 838   Other: 349   Non-hispanic:3268
##                                 White:3188
##
##
##
##
##          marstatus           lfstatus     ottipcomm         hourly
##  Divorced      : 707   Employed  :2809   No  :2341   Hourly    :1839
##  Married       :2377   Not in LF :1098   Yes : 468   Non-hourly: 970
##  Never married: 853   Unemployed: 106   NA's:1204   NA's      :1204
##  Widowed       :  76
##
##
```

```
##
##      unionstatus
##   Non-union:2533
##   Union    : 276
##   NA's     :1204
##
##
##
##
```

The `str(cps)` *command shows the structure of the* `cps` *data frame, indicating the number of observations and variables and, for each variable, the variable type and the first few observed values from the data. The* `summary(cps)` *command shows more detailed information about each variable, with category counts shown for categorical (factor) variables and descriptive statistics shown for numerical variables; for variables with missing values, the number of missing values is indicated in the row labeled* `NA's`*. For example, the* `unionstatus` *variable has 2,533 observations in the* `Non-union` *category, 276 observations in the* `Union` *category, and 1,204 with a missing (*`NA`*) value.*

We can also get the summary statistics for a single variable by specifying that variable, rather than the whole data frame, as the argument for the `summary` *function:*

```
summary(cps$earnwk)

##     Min. 1st Qu.  Median    Mean 3rd Qu.    Max.    NA's
##     12.0   520.0   770.0   971.2  1193.6  8779.7    1204
```

Example 6.2 (Monthly stock returns) *The **sp500** data set consists of 364 monthly observations (January 1991 through April 2021) for a set of 266 individual stocks, each of which is part of the S&P 500 stock market index.*[1] *Each variable in the data set corresponds to a single company, with the variable name being the company's stock ticker. For example, the variable names for AT&T and Bank of America are T and BAC, respectively. For any given stock, the observations constitute a time series with a sample size of $n = 364$. The data are all numerical, with each observation representing a monthly return for a given stock. The monthly return for month m is defined as*

$$return_m = \frac{price_m - price_{m-1}}{price_{m-1}},$$

where $price_m$ is the price at the end of month m and $price_{m-1}$ is the price at the end of the previous month $m - 1$. This variable is continuous with possible values in $[-1, \infty)$, where the -1 corresponds to $price_m$ being 0.[2] *The variable values are unitless here. Even though stock prices $price_m$ and $price_{m-1}$ have monetary units (dollars), the formula given here indicates that the monetary units in the numerator cancel those in the denominator, leaving no units for $return_m$.*

*After loading the **sp500** data set in R, we can use the* `head` *function to display the first few observations for T and BAC, along with the dates:*

```
head(sp500[,c("Date","T","BAC")])

##         Date           T          BAC
## 1 1991-02-01 0.041747550  0.03999942
```

```
## 2 1991-03-01  0.044186174  0.18376090
## 3 1991-04-01 -0.048997714  0.07811400
## 4 1991-05-01 -0.017966258  0.14237314
## 5 1991-06-01  0.033815331 -0.15133543
## 6 1991-07-01  0.009346888 -0.01935816
```

The monthly return for AT&T in January 1991 was approximately 4.17%, *and the monthly return for Bank of America was approximately* 4.00%. *During the first six months, AT&T and Bank of America each had four positive monthly returns and two negative monthly returns, although the timing of the negative-return months was different for the two companies.*

In the rest of this chapter, we discuss several descriptive statistics and data visualization options for different types of data, including the following:

- Categorical data: Sample proportion, bar charts
- Discrete or continuous numerical data: Histograms, measures of location (sample mean, sample median, sample quantiles), box plots, measures of dispersion (interquartile range (IQR), sample variance, sample standard deviation)

6.2 Categorical Data: Sample Proportions and Bar Charts

Consider a categorical variable x, with sample $\{x_1, x_2, \ldots, x_n\}$, where there are C possible categories for x. The sample can be described by the count of observations within each category $c \in \{1, 2, \ldots, C\}$ or, equivalently, the fraction of observations within each category $c \in \{1, 2, \ldots, C\}$. These quantities are known as the **sample counts** and **sample proportions**, respectively.

Definition 6.2 *For a categorical variable x, the* ***sample count*** *associated with category c is the number of observations in category c:*

$$\text{sample count for category } c = \sum_{i=1}^{n} 1(x_i = c).$$

In this definition, the function $1(x_i = c)$ *has the value* 1 *when* $x_i = c$ *(the ith observation is in category c) and the value* 0 *when* $x_i \neq c$ *(the ith observation is not in category c). This function is an example of an* indicator function, *which is used elsewhere in the book. More generally, an indicator function* $1(E)$ *equals* 1 *if event E is true and* 0 *if event E is not true.*

Definition 6.3 *For a categorical variable x, the* ***sample proportion*** *associated with category c is the fraction or percentage of observations in category c:*

$$\text{sample proportion for category } c = \frac{\sum_{i=1}^{n} 1(x_i = c)}{n} = \frac{1}{n} \sum_{i=1}^{n} 1(x_i = c).$$

Since the C categories of x are disjoint and exhaustive, every x_i is in one and only one category, which immediately implies what is stated in proposition 6.1.

Proposition 6.1 *For any sample* $\{x_1, x_2, \ldots, x_n\}$ *of a categorical variable x, the sum of the sample counts is n and the sum of the sample proportions is* 1.

Similar to the discussion of categorical variables in section 5.2.2, the sample proportion of any category c can be inferred if the sample proportions of the other $C - 1$ categories are known since the proportions sum to 1. In the case of a binary categorical variable, the sample proportion of one of the two categories completely summarizes the observed data.

Example 6.3 (Labor force data) *Let's focus on the labor force status (lfstatus) variable from the **cps** data. The sample counts for the lfstatus categories were already seen in example 6.1 as part of the output from the* `summary(cps)` *command. Another method to directly tabulate the sample counts for a categorical variable, which works even when there are more categories than will fit in the* `summary` *output, is to use the* `table` *function:*

```
table(cps$lfstatus)

##
##   Employed  Not in LF Unemployed
##       2809       1098        106

table(cps$lfstatus)/nrow(cps)

##
##   Employed  Not in LF Unemployed
## 0.69997508 0.27361077 0.02641415
```

The first `table` *command provides the sample counts. The second command, which divides by the sample size (i.e., the number of rows in the data set, given by* `nrow(cps)`*), provides the sample proportions. The sample counts and sample proportions for the three categories are provided more neatly in the following table:*

lfstatus Category	*Sample Count*	*Sample Proportion*
Employed	2,809	0.7000
Unemployed	106	0.0264
Not in LF	1,098	0.2736
Total	4,013	1.0000

70.00% *of individuals in the sample are employed, while* 2.64% *are unemployed and* 27.36% *are not in the labor force. (The* 2.64% *value is not what economists would consider the "unemployment rate" since its denominator includes individuals who are not in the labor force. The unemployment rate for this sample is* $\frac{106}{(106+2809)} \approx 3.64\%$.*)*

The sample counts or sample proportions of a categorical variable can be visually portrayed through the use of a **bar chart** with one bar for each category, with a total of C bars for a variable with C categories. The height of each bar corresponds to either the sample count or the sample proportion of that category, depending on which bar chart is used. Likewise, the y-axis for the bar chart corresponds to count values or proportion values, depending on which bar chart we use.[3] As illustrated in figure 6.1, the two bar charts look identical except that the y-axis scale is different.

Example 6.4 (Labor force data) *Continuing example 6.3, the two bar charts in figure 6.1 provide a visual description of the lfstatus variable, where the heights are the sample counts*

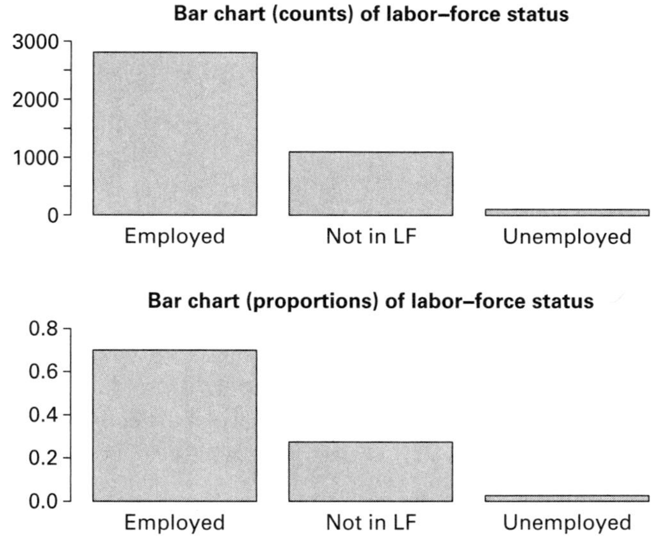

Figure 6.1
Bar charts of labor-force status (CPS data)

in the first bar chart and are the sample proportions in the second bar chart. Note that lfstatus is an unordered categorical variable. Since there is no natural ordering of the three categories, it doesn't really matter in which order the three bars are displayed. If we had an ordered categorical variable instead, we would want to have the bars displayed according to the arrangement of the categories, either in ascending or descending order.

The R code to create figure 6.1 uses the graphing function `barplot`*:*

```
# graph-display format (two rows, one column)
par(mfrow = c(2,1))

# bar chart with sample counts
barplot(table(cps$lfstatus), ylim=c(0,3000),
        main="Bar chart (counts) of labor-force status")

# bar chart with sample proportions
barplot(table(cps$lfstatus)/nrow(cps), ylim=c(0,0.8),
        main="Bar chart (proportions) of labor-force status")
```

For the first `barplot` *command, the first argument is* `table(cps$lfstatus)` *(the table of sample counts), the second argument (*`ylim=c(0,3000)`*) specifies the lower and upper limits to be used for the y-axis, and the third argument (*`main`*) provides a title for the chart. The second* `barplot` *command is similar, except that the table of sample proportions is the first argument and the values of the* `ylim` *and* `main` *are different. The* `ylim` *argument is optional, and R uses its default upper/lower limits for the y-axis if it is omitted. The formatting command* `par(mfrow = c(2,1))` *is used before the bar charts are produced. This command provides a convenient way to display multiple graphs simultaneously. The* `c(2,1)` *can be changed to accommodate a different number of rows and columns to be displayed. Here, there are* `2` *rows and* `1` *column, leading to the first bar chart being displayed* *on top of the second bar chart. If there were six graphs to be displayed in three rows of two graphs each, the appropriate command is* `par(mfrow = c(3,2))`*. The graphs are displayed from left to right, starting with the first row and continuing left to right for subsequent rows.*

6.3 Numerical Data: Histograms

For a numerical variable, a **histogram** is a common method of visually representing the sample distribution of the observed values. A histogram is nothing more than a collection of rectangles, similar to the bars in a bar chart. Each rectangle corresponds to a **bin** of possible values, where the width of the rectangle is known as a **bin width** and the height of the rectangle indicates how many of the variable's values are within that bin. To construct the histogram, the bins need to be specified, which involves specifying the bin width and the starting/ending values for the bins. To illustrate how this works, we start with a simple example involving a discrete numerical variable.

Example 6.5 (Labor force data) *Consider the age variable from the* **cps** *data, which is a discrete numerical variable that can take on the integer values* $30, 31, \ldots, 59$*. Figure 6.2 shows four histograms for the age variable. The top two look identical except for the y-axis, with the one on the left having "frequency" (counts) of observations and the one on the right having the "density" of observations. Both of these histograms have 30 bins, each with a bin width of 1 year. The bins themselves are* $(29.5, 30.5]$, $(30.5, 31.5]$, $(31.5, 32.5]$, *and so on through* $(58.5, 59.5]$*. Since age is only integer-valued, the* $(29.5, 30.5]$ *bin contains observations with age* $= 30$*, the* $(30.5, 31.5]$ *bin contains observations with age* $= 31$*, and so on through the* $(58.5, 59.5]$ *bin which contains observations with age* $= 59$*. Looking at these two histograms, the two most observed age values are age* $= 56$ *and age* $= 59$*, and the two least observed age values are age* $= 38$ *and age* $= 44$*. For the "frequency" histogram on the top left, the height of each rectangle is the count of observations within the corresponding bin or, equivalently, the count of observations with that specific age value. For the "density" histogram on the top right, the height of each rectangle turns out to be the proportion of observations in the associated bin or, equivalently, the proportion of observations with that specific age value; as we'll see, these heights are proportions for this histogram since the bin widths are exactly equal to 1.*

The bottom two histograms use a bin width of 2 *years, with the* 15 *bins defined as* $(29.5, 31.5]$, $(31.5, 33.5]$, $(33.5, 35.5]$, *and so on through* $(57.5, 59.5]$*. The first bin contains the observations with age* $= 30$ *or age* $= 31$*, the second bin contains the observations with age* $= 32$ *or age* $= 33$*, and so on through the last bin, which contains the observations with age* $= 58$ *or age* $= 59$*. For the bottom-left "frequency" histogram, the height of each rectangle still corresponds to a count of the observations within the associated bin. Since the wider* 2*-year bins naturally contain more observations than the* 1*-year bins, the scale of the y-axis is considerably larger. For instance, the height of the first rectangle for the* 2*-year bin width is equal to the sum of the first two rectangles for the* 1*-year bin width histogram above it. For the bottom-right "density" histogram, the scale of the y-axis is quite similar to the "density" histogram with* 1*-year bins (top right). The "density" values on the y-axis lead to the area of each rectangle to be equal to the proportion or fraction of observations within the*

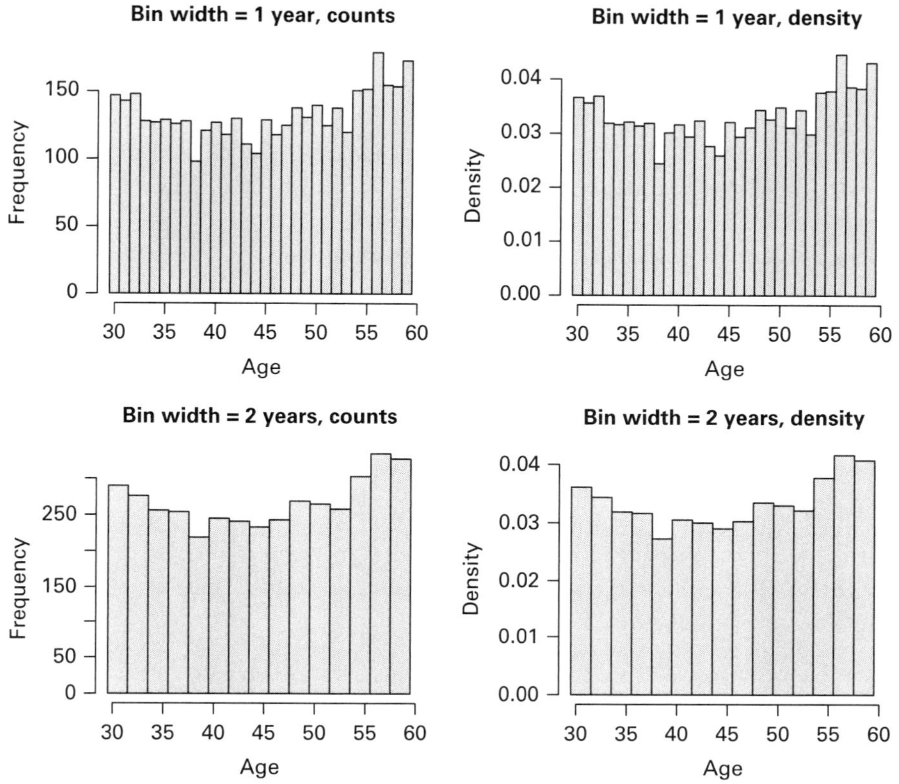

Figure 6.2
Histograms of age (CPS data)

associated bin. For example, the height ("density") of the first bin is approximately 0.036, *so the area of the rectangle is approximately* $(0.036)(2) = 0.072$, *with roughly* 7.2% *of the observations having* $age = 30$ *or* $age = 31$.

The R code to create figure 6.2 uses the function `hist`*:*

```
# graph-display format (two rows, two columns)
par(mfrow = c(2,2))

# histogram of age with bin width = one year, counts
hist(cps$age, breaks=seq(29.5,59.5,1),
     main="Bin width = 1 year, counts", xlab="Age")

# histogram of age with bin width = one year, density
hist(cps$age, breaks=seq(29.5,59.5,1), freq=FALSE,
     main="Bin width = 1 year, density", xlab="Age")

# histogram of age with bin width = one year, counts
hist(cps$age, breaks=seq(29.5,59.5,2),
     main="Bin width = 2 years, counts", xlab="Age")

# histogram of age with bin width = one year, density
hist(cps$age, breaks=seq(29.5,59.5,2), freq=FALSE,
     main="Bin width = 2 years, density", xlab="Age")
```

The first argument for the `hist` *function is the variable of interest (here,* `cps$age`*). The optional* `breaks` *argument allows the user to directly specify the starting/ending values for the bins. For the histograms in the top row, the* 1*-year bins are specified using the vector* `seq(29.5,59.5,1)`*; for the histograms in the bottom row, the* 2*-year bins are specified using the vector* `seq(29.5,59.5,2)`*. The optional argument* `freq` *indicates whether the y-axis of the histogram should display counts (when* `freq` *is* `TRUE`*, which is the default) or densities (when* `freq` *is* `FALSE`*). The* `main` *argument provides a title for the histogram, and the* `xlab` *argument specifies the text to be shown on the x-axis (here specified as* `Age` *rather than the default value, which is* `cps$age`*).*

The properties of the histograms described in example 6.5 are stated in propositions 6.2 and 6.3.

Proposition 6.2 *For a frequency or count histogram, the height of any rectangle is the number of observations within the associated bin. The sum of the heights of all the rectangles is equal to sample size n.*

Proposition 6.3 *For a density histogram, the area of any rectangle (height times bin width) is the proportion or fraction of observations within the associated bin. The sum of the areas of all of the rectangles is equal to* 1*, or* 100%*.*

From this point forward, we focus on density histograms since they have a direct relationship with the probability distributions introduced in chapters 8 and 10.

While example 6.5 considers a discrete numerical variable, example 6.6 considers a continuous numerical variable and discusses the choice of the bin width or the number of bins in more detail.

Example 6.6 (Labor force data) *Consider the weekly earnings (earnwk) variable from the* ***cps*** *data, which is a continuous, or at least approximately continuous, numerical variable. The earnwk variable has nonmissing values for the* 2,809 *employed individuals in the sample. Figure 6.3 shows six different density histograms for earnwk, with the number of bins specified as* 10, 20, 50, 100, 200*, and* 500*. These six histograms illustrate an inherent tradeoff when choosing the number of bins. If the number of bins is too small (or, equivalently, the bin width too large), the histogram will be less noisy but might miss key aspects of the shape of the distribution of the data. The top-left histogram in figure 6.3, with* 10 *bins, completely misses the "hump" in the distribution of the observed weekly earnings. This hump, located just below* \$1,000 *per week, first becomes evident in the* 20*-bin histogram in figure 6.3, and even more so in the histograms with* 50 *bins and* 100 *bins. As the number of bins gets even larger, however, the histograms display less smoothness and more noise. The two histograms with* 200 *bins and* 500 *bins in figure 6.3 have rectangle heights that jump up and down. This lack of smoothness can be exacerbated by variables where data may be bunched at round numbers, as is the case for the earnwk variable, where round numbers like* \$1,000 *or* \$2,000 *are more likely to be reported than, say,* \$1,019 *or* \$1,992*. When the number of bins is very large (or the bin width very small), the histogram is unable to smooth out this type of bunching.*

Here is the R code used to create figure 6.3:

```
# create subsample of the data of only "Employed" individuals
cpsemployed <- cps[cps$lfstatus=="Employed",]

# show the number of employed individuals
nrow(cpsemployed)

## [1] 2809

# graph-display format (three rows, two columns)
par(mfrow = c(3,2))

# histogram of weekly earnings with 10 bins (density on y-axis)
hist(cpsemployed$earnwk, breaks=10, freq=FALSE, main="Weekly earnings, 10 bins",
     xlab="Weekly earnings($)")

# histogram of weekly earnings with 20 bins (density on y-axis)
hist(cpsemployed$earnwk, breaks=20, freq=FALSE, main="Weekly earnings, 20 bins",
     xlab="Weekly earnings($)")

# histogram of weekly earnings with 50 bins (density on y-axis)
hist(cpsemployed$earnwk, breaks=50, freq=FALSE, main="Weekly earnings, 50 bins",
     xlab="Weekly earnings($)")

# histogram of weekly earnings with 100 bins (density on y-axis)
hist(cpsemployed$earnwk, breaks=100, freq=FALSE, main="Weekly earnings, 100 bins",
     xlab="Weekly earnings($)")

# histogram of weekly earnings with 200 bins (density on y-axis)
hist(cpsemployed$earnwk, breaks=200, freq=FALSE, main="Weekly earnings, 200 bins",
     xlab="Weekly earnings($)")

# histogram of weekly earnings with 500 bins (density on y-axis)
hist(cpsemployed$earnwk, breaks=500, freq=FALSE, main="Weekly earnings, 500 bins",
     xlab="Weekly earnings($)")
```

Since weekly earnings are observed only for employed individuals, the relevant rows of the data frame `cps` *are first selected. Specifically,* `cpsemployed` *is created as a new data frame, which consists of the rows of* `cps` *for which* `cps$lfstatus=="Employed"` *is* `TRUE`. *The* `nrow` *command confirms that the number of employed individuals is* 2,809. *The six* `hist` *commands use the* `breaks` *argument, but here, a value is provided for* `breaks`, *which specifies the number of bins in the histogram. This use of* `breaks` *is in contrast to example 6.5, where a vector was provided for* `breaks`, *corresponding to the starting/ending values for the bins rather than the number of bins.*

6.3.1 Bin Width Choice and Density Curves

Given that the bin width can dramatically affect how a histogram looks, how should the bin width be chosen? We don't want a bin width that is too small, nor do we want a bin width that is too large. Most statistical packages offer the option to automatically choose a bin width for a histogram. Different statistical packages may use different rules of thumb for calculating the bin width, but in practice, the overall look of the histogram should not vary much from package to package. One rule of thumb that is sometimes used is the *Freedman-Diaconis rule*, which calculates the bin width for variable x as follows:

$$\text{bin width} = 2 \cdot \frac{IQR_x}{\sqrt[3]{n}},$$

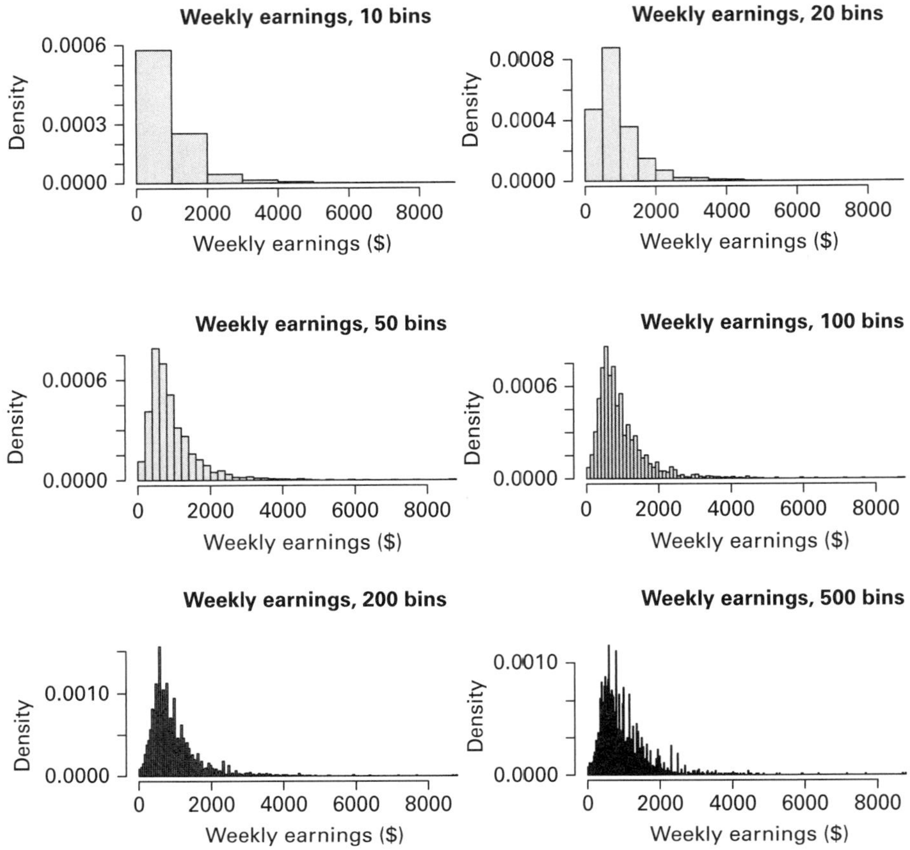

Figure 6.3
Histograms of weekly earnings (CPS data)

where IQR_x is the IQR of x, a measure that we will discuss further in section 6.5, and $\sqrt[3]{n}$ is the cube root of sample size n. The number of bins for the histogram in this case is

$$\frac{x_{max} - x_{min}}{2 \cdot \frac{IQR_x}{\sqrt[3]{n}}} \text{ rounded to the nearest integer,}$$

where x_{max} and x_{min} are the maximum and minimum x-values in the sample, respectively. For the *earnwk* variable in example 6.6, the Freedman-Diaconis rule yields a choice of 92 bins, based upon a calculated bin width of 95.48 since $earnwk_{max} = 8779.73$ and $earnwk_{min} = 12$. Of the histograms in figure 6.3, this choice yields a histogram similar to the one with 100 bins. Figure 6.4 shows the *earnwk* histogram with the 92 bins from the Freedman-Diaconis rule. In addition, it features a **density curve** overlying the histogram. Most statistical packages, including *R*, offer the option to draw a smooth density curve either on its own or along with a histogram. In figure 6.4, the density curve roughly passes through the tops of the histogram rectangles, but it does so in a smooth fashion rather than jumping from flat step to flat step as the histogram does. While a formal discussion of density curves is beyond the scope of this book, a density curve (i) can be quite useful as a descriptive visual

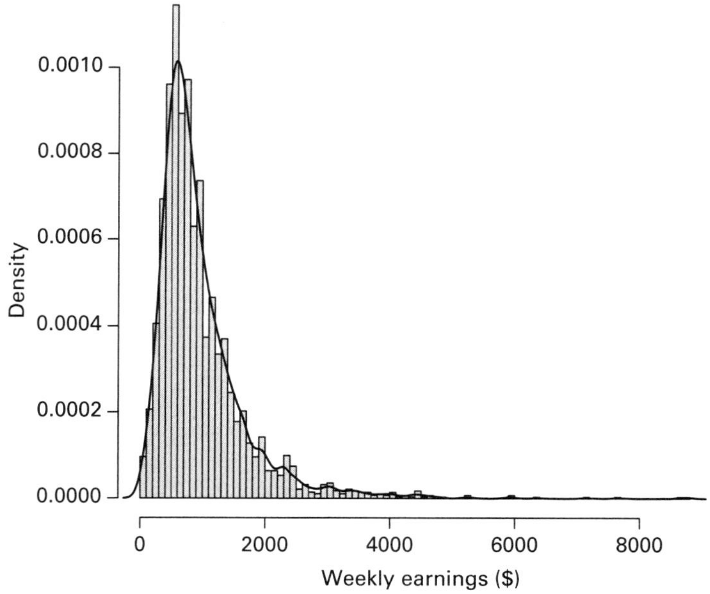

Figure 6.4
Histogram of weekly earnings with a Freedman-Diaconis bin width and density curve

for a numerical variable and (ii) has a direct relationship to the probability distributions introduced in chapters 8 and 10.

Here is the *R* code used to create figure 6.4:

```
# histogram of weekly earnings with 92 bins, with estimated density overlaid
hist(cpsemployed$earnwk, breaks=92, freq=FALSE, main="", xlab="Weekly earnings($)")
lines(density(cpsemployed$earnwk), lwd=2)
```

The `hist` command is similar to those seen in example 6.6, here with `breaks=92` specified. The `lines` command, with `density(cpsemployed$earnwk)` specified as its first argument, draws the density curve *on the same graph* as the original histogram. The optional line width argument `lwd=2` specifies a slightly thicker line for the density curve compared to the default value of `lwd=1`.

6.4 Numerical Data: Measures of Location

6.4.1 Sample Mean and Sample Median

For a numerical variable, the two most common measures of central tendency are the **sample mean** and the **sample median**. Most readers are probably already familiar with the sample mean (or the sample average), which is formally defined in definition 6.4.

Definition 6.4 *The* ***sample mean*** *or* ***sample average*** *of observations* $x_1, x_2, \ldots, x_n$*, denoted as* $\bar{x}$*, is*

$$\bar{x} = \frac{x_1 + x_2 + \cdots + x_n}{n} = \frac{1}{n}\sum_{i=1}^{n} x_i.$$

The sample mean depends on the value of every observation in the sample. As such, the sample mean can be sensitive to unusually small or unusually large values of x, sometimes known as **outliers**.

The sample median is a descriptive statistic used to describe the center of the sample. The sample median depends only on the relative order of the observations and, specifically, the observation values directly in the middle of the sample. The sample median is not affected by the values of outliers.

Definition 6.5 *The* ***sample median*** *of observations* $x_1, x_2, \ldots, x_n$, *denoted as* $\tilde{x}_{1/2}$ *or* $\tilde{x}_{0.5}$, *is the value for which at least half the observations are* $\leq \tilde{x}_{1/2}$ *and at least half the observations are* $\geq \tilde{x}_{1/2}$.

Unlike the sample mean, the sample median does not have a closed-form formula.[4]

To determine the sample median $\tilde{x}_{1/2}$, the following procedure can be used:

1. Sort the observations from lowest to highest. There may be some repeated values ("ties").
2. If n is odd, $\tilde{x}_{1/2}$ is the middle observation in the sorted sample. If n is even, $\tilde{x}_{1/2}$ is the average of the two middle observations in the sorted sample.

Example 6.7 *Suppose that* $n = 7$, *and the sample for* x *is* $\{4, 3, 8, 12, 0, 10, 5\}$. *The sample mean is*

$$\bar{x} = \frac{1}{7}(4 + 3 + 8 + 12 + 0 + 10 + 5) = 6.$$

The sorted sample is $\{0, 3, 4, 5, 8, 10, 12\}$. *Since* n *is odd, the sample median is* $\tilde{x}_{1/2} = 5$. *What would happen if the* $x_4 = 12$ *observation were instead* $x_4 = 26$? *The sample mean would increase to* 8, *showing how the sample mean can be affected by an outlier. In contrast, the sample median is unchanged since the middle value in the sorted list of values* $\{0, 3, 4, 5, 8, 10, 26\}$ *is still* 5.

Example 6.8 *Suppose that* $n = 8$, *and the sample for* x *is* $\{4, 3, 8, 12, 0, 10, 5, 14\}$. *The sample mean is*

$$\bar{x} = \frac{1}{8}(4 + 3 + 8 + 12 + 0 + 10 + 5 + 14) = 7.$$

The sorted sample is $\{0, 3, 4, 5, 8, 10, 12, 14\}$. *Since* n *is even, the sample median is* $\tilde{x}_{1/2} = \frac{5+8}{2} = 6.5$.

Both the sample mean and the sample median have the same units as the underlying x variable. For instance, if the x variable is measured in dollars, the sample mean $\bar{x}$ and sample median $\tilde{x}_{1/2}$ are also measured in dollars.

Example 6.9 (Labor force data) *Examples 6.5 and 6.6 considered histograms for the age and earnwk variables from the* **cps** *data. The sample average and the sample median can be calculated in R using the* `mean` *and* `median` *functions, respectively.*

```
mean(cps$age)

## [1] 45.0167
```

```
median(cps$age)

## [1] 45

mean(cps$earnwk, na.rm = TRUE)

## [1] 971.1785

median(cps$earnwk, na.rm = TRUE)

## [1] 770
```

Since `earnwk` *has missing values for nonemployed individuals, we specify the optional argument* `na.rm = TRUE` *for the* `mean` *and* `median` *functions to ignore the missing values. Alternatively, the* `cpsemployed` *data frame from example 6.6 could be used; for instance, the command* `mean(cpsemployed$earnwk)` *would give the same result as* `mean(cps$earnwk, na.rm = TRUE)`.

*For the age variable, the sample mean (*44.64 *years) and the sample median (*45 *years) are quite close to each other, which is expected from the fairly symmetric histograms in example 6.5. For the earnwk variable, it's a very different story, with the sample mean of weekly earnings (*$971.18*) much larger than the sample median (*$770*). Figure 6.5 shows the same histogram as figure 6.4, but now with the sample mean and sample median indicated on the graph. This histogram exhibits a long* right tail, *with some very large weekly earnings values observed in the right tail. There is no long* left tail *in the histogram since earnwk must be positive. The right-tail values cause the sample mean to be larger than the sample median. Since the sample mean depends on all observations, the large weekly earnings values in the right tail effectively pull the sample mean to the right. On the other hand, the sample median does not increase due to the very large right-tail values. Even if all the weekly earnings values in the right tail were instead equal to* 2,000, *the sample median would be unchanged, as there would still be* 50% *of observations below* 770 *and* 50% *of observations above* 770.

Here is the R code used to create figure 6.5:

```
# histogram of weekly earnings with 92 bins, with estimated density overlaid
hist(cpsemployed$earnwk, breaks=92, freq=FALSE, main="", xlab="Weekly earnings($)")
lines(density(cpsemployed$earnwk), lwd=2)
abline(v=mean(cpsemployed$earnwk), lwd=2, lty=3)
abline(v=median(cpsemployed$earnwk), lwd=2, lty=2)
legend("topright", legend=c("Sample median","Sample mean"), lty=c(2,3), lwd=c(2,2))
```

The `hist` *and* `lines` *commands are identical to those used for figure 6.4. The two* `abline` *commands add vertical lines, using the optional argument* `v`, *at the sample mean and the sample median of earnwk. The vertical lines for the sample mean and the sample median are dotted and dashed, respectively. Finally, the* `legend` *command shows how a legend can be added to a graph. The first argument indicates where the legend appears, and the remaining arguments specify the legend text (*`legend`*), line type (*`lty`*), and line width (*`lwd`*). Refer to the* `legend` *documentation in R for more details on the available options.*

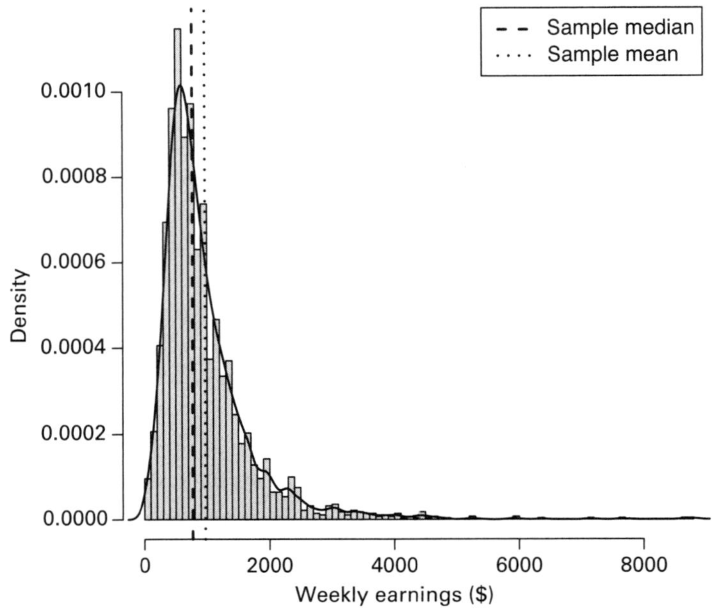

Figure 6.5
Right-skewness of the distribution of weekly earnings

When a histogram is characterized by a long right tail, like weekly earnings in example 6.9, the variable is said to have a **right-skewed distribution**, as specified in definition 6.6.

Definition 6.6 *A variable has a **right-skewed distribution** if there is a longer tail on the right side of its distribution than on the left side of its distribution, as exhibited by a histogram or density curve.*

A variable with a right-skewed sample distribution usually has its sample mean greater than its sample median. Many economic variables naturally have right-skewed distributions, including earnings or wealth for a sample of individuals, sales or profits for a sample of firms, and gross domestic product (GDP) for a sample of countries. While **left-skewed distributions** are less common in economics, we provide a formal definition in definition 6.7 in the interest of completeness.

Definition 6.7 *A variable has a **left-skewed distribution** if there is a longer tail on the left side of its distribution than on the right side of its distribution, as exhibited by a histogram or density curve.*

A variable with a left-skewed sample distribution usually has its sample mean less than its sample median.

Unlike the weekly earnings variable in example 6.9, the age variable exhibits neither a right-skewed distribution nor a left-skewed distribution. In fact, the histogram for the age variable looks approximately flat on both sides of the center of its distribution. This type of distribution is said to be an **approximately symmetric distribution**, as in definition 6.8.

Definition 6.8 *A variable has an* ***approximately symmetric distribution*** *if the shape of the distribution to the left of the sample median is approximately a mirror image of the shape of the distribution to the right of the sample median. If a variable does not have an approximately symmetric distribution, the variable is said to have an* ***asymmetric distribution****.*

A variable with an approximately symmetric sample distribution has a sample mean that is close to the sample median, meaning that either provides a good measure of the center of the sample distribution. A sample with either a right-skewed distribution or a left-skewed distribution must be an asymmetric distribution, as the presence of a longer tail on either the right or the left means that the distribution cannot possibly appear to have mirror images on the two sides of the sample median.

Example 6.10 (Monthly stock returns) *Example 6.2 introduced the monthly stock return data set* ***sp500****. Focusing again on the monthly returns for AT&T and Bank of America, the sample means and sample medians can be calculated with the* `mean` *and* `median` *functions in R. Alternatively, the sample mean and sample median are part of the output provided by the* `summary` *function applied to a numerical variable.*

```
mean(sp500$T)

## [1] 0.008239883

median(sp500$T)

## [1] 0.007072945

mean(sp500$BAC)

## [1] 0.01294691

median(sp500$BAC)

## [1] 0.014761

summary(sp500[,c("T","BAC")])

##        T                  BAC
##  Min.   :-0.191214   Min.   :-0.52203
##  1st Qu.:-0.022352   1st Qu.:-0.03675
##  Median : 0.007073   Median : 0.01476
##  Mean   : 0.008240   Mean   : 0.01295
##  3rd Qu.: 0.047903   3rd Qu.: 0.06571
##  Max.   : 0.276617   Max.   : 0.72658
```

The mean monthly return for AT&T is 0.824%*, and the median monthly return is* 0.707%*. Both measures are larger for the Bank of America, whose monthly returns have a sample mean of* 1.295% *and a sample median of* 1.476%*. Figure 6.6 shows histograms and density curves for the T and BAC variables. Both distributions exhibit a bell curve and look approximately symmetric. That said, the histograms do not look identical to each other, as the BAC distribution has many more monthly returns that are larger in magnitude (i.e., either large positive returns or large negative returns). This feature is discussed further when the dispersion of distributions is explored in section 6.5.*

Here is the R code used to create figure 6.6:

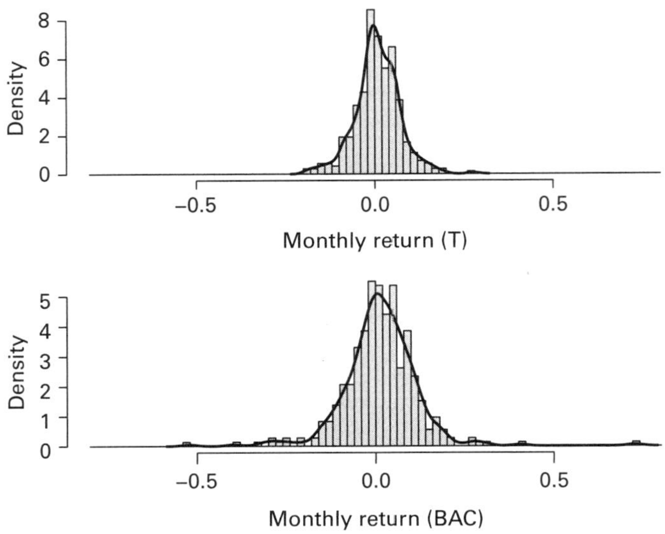

Figure 6.6
Histograms of monthly stock returns

```
# graph-display format (two rows, one column)
par(mfrow = c(2,1))

# histograms of T and BAC returns, with density curves
hist(sp500$T, freq=FALSE, breaks=seq(-0.8,0.8,0.02), main="",
     xlab="Monthly return (T)")
lines(density(sp500$T), lwd=2)

hist(sp500$BAC, freq=FALSE, breaks=seq(-0.8,0.8,0.02), main="",
     xlab="Monthly return (BAC)")
lines(density(sp500$BAC), lwd=2)
```

The use of the `hist` *and* `lines` *functions to draw the histograms and density curves is similar to the examples given previously. For both histograms, the vector for the* `breaks` *argument is specified to be the same (*`seq(-0.8,0.8,0.02)`*) for ease of comparison.*

6.4.2 Sample Quantiles

The sample median, introduced in section 6.4.1, provides a measure of the middle of a variable's sample distribution. Specifically, the sample median is the value $\tilde{x}_{1/2}$ for which half the observations are $\leq \tilde{x}_{1/2}$ and half the observations are $\geq \tilde{x}_{1/2}$. While the sample median is interesting, we might want to look at other parts of the sample distribution. For instance, if we want to know what the large values of variable x might be, the sample median doesn't provide a useful measure. Instead, we might want to find the value for which only 10% of observations are above that value.

To generalize the idea of a sample median to other parts of the sample distribution, **sample quantiles** are defined as in definition 6.9.

Definition 6.9 *For any q where $0 < q < 1$, the* ***sample quantile*** *$\tilde{x}_q$ is a value for which at least $(100q)\%$ of the observations are below $(\leq)$ $\tilde{x}_q$ and at least $(100-100q)\%$ of* *the observations are above (≥) $\tilde{x}_q$. The sample median $\tilde{x}_{1/2}$ is a special case of the sample quantile where $q = 1/2$. Other commonly used sample quantiles are* ***sample quartiles****, corresponding to $q \in \{0.25, 0.50, 0.75\}$, and* ***sample deciles****, corresponding to $q \in \{0.1, 0.2, 0.3, 0.4, 0.5, 0.6, 0.7, 0.8, 0.9\}$.*

As an example, for $q = 0.9$, the sample quantile $\tilde{x}_{0.9}$ is the value for which at least 90% of the observations are less than or equal to $\tilde{x}_{0.9}$ and at least 10% of the observations are greater than or equal to $\tilde{x}_{0.9}$. We call $\tilde{x}_{0.9}$ the sample 90% quantile. For the sample quartiles, the 25% quantile is the *first quartile* or *lower quartile*, and the 75% quantile is the *third quartile* or *upper quartile*. For the sample deciles, the 10% quantile is the *first decile*, the 20% quantile is the *second decile*, and so on. Like the sample mean and sample median, any sample quantile $\tilde{x}_q$ has the same units as the underlying x variable.

Most statistical packages have functions to calculate sample quantiles, and in practice, we rely upon the statistical package to calculate these quantiles. There are several alternative algorithms to calculate sample quantiles, meaning that one statistical package might give a slightly different answer than another statistical package. In reasonably sized samples, these small differences will not be practically meaningful.

Here is a procedure that generalizes the procedure used previously for calculating a sample median and can be used to manually compute the sample quantile for any value q between 0 and 1:

1. Sort the observations from lowest to highest. There may be some repeated values ("ties"), which is fine.
2. If nq is an integer, then $\tilde{x}_q$ is the average of the nqth value and the $(nq+1)$th value in the sorted sample. If nq is not an integer, then $\tilde{x}_q$ is the $\lceil nq \rceil$th value in the sorted sample, where $\lceil nq \rceil$ is the smallest integer larger than nq.

Consider a sample with sample size $n = 50$. First, the 50 observations are sorted in ascending order, from lowest to highest. For the 10% quantile ($q = 0.1$), $nq = 5$ is an integer, so $\tilde{x}_{0.1}$ is the average of the 5th and 6th values of the sorted sample. For the 25% quantile ($q = 0.25$), $nq = 12.5$ is not an integer, so $\tilde{x}_{0.25}$ is equal to the 13th value of the sorted sample since $\lceil nq \rceil = \lceil 12.5 \rceil = 13$.

The *R* function `quantile` calculates sample quantiles:

- `quantile(x, probs = ...)`: Returns sample quantiles of vector `x`, where the quantiles returned are specified by the `probs` argument. For example, `quantile(x, probs = c(0.25,0.75))` returns the sample 25% and 75% quantiles.

Example 6.11 (Labor force data) *Sample quantiles provide a more complete description of the distribution of weekly earnings (earnwk) from the* **cps** *data. The following R output shows the sample quartiles and sample deciles of earnwk:*

```
# sample quartiles of weekly earnings
quantile(cpsemployed$earnwk, probs = c(0.25,0.50,0.75))

##    25%    50%    75%
##  520.0  770.0 1193.6
```

```
# sample quartiles of weekly earnings
quantile(cpsemployed$earnwk, probs = seq(0.1,0.9,0.1))

##     10%     20%     30%     40%     50%     60%     70%     80%     90%
##  354.56  480.00  576.00  670.00  770.00  900.00 1080.00 1346.00 1750.00
```

Interpreting the values for the sample quantiles is straightforward. For instance, for the sample 70% *quantile* ($\tilde{x}_{0.7} = 1{,}080$)*, approximately* 70% *of the sample observations are below* 1,080 *and approximately* 30% *are above* 1,080.

In figure 6.7, five different quantiles ($q = 0.1, 0.25, 0.5, 0.75, 0.9$) *are shown on the histogram. It helps to visualize where the sample quantiles lie along the distribution. Due to the right skewness of earnwk, the 75% and 90% quantile values are pulled to the right. The distance between the 75% quantile and the sample median (50% quantile) is larger than the distance between the 25% quantile and the sample median, and the distance between the 90% quantile and the sample median is much larger than the distance between the 10% quantile and the sample median.*

As this example illustrates, the skewness of a sample distribution has implications for the sample quantile values. For a right-skewed distribution, the distance between the 75% quantile and the sample median ($\tilde{x}_{0.75} - \tilde{x}_{0.5}$) *would be expected to be larger than the distance between the 25% quantile and the sample median* ($\tilde{x}_{0.5} - \tilde{x}_{0.25}$) *and similarly for higher quantiles like the 90% quantile* ($\tilde{x}_{0.9} - \tilde{x}_{0.5}$ *larger than* $\tilde{x}_{0.5} - \tilde{x}_{0.1}$) *and the 95% quantile* ($\tilde{x}_{0.95} - \tilde{x}_{0.5}$ *larger than* $\tilde{x}_{0.5} - \tilde{x}_{0.05}$)*. In contrast, for a sample distribution that is approximately symmetric, the distance between the 75% quantile and the sample*

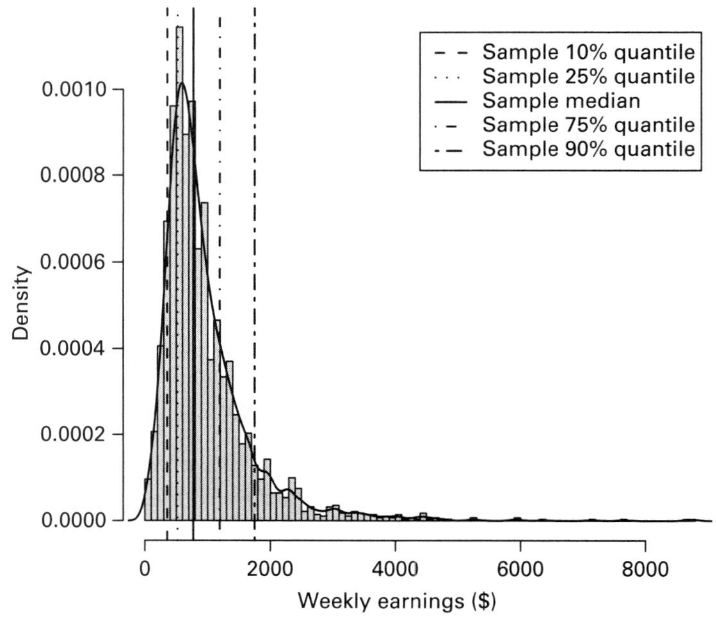

Figure 6.7
Sample quantiles of the weekly earnings distribution

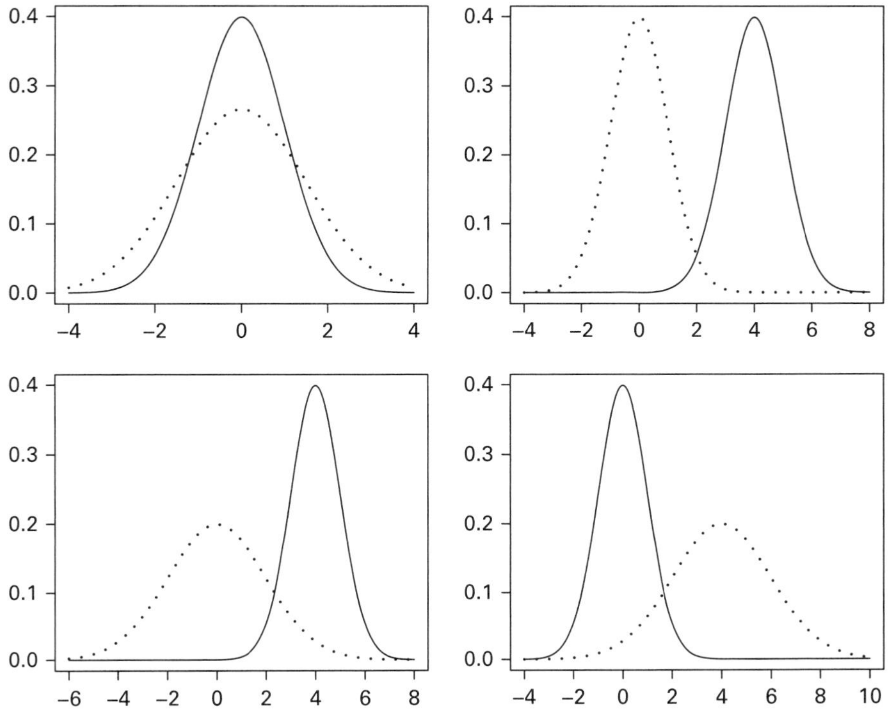

Figure 6.8
Location and dispersion of variable distributions

median would be expected to be similar to the distance between the 25% quantile and the sample median and similarly for higher quantiles like the 90% quantile ($\tilde{x}_{0.9} - \tilde{x}_{0.5}$ similar to $\tilde{x}_{0.5} - \tilde{x}_{0.1}$) and the 95% quantile ($\tilde{x}_{0.95} - \tilde{x}_{0.5}$ similar to $\tilde{x}_{0.5} - \tilde{x}_{0.05}$). As an example, the age variable from the labor-force data is approximately symmetric and has a sample median of 45. *The* 25% *and* 75% *sample quantiles are* 37 *and* 53, *respectively, which are equidistant (8 years) from the sample median. The* 10% *and* 90% *sample quantiles are* 32 *and* 57, *respectively, which are nearly equidistant (13 years and 12 years, respectively) from the sample median.*

6.5 Numerical Data: Measures of Dispersion

We now introduce descriptive statistics that measure the dispersion of a sample distribution. In cases where the dispersion measures for multiple variables are comparable, we can determine which variables exhibit more dispersion or variation.

Let's consider some simple examples of how variables can differ in terms of their location and/or dispersion. In the four graphs shown in figure 6.8, hypothetical sample distributions for two different variables are depicted as a solid, bell-shaped density curve and a dotted, bell-shaped density curve. In the top-left graph, the two variables have the same central location (sample median), but the dotted distribution has longer (and thicker) left and right tails, exhibiting more dispersion than the solid distribution. The solid distribution has more observations closer to the center and fewer observations in the tails. In the top-right graph, the two variables have the same dispersion since their shapes are identical, but their locations are different, with the solid distribution having a higher sample median than the dotted distribution. In the bottom-left graph, the solid distribution has a higher sample median and less dispersion than the dotted distribution. Finally, in the bottom-right graph, the solid distribution has a lower sample median and less dispersion than the dotted distribution.

While histograms and density curves provide a way to visually compare the dispersion of different variables, it is also important to have numerical descriptive statistics to characterize the dispersion of variables. Section 6.5.1 introduces the IQR and a descriptive visual, known as a *box plot*, that is based in part on the IQR. Section 6.5.2 introduces the sample standard deviation and sample variance descriptive statistics.

6.5.1 Interquartile Range and Box Plots

A simple way to describe the dispersion of a sample is to look at the **range** of observed values, as in definition 6.10.

Definition 6.10 *The* ***range*** *of observations* $x_1, x_2, \ldots, x_n$ *is*

$$x_{max} - x_{min},$$

where x_{max} *is the maximum value in the sample and* x_{min} *is the minimum value in the sample.*

Unfortunately, the range is not a very useful dispersion measure since it is so sensitive to the minimum and maximum values. Just a single outlier can dramatically affect the range. For instance, the sample $\{4, 7, 2, 9, 11, 1\}$ has a range of 10, while the same sample with an outlier of 42 rather than the 11 value, such that the sample is $\{4, 7, 2, 9, 42, 1\}$, has a range of 41.

To address this drawback of the range, we can instead look at the difference between certain sample quantiles of the distribution. While the minimum and maximum values are too extreme, the difference between the sample 90% quantile and the sample 10% quantile ($\tilde{x}_{0.9} - \tilde{x}_{0.1}$), or the difference between sample 80% quantile and the sample 20% quantile ($\tilde{x}_{0.8} - \tilde{x}_{0.2}$), could be used as alternative measures of dispersion. A frequently used descriptive statistic, known as the **interquartile range** (IQR), is based on the difference between the sample 75% quantile and the sample 25% quantile or, equivalently, the difference between the third quartile and the first quartile.

Definition 6.11 *The* ***interquartile range (IQR)*** *of observations* $x_1, x_2, \ldots, x_n$*, denoted as* IQR_x*, is*

$$IQR_x = \tilde{x}_{0.75} - \tilde{x}_{0.25}.$$

Since sample quantiles have the same units as the underlying x variable, the units of IQR_x are also the same as the units of the underlying x variable. With 25% of the observations being below the first quartile $\tilde{x}_{0.25}$ and 75% of the observations being below the third quartile $\tilde{x}_{0.75}$, approximately 50% of the observations are between $\tilde{x}_{0.25}$ and $\tilde{x}_{0.75}$. The IQR statistic gives the width of this range, between the first and third quartiles, which covers approximately 50% of the observations. Recall that IQR_x was used for the Freedman-Diaconis method for choosing a histogram bin width in section 6.3.1. The Freedman-Diaconis bin width choice is $2 \cdot \frac{IQR_x}{\sqrt[3]{n}}$, so the bin width choice is directly proportional to IQR_x. Therefore, for the same sample size n, if one variable has an IQR that is twice as large as another variable, that variable would have a bin width that is twice as large as the other variable. The relationship between the bin width and the IQR seems sensible, as it ensures that different variables have comparable numbers of observations within their histogram bins.

Example 6.12 (Labor force data) *For the weekly earnings (earnwk) variable from the* ***cps*** *data, the first and third quartiles of the sample are* 520 *and 1,194, respectively, so approximately half the sample has earnwk values between* 520 *and 1,194. The IQR for earnwk is* $1{,}194 - 520 = 674$ *dollars. For the age variable, the IQR is* $53 - 37 = 16$ *years. It does not make sense to directly compare the IQR values for earnwk and age since they have different units. That is, it would not be appropriate to say that the earnwk variable exhibits more dispersion than age. In other situations, when two variables have the same units, it can make sense to directly compare the IQR values.*

The IQR can be calculated in R directly using the `IQR` *function or, alternatively, as the difference between the 75% and 25% sample quantiles.*

```
# interquartile range of weekly earnings
IQR(cps$earnwk, na.rm = TRUE)

## [1] 673.6

# interquartile range of weekly earnings
IQR(cps$age)

## [1] 16
```

A useful descriptive visual based upon the IQR is a **box plot**. While there are several variants of the box plot, two alternative versions are considered here:

- *Box plot with whiskers at the minimum and maximum*: The box extends from the sample 25% quantile (first quartile) to the sample 75% quantile (third quartile), with the sample median indicated by a line within the box. The whiskers are indicated by lines at the minimum value and the maximum value in the sample.
- *Box plot with whiskers and outliers*: The box extends from the sample 25% quantile (first quartile) to the sample 75% quantile (third quartile), with the sample median indicated by a line within the box. The upper whisker is indicated by a line at the minimum of the following two values: x_{max} and $\tilde{x}_{0.75} + 1.5IQR_x$. The lower whisker is indicated by a line at the maximum of the following two values: x_{min} and $\tilde{x}_{0.25} - 1.5IQR_x$. The outliers, observations that are either above the upper whisker or below the lower whisker, are indicated by dots or circles.

The second version is usually preferred by practitioners since the first version, like the range descriptive statistic, is too sensitive to the minimum and maximum values. The best way to fully understand how these box plots are constructed is to consider example 6.13.

Example 6.13 (Labor force data) *For the weekly earnings (earnwk) variable from the* ***cps*** *data, figure 6.9 shows the two versions of the box plot described here. The one on the left is*

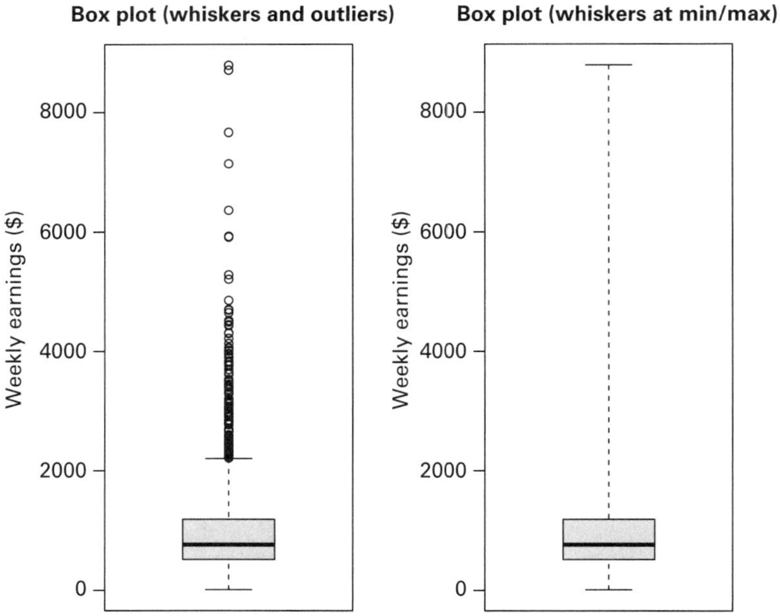

Figure 6.9
Box plots of weekly earnings (CPS data)

the box plot with whiskers and outliers, and the one on the right is the box plot with whiskers at the minimum and maximum. Both box plots have the same box, extending from the first quartile (520) to the third quartile (1,194) with a line indicating the sample median (770). The height of each box is the IQR value of 674. The box plot on the right has the lower whisker at the minimum value (12) and the upper whisker at the maximum value (8,780). For the box plot on the left, the lower whisker is indicated by a line at

$$max(x_{min}, \tilde{x}_{0.25} - 1.5IQR_x) = max(12, 520 - (1.5)(674)) = max(12, -491) = 12,$$

and the upper whisker is indicated by a line at

$$min(x_{max}, \tilde{x}_{0.75} + 1.5IQR_x) = min(8780, 1194 + (1.5)(674)) = min(8780, 2205) = 2205.$$

Finally, the box plot on the left has outliers represented by circles. In this case, all the outliers are above the upper whisker, as there can be no observations below the lower whisker, which is located at the minimum value of the sample. This box plot offers another visual confirmation of the right skewness of the earnwk variable. The distance from the sample median to the third quantile is larger than the distance from the first quartile to the sample median, and a long right tail of outliers appears above the upper whisker, with no such left tail of outliers below the lower whisker.

The R code to create figure 6.9 uses the function `boxplot` *as follows:*

```
# graph-display format (one row, two columns)
par(mfrow = c(1,2))
```

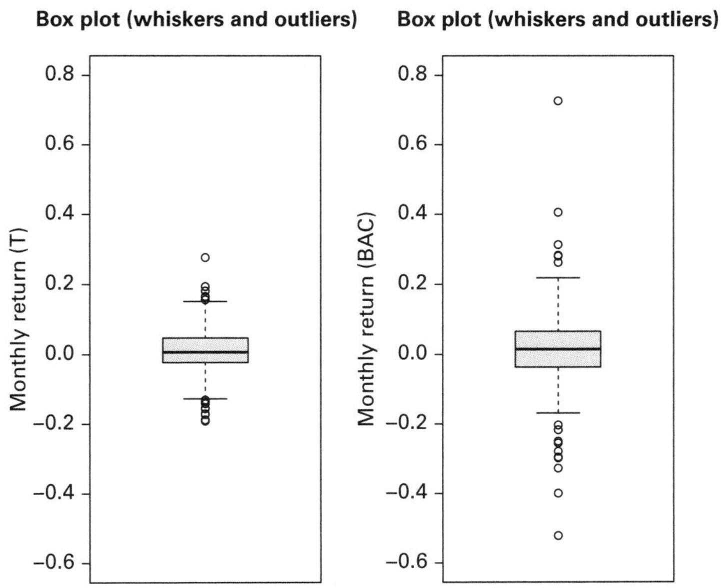

Figure 6.10
Box plots of monthly stock returns

```
# box plot with whiskers and outliers (the default in R)
boxplot(cpsemployed$earnwk, ylab="Weekly earnings",
        main="Box plot (whiskers and outliers)")

# box plot with whiskers at min and max values
boxplot(cpsemployed$earnwk, range=0, ylab="Weekly earnings",
        main="Box plot (whiskers at min/max)")
```

The default for the `boxplot` *function is to have whiskers and outliers displayed. The first* `boxplot` *command creates this box plot for the* `cpsemployed$earnwk` *variable. The second* `boxplot` *command creates a box plot with whiskers at the minimum and maximum by specifying the optional argument* `range=0`.

Example 6.14 (Monthly stock returns) *As in example 6.10, we focus on the monthly returns for AT&T and Bank of America from the* ***sp500*** *data set. Figure 6.10 shows the box plots, with whiskers and outliers, for T and BAC, drawn with the same y-axis (extending from* −0.6 *to* 0.8*) for ease of comparison. Here is the R code to create figure 6.10:*

```
# graph-display format (one row, two columns)
par(mfrow = c(1,2))

# box plot (whiskers and outliers) for AT&T
boxplot(sp500$T, ylab="Monthly return (T)", main="Box plot (whiskers and outliers)",
        ylim=c(-0.6,0.8))

# box plot (whiskers and outliers) for Bank of America
boxplot(sp500$BAC, ylab="Monthly return (BAC)",
        main="Box plot (whiskers and outliers)",ylim=c(-0.6,0.8))
```

*Neither box plot shows strong evidence of right or left skewness, with the first and third quartiles and the lower and upper whiskers fairly equidistant to the sample medians. In this example, a direct comparison of the distributions of T and BAC is possible since they are both unitless. As already seen in the histograms from example 6.10, the distribution of BAC (i.e., Bank of America's monthly stock returns) exhibits more dispersion than the distribution of T (i.e., AT&T's monthly stock returns). The box extends farther in both directions for BAC than for T, corresponding to a larger IQR value (*0.1025*) for BAC than the IQR value (*0.0703*) for T. BAC also has more extreme outliers in both directions than T does. In fact, BAC has six observed returns greater than* 0.3 *in magnitude (either below* -0.3 *or above* $+0.3$*), whereas T has no such observed returns greater than* 0.3 *in magnitude.*

6.5.2 Sample Variance and Sample Standard Deviation

We now introduce descriptive measures of dispersion based upon the **deviations from mean** in the sample.

Definition 6.12 *The* ***deviation from mean*** *for the ith observation of the x variable is* $x_i - \bar{x}$.

Example 6.15 *Suppose* $n = 7$, *and the sample for x is* $\{4, 3, 8, 12, 0, 10, 5\}$. *The sample mean is* $\bar{x} = 6$. *The deviations from mean for each of the observations are as follows:*

i	1	2	3	4	5	6	7
x_i	4	3	8	12	0	10	5
$x_i - \bar{x}$	−2	−3	2	6	−6	4	−1

The sum of the deviations from mean in example 6.15 is exactly zero, which is a general result for any sample, as given in proposition 6.4.

Proposition 6.4 *The sum of the deviations from mean,* $\sum_{i=1}^{n}(x_i - \bar{x})$, *is equal to zero. The average of the deviations from mean,* $\frac{1}{n}\sum_{i=1}^{n}(x_i - \bar{x})$, *is equal to zero.*

This proposition holds since

$$\sum_{i=1}^{n}(x_i - \bar{x}) = (x_1 - \bar{x}) + (x_2 - \bar{x}) + \cdots + (x_n - \bar{x}) = \left(\sum_{i=1}^{n} x_i\right) - n\bar{x} = n\bar{x} - n\bar{x} = 0.$$

Proposition 6.4 implies that it is pointless to use the average deviation from mean as a measure of dispersion since its value is always zero. The negative and positive deviations cancel each other out in the summation. To avoid this canceling out, the key is to treat positive and negative deviations in the same way, viewing the deviation from the mean as a positive distance, regardless of whether an observation is above or below the mean. One dispersion measure based upon this idea is the **sample mean absolute deviation**:[5]

Definition 6.13 *The* ***sample mean absolute deviation*** *of observations* $x_1, x_2, \ldots, x_n$, *denoted as* MAD_x, *is*

$$MAD_x = \frac{1}{n}\sum_{i=1}^{n}|x_i - \bar{x}|.$$

The MAD_x descriptive statistic is interpreted as the average distance that sample observations are from the sample mean. MAD_x is always nonnegative, and it's strictly positive unless all x_i values are the same. The units of MAD_x are the same as the units of x, making it convenient for interpretation.

Example 6.16 *Continuing example 6.15, the absolute deviation values are added to the table:*

i	1	2	3	4	5	6	7
x_i	4	3	8	12	0	10	5
$x_i - \bar{x}$	−2	−3	2	6	−6	4	−1
$\lvert x_i - \bar{x}\rvert$	2	3	2	6	6	4	1

Then, $MAD_x = \frac{1}{7}(2+3+2+6+6+4+1) = \frac{24}{7}$. *The average distance that an observation is from the sample mean is* $\frac{24}{7}$.

Example 6.17 (Monthly stock returns) *The* MAD_x *values for the AT&T monthly stock returns and the Bank of America monthly stock returns can be calculated in R:*

```
# MAD for AT&T
mean(abs(sp500$T-mean(sp500$T)))

## [1] 0.04708729

# MAD for Bank of America
mean(abs(sp500$BAC-mean(sp500$BAC)))

## [1] 0.07205011
```

Since the two variables are unitless, the MAD_x *statistics are also unitless and can be directly compared. On average, the BAC monthly returns are farther from their sample mean than the T returns. The average distance of the T monthly returns to their sample mean is* 0.04709 *or* 4.709%, *and the average distance of the Bank of America monthly returns to their sample mean is* 0.07205 *or* 7.205%. *The sample dispersion is much larger for BAC, with its* MAD_x *value* 53% *larger than the* MAD_x *value for T.*

An alternative dispersion measure based upon the deviations from mean is the **sample variance**, where the squared distance $(x_i - \bar{x})^2$ is used instead of the absolute distance $|x_i - \bar{x}|$, as in definition 6.14.

Definition 6.14 *The* ***sample variance*** *of observations* $x_1, x_2, \ldots, x_n$, *denoted as* s_x^2, *is*

$$s_x^2 = \frac{1}{n-1}\sum_{i=1}^{n}(x_i - \bar{x})^2.$$

The sample variance s_x^2 has a $\frac{1}{n-1}$ scaling, rather than the $\frac{1}{n}$ scaling previously seen for averages. Even with the small difference introduced by the $\frac{1}{n-1}$ scaling, the sample variance s_x^2 can be interpreted as (approximately) the average squared distance that the sample observations are from the sample mean. Like MAD_x, the sample variance s_x^2 is always non-negative, and it's strictly positive unless all x_i values are the same. Unlike MAD_x, however, the units of s_x^2 are not the same as the units of x. Instead, the units of s_x^2 are the units of x squared since each $(x_i - \bar{x})^2$ term has those units. If the units of x are dollars, the units of

s_x^2 are dollars squared; if the units of x are years, the units of s_x^2 are years squared; and it is similar for other variables. This feature of the sample variance makes interpretation problematic since it's hard to explain or comprehend a descriptive statistic that is measured in (units of x)^2 (e.g., dollars squared, years squared, etc.). As a result, it is often more useful to calculate and report the **sample standard deviation**, which is obtained by taking the square root of the sample variance and therefore has the same units as the original x variable, as in definition 6.15.

Definition 6.15 *The* ***sample standard deviation*** *of observations* $x_1, x_2, \ldots, x_n$*, denoted as* s_x*, is*

$$s_x = \sqrt{s_x^2} = \sqrt{\frac{1}{n-1}\sum_{i=1}^{n}(x_i - \bar{x})^2}.$$

While the sample standard deviation s_x is in the units of the x variable, it does *not* have a simple interpretation in the same way that MAD_x does. Recall that the MAD_x can be interpreted as the average distance of sample observations from their mean. Unfortunately, due to the presence of the square root in the definition of s_x, the sample standard deviation is not an average of some interesting underlying quantity. As seen later in this book, the meaning of the sample standard deviation depends on the specific underlying distribution of the variable; for instance, the sample standard deviation has a particularly interesting interpretation in the case of a variable that has a normal distribution. For now, the sample standard deviation should be thought of as an alternative dispersion measure.

Example 6.18 *Continuing example 6.16, the squared deviation values are added to the table:*

i	1	2	3	4	5	6	7
x_i	4	3	8	12	0	10	5
$x_i - \bar{x}$	−2	−3	2	6	−6	4	−1
$\lvert x_i - \bar{x}\rvert$	2	3	2	6	6	4	1
$(x_i - \bar{x})^2$	4	9	4	36	36	16	1

The sample variance is

$$s_x^2 = \frac{1}{7-1}(4+9+4+36+36+16+1) = \frac{106}{6} = \frac{53}{3},$$

and the sample standard deviation is $s_x = \sqrt{s_x^2} = \sqrt{\frac{53}{3}} \approx 4.20.$

Example 6.19 (Monthly stock returns) *Continuing example 6.17, the sample variances and sample standard deviations for the AT&T monthly stock returns and the Bank of America monthly stock returns can be calculated in R using the* `var` *and* `sd` *functions, respectively:*

```
# sample variance and sample standard deviation for AT&T
var(sp500$T)

## [1] 0.003977273

sd(sp500$T)
```

```
## [1] 0.06306562

# sample variance and sample standard deviation for Bank of America
var(sp500$BAC)

## [1] 0.01108884

sd(sp500$BAC)

## [1] 0.1053035
```

*As with MAD_x, the sample standard deviation s_x indicates that BAC has more dispersion than T. The standard deviation of BAC (*0.10530*) is approximately* 67% *larger than the standard deviation of T (*0.06307*).*

Example 6.20 (Union versus nonunion wages) *Suppose that we are interested in a comparison of weekly earnings for union workers versus nonunion workers. Using the* **cps** *data, we construct two* **subsamples** *based upon the unionstatus variable, one with the* 276 *union workers and one with the* 2,533 *nonunion workers. Focusing on weekly earnings ($x =$ earnwk), the following table provides descriptive statistics for the union and nonunion subsamples:*

Sample	n	$\bar{x}$	MAD_x	s_x^2	s_x
Union workers	276	1,197.7	532.4	518,378.8	720.0
Nonunion workers	2,533	946.5	488.8	562,120.3	749.7

The sample mean of weekly earnings is roughly \$250 *higher for union workers (*\$1,198*) than for nonunion workers (*\$947*). The dispersion measures provide mixed evidence on the relative dispersion of the distribution of union weekly earnings versus the distribution of nonunion weekly earnings. The MAD_x measure suggests slightly more dispersion for union workers, whereas the s_x measure suggests slightly more dispersion for nonunion workers. To see whether the histograms for the two subsamples provide any further evidence of their relative dispersion, figure 6.11 plots the two histograms and density curves, using the same x-axis for ease of comparison. It's clear why union workers have higher average weekly earnings, as there is a much higher proportion of observations with earnwk > 1000 in the top histogram compared to the bottom histogram. But, consistent with the descriptive statistics, it's unclear from these histograms whether earnings are more dispersed in one versus the other.*

Example 6.21 (Male versus female wages) *Suppose that instead we are interested in a comparison of weekly earnings for male workers versus female workers. The approach is similar to example 6.20, except we construct two subsamples based on gender—one with the* 1,501 *male workers and one with the* 1,308 *female workers. Again focusing on weekly earnings ($x =$ earnwk), the following table provides the descriptive statistics for the two subsamples:*

Sample	n	$\bar{x}$	MAD_x	s_x^2	s_x
Male workers	1,501	1,117.3	529.9	610,217.8	781.2
Female workers	1,308	803.5	415.1	457,066.4	676.1

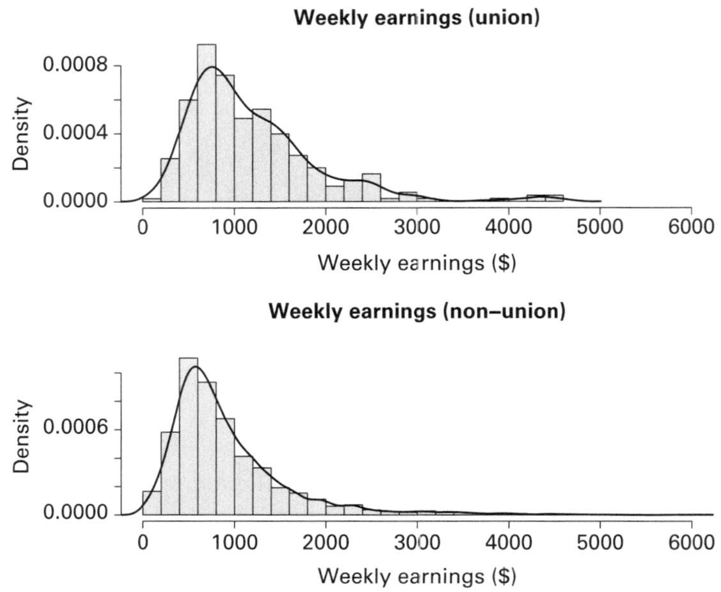

Figure 6.11
Histograms of weekly earnings for union and nonunion subsamples

The average weekly earnings for male workers (\$1,117) is over \$300 higher than the average weekly earnings for female workers (\$804). The MAD_x and s_x statistics both provide evidence that the distribution of male weekly earnings is more dispersed than the distribution of female weekly earnings. Figure 6.12, which shows the histograms and density curves for the two subsamples, indicates the higher dispersion for male weekly earnings distribution is a result of its thicker and longer right tail compared to the female weekly earnings distribution.

When $x \in \{0, 1\}$ is a binary or indicator variable, the sample variance and sample standard deviation take a particularly simple form. For a binary x variable, the sample mean $\bar{x} = \frac{1}{n}\sum_{i=1}^{n} x_i$ is the sample proportion of observations with $x_i = 1$ since the $x_i = 0$ observations do not contribute to the summation. As an example, let's say that x is an indicator of whether a worker is in a union, with 1 indicating a union worker and 0 indicating a nonunion worker. For the sample considered in example 6.20, we have $\bar{x} = \frac{276}{2809} \approx 0.098$ or approximately 9.8% of workers in a union. Proposition 6.5 indicates that the sample variance of a binary x variable depends only on $\bar{x}$.

Proposition 6.5 *If $x \in \{0, 1\}$ is a binary variable, the sample variance of x is*

$$s_x^2 = \frac{n}{n-1}\bar{x}(1-\bar{x}),$$

and the sample standard deviation of x is

$$s_x = \sqrt{s_x^2} = \sqrt{\frac{n}{n-1}\bar{x}(1-\bar{x})}.$$

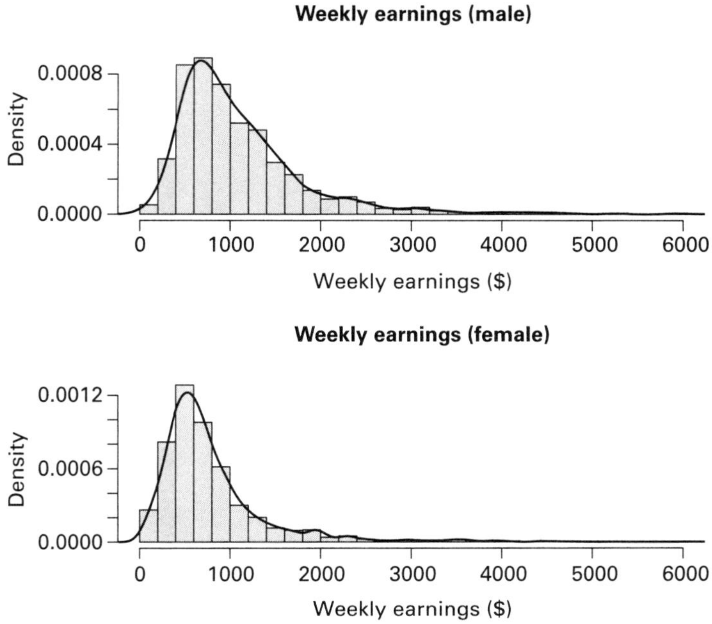

Figure 6.12
Histograms of weekly earnings for male and female subsamples

The proof of this proposition is left as an exercise at the end of this chapter (exercise 6.10). For the union example,

$$s_x^2 = \frac{2809}{2808} \cdot \frac{276}{2809} \cdot \frac{2533}{2809} \approx 0.0886 \text{ and } s_x = \sqrt{\frac{2809}{2808} \cdot \frac{276}{2809} \cdot \frac{2533}{2809}} \approx 0.2977.$$

In a sense, the sample variance doesn't provide any additional information (beyond $\bar{x}$) about the binary variable x since it's directly a function of $\bar{x}$. Once the sample proportion of 1s is known, it completely determines both the sample mean and the sample variance of the binary variable.

6.6 Modal Outcomes

In certain instances, we might be interested in knowing the most likely outcome for a given variable. None of the descriptive statistics discussed to this point are helpful for determining the most likely outcome. Instead, a new descriptive statistic is needed.

Definition 6.16 *The **sample mode** or **modal outcome** of observations $x_1, x_2, \ldots, x_n$ is the value that occurs most often. It is possible that there is more than one sample mode or modal outcome, which happens when two or more outcomes are tied for being the most likely.*

When x is a categorical variable, the sample mode is the category that occurs most often in the sample. For numerical variables, the sample mode is generally most useful for discrete variables, and perhaps for continuous variables which have focal responses/values. For

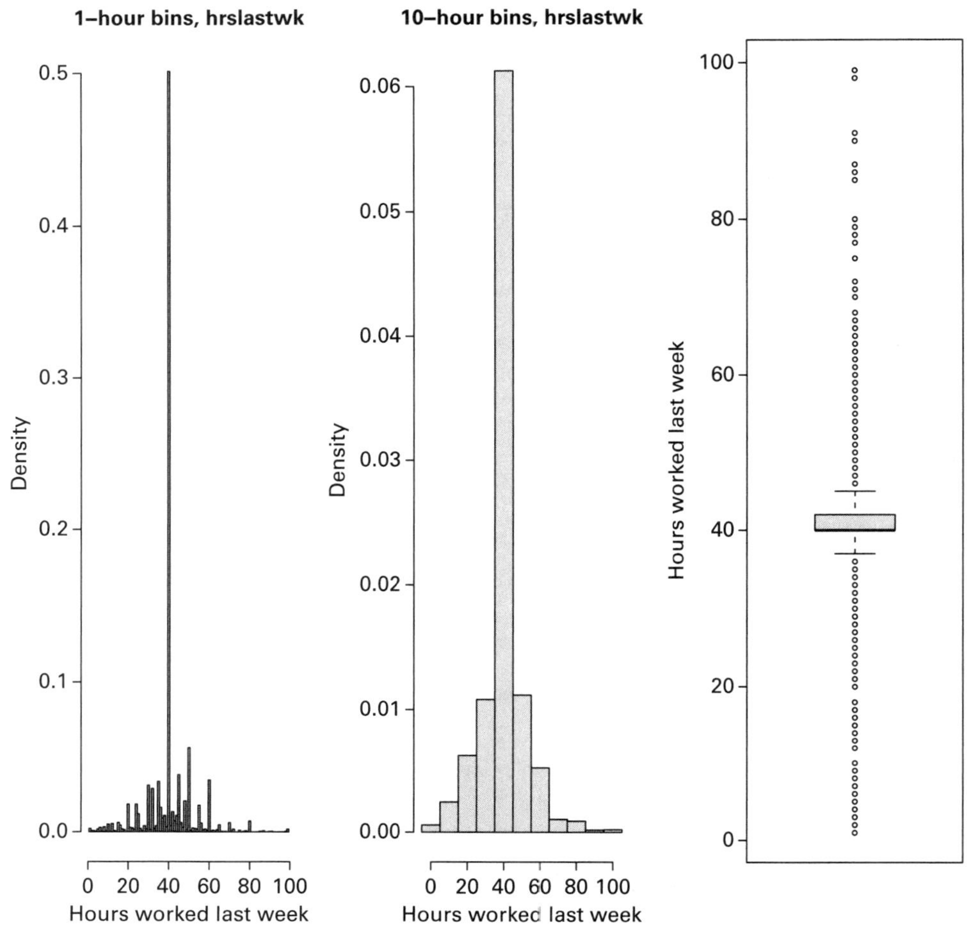

Figure 6.13
Histograms and box plots of weekly hours worked (CPS data)

a continuous numerical variable, even if the sample mode is not useful (e.g., if most of or all the values are distinct), we may refer to a distribution as a *unimodal distribution* if its histogram or density curve exhibits only one "hump," a *bimodal distribution* if its histogram or density curve exhibits only two "humps," and so on. For example, the various distributions of weekly earnings from the **cps** data have been unimodal distributions with a single hump.

Example 6.22 (Labor force data) *Consider the distribution of the hrslastwk variable (hours worked last week) for the sample of* 2,809 *employed individuals from the* **cps** *data. Figure 6.13 provides three ways of looking at the distribution of hrslastwk: a histogram with 1-hour bin widths, a histogram with* 10*-hour bin widths, and a box plot with whiskers and outliers. We use the following R code, specifying the* `breaks` *argument to be* `seq(0.5,99.5,1)` *for the 1 hour bin-width histogram and* `seq(-5,105,10)` *for the* 10*-hour bin-width histogram:*

```
# graph-display format (one row, three columns)
par(mfrow = c(1,3))

# histogram with bin widths of one hour
hist(cpsemployed$hrslastwk, breaks=seq(0.5,99.5,1), freq=FALSE,
     main="1-hour bins, hrslastwk", xlab="Hours worked last week")

# histogram with bin widths of ten hours
hist(cpsemployed$hrslastwk, breaks=seq(-5,105,10), freq=FALSE,
     main="10-hour bins, hrslastwk", xlab="Hours worked last week")

# box plot with whiskers and outliers
boxplot(cpsemployed$hrslastwk, ylab="Hours worked last week")
```

Since hrslastwk is integer-valued, the histogram with 1-hour bin widths has a rectangle corresponding to every possible value of hrlastwk. A huge spike at hrslastwk $=40$ *is evident. In the sample,* 40 *hours is the modal outcome with* 1,408 *individuals reporting* 40 *hours of work, representing just over half the sample. Some other less pronounced spikes can be seen, with the next most common values being* 50 *hours (*158 *individuals),* 45 *hours (*106 *individuals),* 60 *hours (*96 *individuals), and* 35 *hours (*93 *individuals). Perhaps not surprisingly, these common outcomes are at round numbers, which can be a feature of survey variables when individuals are asked for their recollection of a particular activity. The second histogram, with* 10*-hour bin widths, smoothes away each of the spikes seen in the first histogram. This histogram still makes it clear that the vast majority of the individuals work around* 40 *hours per week. Finally, note the somewhat strange box in the box plot on the right. The line indicating the sample median is at the bottom of the box, which also represents the lower quartile. The lower quartile and the median of the sample are both equal to* 40 *hours here because there are so many individuals with that value in the data set.*

6.7 Linear Transformations of Univariate Data

This section considers **linear transformations** of univariate variables and data. As the terminology suggests, we start with an original x variable and take a linear transformation or function of that x variable to create a new variable. That same linear transformation or function is applied to the realized observations of the x variable in the sample.

Definition 6.17 *If a and b are known constants, the variable $y=a+bx$ is a **linear transformation** of the x variable. Moreover, the values $y_i=a+bx_i$ (for $i=1,\ldots,n$) are **linear transformations** of the sample observations $\{x_1,\ldots,x_n\}$.*

In some cases, a linear transformation changes the units of a variable, as in examples 6.23 and 6.24. But linear transformations are much more general and can be used to create new variables of interest, as seen in examples 6.25 and 6.26.

Example 6.23 (Height) *If x is height in inches, the variable y that measures height in feet is a linear transformation of x with $a=0$ and $b=\frac{1}{12}$:*

$$x = \textit{height in inches}$$

$$y = \textit{height in feet} = \frac{1}{12}x.$$

Example 6.24 (Earnings) *If x is weekly earnings, the variable y that measures annualized earnings is a linear transformation of x with $a=0$ and $b=52$:*

$$x = \textit{weekly earnings}$$

$$y = \textit{annualized earnings} = 52x.$$

The units of x are dollars per week, and the units of y are dollars per year.

Example 6.25 (Nonworking hours) *If x is the number of hours worked last week, the variable y that measures the number of nonworking hours last week is a linear transformation of x with $a=168$ and $b=-1$:*

$$x = \textit{hours worked last week}$$

$$y = \textit{nonworking hours last week} = (24)(7) - x = 168 - x.$$

Example 6.26 (Website profits) *Suppose that x is the number of widgets purchased at a website on a given day. If the price of a widget is p, the daily fixed cost for the website is f, and the marginal cost of a widget is c, the variable y that measures the website's daily profit is a linear transformation of x with $a=-f$ and $b=p-c$:*

$$x = \textit{daily purchases of widgets}$$

$$y = \textit{daily profit} = -f + (p-c)x.$$

The units of y are dollars if the units of f are dollars and the units of p and c are dollars per widget.

For a linear transformation y of variable x, the location and/or dispersion of the distribution of the new y_i observations may differ from the distribution of the original x_i observations. Figure 6.14 considers four examples to illustrate this point. For each of the four graphs shown, the same density curve is used for the hypothetical sample associated with the x variable. The distribution for x is bell-shaped, centered around 2, and has nearly all its observations between 0 and 4. The top-left graph shows the density curve for $y=2x$, a linear transformation with $a=0$ and $b=2$. The center of the y density curve is shifted to the right. The center appears to be located at 4, which is equal to b (2) times the center of the x density curve (2). The dispersion is also greater for the y density curve, which occurs since the b value is greater than 1 and therefore increases the scale of the x observations. Moving next to the top-right graph, we have a density curve for $y=0.5x$, a linear transformation with $a=0$ and $b=0.5$. Again, the central location of the y density curve differs from that of the x density curve, but this time it is smaller and is located around 1, which is b (0.5) times the center of the x density curve (2). The dispersion of the y density curve is now less than that of the x density curve, which happens since the b value is less than 1 and has the effect of decreasing the scale of the observations. For the bottom-left graph, we have the density curve for $y=3+x$, a linear transformation with $a=3$ and $b=1$. In this case, the shape of the y density curve looks identical to that of the x density curve, but it is shifted over to the right by 3 units, corresponding to the value of a. As the shapes of the two density curves are identical, their dispersions are also the same, which occurs since b is exactly equal to 1. Finally, for the bottom-right graph, we have the density curve for $y=3+0.5x$, a linear

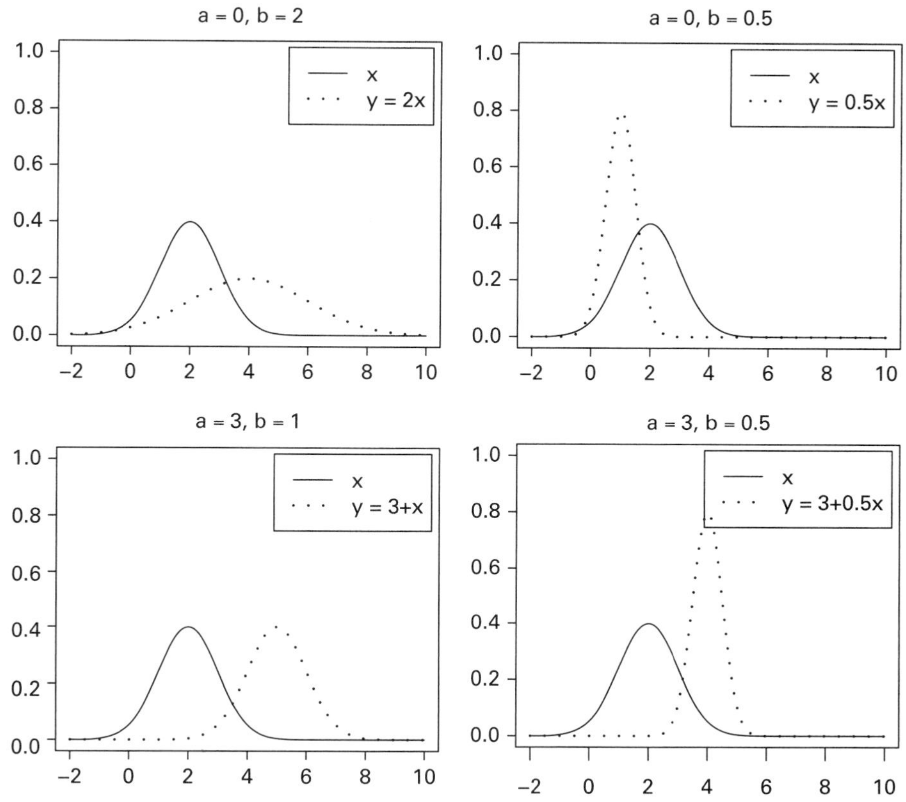

Figure 6.14
Linear transformations of a variable

transformation with $a = 3$ and $b = 0.5$. There is less dispersion for the y density curve, again due to b being less than 1. The shape of the y density curve in this case looks identical to the graph above it (where $y = 0.5x$), but shifted to the right by $a = 3$ units. The central location of the y density is affected by both the a value (a shift of 3 to the right) and the b value (a shift of 1 to the left due to scaling the x variable by 0.5), resulting in an overall shift of 2 units (4 as compared to the central location of the x density at 2).

To formalize some of the properties exhibited in figure 6.14, proposition 6.6 states how the various descriptive statistics for location and dispersion are affected by a linear transformation:

Proposition 6.6 *If a and b are known constants and $y = a + bx$ is a linear transformation of x, the descriptive statistics for the sample $\{y_1, y_2, \ldots, y_n\}$ have the following relationships to the descriptive statistics for the sample $\{x_1, x_2, \ldots, x_n\}$:*

i. *(sample mean)* $\bar{y} = a + b\bar{x}$
ii. *(sample variance)* $s_y^2 = b^2 s_x^2$
iii. *(sample standard deviation)* $s_y = |b| s_x$
iv. *(sample quantiles)*[6] $\tilde{y}_q = a + b\tilde{x}_q$ *if* $b \geq 0$

v. *(sample IQR)*[7] $IQR_y = |b| IQR_x$ *if* $b \geq 0$

vi. *(sample MAD)* $MAD_y = |b| MAD_x$

It is useful to show each of these properties. For the sample mean in property (i),

$$\bar{y} = \frac{1}{n}\sum_{i=1}^{n} y_i = \frac{1}{n}\sum_{i=1}^{n}(a + bx_i) = \frac{1}{n}\left(\sum_{i=1}^{n} a + \sum_{i=1}^{n} bx_i\right) = \frac{1}{n}(na + bn\bar{x}) = a + b\bar{x}.$$

Here, $\bar{y}$ is a linear function of $\bar{x}$, with the same a and b constants as the original linear transformation.

For the sample variance in property (ii),

$$s_y^2 = \frac{1}{n-1}\sum_{i=1}^{n}(y_i - \bar{y})^2 = \frac{1}{n-1}\sum_{i=1}^{n}(a + bx_i - (a + b\bar{x}))^2$$

$$= \frac{1}{n-1}\sum_{i=1}^{n}((bx_i - b\bar{x})^2) = b^2 \cdot \frac{1}{n-1}\sum_{i=1}^{n}(x_i - \bar{x})^2 = b^2 s_x^2.$$

When the x variable is scaled by b in linear transformation y, the sample variance of y gets scaled by b^2. The additive constant a drops out of the sample variance s_y^2 expression, which makes intuitive sense since the constant a shifts the location of the original x distribution but does not change its shape or dispersion. The sign of b doesn't matter here, so, for example, the sample variance of $y = 1 + 2x$ is the same as the sample variance of $y = 1 - 2x$, both being four times the sample variance of x. The magnitude of b dictates whether the sample variance of y is smaller or larger than the sample variance of x. From the sample variance result ($s_y^2 = b^2 s_x^2$),

$$s_y^2 > s_x^2 \text{ if } |b| > 1,\ s_y^2 < s_x^2 \text{ if } |b| < 1, \text{ and } s_y^2 = s_x^2 \text{ if } |b| = 1.$$

For the sample standard deviation in property (iii),

$$s_y = \sqrt{s_y^2} = \sqrt{b^2 s_x^2} = |b| s_x,$$

since $\sqrt{b^2} = |b|$. As with the sample variance, the relative sizes of the sample standard deviations of x and y depend on the magnitude of b:

$$s_y > s_x \text{ when } |b| > 1,\ s_y < s_x \text{ when } |b| < 1, \text{ and } s_y = s_x \text{ when } |b| = 1.$$

For the sample quantiles in property (iv), showing that $\tilde{y}_q = a + b\tilde{x}_q$ when $b \geq 0$ is straightforward. Let's say that the sample of x-values have been sorted, from lowest to highest, to calculate a sample quantile $\tilde{x}_q$. When b is positive, if the sample of y-values is sorted, from lowest to highest, the sorted y-values will have the same exact ordering as the sorted x-values. Therefore, when the algorithm from section 6.4.2 for calculating a sample quantile is applied, the sample quantile of y will be $\tilde{y}_q = a + b\tilde{x}_q$. As a special case, the sample medians are related by the equation $\tilde{y}_{0.5} = a + b\tilde{x}_{0.5}$.

For the sample IQR in property (v), when $b \geq 0$, the result follows from the result for sample quantiles. Since $\tilde{y}_{0.25} = a + b\tilde{x}_{0.25}$ and $\tilde{y}_{0.75} = a + b\tilde{x}_{0.75}$,

$$IQR_y = \tilde{y}_{0.75} - \tilde{y}_{0.25} = a + b\tilde{x}_{0.75} - (a + b\tilde{x}_{0.25}) = b(\tilde{x}_{0.75} - \tilde{x}_{0.25}) = bIQR_x.$$

As with the sample variance and sample standard deviation, the additive constant a has no effect on the IQR dispersion measure. While a may shift the location of the distribution, it has no effect on the difference between the quantiles within the distribution. Instead, the sample IQR of y is just b times the sample IQR of x, with the same scaling (b) as for the sample standard deviation.

For the sample MAD in property (vi),

$$MAD_y = \frac{1}{n}\sum_{i=1}^{n} |y_i - \bar{y}| = \frac{1}{n}\sum_{i=1}^{n} |a + bx_i - (a + b\bar{x})|$$

$$= \frac{1}{n}\sum_{i=1}^{n} |b(x_i - \bar{x})| = |b|\frac{1}{n}\sum_{i=1}^{n} |x_i - \bar{x}| = |b| MAD_x.$$

The additive constant a does not affect MAD_y, and the scaling constant b affects MAD_y in the same way as for the sample standard deviation and the sample IQR.

Example 6.27 (Height) *Example 6.23 had x = height in inches and $y = \frac{1}{12}x$ = height in feet. For a sample of heights, proposition 6.6 implies*

$$\bar{y} = \frac{1}{12}\bar{x},\ \tilde{y}_{0.5} = \frac{1}{12}\tilde{x}_{0.5},\ s_y^2 = \frac{1}{144}s_x^2,\ \text{and } s_y = \frac{1}{12}s_x.$$

Example 6.28 (Website profits) *In example 6.26, the website had daily widget sales x and daily profits $y = -f + (p - c)x$, where f is the fixed daily cost, p is the widget price, and c is the marginal cost of producing each widget. For a sample of daily sales x, from which daily profits y are derived,*

$$\bar{y} = -f + (p - c)\bar{x},\ \tilde{y}_{0.5} = -f + (p - c)\tilde{x}_{0.5}\ (\text{if } p > c),\ s_y^2 = (p - c)^2 s_x^2,\ \text{and } s_y = |p - c| s_x.$$

Example 6.29 (Earnings) *For example 6.24, with x being weekly earnings and $y = 52x$ being annualized earnings, let's consider the descriptive statistics associated with the actual weekly earnings (earnwk) variable from the **cps** data:*

	$\bar{x}$	s_x^2	s_x	$\tilde{x}_{0.5}$	IQR_x	MAD_x
earnwk (x)	971.2	563,227.1	750.48	770	673.6	497.8

	$\bar{y}$	s_y^2	s_y	$\tilde{y}_{0.5}$	IQR_y	MAD_y
earnyr (y)	50,501	1,522,966,024	39,025.2	40,040	35,027.2	25,886.1

Each quantity in the annualized earnings (y) table is equal to 52 times the corresponding quantity in the weekly earnings (x) table, except for the sample variance, which is $52^2 =$ 2,704 times larger.

Example 6.30 (Nonworking hours) *For example 6.25, with x being the number of hours worked last week and $y = 168 - x$ being the number of nonworking hours last week, let's consider x = hrslastwk from the **cps** data. Figure 6.15 shows two histograms with 1-year bin widths, the first for the x-values and the second for the $y = 168 - x$-values. The y distribution is an exact mirror image of the x distribution since $b = -1$. The y distribution is also shifted by the large additive constant $a = 168$. Taking the mirror image of the distribution doesn't affect dispersion, as implied by proposition 6.6 when $|b| = 1$. For instance, the sample standard deviations s_x and s_y are both approximately 11.28, and the sample MAD statistics MAD_x and MAD_y are both approximately 6.53. Since $x + y = 168$, the sample mean of x ($\bar{x} \approx 40.3$) and the sample mean of y ($\bar{y} \approx 127.7$) sum to 168.*

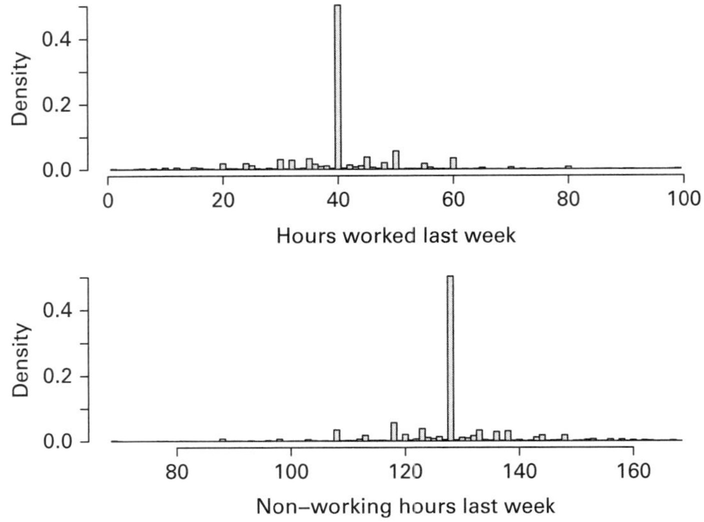

Figure 6.15
Histograms of weekly hours worked and nonworking hours

6.7.1 Standardizing a Variable

Since it can be difficult to compare variables that have different means, different variances, and even different units, a commonly used linear transformation is one that standardizes the values observed in a sample. Specifically, for a sample $\{x_1, x_2, \ldots, x_n\}$ with sample mean $\bar{x}$ and sample standard deviation s_x, the **standardized values**

$$y_i = \frac{x_i - \bar{x}}{s_x}, \text{ for } i = 1, 2 \ldots, n,$$

are obtained by "de-meaning" the observations (i.e., subtracting the sample mean $\bar{x}$) and then dividing by the sample standard deviation s_x. The standardized value y_i is the number of sample standard deviations that x_i is above or below the sample mean $\bar{x}$. For instance, $y_i = 1.5$ when $x_i = \bar{x} + 1.5s_x$ (i.e., x_i is 1.5 sample standard deviations above the sample mean), and $y_i = -0.8$ when $x_i = \bar{x} - 0.8s_x$ (i.e., x_i is 0.8 sample standard deviations below the sample mean). The standardized value y_i is unitless since the units of the numerator and the units of the denominator are both the units of x. This standardization is a special case of the linear transformation, with $a = -\frac{\bar{x}}{s_x}$ and $b = \frac{1}{s_x}$, since y_i can be rewritten

$$y_i = -\frac{\bar{x}}{s_x} + \frac{1}{s_x} x_i \text{ for } i = 1, 2 \ldots, n.$$

Interestingly, regardless of the original sample $\{x_1, x_2, \ldots, x_n\}$, the sample of standardized values, $\{y_1, y_2, \ldots, y_n\}$, always has sample mean $\bar{y} = 0$, since

$$\bar{y} = -\frac{\bar{x}}{s_x} + \frac{1}{s_x}\bar{x} = 0,$$

and sample standard deviation $s_y = 1$, since

$$s_y = \frac{1}{s_x} s_x = 1.$$

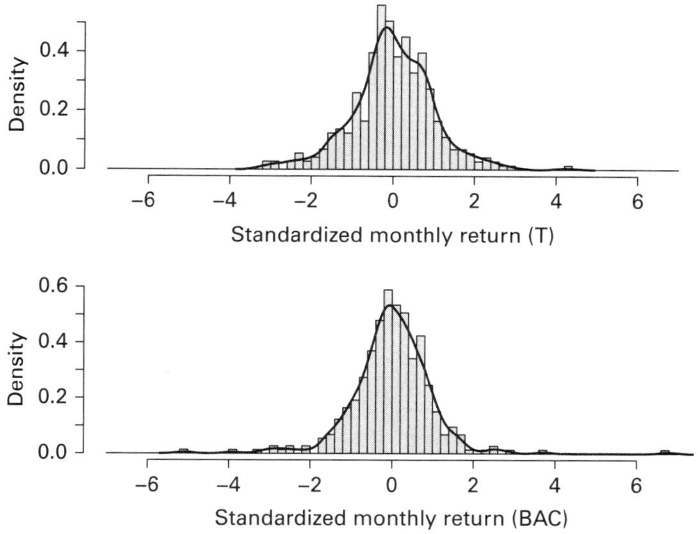

Figure 6.16
Histograms of standardized monthly stock returns

Example 6.31 (Monthly stock returns) *Returning to the AT&T monthly stock returns and the Bank of America monthly stock returns from the* ***sp500*** *data set, the following R code standardizes the variables for T and BAC:*

```
T_std <- (sp500$T-mean(sp500$T))/sd(sp500$T)
BAC_std <- (sp500$BAC-mean(sp500$BAC))/sd(sp500$BAC)

summary(T_std)

##    Min. 1st Qu.  Median    Mean 3rd Qu.    Max.
## -3.1626 -0.4851 -0.0185  0.0000  0.6289  4.2555

sd(T_std)

## [1] 1

summary(BAC_std)

##     Min.  1st Qu.   Median     Mean  3rd Qu.     Max.
## -5.08030 -0.47193  0.01723  0.00000  0.50103  6.77693

sd(BAC_std)

## [1] 1
```

The summary statistics for the two standardized variables confirm that the sample means equal zero and the sample standard deviations equal 1*. Figure 6.16 shows the histograms and density curves for the standardized variables. Compared to the distributions of the original variables T and BAC in figure 6.6, where BAC has visibly more dispersion than T, the dispersion of the two standardized variables in figure 6.16 is quite similar due to the division by their respective standard deviations. The distributions are also centered around zero due to the de-meaning. And, for both standardized variables, a very large proportion of the observations are between* -2 *(two sample standard deviations below the sample mean) and*

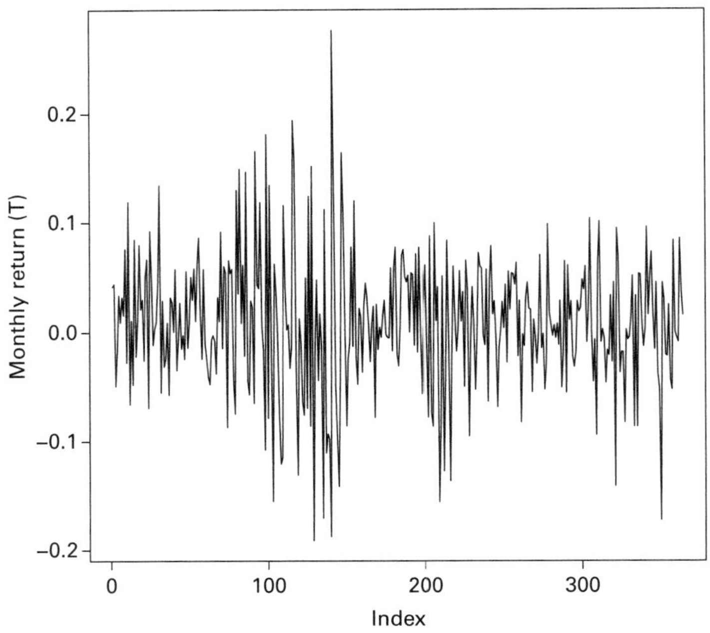

Figure 6.17
Time-series plot of monthly stock returns

2 *(two sample standard deviations above the sample mean). As we'll see in chapter 11, this property is to be expected for variables with distributions that are approximately bell-shaped.*

6.8 Time-Series Plots

For time-series data, there is a natural chronological ordering to the observations. Generally, the data are stored with the earliest observation in the first row, the next observation (in time) in the second row, and so on through the latest observation in the last row. A **time-series plot** graphs a time-series variable (y-axis) against time (x-axis), which is useful for visualizing how a univariate time-series variable behaves over time.

The simplest way in *R* to draw a time-series plot for a single variable is to use the `plot` function. For instance, here is the code that generates the time-series plot of AT&T monthly returns (T) from the **sp500** data set shown in figure 6.17:

```
plot(sp500$T, type="l", ylab="Monthly return (T)")
```

The optional argument `type="l"` provides a line plot rather than a scatterplot. With only a single variable specified in the first argument to `plot`, *R* automatically plots the monthly return data against the row number or, as specified on the x-axis label, the "Index."

To directly incorporate information about actual dates or times associated with the observations, an alternative approach in *R* is to create a time-series object and then draw the time-series plot:

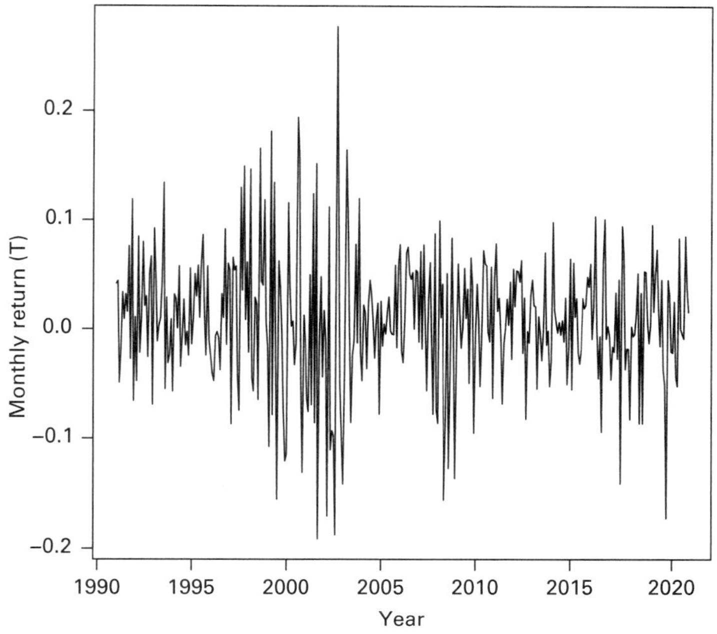

Figure 6.18
Time-series plot of monthly stock returns

```
# create a time-series object for AT&T monthly returns
ts_T <- ts(sp500$T, start=c(1991,1), frequency=12)

plot(ts_T, ylab="Monthly return (T)", xlab="Year")
```

The `ts` function creates the time-series object based on the variable specified by the first argument `sp500$T`. The optional argument `start=c(1991,1)` specifies that the time series begins in the first month of the year 1991, and the optional argument `frequency=12` specifies that the observations are monthly (i.e., at a frequency of 12 per year). The resulting time-series plot, shown in figure 6.18, looks identical to figure 6.17, as it should, but with only one difference: the date values appear on the *x*-axis, now labeled "Year." The dispersion of the AT&T monthly returns is particularly high between 1998 and 2004. This feature of the data is missed by either the histogram or box plot since neither of those visualizations incorporates the time dimension.

We can also graph multiple time-series plots at once. Two approaches are considered: one in which the time-series plots are shown on the same graph and one in which the time-series plots are vertically stacked. The following *R* code yields figure 6.19, with the time-series plots for both AT&T monthly returns and Bank of America monthly returns shown on the same graph:

```
# create the time-series objects for monthly returns
ts_T <- ts(sp500$T, start=c(1991,1), frequency=12)
ts_BAC <- ts(sp500$BAC, start=c(1991,1), frequency=12)
```

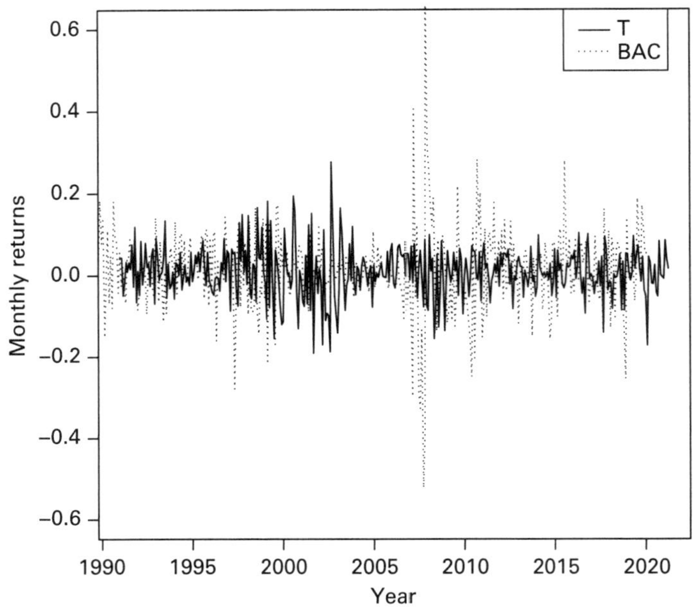

Figure 6.19
Time-series plots of monthly stock returns

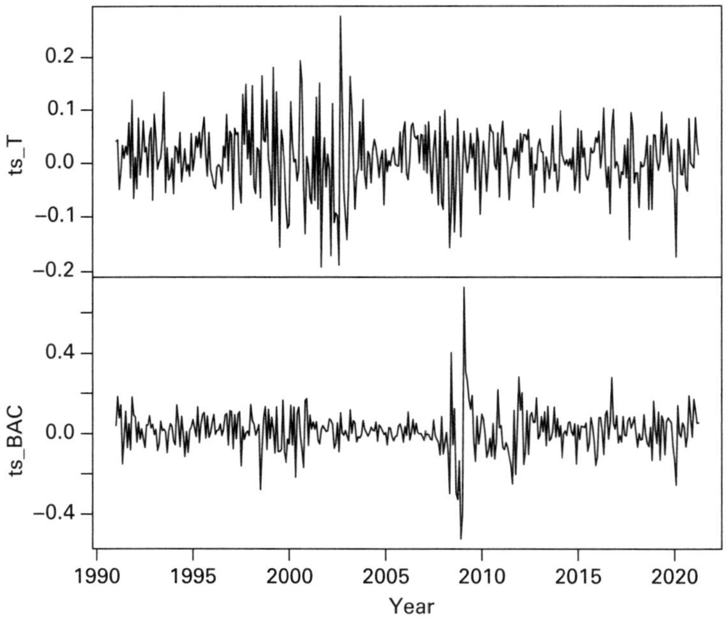

Figure 6.20
Time-series plots of monthly stock returns

```
plot(ts_T, ylab="Monthly returns",  xlab="Year", ylim=c(-0.6,0.6))
lines(ts_BAC, lty=3)
legend("topright", c("T","BAC"), lty=c(1,3))
```

The `plot` function draws the AT&T time-series plot with a solid line. The `lines` function draws the Bank of America time-series plot as a dotted line and does so on the same graph since `lines` is used instead of `plot`. The `legend` function creates the legend in the top-right corner of the graph. The greater dispersion of Bank of America's returns are evident in figure 6.19, with the dotted time-series plot having more pronounced spikes. The largest of these spikes occurred during the financial crisis that occurred in the years before 2010.

Including multiple time-series plots on the same graph can make it difficult to distinguish the behavior of the variables. The following *R* code instead separately graphs the two time-series plots for *T* and *BAC* and stacks them vertically, as shown in figure 6.20:

```
plot(cbind(ts_T,ts_BAC), main="",xlab="Year")
```

The `cbind` function combines the two time-series objects `ts_T` and `ts_BAC` into a single data frame, and `plot` vertically stacks the time-series plots for the variables in the resulting data frame. This approach generalizes to more time-series variables, with additional variables included in the `cbind` function. Alternatively, the first argument can be supplied directly as a data frame that consists of time-series objects.

Notes

1. The Standard and Poor's 500 (S&P 500) index, is a stock market index that tracks the stock performance of 500 of the largest companies listed on US stock exchanges, including the New York Stock Exchange and the Nasdaq. There are fewer than 500 stocks in the **sp500** data set since we have only included companies that were part of the S&P 500 index for the full time period between January 1991 and April 2021.
2. Negative stock prices are not possible. There are other investments, like short sales of stocks, for which returns can be negative and arbitrarily large in magnitude.
3. A *pie chart* is an alternative descriptive visual that can be used, also using either count values or proportion values, where the sizes of the pie wedges correspond to the sample proportions. An advantage of the bar chart is that it's easier to visually compare bar heights rather than the size of pie wedges.
4. Unlike the sample mean, the sample median may not be uniquely defined when n is even. In the algorithm for calculating the sample median for even n, choosing *any* value in between the two middle points would satisfy the formal definition.
5. An alternative measure is the *sample median absolute deviation*, $\frac{1}{n}\sum_{i=1}^{n}|x_i - \tilde{x}_{0.5}|$, calculated with the `mad` function in *R*.
6. The result for negative b is more subtle. For $b < 0$, we have $\tilde{y}_q \approx a + b\tilde{x}_{1-q}$, where we use an approximation sign ($\approx$) since we don't get exact equality based upon the procedure given in section 6.4.2 for calculating sample quantiles. When b is negative, if we sort the sample of y-values from lowest to highest, the sorted values will be in exactly the *reverse* order of the sorted x-values. As an example, if $n = 101$ and $y = -x$, the sample 25% quantile of y is the 26th value from sorted y-values, which corresponds to the 75th value from the sorted x-values (which is $\tilde{x}_{0.74}$); so $\tilde{y}_{0.25} = a + b\tilde{x}_{0.74} \approx a + b\tilde{x}_{0.75} = a + b\tilde{x}_{1-0.25}$.
7. The result for negative b ($b < 0$) is approximately $IQR_y \approx |b| IQR_x$, again due to the procedure given in section 6.4.2.

Exercises

1. Use the **cps** data set for this question.
   a. Provide a table of the marital status (*marstatus*) categorical variable.
   b. For a bar chart of *marstatus* with sample proportions on the y-axis, what is the height of the "Divorced" bar?
   c. Conditional on someone being not "Married," what is the sample proportion that are "Divorced"?
   d. Consider the indicator variable *notmarried*, which is equal to 0 if the variable value is "Married" and 1 otherwise. What is the sample mean of *notmarried*?
2. For a sample with five observations ($n = 5$), the deviations-from-mean for the first four observations are -1.1, 2.3, -0.8, and 0.2.
   a. What is the deviation from the mean for the fifth observation?
   b. What is the sample variance?
3. For the **cps** data set, there are 4,013 observations on adults aged 30 to 59. One of the variables is the number of children (*ownchild*), with the following table describing how many observations take on each of the possible values $0, 1, 2, 3, 4, 5, 6, 7$:

ownchild (# of children)	0	1	2	3	4	5	6	7
# of observations	2,432	654	584	233	81	21	5	3

   For this question, use only the numbers in the table, not the actual **cps** data set; you can use *R* for mathematical calculations, but don't use the built-in *R* functions for descriptive statistics.
   a. What is the sample average of *ownchild*?
   b. What is the sample median of *ownchild*?
   c. What is the sample 90% quantile of *ownchild*?
   d. Is the distribution of *ownchild* symmetric, left-skewed, or right-skewed?
   e. What is the sample mean absolute deviation of *ownchild*?
   f. What is the sample variance of *ownchild*?
   g. What is the sample standard deviation of *ownchild*?
4. Your classmate surveys 223 people and asks each person how many times they ate out at a restaurant in a given month. The minimum response is 0, and the maximum response is 30. Rather than providing the counts for every possible integer between 0 and 30, your classmate summarizes the responses as follows:

# of times eating out in a month	0	1–10	11–20	21–30
# of observations	25	125	49	24

   Let x be the variable representing the number of times that a person eats out in a month.
   a. What can you say about the sample median of x?

b. What is the highest possible value of the sample mean of x? What is the lowest possible value of the sample mean of x?

c. What is the sample proportion of individuals for whom $x \geq 11$?

d. Is there anything that can be said about the modal outcome for x?

5. The monthly rainfall x (in inches) in Statsville from January 2023 through December 2023 is:

   1.89, 1.99, 2.14, 2.51, 5.03, 3.81, 1.97, 2.31, 2.91, 3.97, 2.68, 2.44.

   a. What is the sample mean absolute deviation (MAD_x) of monthly rainfall?

   b. What is the sample variance (s_x^2) of monthly rainfall?

   c. What is the sample standard deviation (s_x) of monthly rainfall?

   d. If y is monthly rainfall measured in feet, what are MAD_y, s_y^2, and s_y?

6. For the **cps** data set, here are summary statistics of the hourly wage last week (*wagehr*). The variable *wagehr* is missing for the 2,174 individuals who are not hourly employees, so it has numeric values for the 1,839 employed individuals paid on an hourly basis.

```
summary(cps$wagehr)

##    Min. 1st Qu.  Median    Mean 3rd Qu.    Max.    NA's
##    1.01   12.78   16.41   18.60   22.00   90.00    2174
```

   a. Is the distribution of *wagehr* likely to be symmetric, left-skewed, or right-skewed?

   b. If you add one observation to this data for a worker with an hourly wage of \$16.41, how would the sample median change? How would the sample mean change?

   c. What is the sample IQR of *wagehr*?

   d. What are the values of the lower whisker and the upper whisker for the box plot of *wagehr*?

7. Use the **website** data set for this question. This time-series data set contains the arrival times of website visitors during 1 hour. The variable *arrival* (in minutes, with values between 0 and 60) indicates when a visitor arrives at the website. The variable *time_since_last* (in minutes) indicates the amount of time since the last visitor's arrival.

   a. Use `hist` to draw a histogram of *arrival* with densities on the y-axis and without specifying the width or number of bins. Does the distribution of *arrival* appear to be approximately symmetric, left-skewed, or right-skewed?

   b. Draw a density histogram of *arrival* with 60 bins, corresponding to the first minute, the second minute, and so on.

   c. Draw a density histogram of *time_since_last* using the Freedman-Diaconis rule. Does *time_since_last* appear to be symmetric, left-skewed, or right-skewed?

d. Add a density curve to the histogram in (c). When using the `density` function, specify the optional argument `na.rm=TRUE` since the first observation of *time_since_last* is missing.

e. Draw a box plot with whiskers and outliers for *time_since_last*.

f. Using other *R* functions as necessary, what are the values of the following features of the box plot from (e): (i) lower whisker, (ii) upper whisker, (iii) bottom of the box, (iv) top of the box, (v) median (line within the box), and (vi) the number of points (outliers) outside the whiskers.

g. If the time-since-last-arrival were measured in seconds rather than minutes, how would the box plot in (e) change? Answer this part without actually drawing a new box plot.

8. Use the **metricsgrades** data set for this question. These data come from a graduate econometrics course with 68 students, containing the following variables:

total = overall composite course grade (out of 100 points)
gre_quant = score on GRE quantitative test (out of 170 points)
gre_verbal = score on GRE (English) verbal test (out of 170 points)
domestic = 1 if domestic (US) student, 0 otherwise

a. What is the sample mode of *gre_quant*?

b. Draw separate frequency histograms of *gre_quant* for domestic students and nondomestic students. For ease of comparison, use the `xlim` and `ylim` options to specify the same scales for both axes.

c. Draw separate frequency histograms of *gre_verbal* for domestic students and nondomestic students. For ease of comparison, use the `xlim` and `ylim` options to specify the same scales for both axes.

d. Compare the following descriptive statistics of *gre_quant* for the subsamples of domestic students and nondomestic students: (i) sample mean, (ii) sample median, (iii) sample 75% quantile, and (iii) sample IQR.

e. Create a standardized version of *gre_quant* called *gre_quant_std*, using the sample mean and sample standard deviation of the full sample. Confirm that *gre_quant_std* has the sample mean zero and the sample standard deviation 1.

f. What are the sample mean, the sample standard deviation, and the sample IQR of *gre_quant_std* for the subsamples of domestic students and nondomestic students?

9. Logan's Lemonade sells lemonade for \$6 per cup each Sunday at a local farmer's market. The number of cups sold each Sunday is represented by variable x. Over 16 weeks, the descriptive statistics for x are $\bar{x} = 120$ and $s_x = 20$.

a. What is the sample mean and the sample standard deviation of Logan's Lemonade's daily revenues (price times quantity sold)?

b. The fee to have a booth at the farmer's market is \$100 each Sunday. If it also costs Logan's Lemonade \$1 per cup to make the lemonade, what are the sample mean and the sample standard deviation of Logan's Lemonade's daily profits (revenues minus costs)?

c. If you knew the value of the sample median $\tilde{x}_{0.5} = x^*$, how would the sample median of daily revenues be related to x^*? How about the sample median of daily profits?

10. Suppose that $x \in \{0, 1\}$ is a binary variable.

a. *Show that the sample variance of x is

$$s_x^2 = \frac{n}{n-1}\bar{x}(1-\bar{x}).$$

(Hint: Use the following facts: (i) $x_i^2 = x_i$ if $x_i \in \{0, 1\}$ and (ii) $\sum_{i=1}^{n} x_i = n\bar{x}$.)

b. In a sample of 200 individuals, 13 had an emergency-room visit in the last year. If variable x is an indicator of an emergency-room visit (equal to 1 if a visit occurred, 0 if not), what is the sample standard deviation of x?

c. *Provide a formula for the sample mean absolute deviation of x (MAD_x) in terms of $\bar{x}$. The formula should not contain any individual x_i values.

11. The **sample skewness** g_x is a descriptive statistic that measures the skewness of a sample distribution, defined as

$$g_x = \frac{\frac{1}{n}\sum_{i=1}^{n}(x_i - \bar{x})^3}{(s_x)^3}.$$

Positive values of g_x are associated with right-skewed sample distributions, with $g_x > 1$ considered highly right-skewed. Negative values of g_x are associated with left-skewed sample distributions, with $g_x < -1$ considered highly left-skewed. Values of g_x closer to zero are associated with sample distributions that are neither left-skewed nor right-skewed.

a. What are the units of g_x?

b. Write an *R* function `skewness` that takes vector `x`, containing the sample, as its only argument and returns the sample skewness.

c. For the **cps** data set, use the `skewness` function to calculate the sample skewness of the *earnwk* and *age* variables for the subsample of employed individuals.

d. How would the sample skewness of *earnwk* change if weekly earnings were measured in thousands of dollars rather than dollars?

12. Use the **bitcoin** data set for this question. This data set consists of daily prices and returns for the Bitcoin cryptocurrency between January 1, 2020 and December 31, 2021. There are 731 observations on the three price variables (*high* = daily high price, *low* = daily low price, *close* = end-of-day price) and 730 observations on the daily return (*return*).

a. Draw a time-series plot of the closing daily price for the full time series.

b. For the first 60 observations through February 29, 2020, draw a time-series plot with the three price variables (*high*, *low*, *close*) on the same plot. Draw *high* and *low* as solid lines and *close* as a dotted line.

c. Draw a time-series plot of the daily returns for the full time series.

d. The time-series plot in (c) should indicate periods of low variance and periods of high variance. Look at the graph, pick one low-variance range and one high-variance range, and then confirm what you see by calculating the sample standard deviations for the two ranges that you've identified.

13. Use the **inflation** data set for this question. This panel data set consists of annual inflation rates for 45 countries over the 10-year period between 2010 and 2019. The variables are *country* (a categorical (factor) variable with a three-character abbreviation), *year* (values between 2010 and 2019), and *inflation* (in percentage points; e.g., $inflation = 3$ means 3% annual inflation).

    a. How many observations are associated with deflation (a *negative* inflation rate)?

    b. How many countries experience deflation at some point during this 10-year period?

    c. Which countries have the lowest and highest average inflation over this 10-year period?

    d. Which countries have the lowest and highest standard deviation of inflation over this 10-year period?

    e. Draw a time-series plot for the inflation rate of the US (*USA*).

    f. Draw a time-series plot that has the inflation rates for the US (*USA*), Canada (*CAN*), and Mexico (*MEX*) on the same plot. Make the three lines a different style and/or color to differentiate them, and make sure that the y-axis range allows all three lines to be completely visible.

7

Descriptive Statistics and Visuals: Bivariate Data

Chapter 6 introduced descriptive statistics and visuals for univariate data. In this chapter, the focus shifts to the case of data on two variables, known as **bivariate data**, with the introduction of descriptive statistics and visuals that describe the relationship between the two variables. Consider a situation where two variables x and y are observed as a collection of pairs:

$$\{(x_1, y_1), (x_2, y_2), \ldots, (x_n, y_n)\},$$

or, more concisely, $\{(x_i, y_i)\}_{i=1}^n$. As before, n denotes the sample size.

Example 7.1 (Education and earnings) *If we are interested in the relationship between earnings and educational attainment, data can be collected on a sample of workers, with x being the years of educational attainment and y being weekly (or annual) earnings. A positive association is expected between these two variables.*

Example 7.2 (Monthly stock returns) *Chapter 6 considered several examples using the monthly stock return data from **sp500**, specifically for the case of AT&T (stock ticker T) and Bank of America (stock ticker BAC). Suppose that we are instead interested in the relationship between monthly stock returns for two companies in the same industry (in this case, the home improvement industry). In the data set, there are monthly stock returns for both Home Depot (stock ticker HD) and Lowe's (stock ticker LOW), so let x be the monthly stock return for HD and let y be the monthly stock return for LOW. For these two companies, there are different factors that might affect the relationship of their stock returns. On the one hand, since they are competitors in the industry, it might be expected that one company does worse when the other company does better. On the other hand, since both companies are affected by the same macroeconomic conditions (e.g., home construction levels), it might be expected that both companies do better (or worse) than usual at the same time.*

Examples 7.1 and 7.2 involve numerical data for the variables being considered. While this chapter focuses primarily on descriptive statistics and visualization for numerical variables, section 7.1 considers categorical variables before numerical variables are covered in section 7.2.

7.1 Categorical Variables

In this section, two cases are considered: (1) bivariate data, where both variables are categorical; and (2) bivariate data, where one variable is categorical and one variable is numerical.

7.1.1 Bivariate Data with Two Categorical Variables

With two categorical variables, bivariate data can be completely characterized by a table with **joint sample counts** or **joint sample proportions**. This approach is quite similar to the treatment of a single categorical variable in section 6.2. The same level of formality is not used here since the bivariate data case is a special case of what was done in that section. If x is a categorical variable with C_x disjoint and exhaustive categories and y is a categorical variable with C_y disjoint and exhaustive categories, the number of categories for the pair of variables is C_xC_y. Any (x_i, y_i) pair in the observed sample must be in one (and only one) of the C_xC_y categories. As such, the sum of the joint sample counts over all the categories equals the sample size n, and the sum of the joint sample proportions equals 1.

Example 7.3 (Race and labor-force status) *We consider the categorical variables race and lfstatus from the* **cps** *data set, which allow us to examine whether there is a relationship between race and labor-force status. race has three categories ("Black," "White," and "Other") and lfstatus has three categories ("Employed," "Unemployed," and "Not in LF"), so the total number of joint categories is nine. To completely describe the joint sample distribution, a table of joint sample counts and a table of joint sample proportions can be created in R:*

```
# table of joint sample counts for labor-force status and race
table(cps$lfstatus, cps$race)

##
##              Black Other White
##   Employed     324   241  2244
##   Not in LF    136    99   863
##   Unemployed    16     9    81

addmargins(table(cps$lfstatus, cps$race))

##
##              Black Other White  Sum
##   Employed     324   241  2244 2809
##   Not in LF    136    99   863 1098
##   Unemployed    16     9    81  106
##   Sum          476   349  3188 4013

# table of joint sample proportions for labor-force status and race
addmargins(table(cps$lfstatus, cps$race))/nrow(cps)

##
##                    Black       Other       White         Sum
##   Employed   0.080737603 0.060054822 0.559182656 0.699975081
##   Not in LF  0.033889858 0.024669823 0.215051084 0.273610765
##   Unemployed 0.003987042 0.002242711 0.020184401 0.026414154
##   Sum        0.118614503 0.086967356 0.794418141 1.000000000
```

The command `table(cps$lfstatus, cps$race)` *provides the joint sample counts, while passing* `table(cps$lfstatus, cps$race)` *as an argument to the function* `addmargins` *also provides row totals and column totals (labeled by* `Sum`*). The table of joint sample proportions, created by the second* `addmargins` *command, divides each count by the sample size (*`nrow(cps)`*, which is* 4,013*).*

Looking at the top-left element of each table, there are 324 *individuals who are black* and *employed, representing* 0.0807*, or* 8.07%*, of the sample. The row and column totals make it easy to say something about either of the individual variables lfstatus and race. For instance, looking at the row labeled* `Sum`*, there are* 3,188 *white individuals, representing* 79.45% *of the sample.*

Unfortunately, these tables do not make it easy to compare the labor-force statuses of the different racial groups since there are different numbers of individuals in the three categories. Similarly, these two tables do not make it easy to compare the racial breakdown of the labor-force statuses since there are different numbers of individuals in the three categories. To facilitate such comparisons, we introduce a different table containing conditional sample proportions. *The idea is to calculate the sample proportions of one categorical variable given the value of the other categorical variable. As an example, the function* `prop.table` *can be used in R to create a table of the conditional sample proportions of labor-force status, where we condition upon the value of the racial category:*

```
# table of sample proportions of labor-force status conditional on race
prop.table(table(cps$lfstatus, cps$race), margin=2)

##
##                  Black      Other      White
##   Employed   0.68067227 0.69054441 0.70388959
##   Not in LF  0.28571429 0.28366762 0.27070263
##   Unemployed 0.03361345 0.02578797 0.02540778
```

The original table, given by `table(cps$lfstatus, cps$race)`*, is passed as the first argument to the* `prop.table` *function. The* `margin` *argument indicates which variable should be conditioned on, with* `margin=2` *specifying that the second variable (*`cps$race`*) is the one being conditioned on here. It can be verified that the sum of the proportions in each column equals* 1*. The values in this table could have been calculated directly from the information in the joint sample count table. For instance, the "Black" and "Employed" table element, which is the proportion of black individuals who are employed, is obtained by dividing the joint sample count of* 324 *by the total for the "Black" column (*476*), which yields* $\frac{324}{476} \approx 0.6807$*. From this table, we see that the sample proportion of black individuals who are employed (*0.6807*) is less than the sample proportion of white individuals who are employed (*0.7039*). There is also a lower sample proportion of white individuals who are not in the labor force (*0.2707*), as compared to either black individuals (*0.2857*) or other-race individuals (*0.2837*).*

By changing the `margins` *argument for the* `prop.table` *function, we can create a table of conditional sample proportions of the racial categories, where we condition on labor-force status:*

```
# table of sample proportions of race conditional on labor-force status
prop.table(table(cps$lfstatus, cps$race), margin=1)

##
##                  Black      Other      White
##   Employed   0.11534354 0.08579566 0.79886080
##   Not in LF  0.12386157 0.09016393 0.78597450
##   Unemployed 0.15094340 0.08490566 0.76415094
```

The `margin=1` *argument specifies that we condition on the first variable (*`cps$lfstatus`*). It can be verified that the sum of the proportions in each row is equal to 1. The values in this table could have been calculated directly from the joint sample count table, by dividing any of the joint sample counts by its associated row total. For instance, the "Black" and "Employed" table element, which is the proportion of employed individuals who are black, is obtained by dividing the joint sample count of* 324 *by the total for the "Employed" row (*2,809*), which yields* $\frac{324}{2809} \approx 0.1153$. *The proportion of white individuals among employed individuals (*0.7989*) is larger than the proportion of white individuals among unemployed workers (*0.7642*), whereas the proportion of black individuals among employed individuals (*0.1153*) is smaller than the proportion of black individuals among unemployed individuals (*0.1509*).*

Building on the idea of using conditional sample proportions as a descriptive tool, as in example 7.3, we can do something similar graphically by providing a descriptive visual (e.g., a bar chart) of one variable conditioned on the categorical value of the other variable.

Example 7.4 (Race and labor-force status) *To visualize the association between race and labor-force status, bar chart versions of the two conditional sample proportions tables from example 7.3 can be created. Figure 7.1 shows bar charts of labor-force status given race. Consistent with example 7.3, the bar chart indicates that the proportion of employed individuals is highest among white individuals and lowest among black individuals, whereas the proportion of not-in-labor-force individuals is lowest among white individuals. Here is the R code used to create figure 7.1:*

```
# create sample count table for race and labor-force status variables
tbl_racelf <- table(cps$race, cps$lfstatus)

# barplot with categories on x-axis based upon columns (lfstatus) of the table
barplot(prop.table(tbl_racelf, margin=1), ylim=c(0,0.8),
        col=c("gray30","gray50","gray70"),
        legend.text=rownames(tbl_racelf), beside=TRUE, main="")
```

When a table is the first argument of the `barplot` *function, the bars are grouped into categories on the x-axis corresponding to the columns of the table. The table* `tbl_racelf` *has been specified to have labor-force status as the columns. The additional argument* `beside=TRUE` *causes the bars for each of the racial categories to be displayed side by side, and the* `legend.text` *argument specifies the racial categories, which are given by the vector returned by* `rownames(tbl_racelf)`.

Figure 7.2 shows bar charts of the racial categories given labor-force status. Consistent with example 7.3, the bar charts indicate that the proportion of white individuals is

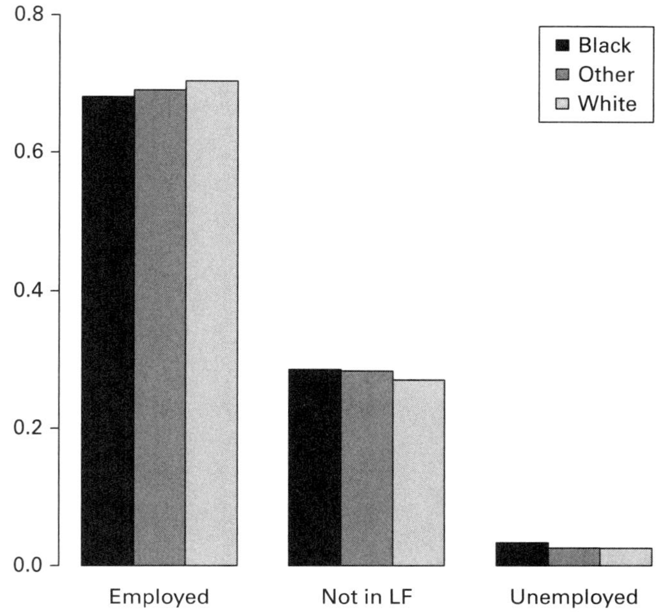

Figure 7.1
Labor-force status proportions by race

highest within the group of employed individuals and lowest within the group of unemployed individuals, whereas the reverse is true for black individuals. Here is the R code used to create figure 7.2:

```
# create sample count table table for race and labor-force status variables
tbl_lfrace <- table(cps$lfstatus, cps$race)

# barplot with categories on x-axis based upon columns (race) of the table
barplot(prop.table(tbl_lfrace, margin=1), ylim=c(0,0.8),
        col=c("gray30","gray50","gray70"),
        legend.text=rownames(tbl_lfrace), beside=TRUE,
        args.legend = list(x="topleft",inset=0.01), main="")
```

The primary difference from the previous code is that the rows and columns of the created table (`tbl_lfrace`) are labor-force status and race, respectively. The position of the legend is also specified since the default position would interfere with the displayed bars.

7.1.2 Bivariate Data with One Categorical Variable and One Numerical Variable

To assess the relationship between a categorical variable and numerical variable, we can examine how the descriptive statistics and distribution of the numerical variable vary over different categories of the categorical variable. With descriptive statistics, the easiest approach is to report descriptive statistics of the numerical variable for each possible value of the categorical variable. This approach is equivalent to breaking the full sample into different subsamples, each of which corresponds to one of the categories for the categorical variable.

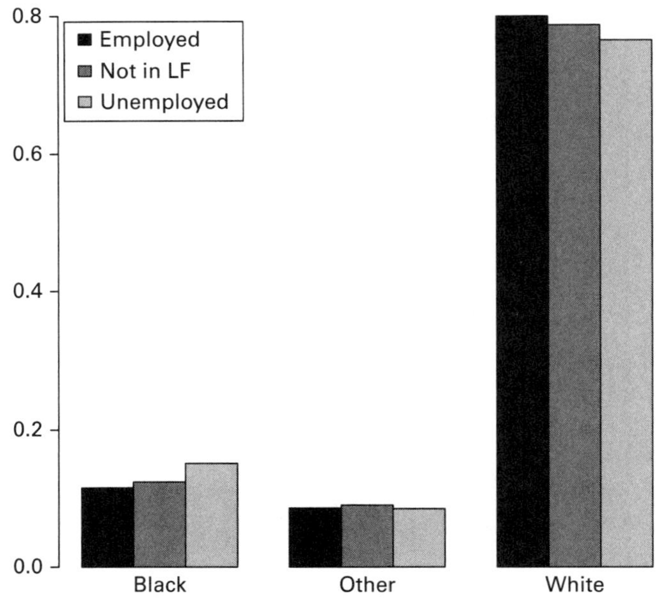

Figure 7.2
Race proportions by labor-force status

Example 7.5 (Race and earnings) *Let's examine the relationship between weekly earnings (earnwk) and racial category (race) from the* **cps** *data set. The full sample of employed individuals has* $n = 2{,}809$*, which is broken into three subsamples for race = "Black" (*324 *observations), race = "White" (*2,244 *observations), and race = "Other" (*241 *observations). For instance, if* `cpsemployed` *is a data frame in R for the* 2,809 *employed individuals, the subsample of black employed individuals is the data frame* `cpsemployed[cpsemployed$race=="Black",]`*. The following table shows several descriptive statistics for weekly earnings (sample mean, sample median, sample standard deviation, and sample interquartile range (IQR)) in each of the three subsamples:*

race	n	$\bar{x}$	$\tilde{x}_{0.5}$	s_x	IQR_x
Black	324	765	640	491	450
White	2,244	1,002	800	780	677
Other	241	962	730	718	685

In comparing white workers to black workers, earnwk has a larger sample mean and sample median for white workers. On average, white workers earn \$1,002 *per week, and black workers earn* \$765 *per week. Both dispersion measures (*s_x *and* IQR_x*) indicate that the distribution of earnings for white workers exhibits considerably more dispersion than the distribution of earnings for black workers. The sample IQR for white workers (*\$677*) is* 50.4*% larger than the sample IQR for black workers (*\$450*).*

The built-in *R* function `tapply` provides one simple way to calculate descriptive statistics of variables for the different subsamples based upon another categorical variable. Here is the basic syntax for the `tapply` function:

- `tapply(x, catvar, func, na.rm=FALSE)`: Outputs the results from applying the function `func` to the vector or data frame `x` for the subsamples based on the data frame factor (categorical) variable `catvar`. The optional argument `na.rm=TRUE` should be specified when missing values must be removed for the data frame and function being used.

Here is the code that produces the descriptive statistics in example 7.5:

```
# summary statistics of weekly earnings by racial category
tapply(cpsemployed$earnwk, cpsemployed$race, summary)

## $Black
##    Min. 1st Qu.  Median    Mean 3rd Qu.    Max.
##    19.0   461.4   640.0   765.0   911.8  3312.5
##
## $Other
##    Min. 1st Qu.  Median    Mean 3rd Qu.    Max.
##    12.0   520.0   730.0   961.9  1205.4  4431.0
##
## $White
##    Min. 1st Qu.  Median    Mean 3rd Qu.    Max.
##      15     540     800    1002    1217    8780

# sample standard deviation of weekly earnings by racial category
tapply(cpsemployed$earnwk, cpsemployed$race, sd)

##    Black    Other    White
## 491.0571 718.0332 779.8147

# sample IQR of weekly earnings by racial category
iqrvec <- tapply(cpsemployed$earnwk, cpsemployed$race, IQR)
iqrvec["White"]

##   White
## 676.875
```

Had we used the original data frame `cps`, which contains missing values for the weekly earnings variable, rather than the `cpsemployed` data frame, the optional argument `na.rm=TRUE` would have been required for the last two uses of the `tapply` function. For example, the sample standard deviation command would be `tapply(cps$earnwk, cps$race, sd, na.rm=TRUE)`. For the IQR calculations, the results are stored in the vector `iqrvec`, and the specific IQR value for the white subsample of workers is obtained by referring to the `"White"` index within square brackets.

In the same way that descriptive statistics can be applied to different subsamples based upon the categorical values, descriptive visuals can also be applied to these subsamples. For example, histograms, density curves, and/or box plots can be drawn and compared for the different subsamples.

Example 7.6 (Race and earnings) *Figure 7.3 shows a box plot of weekly earnings (earnwk) for each of the three subsamples. From left to right, the box plots correspond to black workers, other-race workers, and white workers. Comparing the box plots for black and white workers, there is a higher sample median for white workers, a larger IQR (height of the box) for white workers, and a more pronounced right skew for white workers. Figure 7.3 is particularly easy to create in R:*

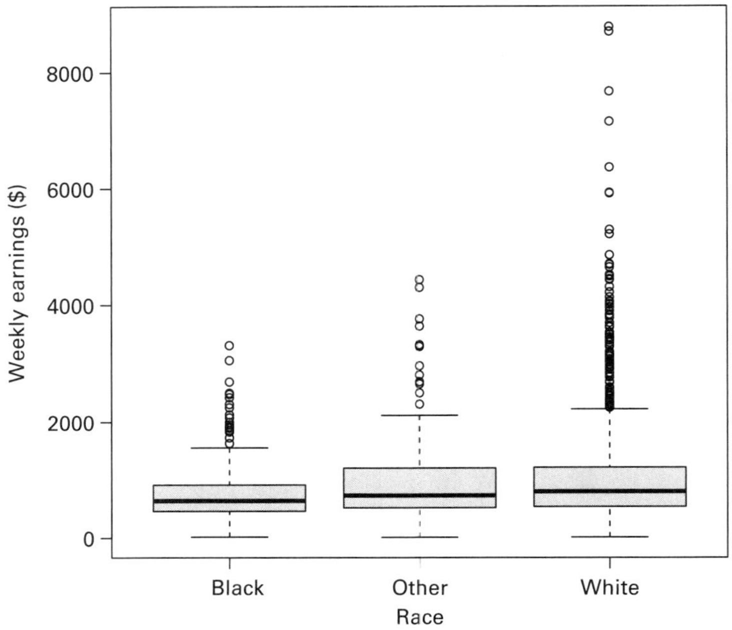

Figure 7.3
Box plots of weekly earnings by race

```
# box plot of weekly earnings, by race
boxplot(cps$earnwk~cps$race, ylab="Weekly earnings($)", xlab="Race")
```

The syntax here for the first argument of the `boxplot` *function has a tilde character (~) between two variables, where the first variable (*`cps$earnwk`*) is the variable of interest for the box plot and the second variable (*`cps$race`*) is a categorical variable used to split the sample into subsamples. For a similar figure to be drawn based on gender rather than race,* `cps$earnwk~cps$gender` *would be the first argument (and* `xlab` *should be appropriately changed).*

Histograms and density curves can be used as alternatives to box plots. As an example, figure 7.4 shows the density curves of the earnwk variable for the three racial categories. To easily compare the distributions, the density curves are drawn on the same graph, with the same x-axis and y-axis. This graph tells much the same story as figure 7.3. All three earnings distributions have a unimodal shape, with the earnings distribution for white workers exhibiting a much thicker right tail than the earnings distribution for black workers. This thicker right tail explains the higher dispersion statistics seen in example 7.5. The hump for the black-worker earnings distribution is considerably higher and peaks at a lower earnings level. A comparison of the curves indicates that a much larger proportion of black workers have weekly earnings below \$1,000 *than either white workers or other-race workers.*

Here is the R code to create figure 7.4:

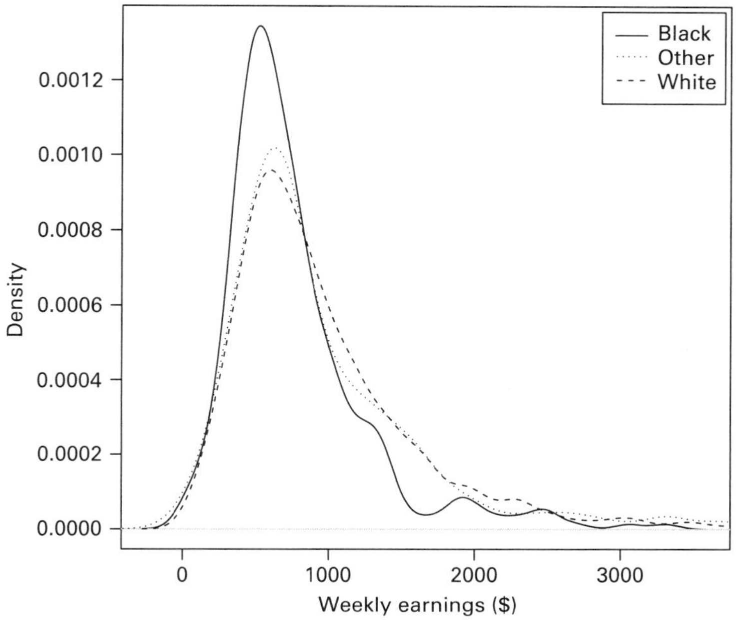

Figure 7.4
Density of weekly earnings by race

```
# create weekly earnings vectors for the three subsamples, by race
earnwk_black <- cpsemployed[cpsemployed$race=="Black","earnwk"]
earnwk_other <- cpsemployed[cpsemployed$race=="Other","earnwk"]
earnwk_white <- cpsemployed[cpsemployed$race=="White","earnwk"]

# first plot the density of weekly earnings for black individuals
plot(density(earnwk_black), main="", xlab="Weekly earnings($)")

# overlay the density of weekly earnings for other-race individuals
lines(density(earnwk_other), lty=3)

# overlay the density of weekly earnings for white individuals
lines(density(earnwk_white), lty=2)

# draw the legend
legend("topright", legend=c("Black","Other","White"), lty=c(1,3,2), inset=0.01)
```

First, vectors containing the earnwk values for each of the three subsamples are created. Then, the three density curves are plotted on the same graph, with each having a different line type, with `lty=1` *(the default) being a solid line for black individuals,* `lty=3` *being a dotted line for other-race individuals, and* `lty=2` *being a dashed line for white individuals.*

7.2 Numerical Data: Scatter Plots, Sample Covariance, and Sample Correlation

This section considers bivariate data with observations on two numerical variables. First, section 7.2.1 introduces scatter plots as a descriptive visual for bivariate numerical data. Then, section 7.2.2 introduces two descriptive statistics, the sample covariance and the sample correlation, which measure the association between two numerical variables. Finally, section 7.2.4 considers how the linear transformations of variables, seen previously in section 6.7, affect these new descriptive statistics, and it also examines the properties associated with a linear combination of variables, in which more than one variable is part of the linear function.

7.2.1 Scatter Plots

The most common visualization tool for bivariate numerical data is a **scatter plot**, where the values of the y variable are plotted against the values of the x variable. Specifically, for each observed (x_i, y_i) pair in the sample, a dot or circle is drawn such that the x_i value is read off the x-axis when a vertical line is drawn through the point (x_i, y_i), and the y_i value is read off the y-axis when a horizontal line is drawn through the point (x_i, y_i). The "scatter" in the term "scatter plot" refers to the fact that the full sample of n points looks like they are scattered on the graph. If it's the case that no two observations share both the same x- and y-values, it may be possible to see all n points on the scatter plot. If there are "tied" pairs, with two or more observations sharing the same (x, y) values, some of the dots or circles will lie on top of each other.

Example 7.7 (Education and earnings) *For the sample of employed individuals ($n = 2{,}809$) from the **cps** data set, let $x = educ$ (educational attainment, in years) and $y = earnwk$ (weekly earnings, in dollars). Figure 7.5 shows a scatter plot of $y = earnwk$ versus $x = educ$. The plot provides some interesting insight into the relationship between earnings and education. First, the higher earnings values are almost exclusively seen at higher education levels, primarily for $educ \geq 12$, which indicates a type of positive association between education and earnings. Second, the dispersion of the earnings values also increases at higher education levels. For example, thinking about all the observations along the vertical line at $educ = 12$ (i.e., the observations for all workers with exactly 12 years of education), their dispersion is the variation that is seen in the earnings values along that vertical line. The dispersion at $educ = 12$ is clearly greater than the dispersion of earnings at lower education values (e.g., if we look at the points along the vertical at $educ = 10$ or along the vertical at $educ = 9$). The R code for figure 7.5 uses the function* `plot` *with the x variable (*`cps$educ`*) as the first argument and the y variable (*`cps$earnwk`*) as the second argument:*[1]

```
# scatter plot of weekly earnings versus years of education
plot(cps$educ, cps$earnwk, xlab="Education", ylab="Weekly earnings($)")
```

*In addition to a standard scatter plot for two variables, R can draw multiple scatter plots simultaneously for a set of three or more variables. This **expanded scatter plot** provides a scatter plot for every possible pair of variables from among the set of variables specified. To see how the expanded scatter plot works in practice, let's add an additional variable (hours worked last week (hrslastwk)) to the two variables (educ and earnwk) already specified. Figure 7.6 shows the expanded scatter plot for these three variables. There are six different plots since there are $P_{3,2} = (3)(2)$ ways to choose a pair of two variables from a set of three variables. (In general, for a set of k variables, an expanded scatter plot has $P_{k,2} = (k)(k-1)$ scatter plots.) The scatter plot in the middle of the top row is a plot of educ*

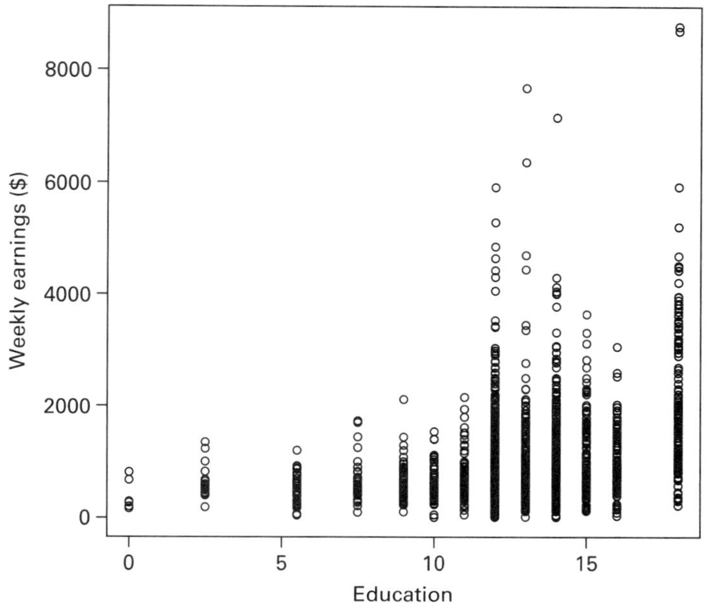

Figure 7.5
Scatter plot of weekly earnings versus years of education

versus earnwk, whereas the scatter plot on the left of the second row is a plot of earnwk versus educ, essentially reversing the roles of the x and y variables. Similarly, the scatter plot on the right of the second row is a plot of earnwk versus hrslaswk, whereas the scatter plot in the middle of the bottom row is a plot of hrslastwk versus earnwk. These two earnings and hours worked plots indicate a positive relationship between earnings and hours worked, which is perhaps unsurprising. In addition to earnings being more likely to be higher for larger hours worked, the dispersion of earnings also increases at higher levels of hours worked.

The R code to create the expanded scatter plot in figure 7.6 is rather simple:

```
# expanded scatter plot, with weekly earnings, years of education,
# and weekly hours worked
plot(cps[,c("educ","earnwk","hrslastwk")])
```

Here, the function `plot` *takes a data frame as its first and only argument, and in this case, the three columns for the variables of interest are selected.*

Example 7.8 (Monthly stock returns) *Continuing example 7.2, let's look at the relationship between the monthly stock returns for Home Depot and Lowe's. Figure 7.7 shows a scatter plot of LOW (y-axis) versus HD (x-axis).*

```
# scatter plot of Lowe's returns versus Home Depot returns
plot(sp500$HD, sp500$LOW, xlab="Monthly returns (HD)", ylab="Monthly returns (LOW)")
```

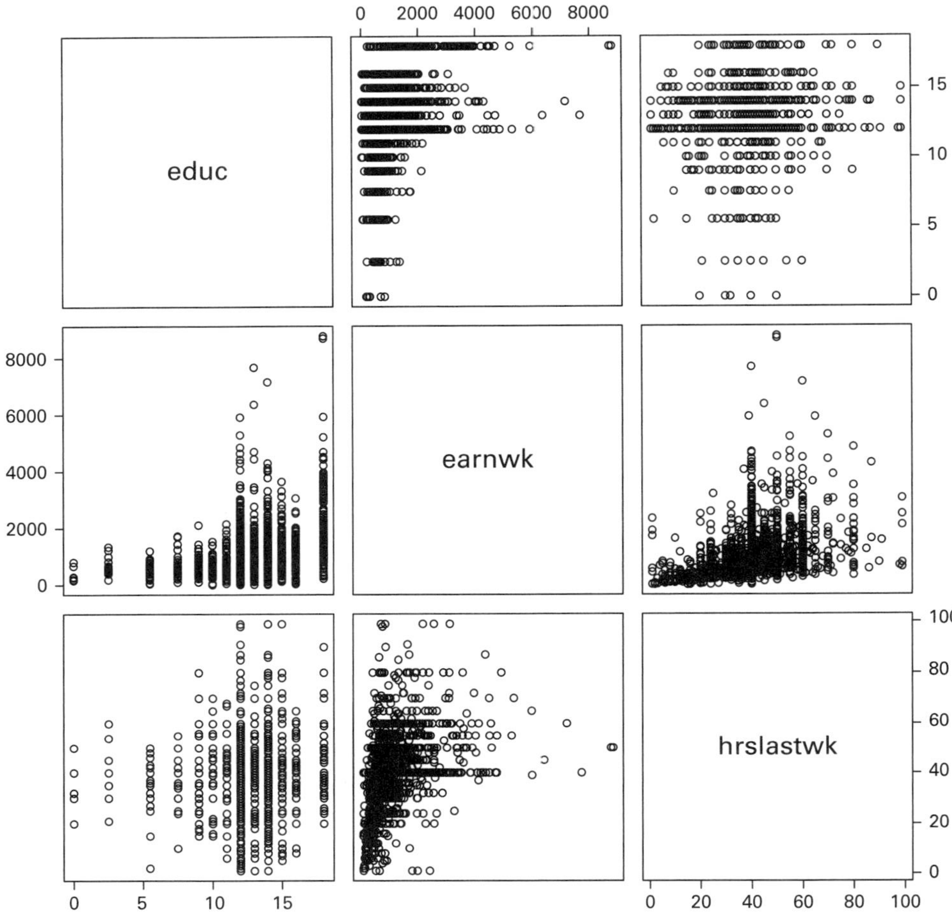

Figure 7.6
Expanded scatter plot of weekly earnings, education, and weekly hours worked

The plot indicates a positive association between HD and LOW, as the cloud of points are roughly contained within an oval stretching from the bottom left to the upper right of the plot. It is more likely to see low values of LOW when HD is low and high values of LOW when HD is high. This positive association supports the idea that the two companies' monthly returns tend to move in the same direction due to common macroeconomic conditions affecting their industry.

An expanded scatter plot can include more variables, so let's add two more stocks: Bank of America (stock ticker BAC) and Wells Fargo (stock ticker WFC), both in the banking industry. Figure 7.8 shows the expanded scatter plot with four stocks (HD, LOW, BAC, WFC).

```
# expanded scatter plot of monthly returns
plot(sp500[,c("HD","LOW","BAC","WFC")])
```

The two BAC-WFC plots, located at the bottom right of the figure, indicate a positive association similar to the one for HD and LOW. BAC tends to be higher when WFC is higher,

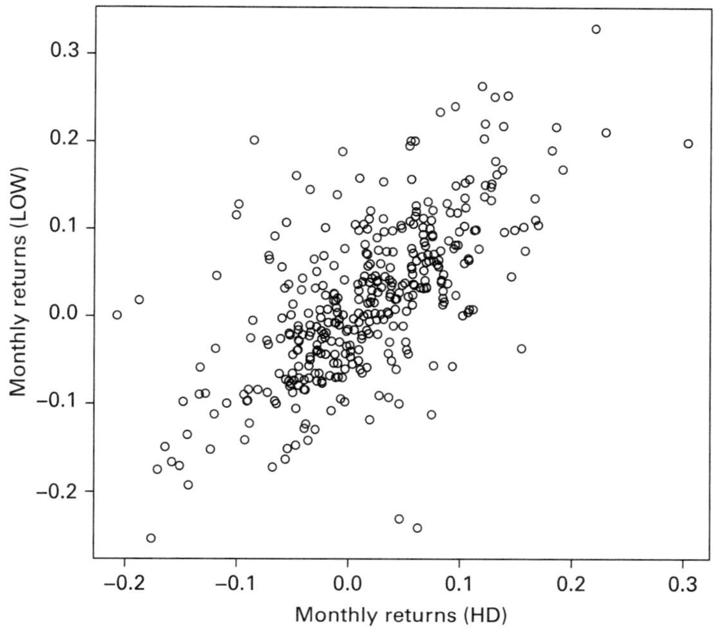

Figure 7.7
Scatter plot of Lowe's monthly returns versus Home Depot's monthly returns

and BAC tends to be lower when WFC is lower. For some of the other plots, specifically the ones across industries (e.g., one stock from home improvement (HD or LOW) versus one stock from banking (BAC or WFC)), there are also positive relationships, but not as strong as either the HD-LOW relationship or the BAC-WFC relationship. For example, in the plot of Lowe's (LOW) versus Bank of America (BAC), located in the third box of the second row, the cloud of points has a roughly oval shape that is slightly tilted to the right from vertical. The slight tilt indicates a postive relationship, but the tilt is not as strong as that seen in, for example, the plot of Lowe's (LOW) versus Home Depot (HD), located in the first box of the second row.

7.2.2 Sample Covariance and Sample Correlation

Scatter plots provide a visualization of the relationship between two numerical variables. To quantify this relationship, this section introduces the **sample covariance** and **sample correlation** descriptive statistics. In the monthly stock return example, for instance, it would be useful to quantify that the positive relationship between the returns of Lowe's and Home Depot is *stronger* than the positive relationship between Lowe's and Bank of America.

Both the sample covariance and the sample correlation provide a measure of the *linear* association between two numerical variables. The definition of the sample covariance is provided first, in definition 7.1.

Definition 7.1 *For observations* $(x_1, y_1), (x_2, y_2), \ldots, (y_n, x_n)$, *the* ***sample covariance*** *between x and y, denoted as* s_{xy}, *is*

$$s_{xy} = \frac{1}{n-1} \sum_{i=1}^{n} (x_i - \bar{x})(y_i - \bar{y}).$$

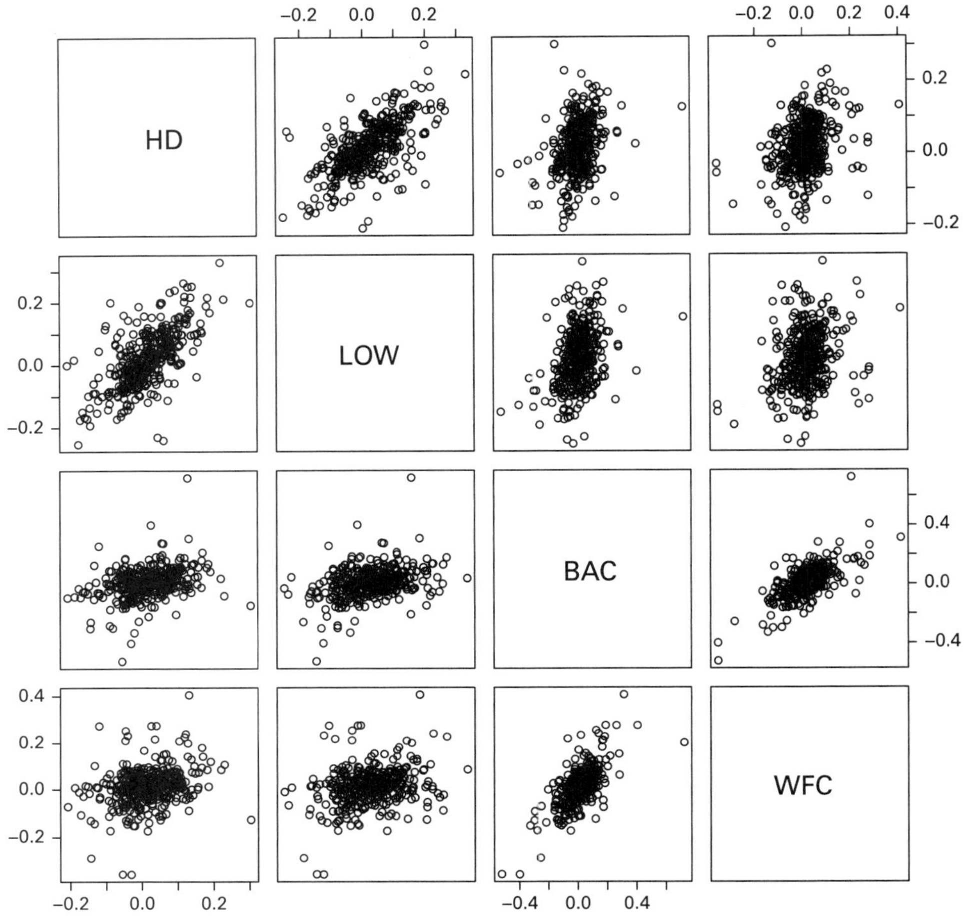

Figure 7.8
Expanded scatter plot of monthly stock returns

The $\frac{1}{n-1}$ scaling for the sample covariance is the same as the scaling for the sample variance s_x^2. The units of the sample covariance s_{xy} are the units of x times the units of y. For example, if x is education (in years) and y is weekly earnings (in dollars), the units of s_{xy} are years × dollars. The sample variance is a special case of the sample covariance, which is obtained by taking the covariance of variable x with *itself*:

$$s_{xx} = \frac{1}{n-1}\sum_{i=1}^{n}(x_i - \bar{x})(x_i - \bar{x}) = \frac{1}{n-1}\sum_{i=1}^{n}(x_i - \bar{x})^2 = s_x^2.$$

Also, the ordering of the variables x and y does not matter: $s_{xy} = s_{yx}$ since $(x_i - \bar{x})(y_i - \bar{y}) = (y_i - \bar{y})(x_i - \bar{x})$ for each i.

There are four types of contributions to the summation $\sum_{i=1}^{n}(x_i - \bar{x})(y_i - \bar{y})$ in the sample covariance s_{xy}:

1. Observation (x_i, y_i), with x_i and y_i above their respective means:

$$x_i > \bar{x} \text{ and } y_i > \bar{y} \text{ implies } (x_i - \bar{x})(y_i - \bar{y}) > 0.$$

2. Observation (x_i, y_i), with x_i and y_i below their respective means:

$$x_i < \bar{x} \text{ and } y_i < \bar{y} \text{ implies } (x_i - \bar{x})(y_i - \bar{y}) > 0.$$

3. Observation (x_i, y_i), with x_i above its mean and y_i below its mean:

$$x_i > \bar{x} \text{ and } y_i < \bar{y} \text{ implies } (x_i - \bar{x})(y_i - \bar{y}) < 0.$$

4. Observation (x_i, y_i), with x_i below its mean and y_i above its mean:

$$x_i < \bar{x} \text{ and } y_i > \bar{y} \text{ implies } (x_i - \bar{x})(y_i - \bar{y}) < 0.$$

The first two types of observations, with x_i and y_i either both above their means or both below their means, lead to positive contributions to the sample covariance. The second two types of observations, with x_i and y_i on opposite sides of their means, lead to negative contributions to the sample covariance. The sample covariance involves a combination of all four types of observations, so its sign depends on the relative magnitudes of the $(x_i - \bar{x})(y_i - \bar{y}) > 0$ contributions from (1) and (2) versus the magnitudes of the $(x_i - \bar{x})(y_i - \bar{y}) < 0$ contributions from (3) and (4).

Generally, there is a *positive* sample covariance when larger values of x tend to be associated with larger values of y and smaller values of x tend to be associated with smaller values of y. In terms of the sample averages, the sample covariance is positive when the x- and y-values are more likely to be on the same side of $\bar{x}$ and $\bar{y}$, respectively, than they are to be on opposite sides. On the other hand, there is a *negative* sample covariance when larger values of x tend to be associated with smaller values of y and smaller values of x tend to be associated with larger values of y. In terms of the sample averages, the sample covariance is negative when the x- and y-values are more likely to be on the opposite sides of $\bar{x}$ and $\bar{y}$, respectively, than they are to be on the same side.

Example 7.9 *Consider the following bivariate data with seven observations ($n = 7$):*

$$\{(x_i, y_i)\}_{i=1}^{7} = \{(4,8), (3,6), (8,10), (12,1), (0,15), (10,3), (5,6)\}.$$

Figure 7.9 shows a scatter plot of these data, indicating a clear negative relationship between x and y. The following table provides a detailed calculation of $(x_i - \bar{x})(y_i - \bar{y})$ for each observation:

i	1	2	3	4	5	6	7
x_i	4	3	8	12	0	10	5
y_i	8	6	10	1	15	3	6
$x_i - \bar{x}$	−2	−3	2	6	−6	4	−1
$y_i - \bar{y}$	1	−1	3	−6	8	−4	−1
$(x_i - \bar{x})(y_i - \bar{y})$	−2	+3	+6	−36	−48	−16	+1

There are three positive contributions and four negative contributions to the sample covariance, with the negative contributions considerably larger in magnitude. In particular, the (12, 1) *point and the* (0, 15) *point have contributions of* −36 *and* −48*, respectively, since their x and y values are very far away from their respective means. The* (12, 1) *point is a type (3) observation with x above its mean and y below its mean, while the* (0, 15) *point is a type (4) observation with x below its mean and y above its mean. On the scatter plot, these*

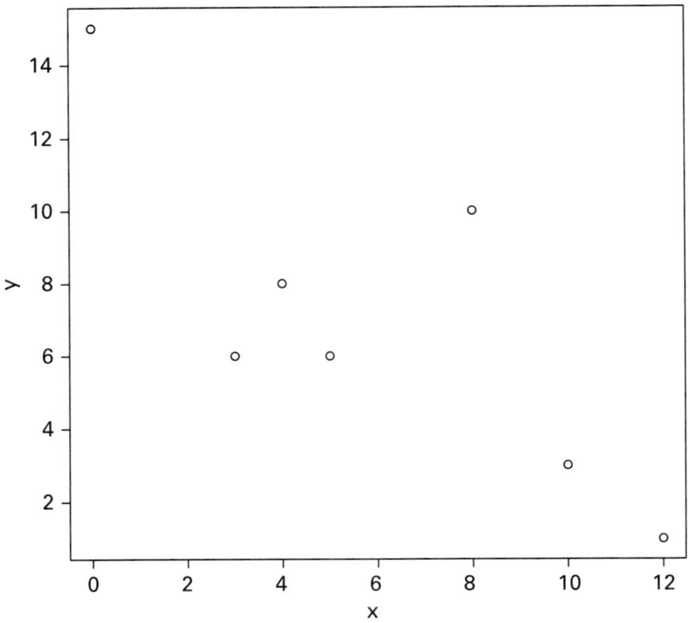

Figure 7.9
Scatter plot of a seven-observation sample

two points are the ones in the lower-right corner and the upper-left corner, respectively. Summing the values in the table's bottom row and dividing by $n-1$*, the sample covariance is*

$$s_{xy} = \frac{1}{7-1}(-2+3+6-36-48-16+1) = \frac{1}{6}(-92) = -\frac{46}{3} \approx -15.33.$$

Example 7.10 (Monthly stock returns) *In example 7.8, the positive association between the monthly stock returns of Home Depot (HD) and Lowe's (LOW) is evident from the scatter plot in figure 7.7. The sample covariance is positive, with* $s_{HD,LOW} = 0.004378$ *calculated in R using the* `cov` *function with the two variables as arguments:*

```
cov(sp500$HD, sp500$LOW)

## [1] 0.004378172
```

*To understand why the sample covariance is positive, figure 7.10 redraws the scatter plot of LOW versus HD with a horizontal line at the sample mean of LOW (*0.02029*) and a vertical line at the sample mean of HD (*0.01668*).*

```
# scatter plot with horizontal/vertical lines drawn at the sample means
plot(sp500$HD, sp500$LOW, main="", xlab="Monthly returns (HD)",
     ylab="Monthly returns (LOW)")
abline(h=mean(sp500$LOW))
abline(v=mean(sp500$HD))
```

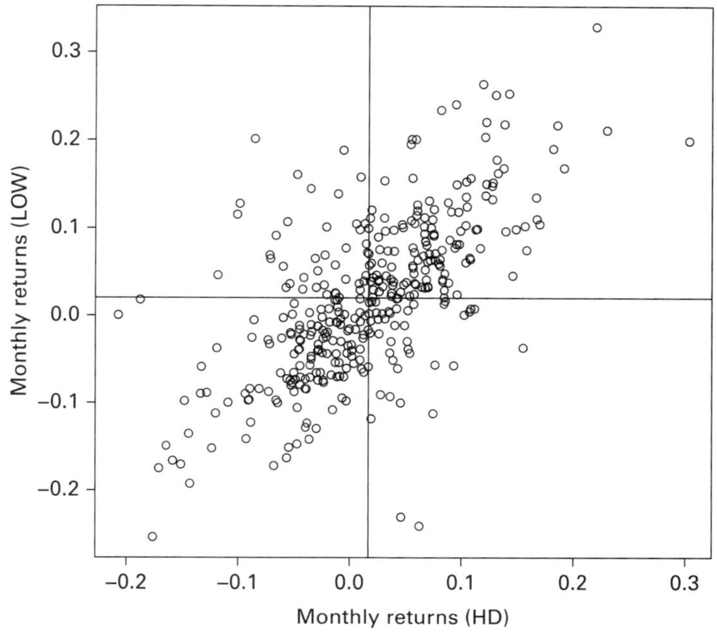

Figure 7.10
Scatter plot of LOW versus HD, with lines at sample means

The positive contributions to the sample covariance come from the observations in the upper-right quadrant (type (1) observations) and the lower-left quadrant (type (2) observations). The negative contributions to the sample covariance come from the observations in the lower-right quadrant (type (3) observations) and the upper-left quadrant (type (4) observations). There are many more observations in the upper-right and lower-left quadrants (positive contributions) than there are in the lower-right and upper-left quadrants (negative contributions). To be more precise, there are 139 *points of type (1),* 141 *points of type (2),* 43 *points of type (3), and* 41 *points of type (4). The sum of the* $(x_i - \bar{x})(y_i - \bar{y})$ *terms for the four types of observations are* 0.933703 *(type (1)),* 0.842564 *(type (2)),* −0.075304 *(type (3)), and* −0.111687 *(type (4)). Adding these four values together and dividing by* $n - 1 = 363$ *yields the sample covariance* $s_{HD,LOW} = 0.004378$.

Example 7.11 (Education and earnings) *In example 7.7, the positive relationship between weekly earnings and educational attainment was evident in the scatter plot of earnwk versus educ (figure 7.5). As in example 7.10, the scatter plot can be redrawn with lines at the sample means of the two variables. This scatter plot is shown in figure 7.11, with a horizontal line at the sample mean of earnwk (*$971.18*) and a vertical line at the sample mean of educ (*12.82 *years). For these data, there are* 600 *upper-right quadrant (type (1)) observations,* 1,113 *lower-right quadrant (type (2)) observations,* 702 *lower-right quadrant (type (3)) observations, and* 394 *upper-left quadrant (type (4)) observations, with overall contributions to the sum of* $(x_i - \bar{x})(y_i - \bar{y})$ *given by* 1,388,551, 811,985, −335,102, *and* −218,892, *respectively. In this example, it is not just the number of upper-right quadrant (type (1)) observations that influences the sample covariance but also the large magnitude of the*

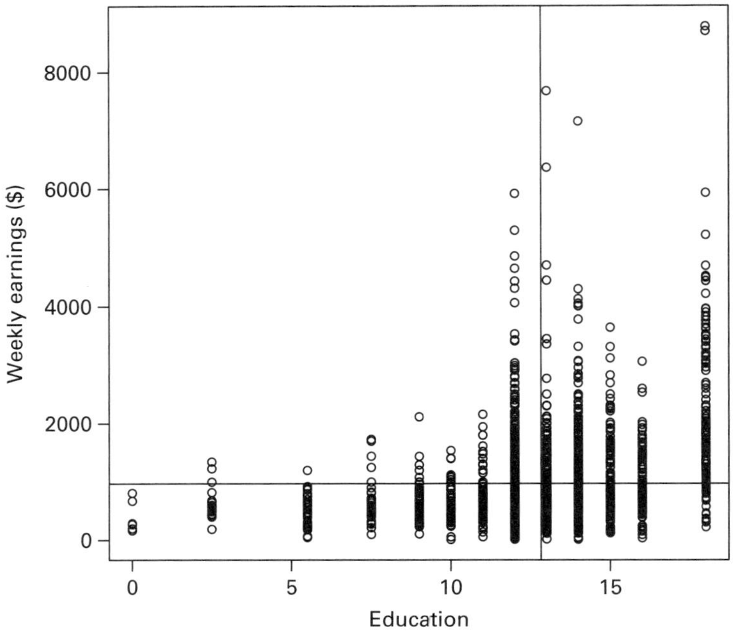

Figure 7.11
Scatter plot of weekly earnings versus education, with lines at sample means

$(x_i - \bar{x})(y_i - \bar{y})$ contributions caused by the considerable number of points that have earnwk (y) values far above the sample mean of earnwk ($\bar{y}$). The resulting sample covariance is $s_{educ,earnwk} = 586.375$.

```
cov(cpsemployed$educ, cpsemployed$earnwk)

## [1] 586.3751
```

The units of the sample covariance $s_{educ,earnwk}$ are years × dollars or, equivalently, dollars × years. These strange units make it very difficult to interpret what the numerical value 586.375 *means. While the positive covariance does reflect the positive relationship between earnwk and educ, the numerical value of the sample covariance may not be an ideal statistic for quantifying this relationship.*

As seen in example 7.11, the sample covariance value can be difficult to interpret, even if the sign of the sample covariance indicates a positive or negative relationship between two variables. The problem involves the units of the sample covariance, which are the units of x times the units of y. It would be more useful if we had a descriptive statistic that could be compared for different pairs of bivariate data. For example, is the relationship between *educ* and *earnwk* stronger than the relationship between *hrslastwk* and *earnwk*? The sample covariance is not useful for this purpose since the units of the first sample covariance are years × dollars and the units of the second sample covariance are hours × dollars. Even after the sample covariances $s_{educ,earnwk}$ and $s_{hrslastwk,earnwk}$ are calculated, their values are not comparable due to the different units.

To address this undesirable feature of the sample covariance, we introduce the sample correlation, a descriptive statistic that also measures the linear association between two variables, but in a way that is comparable across different pairs of variables.

Definition 7.2 *For observations* $(x_1, y_1), (x_2, y_2), \ldots, (y_n, x_n)$, *the* ***sample correlation*** *between x and y, denoted as* r_{xy}, *is*

$$r_{xy} = \frac{s_{xy}}{s_x s_y},$$

where $s_{xy} = \frac{1}{n-1}\sum_{i=1}^{n}(x_i - \bar{x})(y_i - \bar{y})$, $s_x = \sqrt{\frac{1}{n-1}\sum_{i=1}^{n}(x_i - \bar{x})^2}$, *and* $s_y = \sqrt{\frac{1}{n-1}\sum_{i=1}^{n}(y_i - \bar{y})^2}$.

The sample correlation between x and y is the sample covariance between x and y divided by the product of the standard deviations of x and y. Importantly, since the units of s_x are the units of x and the units of s_y are the units of y, the units of the numerator (units of $x \times$ units of y) cancel the units of the denominator, leading to the sample correlation r_{xy} being *unitless*. This fact and additional properties about the sample correlation are stated in proposition 7.1.

Proposition 7.1 (Properties of the sample correlation) *The sample correlation* r_{xy} *has the following properties:*

i. r_{xy} *is unitless;*

ii. *the sign of the sample correlation is the same as the sign of the sample covariance,*

$$sign(r_{xy}) = sign(s_{xy});$$

iii. $r_{xx} = 1$;

iv. $-1 \leq r_{xy} \leq 1$.

Property (i) has already been discussed. For property (ii), note that both $s_x > 0$ and $s_y > 0$ since standard deviations are positive. Therefore, the denominator in $\frac{s_{xy}}{s_x s_y}$ is also positive, meaning that the sign of r_{xy} is the same as the sign of the numerator, s_{xy}. Property (iii) says that the correlation of variable x with itself is exactly equal to 1. For $r_{xx} = \frac{s_{xx}}{s_x s_x}$, the numerator and denominator are both equal to the sample variance s_x^2, yielding $r_{xx} = 1$. Finally, property (iv) says that the magnitude of the sample correlation must be less than or equal to 1; that is, $|r_{xy}| \leq 1$ or, equivalently, $-1 \leq r_{xy} \leq 1$. The proof of this property is beyond the scope of this book, but the property makes sense if two extreme cases are considered. Intuitively, the strongest positive correlation should be between variable x and itself, in which case the sample correlation is equal to 1 from property (iii). Likewise, the strongest negative correlation should be between variable x and its negative $(-x)$, in which case the sample correlation is equal to -1.[2] As property (iv) states, any other sample correlation has a value between the two extremes of -1 and 1.

To see how sample correlation values relate to the bivariate association in scatter plots, figure 7.12 shows a set of six scatter plots, each with a different r_{xy} value. The top row of three scatter plots has $r_{xy} = 0.4$, $r_{xy} = 0.8$, and $r_{xy} = 1$, and the bottom row of three scatter plots has $r_{xy} = -0.4$, $r_{xy} = -0.8$, and $r_{xy} = 0$. The $r_{xy} = 1$ scatter plot indicates a perfect

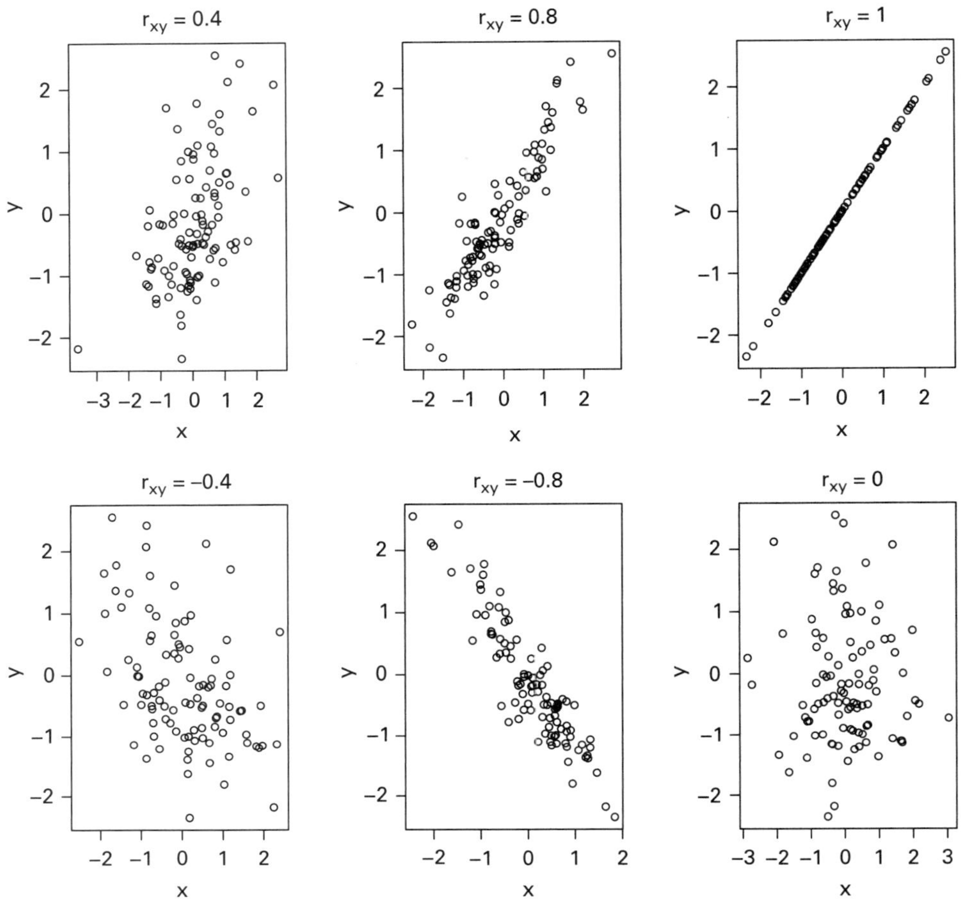

Figure 7.12
Scatter plots for different sample correlations

linear and positive relationship between x and y. In this case, a positive-sloped line can be drawn through all the points. For $r_{xy} = 0.4$ and $r_{xy} = 0.8$, both plots reveal a positive relationship between x and y, but the larger correlation ($r_{xy} = 0.8$) appears to describe a stronger positive relationship, in the sense that the cloud of points is closer to the extreme of a linear relationship compared with the smaller correlation ($r_{xy} = 0.4$). The comparison between the $r_{xy} = -0.4$ and $r_{xy} = -0.8$ scatter plots is similar. Both indicate a negative relationship between x and y, with the $r_{xy} = -0.8$ plot indicating a stronger negative relationship as the cloud of points is closer to the extreme of a negative linear relationship compared with $r_{xy} = -0.4$. Finally, the $r_{xy} = 0$ scatter plot presents a case where there is no evident relationship between x and y. The cloud of points in this case does not show a tendency to be either upward sloping or downward sloping.

Before considering empirical examples involving the sample correlation, it's important to discuss one potential pitfall associated with the sample correlation measure. As stated previously, the sample correlation is only meant to measure the linear association between two variables. As such, a sample correlation of zero means that there is no linear

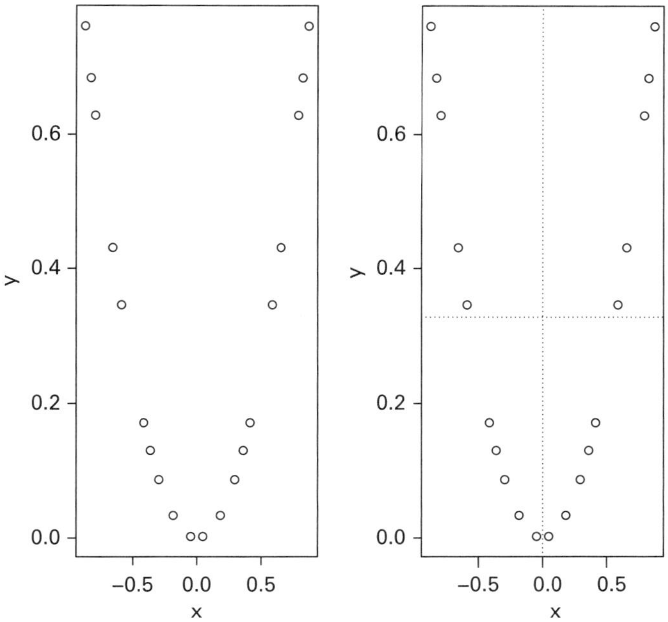

Figure 7.13
Scatter plot for a sample with zero sample correlation

relationship between two variables, but it does *not* necessarily mean that there is no relationship whatsoever between the two variables. Figure 7.13 pictures a rather extreme example, with a scatter plot of points that lie exactly along a parabola. For these bivariate data, there is a perfect *nonlinear* relationship between x and y. However, it turns out that the sample correlation between the two variables is exactly equal to zero. The scatter plot on the right, with horizontal and vertical lines drawn at the sample means, indicates why this is the case. The positive contributions to the sample covariance from the upper-right quadrant are canceled out exactly by the negative contributions from the upper-left quadrant, and similarly the positive contributions from the lower-left quadrant are canceled out exactly by the negative contributions from the lower-right quadrant. The resulting sample covariance is zero, meaning that the sample correlation is also zero.

There's no way to avoid this feature of the sample correlation, as the sample correlation is specifically designed to measure the linear association between two variables. However, this nonlinear scatter plot example highlights the importance of using *both* descriptive visuals, like scatter plots, and descriptive statistics, like sample correlations, when examining the association between variables.

With the sample correlation measure now in our descriptive statistics tool kit, we revisit the previous examples.

Example 7.12 (Education and earnings) *In example 7.11, the sample covariance of educational attainment and weekly earnings was* $s_{educ,earnwk} = 586.375$. *The sample standard deviations of educ and earnwk,* $s_{educ} = 2.4030$ *and* $s_{earnwk} = 750.4846$, *are needed* *to calculate the sample correlation* $r_{educ,earnwk}$:

$$r_{educ,earnwk} = \frac{s_{educ,earnwk}}{s_{educ}s_{earnwk}} = \frac{586.375}{(2.4030)(750.4846)} \approx 0.325.$$

The sample correlation is calculated in R with the `cor` *function, with the two variables as arguments:*

```
cor(cpsemployed$educ, cpsemployed$earnwk)

## [1] 0.3251519
```

Since the sample correlation is unitless, it doesn't matter which units are used for educ or earnwk. For instance, if earnings were measured in thousands of dollars rather than dollars (i.e., $earnwkthous = earnwk/1000$*),*

$$r_{educ,earnwkthous} = \frac{s_{educ,earnwkthous}}{s_{educ}s_{earnwkthous}} = \frac{s_{educ,earnwk} \cdot \frac{1}{1000}}{s_{educ}(s_{earnwk} \cdot \frac{1}{1000})} = r_{educ,earnwk}.$$

Similarly, if education were measured in months rather than years (i.e., $educmonths = 12educ$*),*

$$r_{educmonths,earnwk} = \frac{s_{educmonths,earnwk}}{s_{educmonths}s_{earnwk}} = \frac{s_{educ,earnwk} \cdot 12}{(s_{educ} \cdot 12)s_{earnwk}} = r_{educ,earnwk}.$$

With the sample correlation being unitless, the sample correlation $r_{educ,earnwk}$ *can also be compared to the sample correlation between earnwk and some other variable. For instance, for the relationship between weekly earnings and hours worked, the sample correlation is* $r_{hrslastwk,earnwk} \approx 0.368$:

```
cor(cpsemployed$hrslastwk, cpsemployed$earnwk)

## [1] 0.3681961
```

A comparison of this value with the $r_{educ,earnwk}$ *value (0.325) indicates that the two sample correlations are quite similar, with some evidence that the positive relationship is slightly larger for earnwk and hrslastwk than it is for earnwk and educ.*

Example 7.13 (Monthly stock returns) *Example 7.10 considered the relationship between monthly stock returns for Home Depot and Lowe's by drawing a scatter plot and providing the sample covariance* $s_{HD,LOW}$*. The sample correlation between HD and LOW is*

$$r_{HD,LOW} = \frac{s_{HD,LOW}}{s_{HD}s_{LOW}} = \frac{0.004378}{(0.073706)(0.091598)} \approx 0.648.$$

```
cor(sp500$HD, sp500$LOW)

## [1] 0.6484954
```

A sample correlation of 0.648 seems pretty high, but how does it compare to the sample correlations for other pairs of stocks? Since the sample correlations for other pairs of stocks can be directly compared to $r_{HD,LOW}$*, we can look at the sample correlations among a* *larger group of stocks. Previously, Bank of America and Wells Fargo were considered in the expanded scatter plot of example 7.8. Let's add two more companies to the mix, specifically Marathon Oil (stock ticker MRO) and ConocoPhillips (stock ticker COP), both of which are in the oil industry.*

*A **correlation matrix** can be created in R using the* `cor` *function and a single argument that contains a data frame with the desired variables as follows:*

```
cor(sp500[,c("HD","LOW","BAC","WFC","MRO","COP")])

##            HD       LOW       BAC       WFC       MRO       COP
## HD  1.0000000 0.6484954 0.3311974 0.2803797 0.1887287 0.2145689
## LOW 0.6484954 1.0000000 0.3566068 0.2619187 0.1811412 0.2560374
## BAC 0.3311974 0.3566068 1.0000000 0.6919604 0.3313437 0.3392213
## WFC 0.2803797 0.2619187 0.6919604 1.0000000 0.3793915 0.3960973
## MRO 0.1887287 0.1811412 0.3313437 0.3793915 1.0000000 0.7709914
## COP 0.2145689 0.2560374 0.3392213 0.3960973 0.7709914 1.0000000
```

This correlation matrix provides all possible pairwise sample correlations of the monthly returns for six stocks (HD, LOW, BAC, WFC, MRO, and COP). Interestingly, all the sample correlations are positive, which is consistent with the idea that all six companies are affected by overall macroeconomic conditions, especially since the time frame here (over 20 years) is quite long. There are a lot of duplicate values in the correlation matrix since the sample correlation of x and y (r_{xy}) is the same as the sample correlation of y and x (r_{yx}). For instance, $r_{BAC,LOW} = r_{LOW,BAC} = 0.357$ appears both as the third value in the second row and the second value in the third row. To make the correlation matrix easier to read, it can be shown by reporting only the upper-right triangle of sample correlations (above the diagonal) or the lower-left triangle of correlations (below the diagonal). Here is a version of the correlation matrix that provides the upper-right triangle of sample correlations:

	HD	*LOW*	*BAC*	*WFC*	*MRO*	*COP*
HD	1.000	0.648	0.331	0.280	0.189	0.215
LOW		1.000	0.357	0.262	0.181	0.256
BAC			1.000	0.692	0.331	0.339
WFC				1.000	0.379	0.396
MRO					1.000	0.771
COP						1.000

The three highest values in the correlation matrix are $r_{MRO,COP} = 0.771$, $r_{BAC,WFC} = 0.692$, and $r_{HD,LOW} = 0.648$. Not coincidentally, these three stock pairs correspond to the three pairs of companies that are in the same industry, with MRO and COP in the oil industry, BAC and WFC in the banking industry, and HD and LOW in the home improvement industry. Looking at COP, for instance, its sample correlation with MRO (0.771) is much larger than its sample correlations with the other four companies, which range from 0.215 to 0.396. Similarly, for HD, its sample correlation with LOW (0.648) is much larger than its sample correlations with the other four companies, which range from 0.189 to 0.331. The four lowest values in the correlation matrix are $r_{LOW,MRO} = 0.181$, $r_{HD,MRO} = 0.189$, $r_{HD,COP} = 0.215$, and $r_{LOW,COP} = 0.256$, suggesting that the shocks affecting the home *improvement and oil industries are less related to each other than the shocks affecting other pairs of the three industries.*

It turns out that we can say more about the interpretation of the magnitude of the sample correlation r_{xy} when x and y have a *linear relationship*. In chapter 17, we formally introduce a model that describes such a linear relationship. For now, we can think of a linear relationship as holding if we can imagine being able to draw a straight line through the middle of the cloud of points in the scatterplot of x and y. In this case, we can make statements about how the sample average of one of the variables, such as y, is expected to change when the other variable, such as x, changes.

Let's think about a one-standard-deviation (s_x) change in the x variable, from some value x^* to $x^* + s_x$. Then the sample average of y values associated with x values near $x^* + s_x$ will be approximately $|r_{xy}|s_y$ higher (if r_{xy} is positive) or $|r_{xy}|s_y$ lower (if r_{xy} is negative) than the sample average of y-values associated with x-values near x^*. Put another way, when x increases by one standard deviation, the approximate change in the associated sample average of y-values is r_{xy} standard deviations. Similarly, thinking about a one-standard-deviation (s_y) change in the y variable, from some value y^* to $y^* + s_y$, the sample average of x-values associated with y-values near $y^* + s_y$ will be approximately $|r_{xy}|s_x$ higher (if r_{xy} is positive) or lower (if r_{xy} is negative) than the sample average of x-values associated with y-values near y^*.

As an example, we consider the relationship between the Home Depot (*HD*) and Lowe's (*LOW*) monthly returns from the **sp500** data set. The sample correlation is $r_{HD,LOW} \approx 0.648$, and the two sample standard deviations are $s_{HD} \approx 0.0737$ and $s_{LOW} \approx 0.0916$. Then, when we think about the monthly return of *HD* increasing by 0.0737 or 7.37% (one standard deviation), the sample average of *LOW* monthly returns should increase by approximately $r_{HD,LOW}s_{LOW} = (0.648)(0.0916) \approx 0.0594$, or 5.94%. Similarly, when we think about the monthly return of *LOW* increasing by 0.0916, or 9.16%, the sample average of *HD* monthly returns should increase by approximately $r_{HD,LOW}s_{HD} = (0.648)(0.0737) \approx 0.0478$, or 4.78%.

In the special case where both x and y are standardized variables, with $s_x = s_y = 1$, the interpretation of r_{xy} is even more direct. For instance, a change from x^* to $x^* + 1$ is a one-standard-deviation change. The approximate change in the associated sample average of y-values, comparing those associated with $x \approx x^* + 1$ to those associated with $x \approx x^*$, is just r_{xy} itself.

7.2.3 Sample Correlation When One Variable Is Binary

This section briefly considers a specific case in which one of the two variables in the bivariate data is a binary variable or an indicator variable. Without loss of generality, let's say that x is a binary variable for which observations x_i can only have the values 0 and 1. In this case, proposition 7.2 shows an interesting relationship between the sample covariance/correlation of x with another y variable and how the sample mean of y_i observations varies between the subsample having $x_i = 0$ and having $x_i = 1$.

Proposition 7.2 *If x is a binary variable, the sample covariance and sample correlation with y simplify to*

$$s_{xy} = s_x^2(\bar{y}_1 - \bar{y}_0) \text{ and } r_{xy} = \frac{s_x}{s_y}(\bar{y}_1 - \bar{y}_0),$$

where $\bar{y}_0$ *is the average of* y_i *observations when* $x_i = 0$ *and* $\bar{y}_1$ *is the average of* y_i *observations when* $x_i = 1$.

An immediate implication of proposition 7.2 is that the signs of the sample covariance and the sample correlation are the same as the sign of $\bar{y}_1 - \bar{y}_0$. When $\bar{y}_1 > \bar{y}_0$, which occurs when the sample average of y in the $x=1$ subsample is greater than the sample average of y in the $x=0$ subsample, both s_{xy} and r_{xy} are positive. Likewise, when $\bar{y}_1 < \bar{y}_0$, both s_{xy} and r_{xy} are negative. This result makes sense since $\bar{y}_1 > \bar{y}_0$ means that y tends to be higher for higher x-values (1 here) than for lower x-values (0 here), indicating a positive relationship.

Example 7.14 (Earnings and union status) *Consider the sample of* 2,809 *employed individuals from the* **cps** *data set. Let x be a binary variable based upon the unionstatus categorical variable, with* $x = 1$ *if unionstatus = "Union" and* $x = 0$ *if unionstatus = "Non-union," and let* $y = $ *earnwk be the earnings last week.*

```
# sample mean of weekly earnings for union employees
mean(cpsemployed[cpsemployed$unionstatus=="Union","earnwk"])

## [1] 1197.654

# sample mean of weekly earnings for nonunion employees
mean(cpsemployed[cpsemployed$unionstatus=="Non-union","earnwk"])

## [1] 946.5013

# sample correlation between union indicator and weekly earnings
union_var <- ifelse(cpsemployed$unionstatus=="Union",1,0)
cor(union_var, cpsemployed$earnwk)

## [1] 0.0996309
```

Among the 276 *union workers with* $x = 1$, *the sample mean of weekly earnings is* \$1,198 ($\bar{y}_1 = 1{,}198$). *Among the* 2,533 *nonunion workers with* $x = 0$, *the sample mean of weekly earnings is* \$947 ($\bar{y}_0 = 947$). *Since* $\bar{y}_1 > \bar{y}_0$, *the sample correlation between x and y is positive, with* $r_{xy} \approx 0.100$.

Proposition 7.2 holds if y is also a binary variable. In that case, $\bar{y}_1$ is the proportion of the $x = 1$ subsample with $y = 1$, and $\bar{y}_0$ is the proportion of the $x = 0$ subsample with $y = 0$. The sample correlation between x and y is positive if the proportion of $y = 1$ observations is higher in the $x = 1$ subsample than it is in the $x = 0$ subsample, and negative otherwise.

7.2.4 Linear Transformations

Section 6.7 introduced linear transformations of variables in the context of univariate data and described how descriptive statistics were affected by taking a linear transformation of the data. Having now introduced descriptive statistics for measuring linear association between two variables, we can provide results for the sample covariance and the sample correlation after taking linear transformations of two different variables.

For the known constants a, b, c, and d, consider the following linear transformations of the x and y variables:

$$v = a + bx \quad \text{and} \quad w = c + dy.$$

From proposition 6.6, the sample variances and the sample standard deviations of v and w are directly related to those of x and y, respectively, as follows:

$$s_v^2 = b^2 s_x^2,\ s_v = |b| s_x,\ s_w^2 = d^2 s_y^2,\ s_w = |d| s_y.$$

The sample covariance between v and w is

$$\begin{aligned} s_{vw} &= \frac{1}{n-1}\sum_{i=1}^{n}(v_i - \bar{v})(w_i - \bar{w}) \\ &= \frac{1}{n-1}\sum_{i=1}^{n}(a + bx_i - (a + b\bar{x}))(c + dy_i - (c + d\bar{y})) \\ &= \frac{1}{n-1}\sum_{i=1}^{n}(b(x_i - \bar{x}))(d(y_i - \bar{y})) \\ &= bd\frac{1}{n-1}\sum_{i=1}^{n}(x_i - \bar{x})(y_i - \bar{y}) \\ &= bds_{xy}, \end{aligned}$$

where the second equality follows from $\bar{v} = a + b\bar{x}$ and $\bar{w} = c + d\bar{y}$ (proposition 6.6).

Then the sample correlation between v and w is

$$r_{vw} = \frac{s_{vw}}{s_v s_w} = \frac{bds_{xy}}{(|b|s_x)(|d|s_y)} = \frac{bd}{|b||d|}\cdot\frac{s_{xy}}{s_x s_y} = \frac{bd}{|b||d|}r_{xy}.$$

The expression $\frac{bd}{|b||d|}$ is equal to 1 if b and d have the same sign, which occurs when b and d are both positive or both negative; and -1 if b and d have oppositive signs. Therefore, $r_{vw} = r_{xy}$ when b and d have the same sign, and $r_{vw} = -r_{xy}$ when b and d have opposite signs. The additive constants a and c do not affect either s_{vw} or r_{vw}.

These results for s_{vw} and r_{vw} are summarized in proposition 7.3.

Proposition 7.3 *Suppose that a, b, c, and d are known constants, $v = a + bx$ is a linear transformation of x, and $w = c + dy$ is a linear transformation of y. The sample covariance and sample correlation for the sample $\{(v_1, w_1), (v_2, w_2), \ldots, (v_n, w_n)\}$ have the following relationships to the sample covariance and sample correlation for the sample $\{(x_1, y_1), (x_2, y_2), \ldots, (x_n, y_n)\}$:*

i. *(sample covariance) $s_{vw} = bds_{xy}$. The sign of s_{vw} is the same as the sign of s_{xy} when b and d are both positive or both negative, and the sign of s_{vw} is opposite the sign of s_{xy} when b and d are opposite signs.*

ii. *(sample correlation)* $r_{vw} = \begin{cases} r_{xy} & \text{if } b \text{ and } d \text{ are both positive or both negative} \\ -r_{xy} & \text{if } b \text{ and } d \text{ are opposite signs.} \end{cases}$

iii. *(transforming one variable and not the other)* $s_{vy} = bs_{xy}$ *and* $r_{vy} = \begin{cases} r_{xy} & \text{if } b \geq 0 \\ -r_{xy} & \text{if } b < 0. \end{cases}$

Property (iii), where one variable ($v = a + bx$) is transformed but the other is not, is a special case of transforming both variables with $c = 0$ and $d = 1$.

In a situation where v and w are constructed by simply changing the units of x and y, in which case b and d would be positive numbers, the sign of s_{vw} is the same as the sign of s_{xy}, but the magnitude of s_{vw} is scaled by bd. In this case, $r_{vw} = r_{xy}$, meaning that the sample correlation is not affected by the units of the two variables.

Example 7.15 (Earnings and height) *Suppose that x denotes weekly earnings and y denotes height in inches, with sample covariance s_{xy} and sample correlation r_{xy}. If v is annualized earnings ($v = 52x$) and w is height in feet ($w = \frac{1}{12}y$), the sample covariance s_{vw} and the sample correlation r_{vw} are*

$$s_{vw} = (52)\left(\frac{1}{12}\right)s_{xy} = \frac{13}{3}s_{xy} \quad \text{and} \quad r_{vw} = r_{xy}.$$

Example 7.16 (Earnings and education) *Suppose that x denotes weekly earnings and y denotes educational attainment in years. If x is transformed into annualized earnings, so that $v = 52x$, the sample covariance between annualized earnings and education (in years) is $s_{vy} = 52s_{xy}$, and the sample correlation is $r_{vy} = r_{xy}$. For the* **cps** *data set, the sample covariance and sample correlation are $s_{xy} = 586.3751$ and $r_{xy} = 0.325$, so that $s_{vy} = 30{,}491.51$ and $r_{vy} = 0.325$.*

Example 7.17 (Website profits) *Recall the website profit example considered in example 6.26, where x is the daily widget purchases, p is the price per widget, f is the website's daily fixed cost, and c is the marginal cost per widget. Let's assume that $p > c$, which means that the price for a widget is greater than the marginal cost of a widget. The daily profit is $y = -f + (p - c)x$. If we have data on both x and y, the sample covariance is $s_{xy} = (p - c)s_{xx} = (p - c)s_x^2$, and the sample correlation is 1 since $(p - c) > 0$. There is a perfect linear relationship between x and y here since y is defined to be a linear function of x, based on the three constants (f, p, c).*

Suppose we also have data on variable u measuring the daily number of unique visitors to the website. A positive sample correlation is expected between u and x (more likely to have more purchases on days with more visitors and fewer purchases on days with fewer visitors), but it would not be a perfect linear relationship. In thinking about the relationship between daily visitors and daily profits, how would the sample covariance s_{uy} and the sample correlation r_{uy} compare to the sample covariance s_{ux} and the sample correlation r_{ux}? Since $s_{uy} = (p - c)s_{ux}$ and $r_{uy} = r_{ux}$ (since $(p - c) > 0$), the sample correlation between visitors and profits is the same as the sample correlation between visitors and purchases. The equality $r_{uy} = r_{ux}$ arises since profits have a perfect linear relationship with purchases.

7.2.5 Linear Combination of Two Variables

For the linear transformations considered thus far, only a single variable has been part of any linear transformation. For example, $v = a + bx$ is a linear function of the single variable x. In considering samples with more than one variable, we can generalize the types of linear functions of variables that can be analyzed. When a linear function contains two or more variables, it is known as a **linear combination** of the variables.

We first focus on a linear combination of *two* variables. For the known constants k, a, and b, the variable

$$v = k + ax + by$$

is a linear combination of x and y. Generally, variables x and y will be in the same units, or at least scaled versions of the same units, so it makes sense to take a combination of them. For instance, if a company has two separate products, with x representing the profit for one product and y the profit for the other, it makes perfect sense to take a linear combination (e.g., $v = x + y$). On the other hand, if x represents earnings (in dollars) and y represents height (in inches), it's hard to imagine why we would want to take a linear combination of the two variables; of course, it's possible to mathematically construct something like $v = x + y$ even in this case, but what does $x + y$ mean?

Proposition 7.4 provides results for the descriptive statistics of the linear combination $v = k + ax + by$.

Proposition 7.4 *Suppose that k, a, and b are known constants and $v = k + ax + by$ is a linear combination of x and y. The descriptive statistics for the sample $\{v_1, v_2, \ldots, v_n\}$ have the following relationships to the descriptive statistics for the sample $\{(x_1, y_1), (x_2, y_2), \ldots, (x_n, y_n)\}$:*

i. *(sample mean)* $\bar{v} = k + a\bar{x} + b\bar{y}$
ii. *(sample variance)* $s_v^2 = a^2 s_x^2 + b^2 s_y^2 + 2ab s_{xy}$, *with the following special cases:*
   (no correlation between x and y) for $s_{xy} = 0$: $s_v^2 = a^2 s_x^2 + b^2 s_y^2$
   (sum of two variables) for $v = x + y$: $s_v^2 = s_x^2 + s_y^2 + 2s_{xy}$
   (difference of two variables) for $v = x - y$: $s_v^2 = s_x^2 + s_y^2 - 2s_{xy}$
iii. *(sample standard deviation)* $s_v = \sqrt{s_v^2} = \sqrt{a^2 s_x^2 + b^2 s_y^2 + 2ab s_{xy}}$

Similar to linear transformations, the additive constant, here denoted by k, only plays a role in the sample mean, as it may affect the location of the resulting sample distribution of v but does not affect the dispersion. In property (ii), the sample variance of the linear combination v has three components, with the $a^2 s_x^2$ term coming from the sample variance of x, the $b^2 s_y^2$ term coming from the sample variance of y, and the $2ab s_{xy}$ term coming from the sample covariance between x and y. The first two terms are always positive, and the sign of the third term depends upon the signs of a, b, and s_{xy}. In the case of no sample correlation between x and y, the third term drops out, so the sample variance of the linear combination is just a sum of the scaled sample variances of x and y.

To develop more intuition for the results in property (ii), let's first focus on the case where $v = x + y$ is the sum of the two variables, for which we have $s_v^2 = s_x^2 + s_y^2 + 2s_{xy}$. If x and y are uncorrelated in the sample ($s_{xy} = r_{xy} = 0$), the sample variance of v is the sum of the sample variances of x and y. In other words, the variation contained in the sum of the two variables is equal to the sum of the variations of each variable individually. What if x and y are correlated ($s_{xy} \neq 0$)? Let's think about the positive correlation case first ($r_{xy} > 0$, so $s_{xy} > 0$). The sample variance s_v^2 depends upon how far the $x_i + y_i$ values are from their sample mean $\bar{x} + \bar{y}$. When x_i is very far above its sample mean $\bar{x}$, the positive correlation between x and y implies that we are more likely to see a y_i-value also above its sample mean $\bar{y}$. Similarly, when x_i is very far below its sample mean $\bar{x}$, the positive correlation between x and y implies that we are more likely to see a y_i-value also below its sample mean $\bar{y}$. This relationship causes more extreme values for $(x_i + y_i - (\bar{x} + \bar{y}))^2$ than we'd have if there were no correlation between x and y. As a result, the positive correlation between the two variables causes the dispersion in $x + y$ to increase, as the individual variables x and y move in the same direction.

We can also see this phenomenon graphically. Figure 7.14 shows three cases: a positive correlation between x and y (top row), zero correlation between x and y (middle row), and a negative correlation between x and y (bottom row). The left plot in each row is a scatter plot of y versus x, and the right plot in each row is a histogram of $v = x + y$. To make the three cases directly comparable to each other and to focus on the role of r_{xy}, the sample means $\bar{x}$ and $\bar{y}$ and the sample variances s_x^2 and s_y^2 have been chosen to be the same across all three cases. The axes for the graphs are also exactly the same for each of the three cases. The scatter plots have downward-sloping lines added to them, where each of these dotted lines corresponds to a different value of $x + y$. Specifically, the seven lines drawn on each scatter plot correspond to the lines $x + y = -3$, $x + y = -2$, $x + y = -1$, $x + y = 0$, $x + y = 1$, $x + y = 2$, and $x + y = 3$. Looking at the top scatter plot, when x is above 1, the $x + y$ values are almost all above 1 (i.e., almost all of the points corresponding to $x > 1$ are above the diagonal line corresponding to $x + y = 1$). Similarly, when x is below -1, the $x + y$ values are almost all below -1. In contrast, for the scatter plot in the zero correlation case (second row), when x is above 1, several of the $x + y$ values are below 1 and some are even below the sample mean $\bar{x} + \bar{y} = 0$; and when x is below -1, several of the $x + y$ values are above 1 and some are even above the sample mean $\bar{x} + \bar{y} = 0$. This comparison indicates that the dispersion of the $x + y$ values is greater in the positive correlation case since these values tend to be farther from their mean than in the zero correlation case. The histograms in the first two rows confirm this fact, as the histogram of $x + y$ in the positive correlation case has thicker left and right tails and a lower density in the middle compared to the histogram of $x + y$ in the zero correlation case.

How about the negative correlation case (third row) in figure 7.14? We get something very different from the positive correlation case. In the histogram, there is very little dispersion in $x + y$ compared to the dispersion for the zero correlation case. In the scatter plot, the points are within a pretty narrow band of the possible $x + y$ lines, with almost all of them lying between the $x + y = -1$ and $x + y = 1$ lines. When x_i is above its mean, it's more likely that the y_i-value is below its mean, so there are counteracting effects on the sum $x_i + y_i$, making it more likely that it is *closer* to its mean $\bar{x} + \bar{y}$ than in the case with zero correlation. Similarly, when x_i is below its mean, it's more likely that the y_i-value is above its mean, again making it more likely that $x_i + y_i$ is closer to its mean $\bar{x} + \bar{y}$ than in the case with zero correlation. From the formula $s_v^2 = s_x^2 + s_y^2 + 2s_{xy}$, a negative correlation ($s_{xy} < 0$) leads to $s_v^2 < s_x^2 + s_y^2$. The variance of the sum of the variables is less than the sum of the variances of the variables in this case. The reduction in variance arises precisely because of the negative relationship

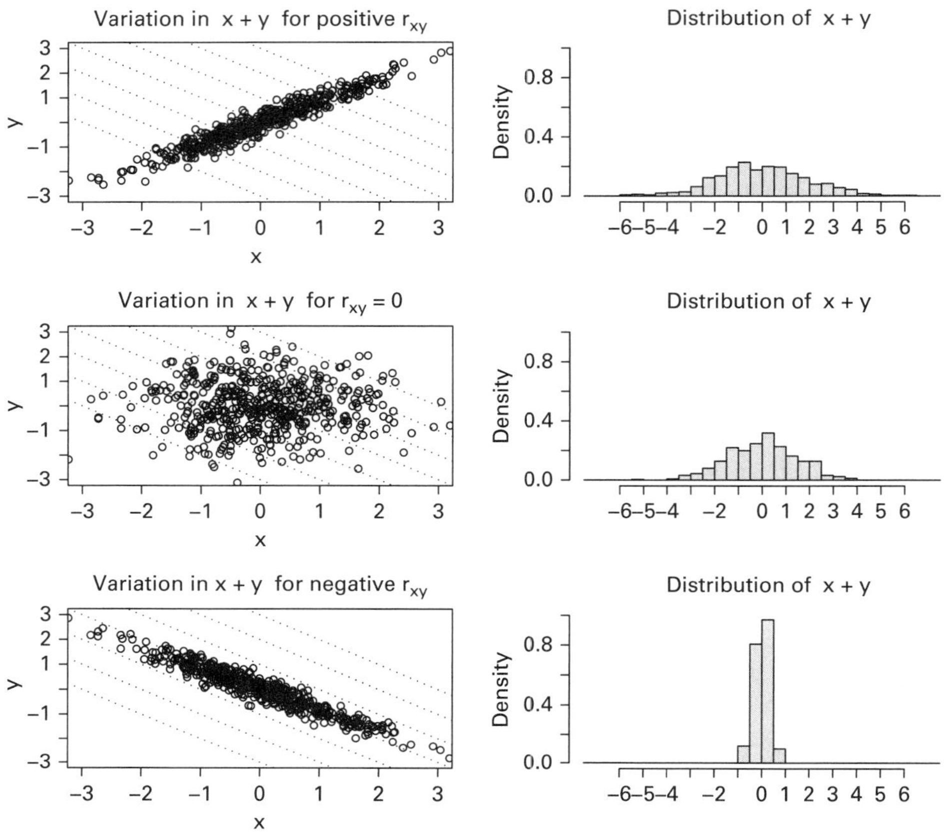

Figure 7.14
Variation in $x+y$ for different sample correlations

between x and y, where the tendency of the variables to be on opposite sides of their means also leads to a tendency for $x+y$ to be closer to its mean.

Next, we discuss the other result in property (ii), which involves the sample variance of the difference of two variables. When $v=x-y$, the resulting sample variance is $s_v^2 = s_x^2+s_y^2-2s_{xy}$. In the case of no correlation between x and y ($s_{xy}=0$), we have $s_v^2=s_x^2+s_y^2$, so the sample variance of the difference of two variables is the sum of the variances of the two variables. When there is correlation, the reasoning for the presence of the $-2s_{xy}$ is similar to that given previously. Let's start with the positive correlation ($s_{xy}>0$) case. When x and y are positive correlated, their values tend to be on the same side of their respective means, so the difference $x-y$ tends to be smaller in magnitude than it would be in the case of zero correlation. The positive relationship between x and y leads to counteracting effects for $x-y$, in contrast to the exaggerating effects for $x+y$, and causes the sample variance s_v^2 to be lower, due to the $-2s_{xy}$ term, than it would be in the zero correlation case. On the other hand, when x and y are negatively correlated, they tend to be on opposite sides of their respective means, which means that the difference $x-y$ tends to be larger in magnitude compared to the case of zero correlation. The negative relationship between x and y increases the dispersion of $x-y$, with the term $-2s_{xy}$ being positive since $s_{xy}<0$.

Example 7.18 (Earnings of siblings) *Suppose that we have data on the earnings of a sample of adult siblings, where x is weekly earnings for the older sibling and y is weekly earnings for the younger sibling. For these variables, a positive correlation ($r_{xy} > 0$) is expected due to common factors like similar parenting, similar educational background, and similar socioeconomic background. Thus, the variance of their combined weekly earnings, $v = x + y$, is larger than the sum of the variances of their individual weekly earnings: $s_v^2 = s_{x+y}^2 > s_x^2 + s_y^2$. The positive correlation between the siblings' earnings leads to a greater dispersion in the sum of their earnings. Compared to two randomly chosen individuals from the population (i.e., nonsiblings), in which case there is zero correlation, there is a larger variance for the sum of the earnings of adult siblings.*

How about the average *of the siblings' weekly earnings? For $v = \frac{x+y}{2} = \frac{1}{2}x + \frac{1}{2}y$, the sample variance of v is*

$$s_v^2 = s^2_{\frac{1}{2}x+\frac{1}{2}y} = \frac{1}{4}s_x^2 + \frac{1}{4}s_y^2 + 2\left(\frac{1}{2}\right)\left(\frac{1}{2}\right)s_{xy} = \frac{1}{4}s_x^2 + \frac{1}{4}s_y^2 + \frac{1}{2}s_{xy}.$$

How about the difference *between the siblings' weekly earnings, such as $v = x - y$ (older sibling's wages minus younger sibling's wages)? For $v = x - y$, the sample variance of v is*

$$s_v^2 = s_{x-y}^2 = s_x^2 + s_y^2 - 2s_{xy},$$

with the positive correlation in their wages leading to a decreased dispersion in the difference compared to the zero correlation case.

Example 7.19 (Two-stock portfolio) *Suppose that we have data on the returns for two stocks, with x denoting the return for stock A and y denoting the return for stock B. These could be monthly returns, annual returns, or returns for some other time interval. We won't specify the time interval for now, in the interest of keeping the discussion general. By applying linear combinations, these data can be used to see how a particular two-stock portfolio would have performed over the same time period. Specifically, consider the two-stock portfolio where fraction a is invested in stock A and the remainder, $(1 - a)$, is invested in stock B, with a being a constant between 0 and 1. Then the return on the two-stock portfolio is the linear combination*

$$v = ax + (1 - a)y.$$

Applying proposition 7.4, the sample mean and the sample variance of the return on the two-stock portfolio are

$$\bar{v} = a\bar{x} + (1 - a)\bar{y}$$

and

$$s_v^2 = a^2 s_x^2 + (1 - a)^2 s_y^2 + 2a(1 - a)s_{xy}.$$

The sample mean of the portfolio return is a weighted average of the sample means for the two stocks, with weight a placed on stock A's sample mean and weight $1 - a$ placed on stock B's sample mean. The sample variance s_v^2 is a measure of the risk associated with the portfolio since it tells us how much variability is in the portfolio's observed returns. Part of this sample variance comes from the sample variances of the individual stocks, reflected by the $a^2 s_x^2$ and $(1 - a)^2 s_y^2$ terms in the s_v^2 formula. But there is also a third term, $2a(1 - a)s_{xy}$. If the stocks' returns are positive correlated, the $2a(1 - a)s_{xy}$ term is positive since $s_{xy} > 0$

(a and $1-a$ are also positive since $0<a<1$). The stock returns move together in this case, with the returns tending to be on the same side of their respective means, leading to an increased variance or risk of the two-stock portfolio relative to the case of no correlation. On the other hand, if the stocks' returns are negative correlated, the $2a(1-a)s_{xy}$ term is negative and leads to a decreased variance or risk of the two-stock portfolio, relative to the case of no correlation.

Interestingly, even in the case of zero correlation between the stocks' returns ($s_{xy}=0$), there is a diversification effect that affects the variance or risk of the two-stock portfolio. Using the fact that $k^2<k$ for any positive constant $k<1$, we can show this effect as follows:

$$s_v^2 = a^2 s_x^2 + (1-a)^2 s_y^2 < a s_x^2 + (1-a) s_y^2 \le a \max(s_x^2, s_y^2) + (1-a) \max(s_x^2, s_y^2)$$
$$= \max(s_x^2, s_y^2),$$

which means that s_v^2 is less than the maximum of the two sample variances s_x^2 and s_y^2. For this case of zero correlation, the same must then also be true for the standard deviations: $s_v < max(s_x, s_y)$.

Let's consider real-world examples of two-stock portfolios from the **sp500** *data set. The following table shows sample means and standard deviations for three stocks (Bank of America (BAC), Wells Fargo (WFC), and ConocoPhillips (COP)) and also two different two-stock portofolios: one equally weighted ($a=1-a=0.5$) between BAC and WFC and one equally weighted ($a=1-a=0.5$) between COP and WFC:*

	$\bar{x}$	s_x
BAC	0.01295	0.10530
WFC	0.01351	0.08157
COP	0.01093	0.08173
$1/2BAC+1/2WFC$	0.01323	0.08607
$1/2COP+1/2WFC$	0.01222	0.06822

For the BAC-WFC portfolio, we can confirm that for the sample means, $0.01323 = (0.5)(0.01295)+(0.5)(0.01351)$. *For the sample standard deviation, using the sample correlation $r_{BAC,WFC}=0.692$ (example 7.13) and the fact that $s_{BAC,WFC}=r_{BAC,WFC}s_{BAC}s_{WFC}$, we have*

$$s^2_{\frac{1}{2}BAC+\frac{1}{2}WFC} = 0.25 s^2_{BAC} + 0.25 s^2_{WFC} + 0.5 s_{BAC,WFC} = (0.25)(0.10530)^2$$
$$+ (0.25)(0.08157)^2 + (0.5)(0.692)(0.10530)(0.08157) \approx 0.0074073.$$

The standard deviation is $s_{\frac{1}{2}BAC+\frac{1}{2}WFC} = \sqrt{0.0074073} \approx 0.08607$. The calculations are similar for $s_{\frac{1}{2}COP+\frac{1}{2}WFC}$. The sample standard deviation of the COP-WFC portfolio is considerably lower than the sample standard deviation of the BAC-WFC portfolio, which occurs for two reasons: (1) COP has a lower variance than BAC, and (2) the sample correlation between COP and WFC is lower than the sample correlation between BAC and WFC.

Other weights can be considered, in addition to the $a=0.5$ choice given here. Figure 7.15 shows the sample means and sample standard deviations for different weights for the

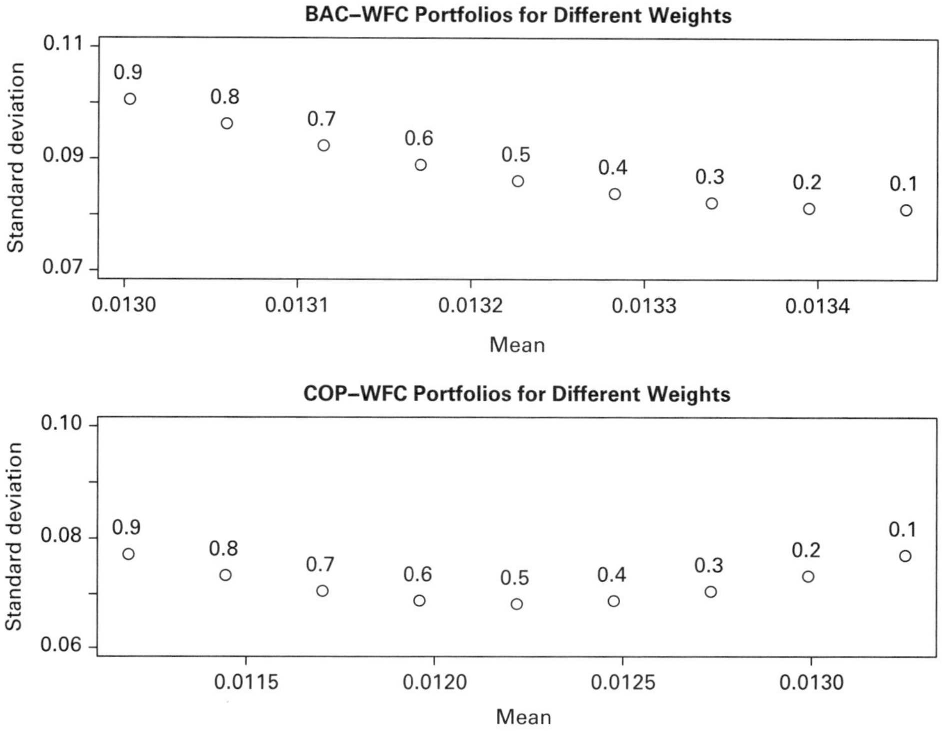

Figure 7.15
Means and standard deviations for weighted two-stock portfolios

two stocks. The top graph plots the sample standard deviation versus the sample mean for the BAC-WFC portfolio with possible weights $a \in \{0.1, 0.2, \ldots, 0.9\}$, *where a is the weight on BAC. The bottom graph plots the sample standard deviation versus the sample mean for the COP-WFC with possible weights* $a \in \{0.1, 0.2, \ldots, 0.9\}$, *where a is the weight on COP.*

7.2.6 Linear Combination of More Than Two Variables

What if a linear combination consists of *more than two* variables? The results of proposition 7.4 can be generalized to additional variables. First, consider a linear combination of three variables x, y, and z, with

$$v = k + ax + by + cz.$$

The observed sample is $\{(x_1, y_1, z_1), (x_2, y_2, z_2), \ldots, (x_n y_n, z_n)\}$, with corresponding linear combinations $\{v_1, v_2, \ldots, v_n\}$. One can show that

$$\bar{v} = k + a\bar{x} + b\bar{y} + c\bar{z},$$

analogous to property (i) of proposition 7.4. For the sample variance s_v^2, the expression is more complicated:

$$s_v^2 = a^2 s_x^2 + b^2 s_y^2 + c^2 s_z^2 + 2ab s_{xy} + 2ac s_{xz} + 2bc s_{yz}.$$

The sample variance of v involves the variances of the three variables individually, through the $a^2s_x^2 + b^2s_y^2 + c^2s_z^2$ component, but it also involves each of the possible pairwise covariances among the three variables. With three variables, there are $\binom{3}{2} = 3$ different pairs of variables and, thus, three different covariances.

These results can be generalized to an even larger number of variables in the linear combination. Proposition 7.5 provides the general results for the case of $m \geq 2$ variables.

Proposition 7.5 *Suppose k and $a_1, a_2, \ldots, a_m$ are known constants, and*

$$v = k + a_1x_1 + a_2x_2 + \cdots + a_mx_m = k + \sum_{j=1}^{m} a_jx_j$$

is a linear combination of the m variables $x_1, x_2, \ldots, x_m$. The descriptive statistics for the sample $\{v_1, v_2, \ldots, v_n\}$ have the following relationships to the descriptive statistics for the sample of observations for the variables $x_1, x_2, \ldots, x_m$:

i. *(sample mean)* $\bar{v} = k + a_1\overline{x_1} + a_2\overline{x_2} + \cdots + a_m\overline{x_m} = k + \sum_{j=1}^{m} a_j\overline{x_j}$

ii. *(sample variance)* $s_v^2 = \sum_{j=1}^{m} a_j^2 s_{x_j}^2 + 2\sum_{j=1}^{m-1}\sum_{\ell=j+1}^{m} a_j a_\ell s_{x_j x_\ell}$

iii. *(sample standard deviation)* $s_v = \sqrt{s_v^2} = \sqrt{\sum_{j=1}^{m} a_j^2 s_{x_j}^2 + 2\sum_{j=1}^{m-1}\sum_{\ell=j+1}^{m} a_j a_\ell s_{x_j x_\ell}}$

Property (i) remains simple, with the sample mean of v having the same linear relationship with the sample means of the variables as in the original linear combination. As seen previously for the case of three variables ($m = 3$), property (ii) says that the sample variance of the linear combination v involves each of the sample variances of the m variables through the $\sum_{j=1}^{m} a_j^2 s_{x_j}^2$ summation term, but also all possible pairwise covariances among the m variables through the double summation term. There are $\binom{m}{2}$ terms in the double summation, corresponding to the $\binom{m}{2}$ possible variable pairs among the m variables. Finally, property (iii) says that the sample standard deviation, as always, is equal to the square root of the sample variance.

7.3 Correlation Is Not Causation

In describing the sample covariance or the sample correlation between two variables x and y, we have been careful to use phrases like "the positive association between x and y" or "the positive relationship between x and y" rather than saying "x causes y to be larger" or "x has a positive effect on y." It is important to stress that there is no way to tell whether there is causality between x and y, with x causing y or vice versa, just by looking at a sample of bivariate data. To establish a causal relationship, additional information about how the data are generated is required.

Consider the example of education and earnings. While there is a positive relationship between education and earnings, as measured by their positive correlation, we would not necessarily want to conclude that higher education leads to higher earnings. There may be some other factor that is related to both of these variables and, therefore, may be causing the observed positive correlation. For instance, parental income is positively related to an individual's educational attainment (e.g., more resources to go to college) and positively related to an individual's earnings (e.g., through a parent's professional connections). Even in the absence of any causal relationship between education and earnings, this outside factor would cause a positive relationship between the variables. Another possible outside factor is whether an individual has a "go-getter" attitude. Even though this factor is not something observed in a data set, we can still think about what it implies for the causality argument. If a go-getter is both more likely to attain more education and to strive for higher earnings (and a nongo-getter is less likely to do either), this outside factor would also lead to a positive correlation between education and earnings in the sample, even if there were no causal relationship.

The type of problem described for the education-earnings relationship is the norm rather than the exception with observational data. It is often difficult to establish causality between two variables unless we can rule out the types of outside related factors discussed here. While a complete treatment of causal inference is beyond the scope of this book, we can discuss one case in which causality *can* be reasonably established. Looking back at example 2.1, we have a situation in which one of the variables is randomly created through an experiment. Recall that Widgets.com randomized whether a user received e-mail A, e-mail B, or no e-mail at all. Suppose that we consider only the users who receive e-mail A or e-mail B and collect the resulting data on purchases. If $x=0$ for e-mail A recipients, $x=1$ for e-mail B recipients, and y denotes the purchases, a positive correlation between x and y indicates that average purchases among e-mail B recipients are higher than among e-mail A recipients. Moreover, we could argue here that the observed positive relationship is causal because of the randomization used to create the x variable. The randomization rules out the possibility of some outside factor being related to both x and y. Since x is generated randomly, with no regard to any other factor like age or income, there should not be any relationship between x and an outside factor.

The value of randomization for establishing causality explains why randomized **A/B testing** is so prevalent in many areas, including medical research, business improvements in the tech industry, and field experiments in economic research. In medical research, for instance, the standard protocol is to randomly assign one subset of individuals to a treatment group (i.e., receiving a new medicine or a new treatment) and another subset of individuals to a control group (i.e., not receiving the medicine/treatment). The randomization is important since we don't want the patients to be able to choose whether to receive the treatment, as that choice itself could be related to many outside factors whereas the randomization, by construction, is not.

Notes

1. Alternatively, a single argument can be used for `plot` if it consists of a data frame with the x variable as the first column and the y variable as the second column. For figure 7.5, the appropriate argument would be `cps[,c("educ","wagehr")]`.
2. The sample correlation between x and $-x$ equals -1 since $r_{x,-x}=\frac{s_{x,-x}}{s_x s_{-x}}=\frac{-s_{x,x}}{s_x s_x}=\frac{-s_x^2}{s_x^2}=-1$, using the facts that $s_{x,-x}=-s_{x,x}$ and $s_{-x}=s_x$.

Exercises

1. Use the **cps** data set for this question, focusing on the subsample of 2,809 employed individuals.
   a. Provide a table of the joint sample counts for *hourly* (rows) and *race* (columns).
   b. Provide a table of the joint sample proportions for *hourly* (rows) and *race* (columns).
   c. Same as (b), but condition on *race* so that the column values sum to 1. What does the table say about the relationship between race and being paid hourly?
   d. Draw a graph similar to figure 7.1 with a bar plot of hourly-wage status by race. Compared to figure 7.1, this graph will have only two categories ("Hourly" and "Nonhourly") on the x-axis.
   e. Draw box plots of weekly earnings (*earnwk*) by hourly-wage status.
   f. Draw the two densities of weekly earnings (*earnwk*), one for hourly workers and one for nonhourly workers, on the same graph. Use different line styles or colors to differentiate the two densities.
   g. Based on (e) and (f), how does the sample mean of weekly earnings for hourly workers compare to that of nonhourly workers? How about the sample standard deviation? How about the sample 90% quantile?
2. Use the **sp500** data set for this question. The variable *IDX* represents the monthly return for the overall stock market (as measured by the S&P 500 index).
   a. Use the command `sp500$mkt <- ifelse(sp500$IDX>=0, "Up", "Down")` to create a categorical variable *mkt* whose value is `Up` when the market has a positive monthly return and `Down` when the market has a negative monthly return.
   b. Provide a table of *mkt*. What proportion of observations have *mkt* equal to `Up`?
   c. Use the `summary` and `sd` functions to provide summary statistics of Apple's monthly returns (*AAPL*).
   d. Repeat (c) on the two subsamples of observations corresponding to *mkt* = `Up` and *mkt* = `Down`, using the `tapply` function. How do the sample means and the sample standard deviations compare to each other for the two subsamples?
   e. On the same graph, draw two densities of *AAPL*, one for the *mkt* = `Up` subsample and one for the *mkt* = `Down` subsample. Use different line styles or colors to differentiate between the two densities.
   f. Defining the binary variable *mktup* to be 1 if *mkt* = `Up` and 0 if *mkt* = `Down`, use proposition 7.2 to calculate the sample correlation between *AAPL* and *mktup* based on the sample means and the sample standard deviations of *AAPL* and *mktup*. The necessary information is available from the answers to (b), (c), and (d).
3. For a sample size of six observations and two variables x and y, draw a possible scatter plot of data for which $(x_i - \bar{x})(y_i - \bar{y})$ is positive for every $i = 1, \ldots, 6$, but x and y are not perfectly correlated.
4. Use the data set **auctions** for this question. The data set consists of 684 eBay auctions for Apple iPod Mini devices in June and July 2006. The binary variables *new*, *used*,

and *refurb* indicate the condition of the device (e.g., a new device has $new = 1$ and $used = refurb = 0$). For this question, focus only on the subsample of 624 auctions of used and new devices, so drop those with $refurb = 1$.

a. Draw side-by-side box plots of auction sales prices (*finalprice*, in dollars) for used devices and new devices.

b. What is the sample correlation between *finalprice* and *new*?

c. Draw a scatter plot of *finalprice* versus the number of bidders (*bidders*).

d. What is the sample correlation between *finalprice* and *bidders*?

e. For the plot in (c), does the variability of *finalprice* change for larger values of *bidders*?

f. For the plot in (c), does the average *finalprice* appear to always increase as the value of *bidders* increases? Explain.

5. For a sample of firms, there are data on $x =$ electricity purchased (in kilowatt-hours, or kwh) and $y =$ firm revenues (in dollars). What are the units of the following?

a. Sample average of electricity purchased

b. Sample variance of firm revenues

c. Sample covariance between electricity purchased and firm revenues

d. Sample correlation between electricity purchased and firm revenues

6. Last week, Widgets.com had daily sales of 24, 13, 19, 21, 12, 28, and 18 widgets.

a. What is the sample median of daily widget sales?

b. What is the sample mean of daily widget sales?

c. What is the sample standard deviation of daily widget sales?

d. If daily fixed cost is \$100 and profit margin per widget is \$20, daily profits are given by $-100 + 20\,widgets$. Using the results for linear transformations of univariate data, what are the sample median, the sample mean, and the sample standard deviation of daily profits?

e. What is the sample correlation between daily widget sales and daily profits?

f. What is the sample covariance between daily widget sales and daily profits?

7. A survey of chief executive officers collected data on $x =$ salary (in thousands of dollars) and $y =$ education (in years) for each person. The sample has $\bar{x} = 200$, $s_x = 30$, $\bar{y} = 17$, and $s_y = 3$. The sample covariance between x and y is $s_{xy} = 50$.

a. What is the sample correlation between salary and education?

b. What is the sample covariance between salary in dollars (not thousands of dollars) and education?

c. What is the sample correlation between salary in dollars and education?

8. You have data on the monthly returns of two stocks A and B, given by the variables x and y, respectively. The sample variance for stock A is 0.006 ($s_x^2 = 0.006$), and the sample variance for stock B is 0.008 ($s_x^2 = 0.008$).

a. What must be true about the sample correlation between x and y for the average return $\frac{1}{2}(x+y)$ to have a sample variance less than or equal to 0.007?

b. What must be true about the sample correlation between x and y for the average return $\frac{1}{2}(x+y)$ to have a sample variance less than or equal to 0.006?

c. What must be true about the sample correlation between x and y for the difference in returns $(x-y)$ to have a sample variance less than 0.012?

9. You have data for 1,230 individuals, including their education (in years, denoted as *educ*) as well as the education, in years, for each individual's mother (*motheduc*) and father (*fatheduc*). The sample correlation matrix is:

	educ	*motheduc*	*fatheduc*
educ	1.000	0.452	0.440
motheduc		1.000	0.599
fatheduc			1.000

The sample variance of *educ* is 5.543, the sample variance of *motheduc* is 5.190, and the sample variance of *fatheduc* is 10.653. The sample covariance between *motheduc* and *fatheduc* is 4.454.

a. What is the sample covariance between *educ* and *motheduc*? What are the units?

b. What is the sample variance of the sum of *motheduc* and *fatheduc*?

c. What is the sample variance of the average of *motheduc* and *fatheduc*?

d. What is the sample variance of the difference of *motheduc* and *fatheduc*?

e. Explain why the sample variance in (b) is greater than the sample variance in (d).

f. What additional information is needed to calculate the sample variance of the average of all three education variables (*educ*, *motheduc*, and *fatheduc*)?

10. Use the **exams** data set for this question.

a. Draw a scatter plot of the second exam score (*exam*2) versus the first exam score (*exam*1), with lines drawn through the sample means of the two variables. Specify the range of both axes as being between 0 and 100.

b. Using the appropriate *R* commands (rather than counting points in the scatter plot), how many points are in each of the four quadrants of the scatter plot in (a)?

c. What is the sample correlation between *exam*1 and *exam*2?

d. Based on (c), which of the following would have a greater sample standard deviation: the sum of the two exam scores or the difference of the two exam scores? Answer without using *R*.

e. If you standardize both *exam*1 and *exam*2 (by de-meaning and dividing by the sample standard deviation), what would be the sample correlation between the two standardized exam scores? Answer without using *R*.

f. If you standardize both *exam*1 and *exam*2 as in (e), what would be the sample variance of the sum of the two standardized exam scores? Answer without using *R*.

g. Create two new variables *exam1_std* and *exam2_std* with the standardized exam scores. Suppose that the instructor calculates a composite score (*score*) as the sum of 0.75 times the higher standardized exam score and 0.25 times the lower standardized exam score. Create the new variable *score*. What is the sample mean and the sample standard deviation of the composite scores? If the instructor would like the top 20 students in the class to get an A, what would be the appropriate cutoff for the composite scores?

11. a. *For two variables x and y, show that the sample covariance of x and y is

$$s_{xy} = \frac{n}{n-1}\left(\overline{xy} - \bar{x}\bar{y}\right),$$

where $\overline{xy} = \frac{1}{n}\sum_{i=1}^{n} x_i y_i$ is the sample average of the $x_i y_i$ values.

b. Suppose that $x \in \{0, 1\}$ and $y \in \{0, 1\}$ are binary variables. Using the result in (a) and the fact that a binary x has sample variance $s_x^2 = \frac{n}{n-1}\bar{x}(1-\bar{x})$, provide a formula for r_{xy} in terms of $\bar{x}$, $\bar{y}$, and $\overline{xy}$.

c. A city has two newspapers, the *Daily Bugle* and the *Daily Planet*. The binary variable x indicates whether a local company advertises in the *Daily Bugle* in a given year (1 means yes, 0 means no), and the binary variable y indicates whether a local company advertises in the *Daily Planet* in a given year (1 means yes, 0 means no). The following table describes the advertising behavior of a sample of 80 local companies in a given year:

		y = 0	y = 1
x	0	24	28
	1	18	10

Using the result from (b), what is the sample correlation between x and y?

12. Use the **sp500** data set for this question. If the data are not already visible in the top-left window of *RStudio*, use the command `View(sp500)`.

a. First, focus on the first 20 stocks that appear in the spreadsheet. Ignore *IDX*, which is in the second column, so the first 20 stocks are given by the stock tickers *AAPL* through *APA*. Output the descriptive statistics for these 20 stocks using the command `summary(sp500[,3:22])`. What is the sample mean for *AMD*?

b. Use the command `sapply(sp500[,3:22], sd)` to calculate sample standard deviations for the stocks in (a). The `sapply` function applies the `sd` function to each of the columns of the first argument `sp500[,3:22]`. Which stock has the highest sample standard deviation? Which stock has the lowest sample standard deviation?

c. Create a new variable that contains the monthly returns for a portfolio with equal (1/2) weights on the first two stocks (*AAPL*, *ABMD*). What are the sample mean and the sample standard deviation for this two-stock portfolio?

d. Create a new variable that contains the monthly returns for a portfolio with equal (1/3) weights on the first three stocks (*AAPL*, *ABMD*, and *ABT*). What are the sample mean and the sample standard deviation for this three-stock portfolio?

e. *Continue this process to get an equally weighted 4-stock portfolio with the first 4 stocks, an equally weighted 5-stock portfolio with the first 5 stocks, and so on through an equally weighted 20-stock portfolio with the first 20 stocks. For each portfolio, calculate the sample mean and the sample standard deviation. Then, make two plots: (i) sample mean versus the number of stocks (ranging from 2 to 20) in the portfolio and (ii) sample standard deviation versus the number of stocks (ranging from 2 to 20) in the portfolio. (Hint: A useful function is `rowMeans`, which creates a vector that averages across columns of a data frame. For example, the command `portfolio <- rowMeans(sp500[,3:8])` creates a `portfolio` variable corresponding to an equally weighted portfolio consisting of the first six stocks, as shown in columns 3 through 8.)

f. *Rather than using the first 20 stocks as in (e), instead choose 20 stocks randomly *without replacement* from the full set of stocks, contained in columns 3 through 268 of the data frame. Calculate the sample mean and the sample standard deviation for the equally weighted 20-stock portfolio. Use a loop to do this 1,000 times, randomly picking 20 stocks each time, and store the sample means and the sample standard deviations along the way. Plot the histogram and/or smoothed density of the 1,000 sample means. Plot the histogram and/or smoothed density of the 1,000 sample standard deviations.

8

Discrete Random Variables

Chapter 5 introduced the concept of numerical variables, including both discrete numerical variables and continuous numerical variables. Then, chapter 6 introduced several descriptive statistics for summarizing numerical variables. At this point, we take a step back and, applying the concepts of probability theory from chapter 2, formally model the process by which a numerical variable arises and is observed. This chapter provides the theoretical framework for discrete numerical variables, and chapter 10 introduces the theoretical framework for continuous numerical variables. While the approaches for the two types of variables have similarities, the mathematics are sufficiently different that it is useful to consider them separately. For instance, while discrete summations are used for much of the analysis in the discrete case, integration is necessary for much of the analysis in the continuous case. Although categorical variables are not covered explicitly in this chapter, we have already seen that categorical variables can be represented by one or more discrete binary numerical variables, so the framework discussed in this chapter is applicable.

8.1 Using Sample Proportions to Calculate Descriptive Statistics

Example 8.1 (Car repairs) *Suppose that a car dealership sells n new cars and keeps track of how often they need to repair those cars during the three-year warranty period. The sample proportions and sample counts for the possible values,* $\{0, 1, 2, 3, 4\}$, *of the repair variable x are provided in the following table:*

# of repairs on a car (x)	0	1	2	3	4
Proportion of cars	0.26	0.34	0.24	0.12	0.04
Number of cars	$0.26n$	$0.34n$	$0.24n$	$0.12n$	$0.04n$

For example, if $n = 1{,}000$, *there would be* 260 *cars with zero repairs,* 340 *cars with one repair, and so on.*

Thinking about the sample mean of x for n cars, there are two alternative ways to calculate it. The first method uses the observation values themselves, as in the original definition of $\bar{x}$:

$$\bar{x} = \frac{1}{n}\sum_{i=1}^{n} x_i = \frac{1}{n}(0 \times 0.26n + 1 \times 0.34n + 2 \times 0.24n + 3 \times 0.12n + 4 \times 0.04n)$$

$$\implies \bar{x} = 0 \times 0.26 + 1 \times 0.34 + 2 \times 0.24 + 3 \times 0.12 + 4 \times 0.04 = 1.34.$$

The second method directly uses the sample proportions, as suggested by the expression for $\bar{x}$ *after the* $\implies$ *above. The values* 0.26, 0.34, 0.24, 0.12, *and* 0.04 *are, respectively, the sample proportions corresponding to* $x=0$, $x=1$, $x=2$, $x=3$, *and* $x=4$. *The expression for* $\bar{x}$ *is, therefore, a summation of the products of sample proportion times variable value, taken over all possible values of the variable.*

Similarly, the sample variance of the number of repairs (per car) is

$$s_x^2 = \frac{1}{n-1}\sum_{i=1}^{n}(x_i - \bar{x})^2$$

$$= \frac{1}{n-1}((0-1.34)^2 \times 0.26n + (1-1.34)^2 \times 0.34n + (2-1.34)^2 \times 0.24n$$
$$+ (3-1.34)^2 \times 0.12n + (4-1.34)^2 \times 0.04n)$$
$$= \frac{n}{n-1}((0-1.34)^2 \times 0.26 + (1-1.34)^2 \times 0.34 + (2-1.34)^2 \times 0.24$$
$$+ (3-1.34)^2 \times 0.12 + (4-1.34)^2 \times 0.04)$$
$$\approx (0-1.34)^2 \times 0.26 + (1-1.34)^2 \times 0.34 + (2-1.34)^2 \times 0.24$$
$$+ (3-1.34)^2 \times 0.12 + (4-1.34)^2 \times 0.04 = 1.2244.$$

The last line has an approximation sign due to the $\frac{1}{n-1}$ *scaling in the* s_x^2 *formula, leading to a scaling by* $\frac{n}{n-1}$ *rather than* $\frac{n}{n}$. *As with the sample mean, the last line of the calculation for* s_x^2 *indicates that sample proportions can be used directly to calculate the sample variance. In this case, the sample variance is approximately a summation of the products of sample proportion times the squared difference between the variable value and the sample mean, taken over all possible values of the variable.*

To generalize the ideas from example 8.1, some additional notation is required. Suppose that x is a discrete variable with K possible outcomes, where K can be infinite. The possible outcomes are

$$x_1^*, x_2^*, \ldots, x_K^*$$

in the case of finite K, and

$$x_1^*, x_2^*, \ldots, x_k^*, \ldots$$

in the case of $K = \infty$. The sample proportions associated with these possible outcomes, listed in order, are

$$p_1, p_2, \ldots, p_K$$

in the case of finite K, and

$$p_1, p_2, \ldots, p_k, \ldots$$

in the case of $K=\infty$. Finally, the notation $\sum_k$, which is a summation over k, represents $\sum_{k=1}^{K}$ in the case of finite K and $\sum_{k=1}^{\infty}$ in the case of infinite K.

Proposition 8.1 *Suppose that x is a discrete variable with K possible outcomes, where K may be infinite. The sample mean $\bar{x}$ and the sample variance s_x^2 are related to the sample proportions as follows:*

$$\bar{x}=\sum_k x_k^* p_k$$

and

$$s_x^2=\frac{n}{n-1}\sum_k (x_k^*-\bar{x})^2 p_k \approx \sum_k (x_k^*-\bar{x})^2 p_k.$$

8.2 Random Variables and Discrete Random Variables

Recall the sample space $\mathcal{S}$ introduced in chapter 2. The main theoretical construct used to model variables is known as a **random variable** and is formally defined as in definition 8.1.

Definition 8.1 *A* ***random variable*** *X is a function that maps each outcome of sample space $\mathcal{S}$ to a number.*

While definition 8.1 indicates that X is a function, it is standard to suppress the argument when referring to the random variable X, in the interest of brevity. If e is a simple event (outcome) from $\mathcal{S}$, it is more accurate to say that $X(e)$ is some numerical value. Usually, however, the argument e is suppressed, with X instead of $X(e)$ used to denote the random variable. The standard convention is to use a capital letter, like X, to denote a random variable.

For modeling discrete variables, a specific type of random variable known as a **discrete random variable** is used, as in definition 8.2.

Definition 8.2 *A* ***discrete random variable*** *is a random variable with a finite or countable set of possible values.*

The notation for the possible values, referred to in definition 8.2, is the same as introduced in section 8.1, with $x_1^*, x_2^*, \ldots, x_K^*$ for finite K and $x_1^*, x_2^*, \ldots, x_k^*, \ldots$ for infinite K.

Here are some examples of discrete random variables:

- Coin toss: $\mathcal{S}=\{H,T\}$, with $X=1$ if H and $X=0$ if T
- Roll of a die: $\mathcal{S}=\{1,2,3,4,5,6\}$, with X equal to the outcome
- Three website visitors, purchase (Y) or not (N):

$$\mathcal{S}=\{YYY, YYN, YNY, YNN, NYY, NYN, NNY, NNN\}$$

  - Total number of purchases X = the number of purchases in the three-visitor sequence
  - Indicator of at least two purchases T = 1 if the number of purchases is ≥ 2 and 0 otherwise
- Website visitors until first purchase: $\mathcal{S}=\{Y, NY, NNY, NNNY, \ldots\}$, with corresponding X values 1, 2, 3, 4, ...

- Number of firm patents: $\mathcal{S} = \{0, 1, 2, 3, \ldots\}$, with X equal to the outcome
- Income threshold: $\mathcal{S}$ containing all possible annual income values, with $X = 1$ if outcome > \$100,000 and 0 otherwise

8.2.1 A Random Variable and the Population

Recalling the concepts of an "experiment" and the "population" from earlier chapters, the experiment associated with the random variable X can be thought of as a draw of an observed variable x from the population. The lowercase x notation denotes the observed variable, a value drawn from the population, and the uppercase notation X denotes the random variable describing the population. An observed sample $\{x_1, x_2, \ldots, x_n\}$ is the result of n repetitions of the experiment, with each x_i generated by the experiment of drawing an observation from the population associated with random variable X. In the example involving the number of firm patents, the population can be thought of as the number of patents for all possible firms in existence. An observation x_i drawn from that population is associated with a specific firm, and its value is one of the possible values $x_k^* \in \{0, 1, 2, 3, \ldots\}$ associated with the random variable X.

We introduce tools to characterize a discrete random variable X and, therefore, the underlying population for the variable that X represents. In the case of a discrete random variable X, we need to specify how likely each of the potential values of x_k^* is. This characterization involves specifying the probability of drawing each x_k^* value from the population associated with X. In the firm patent example, we need to specify the probabilities for $x_k^* \in \{0, 1, 2, 3, \ldots\}$, which are $P(X = 0)$, $P(X = 1)$, $P(X = 2)$, $P(X = 3)$, and so on. Whether these are *known* probabilities (i.e., known to the researcher) depends on the example. In the coin toss and die roll examples, the probabilities of the possible outcomes are known, so long as the coin and die are fair. However, in the firm patent example, the outcome probabilities are likely unknown, and we can only guess or estimate what they are after observing a sample from the population. In fact, for most cases of interest where observational data are analyzed, the actual probabilities associated with the possible outcomes are unknown. Despite that reality, it is very powerful to model the probabilities of those outcomes, providing the theoretical underpinning for the statistical analysis of discrete variables.

8.2.2 Probability Mass Function

To completely describe a discrete random variable X, the probability $P(X = x_k^*)$ is specified for every possible value x_k^*. These probabilities are collectively known as the **probability mass function (pmf)** of the random variable, as in definition 8.3:

Definition 8.3 *The* ***probability mass function (pmf)*** *of a discrete random variable X, denoted as $p_X(\cdot)$, gives the probability of each possible value of X:*

$$p_X(x_k^*) = P(X = x_k^*).$$

For each possible value x_k^*, the probability $p_X(x_k^*)$ can be determined as follows:

1. Find all of the outcomes in $\mathcal{S}$ for which $X = x_k^*$:

$$A = \{e : X(e) = x_k^*\}.$$

2. Determine $p_X(x_k^*) = P(X = x_k^*) = P(A)$.

From the Axioms of Probability (section 2.4), the pmf satisfies the following properties:

$$0 \leq p_X(x_k^*) \leq 1 \text{ for any possible value } x_k^*$$

and

$$\sum_k p_X(x_k^*) = 1.$$

That is, the pmf probabilities are all positive and sum to 1.

Here are the pmfs for four of the examples considered here:

- Coin toss: For a fair coin, $p_X(1) = P(H) = 0.5$ and $p_X(0) = P(T) = 0.5$.
- Roll of a die: For a fair die, $p_X(x_k^*) = \frac{1}{6}$ for each $x_k^* \in \{1, 2, 3, 4, 5, 6\}$.
- Three website visitors, purchase (Y) or not (N); X = total number of purchases: Assume that the purchase probability is 20% (0.2) for each visitor and the purchase behavior by each visitor is independent of other visitors. Then the pmf is

$$p_X(0) = P(NNN) = 0.8^3 = 0.512,$$

$$p_X(1) = P(YNN \cup NYN \cup NNY) = 3(0.2)(0.8)(0.8) = 0.384,$$

$$p_X(2) = P(YYN \cup YNY \cup NYY) = 3(0.2)(0.2)(0.8) = 0.096,$$

and

$$p_X(3) = P(YYY) = 0.2^3 = 0.008.$$

- Three website visitors, purchase (Y) or not (N); X = website visitors until first purchase. Assume that the purchase probability is 20% (0.2) for each visitor and the purchase behavior by each visitor is independent of other visitors. Then, the pmf is:

$$p_X(1) = 0.2,$$

$$p_X(2) = (0.8)(0.2) = 0.16,$$

$$p_X(3) = (0.8)(0.8)(0.2) = 0.128,$$

and so on, with the general formula

$$p_X(x_k^*) = (0.8)^{k-1}(0.2).$$

For these four examples, figure 8.1 provides a graph of each of the four pmfs. Each graph has probability values on the y-axis and possible x_k^* values on the x-axis, with each vertical line indicating the probability associated with any given possible value x_k^*. For instance, the pmf for the coin toss, in the top-left graph, has vertical lines drawn at $x = 0$ and $x = 1$, each with height 0.5, corresponding to the equal probabilities of tails ($X = 0$) and heads ($X = 1$). Similarly, the pmf for the die roll, in the top-right graph, has vertical lines with heights of 1/6 at each of the six possible values. In the lower-right graph for the fourth example (website visitors until first purchase), note that the x-axis only extends to a maximum of $x_k^* = 20$, even though K is infinite. Although the x-axis could be extended more, the

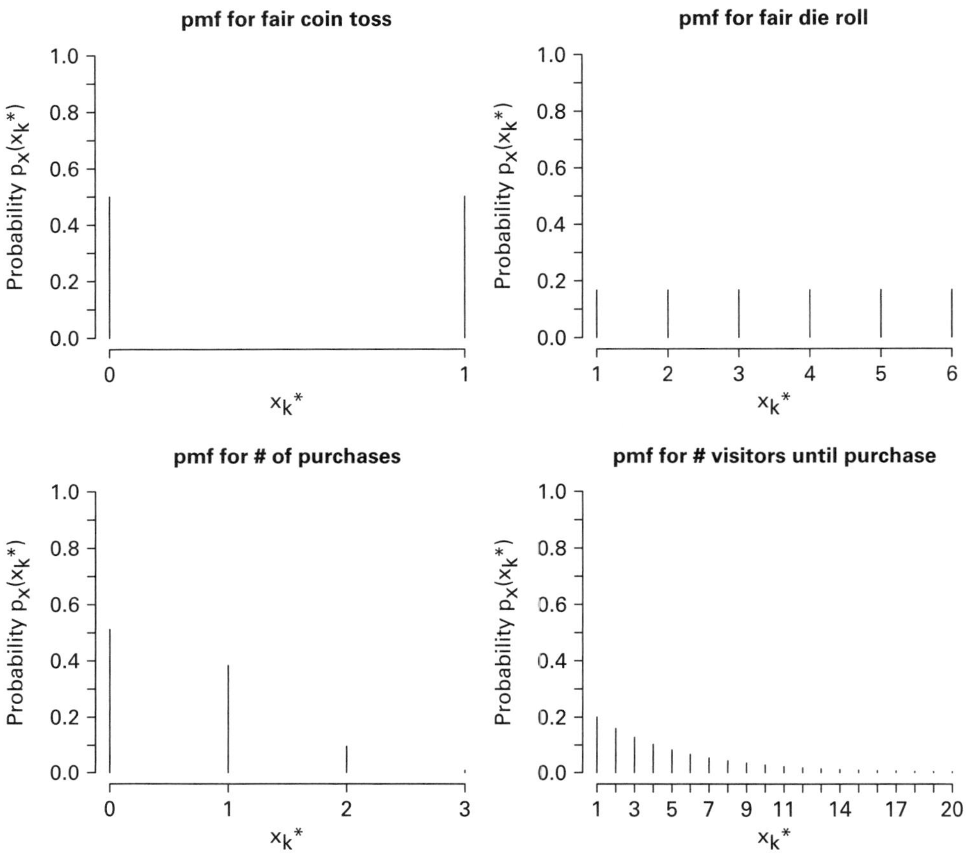

Figure 8.1
The pmfs for four examples

probability values become very close to zero as x_k^* gets larger; for example, at $x_k^* = 20$, the probability is $p_X(20) \approx 0.00288$, or 0.288%.

8.2.3 Cumulative Distribution Function

The pmf $p_X(\cdot)$ completely characterizes the discrete random variable X since it specifies the probability associated with every possible value of X. This section introduces another convenient way to characterize the discrete random variable X, known as the **cumulative distribution function (cdf)**. It may seem redundant to introduce another approach when the pmf already offers a complete characterization of X. However, the cdf provides a unifying approach that can be used for both discrete random variables and continuous random variables, whereas the pmf is useful only for the case of discrete random variables. The cdf of a discrete random variable is defined formally as in definition 8.4.

Definition 8.4 *The **cumulative distribution function (cdf)** of a discrete random variable X, denoted as $F_X(\cdot)$, gives the probability that X is less than or equal to any argument*

x_0 *of* $F_X(\cdot)$:

$$F_X(x_0) = P(X \le x_0) = \sum_{x_k^* \le x_0} p_X(x_k^*).$$

The generic argument x_0 can take any possible value on the real line. Although x_0 *can* equal a possible value x_k^*, the definition also allows for x_0 values that are between possible values of the random variable. The cdf has the following properties:

$$0 \le F_X(x_0) \le 1 \text{ for every } x_0$$

and

$$x_0 < x_1 \implies F_X(x_0) \le F_X(x_1).$$

The first property follows directly from the fact that that $F_X(x_0) = P(X \le x_0)$ is a probability and therefore must be between 0 and 1 (inclusive). The second property says that the cdf is a weakly increasing function. For $x_0 < x_1$, this property holds since

$$F_X(x_1) = P(X \le x_1) = P(X \le x_0 \text{ or } x_0 < X \le x_1) \ge P(X \le x_0) = F_X(x_0).$$

Let's consider the four examples from section 8.2.2 to illustrate the concept of cdfs.

Example 8.2 (Coin toss) *For a fair coin, the pmf and the cdf at* 0 *and* 1 *are given in the following table.*

x_k^*	$p_X(x_k^*)$	$F_X(x_k^*)$
0	0.5	0.5
1	0.5	1

How about the cdf $F_X(x_0)$ *for other values of* x_0? *For* $x_0 = -0.4$, $F_X(-0.4) = P(X \le -0.4) = 0$ *since there are no possible values of* X *below* -0.4. *The same is true of any negative* x_0 *value, so* $F_X(x_0) = 0$ *when* $x_0 < 0$. *This means that the* $F_X(\cdot)$ *function jumps from* 0 *to* 1 *exactly at the point* $x_0 = 0$. *For* $x_0 = 0.6$ *(a point between* 0 *and* 1*),* $F_X(0.6) = P(X \le 0.6) = P(X = 0) = 0.5$. *The same is true of any* x_0 *that is strictly between* 0 *and* 1, *meaning the* $F_X(\cdot)$ *function jumps from* 0.5 *to* 1 *exactly at the point* $x_0 = 1$. *Finally, for any* x_0 *value greater than* 1, $P_X(x_0) = P(X \le x_0) = P(X \le 1) = 1$.

Taking these results together, the cdf is a step function, *as shown in figure 8.2. The graph has been drawn with the x-axis extending from* -1 *to* 2, *but it should be understood that the cdf extends to the left forever (with a value of* 0*) and to the right forever (with a value of* 1*). The solid lines indicate the cdf value at any given* x_0 *value. There is a closed dot and an open dot at the* x_0 *values where the function jumps up. The closed dot is the cdf value at the corresponding point. For instance, at* $x_0 = 0$, *the cdf value is* $F_X(0) = 0.5$, *represented by the closed dot and* not *the open dot. Similarly, at* $x_0 = 1$, *the cdf value is* $F_X(1) = 1$, *represented by the closed dot, not the open dot.*

Example 8.3 (Six-sided die) *For a fair die, the pmf and the cdf for the possible outcome values are given in the following table.*

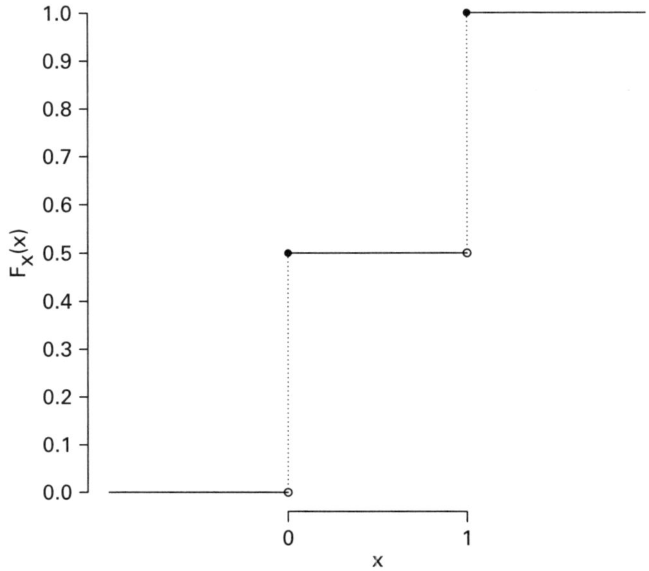

Figure 8.2
The cdf for a fair coin toss

x_k^*	$p_X(x_k^*)$	$F_X(x_k^*)$
1	1/6	1/6
2	1/6	2/6 = 1/3
3	1/6	3/6 = 1/2
4	1/6	4/6 = 2/3
5	1/6	5/6
6	1/6	1

For any number x_0 in between these outcomes, the cdf $F_X(x_0)$ is the $F_X(x_k^*)$ for the lowest x_k^* that is greater than x_0. For instance, if $x_0 = 3.7$, the cdf is $F_X(3.7) = P(X \leq 3.7) = P(X \leq 3) = 1/2$. Also, $F_X(x_0) = 0$ for $x_0 < 1$, and $F_X(x_0) = 1$ for $x_0 > 6$. Figure 8.3 shows the cdf for the fair die roll. Again, the closed dots indicate the cdf values at the x_0 points where the cdf jumps up, and the cdf extends forever to the left (with value 0) and to the right (with value 1).

Example 8.4 (Three website visitors) *For the example with three website visitors and independent purchase probabilities of 0.2, the pmf and cdf of X = number of purchases are given by the following table:*

x_k^*	$p_X(x_k^*)$	$F_X(x_k^*)$
0	0.512	0.512
1	0.384	0.896
2	0.096	0.992
3	0.008	1

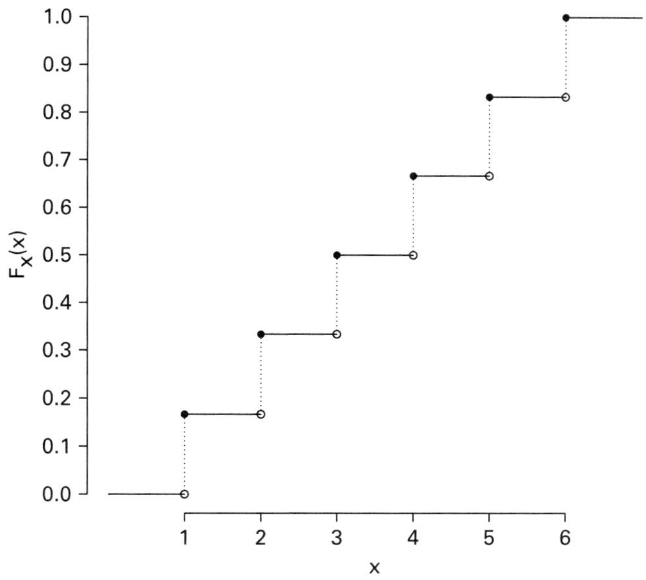

Figure 8.3
The cdf for a fair die roll

We determine the cdf values for other x_0 *values in the same way as the previous two examples. Figure 8.4 graphs the cdf for this example. The step function in this case has steps at* 0 *(for* $x_0 < 0$*),* 0.512 *(for* $0 \leq x_0 < 1$*),* 0.896 *(for* $1 \leq x_0 < 2$*),* 0.992 *(for* $2 \leq x_0 < 3$*), and* 1 *(for* $x_0 \geq 3$*).*

Example 8.5 (Website visitors until first purchase) *For the example with continual website visitors and independent purchase probabilities of* 0.2*, the pmf and cdf of* X = *number of visitors until first purchase are given by the following table:*[1]

x_k^*	$p_X(x_k^*)$	$F_X(x_k^*)$
1	0.2	0.2
2	$(0.8)(0.2) = 0.16$	0.36
3	$(0.8)^2(0.2) = 0.128$	0.488
$\vdots$	$\vdots$	$\vdots$
k	$(0.8)^{k-1}(0.2)$	$\sum_{j=1}^{k}(0.8)^{j-1}(0.2) = \frac{1-(0.8)^k}{1-0.8}(0.2) = 1-(0.8)^k$
$\vdots$	$\vdots$	$\vdots$

The cdf values for other x_0 *values can be determined in the same way as example 8.4. For example, for* $x_0 = 2.8$ *(between* 2 *and* 3*), the cdf is* $F_X(2.8) = F_X(2) = 0.36$*. As k increases, the cdf value gets closer and closer to 1. Mathematically,* $F_X(x_k^*) = 1 - (0.8)^k$*, and as* $k \to \infty$*,* $(0.8)^k \to 0$ *so that* $F_X(x_k^*) \to 1$*. The cdf never reaches 1, but its value gets closer and closer to 1 for larger k.*

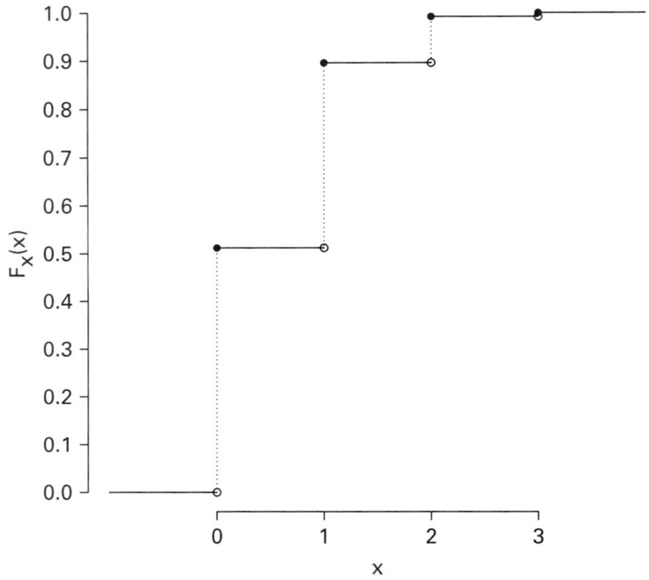

Figure 8.4
The cdf for number of purchases

8.2.4 Probability of Lying in an Interval

The cdf is a useful tool to determine the probability that a random variable X lies in a particular interval. Since discrete random variables are being considered, we need to be careful about how the end points of the interval are handled. Specifically, when specifying the interval of interest, are the lower end point and/or the upper end point included as values to consider? In mathematical terms, the inclusion of an end point depends on whether we use a strict inequality ($<$ or $>$) or a weak inequality ($\leq$ or $\geq$) for the interval.

To see why the form of the inequality matters, consider the probability that $P(X \leq c)$ versus $P(X < c)$ for some number c. The first probability, $P(X \leq c)$, is equal to $F_X(c)$ by the definition of the cdf. The second probability, however, is not necessarily equal to $F_X(c)$. If c is a possible value for X, then $P(X < c)$ does not include the probability $P(X = c)$, whereas $F_X(c) = P(X \leq c)$ does include $P(X = c)$. Instead, $P(X < c)$ is equal to the cdf evaluated at the largest possible x_k^* value that is less than c. If the notation c^- is used to denote the largest possible x_k^* value that is less than c, then $P(X < c) = F_X(c^-)$. Using this idea, proposition 8.2 indicates how the probability of X being in an interval is related to the cdf of X.

Proposition 8.2 *Suppose that a and b are two numbers, with $a \leq b$. If X is a discrete random variable with possible values $\{x_k^*\}$, then*

$$P(a < X \leq b) = P(X \leq b) - P(X \leq a) = F_X(b) - F_X(a)$$

$$P(a \leq X \leq b) = P(X \leq b) - P(X < a) = F_X(b) - F_X(a^-)$$

$$P(a < X < b) = P(X < b) - P(X \leq a) = F_X(b^-) - F_X(a)$$

$$P(a \leq X < b) = P(X < b) - P(X < a) = F_X(b^-) - F_X(a^-),$$

where a^- is the largest possible x_k^* value that is less than a, and b^- is the largest possible x_k^* value that is less than b.

There are four cases for the form of the interval since both inequalities in the probability expression can be either weak ($\leq$) or strict ($<$). For the first result, $P(X \leq b) = P(X \leq a) + P(a < X \leq b)$, which implies that $P(a < X \leq b) = P(X \leq b) - P(X \leq a)$. A similar decomposition of $P(X \leq b)$ or $P(X < b)$ for the other three cases leads to the results given in proposition 8.2.

To see how these interval probabilities work in practice, consider example 8.6.

Example 8.6 (Website visitors until first purchase) *In example 8.5, with independent purchase probabilities of* 0.2*, the cdf was provided for X = number of visitors until first purchase. The probability that X is greater than* 3 *and less than or equal to* 6 *is*

$$P(3 < X \leq 6) = p_X(4) + p_X(5) + p_X(6) = F_X(6) - F_X(3) \approx 0.2499.$$

The probability that X is greater than or equal to 3 *and less than or equal to* 6 *is*

$$P(3 \leq X \leq 6) = p_X(3) + p_X(4) + p_X(5) + p_X(6) = F_X(6) - F_X(3^-)$$
$$= F_X(6) - F_X(2) \approx 0.3779.$$

The difference in these two probabilities arises since the first probability has a strict inequality for the interval's lower end point while the second probability has a weak inequality.

8.3 Population-Descriptive Statistics

8.3.1 Population Mean or Expected Value

Proposition 8.1 stated the formula for the sample mean of a discrete variable x in terms of sample proportions:

$$\bar{x} = \sum_k x_k^* p_k,$$

where x_k^* are the possible values of variable x and p_k are the associated observed sample proportions. The sample mean is a function of the realized observations or, equivalently, the observed sample proportions. This section introduces the concept of a mean or average associated with a random variable X. Unlike the sample mean, the mean of the random variable X is *not* a function of the sample, but rather a function of the true outcome probabilities associated with the population. To stress this distinction, the term **population mean** or **population average** is used for the mean of X, as in definition 8.5.

Definition 8.5 *The* ***population mean*** *(or* ***population average*** *or* ***expected value****) of a discrete random variable X, denoted as μ_X or E(X), is*

$$\mu_X = E(X) = \sum_k x_k^* p_X(x_k^*).$$

The population mean μ_X or expected value $E(X)$ is a weighted average of the possible x_k^*, where the weights are the pmf probabilities $p_X(x_k^*) = P(X = x_k^*)$. Notice the similarity between $\sum_k x_k^* p_k$ (for the sample mean) and $\sum_k x_k^* p_X(x_k^*)$ (for the population mean), both of which are weighted averages of the x_k^* values, with the weights being the sample proportions for the sample mean and the true outcome probabilities for the population mean. Recall the frequentist interpretation of probabilities discussed in section 2.3, where we viewed the probability of an outcome as the long-run frequency or proportion of the outcome being observed over repeated experiments. Here, for the thought experiment of taking many repeated draws from the population, the probability $p_X(x_k^*)$ is the number that the sample proportion p_k approaches as the sample size n gets arbitrarily large. Since this relationship holds for each probability $p_X(x_k^*)$ and each p_k, the sample mean $\bar{x}$ should also approach the population mean μ_X as the sample size n gets arbitrarily large. This result, known as the Law of Large Numbers (LLN), is formalized in chapter 13, but it is worthwhile to introduce the intuition here. To summarize, for any given sample, it is generally the case that $p_k \neq p_X(x_k^*)$ for each of the possible outcomes and $\bar{x} \neq \mu_X$, either of which would happen only by chance, but as sample size increases, the sample proportion p_k gets closer to the probability $p_X(x_k^*)$ for each possible outcome and $\bar{x}$ gets closer to μ_X.

Example 8.7 (Union status) *Let X be a binary random variable representing union status, equal to 1 for a union worker and 0 for a nonunion worker. The population consists of the union status $x \in \{0, 1\}$ for all possible workers. The population mean is*

$$\mu_X = E(X) = 0 \times p_X(0) + 1 \times p_X(1) = p_X(1).$$

The population mean is equal to the probability of union status in the population, $p_X(1) = P(X = 1)$. As there is nothing special about union status in this example, $\mu_X = P(X = 1)$ holds for any binary random variable $X \in \{0, 1\}$.

Example 8.8 (Six-sided die) *For $X \in \{1, 2, 3, 4, 5, 6\}$ being the outcome of a fair die roll, the pmf probabilities are $p_X(x_k^*) = \frac{1}{6}$ for each outcome. The population mean or expected value of X is*

$$\mu_X = E(X) = 1 \times \frac{1}{6} + 2 \times \frac{1}{6} + 3 \times \frac{1}{6} + 4 \times \frac{1}{6} + 5 \times \frac{1}{6} + 6 \times \frac{1}{6} = \frac{21}{6} = 3.5.$$

Example 8.9 (Three website visitors) *For the example with three website visitors and independent purchase probabilities of 0.2, example 8.4 provided the pmf of X = number of purchases. The following table lists the pmf values $p_X(x_k^*)$ for $x_k^* \in \{0, 1, 2, 3\}$ and calculates the $x_k^* p_X(x_k^*)$ terms in the summation for μ_X:*

x_k^*	$p_X(x_k^*)$	$x_k^* p_X(x_k^*)$
0	0.512	$(0)(0.512) = 0$
1	0.384	$(1)(0.384) = 0.384$
2	0.096	$(2)(0.096) = 0.192$
3	0.008	$(3)(0.008) = 0.024$

Then the population mean or expected value of X is

$$\mu_X = E(X) = \sum_k x_k^* p_X(x_k^*) = 0 + 0.384 + 0.192 + 0.024 = 0.6.$$

While we have been able to calculate the population mean in the examples given here, there is no guarantee that the population mean or expected value is well defined (i.e., a finite number), as example 8.10 illustrates.

Example 8.10 (Infinite expected value) *Suppose that $p_X(x) = \frac{1}{x}$ for any x that is a power of 2—that is, $x \in \{2, 4, 8, \dots, 2^k, \dots\}$. These outcomes and probabilities constitute a valid pmf since $\sum_{k=1}^{\infty} \frac{1}{2^k} = \frac{1/2}{1-1/2} = 1$, but the expected value of X is infinite since*

$$\mu_X = \sum_{k=1}^{\infty} \left(2^k\right)\left(\frac{1}{2^k}\right) = \sum_{k=1}^{\infty} 1 = \infty.$$

8.3.2 Population Variance and Population Standard Deviation

This section introduces the **population variance** of a discrete random variable X. For the sample variance of a discrete variable x, proposition 8.1 stated the formula in terms of sample proportions:

$$s_x^2 = \frac{n}{n-1} \sum_k (x_k^* - \bar{x})^2 p_k \approx \sum_k (x_k^* - \bar{x})^2 p_k.$$

The sample variance is approximately a weighted average of the $(x_k^* - \bar{x})^2$ values, where the weights are the associated sample proportions of the outcomes. For the population variance, there are two main differences from the sample variance formula. First, as with the population mean, the weights are the probabilities of the possible outcomes rather than the sample proportions. Second, the sample mean, which appears in each of the $(x_k^* - \bar{x})^2$ expressions, is replaced by the population mean, μ_X. The formal definition of the population variance is given in definition 8.6.

Definition 8.6 *The **population variance** of a discrete random variable X, denoted as σ_X^2 or $Var(X)$, is*

$$\sigma_X^2 = Var(X) = E[(X - \mu_X)^2] = \sum_k (x_k^* - \mu_X)^2 p_X(x_k^*).$$

The sample variance is a weighted average of the squared difference of possible outcomes from the sample mean, and the population variance is a weighted average of the squared difference of possible outcomes from the population mean. The weights for the sample variance are the sample proportions, and the weights for the population variance are the probabilities $p_X(x_k^*) = P(X = x_k^*)$. For a given sample, the sample proportions generally differ from the pmf probabilities ($p_k \neq p_K(x_k^*)$), leading to a sample mean that differs from the population mean ($\bar{x} \neq \mu_X$) and a sample variance that differs from the population variance ($s_x^2 \neq \sigma_X^2$). That said, if we conduct the thought experiment of taking many repeated draws from the population, the sample variance s_x^2 should get closer to the population variance σ_X^2 as the sample size n gets larger and larger. The reason here is that, as the sample size n becomes very large, (1) the sample proportions p_k get closer to the pmf probabilities $p_X(x_k^*)$ and (2) the sample mean $\bar{x}$ gets closer to the population mean μ_X, which, taken together, imply that each $(x_k^* - \bar{x})^2 p_k$ term in s_x^2 gets closer to each $(x_k^* - \mu_X)^2 p_X(x_k^*)$ term in σ_X^2.

We also define the **population standard deviation** associated with the random variable X in definition 8.7.

Definition 8.7 *The* ***population standard deviation*** *of a discrete random variable X, denoted as σ_X or $sd(X)$, is*

$$\sigma_X = sd(X) = \sqrt{\sigma_X^2} = \sqrt{\sum_k (x_k^* - \mu_X)^2 p_X(x_k^*)}.$$

The population standard deviation σ_X differs from the sample standard deviation s_x in the same way that the population variance σ_X^2 differs from the sample variance s_X^2. Similar to the variance measures, the sample standard deviation s_x is expected to get closer to the population standard deviation σ_X as the sample size n gets larger and larger.

Example 8.11 (Union status) *In example 8.7, where $X \in \{0, 1\}$ is a binary random variable representing union status (1 for union, 0 for nonunion), the population mean was shown to be $\mu_X = p_X(1)$. Using this result and the fact that $p_X(0) = 1 - p_X(1)$, the population variance is*

$$\begin{aligned}
\sigma_X^2 = Var(X) &= (0 - \mu_X)^2 p_X(0) + (1 - \mu_X)^2 p_X(1) \\
&= (0 - p_X(1))^2 p_X(0) + (1 - p_X(1))^2 p_X(1) \\
&= (0 - p_X(1))^2 (1 - p_X(1)) + (1 - p_X(1))^2 p_X(1) \\
&= (p_X(1)^2 + (1 - p_X(1)) p_X(1))(1 - p_X(1)) \\
&= p_X(1)(1 - p_X(1)).
\end{aligned}$$

For this binary X, the population variance is $p_X(1)(1 - p_X(1))$, and the population standard deviation is

$$\sigma_X = sd(X) = \sqrt{p_X(1)(1 - p_X(1))}.$$

As in example 8.7, there's nothing special about union status here, so these are general results for a binary random variable $X \in \{0, 1\}$. These results will be revisited when binary random variables are discussed further in chapter 9.

Example 8.12 (Six-sided die) *In example 8.8, the population mean was shown to be $\mu_X = 3.5$ for the random variable $X \in \{1, 2, 3, 4, 5, 6\}$ as the outcome of a fair die roll. Since the probability of each outcome is $\frac{1}{6}$, we have*

$$\begin{aligned}
\sigma_X^2 = Var(X) &= \textstyle\sum_{x_k^*=1}^{6} (x_k^* - 3.5)^2 p_X(x_k^*) \\
&= \textstyle\sum_{x_k^*=1}^{6} (x_k^* - 3.5)^2 \left(\frac{1}{6}\right) \\
&= ((-2.5)^2 + (-1.5)^2 + (-0.5)^2 + (0.5)^2 + (1.5)^2 + (2.5)^2) \left(\frac{1}{6}\right) \\
&= (17.5) \left(\frac{1}{6}\right) = \frac{35}{12} \approx 2.9167.
\end{aligned}$$

The population standard deviation is $\sigma_X = \sqrt{\frac{35}{12}} \approx 1.7078$. To see how the sample descriptive statistics relate to the population descriptive statistics as the sample size n increases, figure 8.5 shows the results from a computer simulation of rolling a fair die 5,000 times. After each die roll, the following four quantities are updated: (1) sample proportion of 6 being the outcome of a roll (top-left graph), (2) sample mean of the die rolls (top-right graph), (3) sample variance of the die rolls (bottom-left graph), and (4) sample standard deviation of the die rolls (bottom-right graph). The figure provides plots of these four quantities as the number of tosses n (along the x-axis) increases to 5,000. For comparison purposes,

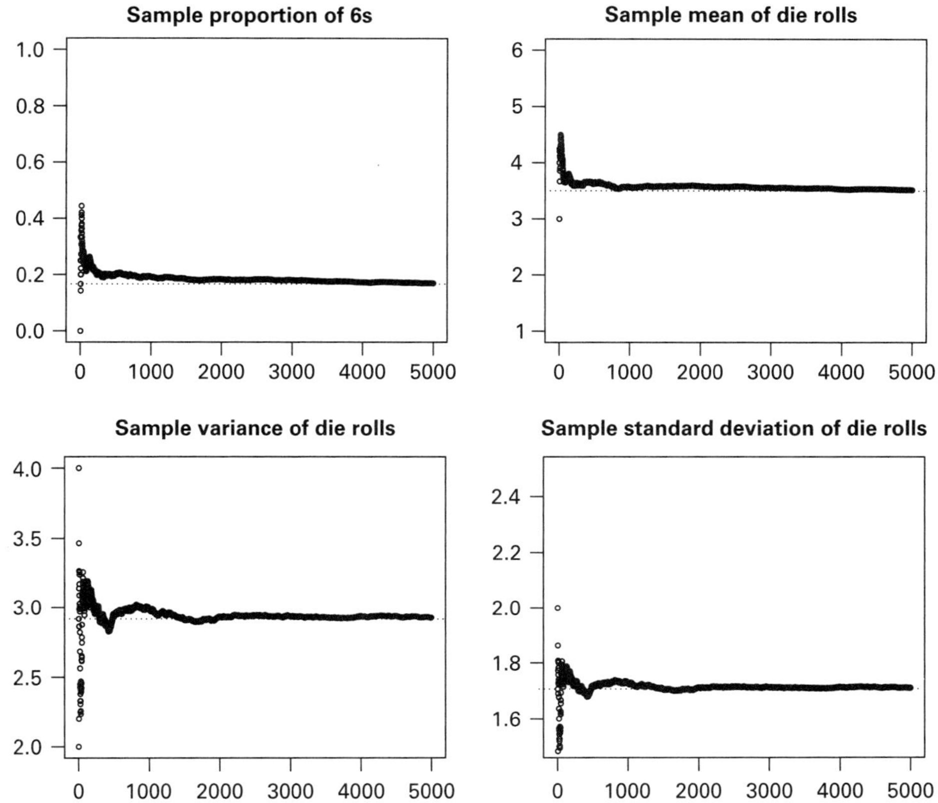

Figure 8.5
Descriptive statistics for 5,000 simulated die rolls

the corresponding population statistics are drawn as horizontal dotted lines on each plot. These values are the probability of a 6 ($p_X(6) = \frac{1}{6}$), the population mean ($\mu_X = 3.5$), the population variance ($\sigma_X^2 = \frac{35}{12}$), and the population standard deviation ($\sigma_X = \sqrt{\frac{35}{12}}$). As evident in the four graphs, the sample descriptive statistic in each case gets very close to its corresponding population statistic as n increases toward 5,000.

Here is the R code to create figure 8.5:

```
set.seed(1234)

# initialize the number of simulations
num_simulations <- 5000

# roll a fair six-sided die for all simulations
dierolls <- sample(1:6, num_simulations, replace=TRUE)

# graph-display format (two rows, two columns)
par(mfrow=c(2,2))

# plot the sample frequency of a "6" being rolled
plot(cumsum(dierolls==6)/(1:num_simulations), main="Sample proportion of 6s",
```

```
        xlab="", ylab="", ylim=c(0,1), cex=0.5)
abline(h=1/6,lty=3)

# plot the sample mean of die rolls
plot(cumsum(dierolls)/(1:num_simulations), main="Sample mean of die rolls",
        xlab="", ylab="", ylim=c(1,6), cex=0.5)
abline(h=3.5,lty=3)

# plot the sample variance of die rolls
tempvar <- rep(0,num_simulations)
for (i in 2:num_simulations) {
  tempavg <- cumsum(dierolls)[i]/i
  tempvar[i] <- sum((dierolls[1:i]-tempavg)^2)/(i-1)
}
plot(tempvar[2:num_simulations], main="Sample variance of die rolls",
        xlab="", ylab="", ylim=c(2,4), cex=0.5)
abline(h=35/12,lty=3)

# plot the sample standard deviation of die rolls
plot(sqrt(tempvar[2:num_simulations]),
     main="Sample standard deviation ofdie rolls",
     xlab="", ylab="", ylim=c(1.5,2.5), cex=0.5)
abline(h=sqrt(35/12),lty=3)
```

8.4 Multiple Discrete Random Variables

This section considers a situation in which there are two or more discrete random variables associated with a population. In the case of two discrete random variables, we will generally use X and Y to denote the variables.

Example 8.13 (Phone and computer ownership) *Suppose a random individual drawn from a population $\mathcal{S}$ of individuals is surveyed to determine the number of phones and computers that they own. Let e be the identity of the individual chosen and surveyed. The underlying discrete random variables are $X = X(e)$ = number of phones owned and $Y = Y(e)$ = number of computers owned.*

Example 8.14 (Patents and R&D staff) *Suppose a random firm drawn from a population $\mathcal{S}$ of firms is surveyed to determine the number of patents awarded (last year) and the size of the firm's R&D staff. Let e be the identity of the firm chosen and surveyed. The underlying discrete random variables are $X = X(e)$ = number of patents awarded last year and $Y = Y(e)$ = size of R&D staff.*

8.4.1 Joint Probability Mass Function and Joint Cumulative Distribution Function

Let K and L denote the number of possible outcomes for X and Y, respectively, where K and/or L may be infinite. The possible outcomes for X are $\{x_1^*, x_2^*, \ldots, x_K^*\}$ if K is finite and $\{x_1^*, x_2^*, \ldots, x_k^*, \ldots\}$ if K is infinite. The possible outcomes for Y are $\{y_1^*, y_2^*, \ldots, y_L^*\}$ if L is finite and $\{y_1^*, y_2^*, \ldots, y_\ell^*, \ldots\}$ if L is infinite. The concept of the pmf can be extended to a **joint probability mass function (joint pmf)** as given in definition 8.8.

Definition 8.8 *The **joint probability mass function (joint pmf)** of two discrete random variables X and Y, denoted as $p_{XY}(\cdot,\cdot)$, gives the joint probability*

$$p_{XY}(x_k^*, y_\ell^*) = P(X = x_k^* \cap Y = y_\ell^*) = P(X = x_k^*,\ Y = y_\ell^*)$$

for every possible outcome x_k^* for X and every possible outcome y_ℓ^* for Y.

The collection of possible (x_k^*, y_ℓ^*) values is a set of disjoint and exhaustive outcomes, meaning that the joint pmf satisfies the following properties:

$$0 \leq p_{XY}(x_k^*, y_\ell^*) \leq 1 \text{ for any possible outcome pair } (x_k^*, y_\ell^*)$$

and

$$\sum_{(k,\ell)} p_{XY}(x_k^*, y_\ell^*) = \sum_k \sum_\ell p_{XY}(x_k^*, y_\ell^*) = 1.$$

The joint pmf probabilities are each between 0 and 1 (inclusive) and sum to 1. The summations in the expression above are written in two equivalent ways: (1) $\sum_{(k,\ell)}$ is a summation over all possible pairs (k, ℓ), and (2) $\sum_k \sum_\ell$ is a double summation, with the inner summation taken over possible ℓ values and the outer summation taken over possible k values.

If X and Y are both *finite* discrete random variables, the number of possible joint outcomes is KL. Following the approach that was taken for probability tables in section 3.3, the joint pmf can be represented by the following table:

	y_1^*	y_2^*	$\cdots$	y_L^*
x_1^*	$p_{XY}(x_1^*, y_1^*)$	$p_{XY}(x_1^*, y_2^*)$	$\cdots$	$p_{XY}(x_1^*, y_L^*)$
x_2^*	$p_{XY}(x_2^*, y_1^*)$	$p_{XY}(x_2^*, y_2^*)$	$\cdots$	$p_{XY}(x_2^*, y_L^*)$
$\vdots$	$\vdots$	$\vdots$	$\ddots$	$\vdots$
x_K^*	$p_{XY}(x_K^*, y_1^*)$	$p_{XY}(x_K^*, y_2^*)$	$\cdots$	$p_{XY}(x_K^*, y_L^*)$

Example 8.15 (Phone and computer ownership) *Continuing example 8.13, where X = number of phones owned and Y = number of computers owned, let's assume that no individual in the population owns more than two phones or more than two computers, so $K = 2$ and $L = 2$. Suppose the joint pmf is given by the following probability table:*

		Y (# computers) 0	1	2
	0	0.06	0.03	0.01
X (# phones)	1	0.22	0.48	0.05
	2	0.02	0.04	0.09

The joint pmf values in this table are the probabilities associated with the population, not *sample proportions based on an observed sample. For example, $p_{XY}(0, 1) = 0.03$ indicates there is a 3% probability that an individual drawn from the population doesn't own a phone and owns one computer. Similarly, $p_{XY}(1, 2) = 0.05$ indicates there is a 5% probability that an individual drawn from the population owns one phone and two computers.*

The concept of the cdf can also be extended to the case of two discrete random variables, as in definition 8.9.

Definition 8.9 *The **joint cumulative distribution function (joint cdf)** of two discrete random variables, denoted as $F_{XY}(\cdot, \cdot)$, gives the probability that* both *X and Y are less than* *or equal to their corresponding arguments:*

$$F_{XY}(x_0, y_0) = P(X \le x_0 \cap Y \le y_0) = P(X \le x_0,\ Y \le y_0).$$

The joint cdf has the following properties:

$$0 \le F_{XY}(x_0, y_0) \le 1 \text{ for every } x_0 \text{ and } y_0$$

and

$$x_0 < x_1 \implies F_{XY}(x_0, y_0) \le F_{XY}(x_1, y_0) \text{ and } y_0 < y_1 \implies F_{XY}(x_0, y_0) \le F_{XY}(x_0, y_1).$$

The first property follows from the fact that $F_{XY}(x_0, y_0)$ is a probability. The first part of the second property says that the joint cdf is weakly increasing in its first argument; that is, if y_0 is held fixed, the value of the joint cdf weakly increases as x_0 increases. Similarly, the second part says that the joint cdf is weakly increasing in its second argument, so holding x_0 fixed, the value of the joint cdf weakly increases as y_0 increases.

Example 8.16 (Phone and computer ownership) *Using the probability table in example 8.15, the joint cdf can be calculated at any specified arguments. For example, with $x_0 = 2$ and $y_0 = 1$, the joint cdf $F_{XY}(2, 1) = P(X \le 2,\ Y \le 1)$ is the sum of all the probabilities from the first two columns, which is 0.85. If y_0 is held fixed at $y_0 = 1$, focusing on the subpopulation of individuals who own one computer, the joint cdf values for the possible values of X are $F_{XY}(0, 1) = 0.09$, $F_{XY}(1, 1) = 0.79$, and $F_{XY}(2, 1) = 0.85$. These joint cdf values increase as x_0 increases with y_0 fixed.*

8.4.2 Marginal and Conditional Probability Mass Functions

Using the concepts of marginal probabilities and conditional probabilities introduced in chapter 3, **marginal probability mass functions** and **conditional probability mass functions** can be inferred from the joint pmfs of X and Y. For the random variable X, the marginal pmf is given by $p_X(x_k^*)$ for all possible values x_k^*. Similarly, for the random variable Y, the marginal pmf is given by $p_Y(y_\ell^*)$ for all possible values y_ℓ^*. By applying the Law of Total Probability, the marginal pmf is related to the joint pmf as given in proposition 8.3.

Proposition 8.3 *The marginal pmfs of X and Y are, respectively,*

$$p_X(x_k^*) = \sum_\ell p_{XY}(x_k^*, y_\ell^*)$$

and

$$p_Y(y_\ell^*) = \sum_k p_{XY}(x_k^*, y_\ell^*).$$

When K and L are finite, in which case the joint pmf can be specified as a probability table, the $p_Y(x_k^*)$ values are the row totals of the joint probabilities and the $p_Y(y_\ell^*)$ values are the column totals of the joint probabilities.

The conditional probability mass functions (conditional pmfs) are defined as in definition 8.10.

Definition 8.10 *The **conditional probability mass function (conditional pmf)** of X given Y, denoted as $p_{X|Y}(\cdot|\cdot)$, is*

$$p_{X|Y}(x_k^*|y_\ell^*) = P(X = x_k^*|Y = y_\ell^*) = \frac{p_{XY}(x_k^*, y_\ell^*)}{p_Y(y_\ell^*)}.$$

Similarly, the conditional pmf of Y given X, denoted as $p_{Y|X}(\cdot|\cdot)$, *is*

$$p_{Y|X}(y_\ell^*|x_k^*) = P(Y = y_\ell^*|X = x_k^*) = \frac{p_{XY}(x_k^*, y_\ell^*)}{p_X(x_k^*)}.$$

For the probability table when K and L are finite, the conditional probability $p_{X|Y}(x_k^*|y_\ell^*)$ is focused on the joint probabilities in the column corresponding to $Y = y_\ell^*$, and the conditional probability $p_{Y|X}(y_\ell^*|x_k^*)$ is focused on the joint probabilities in the row corresponding to $X = x_k^*$. The conditional pmf is itself a pmf, having the properties that its probabilities are positive and add up to 1. Therefore, a conditional cdf can be defined based upon the conditional pmf's probabilities as in definition 8.11.

Definition 8.11 *The* ***conditional cumulative distribution function (conditional cdf)*** *of X given Y, denoted as* $F_{X|Y}(\cdot|\cdot)$, *gives the probability that X is less than or equal to any argument* x_0 *conditional on* $Y = y_\ell^*$:

$$F_{X|Y}(x_0|y_\ell^*) = P(X \le x_0|Y = y_\ell^*) = \sum_{x_k^* \le x_0} p_{X|Y}(x_k^*|y_\ell^*).$$

Similarly, the conditional cdf of Y given X, denoted as $F_{Y|X}(\cdot|\cdot)$, *gives the probability that Y is less than or equal to any argument* y_0 *conditional on* $X = x_k^*$:

$$F_{Y|X}(y_0|x_k^*) = P(Y \le y_0|X = x_k^*) = \sum_{y_\ell^* \le y_0} p_{Y|X}(y_\ell^*|x_k^*).$$

Example 8.17 (Phone and computer ownership) *The joint probability table from example 8.15 is replicated here, with the marginal pmfs for X and Y now included:*

		Y (# computers)			
		0	1	2	$p_X(x)$
	0	0.06	0.03	0.01	0.10
X (# phones)	1	0.22	0.48	0.05	0.75
	2	0.02	0.04	0.09	0.15
	$p_Y(y)$	0.30	0.55	0.15	

The marginal pmfs of X, given by the $p_X(x)$ *values, are obtained by summing the probabilities in each row. The marginal pmfs of Y, given by the* $p_Y(y)$ *values, are obtained by summing the probabilities in each column. For the marginal pmfs, the sum of the* $p_X(x)$ *probabilities equals 1 and the sum of the* $p_Y(y)$ *probabilities also equals* 1.

Focusing on the $Y = 1$ *column (individuals who own one computer), the conditional pmf is*

$$p_{X|Y}(0|1) = \frac{p_{XY}(0,1)}{p_Y(1)} = \frac{0.03}{0.55} = \frac{3}{55},$$

$$p_{X|Y}(1|1) = \frac{p_{XY}(1,1)}{p_Y(1)} = \frac{0.48}{0.55} = \frac{48}{55},$$

and

$$p_{X|Y}(2|1) = \frac{p_{XY}(2,1)}{p_Y(1)} = \frac{0.04}{0.55} = \frac{4}{55}.$$

For instance, the probability than an individual owns two phones given that the person owns one computer is $p_{X|Y}(2|1) = \frac{4}{55}$ *or approximately* 7.27%. *The associated conditional cdf is*

$$F_{X|Y}(0|1) = p_{X|Y}(0|1) = \frac{3}{55},$$

$$F_{X|Y}(1|1) = p_{X|Y}(0|1) + p_{X|Y}(1|1) = \frac{3}{55} + \frac{48}{55} = \frac{51}{55},$$

and

$$F_{X|Y}(2|1) = p_{X|Y}(0|1) + p_{X|Y}(1|1) + p_{X|Y}(2|1) = 1.$$

For instance, the probability that an individual owns (strictly) less than two phones, given that the person owns one computer is $F_{X|Y}(1|1) = \frac{51}{55}$ *or approximately* 92.73%.

Similarly, conditioning on the number of phones owned involves focusing on a specific row. For the $X = 0$ *row (individuals who do not own a phone), the conditional pmf is*

$$p_{Y|X}(0|0) = \frac{p_{XY}(0,0)}{p_X(0)} = \frac{0.06}{0.10} = 0.6,$$

$$p_{Y|X}(1|0) = \frac{p_{XY}(0,1)}{p_X(0)} = \frac{0.03}{0.10} = 0.3,$$

and

$$p_{Y|X}(2|0) = \frac{p_{XY}(0,2)}{p_X(0)} = \frac{0.01}{0.10} = 0.1.$$

The probability that an individual owns one computer given that the person does not own a phone is $p_{Y|X}(1|0) = 0.3$, *or* 30%. *This conditional probability is considerably less than the unconditional probability that an individual from the population owns one computer, which is* $p_Y(1) = 0.55$, *or* 55%. *This difference tells us that knowing* $X = 0$ *provides additional information about the probability that* $Y = 1$.

As mentioned previously, each conditional pmf is itself a pmf. That is, fixing $Y = y_\ell^*$ for some possible outcome y_ℓ^*, the collection of $p_{X|Y}(x_k^*|y_\ell^*)$ probabilities constitutes a pmf for the x_k^* values. Similarly, fixing $X = x_k^*$ for some possible outcome x_k^*, the collection of $p_{Y|X}(y_\ell^*|x_k^*)$ probabilities constitutes a pmf for the y_ℓ^* values. In example 8.17, the conditional pmf values sum to 1 ($\frac{3}{55} + \frac{48}{55} + \frac{4}{55} = 1$, conditioning on $Y = 1$; and $0.6 + 0.3 + 0.1 = 1$, conditioning on $X = 0$), as is expected for pmfs.

Since a conditional pmf is just a special case of a pmf, it is natural to introduce conditional versions of the population descriptive statistics, including the mean, variance, and standard deviation, as in definitions 8.12, 8.13, and 8.14.

Definition 8.12 *The* ***population conditional mean*** *or* ***conditional expectation*** *of* X *given* $Y = y_\ell^*$, *denoted as* $\mu_{X|Y=y_\ell^*}$, *is*

$$\mu_{X|Y=y_\ell^*} = E(X|Y = y_\ell^*) = \sum_k x_k^* p_{X|Y}(x_k^*|y_\ell^*).$$

Definition 8.13 *The* ***population conditional variance*** *of X given* $Y = y^*_\ell$, *denoted as* $\sigma^2_{X|Y=y^*_\ell}$, *is*

$$\sigma^2_{X|Y=y^*_\ell} = Var(X|Y = y^*_\ell) = \sum_k (x^*_k - \mu_{X|Y=y^*_\ell})^2 p_{X|Y}(x^*_k|y^*_\ell).$$

Definition 8.14 *The* ***population conditional standard deviation*** *of X given* $Y = y^*_\ell$, *denoted as* $\sigma_{X|Y=y^*_\ell}$, *is*

$$\sigma_{X|Y=y^*_\ell} = sd(X|Y = y^*_\ell) = \sqrt{\sigma^2_{X|Y=y^*_\ell}} = \sqrt{\sum_k (x^*_k - \mu_{X|Y=y^*_\ell})^2 p_{X|Y}(x^*_k|y^*_\ell)}.$$

These definitions are stated for the conditional distribution of X given Y. For the conditional distribution of Y given X, the same definitions can be used with the roles of X and Y reversed. For example, the population conditional mean or conditional expectation of Y given $X = x^*_k$ is

$$\mu_{Y|X=x^*_k} = E(Y|X = x^*_k) = \sum_\ell y^*_\ell p_{Y|X}(y^*_\ell|x^*_k).$$

Example 8.18 (Phone and computer ownership) *In example 8.17, the conditional pmf associated with phone ownership (X) given that an individual owned one computer* ($Y = 1$) *is* $p_{X|Y}(0|1) = \frac{3}{55}$, $p_{X|Y}(1|1) = \frac{48}{55}$, *and* $p_{X|Y}(2|1) = \frac{4}{55}$. *Then, using definitions 8.12, 8.13, and 8.14, the population conditional mean of X given* $Y = 1$ *is*

$$\mu_{X|Y=1} = E(X|Y = 1) = 0 \times \frac{3}{55} + 1 \times \frac{48}{55} + 2 \times \frac{4}{55} = \frac{56}{55} \approx 1.018,$$

the population conditional variance of X given $Y = 1$ *is*

$$\sigma^2_{X|Y=1} = Var(X|Y = 1) = \left(0 - \frac{56}{55}\right)^2 \times \frac{3}{55} + \left(1 - \frac{56}{55}\right)^2 \times \frac{48}{55} + \left(2 - \frac{56}{55}\right)^2 \times \frac{4}{55} \approx 0.1269,$$

and the population conditional standard deviation of X given $Y = 1$ *is*

$$\sigma_{X|Y=1} = sd(X|Y = 1) \approx \sqrt{0.1269} \approx 0.3563.$$

These population descriptive statistics describe the number of phones owned (X) in the subpopulation of individuals who own one computer. The unconditional population mean of X is $\mu_X = (0)(0.10) + (1)(0.75) + (2)(0.15) = 1.05$, *and the unconditional population variance of X is* $\sigma^2_X = (0 - 1.05)^2(0.10) + (1 - 1.05)^2(0.75) + (2 - 1.05)^2(0.15) = 0.2475$. *Therefore, the conditional distribution of X given* $Y = 1$ *has a slightly lower population mean and a much lower population variance, indicating that knowing* $Y = 1$ *provides useful information about the distribution of X.*

Similarly, we can find the population descriptive statistics associated with the conditional pmf of Y given $X = 0$ *found in example 8.17, which is* $p_{Y|X}(0|0) = 0.6$, $p_{Y|X}(1|0) = 0.3$, *and* $p_{Y|X}(2|0) = 0.1$. *The population conditional mean of Y given* $X = 0$ *is*

$$\mu_{Y|X=0} = E(Y|X = 0) = 0 \times 0.6 + 1 \times 0.3 + 2 \times 0.1 = 0.5,$$

the population conditional variance of Y given X = 0 is

$$\sigma^2_{Y|X=0} = Var(Y|X=0) = (0-0.5)^2 \times 0.60 + (1-0.5)^2 \times 0.30 + (2-0.5)^2 \times 0.10 = 0.45,$$

and the population conditional standard deviation of Y given X = 0 is

$$\sigma_{Y|X=0} = sd(Y|X=0) = \sqrt{0.45} \approx 0.6708.$$

For comparison, the unconditional population mean of Y is $\mu_Y = (0)(0.30) + (1)(0.55) + (2)(0.15) = 0.85$, *and the unconditional population variance of Y is* $\sigma^2_Y = (0-0.85)^2(0.30) + (1-0.85)^2(0.55) + (2-0.85)^2(0.15) = 0.4275$. *Therefore, knowing that* $X = 0$ *provides useful information about the distribution of Y.*

8.4.3 Population Covariance and Population Correlation

This section introduces population analogs to the sample covariance and the sample correlation, which were introduced in section 7.2.2 to measure the relationship between two numerical variables. We introduce the **population covariance** and the **population correlation**, which are descriptive statistics measuring the relationship between two random variables in the population. We continue to assume that both X and Y are discrete random variables.

Recall the sample covariance formula for numerical variables x and y:

$$s_{xy} = \frac{1}{n-1} \sum_{i=1}^{n} (x_i - \bar{x})(y_i - \bar{y}).$$

The sample covariance can be expressed in terms of sample proportions, in this case the *joint* sample proportions for the possible joint outcomes (x^*_k, y^*_ℓ):

$$s_{xy} = \frac{n}{n-1} \sum_{(k,\ell)} (x^*_k - \bar{x})(y^*_\ell - \bar{y}) p_{k\ell} \approx \sum_{(k,\ell)} (x^*_k - \bar{x})(y^*_\ell - \bar{y}) p_{k\ell},$$

where p_{kl} is the sample proportion of observations having $x = x^*_k$ and $y = y^*_\ell$. For the analogous population-descriptive statistic, the population covariance, the sample proportions p_{kl} are replaced by the joint probabilities $p_{XY}(x^*_k, y^*_\ell)$, and the sample means $\bar{x}$ and $\bar{y}$ are replaced by the population means μ_X and μ_Y as in definition 8.15.

Definition 8.15 *The* ***population covariance*** *between two discrete random variables X and Y, denoted as* σ_{XY}, *is*

$$\sigma_{XY} = Cov(X, Y) = \sum_{(k,\ell)} (x^*_k - \mu_X)(y^*_\ell - \mu_Y) p_{XY}(x^*_k, y^*_\ell).$$

If we conduct the thought experiment of taking many repeated draws from the population, the sample covariance s_{xy} should get close to the population covariance σ_{XY} as the sample size n gets larger. We expect this to occur since (1) the joint probability $p_{XY}(x^*_k, y^*_\ell)$ is the long-run frequency that the sample proportions p_{kl} approach for large n and (2) the sample means $\bar{x}$ and $\bar{y}$ approach the population means μ_X and μ_Y for large n.

It is important to stress the difference between the sample covariance and the population covariance. While both are measures of linear association, the sample covariance measures that association between x and y in the observed sample, and the population covariance measures that association between X and Y in the population. For the population covariance, the terms $(x_k^* - \mu_X)(y_\ell^* - \mu_Y)p_{XY}(x_k^*, y_\ell^*)$ in the summation have positive contributions for possible outcomes x_k^* and y_ℓ^* on the same side of their respective population means, μ_X and μ_Y, and negative contributions for possible outcomes x_k^* and y_ℓ^* on opposite sides of their respective population means. The overall population covariance depends on the magnitude and weighting of each term, where the weights are the joint probabilities $p_{XY}(x_k^*, y_\ell^*)$. The sign of the population covariance indicates, for a joint draw of X and Y from the population, whether the X and Y draws tend to be on the same side of their population means (a positive covariance) or on opposite sides of their population means (a negative covariance).

The units of the population covariance σ_{XY} are the units of X times the units of Y. Like the sample covariance, these units make the population covariance difficult to interpret. As before, it is useful to have a unitless measure of linear association, this time in the population, that is easier to interpret. With the population covariance defined, the population correlation is defined analogously to the sample correlation, as in definition 8.16.

Definition 8.16 *The* ***population correlation*** *between two discrete random variables X and Y, denoted as ρ_{XY}, is*

$$\rho_{XY} = Corr(X, Y) = \frac{\sigma_{XY}}{\sigma_X \sigma_Y}.$$

The population covariance and correlation have properties analogous to those seen for the sample covariance and correlation (given in proposition 7.1).

Proposition 8.4 *For random variables X and Y, the population covariance σ_{XY} and the population correlation ρ_{XY} satisfy the following properties:*

i. ρ_{XY} *is unitless.*

ii. *The sign of the population correlation is the same as the sign of the population covariance,*

$$sign(\rho_{XY}) = sign(\sigma_{XY}).$$

iii. $\rho_{XX} = 1.$

iv. $-1 \leq \rho_{XY} \leq 1.$

Proposition 8.4 does *not* assume that X and Y are discrete random variables. These properties hold more generally for any type of numerical random variables X and Y, so there is no reason to restrict our discussion to the case of discrete random variables.

Example 8.19 (Phone and computer ownership) *With X = number of phones owned and Y = number of computers owned, recall the joint pmf from example 8.15:*

		Y (# computers)		
		0	1	2
X (# phones)	0	0.06	0.03	0.01
	1	0.22	0.48	0.05
	2	0.02	0.04	0.09

We have already seen that the population means are $\mu_X = 1.05$ *and* $\mu_Y = 0.85$*. Thus, the population covariance is*

$$\begin{aligned}\sigma_{XY} = & (0-1.05)(0-0.85)(0.06)+(0-1.05)(1-0.85)(0.03)\\ & +(0-1.05)(2-0.85)(0.01)+(1-1.05)(0-0.85)(0.22)\\ & +(1-1.05)(1-0.85)(0.48)+(1-1.05)(2-0.85)(0.05)\\ & +(2-1.05)(0-0.85)(0.02)+(2-1.05)(1-0.85)(0.04)\\ & +(2-1.05)(2-0.85)(0.09)=0.1275.\end{aligned}$$

The population correlation is

$$\rho_{XY} = \frac{\sigma_{XY}}{\sigma_X \sigma_Y} = \frac{0.1275}{\sqrt{0.2475}\sqrt{0.4275}} \approx 0.392,$$

where σ_X *and* σ_Y *were calculated in example 8.17. The population correlation* $\rho_{XY} \approx 0.392$ *indicates a positive relationship between the random variables* X *and* Y*.*

Example 8.20 (Stock price up or down?) *Suppose that* X *and* Y *are defined as binary random variables based on the monthly returns of two underlying stocks, A and B:*

$$X = \begin{cases} 1 & \textit{if stock A's price goes up during the month (positive return)} \\ 0 & \textit{if stock A's price goes down or stays the same during the month} \\ & \textit{(nonpositive return)} \end{cases}$$

and

$$Y = \begin{cases} 1 & \textit{if stock B's price goes up during the month (positive return)} \\ 0 & \textit{if stock B's price goes down or stays the same during the month} \\ & \textit{(nonpositive return).} \end{cases}$$

We want to know whether there is a relationship between X *and* Y *in the population. Let's assume that the joint pmf of* X *and* Y *is given by the following probability table:*

		Y = 0	Y = 1
X	0	0.30	0.15
	1	0.10	0.45

Since both X *and* Y *are binary variables, their population means and variances are*

$$\mu_X = p_X(1) = 0.55,\ \sigma_X^2 = p_X(1)(1 - p_X(1)) = 0.2475,$$

and

$$\mu_Y = p_Y(1) = 0.60,\ \sigma_Y^2 = p_Y(1)(1 - p_Y(1)) = 0.24.$$

The population covariance between X *and* Y *is*

$$\begin{aligned}\sigma_{XY} = & (0-0.55)(0-0.60)(0.30)+(0-0.55)(1-0.60)(0.15)\\ & +(1-0.55)(0-0.60)(0.10)+(1-0.55)(1-0.60)(0.45)=0.12,\end{aligned}$$

and the population correlation between X and Y is

$$\rho_{XY} = \frac{\sigma_{XY}}{\sigma_X \sigma_Y} = \frac{0.12}{\sqrt{0.2475}\sqrt{0.24}} \approx 0.492.$$

There is a positive relationship between X and Y, as it is much more likely that X and Y will both be above their population means or both be below their population means (75%), as opposed to being on oppositive sides of their population means (25%). When $X=1$, it is much more likely that $Y=1$ than in the overall population; the conditional probability $p_{Y|X}(1|1) = \frac{0.45}{0.55} \approx 0.818$ compared to the unconditional probability $p_Y(1) = 0.60$. Similarly, when $X=0$, it is much more likely that $Y=0$ than in the overall population; the conditional probability $p_{Y|X}(0|0) = \frac{0.30}{0.45} \approx 0.667$ compared to the unconditional probability $p_Y(0) = 0.40$.

8.4.4 Independence of Discrete Random Variables

In examples 8.19 and 8.20, knowing the value of one random variable provided information about the distribution of the other random variable. The association between the two random variables, given by a nonzero population covariance/correlation, indicates a type of dependence between the two random variables. In this section, we formalize the concept of whether two random variables depend on each other in any way. Specifically, two discrete random variables X and Y are said to be **independent** when they have no relationship to each other and **dependent** when they do have a relationship, as described in definition 8.17.

Definition 8.17 *Two discrete random variables X and Y are* ***independent*** *if and only if*

$$p_{XY}(x_k^*, y_\ell^*) = p_X(x_k^*) p_Y(y_\ell^*) \quad \textit{for every possible outcome pair } (x_k^*, y_\ell^*)$$

or, equivalently,

$$F_{XY}(x_k^*, y_\ell^*) = F_X(x_k^*) F_Y(y_\ell^*) \quad \textit{for every possible outcome pair } (x_k^*, y_\ell^*).$$

If this equality fails for any (x_k^*, y_ℓ^*), then the discrete random variables X and Y are ***dependent****.*

Independent discrete random variables X and Y have joint probabilities $p_{XY}(x_k^*, y_\ell^*)$ equal to the product of the marginal probabilities $p_X(x_k^*)$ and $p_Y(y_\ell^*)$ for every possible outcome pair (x_k^*, y_ℓ^*). This concept of independence is closely related to the independence of events discussed in chapter 3–specifically, that two events are independent if and only if their joint probability equals the product of the marginal probabilities of the two events (described in proposition 3.6). Here, any given outcome of X being observed can be thought of as an event and, likewise, any given outcome of Y being observed can be thought of as an event. Framed in this way, definition 8.17 is equivalent to any possible outcome (event) associated with X being independent of any possible outcome (event) associated with Y, in the sense discussed in chapter 3.

To show that two discrete random variables are dependent, it is only necessary to show that there is one outcome pair (x_k^*, y_ℓ^*) for which $p_{XY}(x_k^*, y_\ell^*) \neq p_X(x_k^*) p_Y(y_\ell^*)$.

Example 8.21 (Phone and computer ownership) *Continuing example 8.19, are X (number of phones owned) and Y (number of computers owned) independent? The answer is no since, for instance,* $p_{XY}(0,0) = 0.06$ *and* $p_X(0)p_Y(0) = (0.10)(0.30) = 0.03$ *are not equal. Other joint probabilities could be checked, but it's sufficient to find just one case where* $p_{XY}(x_k^*, y_\ell^*) \neq p_X(x_k^*)p_Y(y_\ell^*)$ *to show that X and Y are dependent.*

Example 8.22 (Stock price up or down?) *Continuing example 8.20, are X (binary variable indicating positive return of stock A) and Y (binary variable indicating positive return of stock B) independent? The answer is no since* $p_{XY}(0,0) = 0.30$ *and* $p_X(0)p_Y(0) = (0.45)(0.40) = 0.18$ *are not equal.*

At times throughout this book, we have made the implicit assumption that certain random variables are independent of each other. For example, for two coin tosses that have nothing to do with each other, the tosses can be thought of as independent random variables. To be more precise, if $X = 1$ if the first coin is heads and 0 if the first coin is tails and $Y = 1$ if the second coin is heads and 0 if the second coin is tails, the following joint probability table represents the case when X and Y are independent:

		Y = 0	Y = 1
X	0	0.25	0.25
	1	0.25	0.25

The joint probabilities are products of the respective marginal probabilities since each marginal probability (of heads or tails for either toss) equals 0.5. Similarly, for two fair die rolls $X \in \{1,2,3,4,5,6\}$ and $Y \in \{1,2,3,4,5,6\}$ that are independent, the joint probability of any of the 36 possible outcomes (x, y) is $\frac{1}{36}$ since each marginal probability for X equals $\frac{1}{6}$ and each marginal probability for Y equals $\frac{1}{6}$. Here is another example of two independent discrete random variables.

Example 8.23 (Website purchases) *For two visitors to a website, let* $X = 1$ *if the first visitor makes a purchase and* 0 *otherwise and* $Y = 1$ *if the second visitor makes a purchase and* 0 *otherwise. If the marginal probability of purchase is* 0.2 *for both individuals, the probability table in the case of independent X and Y (i.e., their purchase behaviors are not related) is*

		Y = 0	Y = 1
X	0	$(0.8)(0.8) = 0.64$	$(0.8)(0.2) = 0.16$
	1	$(0.2)(0.8) = 0.16$	$(0.2)(0.2) = 0.04$

An alternative way to show that two discrete random variables are dependent is to consider the population covariance or population correlation. If the population covariance/correlation is nonzero, there is some linear association between the two random variables, meaning that they must be dependent. Proposition 8.5 formally states this result.

Proposition 8.5 *If the discrete random variables X and Y are independent, the population covariance* σ_{XY} *and the population correlation* ρ_{XY} *equal zero* ($\sigma_{XY} = \rho_{XY} = 0$).

Equivalently, if X and Y have a nonzero population covariance or correlation, X and Y are dependent.

The reverse is not necessarily true: having a population covariance/correlation between X and Y equal to zero does *not* necessarily imply independence. As with sample covariance/correlation, these population statistics only measure the linear relationship between two random variables, so it's possible that there is some nonlinear dependence between the two random variables even when the covariance/correlation equals zero. Example 8.24 shows such a situation, where X and Y are dependent even though the population covariance is zero.

Example 8.24 *Consider the following probability table for* $X \in \{0, 1\}$ *and* $Y \in \{-1, 0, 1\}$*:*

		Y = −1	Y = 0	Y = 1	$p_X(x)$
X	0	0	0.5	0	0.5
	1	0.25	0	0.25	0.5
	$p_Y(y)$	0.25	0.5	0.25	

X and Y are dependent here since none of the joint probabilities are equal to the product of their respective marginal probabilities. For example, $p_{XY}(0, 0) = 0$ *and* $p_X(0)p_Y(0) = (0.5)(0.25) = 0.125$*. The population covariance is zero since* $\mu_X = 0.5$*,* $\mu_Y = 0$*, and*

$$\sigma_{XY} = (0 - 0.5)(0 - 0)(0.5) + (1 - 0.5)(-1 - 0)(0.25) + (1 - 0.5)(1 - 0)(0.25) = 0.$$

Another way to think about independence is to ask whether knowing the realized draw of a random variable X provides any useful information about the distribution of Y. If X and Y are independent, knowledge of the value of X should provide no information about the distribution of Y, and vice versa. Therefore, an equivalent way to establish that X and Y are independent is to show that the conditional distribution of $Y|X$ is the same for all possible values of X or, equivalently, that the conditional distribution of $X|Y$ is the same for all possible values of Y. If the conditional distribution of one random variable changes for different values of the other random variable, then the two random variables are dependent. This result is stated in proposition 8.6:

Proposition 8.6 *The discrete random variables X and Y are independent if and only if the following are true:*

- *For any possible value* y_ℓ^**, the conditional distribution of X given* $Y = y_\ell^*$ *is the same as the unconditional marginal distribution of X.*
- *For any possible value* x_k^**, the conditional distribution of Y given* $X = x_k^*$ *is the same as the unconditional marginal distribution of Y.*

Based on proposition 8.6, X and Y can be shown to be dependent by finding one case where a conditional distribution differs from the marginal distribution, as in example 8.25.

Example 8.25 (Phone and computer ownership) *Recall the probability table from example 8.15, with X = number of phones owned and Y = number of computers owned:*

		Y (# computers)		
		0	1	2
	0	0.06	0.03	0.01
X (# phones)	1	0.22	0.48	0.05
	2	0.02	0.04	0.09

We can verify that the conditional distribution of $Y|X$ depends on the value of X:

$$p_{Y|X}(0|0)=0.6,\ p_{Y|X}(1|0)=0.3,\ p_{Y|X}(2|0)=0.1 \text{ (conditional on } X=0)$$

$$p_{Y|X}(0|1)=\frac{22}{75},\ p_{Y|X}(1|1)=\frac{48}{75},\ p_{Y|X}(2|1)=\frac{5}{75} \text{ (conditional on } X=1)$$

$$p_{Y|X}(0|2)=\frac{2}{15},\ p_{Y|X}(1|2)=\frac{4}{15},\ p_{Y|X}(2|2)=\frac{9}{15} \text{ (conditional on } X=2).$$

Each of these conditional distributions differs from the marginal distribution of Y:

$$p_Y(0)=0.30,\ p_Y(1)=0.55,\ p_Y(2)=0.15.$$

Again, it is sufficient to show that just one of the conditional distributions differs from the marginal distribution, even though we have shown it for all three. Similarly, the conditional distribution of $X|Y$ depends on the value of Y, which is left for the reader to verify.

Example 8.26 *Consider the following probability table for $X \in \{0,1\}$ and $Y \in \{-1,0,1\}$, with the marginal probabilities also calculated:*

		Y			
		-1	0	1	$p_X(x)$
X	0	7/36	2/36	3/36	12/36
	1	14/36	4/36	6/36	24/36
	$p_Y(y)$	21/36	6/36	9/36	

The conditional distributions of $Y|X$ are the same as the marginal distribution of Y:

$$p_{Y|X}(-1|0)=\frac{7}{12},\ p_{Y|X}(0|0)=\frac{2}{12},\ p_{Y|X}(1|0)=\frac{3}{12} \text{ (conditional on } X=0)$$

$$p_{Y|X}(-1|1)=\frac{14}{24},\ p_{Y|X}(0|1)=\frac{4}{24},\ p_{Y|X}(1|1)=\frac{6}{24} \text{ (conditional on } X=1).$$

The probabilities in the $X=1$ row are all proportional to the corresponding probabilities in the $X=0$ row; specifically, each one is 2 times the probability of the corresponding probability in the $X=0$ row. This proportionality is what leads the conditional pmfs to be unchanged, and therefore equal to the marginal distribution function. Similarly, the conditional distributions of $X|Y$ are the same as the marginal distribution of X, which can be verified by the reader. Therefore, in this case, the random variables X and Y are independent.

The concept of independence can be generalized to additional random variables, with the intuition remaining the same. When random variables have no relationship with each other, they are independent; otherwise, they are dependent. Definition 8.17, which considered the case of two discrete random variables, can be extended to the general case of multiple discrete random variables as in definition 8.18.

Definition 8.18 *The m discrete random variables $X_1, X_2, \ldots, X_m$, where $m \geq 2$, are* ***independent*** *if and only if*

$$P(X_1 = x_{k_1}^*, X_2 = x_{k_2}^*, \ldots, X_m = x_{k_m}^*) = P(X_1 = x_{k_1}^*)P(X_2 = x_{k_2}^*) \cdots P(X_m = x_{k_m}^*)$$

for any possible joint outcome $(x_{k_1}^*, x_{k_2}^*, \ldots, x_{k_m}^*)$ of $(X_1, X_2, \ldots, X_m)$. If this equality fails for any joint outcome $(x_{k_1}^*, x_{k_2}^*, \ldots, x_{k_m}^*)$, then the discrete random variables $X_1, X_2, \ldots, X_m$ are ***dependent***.

Using the notation $p_{X_1X_2\cdots X_m}(x_{k_1}^*, x_{k_2}^*, \ldots, x_{k_m}^*)$ to denote the joint pmf of $(x_{k_1}^*, x_{k_2}^*, \ldots, x_{k_m}^*)$, the definition can be restated as the discrete random variables $X_1, X_2, \ldots, X_m$ being independent if and only if

$$p_{X_1X_2\cdots X_m}(x_{k_1}^*, x_{k_2}^*, \ldots, x_{k_m}^*) = p_{X_1}(x_{k_1}^*)p_{X_2}(x_{k_2}^*) \cdots p_{X_m}(x_{k_m}^*)$$

for any possible joint outcome $(x_{k_1}^*, x_{k_2}^*, \ldots, x_{k_m}^*)$ of $(X_1, X_2, \ldots, X_m)$. Equivalently, the definition could be stated in terms of cdfs rather than pdfs, with the discrete random variables $X_1, X_2, \ldots, X_m$ being independent if and only if

$$F_{X_1X_2\cdots X_m}(x_{k_1}^*, x_{k_2}^*, \ldots, x_{k_m}^*) = F_{X_1}(x_{k_1}^*)F_{X_2}(x_{k_2}^*) \cdots F_{X_m}(x_{k_m}^*)$$

for any possible joint outcome $(x_{k_1}^*, x_{k_2}^*, \ldots, x_{k_m}^*)$ of $(X_1, X_2, \ldots, X_m)$.

As with two discrete random variables, the independence of multiple random variables is characterized by the joint probability of an outcome being equal to the product of the marginal probabilities. It becomes more difficult to check independence when the number of random variables increases, as there are more joint probabilities to consider. That said, the most important use of this concept is to apply the fact that the joint probability is equal to the product of the marginal probabilities when it is either known or assumed that a collection of random variables are independent.

8.5 Linear Transformations

8.5.1 Linear Transformations of One Random Variable

Section 6.7 considered linear transformations of univariate variables and the descriptive statistics of those linear transformations. In the same way that a linear transformation $y = a + bx$ of a discrete numerical variable x can be constructed, a linear transformation $Y = a + bX$ of a discrete random variable X can also be constructed. The population descriptive statistics for the resulting discrete random variable Y are analogous to those from proposition 6.6 and are summarized in the following table. We hold off on formally presenting these population results as a proposition since the same results also hold for continuous random variables. It is more convenient to state the proposition later in this book (see chapter 10) for all random variables rather than specifically for discrete random variables.

	Sample	Population
Linear transformation	$y_i = a + bx_i$	$Y = a + bX$
Mean	$\bar{y} = a + b\bar{x}$	$\mu_Y = a + b\mu_X$
Variance	$s_y^2 = b^2 s_x^2$	$\sigma_Y^2 = b^2\sigma_X^2$
Standard deviation	$s_y = \lvert b\rvert s_x$	$\sigma_Y = \lvert b\rvert \sigma_X$

As with sample descriptive statistics, (1) the additive constant a only affects the location (mean) of the distribution and (2) the scaling constant b affects both the location (mean) and the dispersion (variance or standard deviation) of the distribution.

Example 8.27 (Earnings) *In example 6.29, a linear transformation of weekly earnings, given by $y = 52x$, was used to get annualized earnings. That example considered actual data and the descriptive statistics associated with a sample. An analogous transformation, this time in terms of the population rather than the sample, can be done for the associated random variables. If X is a random variable associated with an individual's weekly earnings, where the individual is drawn randomly from the population of possible individuals, the random variable $Y = 52X$ is a random variable representing the individual's annualized earnings. In this case, $a = 0$ and $b = 52$, so that*

$$\mu_Y = 52\mu_X,\ \sigma_Y^2 = (52)^2\sigma_X^2,\ \text{and}\ \sigma_Y = 52\sigma_X.$$

Example 8.28 (Placing a bet) *Suppose that you place a bet on a certain outcome (e.g., a team winning a game, a political candidate winning an election). Let X denote the binary variable indicating whether the event happens ($X = 1$) or doesn't happen ($X = 0$). If you win the bet, you earn w dollars. If you lose the bet, you lose ℓ dollars. Then the net gain Y on the bet is*

$$Y = -\ell + (w + \ell)X,$$

so $Y = w$ when $X = 1$ and $Y = -\ell$ when $X = 0$. Since X is a binary variable, the population mean and population variance of X are $\mu_X = p_X(1)$ and $\sigma_X^2 = p_X(1)(1 - p_X(1))$, respectively. The population mean of Y is

$$\mu_Y = -\ell + (w + \ell)\mu_X = -\ell + (w + \ell)p_X(1).$$

The population variance of Y is

$$\sigma_Y^2 = (w + \ell)^2\sigma_X^2 = (w + \ell)^2 p_X(1)(1 - p_X(1)).$$

The population standard deviation of Y is

$$\sigma_Y = |w + \ell|\sigma_X = (w + \ell)\sqrt{p_X(1)(1 - p_X(1))}.$$

since w and ℓ are both positive.

8.5.2 Linear Transformations of Two Random Variables

Section 6.7 introduced linear transformations for bivariate data, with $v = a + bx$ and $w = c + dy$ as linear transformations of variables x and y. Analogous linear transformations for the case of two discrete random variables X and Y can be constructed as follows:

$$V = a + bX \text{ and } W = c + dY,$$

where a, b, c, and d are known constants. The following table provides the population covariance and correlation of V and W in terms of the population covariance and correlation of X and Y, along with a side-by-side comparison to the results for the analogous sample descriptive statistics from section 6.7:

	Sample	Population
Linear transformations	$v = a + bx,\ w = c + dy$	$V = a + bX,\ W = c + dY$
Covariance	$s_{vw} = bds_{xy}$	$\sigma_{VW} = bd\sigma_{XY}$
Correlation	$r_{vw} = \begin{cases} r_{xy} & \text{if } bd > 0 \\ -r_{xy} & \text{if } bd < 0 \end{cases}$	$\rho_{VW} = \begin{cases} \rho_{XY} & \text{if } bd > 0 \\ -\rho_{XY} & \text{if } bd < 0 \end{cases}$

Example 8.29 (Placing two bets) *Example 8.28 considered a situation in which a bet was placed on the outcome of an event. For this example, let's consider two separate bets on different events, where $X = 1$ if the first event occurs and 0 otherwise and $Y = 1$ if the second event occurs and 0 otherwise. For the first bet, the winnings are w_1 if $X = 1$ and the losses are ℓ_1 if $X = 0$. For the second bet, the winnings are w_2 if $Y = 1$ and the losses are ℓ_2 if $Y = 0$. The net gains on the two bets, V and W respectively, are given by the following linear transformations:*

$$V = -\ell_1 + (w_1 + \ell_1)X \text{ and } W = -\ell_2 + (w_2 + \ell_2)Y.$$

The population covariance between V and W is

$$\sigma_{VW} = (w_1 + \ell_1)(w_2 + \ell_2)\sigma_{XY},$$

and the population correlation between V and W is $\rho_{VW} = \rho_{XY}$ since $w_1 > 0$, $\ell_1 > 0$, $w_2 > 0$, and $\ell_2 > 0$. Perhaps not surprisingly, the sign of the population correlation between the net gains on the two bets is the same as the sign of the population correlation between the two events. If $Y = 1$ is more likely when $X = 1$ as compared to $X = 0$, this correlation is positive. If $Y = 1$ is less likely when $X = 1$ as compared to $X = 0$, this correlation is negative. The equality $\rho_{VW} = \rho_{XY}$ also tells us that, in addition to the signs being equal, the magnitude of the two population correlations is the same. As seen in the previous table, this last finding is general, with linear transformations of two random variables having a population correlation with the same magnitude as the population correlation of the underlying two random variables.

8.6 Linear Combination of Multiple Random Variables

Section 7.2.4 introduced the concept of a linear combination for two variables, where $v = k + ax + by$ for the known constants k, a, and b. In this section, an analogous linear combination is constructed in the context of two random variables. For the known constants k, a, and b, let V denote a random variable that is defined as the following linear combination of the two discrete random variables X and Y:

$$V = k + aX + bY.$$

The following table summarizes the population descriptive statistics for V, along with a side-by-side comparison to the results for the analogous sample descriptive statistics from section 6.7:

	Sample	Population
Linear combination	$v_i = k + ax_i + by_i$	$V = k + aX + bY$
Mean	$\bar{v} = k + a\bar{x} + b\bar{y}$	$\mu_V = k + a\mu_X + b\mu_Y$
Variance	$s_v^2 = a^2 s_x^2 + b^2 s_y^2 + 2ab s_{xy}$	$\sigma_V^2 = a^2\sigma_X^2 + b^2\sigma_Y^2 + 2ab\sigma_{XY}$
Standard deviation	$s_v = \sqrt{s_v^2}$	$\sigma_V = \sqrt{\sigma_V^2}$

Recall that proposition 7.4 included the results for two special cases of the linear combination, the sum of two variables ($v = x + y$) and the difference of two variables ($v = x - y$). In the case of random variables,

$$V = X + Y \Longrightarrow \sigma_V^2 = \sigma_X^2 + \sigma_Y^2 + 2\sigma_{XY},$$

and

$$V = X - Y \Longrightarrow \sigma_V^2 = \sigma_X^2 + \sigma_Y^2 - 2\sigma_{XY}.$$

Example 8.30 (Correlated purchases) *Suppose that a website sells two types of widgets (widget A and widget B) and a visitor to the website is allowed to purchase one of each type of widget. For a given visitor to the website, let $X = 1$ if they buy a widget of type A and $X = 0$ if not, and let $Y = 1$ if they buy a widget of type B and $Y = 0$ if not. The joint pmf for X and Y is given in the following probability table:*

		Y = 0	Y = 1
X	0	0.70	0.05
	1	0.05	0.20

X and Y are positively correlated here, with

$$\begin{aligned}\sigma_{XY} &= (0-0.25)(0-0.25)(0.70) + (0-0.25)(1-0.25)(0.05)\\ &\quad + (1-0.25)(0-0.25)(0.05) + (1-0.25)(1-0.25)(0.20) = 0.1375\\ \rho_{XY} &= \frac{\sigma_{XY}}{\sigma_X\sigma_Y} = \frac{0.1375}{\sqrt{0.25(1-0.25)}\sqrt{0.25(1-0.25)}} = \frac{0.1375}{(0.25)(0.75)} \approx 0.7333.\end{aligned}$$

Letting V denote the total number of widgets purchased by a given website visitor, $V = X + Y$, the population mean of V is $\mu_V = \mu_X + \mu_Y = 0.25 + 0.25 = 0.5$. Due to the positive correlation/covariance of X and Y, the population variance of V, $\sigma_V^2 = \sigma_X^2 + \sigma_Y^2 + 2\sigma_{XY}$, is greater than the sum of the variances of X and Y ($\sigma_X^2 + \sigma_Y^2$). X and Y tend to be on the same sides of their respective population means, which here means that both being 0 or both being 1 is more likely than having one equal to 1 and the other equal to 0. This positive correlation causes the dispersion in $V = X + Y$ to be larger than the dispersion coming from *just the variance of X (σ_X^2) and the variance of Y (σ_Y^2). Since $\sigma_X^2 = p_X(1)(1 - p_X(1))$ and $\sigma_Y^2 = p_Y(1)(1 - p_Y(1))$, the sample variance of V is*

$$\sigma_V^2 = (0.25)(0.75) + (0.25)(0.75) + (2)(0.1375) = 0.65,$$

compared to $\sigma_X^2 + \sigma_Y^2 = 0.375$.

If X and Y were instead negatively correlated, there would be less dispersion in V ($\sigma_V^2 < \sigma_X^2 + \sigma_Y^2$) since X and Y would tend to be on opposite sides of their respective population means.

These results can be generalized to an even larger number of variables in the linear combination. For $m \geq 2$, let $X_1, X_2, \ldots, X_m$ denote random variables. Then, for the constants k and $a_1, a_2, \ldots, a_m$, a linear combination of the random variables is

$$V = k + a_1X_1 + a_2X_2 + \cdots + a_mX_m = k + \sum_{j=1}^{m} a_jX_j.$$

Analogous to proposition 7.5, proposition 8.7 provides general results for the linear combination of $m \geq 2$ random variables. We do not specify that the random variables are discrete, as the properties hold for any types of random variables.

Proposition 8.7 *Suppose that k and $a_1, a_2, \ldots, a_m$ are known constants, and*

$$V = k + a_1X_1 + a_2X_2 + \cdots + a_mX_m = k + \sum_{j=1}^{m} a_jX_j$$

is a linear combination of the m random variables $X_1, X_2, \ldots, X_m$. The population statistics for the random variable V have the following relationships to the population statistics for the random variables $X_1, X_2, \ldots, X_m$:

i. *(population mean)* $\mu_V = k + a_1\mu_{X_1} + a_2\mu_{X_2} + \cdots + a_m\mu_{X_m} = k + \sum_{j=1}^{m} a_j\mu_{X_j}$.

ii. *(population variance)* $\sigma_V^2 = \sum_{j=1}^{m} a_j^2\sigma_{X_j}^2 + 2\sum_{j=1}^{m-1}\sum_{\ell=j+1}^{m} a_ja_\ell\sigma_{X_jX_\ell}$.

iii. *(population standard deviation)* $\sigma_V = \sqrt{\sigma_V^2}$
$= \sqrt{\sum_{j=1}^{m} a_j^2\sigma_{X_j}^2 + 2\sum_{j=1}^{m-1}\sum_{\ell=j+1}^{m} a_ja_\ell\sigma_{X_jX_\ell}}$.

Property (ii) of proposition 8.7 implies that, for random variables that are not independent of each other, the population variance of the linear combination depends on the covariances that exist between any pair of random variables. For a linear combination of $m = 3$ random variables ($V = k + a_1X_1 + a_2X_2 + a_3X_3$), the population variance is

$$\sigma_V^2 = a_1^2\sigma_{X_1}^2 + a_2^2\sigma_{X_2}^2 + a_3^2\sigma_{X_3}^2 + 2a_1a_2\sigma_{X_1X_2} + 2a_1a_3\sigma_{X_1X_3} + 2a_2a_3\sigma_{X_2X_3},$$

which has $\binom{3}{2} = 3$ covariance terms. More generally, the number of covariance terms is $\binom{m}{2}$ for a linear combination of m random variables. As seen in the next section, the population variance σ_V^2 simplifies considerably when $X_1, X_2, \ldots, X_m$ are independent since all the covariance terms are zero, leaving just the sum of the scaled underlying variances.

8.6.1 Linear Combination of Independent Random Variables

If $V = k + aX + bY$ is a linear combination of the *independent* discrete random variables X and Y, the population variance of V simplifies to

$$\sigma_V^2 = a^2\sigma_X^2 + b^2\sigma_Y^2$$

since X and Y being independent implies that $\rho_{XY} = 0$. This result can be generalized to a linear combination of more than two independent discrete random variables. If $X_1, X_2, \ldots, X_m$ are independent random variables, the covariance between any two random variables equals zero, and the population variance (proposition 8.7) simplifies to

$$\sigma_V^2 = a_1^2\sigma_{X_1}^2 + a_2^2\sigma_{X_2}^2 + \cdots + a_m^2\sigma_{X_m}^2 = \sum_{j=1}^{m} a_j^2\sigma_{X_j}^2.$$

Two specific cases of linear combinations that are particularly useful are the sum of independent random variables and the average of independent random variables.[2]

1. Sum of independent random variables: $V = X_1 + X_2 + \cdots + X_m$:

$$\mu_V = \mu_{X_1} + \mu_{X_2} + \cdots + \mu_{X_m} = \sum_{j=1}^{m} \mu_{X_j}$$

$$\sigma_V^2 = \sigma_{X_1}^2 + \sigma_{X_2}^2 + \cdots + \sigma_{X_m}^2 = \sum_{j=1}^{m} \sigma_{X_j}^2.$$

2. Average of independent random variables: $V = \frac{1}{m}(X_1 + X_2 + \cdots + X_m)$:

$$\mu_V = \frac{1}{m}\mu_{X_1} + \frac{1}{m}\mu_{X_2} + \cdots + \frac{1}{m}\mu_{X_m} = \frac{1}{m}\sum_{j=1}^{m} \mu_{X_j}$$

$$\sigma_V^2 = \frac{1}{m^2}\sigma_{X_1}^2 + \frac{1}{m^2}\sigma_{X_2}^2 + \cdots + \frac{1}{m^2}\sigma_{X_m}^2 = \frac{1}{m^2}\sum_{j=1}^{m} \sigma_{X_j}^2.$$

Example 8.31 (Three rolls of a die) *Suppose that a fair die is rolled three times, where X_1, X_2, and X_3 are the random variables associated with the possible outcomes of the three rolls. Assume that the three random variables are independent, meaning that the outcome of each roll has nothing to do with the outcomes of the other rolls. In that case, the random variable representing the* sum *of the three outcomes, $X_1 + X_2 + X_3$, has the population mean*

$$\mu_{X_1} + \mu_{X_2} + \mu_{X_3} = 3.5 + 3.5 + 3.5 = 10.5$$

and the population variance

$$\sigma_{X_1}^2 + \sigma_{X_2}^2 + \sigma_{X_3}^2 = \frac{35}{12} + \frac{35}{12} + \frac{35}{12} = \frac{35}{4}.$$

Since each roll has the same population mean (3.5) and population variance (35/12), the expected value of the sum equals three times the expected value of a single roll, and the population variance of the sum equals three times the population variance of a single roll.

The random variable representing the average *of the three outcomes,* $\frac{1}{3}(X_1 + X_2 + X_3)$, *has the population mean*

$$\frac{1}{3}\mu_{X_1} + \frac{1}{3}\mu_{X_2} + \frac{1}{3}\mu_{X_3} = \frac{1}{3}(3.5 + 3.5 + 3.5) = 3.5$$

and the population variance

$$\frac{1}{9}\sigma^2_{X_1} + \frac{1}{9}\sigma^2_{X_2} + \frac{1}{9}\sigma^2_{X_3} = \frac{1}{9}\left(\frac{35}{12} + \frac{35}{12} + \frac{35}{12}\right) = \frac{35}{36}.$$

The expected value for the average of three die rolls is the same as the expected value of a single roll, and the population variance of the average of three die rolls is $\frac{1}{3}$ *times the population variance of a single roll.*

Example 8.32 (100 coin tosses) *Suppose that a fair coin is tossed* 100 *times, where* $X_1, X_2, \ldots, X_{100}$ *are the random variables associated with the possible outcomes of the tosses. Assume that the coin tosses are all independent of each other. For any given coin toss* X_j*, the population mean is* 0.5*, and the population variance is* $(0.5)(1 - 0.5) = 0.25$. *The random variable representing the sum of the* 100 *outcomes has the population mean*

$$\mu_{X_1} + \mu_{X_2} + \cdots + \mu_{X_{100}} = (100)(0.5) = 50$$

and the population variance

$$\sigma^2_{X_1} + \sigma^2_{X_2} + \cdots + \sigma^2_{X_{100}} = (100)(0.25) = 25.$$

For 100 *coin tosses, the population mean or expected value is* 50 *heads. The population variance is* 100 *times the population variance of an single coin toss.*

The random variable representing the average of the 100 *outcomes has the population mean*

$$\frac{1}{100}(\mu_{X_1} + \mu_{X_2} + \cdots + \mu_{X_{100}}) = \left(\frac{1}{100}\right)(100)(0.5) = 0.5$$

and the population variance

$$\frac{1}{100^2}\sigma^2_{X_1} + \frac{1}{100^2}\sigma^2_{X_2} + \cdots + \frac{1}{100^2}\sigma^2_{X_{100}} = \left(\frac{1}{100^2}\right)(100)(0.25) = 0.0025.$$

The expected value for the average of 100 *tosses is the same as the expected value of a single toss, both equal to* 0.5*, and the population variance of the average of* 100 *tosses is* $\frac{1}{100}$ *times the population variance of a single toss.*

For examples 8.31 and 8.32, the underlying independent random variables $X_1, X_2, \ldots, X_m$ share a common population mean μ_X, equal to 3.5 for the die rolls and 0.5 for the coin tosses, and a common population variance σ^2_X, equal to $\frac{35}{12}$ for the die rolls and 0.25 for the coin tosses. When the independent variables $X_1, X_2, \ldots, X_m$ share a common mean μ_X and a common variance σ^2_X, it follows that the population mean, population variance, and population standard deviation of the sum of the random variables are $m\mu_X$, $m\sigma^2_X$, and $\sqrt{m}\sigma_X$, respectively, and the population mean, population variance, and population standard deviation of the average of the random variables are μ_X, $\frac{\sigma^2_X}{m}$, and $\frac{\sigma_X}{\sqrt{m}}$, respectively.

8.7 Expected Values of Functions of Discrete Random Variables

Sections 8.5 and 8.6 considered linear functions of discrete random variables, with section 8.5 focused on single discrete random variables and section 8.6 focused on multiple discrete random variables. In this section, we consider more general functions that include linear functions as a special case, but also allow nonlinearity. In particular, we discuss how to determine the expected value or population mean of functions of random variables.

Example 8.33 (Office space) *A new employee is hired by a company housed in a large office building. All the offices are square. Let X be a random variable denoting the length of an office wall (in feet). The employee is offered a new office corresponding to a draw of the discrete random variable X, with $p_X(8) = 0.5$, $p_X(10) = 0.3$, and $p_X(12) = 0.2$. What is the expected* area *of the new employee's office? Since the office is square, the office area is the random variable X^2, whose pmf is $p_{X^2}(64) = 0.5$, $p_{X^2}(100) = 0.3$, and $p_{X^2}(144) = 0.2$. The expected value of the office area is then*

$$\mu_{X^2} = E(X^2) = (64)(0.5) + (100)(0.3) + (144)(0.2) = 90.8 \textit{ (square feet)}.$$

As example 8.33 illustrates, there is a two-step approach to calculating the expected value of a function of a discrete random variable. First, apply the function to each possible outcome x_k^* to get the full set of possible outcomes. Second, use the original probabilities $p_X(x_k^*)$ to weight the new set of possible outcomes. This approach is stated as a general result in proposition 8.8:

Proposition 8.8 *For any function $g(x)$ and a discrete random variable X with pmf $p_X(\cdot)$, the population mean or expected value of $g(X)$ is*

$$\mu_{g(X)} = E(g(X)) = \sum_k g(x_k^*) p_X(x_k^*).$$

A special case of proposition 8.8 is the formula for the population variance (definition 8.6), where $g(X) = (X - \mu_X)^2$.

Example 8.34 (Fair-die gamble) *Consider the following gamble based upon the roll of a fair die. A player pays \$2, rolls a fair die whose outcome is the random variable $X \in \{1, 2, 3, 4, 5, 6\}$, and receives a payout of $\sqrt{X}$ dollars. Applying proposition 8.8 with $g(X) = \sqrt{X} - 2$ yields an expected value of net winnings of*

$$\sum_{k=1}^{6} \left(\sqrt{k} - 2\right)\left(\frac{1}{6}\right) = \frac{1}{6}\left(\sqrt{1} + \sqrt{2} + \sqrt{3} + \sqrt{4} + \sqrt{5} + \sqrt{6}\right) - 2 \approx -0.1947.$$

Proposition 8.8 can be generalized to multiple discrete random variables. Proposition 8.9 considers the case of two discrete random variables, where the function of interest now has two arguments (random variables X and Y) and the joint pmf $p_{XY}(\cdot, \cdot)$ is used.

Proposition 8.9 *For any function $g(x, y)$ and discrete random variables X and Y with joint pmf $p_{XY}(\cdot, \cdot)$, the population mean or expected value of $g(X, Y)$ is*

$$\mu_{g(X,Y)} = E(g(X, Y)) = \sum_{(k,\ell)} g(x_k^*, y_\ell^*) p_{XY}(x_k^*, y_\ell^*).$$

A special case of proposition 8.9 is the formula for the population covariance (definition 8.15), where $g(X,Y)=(X-\mu_X)(Y-\mu_Y)$.

Example 8.35 (Two projects) *A firm is undertaking two projects A and B, each of which may succeed or fail. The joint probabilities of success and failure are* $p_{AB}(0,0)$ *(both fail),* $p_{AB}(0,1)$ *(A fails, B succeeds),* $p_{AB}(1,0)$ *(A succeeds, B fails), and* $p_{AB}(1,1)$ *(both succeed). The firm realizes a profit of* K_A *if project A succeeds, a profit of* K_B *if B succeeds, and an additional profit* K_{extra} *if both succeed, where* K_A*,* K_B*, and* K_{extra} *are constants. Then the firm's profit is*

$$g(A,B)=K_AA+K_BB+K_{extra}AB.$$

The additional profit K_{extra} *is realized only when* $AB=1$ *or* $A=B=1$*. Applying proposition 8.9, the expected profit is*

$$E(g(A,B))=(0)p_{AB}(0,0)+(K_B)p_{AB}(0,1)+(K_A)p_{AB}(1,0)+(K_A+K_B+K_{extra})p_{AB}(1,1)$$

or

$$E(g(A,B))=K_A(p_{AB}(1,0)+p_{AB}(1,1))+K_B(p_{AB}(0,1)+p_{AB}(1,1))+K_{extra}p_{AB}(1,1).$$

Notes

1. The result for $F_X(x_k^*)$ simplifies using the formula for the sum of a finite geometric series: $\sum_{j=1}^{k} ar^{j-1}=\frac{a(1-r^k)}{(1-r)}$.
2. The results for the expecation μ_V do not require independence, as μ_V does not depend upon the covariances.

Exercises

1. The pmf of the number of machines X at a given factory that break down in a day is

# of breakdowns (x_k^*)	0	1	2	3
probability ($p_X(x_k^*)$)	0.2	0.5	0.25	0.05

   a. Determine the cdf of the daily breakdowns.
   b. What is the population mean of the daily breakdowns?
   c. What is the population variance of the daily breakdowns?

2. Suppose that you want to "load" a six-sided die so you are more likely to roll a 6.
   a. If you get to choose the probability $p=P(6)$, with the remaining five outcomes $(1,2,3,4,5)$ still having equal probabilities, what choice of p would give you an expected value of 5 for a roll of the six-sided die?
   b. With your choice of p from (a), what is the population variance of the loaded die roll? How does this variance compare to the variance of the fair die roll?
   c. *Determine the population variance of the loaded die roll in terms of p. Using R, define a function that takes p as an argument and returns the population variance. Then, evaluating the population variance for all $p\in\{0,0.01,0.02,\ldots,0.99,1\}$, determine the value of p that maximizes the population variance.

3. A company sells widgets and receives shipments of its widgets from a supplier. Suppose that the true probability of a defect in any given widget is 10%, and assume that the production process is independent for every widget made. Consider picking two widgets at random from a shipment, and let the random variable $X \in \{0, 1, 2\}$ denote the number of defective widgets.
   a. What is the pmf of X?
   b. What is the cdf of X evaluated at 1?
   c. What is the expected value of X?
   d. What is the population standard deviation of X?
4. A venture capital (VC) firm raises funds from investors and makes investments in early-stage companies. One particular VC firm, called Mill VC, invests exactly \$1 million in each company in its portfolio. For each investment, the pmf of the amount X (in millions) that Mill VC receives after 2 years is

x	0	5	10	50
$p_X(x)$	0.90	0.05	0.03	0.02

   Let n be the number of investments made by Mill VC, and assume that the outcomes of Mill VC's investments are independent of each other.
   a. What is the expected profit associated with any single investment?
   b. If $n = 4$, what is the probability that Mill VC's total profit is negative?
   c. For $n = 10$, conduct 10,000 simulations in *R* to approximate (i) the probability that Mill VC's total profit is negative and (ii) Mill VC's average profit. (For each simulation, make 10 independent draws of the discrete random variable X, and calculate the total profit.) Does the simulated average profit make sense given the answer in (a)?
5. An investor is considering three options for a \$1,000 investment, as follows:

$$\textbf{Option 1: } \begin{cases} \text{Profit of \$10,000} & \text{with probability 0.15} \\ \text{Loss of \$1,000} & \text{with probability 0.85.} \end{cases}$$

$$\textbf{Option 2: } \begin{cases} \text{Profit of \$1,000} & \text{with probability 0.50} \\ \text{Profit of \$500} & \text{with probability 0.30} \\ \text{Loss of \$500} & \text{with probability 0.20.} \end{cases}$$

   Option 3: Profit of \$400 with probability 1.
   a. Which option has the highest expected profit?
   b. Which option has the highest population variance?
   c. Thinking about option 1, what probability p on the \$10,000 profit (and $1 - p$ on the \$1,000 loss) would yield an expected profit equal to that of option 3?
   d. If the investor randomly chooses one of the three options (each with probability 1/3), what is the probability that the investor realizes a profit?

e. Let X_2 denoting the random variable associated with the profit or loss associated with option 2. Draw the cdf associated with X_2.

6. The joint pmf of two random variables X and Y is given by the following table:

		Y 1	2	3
	1	0.24	0.12	0.04
X	2	0.12	0.06	0.02
	3	0.24	0.12	0.04

a. Are X and Y independent?

b. What is the expected value of $X+Y$?

c. What is the population variance of $X+Y$?

d. What is the expected value of XY?

e. What is the probability of $X=1$ conditional on $Y=X+1$?

7. For two positive numbers p and q with $p+q<1$, the joint pmf of random variables X and Y is given by the following table:

		Y y_1^*	y_2^*
	x_1^*	p	???
X	x_2^*	q	???

If X and Y are independent, what are the values of the two joint probabilities (in terms of p and q) in the second column?

8. A bank has two car lanes with automatic teller machines (ATMs). The following table provides the joint pmf for the number of cars in each of the two lanes at a given time:

		Cars in Lane 2 (Y) 0	1	2	3
	0	0.05	0.15	0.02	0
Cars in Lane 1 (X)	1	0.15	0.20	0.10	0
	2	0.02	0.10	0.10	0.03
	3	0	0	0.03	0.05

a. What is the probability that exactly one car is in lane 1?

b. What is the probability that exactly one car is in either lane?

c. What is the probability that exactly one car is in both lanes?

d. What is the probability that there are the same number of cars in the two lanes?

e. What is the probability that exactly one car is in lane 1 if two cars are in lane 2?

f. Are X and Y independent?

g. Calculate the population covariance of X and Y.

h. Consider the total number of cars in the two lanes, given by $X + Y$. What is the pmf of $X + Y$? What is $E(X + Y)$? What is $Var(X + Y)$?

i. Consider the difference in the number of cars in the two lanes, given by $X - Y$. What is the pmf of $X - Y$? What is $E(X - Y)$? What is $Var(X - Y)$?

j. Does the difference between $Var(X + Y)$ and $Var(X - Y)$ found in (h) and (i) make sense given the population covariance found in (g)?

9. Consider the population of families in the US having at least two children. The following table describes the probability that a randomly chosen family has a third child, broken down by the number of boys that the family has among its first two children:

# of boys among first two children	0	1	2
Probability of having a third child	0.30	0.25	0.30

Let $B \in \{0, 1, 2\}$ be the random variable representing the number of boys among the first two children. Let T be the random variable indicating whether the family has a third child, with $T = 1$ for families having a third child and $T = 0$ for families not having a third child. Assume the following pmf for B: $p_B(0) = 0.25$, $p_B(1) = 0.50$, and $p_B(2) = 0.25$.

a. Explain why the probabilities in the table do not add up to 1.

b. In terms of B and T, what does the 0.25 value in the table represent?

c. What is the unconditional probability that a family with two children has a third child?

d. Show the joint pmf of B and T in a table.

e. Are B and T independent?

f. What is the population covariance between B and T?

10. You are faced with two possible investment choices, whose annual returns are represented by the random variables X and Y. The possible returns for the first investment (X) are 1%, 2%, 3%, and 4%, and the possible returns for the second investment (Y) are 2%, 3%, and 4%. For simplicity, we omit the "%" sign below. The joint pmf is

		X			
		1	2	3	4
	2	0.10	0.05	0.05	0
Y	3	0.05	0.15	0.05	0.05
	4	0.05	0.05	0.15	0.25

a. What is the joint cdf evaluated at $X = 2$ and $Y = 3$ ($F_{XY}(2, 3)$)?

b. What are the marginal pmfs of X and Y?

c. What is the conditional pmf of X given $Y = 3$?

d. What is the expected value of X?

e. What is the expected value of Y?

f. What is the population variance of X?

g. What is the population variance of Y?

h. What is the population covariance between X and Y?

i. You put \$1,000 into investment X and \$2,000 into investment Y. The trades cost a total of \$20 to execute. Write an expression for the random variable G, which represents your net gain (in dollars) over the next year. (Remember that the returns are in percentages, so, for instance, $X = 1$ corresponds to a 1% return on \$1000, which is $(0.01)(1000) = \$10$.)

j. What is the expected value of G?

k. What is the population variance of G?

11. Companies A and B are in an R&D race to develop a new technology. Each company may develop the technology in one, two, or three years. For each company, the probability that the technology is developed in one year is 20%, the probability that the technology is developed in two years is 50%, and the probability that the technology is developed in three years is 30%. Assume that the companies' efforts are independent of each other; the time that it takes Company B to develop the technology has nothing to do with how long it takes Company A to do so.

Regardless of who develops the technology first, both companies begin commercial production in five years. If one company develops the technology first, its market share is 75% and its competitor's market share is 25%. If the two companies take the same number of years to develop the technology, each has a market share of 50%.

Define the random variables A and B as the number of years that it takes company A and company B to develop the technology, respectively.

a. Find the joint pmf of A and B.

b. Let M denote the eventual market share for company A. What is the pmf of M?

c. Each year of R&D costs the company \$5 million. The total market is worth \$100 million in net revenues (sales minus costs of production), which the two firms divide according to their market shares. Let the random variable R be the profits of company A, equal to net revenues minus the R&D costs.

i. Write R as a function of A and M.

ii. What is the expected value of R (in millions of dollars)?

12. Suppose a fair die is rolled 200 times, with each roll being independent. The rolls are recorded as a 200-character sequence of integers between 1 and 6 (inclusive).

a. What is the probability that the sequence 354 occurs in the first three rolls?

b. *What is the expected value of the number of times that the sequence 354 shows up in the full 200-roll sequence? (Hint: Think about the random variables X_j defined as 1 if 354 shows up starting with the jth roll and 0 otherwise. Then, use the result in proposition 8.7 for the expected value of a linear combination of random variables.)

c. Conduct 10,000 simulations in R to confirm your answer to (b). Each simulation involves rolling a fair die 200 times and counting the number of times that the sequence 354 occurs.

13. The annual income X (in thousands of dollars) in a population of workers has the expected value 70 and population standard deviation 50. Assume that all workers' incomes are independent of each other. For this question, you may assume that X is a discrete random variable.

    a. What is the expected value and population standard deviation for the average of two workers' annual incomes?

    b. What is the expected value and population standard deviation for the difference between two workers' annual incomes?

    c. What is the expected value and population standard deviation for the average of 10 workers' annual incomes?

    d. What is the expected value and population standard deviation for the total of 10 workers' annual incomes?

    e. Assume that every worker in the population works 2,080 hours (52 weeks times 40 hours per week). Now, answer (d) using hourly incomes rather than annual incomes.

14. For the population of graduating economics students, the probability distribution (pmf) associated with X equals the number of job offers that a student receives before graduation is as follows:

x	0	1	2	3	4
$p_X(x)$	0.15	0.40	0.30	0.10	0.05

    a. Use R to determine the expected value of X. (Hint: Create two vectors, one containing the possible outcomes for X and one containing their associated probabilities, and then do the appropriate calculation based on these vectors.)

    b. Use R to determine the population variance of X.

    c. Use R to determine the population standard deviation of X.

    d. Use the `sample` function to create a vector with 10,000 random draws of X.

        i. How do the sample mean, sample variance, and sample standard deviation compare to their population counterparts?

        ii. What proportion of the random draws are less than or equal to 2? How does that compare to $F_X(2)$?

        iii. Draw a histogram or bar chart of the 10,000 random draws.

    e. Suppose that 400 economics students graduate in a given year, and the career services center is interested in T = the *total* number of job offers for the 400 students. You can simulate one draw of T by summing the results of a `sample` call—for instance, something like `sum(sample(...,400,...,replace=TRUE))`, where you fill in the first ... with the outcome vector and the second ... with the probability vector. Try it. What do you get for the draw of T?

    f. Simulate and store the results from 5,000 random draws of T.

        i. Draw a histogram of the 5,000 random draws.

ii. What is the sample average of the draws? How does it compare to the expected value of T? (Note: $E(T)$ should be $400E(X)$, and you found $E(X)$ in (a).)

iii. What is the sample standard deviation of the draws?

iv. Based upon your draws, fill in the blank in the following statement: "I estimate that there's a 90% probability that the total number of job offers for the 400 students is greater than —."

9

Models of Discrete Random Variables

Chapter 8 introduced the concept of a discrete random variable, defined population-descriptive statistics for discrete random variables, and considered the properties of linear transformations and linear combinations involving discrete random variables. This chapter introduces some commonly used models for discrete random variables. Each of these models gives rise to a probability mass function (pmf) for the underlying discrete random variable, where the pmf depends upon the **parameters** of the specific model being used. For instance, for the binary random variable X associated with a website purchase ($X=1$ if purchase, $X=0$ if not), we can formalize a model that assumes there is some (possibly unknown) probability π of a purchase; here, π is the "parameter" of the model, which gives rise to the pmf for X, with $p_X(1)=\pi$ and $p_X(0)=1-\pi$.

9.1 Bernoulli Random Variable

The simplest discrete random variable is a binary random variable X that is often called a **Bernoulli random variable**. The binary X has two possible values: $X=1$ (called a "success") and $X=0$ (called a "failure"). The success probability $p_X(1)=P(X=1)$ completely characterizes the Bernoulli random variable since $P(X=0)=1-P(X=1)$. If the parameter $\pi=p_X(1)$ denotes the success probability, $p_X(0)=1-\pi$. A Bernoulli random variable is defined in definition 9.1.

Definition 9.1 *A random variable X is a **Bernoulli random variable** with parameter π if $X\in\{0,1\}$ and $\pi=p_X(1)=P(X=1)$. We write $X\sim Bernoulli(\pi)$, where "$\sim$" is read "is distributed as."*

Examples 8.7 and 8.11 considered a Bernoulli random variable for union status, where $X=1$ indicated a union worker and $X=0$ a nonunion worker. The results from those examples for the union-status random variable hold for any Bernoulli random variable, as summarized in proposition 9.1.

Proposition 9.1 *If $X\sim Bernoulli(\pi)$,*

$$\mu_X=E(X)=\pi,\ \sigma_X^2=\pi(1-\pi),\ \textit{and}\ \sigma_X=\sqrt{\pi(1-\pi)}.$$

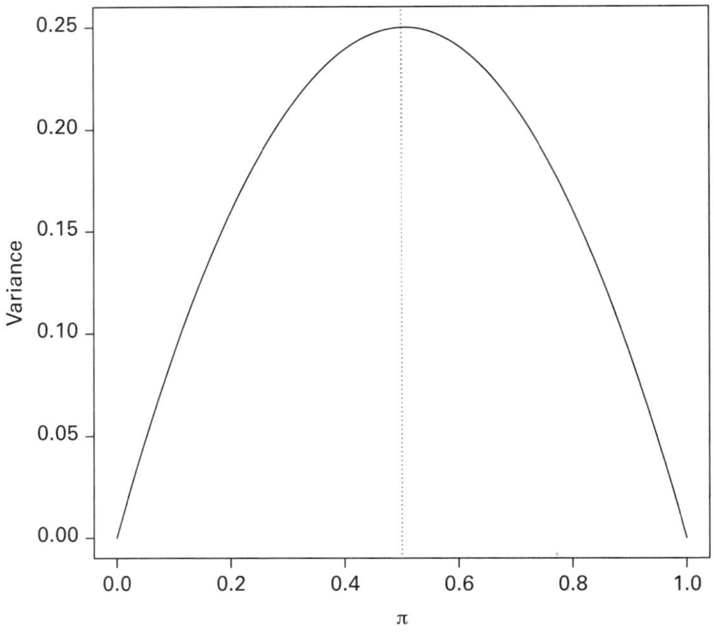

Figure 9.1
Bernoulli variance as a function of π

The population mean μ_X is the average over the 0 and 1 values in the population. Since the 0s contribute nothing to the average, μ_X is equal to the true proportion of 1s in the population, which is $p_X(1)$, or π. For instance, for $\pi = 0.4$, there is a 40% chance of drawing a 1 from the population in a single experiment associated with X, and the population variance and population standard deviation are $\sigma_X^2 = (0.4)(0.6) = 0.24$ and $\sigma_X = \sqrt{0.24}$, respectively.

The sample variance $\sigma_X^2 = \pi(1-\pi)$ is a quadratic function of π, as shown in figure 9.1. The largest possible variance occurs when π equals 0.5, or $\frac{1}{2}$, which is the case for a fair coin toss. Having $\pi = 0.5$ maximizes the variance or noise associated with the experiment of drawing a value of the random variable X from the population. The sample variance, as a function of π, decreases as we move away from the 0.5 value, either to the left toward 0 or to the right toward 1. For $\pi = 0.8$, for instance, there is less uncertainty about X than in the $\pi = 0.5$ case since it is much more likely to see a 1 value than a 0 value. For even larger values, such as $\pi = 0.9$ and $\pi = 0.95$, this uncertainty decreases even more. At the extremes of $\pi = 0$ and $\pi = 1$, there is no uncertainty at all, so the variance is equal to 0, indicating that X is not random in these extreme cases. There is a symmetry in the $\sigma_X^2 = \pi(1-\pi)$ formula, so with respect to the variance, there is no difference between $\pi = 0.8$ and $\pi = 0.2$. For these two π values, the uncertainty or variance of X is the same; the only difference is which of the two outcomes has 80% probability of being drawn. Of course, the population *mean* for these two π values would be different since $\mu_X = \pi$.

Whether the parameter π is known depends on the application. For a coin toss, it makes sense to say that $\pi = 0.5$ is known. For a political poll, π is presumably unknown, and polls are conducted to estimate what π is. In a case where π is unknown, like a political poll, chapter 14 describes how to estimate π and how to measure the uncertainty of that estimate.

9.2 Binomial Random Variable

A **binomial model** is used when repeated (and independent) trials of a Bernoulli random variable are observed, and we are interested in the *total number of successes* in those trials. Let's assume that there are n independent Bernoulli random variables, denoted as $X_1, X_2, \ldots, X_n$, where each random variable X_j has a common success probability equal to π. When the two assumptions of independence and a common pmf hold for discrete random variables, the following terminology describes the random variables.

Definition 9.2 *The discrete random variables $X_1, X_2, \ldots, X_n$ are **independent and identically distributed (i.i.d.)** if (1) $X_1, X_2, \ldots, X_n$ are independent and (2) each X_j has the same pmf.*

When $X_1, X_2, \ldots, X_n$ are i.i.d. *Bernoulli*(π) random variables, the random variable associated with the sum of the values of X_j is a **binomial random variable**, as described in definition 9.3.

Definition 9.3 *A random variable X is a **binomial random variable** with parameters n and π, written $X \sim Binomial(n, \pi)$, if*

$$X = X_1 + X_2 + \cdots + X_n,$$

where $X_1, X_2, \ldots, X_n$ are independent random variables with each $X_j \sim Bernoulli(\pi)$.

The possible outcomes for X are $\{0, 1, \ldots, n\}$, with 0 corresponding to no successes among the n Bernoulli trials and n corresponding to all successes. Since X is a linear combination of independent discrete random variables, proposition 8.7 implies that the population mean of the binomial random variable is

$$\mu_X = \mu_{X_1} + \mu_{X_2} + \cdots + \mu_{X_n} = n\pi,$$

and the population variance of the binomial random variable is

$$\sigma_X^2 = \sigma_{X_1}^2 + \sigma_{X_2}^2 + \cdots + \sigma_{X_n}^2 = n\pi(1 - \pi).$$

Example 9.1 (Website purchases) *Each day, a website records whether each of the first 10 visitors makes a purchase (1) or not (0). Assume that (1) each visitor's behavior is independent of every other visitor and (2) the probability of purchase is 20% for each visitor. Then $X \sim Binomial(10, 0.2)$ is the total number of the first 10 visitors on a given day who make a purchase. The possible values of X are $\{0, 1, 2, \ldots, 10\}$, and its population mean and variance are*

$$\mu_X = n\pi = (10)(0.2) = 2$$

and

$$\sigma_X^2 = n\pi(1 - \pi) = (10)(0.2)(0.8) = 1.6.$$

The population mean and variance do not directly tell us the probabilities of the possible X outcomes. Let's consider the probability of three purchases being made by the first 10 *visitors, which is* $P(X=3)$. *To get* $X=3$, *it must be the case that exactly* 3 *visitors make a purchase and exactly* 7 *visitors do not make a purchase. For a sequence of* 10 *visitors, the number of possible ways to get* 3 *visitors purchasing (Bernoulli* $X_j=1$*) and* 7 *visitors not purchasing (Bernoulli* $X_j=0$*) is* $\binom{10}{3}$. *One of these sequences is*

$$(1,1,1,0,0,0,0,0,0,0),$$

which occurs with probability

$$(0.2)(0.2)(0.2)(0.8)(0.8)(0.8)(0.8)(0.8)(0.8)(0.8)=(0.2)^3(0.8)^7.$$

Similarly, any other sequence with three 1*s and seven* 0*s also occurs with probability* $(0.2)^3(0.8)^7$, *so the total probability of three purchases is*

$$P(X=3)=\binom{10}{3}(0.2)^3(0.8)^7\approx 0.201.$$

How about $P(X=4)$? *There are* $\binom{10}{4}$ *sequences corresponding to* 4 *purchases and* 6 *non-purchases, each with probability* $(0.2)^4(0.8)^6$, *so*

$$P(X=4)=\binom{10}{4}(0.2)^4(0.8)^6\approx 0.088.$$

This idea extends to all possible outcomes in $\{0,1,2,\dots,10\}$, *with*

$$P(X=k)=\binom{10}{k}(0.2)^k(0.8)^{10-k} \text{ for any } k\in\{0,1,2,\dots,10\}.$$

The quantity $\binom{10}{k}$ *is the number of sequences with* k *successes and* $10-k$ *failures out of* 10 *trials, where the probability of any individual sequence is equal to* $(0.2)^k(0.8)^{10-k}$. *For* $P(X=0)$, *note that* $\binom{10}{0}=1$, *corresponding to the one possible sequence (all 0s) for no purchases. And, for* $P(X=10)$, *note that* $\binom{10}{10}=1$, *corresponding to the one possible sequence of all 1s (all* 10 *individuals making a purchase).*

The results derived in example 9.1 can be generalized to any binomial random variable.

Proposition 9.2 *The pmf for* $X\sim Binomial(n,\pi)$ *is*

$$P(X=k)=\binom{n}{k}\pi^k(1-\pi)^{n-k} \text{ for } k\in\{0,1,2,\dots,n\}.$$

The population mean of X is

$$\mu_X=n\pi,$$

the population variance of X is

$$\sigma_X^2=n\pi(1-\pi),$$

and the population standard deviation of X is

$$\sigma_X=\sqrt{n\pi(1-\pi)}.$$

The following *R* functions are useful for working with a binomial random variable:

- `dbinom(x, size, prob)`: Returns the pmf of a *Binomial(*`size`, `prob`*)* random variable evaluated at argument `x`, which may be a single number or a vector.
- `pbinom(x, size, prob)`: Returns the cdf of a *Binomial(*`size`, `prob`*)* random variable evaluated at argument `x`, which may be a single number or a vector.
- `rbinom(n, size, prob)`: Creates a vector of `n` i.i.d. random draws of a *Binomial(*`size`, `prob`*)* random variable.

R has functions like these for other random variables, always using the convention that a function with `d` at the beginning (e.g., `dbinom`) returns a density or pmf, a function with `p` at the beginning (e.g., `pbinom`) returns a cdf, and a function with `r` at the beginning (e.g., `rbinom`) creates random draws of the random variable.

The following *R* code calculates $P(X=3)$ and $P(X=4)$ from example 9.1, where $X \sim Binomial(10, 0.2)$:

```
dbinom(3,10,0.2)

## [1] 0.2013266

dbinom(4,10,0.2)

## [1] 0.08808038
```

Using the vector `0:10`, consisting of all integers between 0 and 10 (inclusive), the complete pmf and cdf can be calculated as follows:

```
dbinom(0:10,10,0.2)

##  [1] 0.1073741824 0.2684354560 0.3019898880 0.2013265920 0.0880803840
##  [6] 0.0264241152 0.0055050240 0.0007864320 0.0000737280 0.0000040960
## [11] 0.0000001024

pbinom(0:10,10,0.2)

##  [1] 0.1073742 0.3758096 0.6777995 0.8791261 0.9672065 0.9936306 0.9991356
##  [8] 0.9999221 0.9999958 0.9999999 1.0000000
```

And the following code simulates 50 i.i.d. draws from the $X \sim Binomial(10, 0.2)$ distribution of example 9.1:

```
set.seed(1234)
rbinom(50,10,0.2)

##  [1] 1 2 2 2 3 2 0 1 2 2 3 2 1 4 1 3 1 1 1 1 1 1 1 0 1 3 2 4 3 0 2 1 1 2 1 3 1 1
## [39] 5 3 2 2 1 2 1 2 2 2 1 3
```

Example 9.2 (Days of stock price increases) *Suppose that whether a stock goes up on a given day is described by a Bernoulli*(0.6) *random variable, where "success" (1) indicates the stock goes up and "failure" (0) indicates that the stock doesn't go up. Moreover, assume that the Bernoulli random variables associated with each day are independent of each other. Under these assumptions, for a sequence of 20 days, what is the probability that the stock*

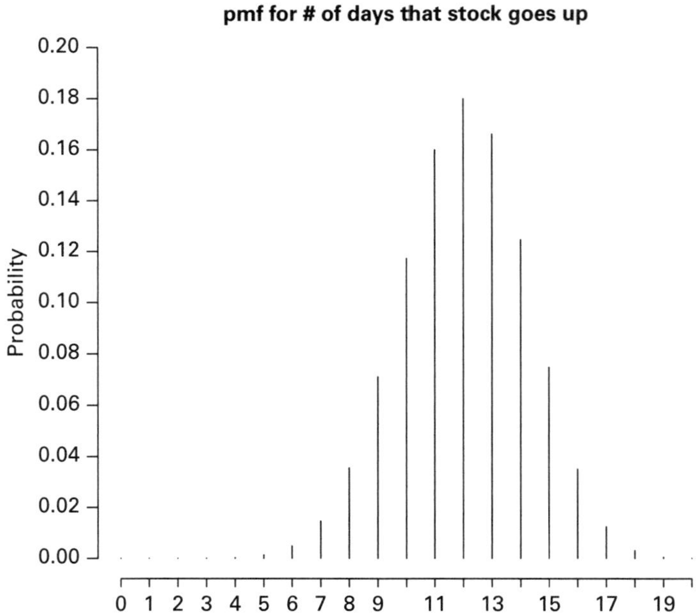

Figure 9.2
The pmf for the stock-increase example

goes up on exactly 12 *of the* 20 *days? In this case,* $X \sim Binomial(20, 0.6)$*, implying*

$$P(X = 12) = \binom{20}{12} 0.6^{12} 0.4^{8} \approx 0.180.$$

The probabilities $P(X = x)$ *for other x values can be similarly calculated. The following R code creates figure 9.2, which graphs the pmf of X. The* `type="h"` *optional argument for the* `plot` *function is used to draw a vertical line from the x-axis to each pmf value:*

```
plot(0:20, dbinom(0:20,20,0.6), type="h", axes=FALSE,
     main="pmf for # of days that stock goes up",xlab="", ylab="Probability",
     xlim=c(0,20), ylim=c(0,0.2))
axis(1, at=0:20)
axis(2, at=seq(0,0.2,0.02))
```

$X = 12$ *is the most likely (modal) outcome here, and the probabilities decrease either to the left or to the right of* 12*. The pmf looks somewhat symmetric, but it is not exactly symmetric around* 12*; for instance,* $P(X = 13)$ *is slightly larger than* $P(X = 11)$.

Interestingly, the pmf in example 9.2 has an approximate bell shape. This particular shape arises due to the parameters of the binomial distribution ($n = 20$, $\pi = 0.6$) being considered. To examine other possible shapes of binomial distributions, figure 9.3 shows pmfs for four different binomial random variables: (1) *Binomial*(10, 0.1) in the upper-left graph, (2) *Binomial*(10, 0.2) in the upper-right graph, (3) *Binomial*(10, 0.5) in the lower-left graph, and (4) *Binomial*(100, 0.1) in the lower-right graph. For the *Binomial*(10, 0.1) random variable, the success probability is so low (10%) that the most likely outcomes are one success and

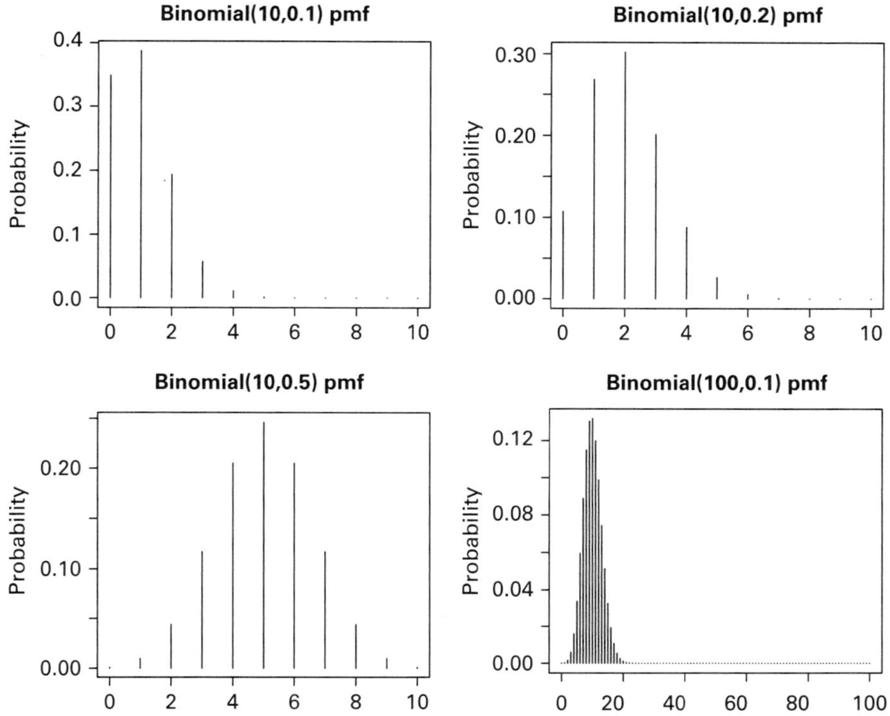

Figure 9.3
The pmfs for binomial random variables

zero successes, with the other outcomes (two or more successes) having declining probabilities. There is no evident bell shape in this graph. For the *Binomial*$(10, 0.2)$ random variable, the bell shape begins to emerge, although there isn't a left tail since the lowest possible outcome is zero successes. For the *Binomial*$(10, 0.5)$ random variable, the pmf has an approximate bell shape and, moreover, it is perfectly symmetric. With a success probability of 0.5, the likelihood of having 4 successes out of 10 trials is the same as having 6 successes out of 10 trials, the likelihood of having 3 successes is the same as having 7 successes, and so on. Finally, for the *Binomial*$(100, 0.1)$ random variable, the pmf has a bell shape even though the success probability is low (10%) and, in fact, it is the same as we had with the *Binomial*$(10, 0.1)$ random variable. The difference here is that the number of trials is much larger (100) than the number of trials (10) for the *Binomial*$(10, 0.1)$ random variable.

9.2.1 The Sample Proportion as a Random Variable

Suppose that rather than the *total* number of successes, we are interested in the *proportion* of successes out of n i.i.d. Bernoulli trials. If X is a binomial random variable, such as $X \sim Binomial(n, \pi)$, define the random variable

$$P = \frac{1}{n}X,$$

which represents the proportion of successes out of n independent $Bernoulli(\pi)$ trials. P is a linear transformation of X, with scaling constant $\frac{1}{n}$, so its population mean, variance, and standard deviation can easily be determined. Specifically, using the results for the binomial random variable ($\mu_X = n\pi$, $\sigma_X^2 = n\pi(1-\pi)$, and $\sigma_X = \sqrt{n\pi(1-\pi)}$),

$$\mu_P = \frac{1}{n}\mu_X = \frac{1}{n}n\pi = \pi,$$

$$\sigma_P^2 = \frac{1}{n^2}\sigma_X^2 = \frac{1}{n^2}n\pi(1-\pi) = \frac{\pi(1-\pi)}{n},$$

and

$$\sigma_P = \frac{1}{n}\sigma_X = \frac{1}{n}\sqrt{n\pi(1-\pi)} = \sqrt{\frac{\pi(1-\pi)}{n}}.$$

The expected value of proportion P is the success probability π, regardless of the number of trials n. The population variance and standard deviation depend on n, with both decreasing as n increases. The dispersion of the proportion random variable P decreases as the number of trials increases, with its pmf becoming more tightly concentrated around the population mean π.

The possible values for P are $\{0, \frac{1}{n}, \frac{2}{n}, \ldots, \frac{n-1}{n}, 1\}$, corresponding to the possible values $\{0, 1, 2, \ldots, n-1, n\}$ for X. Since the pmf for the binomial random variable X is known, the pmf of P follows directly:

$$P\left(P = \frac{k}{n}\right) = P(X = k) = \binom{n}{k}\pi^k(1-\pi)^{n-k} \text{ for each } k \in \{0, 1, 2, \ldots, n\}.$$

Example 9.3 (Political polling) *Suppose that we conduct a poll to see if randomly selected voters favor candidate A or candidate B. Let $X_j = 1$ if a voter from the population favors candidate A and $X_j = 0$ if a voter favors candidate B. Although the true probability $P(X_j = 1)$ would not ordinarily be known, let's assume that we know its value for this example. Specifically, assume that $P(X_j = 1) = 0.52$, so that $X_j \sim Bernoulli(0.52)$. If a poll of 200 voters is conducted, what is the probability that the proportion supporting candidate A is strictly greater than $\frac{1}{2}$? With 200 i.i.d. Bernoulli(0.52) draws and $X \sim Binomial(200, 0.52)$, the probability that the proportion is strictly greater than $\frac{1}{2}$ is equal to the probability that X is strictly greater than 100:*

$$P(P > 0.5) = P(X > 100) = \sum_{k=101}^{200}\binom{200}{k}(0.52)^k(0.48)^{200-k} \approx 0.690.$$

In R, the probability $P(X > 100)$ can be calculated with the `dbinom` *function since $P(X > 100) = 1 - P(X \leq 100)$:*

```
1-pbinom(100,200,0.52)
## [1] 0.6900274
```

Even though the true probability (0.52) of favoring candidate A is greater than $\frac{1}{2}$, there is approximately a 31.0% chance that the observed proportion in the 200-voter poll is less than or equal to $\frac{1}{2}$. What if the poll has more participants? If 1,000 voters were polled

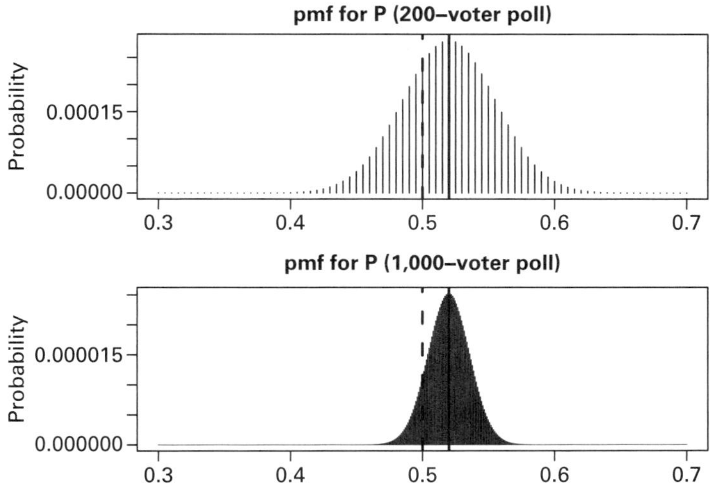

Figure 9.4
The pmfs for voter-poll example

instead of 200 *voters, so* $X \sim Binomial(1000, 0.52)$, *the probability that the proportion is strictly greater than* $\frac{1}{2}$ *equals*

$$P(P > 0.5) = P(X > 500) = \sum_{k=501}^{1000} \binom{1000}{k} (0.52)^k (0.48)^{1000-k} \approx 0.891.$$

```
1-pbinom(500,1000,0.52)

## [1] 0.8914189
```

In this larger poll, there is a considerably lower chance (approximately 10.9%*), that the observed proportion in the poll is less than or equal to* $\frac{1}{2}$. *If we continue to increase the size of the poll, the probability* $P(P > 0.5)$ *gets closer and closer to 1. Figure 9.4 shows the pmfs for the* 200*-voter and* 1,000*-voter polls. For each pmf, a dashed vertical line is drawn at* 0.5*, and a solid vertical line is drawn at* 0.52. *For both pmfs, the solid vertical line is at the center of the distribution. The probability* $P(P > 0.5)$ *corresponds to the sum of the probability values to the right of the dashed vertical line. In comparing the two pmfs, a much larger portion of the distribution is to the right of the* 0.50 *for the* 1,000*-voter poll, corresponding to the* 0.891 *value, compared to the* 200*-voter poll, corresponding to the* 0.690 *value. The dispersion for the* 200*-voter poll is also clearly much larger than the dispersion for the* 1,000*-voter poll since* $\sigma_P^2 = \frac{\pi(1-\pi)}{n} = \frac{(0.52)(0.48)}{n}$.

9.3 Geometric Random Variable

A **geometric model** is used when repeated (and independent) trials of a Bernoulli random variable are observed and we are interested in the *number of failures that occur before the first success*. Rather than fixing the number of trials, the Bernoulli trials $X_1, X_2, \ldots$ are allowed to continue forever since it is not known when the first success will occur.

Definition 9.4 *A **geometric random variable** X with parameter π, written as $X \sim Geo(\pi)$, is the number of failures observed before a success for a sequence of $X_1, X_2, \ldots$ independent random variables with each $X_j \sim Bernoulli(\pi)$.*

The possible outcomes of $X \sim Geo(\pi)$ are $\{0, 1, 2, \ldots\}$. If the first trial is a success, then $X = 0$ since there were no failures before the success. If the second trial is a success, then $X = 1$ since there was one failure before the success. And so on. Since successes are more likely for higher values of π, we should tend to get low values of X if π is high and high values of X if π is low.

Example 9.4 (Visitors before a purchase) *On a given day, the Widgets.com website has a sequence of visitors, each of whom has a 20% probability of purchasing a widget during the visit. The purchase behavior for the sequence of visitors is described by a sequence of i.i.d. Bernoulli(0.2) random variables, with $X_j = 1$ for purchase and $X_j = 0$ for no purchase. Then, the pmf of $X \sim Geo(0.2)$, the number of nonpurchasing individuals that visit the website prior to a purchase being made, is*

$$
\begin{aligned}
P(X=0) &= P(X_1=1) = 0.2 \\
P(X=1) &= P(X_1=0, X_2=1) = (0.8)(0.2) = 0.16 \\
P(X=2) &= P(X_1=0, X_2=0, X_3=1) = (0.8)^2(0.2) = 0.128 \\
&\vdots \\
P(X=k) &= P(X_1=0, X_2=0, \ldots, X_k=0, X_{k+1}=1) = (0.8)^k(0.2) \\
&\vdots
\end{aligned}
$$

The term "geometric" is used for this random variable since each successive probability is equal to the previous probability multiplied by the same constant (0.8 here). The interested reader can use proposition 3.7 to confirm that these probabilities add up to 1.

Proposition 9.3 provides the general form of the pmf of a geometric random variable.

Proposition 9.3 *If $X \sim Geo(\pi)$, then the pmf of X is*

$$P(X=k) = (1-\pi)^k \pi \text{ for } k \in \{0, 1, 2, \ldots\}.$$

The population mean of X is

$$\mu_X = \frac{1-\pi}{\pi},$$

the population variance of X is

$$\sigma_X^2 = \frac{1-\pi}{\pi^2},$$

and the population standard deviation of X is

$$\sigma_X = \frac{\sqrt{1-\pi}}{\pi}.$$

The following *R* functions are useful for working with a geometric random variable:

- `dgeom(x, prob)`: Returns the pmf of a *Geo(*`prob`*)* random variable evaluated at argument `x`, which may be a single number or a vector.
- `pgeom(x, prob)`: Returns the cdf of a *Geo(*`prob`*)* random variable evaluated at argument `x`, which may be a single number or a vector.
- `rgeom(n, prob)`: Creates a vector of `n` i.i.d. random draws of a *Geo(*`prob`*)* random variable.

For the $X \sim Geo(0.2)$ random variable of example 9.4, the pmf and cdf values for $k \in \{0, 1, \cdots, 10\}$ are calculated as follows:

```
dgeom(0:10,0.2)

##  [1] 0.20000000 0.16000000 0.12800000 0.10240000 0.08192000 0.06553600
##  [7] 0.05242880 0.04194304 0.03355443 0.02684355 0.02147484

pgeom(0:10,0.2)

##  [1] 0.2000000 0.3600000 0.4880000 0.5904000 0.6723200 0.7378560 0.7902848
##  [8] 0.8322278 0.8657823 0.8926258 0.9141007
```

Regardless of the value of π, $P(X=0)$ is the largest probability in the pmf. The successive probabilities are strictly decreasing as k gets higher since each one is equal to the previous probability times $(1-\pi)$, which is less than 1. Figure 9.5 shows the pmfs for four geometric random variables: (1) $X \sim Geo(0.1)$ in the upper-left graph, (2) $X \sim Geo(0.2)$ in the upper-right graph, (3) $X \sim Geo(0.5)$ in the lower-left graph, and (4) $X \sim Geo(0.8)$ in the lower-right graph. Each graph has the same y-axis, extending from 0 to 0.8, for ease of comparison.

```
# graph-display format (two rows, two columns)
par(mfrow=c(2,2))

# plots of four different geometric pmf's
plot(0:20, dgeom(0:20,0.1), type="h", main="Geo(0.1) pmf", xlab="",
     ylab="Probability", ylim=c(0,0.8))
plot(0:20, dgeom(0:20,0.2), type="h", main="Geo(0.2) pmf", xlab="",
     ylab="Probability", ylim=c(0,0.8))
plot(0:20, dgeom(0:20,0.5), type="h", main="Geo(0.5) pmf", xlab="",
     ylab="Probability", ylim=c(0,0.8))
plot(0:20, dgeom(0:20,0.8), type="h", main="Geo(0.8) pmf", xlab="",
     ylab="Probability", ylim=c(0,0.8))
```

For $X \sim Geo(0.1)$, the pmf starts at 0.1 and slowly declines with a long right tail, so even very high values for X (i.e., many failures before a success) are possible. For $X \sim Geo(0.2)$, considered in example 9.4, the pmf starts at 0.2 and declines more quickly than when $\pi = 0.1$. The decline in probabilities becomes even more pronounced for $X \sim Geo(0.5)$ and $X \sim Geo(0.8)$. $X \sim Geo(0.5)$ is the random variable describing the number of tails observed before a head is tossed in a sequence of fair coin tosses. When $X \sim Geo(0.8)$, it is very likely that the random variable X has a low value; in this case, $P(X=0) = 0.8$ and

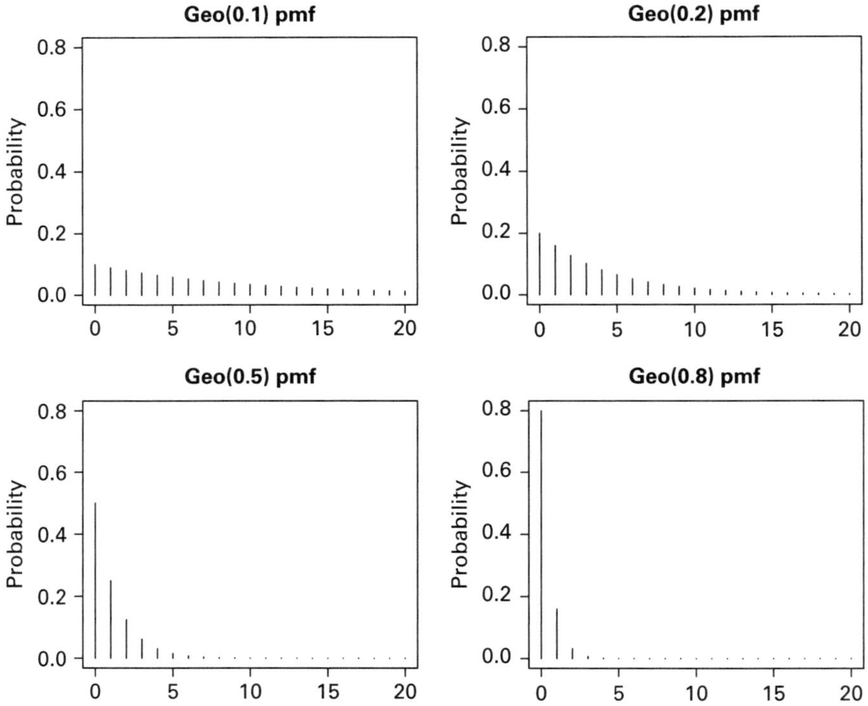

Figure 9.5
The pmfs for geometric random variables

$P(X \leq 2) = 0.8 + (0.2)(0.8) + (0.2)^2(0.8) = 0.992$, so there is only a 0.8% chance of having an X value of 3 or higher.

While we will not prove the results for the population descriptive statistics given in proposition 9.3 here, we can apply the formulas for the population mean, population variance, and population standard deviation. For the website purchase example (example 9.4), where $X \sim Geo(0.2)$, the population mean is $\mu_X = \frac{0.8}{0.2} = 4$ and the population variance is $\sigma_X^2 = \frac{0.8}{0.2^2} = 20$. For the much larger π parameter ($\pi = 0.8$) in the lower-right graph in figure 9.5, the population mean is $\mu_X = \frac{0.2}{0.8} = 0.25$ and the population variance is $\sigma_X^2 = \frac{0.2}{0.8^2} = 0.3125$. Consistent with figure 9.5, the population mean and variance are both much less for $\pi = 0.8$ than for $\pi = 0.2$.

9.4 Negative Binomial Random Variable

The **negative binomial model** generalizes the geometric model described in section 9.3. A geometric random variable is the number of failures before a *single* success. If we are instead interested in the number of failures that occur before *multiple* successes, a negative binomial random variable can be used.

Definition 9.5 *A **negative binomial random variable** X with parameters π and $r \geq 1$, written as $X \sim NegBin(r, \pi)$, is the number of failures observed before r successes for a sequence of $X_1, X_2, \ldots$ independent random variables with each $X_j \sim Bernoulli(\pi)$.*

When $r=1$, a negative binomial random variable is a geometric random variable. That is, $X \sim NegBin(1, \pi)$ is equivalent to $X \sim Geo(\pi)$. The following *R* functions are useful for working with a negative binomial random variable:

- `dnbinom(x, size, prob)`: Returns the pmf of a *NegBin(*`size`, `prob`*)* random variable evaluated at argument `x`, which may be a single number or a vector.
- `pnbinom(x, size, prob)`: Returns the cdf of a *NegBin(*`size`, `prob`*)* random variable evaluated at argument `x`, which may be a single number or a vector.
- `rnbinom(n, size, prob)`: Creates a vector of `n` i.i.d. random draws of a *NegBin(*`size`, `prob`*)* random variable.

Since the geometric random variable is a special case of the negative binomial random variable, `dnbinom(x,1,prob)` is equivalent to `dgeom(x,prob)`, `pnbinom(x,1,prob)` is equivalent to `pgeom(x,prob)`, and `rnbinom(x,1,prob)` is equivalent to `rgeom(x,prob)`.

Example 9.5 (Visitors before three purchases) *Consider the same setup as example 9.4, except that we are now interested in the number of nonpurchasing individuals who visit the website prior to* three *purchases being made, which is the random variable $X \sim NegBin(3, 0.2)$. To have $X=0$, the first three visitors must make a purchase:*

$$P(X=0)=(0.2)^3.$$

For $X=1$, the fourth visitor makes a purchase, with the other two purchases occurring among the first three visitors. In other words, the possible sequences associated with $X=1$ are $(0,1,1,1)$, $(1,0,1,1)$, and $(1,1,0,1)$, so

$$P(X=1)=(3)(0.8)(0.2)^3=0.0192.$$

For $X=2$, the fifth visitor makes a purchase, with the other two purchases occurring among the first four visitors. How many sequences have two purchases and two nonpurchases among the first four visitors? There are $\binom{4}{2}=6$ such sequences, so

$$P(X=2)=\binom{4}{2}(0.8)^2(0.2)^3=0.03072.$$

```
dnbinom(0:2,3,0.2)

## [1] 0.00800 0.01920 0.03072
```

For the general case, for $X=k$, the $(k+3)$th visitor makes a purchase, with the other two purchases occurring among the first $k+2$ visitors. There are $\binom{k+2}{2}$ such sequences, so

$$P(X=k)=\binom{k+2}{2}(0.8)^k(0.2)^3.$$

Figure 9.6 shows the pmf of $X \sim NegBin(3, 0.2)$ in the top graph. The most likely outcomes are roughly between 5 and 10. As a comparison, two other pmfs are shown in the figure. The first is the pmf for a lower purchase probability (10% rather than 20%) in the middle graph,

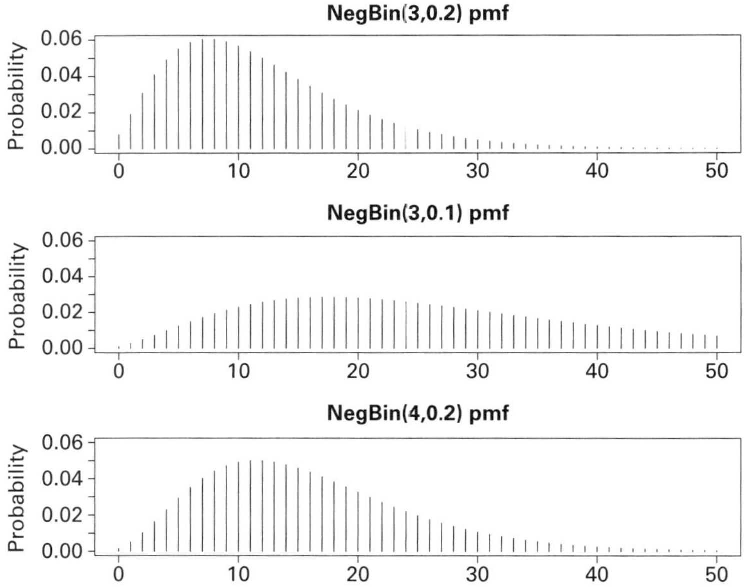

Figure 9.6
The pmfs for negative binomial random variables

$X \sim NegBin(3, 0.1)$. *With the lower success probability, the lower outcomes for X become less likely and the distribution gets stretched out with a very long and thick right tail. The second is the pmf for a larger number of purchases* (4 *rather than* 3) *in the bottom graph,* $X \sim NegBin(4, 0.2)$. *Since more purchases are required, the distribution appears to shift a little to the right and also is a bit lower at its peak relative to the* $NegBin(3, 0.2)$ *pmf.*

Here is the R code to create figure 9.6:

```
# graph-display format (three rows, one column)
par(mfrow=c(3,1))

# plots of three different negative binomial pmf's
plot(0:50, dnbinom(0:50,3,0.2), type="h", main="NegBin(3,0.2) pmf", xlab="",
       ylab="Probability", ylim=c(0,0.06))
plot(0:50, dnbinom(0:50,3,0.1), type="h", main="NegBin(3,0.1) pmf", xlab="",
       ylab="Probability", ylim=c(0,0.06))
plot(0:50, dnbinom(0:50,4,0.2), type="h", main="NegBin(4,0.2) pmf", xlab="",
       ylab="Probability", ylim=c(0,0.06))
```

Let's generalize the approach from example 9.5 to a general $X \sim NegBin(r, \pi)$, for any success probability π and any number of successes r. For $X = k$, it must be the case that (1) there is a total of $k + r$ trials, with k failures and r successes; and (2) the last trial is a success. Combining these two facts, there are k failures and $r - 1$ successes among the first $k + r - 1$ trials. There are $\binom{k+r-1}{r-1}$ possible sequences for which this is true. And the probability of any individual sequence of the $k + r$ trials is $(1 - \pi)^k \pi^r$, corresponding to k failures and r successes. Proposition 9.4 provides the general form of the pmf of a negative binomial random variable.

Proposition 9.4 *If $X \sim NegBin(r, \pi)$, then the pmf of X is*

$$P(X = k) = \binom{k+r-1}{r-1}(1-\pi)^k \pi^r \text{ for } k \in \{0, 1, 2, \ldots\}.$$

The population mean of X is

$$\mu_X = \frac{1-\pi}{\pi} r,$$

the population variance of X is

$$\sigma_X^2 = \frac{1-\pi}{\pi^2} r,$$

and the population standard deviation of X is

$$\sigma_X = \frac{\sqrt{(1-\pi)r}}{\pi}.$$

Both the population mean and the population variance equal r times the population mean and the population variance, respectively, of the $Geo(\pi)$ random variable (proposition 9.3). The reason is that the $X \sim NegBin(r, \pi)$ random variable turns out to be equivalent to the sum of r i.i.d. $Geo(\pi)$ random variables.[1]

For example 9.5, where $X \sim NegBin(3, 0.2)$, proposition 9.4 implies that the population mean is $\mu_X = \frac{0.8}{0.2}(3) = 12$ and the population variance is $\sigma_X^2 = \frac{0.8}{0.2^2}(3) = 60$. Looking at the pmf of X in figure 9.6, the population mean of 12 is a bit to the right of where the most likely modal outcomes of X occur since the long right tail, with large possible values of X, leads to a larger value of μ_X. For the $X \sim NegBin(4, 0.2)$ distribution, the population mean and variance are $\mu_X = 16$ and $\sigma_X^2 = 80$, which are larger than those for $X \sim NegBin(3, 0.2)$.

9.5 Poisson Random Variable

The **Poisson model** models the number of events that occur during a fixed time interval, specifically for a situation in which the possibility of an event occurring at any time during the interval is constant and this possibility is independent of when the last event occurred. The parameter for the Poisson model, denoted as λ, represents the expected value (population average) of the number of events that occur within the fixed time interval.

Definition 9.6 *A random variable X is a* ***Poisson random variable*** *with parameter $\lambda > 0$, written as $X \sim Poisson(\lambda)$, if X represents the number of times that an event occurs within a fixed time interval and the following are true: (1) $\lambda = \mu_X = E(X)$, (2) the average rate at which events occur is constant (depending on λ) and does not depend on whether previous events have occurred, (3) the occurrence of one event does not affect the likelihood that a future event occurs, and (4) two events cannot occur at exactly the same instant.*

The possible outcomes for $X \sim Poisson(\lambda)$ are $\{0, 1, 2, \ldots\}$, with no upper limit imposed on the value of X. The following are two examples of Poisson models.

Example 9.6 (Coffee shop customers) *Consider a random variable X given as the number of customers who arrive at a coffee shop between 10 a.m. and 11 a.m. on a given weekday. The expected value or population mean of X is 20. So, on average, a customer arrives every* *three minutes during the 10 a.m.-to-11 a.m. hour. For the assumptions in definition 9.6 to hold, it must be the case that the likelihood of one customer arriving has nothing to do with other customers arriving and the likelihood of arrival is constant throughout the hour. Then,* $X \sim Poisson(20)$.

Example 9.7 (Research and development and patents) *A firm has a research and development department that does research throughout the year. Every so often, the department makes a discovery that leads to a patent application. Let the random variable X be the number of discoveries (or patent applications) by the firm in a given year. If the expected value of discoveries in a given year is equal to 2 and the research and development process satisfies the assumptions from definition 9.6, then* $X \sim Poisson(2)$.

Proposition 9.5 provides the pmf of a Poisson random variable.

Proposition 9.5 *If* $X \sim Poisson(\lambda)$, *then the pmf of X is*

$$P(X=k)=\frac{e^{-\lambda}\lambda^k}{k!} \quad \text{for } k \in \{0,1,2,3,\ldots\}.$$

The population mean of X is

$$\mu_X=\lambda,$$

the population variance of X is

$$\sigma_X^2=\lambda,$$

and the population standard deviation of X is

$$\sigma_X=\sqrt{\lambda}.$$

Deriving the probabilities associated with the pmf, $P(X=k)=\frac{e^{-\lambda}\lambda^k}{k!}$, is beyond the scope of this book, but we can still apply the results from proposition 9.5 to examples involving Poisson random variables. A Poisson random variable has the interesting property that its population mean is the same as its population variance, with both equaling the parameter λ.

The following *R* functions are useful for working with a Poisson random variable:

- `dpois(x, lambda)`: Returns the pmf of a *Poisson(*`lambda`*)* random variable evaluated at argument `x`, which may be a single number or a vector.
- `ppois(x, lambda)`: Returns the cdf of a *Poisson(*`lambda`*)* random variable evaluated at argument `x`, which may be a single number or a vector.
- `rpois(n, lambda)`: Creates a vector of `n` i.i.d. random draws of a *Poisson(*`lambda`*)* random variable.

Example 9.8 (Research and development and patents) *Let's continue example 9.7, where the number of discoveries X in a given year is a Poisson random variable,* $X \sim Poisson(2)$. *The population mean* μ_X *and population variance* σ_X^2 *both equal* $\lambda=2$, *and the population standard deviation is* $\sigma_X=\sqrt{2}$. *The pmf is*

$$P(X=0)=\frac{e^{-2}2^0}{0!}=e^{-2}\approx 0.135$$

$$P(X=1)=\frac{e^{-2}\,2^1}{1!}=2e^{-2}\approx 0.271$$

$$P(X=2)=\frac{e^{-2}\,2^2}{2!}=2e^{-2}\approx 0.271$$

$$P(X=3)=\frac{e^{-2}\,2^3}{3!}=\frac{4}{3}e^{-2}\approx 0.180$$

$$P(X=4)=\frac{e^{-2}\,2^4}{4!}=\frac{2}{3}e^{-2}\approx 0.090$$

$$\vdots$$

$$P(X=k)=\frac{e^{-2}\,2^k}{k!}$$

$$\vdots$$

The pmf of $X \sim Poisson(2)$ *is shown in the top graph of figure 9.7. The most likely outcomes are* $X=1$ *and* $X=2$*, both with probability* 0.271. *The pmf indicates that large values of X are not likely. For example, the probability of having more than 5 discoveries,* $P(X>5)$*, is approximately* 0.0166, *or* 1.66%. *What would happen if the firm had a more productive research and development department? If the Poisson parameter is increased to* 4, *corresponding to an expected value of 4 discoveries in a given year, the pmf is shown in the bottom graph of figure 9.7. In this case, the most likely outcomes are* $X=3$ *and* $X=4$*, both with*

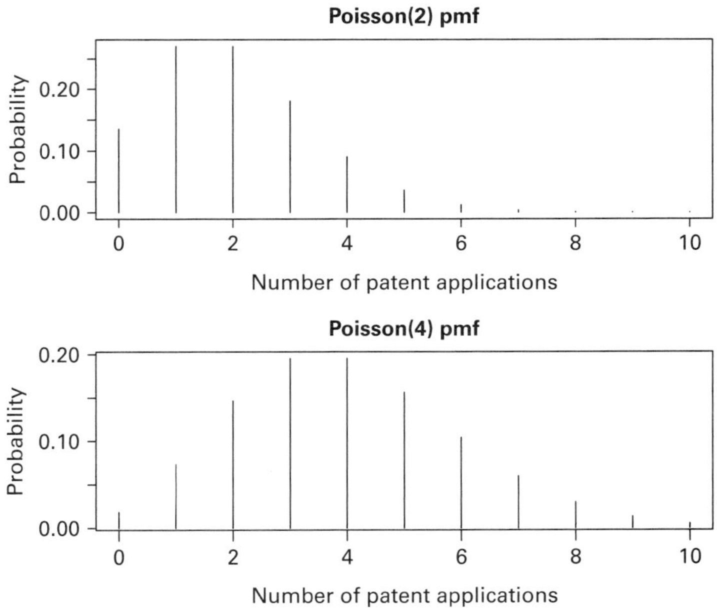

Figure 9.7
The pmfs for Poisson patent example

probability approximately equal to 0.195. *As compared to* $\lambda = 2$, *larger outcomes for* X *are much more likely with* $\lambda = 4$; *for instance,* $P(X > 5)$ *is approximately* 0.215, *or* 21.5%, *and even the probability of* 8 *discoveries,* $P(X = 8)$, *is approximately* 2.98%.

Here is the R code to create figure 9.7:

```
# graph-display format (two rows, one column)
par(mfrow=c(2,1))

# plots of two different Poisson pmf's
plot(0:10, dpois(0:10,2), type="h", main="Poisson(2) pmf",
       xlab="Number of patent applications", ylab="Probability")
plot(0:10, dpois(0:10,4), type="h", main="Poisson(4) pmf",
       xlab="Number of patent applications", ylab="Probability")
```

Example 9.9 (Coffee shop customers) *For the coffee shop example (example 9.6),* $X \sim$ *Poisson*(20) *is the number of customers that arrive at a coffee shop between 10 a.m. and 11 a.m. on a given weekday. Figure 9.8 shows the pmf of* $X \sim$ *Poisson*(20).

```
plot(0:40, dpois(0:40,20), type="h", main="",
     xlab="Customers between 10 a.m. and 11 a.m.", ylab="Probability")
```

For this λ *value, the pmf has a bell shape that peaks at around* $X = 20$; *in fact, the most likely outcomes here are* $X = 19$ *and* $X = 20$, *both of which occur with a probability of approximately* 0.089, *or* 8.9%. *The bell shape of this pmf is quite different from the shape of the pmf for the lower* λ *value of* $\lambda = 2$ *in figure 9.7. In fact, when* λ *is very large, alternative*

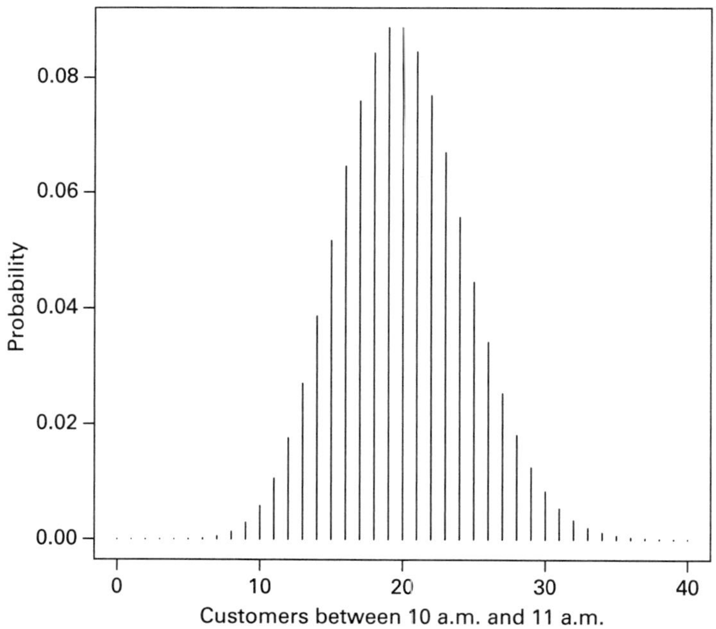

Figure 9.8
The pmfs for Poisson coffee shop example

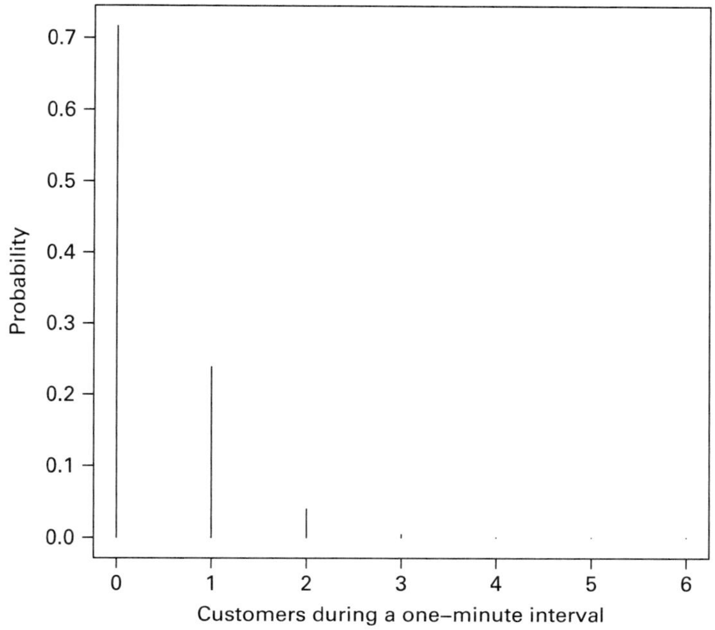

Figure 9.9
The pmf for the Poisson coffee shop example

approaches can be used to model the random variable (e.g., using a normal distribution, which is discussed in chapter 11).

What if we are interested in the number of customers who arrive during a one-minute interval (still at some point between 10 a.m. and 11 a.m. on a weekday) rather than during a one-hour interval? If the expected number of customers during the one-hour interval is 20, the expected number of customers during a one-minute interval must be $\frac{20}{60} = \frac{1}{3}$ *since there is a constant arrival rate. Figure 9.9 shows the pmf for the associated Poisson random variable, which is a* $Poisson(1/3)$ *random variable. The arrival of zero customers is by far the most likely outcome (71.7%), followed by one customer (23.9%), and two customers (4.0%), with these three outcomes accounting for a total of 99.5% in probability. Despite the same underlying process as the one-hour interval, this pmf has a quite different shape due to the much lower* λ *parameter value. With* $\lambda = \frac{1}{3}$*, the arrival of a customer is a more unusual event since it must occur during a one-minute interval compared to the arrival of a customer during the full one-hour interval.*

Note

1. Let Y_1 be the number of failures before the first success, Y_2 be the number of failures after the first success and before the second success, and so on through Y_r being the number of failures after the $(r-1)$th success and before the rth success. With these definitions, $Y_1, Y_2, \ldots, Y_r$ are all geometric random variables $Geo(\pi)$. Moreover, since the underlying Bernoulli trials are independent, it is also the case that the $Y_1, Y_2, \ldots, Y_r$ are independent random variables. Noting that, for $X \sim NegBin(r, \pi)$, we have $X = Y_1 + Y_2 + \cdots + Y_r$. Then, using

the results for a linear combination of independent random variables (proposition 10.13),

$$\mu_X = \mu_{Y_1} + \mu_{Y_2} + \cdots + \mu_{Y_r} = \frac{1-\pi}{\pi} r$$

and

$$\sigma_X^2 = \sigma_{Y_1}^2 + \sigma_{Y_2}^2 + \cdots + \sigma_{Y_r}^2 = \frac{1-\pi}{\pi^2} r.$$

Exercises

1. Let $X = 1$ if a fair die roll results in 5 or 6 and $X = 0$ otherwise.
   a. How would you describe the random variable X?
   b. What is the expected value of X?
   c. What is the population variance of X?
   d. How does the population variance of X compare to the population variance of Y, where $Y = 1$ if a 6 is rolled and $Y = 0$ otherwise?
2. At the Capita County Fair, people can win tickets by playing carnival games. At the end of the fair, the tickets can be turned in for a chance to win prizes. Suppose that (i) there is a 1% chance of winning a prize for each ticket and (ii) all tickets are independent of each other.
   a. If an individual has 10 tickets, what is the expected value of the number of prizes won?
   b. If an individual has 10 tickets, what is the population standard deviation of the number of prizes won?
   c. If an individual has 20 tickets, what is the population standard deviation of the number of prizes won?
3. There are two coins on a table, one of which is fair (50% heads when flipped) and one of which is unfair (60% heads when flipped). You pick one of the two coins at random and flip it 10 times.
   a. What is the probability that the outcome is exactly 4 heads?
   b. What is the probability that the coin is fair, given that the outcome is exactly 4 heads?
4. A politician believes that 30% of all macroeconomists would support a proposal that she wishes to advance. Suppose that this belief is correct and that five macroeconomists are approached at random.
   a. What is the probability that none of the macroeconomists would support the proposal?
   b. What is the probability that exactly three of the macroeconomists would support the proposal?
   c. What is the probability that a majority of the macroeconomists would support the proposal?
   d. Plot the pmf associated with the random variable X being the total number of macroeconomists (out of five) who support the proposal.

5. A recent study found that 13% of individuals in the US are left-handed and 87% are right-handed.
   a. For a group of $n \geq 2$ independent individuals, what is the probability that at least one individual is left-handed? The expression should be a function of n. Evaluate the expression for $n = 3$, $n = 4$, and $n = 5$.
   b. For a group of $n \geq 2$ independent individuals, what is the probability that exactly one individual is left-handed if at least one individual is left-handed? The expression should be a function of n. Evaluate the expression for $n = 3$, $n = 4$, and $n = 5$.
   c. For a group of $n \geq 2$ independent individuals, what is the probability that exactly two individuals are left-handed if at least one individual is left-handed? The expression should be a function of n. Evaluate the expression for $n = 3$, $n = 4$, and $n = 5$.
6. You manage a factory, and due to the complexity of the manufacturing process, there is a 10% probability that any one of your manufactured products will have a defect. You may assume that each product is independent; that is, whether one product is defective is not associated with another product being defective. On any given day, your factory produces 2,000 products. Let the random variable D equal the total number of defective products on a given day.
   a. Plot the pmf of D.
   b. What is the most likely value of D?
   c. What is $P(200 < D \leq 300)$?
   d. What is $P(200 \leq D \leq 300)$?
   e. Find the smallest integer d for which $P(200 - d \leq D \leq 200 + d)$ is greater than 90%. For this value of d, what is $P(200 - d \leq D \leq 200 + d)$? (Hint: Start at $d = 0$ and repeatedly increase d until the probability is above 90%.)
7. A university is experiencing increased enrollment, so it needs to accommodate more students in its classes. The university schedules a 100-student class in a room that seats only 90 students.
   a. On a given day, suppose that each student attends class with an 85% probability and all students can be considered to be independent of each other. What is the probability that all students who attend the class will have a seat?
   b. *Now suppose that the attendance probability π of an individual student is unknown. Limiting the possible values to the set $\pi \in \{0.01, 0.02, \ldots, 0.99, 1.00\}$, what is the largest value of π for which there is at least a 99% probability that all the students who attend the class will have a seat?
8. An entrepreneur is intent on starting a profitable business. The probability that she succeeds in any given year is 15%, and her success in any given year is independent of the outcomes in any previous year. The entrepreneur starts trying in year 1. Let $Y \in \{1, 2, 3, \cdots\}$ be the random variable indicating the year in which the entrepreneur is first successful in starting a profitable business.
   a. Explain why Y is not itself a geometric random variable.
   b. Write Y in terms of a geometric random variable.
   c. Determine $P(4 \leq Y \leq 10)$ using the R function `dgeom`.

d. What is the expected value of Y?

e. What is the population standard deviation of Y?

9. A family really wants to have a daughter, so they decide to keep having children until they have a daughter. Assume that the gender of any given child is independent of the genders of all other children. The probability of having a daughter for any given birth is 48.8% (yes, it is slightly lower than 50%).

   a. What is the probability that the family will have at least two sons before having a daughter?

   b. What is the expected number of sons that the family will have before having a daughter?

   c. If the family wants to have two daughters, what is the probability that the family has at least two sons before having two daughters? What is the expected number of sons that the family will have before having two daughters?

10. A political activist collects petition signatures by going door-to-door in a neighborhood. Suppose that the probability of successfully getting a signature at any given house is 5% and the random variables associated with success at all houses are i.i.d. Let $H \in \{20, 21, 22, 23, \ldots\}$ be the random variable indicating the number of houses visited to obtain 20 signatures.

    a. Explain why H is not itself a negative binomial random variable.

    b. Write H in terms of a negative binomial random variable.

    c. What is the expected value of H?

    d. What is the population standard deviation of H?

    e. Determine $P(300 \leq H \leq 400)$ using the *R* function `pnbinom`.

    f. A more convincing individual has a success probability of 7%. Conduct 10,000 simulations in *R* to approximate the probability that this individual (7% success rate) visits strictly fewer houses than the other individual (5% success rate) to obtain 20 signatures.

11. *A *multinomial distribution* provides a generalization of the binomial distribution that allows more than two outcomes. Specifically, the multinomial distribution is based upon n i.i.d. draws of a discrete random variable with $m \geq 2$ possible outcomes (denoted, without loss of generality, $\{1, 2, \ldots, m\}$) and probabilities $p_1, p_2, \ldots, p_m$ (with $\sum_{j=1}^{m} p_j = 1$). The discrete random variables $X_1, X_2, \ldots, X_m$ correspond to the total number of times that outcome 1 occurs, outcome 2 occurs, and so on through outcome m. The joint pmf is

$$p_{X_1X_2\cdots X_m}(x_1, x_2, \ldots, x_m) = P(X_1 = x_1, X_2 = x_2, \ldots, X_m = x_m)$$
$$= \frac{n!}{x_1! x_2! \cdots x_m!} p_1^{x_1} p_2^{x_2} \cdots p_m^{x_m},$$

where $x_1, x_2, \ldots, x_m$ are nonnegative integers with $x_1 + x_2 + \cdots + x_m = n$.

There are three brands of a certain product (brands 1, 2, and 3). In the population, 15% of consumers prefer brand 1, 25% of consumers prefer brand 2, and 60% of consumers prefer brand 3.

a. If 10 consumers are chosen at random, what is the probability that 2 prefer brand 1, 3 prefer brand 2, and 5 prefer brand 3?
b. If 10 consumers are chosen at random, what is the probability that 5 prefer brand 3 *and* more consumers prefer brand 2 than brand 1?
c. If 10 consumers are chosen at random, what is the expected value of the number of consumers who prefer brand 3? (Hint: Don't use the joint pmf. Instead, consider the binomial distribution based on brand 3 being chosen or not chosen.)
d. Make 10,000 simulated draws of (X_1, X_2, X_3) in *R* for 10 randomly chosen consumers. That is, for each simulation, consider 10 hypothetical consumers, assign them to a preferred brand based upon the probabilities (15%, 25%, and 60%), and let $(X_1, X_2, \textit{and}\, X_3)$ be the overall counts for the three brands.
e. Using the simulated draws from (d), confirm your answer to (b).
f. Using the simulated draws from (d), what is the approximate population correlation between X_1 and X_2? Does the sign of this correlation make sense?

12. A financial company releases a market update e-mail every Friday afternoon. The expected number of typos in a given e-mail is 0.8, and the number of typos follows a Poisson distribution.
    a. What is the probability that a given e-mail has no typos?
    b. What is the probability that a given e-mail has two or more typos?
    c. *Assume that each weekly e-mail can be considered independent. This part considers the number of typos in two consecutive weekly e-mails, denoted as X_1 and X_2.
        i. Use a `while` loop in *R* to calculate the probability that the two e-mails have the *same* number of typos, given by

$$P(X_1 = X_2) = P(X_1 = X_2 = 0) + P(X_1 = X_2 = 1) + P(X_1 = X_2 = 2) + \cdots$$

Since this probability involves an infinite sum, continue looping until the probability $P(X_1 = X_2 = j)$ falls below a very small value, such as 0.000001, to get an accurate approximation of the probability.

        ii. Conduct 10,000 simulations in *R*, taking independent draws of X_1 and X_2 in each simulation, to confirm your answer to part (i) of this question.

13. Consider the number of customers X who arrive at a coffee shop during one minute (within the 10 a.m.-to-11 a.m. hour). Suppose that the store manager knows that the average arrival rate of customers is two per minute. Based upon this information, we can model X as a Poisson random, $X \sim Poisson(2)$, with pmf

$$p_X(x) = \frac{e^{-2} 2^x}{x!} \quad \text{for } x = 0, 1, 2, 3, \ldots$$

a. Using the pmf formula directly, what is the probability that no customers arrive during a given minute?

b. Confirm your answer to (a) by using the *R* function `dpois`.

c. Plot the pmf of X for all values less than or equal to 10.

d. What is the probability that 4 or more customers arrive during a given minute?

e. Simulate 10,000 draws of $X \sim Poisson(2)$ in *R*. What proportion of the 10,000 draws are greater than or equal to 4? Is your answer similar, aside from simulation noise, to the probability found in (d)?

f. A more popular coffee shop on the other side of town has an average of three customers per minute during the 10 a.m.-to-11 a.m. hour, with the number of customers arriving during a given minute (denoted as Y) distributed as a $Poisson(3)$ random variable. Assume that X and Y are independent of each other.

   i. What is the joint probability that, during a given minute, there is one customer who arrives at the the first store (the $\lambda = 2$ one) and one customer who arrives at the second store (the $\lambda = 3$ one)?

   ii. Use simulated draws of X and Y to approximate the probability that more customers arrive at the first store (the $\lambda = 2$ one) than the second store (the $\lambda = 3$ one) during a given minute. Specifically, simulate 10,000 draws of both $X \sim Poisson(2)$ and $Y \sim Poisson(3)$, and calculate the proportion of times that the X draw is strictly greater than the Y draw.

   iii. What is the expected value of the total number of customers arriving at the two stores during a given minute?

   iv. What is the population standard deviation of the total number of customers arriving at the two stores during a given minute?

   v. Use simulated draws of X and Y (again 10,000 for each) to approximate the probability that the total number of customers arriving at the two stores is greater than or equal to 6 during a given minute.

14. *Suppose that $X_1 \sim Poisson(\lambda_1)$ and $X_2 \sim Poisson(\lambda_2)$ are independent random variables. Show that $X_1 + X_2$ is a $Poisson(\lambda_1 + \lambda_2)$ random variable by verifying that

$$P(X_1 + X_2 = k) = \frac{e^{-(\lambda_1+\lambda_2)}\,(\lambda_1+\lambda_2)^k}{k!} \quad \text{for } k \in \{0, 1, 2, 3, \ldots\}.$$

(Hint: Use the "binomial theorem," which is

$$(a+b)^m = \sum_{j=0}^{m} \binom{m}{j} a^j b^{m-j} = \sum_{j=0}^{m} \frac{m!}{j!(m-j)!} a^j b^{m-j}$$

for any $m \in \{0, 1, 2, \ldots\}$ and any constants a and b.)

15. Two employees must separately submit a report to their manager. Let T_1 and T_2 be random variables that represent the number of typos in the two reports. Assume that T_1, the number of typos in the first employee's report, is a $Poisson(1.8)$ random variable, and T_2, the number of typos in the second employee's report, is a $Poisson(2.4)$ random variable. Assume that T_1 and T_2 are independent.

a. What is the joint pmf of T_1 and T_2?

b. What is the probability that there are no typos in the two reports?

c. What is the probability that there are at most a total of two typos in the two reports?

d. Conduct 10,000 simulations in *R* to approximate $P(T_1 = T_2)$, the probability that each report has the same number of typos.

e. Suppose that each employee now must submit *two* reports to the manager. Assume that the reports are all independent, with the same underlying random variables as in the original scenario. Conduct 10,000 simulations in *R* to estimate the probability that the total number of typos for each employee is the same.

f. Repeat (e) for the case where each employee must submit three reports to the manager.

10

Continuous Random Variables

Chapter 8 formalized the concept of a discrete random variable, introducing the probability mass function (pmf) as a complete characterization of the random variable and discussing population quantities like the population mean and population variance. Chapter 9 considered some examples of models that are used in specific situations with discrete random variables. This chapter departs from the case of discrete random variables and formalizes the concept of a **continuous random variable**, where the underlying quantity being measured is a continuous or approximately continuous variable.

10.1 Continuous Random Variables Versus Discrete Random Variables

Recall from definition 8.1 that a random variable X is a function that maps each outcome of sample space $\mathcal{S}$ to a number. Whereas a discrete random variable has a finite or countable set of possible values, a continuous random variable has an uncountable set of possible values. Analogous to the definition of a continuous numerical variable (definition 5.7), a continuous random variable is described as in definition 10.1.

Definition 10.1 *A* ***continuous random variable*** *is a random variable that can take any value on some interval or intervals of the real line, including perhaps the entire real line, and for which the probability of any specific outcome x^* occurring equals zero.*

Before discussing the last part of this definition ("the probability of any specific outcome x^* occurring equals zero"), let's consider a few examples of continuous random variables:

- Weekly earnings: $\mathcal{S}$ contains all possible positive weekly earnings values for employed individuals, and X equals the outcome.
- Monthly stock return:[1] $\mathcal{S}$ contains all possible monthly returns for a given stock, and X equals the outcome.
- Unemployment rate: $\mathcal{S}$ contains all possible unemployment rates, ranging from 0% to 100% (or 0 to 1), and X equals the outcome.

Why does it make sense to say that $P(X = x^*) = 0$ for any specific outcome x^*? Let's think about a random variable X, which is equally likely to be any real number between 0 and 1.

(As will be seen later in this chapter, such an X is called a *uniform random variable*.) In this case, X can be thought of as the limit of a particular sequence of discrete random variables, as follows:

- $X_1 \in \{0.1, 0.2, \cdots, 0.9, 1.0\}$, with 10 equally likely outcomes (each with probability 0.1)
- $X_2 \in \{0.01, 0.02, \cdots, 0.99, 1.00\}$, with 100 equally likely outcomes (each with probability 0.01)
- $X_3 \in \{0.001, 0.002, \cdots, 0.999, 1.000\}$, with 1,000 equally likely outcomes (each with probability 0.001)
- $X_4 \in \{0.0001, 0.0002, \cdots, 0.9999, 1.0000\}$, with 10,000 equally likely outcomes (each with probability 0.0001)

The construction of the X_j random variables can continue for larger j, with the probability of any individual outcome becoming smaller (approaching zero) as j becomes larger. For the random variable X, as the limit of this sequence of X_j random variables, the probability of any specific outcome x^* should equal zero. In fact, if we had $P(X = x^*) = c$ for some positive number $c > 0$, then it must also be the case that *any* other outcome on the [0, 1] interval must also have a probability c of occurring since every value is equally likely. However, there are an infinite number of possible outcomes on the interval [0, 1], meaning that the probabilities of all these outcomes would sum to infinity, violating the Axioms of Probability.

The fact that $P(X = x^*) = 0$ for any specific outcome x^* is a key distinguishing feature of a continuous random variable, in contrast to a discrete random variable, which has positive probabilities for at least two discrete outcomes. For a continuous random variable, rather than analyzing the probabilities of specific outcomes, which all equal zero, the probabilities associated with *intervals* of outcomes are considered.

10.1.1 Mixtures of Discrete and Continuous Random Variables

Before introducing a framework for continuous random variables, it is worth mentioning that some random variables may be a hybrid or *mixture* of a discrete random variable and a continuous random variable. Here are some examples:

- Monthly paycheck savings: $\mathcal{S} = [0, 1]$ contains all possible values corresponding to the fraction of a paycheck that is saved in a given month by a given individual, and X equals the outcome.
- Weekly earnings: $\mathcal{S} = [0, \infty)$ contains all possible weekly earnings values for individuals (*including* those who are unemployed or not in the labor force), and X equals the outcome.
- Annual health-care expenditures: $\mathcal{S} = [0, \infty)$ contains all possible health-care expenditure values for individuals, and X equals the outcome.

For the paycheck savings example, $X = 0$ when an individual has no savings, and $X = 1$ when an individual saves the entire paycheck. These two extremes may have nonzero probabilities and can be considered as discrete outcomes, while anything in between (outcomes strictly between 0 and 1) can be considered continuous outcomes. For the weekly earnings example, the population includes individuals who are unemployed or not in the labor force, which means that individuals can have weekly earnings equal to zero. There is a nonzero probability of zero weekly earnings so long as the probability of being unemployed or not in the labor force is positive. This zero outcome can be considered a discrete outcome, while all other positive outcomes can be considered continuous outcomes. The health-care expenditure example is similar, with the zero outcome having a positive probability associated with individuals in the population who have no health-care expenditures in a given year.

10.2 Probability Density Function

For a discrete random variable, the pmf characterizes the probabilities for all possible outcomes. Since the probabilities of specific outcomes are all zero for a continuous random variable, an alternative function, known as the **probability density function (pdf)**, describes the probabilities associated with any possible *interval* of outcomes.

Definition 10.2 *The* ***probability density function (pdf)*** *of a continuous random variable X is a function $f_X(\cdot)$ such that for any two numbers a and b with $a \le b$,*

$$P(a \le X \le b) = \int_a^b f_X(x)\, dx.$$

This definition places no restriction on the shape of the pdf $f_X(\cdot)$. The fact that any outcome x^* has a zero probability follows directly from definition 10.2 since

$$P(X = x^*) = P(x^* \le X \le x^*) = \int_{x^*}^{x^*} f_X(x)\, dx = 0.$$

As special cases, definition 10.2 also implies

$$P(X \le b) = \int_{-\infty}^{b} f_X(x)\, dx$$

when $a = -\infty$, and

$$P(X \ge a) = \int_{a}^{\infty} f_X(x)\, dx$$

when $b = \infty$.

For a given x^*, it is important to stress that $f_X(x^*)$ is not itself a probability. The probability of any specific x^* occurring equals zero, and the probabilities in definition 10.2 are probabilities of intervals rather than specific values. Since $f_X(x^*)$ is not a probability, it is possible that $f_X(x^*) > 1$. Also, since the probability of any given outcome is zero, the use of strict versus weak inequalities does not matter for the probability of an interval, unlike discrete random variables:

$$P(a \le X \le b) = P(a < X \le b) = P(a \le X < b) = P(a < X < b) \text{ for any } a \le b.$$

This result is easy to show. For the first equality, $P(a \le X \le b) = P(X = a) + P(a < X \le b) = P(a < X \le b)$ since $P(X = a) = 0$. For the second equality, $P(a \le X \le b) = P(a \le X < b) + P(X = b) = P(a \le X < b)$ since $P(X = b) = 0$. And for the third equality, $P(a \le X \le b) = P(X = a) + P(a < X < b) + P(X = b) = P(a < X < b)$ since $P(X = a) = P(X = b) = 0$.

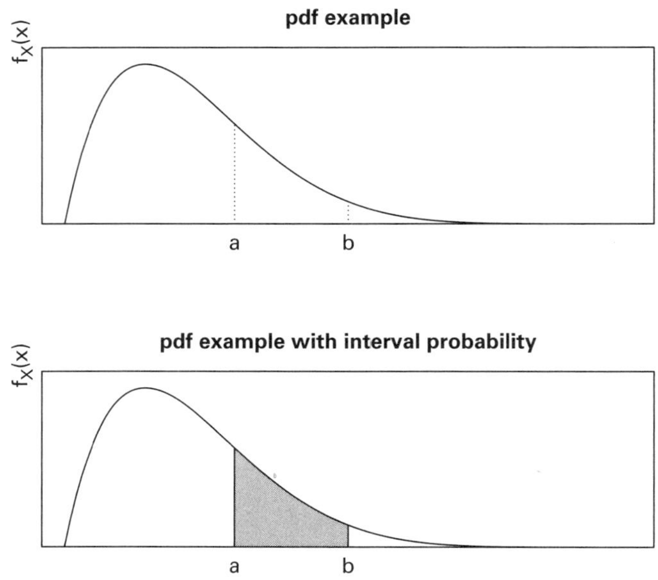

Figure 10.1
An example of a pdf

Figure 10.1 shows an example of a pdf $f_X(\cdot)$, with the top graph showing just the pdf and the bottom graph showing the pdf along with the shaded area between the two values a and b. For any given x^*, the value $f_X(x^*)$ is the height of the pdf function at x^* and can be thought of as the relative likelihood of outcomes being near x^*. In the pdf shown, outcomes near a are more likely than outcomes near b since $f_X(a) > f_X(b)$. Also, the x values that are near the hump of the distribution have the largest pdf values, so an interval of x values near this hump should have a higher probability than an interval (of the same width) away from the hump. The gray shading represents the area under the $f_X(\cdot)$ curve between a and b, and this area is equal to the integral $\int_a^b f_X(x)\,dx$ or, equivalently, the probability $P(a \le X \le b)$. Similarly, the probability $P(X \le a) = \int_{-\infty}^a f_X(x)\,dx$ is equal to the area under the $f_X(\cdot)$ curve and to the left of a, and the probability $P(X \ge b) = \int_b^\infty f_X(x)\,dx$ is equal to the area under the $f_X(\cdot)$ curve and to the right of b.

Proposition 10.1 *For a continuous random variable X, the pdf $f_X(\cdot)$ has the following properties:*

i. $f_X(x) \ge 0$ *for all x*
ii. $\int_{-\infty}^{\infty} f_X(x)\,dx = 1$

Property (i) rules out negative $f_X(x)$ values. This property makes sense since if $f_X(\cdot)$ is negative on any interval, even a tiny one, the associated integral for that interval would be negative, leading to a negative probability. Proposition 10.1 does not restrict $f_X(x) \le 1$ since $f_X(x)$ is not a probability. Property (ii) states that the total probability across all possible outcomes equals 1 since $P(-\infty < X < \infty) = \int_{-\infty}^{\infty} f_X(x)\,dx = 1$. This property is analogous to the property of a discrete random variable that the sum of the probabilities of disjoint and exhaustive outcomes equals 1. With respect to the graph of the pdf $f_X(\cdot)$, property (ii) says

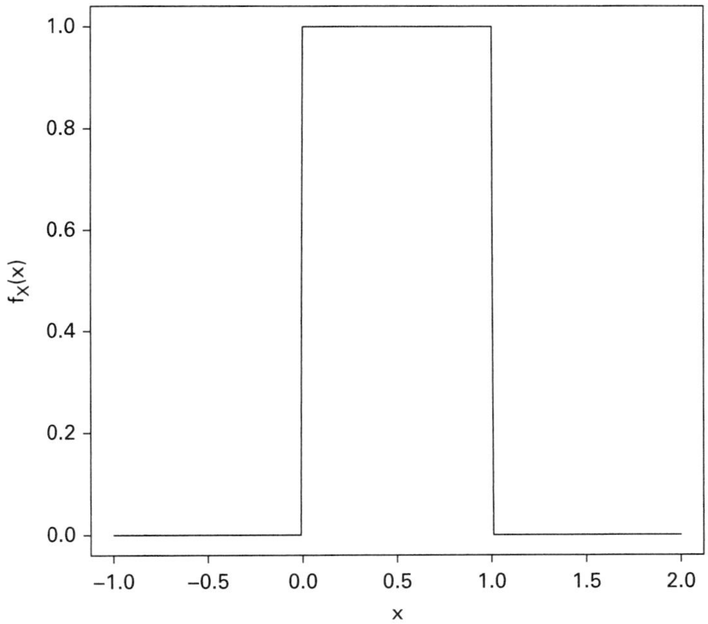

Figure 10.2
The pdf of a $U(0, 1)$ random variable

that the total area under the pdf curve (and above the x-axis) must equal 1. For a given shape of the pdf, this property dictates the scale of the $f_X(\cdot)$ function along the y-axis.

Example 10.1 (Uniform distribution) *Let X be a continuous random variable with pdf*

$$f_X(x) = \begin{cases} 1 & \text{if } 0 \leq x \leq 1 \\ 0 & \text{otherwise.} \end{cases}$$

*In this case, X is said to have a **uniform distribution** between 0 and 1, written as $X \sim U(0, 1)$. Figure 10.2 graphs the pdf curve $f_X(x)$, which has a rectangular shape with height equal to 1 along the $[0, 1]$ interval, which has a width equal to 1. The full area under the pdf is equal to 1, as the area of the rectangle is $1 \times 1 = 1$. Alternatively, it can be verified using the following integral:*

$$\int_{-\infty}^{\infty} f_X(x)\,dx = \int_0^1 f_X(x)\,dx = \int_0^1 1\,dx = x|_0^1 = 1 - 0 = 1.$$

For the uniform distribution, we can easily determine the probability of an interval. For example, $P(0.2 \leq X \leq 0.5)$ can be evaluated by either finding the area under the rectangle between 0.2 and 0.5 or evaluating the integral $\int_{0.2}^{0.5} f_X(x)\,dx = \int_{0.2}^{0.5} 1\,dx$. Using either method, $P(0.2 \leq X \leq 0.5) = 0.3$.

There is nothing special about the end points 0 and 1 in the $U(0, 1)$ random variable, and a uniform random variable can be specified with different end points. For example, if X is a uniform random variable between 5 and 10, written as $X \sim U(5, 10)$, the pdf is a constant value within the $[5, 10]$ interval and zero outside the $[5, 10]$ interval. The constant value can't

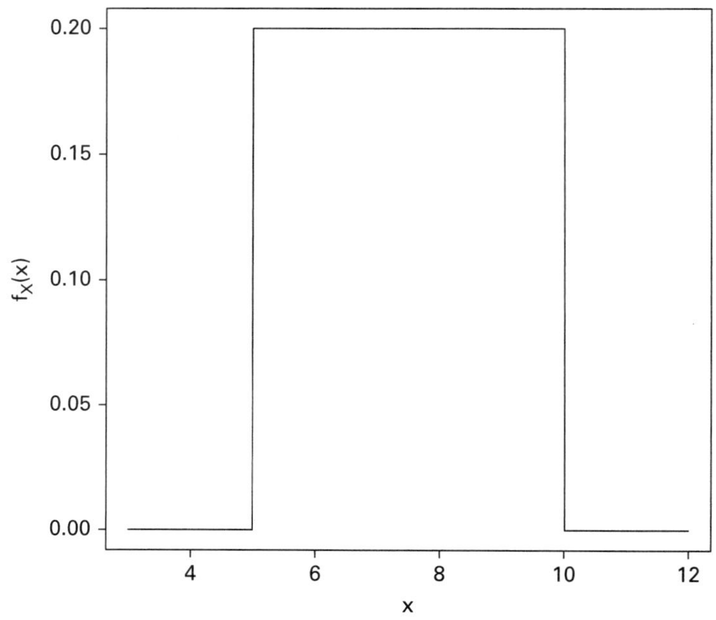

Figure 10.3
The pdf of a $U(5, 10)$ random variable

be equal to 1, *as it is for* $U(0, 1)$, *since the area of the rectangle would be* $1 \times (10 - 5) = 5$. *To have a rectangle area equal to 1, the constant value must be* $\frac{1}{5}$, *or* 0.2, *implying that the pdf is*

$$f_X(x) = \begin{cases} 0.2 & \text{if } 5 \le x \le 10 \\ 0 & \text{otherwise.} \end{cases}$$

Figure 10.3 shows the pdf curve for $X \sim U(5, 10)$. *Probabilities of intervals can be easily determined here as well. For instance,* $P(7 \le X \le 9) = (0.2)(9 - 7) = 0.4$.

More generally, the pdf for a uniform random variable $X \sim U(a, b)$, *where* $a < b$, *is*

$$f_X(x) = \begin{cases} \frac{1}{b-a} & \text{if } a \le x \le b \\ 0 & \text{otherwise.} \end{cases}$$

The height of the pdf, equal to $\frac{1}{b-a}$, *is the value that ensures that the area of the rectangle, which has the width* $b - a$, *is equal to 1.*

The following *R* functions are useful for working with a uniform random variable:

- `dunif(x, min=0, max=1)`: Returns the pdf of a *U(*`min`, `max`*)* random variable evaluated at argument `x`, which may be a single number or a vector. The optional arguments `min` and `max` have default values of `C` and `1`, respectively.
- `punif(x, min=0, max=1)`: Returns the cdf of a *U(*`min`, `max`*)* random variable evaluated at argument `x`, which may be a single number or a vector. The optional arguments `min` and `max` have default values of `0` and `1`, respectively.

- `runif(n, min=0, max=1)`: Creates a vector of `n` i.i.d. random draws of a $U($`min`, `max`$)$ random variable. The optional arguments `min` and `max` have default values of `0` and `1`, respectively.

The `punif` function returns the cdf, which will be discussed in section 10.3. The following code shows examples of `dunif` and `runif` for a $U(5, 10)$ random variable:

```
dunif(4,5,10)

## [1] 0

dunif(6,5,10)

## [1] 0.2

dunif(8,5,10)

## [1] 0.2

set.seed(1234)
runif(20,5,10)

##  [1] 5.568517 8.111497 8.046374 8.116897 9.304577 8.201553 5.047479 6.162753
##  [9] 8.330419 7.571256 8.467956 7.724874 6.413668 9.617167 6.461579 9.186478
## [17] 6.431116 6.334104 5.933614 6.161130
```

Example 10.2 (Triangular distribution) *Let X be a triangular random variable, with the pdf*

$$f_X(x) = \begin{cases} x & \text{if } 0 \le x \le 1 \\ 2-x & \text{if } 1 < x \le 2 \\ 0 & \text{otherwise.} \end{cases}$$

Figure 10.4 shows the pdf curve for X, which has a triangular shape. To confirm that the area under the pdf curve equals 1, the area of a triangle is $\frac{1}{2}$ times the length of the triangle base times the height of the triangle, which here is $\frac{1}{2}(2-0)(1) = 1$.

Probabilities of intervals can be determined from the shape of the pdf in figure 10.4. Alternatively, the integral formula can be used, so, for example, the probability that X is less than 0.5 is

$$P(X < 0.5) = \int_{-\infty}^{0.5} f_X(x)\,dx = \int_0^{0.5} x\,dx = \frac{x^2}{2}\bigg|_0^{0.5} = 0.125.$$

As another (more difficult) example, the probability that X is between 0.5 and 1.3 is

$$\begin{aligned} P(0.5 \le X \le 1.3) &= \textstyle\int_{0.5}^{1.3} f_X(x)\,dx \\ &= \textstyle\int_{0.5}^{1} x\,dx + \int_1^{1.3} (2-x)\,dx \\ &= \frac{x^2}{2}\Big|_{0.5}^{1} + \frac{-(2-x)^2}{2}\Big|_1^{1.3} \\ &= \left(\frac{1}{2} - \frac{1}{8}\right) + \left(\frac{-0.49}{2} + \frac{1}{2}\right) = 0.63. \end{aligned}$$

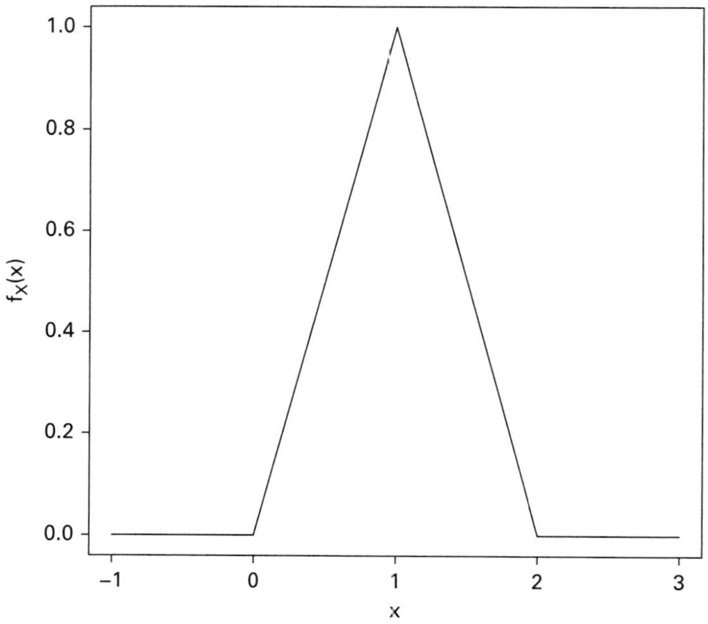

Figure 10.4
The pdf of a triangular random variable

Here is the R code to create figure 10.4:

```
# define a function that returns the pdf of a triangular distribution
tri_pdf <- function(x) {
    return( (x>=0)*(x<=1)*x + (x>1)*(x<=2)*(2-x) )
}

# create a grid of values to evaluate the pdf
temp <- seq(-1,3,0.01)

# graph the pdf
plot(temp, tri_pdf(temp), type='l', main="", ylab=expression(f[X](x)), xlab="x")
```

A new function `tri_pdf` *is defined since R does not have a built-in function for the pdf of a triangular distribution. The function* `tri_pdf` *returns* `x` *when* `x` *is between* `0` *and* `1`, `2-x` *when* `x` *is between* `1` *and* `2`, *and* `0` *otherwise.*

10.3 Cumulative Distribution Function

The pdf $f_X(\cdot)$ completely characterizes the continuous random variable X. As with discrete random variables, the cdf can also be used to characterize a continuous random variable. The advantage of the cdf is that it can be used for *any* random variable X, whether it's discrete, continuous, or even a mix of discrete and continuous. On the other hand, a pmf is used for a discrete random variable, a pdf is used for a continuous random variable, and there's no direct relationship between a pmf and a pdf.

The cdf of a continuous random variable is defined formally as in definition 10.3.

Definition 10.3 *The **cumulative distribution function (cdf)** of a continuous random variable X, denoted as $F_X(\cdot)$, gives the probability that X is less than or equal to any argument x_0 of $F_X(\cdot)$ as follows:*

$$F_X(x_0) = P(X \leq x_0) = \int_{-\infty}^{x_0} f_X(x)\, dx.$$

Like the cdf for a discrete random variable (definition 8.4), the cdf for a continuous random variable equals $P(X \leq x_0)$. Rather than the discrete summation used for a discrete random variable, the cdf of a continuous random variable involves integration of the pdf $f_X(\cdot)$ from $-\infty$ to argument x_0. The properties of the cdf of a continuous random variable are given in proposition 10.2:

Proposition 10.2 *The cdf $F_X(\cdot)$ of a continuous random variable X has the following properties:*

i. $0 \leq F_X(x_0) \leq 1$ *for every* x_0

ii. $x_0 < x_1 \implies F_X(x_0) \leq F_X(x_1)$

iii. *At every x_0 for which the derivative $F'_X(x_0)$ exists, $f_X(x_0) = F'_X(x_0)$*

Property (i) follows from the fact that $F_X(x_0) = P(X \leq x_0)$ is a probability, and therefore must be between 0 and 1 (inclusive). Property (ii) says that the cdf $F_X(\cdot)$ is a weakly increasing function. While it may stay level on certain intervals of the real line, the cdf can never decrease as x increases. Property (iii), which is a restatement of the "first fundamental theorem of calculus," provides an approach to derive the pdf $f_X(\cdot)$ if the cdf $F_X(\cdot)$ is known.

If the cdf $F_X(\cdot)$ is completely known, determining the probabilities of intervals is considerably simplified compared to the case when only the pdf $f_X(\cdot)$ is known. When only $f_X(\cdot)$ is known, we generally must calculate the integral associated with an interval, whereas integration is unnecessary if the cdf $F_X(\cdot)$ is known. Using the cdf, to determine the probability that X is less than value a,

$$P(X \leq a) = P(X < a) = F_X(a).$$

To determine the probability that X is greater than value a,

$$P(X \geq a) = P(X > a) = 1 - F_X(a).$$

And to determine the probability that X is between two values, a and b, where $a < b$,

$$P(a \leq X \leq b) = P(a < X \leq b) = P(a \leq X < b) = P(a < X < b) = F_X(b) - F_X(a).$$

Figure 10.5 shows an example of a pdf $f_X(\cdot)$ and its associated cdf, $F_X(\cdot)$. The top graph shows the pdf along with shading corresponding to the area representing $P(X \leq a)$. The middle graph shows the same pdf along with shading corresponding to the area representing $P(X \leq b)$. The bottom graph shows the cdf associated with the pdf in the other two graphs. The cdf starts at 0 and increases, eventually approaching 1. The probability $P(a \leq X \leq b)$ can be determined by the difference $F_X(b) - F_X(a)$, where the two cdf values are read off the y-axis. In terms of the two pdfs, the probability $P(a \leq X \leq b)$ is the difference between the area to the left of b (middle graph) and the area to the left of a (top graph) or, equivalently, the area under the pdf curve between a and b.

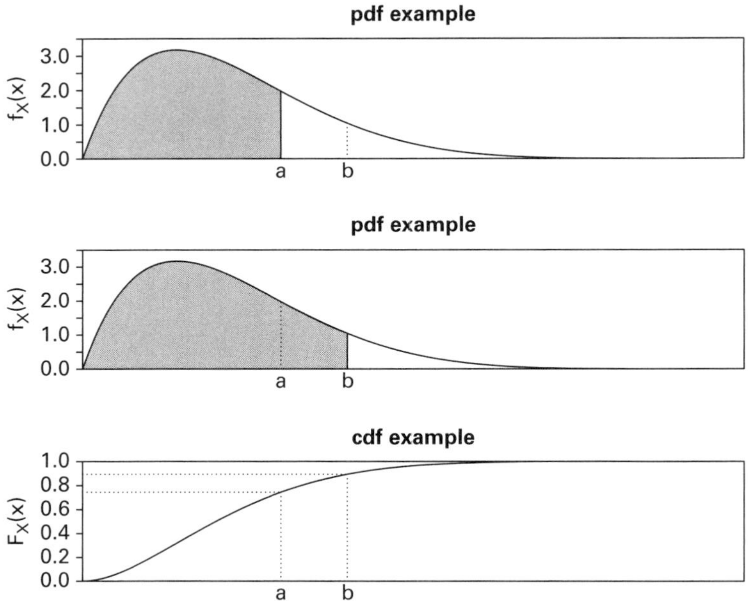

Figure 10.5
Example of a cdf

Example 10.3 (Uniform distribution) *Recall from example 10.1 that the pdf of the uniform random variable $X \sim U(0,1)$ is*

$$f_X(x) = \begin{cases} 1 & \text{if } 0 \leq x \leq 1 \\ 0 & \text{otherwise.} \end{cases}$$

To determine the cdf $F_X(\cdot)$ from the pdf $f_X(\cdot)$, the integral $F_X(x_0) = P(X \leq x_0) = \int_{-\infty}^{x_0} f_X(x)\,dx$ must be evaluated for all possible values of x_0. There are three cases that can be treated separately ($x_0 \leq 0$, $0 < x_0 \leq 1$, and $x_0 > 1$):

$$x_0 \leq 0: F_X(x_0) = \int_{-\infty}^{x_0} f_X(x)\,dx = \int_{-\infty}^{x_0} 0\,dx = 0$$

$$0 < x_0 \leq 1: F_X(x_0) = \int_{-\infty}^{x_0} f_X(x)\,dx = \int_{0}^{x_0} 1\,dx = x|_0^{x_0} = x_0$$

$$x_0 > 1: F_X(x_0) = \int_{-\infty}^{x_0} f_X(x)\,dx = \int_{0}^{1} 1\,dx = x|_0^1 = 1.$$

Putting these results together, the cdf of X is

$$F_X(x) = \begin{cases} 0 & \text{if } x \leq 0 \\ x & \text{if } 0 < x \leq 1 \\ 1 & \text{if } x > 1. \end{cases}$$

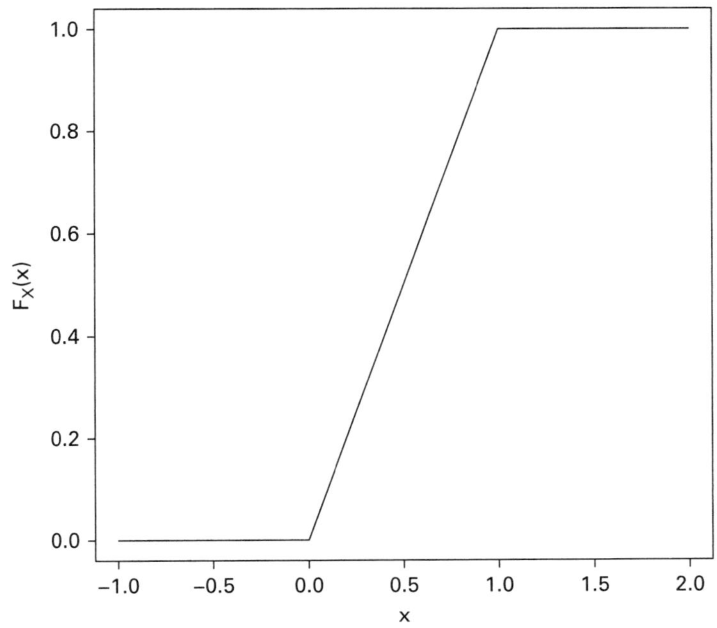

Figure 10.6
The cdf for a $U(0,1)$ random variable

Figure 10.6 shows this cdf for $X \sim U(0,1)$. *The pdf function is a constant on the* $(0,1)$ *interval for the uniform random variable* $U(0,1)$, *which yields a cdf function that is linear on the* $(0,1)$ *interval due to the integration of the pdf.*

In example 10.1, $P(0.2 \leq X \leq 0.5) = 0.3$ *was determined by integrating pdf* $f_X(\cdot)$ *from* 0.2 *to* 0.5. *Using the cdf* $F_X(\cdot)$, $P(0.2 \leq X \leq 0.5) = F_X(0.5) - F_X(0.2) = 0.5 - 0.2 = 0.3$. *The answer can also be verified in R:*

```
punif(0.5)-punif(0.2)

## [1] 0.3
```

Since the default for `punif` *is the* $U(0,1)$ *distribution, we do not specify the end points of the uniform distribution.*

Example 10.4 (Triangular distribution) *From example 10.2, the pdf of the triangular random variable* X *is*

$$f_X(x) = \begin{cases} x & \text{if } 0 \leq x \leq 1 \\ 2 - x & \text{if } 1 < x \leq 2 \\ 0 & \text{otherwise.} \end{cases}$$

To determine the cdf, the same approach as example 10.3 can be used, except that there are now four cases to consider ($x_0 < 0$, $0 \leq x_0 \leq 1$, $1 < x_0 \leq 2$, *and* $x_0 > 2$)*:*

$$x_0 < 0: F_X(x_0) = \int_{-\infty}^{x_0} f_X(x)\,dx = \int_{-\infty}^{x_0} 0\,dx = 0.$$

$$0 \le x_0 \le 1: F_X(x_0) = \int_{-\infty}^{x_0} f_X(x)\,dx = \int_0^{x_0} x\,dx = \left.\frac{x^2}{2}\right|_0^{x_0} = \frac{x_0^2}{2}.$$

$$1 < x_0 \le 2: F_X(x_0) = \int_{-\infty}^{x_0} f_X(x)\,dx = \int_0^1 x\,dx + \int_1^{x_0} (2-x)\,dx = \left.\frac{x^2}{2}\right|_0^1 + \left.\frac{-(2-x)^2}{2}\right|_1^{x_0}$$

$$= 1 - \frac{(2-x_0)^2}{2}.$$

$$x_0 > 2: F_X(x_0) = \int_{-\infty}^{x_0} f_X(x)\,dx = \int_0^1 x\,dx + \int_1^2 (2-x)\,dx = \left.\frac{x^2}{2}\right|_0^1 + \left.\frac{-(2-x)^2}{2}\right|_1^2 = 1.$$

Putting these results together, the cdf is

$$F_X(x) = \begin{cases} 0 & \text{if } x < 0 \\ \frac{x^2}{2} & \text{if } 0 \le x \le 1 \\ 1 - \frac{(2-x)^2}{2} & \text{if } 1 < x \le 2 \\ 1 & \text{if } x > 2. \end{cases}$$

Figure 10.7 shows the cdf for this triangular distribution. The pdf is linear on the $(0,1)$ *and* $(1,2)$ *intervals, which yields a cdf that is quadratic on the* $(0,1)$ *and* $(1,2)$ *intervals due to the integration of the pdf. Here is the R code to create figure 10.7, with a new function* `tri_cdf` *defined for the cdf of the triangular distribution:*

```
# define a function that returns the cdf of a triangular distribution
tri_cdf <- function(x) {
    return( (x>=0)*(x<=1)*(x^2/2) + (x>1)*(x<=2)*(1-((2-x)^2/2)) + (x>2)*1 )
}

# create a grid of values to evaluate the pdf
temp <- seq(-1,3,0.01)

# graph the cdf
plot(temp, tri_cdf(temp), type='l', main="", ylab=expression(F[X](x)), xlab="x")
```

In example 10.2, the probability $P(0.5 \le X \le 1.3)$ *was determined by integrating the pdf* $f_X(\cdot)$ *from* 0.5 *to* 1.3. *Alternatively, using the cdf* $F_X(\cdot)$ *specified here,*

$$P(0.5 \le X \le 1.3) = F_X(1.3) - F_X(0.5) = \left(1 - \frac{(2-1.3)^2}{2}\right) - \frac{0.5^2}{2} = \frac{1.51}{2} - \frac{0.25}{2} = 0.63.$$

```
tri_cdf(1.3)-tri_cdf(0.5)
## [1] 0.63
```

10.3.1 Using *R* to Calculate Integrals

Working out integrals analytically can be difficult. Even for the somewhat simple example of the triangular distribution discussed here, using integration to derive the cdf from the pdf required quite a bit of work. On the other hand, *R* can perform integration with a high

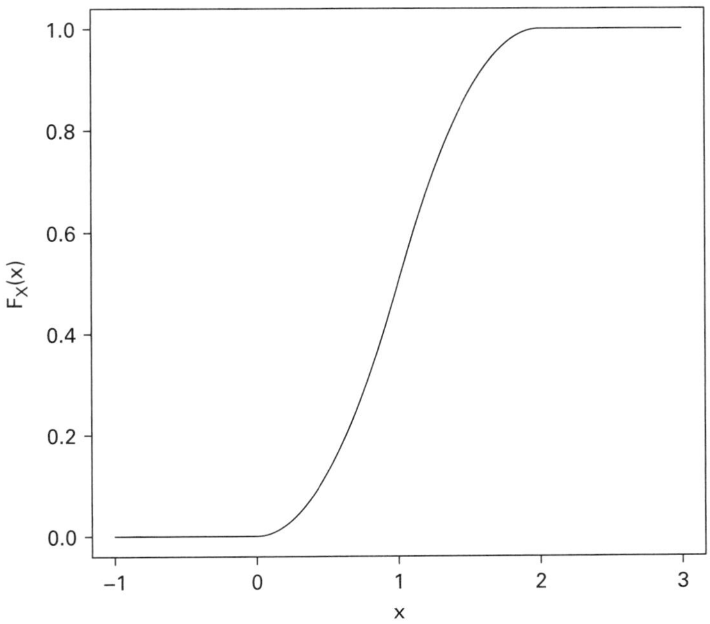

Figure 10.7
The cdf for a triangular random variable

degree of accuracy, even for very complicated functions. This type of integration is often called **numerical integration**, and it only requires knowledge of the underlying function being integrated. The *R* function for numerical integration is `integrate`:

- `integrate(f, lower, upper)`: Returns the integral of function `f` with the lower limit `lower` and upper limit `upper`. A lower limit of negative infinity can be specified by using `lower=-Inf`. An upper limit of positive infinity can be specified by using `upper=Inf`. Function `f` should take a numerical value or vector as its only argument and should return a numerical vector of the same length.

Here are some examples of `integrate` being used to calculate cdfs and probability intervals for the uniform and triangular distribution:

```
integrate(dunif,0.2,0.8)
## 0.6 with absolute error < 0.0000000000000067
integrate(dunif,-Inf,0.8)
## 0.7999995 with absolute error < 0.0000013
integrate(tri_pdf,0.5,1.3)
## 0.63 with absolute error < 0.000000000000007
```

These examples illustrate that the `integrate` function can be used with either a build-in *R* function (like `dunif`) or a user-defined function (like `tri_pdf`). The first `integrate` command integrates the pdf of the $X \sim U(0, 1)$ random variable (`dunif`) from `0.2` to

`0.8`, which is equivalent to $F_X(0.8) - F_X(0.2)$. The second `integrate` command uses a lower limit of $-\infty$, so it is equivalent to the cdf $F_X(0.8)$. The third `integrate` command provides an alternative method of determining $P(0.5 \leq X \leq 1.3)$ when X has a triangular distribution. Rather than determining the cdf function and coding it, as we did with `tri_cdf`, the pdf `tri_pdf` is integrated directly here. For each use of `integrate`, *R* returns the result of the numerical integration along with the phrase "`with absolute error < ...`" Since it's performing numerical integration, the function `integrate` is only approximating the value of the integral, but the *very* small absolute errors indicate that the approximations are very accurate. The desired accuracy can be specified as an argument of `integrate`. For this and other options, refer to the *R* documentation for `integrate`.

As seen in the *R* output, the `integrate` function does more than simply return the numeric value of the integral. If the value of the integral needs to be stored in a variable and/or used in subsequent calculations, it can be accessed directly by appending `$value` after the `integrate` function:

```
integral_value <- integrate(tri_pdf,0.5,1.3)$value
paste("Calculated integral:", round(integral_value,3))

## [1] "Calculated integral: 0.63"
```

10.3.2 The cdf of a Mixture of Discrete and Continuous Outcomes

The cdf is a unifying concept used for both discrete and continuous random variables. In fact, it can be used for *any* random variable, including one with a mixture of discrete outcomes and continuous outcomes. The key idea is that the cdf $F_X(\cdot)$ is always defined as $F_X(x_0) = P(X \leq x_0)$ for every value x_0, as in definition 8.4 for a discrete random variable and definition 10.3 for a continuous random variable. Example 10.5 considers the cdf of a random variable that is a mixture of discrete outcomes and continuous outcomes.

Example 10.5 (Uniform distribution with point masses) *Suppose that the cdf $F_X(\cdot)$ associated with a random variable X is*

$$F_X(x) = \begin{cases} 0 & \text{if } x < 0 \\ 0.3 + 0.5x & \text{if } 0 \leq x < 1 \\ 1 & \text{if } x \geq 1. \end{cases}$$

In this example, it's important to pay close attention to whether an inequality is weak or strict in the specification of $F_X(\cdot)$. The cdf is flat at zero for all $x < 0$, and then at $x = 0$, it jumps to $F_X(0) = 0.3$. Then the cdf is linear on the interval $(0, 1)$, increasing from 0.3 to 0.8 in this interval. Finally, at $x = 1$, the cdf jumps to $F_X(1) = 1$. Figure 10.8 provides a graph of the cdf $F_X(\cdot)$.

The only way for a jump to occur at $x = 0$ is if the event $X = 0$ has positive probability. In this case, we say that there is a "point mass" at $x = 0$, with $P(X = 0) = 0.3$. Similarly, the jump from 0.8 to 1 at $x = 1$ is associated with a "point mass" at $x = 1$, with $P(X = 1) = 1 - 0.8 = 0.2$. For values within the interval $(0, 1)$, the distribution of X is similar to a uniform distribution. This random variable, then, can be thought of as a uniform distribution with point masses at its end points of 0 and 1. Any value within the interval $(0, 1)$ has zero probability, but the discrete outcomes 0 and 1 each have positive probability. The probability

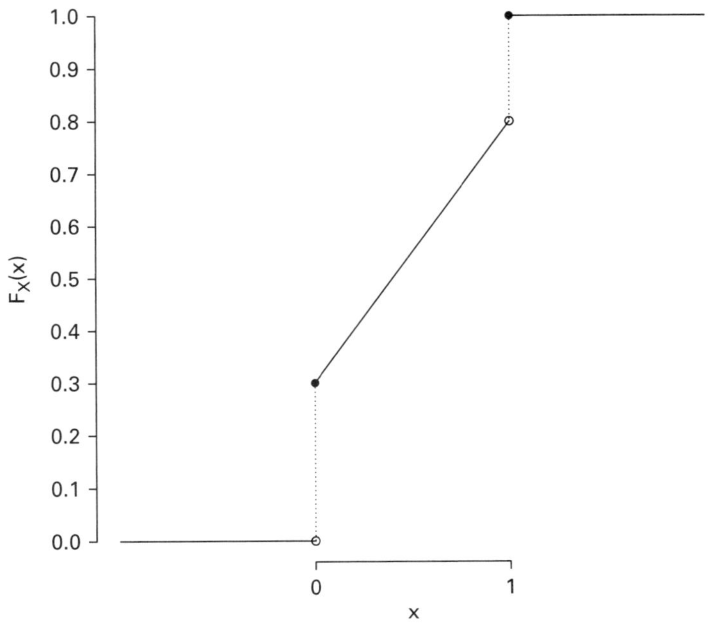

Figure 10.8
The cdf for a mixture random variable

of any interval can be determined with the cdf. For example, the probability that X is less than 0.3 *is* $P(X < 0.3) = F_X(0.3) = 0.45$*, and the probability that X is between* 0.3 *and* 0.9 *is* $F(0.3 < X < 0.9) = F_X(0.9) - F_X(0.3) = 0.75 - 0.45 = 0.30$.

10.4 Population-Descriptive Statistics

10.4.1 Population Quantiles

Section 6.4.2 introduced sample quantiles as descriptive statistics for an observed sample of data. For example, the sample 80% quantile $\tilde{x}_{0.8}$ represents the value for which 80% of the observed data are below $\tilde{x}_{0.8}$. An oft-used special case of the sample quantile is the sample median $\tilde{x}_{0.5}$, the value for which 50% of the observed data are below $\tilde{x}_{0.5}$. This section introduces a similar quantile concept for random variables, but now the quantile describes the distribution of the random variable in the population rather than the distribution of observed data in the sample. The terminology **population quantile**, described in definition 10.4, distinguishes it from the sample quantile.

Definition 10.4 *For any q, with* $0 < q < 1$*, the* ***population quantile*** $\tau_{X,q}$ *of a continuous random variable X is the value for which*

$$P(X \leq \tau_{X,q}) = F_X(\tau_{X,q}) = q.$$

A special case is the ***population median*** *(*$\tau_{X,0.5}$ *or* $\tau_{X,1/2}$*) of a continuous random variable X, with*

$$P(X \leq \tau_{X,0.5}) = F_X(\tau_{X,0.5}) = 0.5.$$

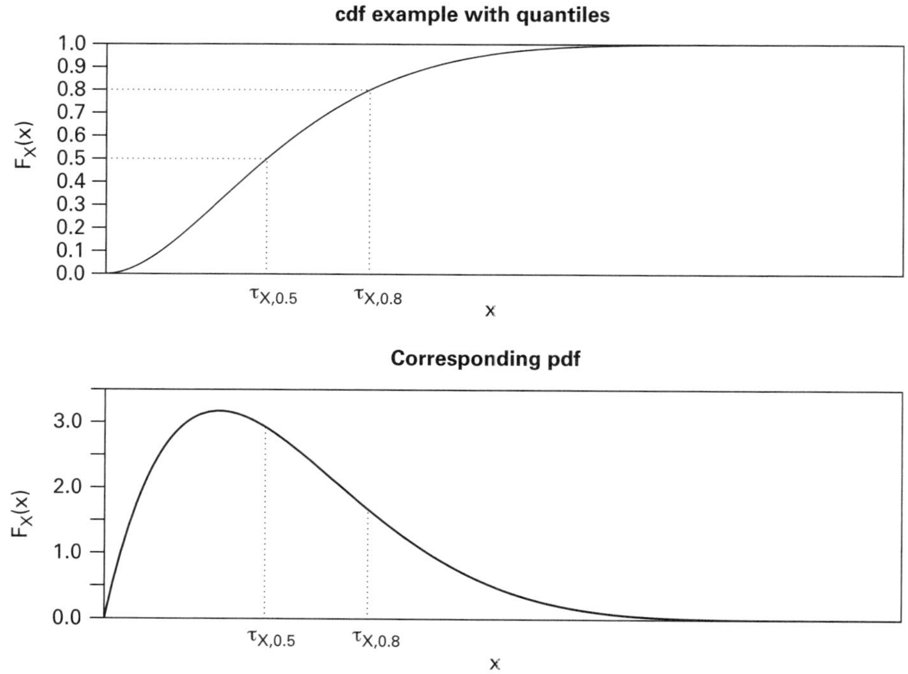

Figure 10.9
Example of population quantiles for a continuous random variable

Let's say that we are interested in the population 80% quantile $\tau_{X,0.8}$ of X, where X is a continuous random variable. $\tau_{X,0.8}$ is the value for which there is an 80% probability that X is less than or equal to $\tau_{X,0.8}$: $P(X \leq \tau_{X,0.8}) = F_X(\tau_{X,0.8}) = 0.8$. If we had a graph of the cdf $F_X(\cdot)$, finding this population 80% quantile would involve drawing a horizontal line at 0.8 and finding the x-value where this line hits the cdf. The top graph of figure 10.9 shows an example of a cdf, where the population 80% quantile ($\tau_{X,0.8}$) and population median ($\tau_{X,0.5}$) are shown as the values where the horizontal lines at 0.8 and 0.5, respectively, cross the cdf curve. The corresponding pdf is shown in the bottom graph of the figure, with the population 80% quantile $\tau_{X,0.8}$ and the population median $\tau_{X,0.5}$ indicated. For the population median $\tau_{X,0.5}$, the area under the pdf to the left of $\tau_{X,0.5}$ equals 0.5. For the population 80% quantile $\tau_{X,0.8}$, the area under the pdf to the left of $\tau_{X,0.8}$ equals 0.8.

Just as the sample interquartile range $IQR_x = \tilde{x}_{0.75} - \tilde{x}_{0.25}$ is defined as the difference between the sample 75% quantile and sample 25% quantile, the **population interquartile range** is defined analogously as the difference between the population 75% quantile and population 25% quantile, as in definition 10.5.

Definition 10.5 *The* ***population interquartile range (population IQR)*** *of a continuous random variable X, denoted as IQR_X, is*

$$IQR_X = \tau_{X,0.75} - \tau_{X,0.25}.$$

Example 10.6 illustrates how population quantiles can be determined analytically when the cdf $F_X(\cdot)$ is known. Since it is easier to determine population quantiles from the cdf, it is recommended to first find it if only the pdf is specified.

Example 10.6 (Triangular distribution) *Continuing example 10.4, recall that the cdf of the triangular random variable X is*

$$F_X(x) = \begin{cases} 0 & \text{if } x < 0 \\ \frac{x^2}{2} & \text{if } 0 \le x \le 1 \\ 1 - \frac{(2-x)^2}{2} & \text{if } 1 < x \le 2 \\ 1 & \text{if } x > 2. \end{cases}$$

Suppose that we are interested in determining the population 40% quantile. We want to find the value $\tau_{X,0.4}$ for which $F_X(\tau_{X,0.4}) = 0.4$. We know we don't get a cdf value equal to 0.4 for $x \le 0$, so we move to the $0 < x \le 1$ interval. Is there an $x \in (0, 1]$ such that $\frac{x^2}{2} = 0.4$? The answer is yes, with $x = \sqrt{0.8}$, so that $\tau_{X,0.4} = \sqrt{0.8} \approx 0.894$.

Suppose that we are interested in determining the population 70% quantile. We want to find the value $\tau_{X,0.7}$ for which $F_X(\tau_{X,0.7}) = 0.7$. We know that we need to move past the $0 < x \le 1$ interval since the maximum cdf value, which occurs at $x = 1$, is only 0.5. Moving to the $1 < x \le 2$ interval, is there an $x \in (1, 2]$ such that $1 - \frac{(2-x)^2}{2} = 0.7$? The answer is yes, with $x = 2 - \sqrt{0.6}$, so that $\tau_{X,0.7} = 2 - \sqrt{0.6} \approx 1.225$.

For commonly used continuous distributions, *R* provides functions to calculate population quantiles. For the uniform distribution, the `qunif` function is available for this purpose:

- `qunif(p, min=0, max=1)`: Returns the population quantiles of a *U*(`min`, `max`) random variable specified by argument `p`, which may be a single number or a vector. The optional arguments `min` and `max` have default values of `0` and `1`, respectively.

R uses the convention of having `q` as the first letter for functions that return population quantiles. The following code returns the population 80% quantile of a $U(0, 1)$ random variable and the population 50% and 75% quantiles of a $U(5, 10)$ random variable:

```
qunif(0.8)

## [1] 0.8

qunif(c(0.50,0.75),5,10)

## [1] 7.50 8.75
```

10.4.2 Population Mean or Expected Value

Section 8.3.1 introduced the population mean for a discrete random variable X, with possible outcomes $\{x_k^*\}_{k=1}^K$ for finite or (countably) infinite K:

$$\mu_X = E(X) = \sum_k x_k^* p_X(x_k^*).$$

The population mean of X is a weighted average of the possible outcomes x_k^*, with weights given by the probabilities of each x_k^* in the population. This definition does not work for a continuous random variable since (1) the number of possible outcomes is uncountable and (2) the probability of any outcome equals zero. Instead, we introduce a version of the population mean that replaces the discrete summation here with an integral and the pmf values with pdf values.

Definition 10.6 *The* ***population mean*** *(or* ***population average*** *or* ***expected value****) of a continuous random variable X, denoted as μ_X or $E(X)$, is*

$$\mu_X = E(X) = \int_{-\infty}^{\infty} x f_X(x)\, dx.$$

The integral in definition 10.6 can be thought of as a summation over *very* small intervals of the x values. Specifically, consider a slice of the pdf function right at the value x, with height $f_X(x)$ and width dx. The area of this slice is $f_X(x)dx$ and can be thought of as the probability that the outcome of X is within the slice. The integral essentially sums over all these little slices, and the expected value provides a weighted average of the x values with weights given by the probabilities $f_X(x)dx$.

Example 10.7 (Uniform distribution) *The pdf of the uniform random variable $X \sim U(0,1)$ is*

$$f_X(x) = \begin{cases} 1 & \text{if } 0 \le x \le 1 \\ 0 & \text{otherwise.} \end{cases}$$

The population mean is

$$\mu_X = \int_{-\infty}^{\infty} x f_X(x)\, dx = \int_0^1 x\, dx = \left.\frac{x^2}{2}\right|_0^1 = 0.5,$$

which is as expected since $X \sim U(0,1)$ is symmetric around the center of the $(0,1)$ interval.

How about the general uniform distribution $X \sim U(a,b)$ for $a < b$? The pdf is

$$f_X(x) = \begin{cases} \frac{1}{b-a} & \text{if } a \le x \le b \\ 0 & \text{otherwise.} \end{cases}$$

Again, due to symmetry around the middle of the (a,b) interval, the population mean is the midpoint $\frac{a+b}{2}$, which can be confirmed as follows:

$$\mu_X = \int_{-\infty}^{\infty} x f_X(x)\, dx = \int_a^b x \frac{1}{b-a}\, dx = \left.\frac{x^2}{2(b-a)}\right|_a^b = \frac{b^2 - a^2}{2(b-a)} = \frac{a+b}{2}.$$

In example 10.7, we referred to the symmetry of the uniform distribution, a concept that we formally define in definition 10.7.

Definition 10.7 *A continuous random variable X is said to have a* ***symmetric distribution*** *or* ***symmetric pdf*** *if there is a midpoint, call it x^*, for which the pdf on one side of x^* is a mirror image of the pdf on the other side of x^*. Mathematically, with midpoint x^*, a symmetric* *distribution is symmetric about x^* and has*

$$f_X(x^* - v) = f_X(x^* + v) \text{ for all } v \geq 0$$

or, equivalently,

$$F_X(x^* - v) = 1 - F_X(x^* + v) \text{ for all } v \geq 0.$$

Symmetric distributions have some nice properties, as described in proposition 10.3.

Proposition 10.3 *A continuous random variable X with a symmetric distribution, for which the pdf is symmetric around midpoint x^*, has the following properties:*

i. *the population mean and the population median are equal to each other and to x^*:*

$$\mu_X = \tau_{X,0.5} = x^*.$$

ii. *for any q, with $0 < q < 0.5$, the population quantiles $\tau_{X,q}$ and $\tau_{X,1-q}$ are equidistant from the midpoint x^*:*

$$|\tau_{X,q} - x^*| = |\tau_{X,1-q} - x^*| \text{ or, equivalently, } |\tau_{X,q} - \tau_{X,0.5}| = |\tau_{X,1-q} - \tau_{X,0.5}|.$$

Both properties should make intuitive sense since the part of the symmetric distribution to the left of its midpoint x^* is a mirror image of the part of the distribution to the right of its midpoint x^*. For the uniform distributions in example 10.7, $X \sim U(0, 1)$ has midpoint $x^* = 0.5$ and $X \sim U(a, b)$ has midpoint $x^* = \frac{a+b}{2}$. For $X \sim U(0, 1)$, the equidistant quantile property (property (ii)) clearly holds; for instance, the population 20% and 80% quantiles, $\tau_{X,0.2} = 0.2$ and $\tau_{X,0.8} = 0.8$, are the same distance from the midpoint $x^* = \tau_{X,0.5} = 0.5$.

As was the case for discrete random variables, it is possible that the population mean or expected value of a continuous random variable is not well defined, as illustrated by example 10.8.

Example 10.8 (Infinite expected value) *Consider the following pdf for X:*

$$f_X(x) = \begin{cases} \frac{1}{x^2} & \text{if } x > 1 \\ 0 & \text{otherwise.} \end{cases}$$

This pdf is valid since $\int_1^\infty \frac{1}{x^2} dx = -\frac{1}{x}\Big|_1^\infty = 0 - (-1) = 1$, but the expected value of X is infinite since

$$\mu_X = \int_1^\infty x \frac{1}{x^2} dx = \int_1^\infty \frac{1}{x} dx = \ln(x)|_1^\infty = \ln(\infty) - 0 = \infty.$$

10.4.3 Population Variance and Population Standard Deviation

Section 8.3.2 introduced the population variance for a discrete random variable X, with possible outcomes $\{x_k^*\}_{k=1}^K$ for finite or (countably) infinite K:

$$\sigma_X^2 = Var(X) = E\left[(X - \mu_X)^2\right] = \sum_k (x_k^* - \mu_X)^2 p_X(x_k^*).$$

For a discrete random variable X, the population variance is a weighted average of the $(x_k^* - \mu_X)^2$ values for all possible outcomes x_k^*, with weights given by the probabilities of each x_k^* in the population. As with the population mean, this definition won't work for a continuous random variable due to the number of outcomes being uncountable and each outcome having a probability of zero. Instead, an analogous definition for a continuous random variable is introduced, replacing the summation with an integral and replacing the pmf values with pdf values.

Definition 10.8 *The* ***population variance*** *of a continuous random variable X, denoted as σ_X^2 or $Var(X)$, is*

$$\sigma_X^2 = Var(X) = E\left[(X-\mu_X)^2\right] = \int_{-\infty}^{\infty} (x-\mu_X)^2 f_X(x)\, dx.$$

The population standard deviation is defined as the square root of the population variance.

Definition 10.9 *The* ***population standard deviation*** *of a continuous random variable X, denoted as σ_X or $sd(X)$, is*

$$\sigma_X = sd(X) = \sqrt{\sigma_X^2} = \sqrt{\int_{-\infty}^{\infty} (x-\mu_X)^2 f_X(x)\, dx}.$$

Example 10.9 (Uniform distribution) *In example 10.7, the population mean of $X \sim U(0,1)$ was determined to be $\mu_X = 0.5$ and, in the general case, the population mean of $X \sim U(a,b)$ was determined to be $\mu_X = \frac{a+b}{2}$. How about their respective population variances and population standard deviations? Starting with $X \sim U(0,1)$, the population variance is*

$$\sigma_X^2 = \int_{-\infty}^{\infty} (x-\mu_X)^2 f_X(x)\, dx = \int_0^1 (x-0.5)^2(1)\, dx = \left.\frac{(x-0.5)^3}{3}\right|_0^1 = \frac{1}{24} - \left(-\frac{1}{24}\right) = \frac{1}{12}$$

and the population standard deviation is

$$\sigma_X = \sqrt{\sigma_X^2} = \sqrt{\frac{1}{12}} = \frac{1}{\sqrt{12}}.$$

For the general case, $X \sim U(a,b)$, the population variance is

$$\begin{aligned}\sigma_X^2 &= \textstyle\int_{-\infty}^{\infty} (x-\mu_X)^2 f_X(x)\, dx \\ &= \textstyle\int_a^b \left(x - \frac{a+b}{2}\right)^2 \left(\frac{1}{b-a}\right) dx \\ &= \left(\frac{1}{b-a}\right) \left.\frac{\left(x-\frac{a+b}{2}\right)^3}{3}\right|_a^b = \left(\frac{1}{b-a}\right)\left(\frac{(b-a)^3}{24} - \left(\frac{(a-b)^3}{24}\right)\right) = \frac{(b-a)^2}{12}\end{aligned}$$

and the population standard deviation is

$$\sigma_X = \sqrt{\sigma_X^2} = \sqrt{\frac{(b-a)^2}{12}} = \frac{b-a}{\sqrt{12}}.$$

Example 10.10 (Triangular distribution) *The triangular distribution (example 10.2) has the pdf*

$$f_X(x) = \begin{cases} x & \text{if } 0 \le x \le 1 \\ 2-x & \text{if } 1 < x \le 2 \\ 0 & \text{otherwise.} \end{cases}$$

From the graph of the pdf $f_X(\cdot)$ in figure 10.4, the triangular distribution appears to be symmetric with midpoint $x^* = 1$. We can confirm that it's a symmetric distribution by checking that $f_X(x^* - v) = f_X(x^* + v)$ for all $v \ge 0$. First, for any v such that $0 \le v \le 1$, $f_X(1-v) = 1 - v$ since $0 \le 1 - v \le 1$ and $f_X(1+v) = 2 - (1+v) = 1 - v$ since $1 \le 1 + v \le 2$. Second, for any $v > 1$, $f_X(1-v) = f_X(1+v) = 0$ since $1 - v < 0$ and $1 + v > 2$. Therefore, X has a symmetric distribution, meaning that its population mean is equal to the midpoint, $\mu_X = x^* = 1$.

Using the fact that $\mu_X = 1$, the population variance of X is

$$\begin{aligned}
\sigma_X^2 &= \int_{-\infty}^{\infty} (x - \mu_X)^2 f_X(x)\,dx \\
&= \int_0^1 (x-1)^2\,(x)\,dx + \int_1^2 (x-1)^2\,(2-x)\,dx \\
&= \int_0^1 (x^3 - 2x^2 + x)\,dx + \int_1^2 (-x^3 + 4x^2 - 5x + 2)\,dx \\
&= \left(\frac{x^4}{4} - \frac{2x^3}{3} + \frac{x^2}{2}\right)\Big|_0^1 + \left(-\frac{x^4}{4} + \frac{4x^3}{3} - \frac{5x^2}{2} + 2x\right)\Big|_1^2 \\
&= \left(\frac{1}{4} - \frac{1}{3} + \frac{1}{2}\right) + \left(-\frac{15}{4} + \frac{28}{3} - \frac{15}{2} + 2\right) = \frac{1}{12} + \frac{1}{12} = \frac{1}{6},
\end{aligned}$$

and the population standard deviation of X is

$$\sigma_X = \sqrt{\sigma_X^2} = \sqrt{\frac{1}{6}} = \frac{1}{\sqrt{6}}.$$

10.4.4 Using Integration in *R* to Calculate Population Statistics

Section 10.3.1 discussed numerical integration in *R*, using the `integrate` function. Since population means, variances, and standard deviations are defined in terms of an integral, we can use numerical integration to calculate their values if the pdf $f_X(\cdot)$ is known, providing an alternative to analytic derivation of the integral. For example, the population mean is equal to $\int_{-\infty}^{\infty} x f_X(x)\,dx$, which can be evaluated by providing a function that calculates $x f_X(x)$ as the first argument of `integrate` and $-\infty$ and ∞ as the second and third arguments. Here is an example for a $U(0, 1)$ random variable:

```
# auxiliary function that returns x*f(x) for a U(0,1) random variable
exp_unif <- function(x) {
    return( x*dunif(x) )
}

# numerical integration to evaluate E(X) for X~U(0,1)
integrate(exp_unif,-Inf,Inf)
```

```
## 0.5 with absolute error < 0.0000000013

integrate(exp_unif,-Inf,Inf)$value

## [1] 0.5
```

The numerical integration gives the correct answer $\mu_X = 0.5$ for $X \sim U(0, 1)$. Two slightly different versions of the integrate command are used. The first, a standard call to integrate, reports the value along with the accuracy. The second, which appends the syntax $value function, provides just the numerical value of the integral. This version is useful when we need to use the numerical value of the integral in further calculations, as seen next for the population variance.

The population variance is equal to $\int_{-\infty}^{\infty}(x-\mu_X)^2 f_X(x)\,dx$, so numerical integration requires a function that returns $(x-\mu_X)^2 f_X(x)$ to be used as the first argument of integrate. Here is an example for a $U(0, 1)$ random variable:

```
# store E(X) for X~U(0,1) in a variable
mean_unif <- integrate(exp_unif,-Inf,Inf)$value

# auxiliary function that returns (x-E(X))^2*f(x) for a U(0,1) random variable
var_unif <- function(x) {
    return( (x-mean_unif)^2*dunif(x) )
}

# numerical integration to evaluate Var(X) for X~U(0,1)
integrate(var_unif,-Inf,Inf)

## 0.08333333 with absolute error < 0.000000046

# numerical integration to evaluate sd(X) for X~U(0,1)
sqrt(integrate(var_unif,-Inf,Inf)$value)

## [1] 0.2886751
```

The numerical integration provides the correct answer $\sigma_X^2 = \frac{1}{12} \approx 0.083333$ for $X \sim U(0, 1)$.

There's nothing inherently special about the uniform distribution in these *R* code examples. To calculate the population mean and population variance for the triangular distribution, for example, the density function tri_pdf that was defined in example 10.2 can be substituted in the code wherever dunif appeared:

```
# auxiliary function that returns x*f(x) for a triangular random variable
exp_tri <- function(x) {
    return( x*tri_pdf(x) )
}

# store and print E(X) for a triangular random variable
mean_tri <- integrate(exp_tri,-Inf,Inf)$value
print(mean_tri)

## [1] 1

# auxiliary function that returns (x-E(X))^2*f(x) for a triangular random variable
var_tri <- function(x) {
    return( (x-mean_tri)^2*tri_pdf(x) )
}
```

```
# numerical integration to evaluate Var(X) for a triangular random variable
integrate(var_tri,-Inf,Inf)

## 0.1666671 with absolute error < 0.000093

# numerical integration to evaluate sd(X) for a triangular random variable
sqrt(integrate(var_tri,-Inf,Inf)$value)

## [1] 0.4082488
```

10.5 Linear Transformations of One Random Variable

Section 8.5.1 considered linear transformations of a discrete random variable. In this section, we generalize the results from that section to all random variables, including some additional results related to quantiles that hold for continuous random variables. The properties of a linear transformation of a random variable are summarized in proposition 10.4:

Proposition 10.4 *If X is a random variable and $Y = a + bX$ for known constants a and b, then:*

i. (population mean) $\mu_Y = a + b\mu_X$.

ii. (population variance) $\sigma_Y^2 = b^2\sigma_X^2$

iii. (population standard deviation) $\sigma_Y = |b|\sigma_X$.

iv. (population quantiles) $\tau_{Y,q} = a + b\tau_{X,q}$ if $b \geq 0$ and X is a continuous random variable; $\tau_{Y,q} = a + b\tau_{X,1-q}$ if $b < 0$ and X is a continuous random variable.

v. (population IQR) $IQR_Y = |b|IQR_X$ if X is a continuous random variable.

Parts (i) through (iii) are identical to the results from section 8.5.1 for discrete random variables. The additive constant a affects the population mean, but not the population variance or standard deviation, whereas the scaling constant b affects all three quantities. Part (iv) states that when b is positive, the population quantiles for the linear combination Y ($\tau_{Y,q}$) are the same linear function of the population quantiles of X ($\tau_{X,q}$); when b is negative, in which case we can think of the distribution being flipping around before being scaled, the linear function is applied to $\tau_{X,1-q}$ rather than $\tau_{X,q}$. Part (v) is analogous to the result for the sample IQR, where we had $IQR_y = |b|IQR_x$ for the linear transformation $y = a + bx$ of variable x.

Example 10.11 (Annualized earnings) *If X represents the weekly earnings for an employed individual from the population, the linear transformation $Y = 52X$ is a random variable that represents that annualized earnings for an employed individual from the population. Similar to the results for descriptive statistics in example 6.29, proposition 10.4 can be applied here to get $\mu_Y = 52\mu_X$, $\sigma_Y^2 = 52^2\sigma_X^2 = 2704\sigma_X^2$, $\sigma_Y = 52\sigma_X$, $\tau_{Y,q} = 52\tau_{X,q}$ for any $q \in (0, 1)$, and $IQR_Y = 52IQR_X$.*

Example 10.12 (Uniform distribution) *Let $X \sim U(0, 1)$ be a uniform random variable on the interval $(0, 1)$. For constants a and b, where $a < b$, if the random variable Y is defined as the linear transformation*

$$Y = a + (b - a)X,$$

the population mean of Y is

$$\mu_Y = a + (b-a)\mu_Y = a + \frac{b-a}{2} = \frac{a+b}{2},$$

the population variance of Y is

$$\sigma_Y^2 = (b-a)^2\sigma_X^2 = \frac{(b-a)^2}{12},$$

and the population standard deviation of Y is

$$\sigma_Y = |b-a|\sigma_X = \frac{b-a}{\sqrt{12}}.$$

Recall from examples 10.7 and 10.9 that these quantities are the same population mean, variance, and standard deviation derived for the $U(a,b)$ distribution. In fact, $Y \sim U(a,b)$ can be shown as follows:

$$F_Y(y) = P(Y \le y) = P(a + (b-a)X \le y) = P\left(X \le \frac{y-a}{b-a}\right) = \frac{y-a}{b-a} \text{ for } y \in (a,b),$$

which implies that

$$f_Y(y) = F_Y'(y) = \frac{1}{b-a} \text{ for } y \in (a,b),$$

by applying part (iii) of proposition 10.2. Thus, the pdf of Y is the same as the pdf of a $U(a,b)$ random variable. This example shows that the linear transformation of a uniform $U(0,1)$ random variable is also a uniform random variable. Using this fact to derive the population mean, variance, and standard deviation is much easier than the work required in example 10.9 to derive those quantities directly. And any uniform random variable can be written as some linear transformation of the $U(0,1)$ random variable. For instance, to get $Y \sim U(-2,5)$, the linear transformation $Y = -2 + 7X$ can be used for $X \sim U(0,1)$.

A specific linear transformation that is often quite useful is the transformation that standardizes a random variable X to have a population mean equal to 0 and a population variance (and population standard deviation) equal to 1. This **standardized random variable** is formally defined as in definition 10.10, where we do not restrict the type of random variable.

Definition 10.10 *For a random variable X, a **standardized random variable** Z is constructed by de-meaning the random variable X and then dividing by its standard deviation:*

$$Z = \frac{X - \mu_X}{\sigma_X}.$$

From this definition, Z is a linear transformation of X with additive constant $a = -\frac{\mu_X}{\sigma_X}$ and scaling constant $b = \frac{1}{\sigma_X}$. Using the results from proposition 10.4,

$$\mu_Z = a + b\mu_X = -\frac{\mu_X}{\sigma_X} + \frac{1}{\sigma_X}\mu_X = 0,$$

$$\sigma_Z^2 = b^2\sigma_X^2 = \left(\frac{1}{\sigma_X}\right)^2 \sigma_X^2 = 1,$$

and

$$\sigma_Z = |b|\sigma_X = \frac{1}{\sigma_X}\sigma_X = 1.$$

Therefore, the standardized random variable Z has a population mean equal to 0 and population variance and standard deviation both equaling 1. Moreover, Z is unitless since both its numerator $X - \mu_X$ and its denominator σ_X are expressed in the units of X. The random variable Z can be interpreted as the number of population standard deviations that X is from its population mean μ_X, with negative Z corresponding to X being below μ_X and positive Z corresponding to X being above μ_X. For instance, $Z = -1.5$ indicates that X is 1.5 standard deviations below μ_X (i.e., $X = \mu_X - 1.5\sigma_X$), and $Z = 2.7$ indicates that X is 2.7 standard deviations above μ_X (i.e., $X = \mu_X + 2.7\sigma_X$).

10.6 Multiple Continuous Random Variables

10.6.1 Joint Probability Density Function and Joint Cumulative Distribution Function

When there are two continuous random variables, their joint distribution can be represented by a **joint probability density function (joint pdf)**, defined as in definition 10.11.

Definition 10.11 *The **joint probability density function (joint pdf)** of continuous random variables X and Y is a function $f_{XY}(\cdot,\cdot)$ such that, for any two numbers a and b with $a \le b$ and any two numbers c and d with $c \le d$,*

$$P(a \le X \le b,\ c \le Y \le d) = \int_c^d \int_a^b f_{XY}(x,y)\,dx\,dy.$$

Whereas the probability of an interval for a single continuous random variable is a single integral, the joint probability of X and Y being in their respective intervals is a double integral. The inner integral integrates over the x values between a and b, while the outer integral integrates over the y values between c and d. Since X and Y are continuous random variables, the joint probability of any specific outcome equals zero; that is, $P(X = x, Y = y) = 0$ for any (x,y). Like the marginal pdf, the joint pdf is nonnegative for all (x,y) and integrates to 1. These properties, along with the relationship of the joint pdf to the marginal pdfs, are stated in proposition 10.5.

Proposition 10.5 *For continuous random variables X and Y, the joint pdf $f_{XY}(\cdot,\cdot)$ has the following properties:*

i. $f_{XY}(x,y) \ge 0$ *for all* (x,y).

ii. $\int_{-\infty}^{\infty}\int_{-\infty}^{\infty} f_{XY}(x,y)\,dx\,dy = 1$.

iii. $f_X(x) = \int_{-\infty}^{\infty} f_{XY}(x,y)\,dy$ *for all* (x,y).

iv. $f_Y(y) = \int_{-\infty}^{\infty} f_{XY}(x,y)\,dx$ *for all* (x,y).

Property (iii) states that the marginal pdf of X, evaluated at x, is obtained by fixing x and integrating the joint pdf over all values of y. This relationship is analogous to the case of discrete random variables, where the marginal pmf of X, evaluated at a possible outcome x_k^*, is obtained by fixing x_k^* and summing the joint pmf over all possible y_ℓ^* values. Similarly, property (iv) states that the marginal pdf of Y, evaluated at y, is obtained by fixing y and integrating the joint pdf over all values of x.

Example 10.13 (Unrelated uniform random variables) *Suppose that X and Y are continuous random variables with the joint pdf*

$$f_{XY}(x,y) = \begin{cases} 1 & \text{if } 0 \le x \le 1 \text{ and } 0 \le y \le 1 \\ 0 & \text{otherwise.} \end{cases}$$

It can be confirmed that the joint pdf integrates to 1:

$$\int_{-\infty}^{\infty}\int_{-\infty}^{\infty} f_{XY}(x,y)\,dx\,dy = \int_0^1\int_0^1 1\,dx\,dy = \int_0^1 1\,dy = 1.$$

The double integral $\int_0^1\int_0^1 1\,dx\,dy$ can be interpreted as the volume of a rectangular solid, with height 1 (the constant $f_{XY}(\cdot,\cdot)$ value) and rectangular sides of 1 (for the range of x) and 1 (for the range of y).

For the marginal pdf of X, when $0 \le x \le 1$,

$$f_X(x) = \int_{-\infty}^{\infty} f_{XY}(x,y)\,dy = \int_0^1 1\,dy = 1,$$

and otherwise (when $x < 0$ or $x > 1$),

$$f_X(x) = \int_{-\infty}^{\infty} f_{XY}(x,y)\,dy = \int_{-\infty}^{\infty} 0\,dy = 0.$$

Therefore, X is a uniform random variable, $X \sim U(0,1)$. Similarly, it can be shown that $Y \sim U(0,1)$.

Example 10.14 (Related uniform random variables) *Suppose that X and Y are continuous random variables with the joint pdf*

$$f_{XY}(x,y) = \begin{cases} 2 & \text{if } 0 \le x \le 0.5 \text{ and } 0 \le y \le 0.5 \\ 2 & \text{if } 0.5 < x \le 1 \text{ and } 0.5 < y \le 1 \\ 0 & \text{otherwise.} \end{cases}$$

It can be confirmed that the joint pdf integrates to 1:

$$\int_{-\infty}^{\infty}\int_{-\infty}^{\infty} f_{XY}(x,y)\,dx\,dy = \int_0^{0.5}\int_0^{0.5} 2\,dx\,dy + \int_{0.5}^1\int_{0.5}^1 2\,dx\,dy = 0.5 + 0.5 = 1.$$

The rectangular solid volume intuition can be used to interpret $\int_0^{0.5}\int_0^{0.5} 2\,dx\,dy$ as the volume of a rectangular solid with dimensions 2, 0.5, and 0.5. The same intuition can be used for $\int_{0.5}^1\int_{0.5}^1 2\,dx\,dy$, again as the volume of a rectangular solid with dimensions 2, 0.5, and 0.5.

For the marginal pdf of X, when $0 \le x \le 0.5$,

$$f_X(x) = \int_{-\infty}^{\infty} f_{XY}(x,y)\,dy = \int_0^{0.5} 2\,dy = 1,$$

and when $0.5 < x \leq 1$*,*

$$f_X(x) = \int_{-\infty}^{\infty} f_{XY}(x,y)\,dy = \int_{0.5}^{1} 2\,dy = 1,$$

and otherwise (when $x < 0$ *or* $x > 1$*),*

$$f_X(x) = \int_{-\infty}^{\infty} f_{XY}(x,y)\,dy = \int_{-\infty}^{\infty} 0\,dy = 0.$$

Therefore, X is a uniform random variable, $U(0,1)$*. Similarly, it can be shown that* $Y \sim U(0,1)$*. Interestingly, the marginal distributions for X and Y here, both* $U(0,1)$*, are identical to those in example 10.13, even though the joint pdf here is different from that example. In this example, the joint pdf implies that X and Y are related to each other, which is not the case in example 10.13. Here, X can only be in the range* $(0,0.5)$ *when Y is in the range* $(0,0.5)$ *and vice versa, and X can only be in the range* $(0.5,1)$ *when Y is in the range* $(0.5,1)$ *and vice versa.*

Definition 10.12 generalizes the definition of the joint cdf, introduced in chapter 8 for discrete random variables (definition 8.9), to all types of random variables.

Definition 10.12 *The* ***cumulative distribution function (joint cdf)*** *of two random variables X and Y, denoted as* $F_{XY}(\cdot,\cdot)$*, gives the probability that* both *X and Y are less than or equal to their corresponding arguments:*

$$F_{XY}(x_0,y_0) = P(X \leq x_0, Y \leq y_0).$$

When X and Y are continuous random variables, the joint cdf can be written in terms of the joint pdf:

$$F_{XY}(x_0,y_0) = \int_{-\infty}^{y_0} \int_{-\infty}^{x_0} f_{XY}(x,y)\,dx\,dy.$$

Regardless of the types of the two random variables X and Y, the joint cdf has the same properties stated in chapter 8:

$$0 \leq F_{XY}(x_0,y_0) \leq 1 \text{ for every } x_0 \text{ and } y_0$$

and

$$x_0 < x_1 \implies F_{XY}(x_0,y_0) \leq F_{XY}(x_1,y_0) \text{ and } y_0 < y_1 \implies F_{XY}(x_0,y_0) \leq F_{XY}(x_0,y_1).$$

The joint cdf is always between 0 and 1 (inclusive) since it is defined as a probability, and the joint cdf is weakly increasing in both of its arguments.

Example 10.15 (Related uniform random variables) *Continuing example 10.14, where X and Y have the joint pdf*

$$f_{XY}(x,y) = \begin{cases} 2 & \textit{if } 0 \leq x \leq 0.5 \textit{ and } 0 \leq y \leq 0.5 \\ 2 & \textit{if } 0.5 < x \leq 1 \textit{ and } 0.5 < y \leq 1 \\ 0 & \textit{otherwise,} \end{cases}$$

the joint cdf is determined by calculating the double integral from definition 10.12 for different regions based on the (x, y) values. The joint cdf, whose derivation is left as an exercise (exercise 10.13), is

$$F_{XY}(x,y) = \begin{cases} 0 & \text{if } x<0 \text{ or } y<0 \\ 2xy & \text{if } 0 \le x \le 0.5 \text{ and } 0 \le y \le 0.5 \\ 0.5+2(x-0.5)(y-0.5) & \text{if } 0.5 < x \le 1 \text{ and } 0.5 < y \le 1 \\ x & \text{if } 0 \le x \le 0.5 \text{ and } y > 0.5 \\ y & \text{if } x > 0.5 \text{ and } 0 \le y \le 0.5 \\ x & \text{if } 0.5 < x \le 1 \text{ and } y > 1 \\ y & \text{if } x > 1 \text{ and } 0.5 < y \le 1 \\ 1 & \text{if } x > 1 \text{ and } y > 1. \end{cases}$$

10.6.2 Conditional Probability Density Functions

This section considers the conditional version of the pdf, which describes the pdf of a continuous random variable X given the outcome of another random variable Y. We do not restrict Y to be continuous, but when it is continuous, we are able to express the **conditional probability density function (conditional pdf)** in terms of the joint pdf function $f_{XY}(x, y)$ from section 10.6.1.

Definition 10.13 *For a continuous random variable X and a random variable Y, the* ***conditional probability density function (conditional pdf)*** *of X given Y, is denoted as $f_{X|Y}(\cdot|\cdot)$. The function $f_{X|Y}(\cdot|y)$ is the pdf associated with X when $Y=y$. When Y is a continuous random variable,*

$$f_{X|Y}(x|y) = \frac{f_{XY}(x,y)}{f_Y(y)} \text{ for all } (x,y) \text{ such that } f_Y(y) > 0.$$

The roles of the two random variables can also be reversed in definition 10.13. That is, if Y is also a continuous random variable, its conditional pdf can be written as

$$f_{Y|X}(y|x) = \frac{f_{XY}(x,y)}{f_X(x)} \text{ for all } (x,y) \text{ such that } f_X(x) > 0.$$

Definition 10.14 *For random variables X and Y, the* ***conditional cumulative distribution function (conditional cdf)*** *of X given Y, denoted as $F_{X|Y}(\cdot|\cdot)$, is*

$$F_{X|Y}(x|y) = P(X \le x | Y = y) \text{ for all } (x,y).$$

If X and Y are both continuous random variables,

$$F_{X|Y}(x|y) = P(X \le x | Y = y) = \frac{\int_{-\infty}^{x} f_{XY}(v,y)\,dv}{f_Y(y)} \text{ for all } (x,y) \text{ such that } f_Y(y) > 0.$$

We consider three examples, the first having a discrete X and continuous Y and the second and third having X and Y both continuous.

Example 10.16 (Data analyst salaries) *A large firm employs many data analysts, some of whom have graduate degrees. Let $X \in \{0, 1\}$ be an indicator of whether a data analyst has* *a graduate degree ($X = 1$ for graduate degree, $X = 0$ for no graduate degree). Suppose the salaries (in thousands of dollars) of data analysts without a graduate degree can be modeled as a $U(60, 100)$ random variable and salaries of data analysts with a graduate degree can be modeled as a $U(90, 210)$ random variable. In this case, X is discrete with two possible values, leading to two conditional pdfs:*

$$f_{Y|X}(y|0) = \begin{cases} \frac{1}{40} & \text{if } 60 \leq y \leq 100 \\ 0 & \text{otherwise;} \end{cases}$$

and

$$f_{Y|X}(y|1) = \begin{cases} \frac{1}{120} & \text{if } 90 \leq y \leq 210 \\ 0 & \text{otherwise.} \end{cases}$$

The associated conditional cdfs are

$$F_{Y|X}(y|0) = \begin{cases} 0 & \text{if } y < 60 \\ \frac{y-60}{40} & \text{if } 60 \leq y \leq 100 \\ 1 & \text{if } y > 100; \end{cases}$$

and

$$F_{Y|X}(y|1) = \begin{cases} 0 & \text{if } y < 90 \\ \frac{y-90}{120} & \text{if } 90 \leq y \leq 210 \\ 1 & \text{if } y > 210. \end{cases}$$

Example 10.17 (Unrelated uniform random variables) *Continuing example 10.13, the conditional pdf of X given Y is defined for y such that $0 \leq y \leq 1$, for which*

$$f_{X|Y}(x|y) = \frac{f_{XY}(x,y)}{f_Y(y)} = \frac{f_{XY}(x,y)}{1} = \begin{cases} 1 & \text{if } 0 \leq x \leq 1 \\ 0 & \text{otherwise.} \end{cases}$$

The second equality follows from the marginal distribution $Y \sim U(0, 1)$, for which $f_Y(y) = 1$ when $0 \leq y \leq 1$. Thus, the conditional distribution of X given $Y = y$, for $0 \leq y \leq 1$, is the uniform distribution $U(0, 1)$, which is the same as the marginal distribution of X.

Similarly, the conditional pdf of Y given X is defined for x such that $0 \leq x \leq 1$, for which

$$f_{Y|X}(y|x) = \frac{f_{XY}(x,y)}{f_X(x)} = \frac{f_{XY}(x,y)}{1} = \begin{cases} 1 & \text{if } 0 \leq y \leq 1 \\ 0 & \text{otherwise.} \end{cases}$$

Thus, the conditional distribution of Y given $X = x$, for $0 \leq x \leq 1$, is the uniform distribution $U(0, 1)$, which is the same as the marginal distribution of Y.

Example 10.18 (Related uniform random variables) *Continuing example 10.14, the conditional pdf of X given Y is defined for y such that $0 \leq y \leq 1$. Recall that the marginal distribution of Y is the $U(0, 1)$ distribution, with $f_Y(y) = 1$ for $0 \leq y \leq 1$. For y values such that $0 \leq y \leq 0.5$, the conditional pdf of X given Y is*

$$f_{X|Y}(x|y) = \frac{f_{XY}(x,y)}{f_Y(y)} = \begin{cases} 2 & \text{if } 0 \leq x \leq 0.5 \\ 0 & \text{otherwise;} \end{cases}$$

and for y values such that $0.5 < y \leq 1$, *the conditional pdf of X given Y is*

$$f_{X|Y}(x|y) = \frac{f_{XY}(x,y)}{f_Y(y)} = \begin{cases} 2 & \text{if } 0.5 < x \leq 1 \\ 0 & \text{otherwise} \end{cases}$$

Thus, the conditional distribution of X given Y is $U(0, 0.5)$ *when* $0 \leq y \leq 0.5$ *and* $U(0.5, 1)$ *when* $0.5 < y \leq 1$. *Similarly, it can be shown that the conditional distribution of Y given X is* $U(0, 0.5)$ *when* $0 \leq x \leq 0.5$ *and* $U(0.5, 1)$ *when* $0.5 < x \leq 1$.

Since a conditional pdf is itself a pdf, it is natural to introduce conditional versions of the population-descriptive statistics, including the mean, variance, and standard deviation.

Definition 10.15 *The* ***population conditional mean*** *or* ***conditional expectation*** *of a continuous random variable X given* $Y = y$, *denoted as* $\mu_{X|Y=y}$, *is*

$$\mu_{X|Y=y} = E(X|Y=y) = \int_{-\infty}^{\infty} x f_{X|Y}(x|y)\, dx.$$

Definition 10.16 *The* ***population conditional variance*** *of a continuous random variable X given* $Y = y$, *denoted as* $\sigma^2_{X|Y=y}$, *is*

$$\sigma^2_{X|Y=y} = Var(X|Y=y) = \int_{-\infty}^{\infty} (x - \mu_{X|Y=y})^2 f_{X|Y}(x|y)\, dx.$$

Definition 10.17 *The* ***population conditional standard deviation*** *of a continuous random variable X given* $Y = y$, *denoted as* $\sigma_{X|Y=y}$, *is*

$$\sigma_{X|Y=y} = sd(X|Y=y) = \sqrt{\sigma^2_{X|Y=y}}.$$

As with the the definitions of the conditional pdf and the conditional cdf, the roles of Y and X can be reversed in these definitions when Y is a continuous random variable.

Example 10.19 (Data analyst salaries) *Continuing example 10.16, where the conditional distribution of salaries for data analysts with no graduate degree* ($Y|X = 0$) *is* $U(60, 100)$ *and the conditional distribution of salaries for data analysts with a graduate degree* ($Y|X = 1$) *is* $U(90, 210)$, *the conditional expectations and conditional variances are*

$$E(Y|X=0) = \frac{60+100}{2} = 80, \quad Var(Y|X=0) = \frac{(100-60)^2}{12} \approx 133.33,$$

$$E(Y|X=1) = \frac{90+210}{2} = 150, \quad Var(Y|X=1) = \frac{(210-90)^2}{12} = 1200.$$

Thus, as expected from a comparison of the two conditional distributions, the expected value and variance of salaries for nongraduate-degree data analysts are both lower than they are for graduate-degree data analysts.

Example 10.20 (Related uniform random variables) *Continuing example 10.14, the conditional expectation of X given* $Y = y$ *is*

$$E(X|Y=y) = \int_0^{0.5} (x)(2)\, dx = 0.25$$

when $0 \le y \le 0.5$, *and*

$$E(X|Y=y) = \int_{0.5}^{1} (x)(2)\,dx = 0.75$$

when $0.5 < y \le 1$. *The conditional variance of* X *given* $Y=y$ *is*

$$Var(X|Y=y) = \int_{0}^{0.5} (x-0.25)^2(2)\,dx = \frac{2}{3}(x-0.25)^3\bigg|_0^{0.5} = \frac{1}{48}$$

when $0 \le y \le 0.5$, *and*

$$Var(X|Y=y) = \int_{0.5}^{1} (x-0.75)^2(2)\,dx = \frac{2}{3}(x-0.75)^3\bigg|_{0.5}^{1} = \frac{1}{48}.$$

In fact, in this simple example, it would have been possible to determine the conditional expectations and conditional variances without calculating the integrals. Since we already knew that the conditional distribution $X|Y=y$ *is* $U(0,0.5)$ *when* $0 \le y \le 0.5$ *and* $U(0.5,1)$ *when* $0.5 < y \le 1$, *the conditional expectations are just the midpoints of the uniform distribution (*0.25 *and* 0.75, *respectively) and the conditional variances are both equal to* $\frac{0.5^2}{12} = \frac{1}{48}$ *from the general variance formula* $\frac{(b-a)^2}{12}$ *for a* $U(a,b)$ *distribution.*

When information is available about the conditional distributions of a random variable, and therefore its conditional expectations, we can determine the unconditional expected value of that random variable. This idea is analogous to the Law of Total Probability from chapter 3, where the unconditional probability is equal to a weighted sum of conditional probabilities. Proposition 10.6 provides several results for expressing the unconditional expected value of a random variable in terms of conditional expectations:

Proposition 10.6 (Expected value in terms of conditional expectations) *Let* Y *be a random variable.*

i. *If* $A_1, A_2, \ldots, A_k$ *are disjoint and exhaustive events, then*[2]

$$E(Y) = \sum_{j=1}^{k} E(Y|A_j)P(A_j).$$

ii. *As a special case of (i), if* X *is a discrete random variable, then*

$$E(Y) = \sum_{k} E(Y|X=x_k^*)p_X(x_k^*).$$

iii. *If* X *is a continuous random variable, then*

$$E(Y) = \int_{-\infty}^{\infty} E(Y|X=x)f_X(x)\,dx.$$

This proposition holds for any type of random variable Y. When conditioning on discrete events, as in part (i), or discrete outcomes of a random variable, as in part (ii), the expected value of Y is a weighted average of the conditional expectations, where the weights are the probabilities of the events/outcomes, similar to the Law of Total Probability. Part (iii) is a continuous version of the weighted summation, where an integral is used and, for the weighting, the pdf of x is used rather than the pmf of x. Interestingly, we can also use part (i) to partition the outcomes of a continuous random variable. For example, if X is a continuous random variable with all real numbers as possible values, we can partition X into negative values ($X < 0$) and nonnegative values ($X \geq 0$). Then, applying part (i) to the events $\{X < 0\}$ and $\{X \geq 0\}$ yields

$$E(Y) = E(Y|X < 0) \cdot P(X < 0) + E(Y|X \geq 0) \cdot P(X \geq 0).$$

Example 10.21 (Data analyst salaries) *For example 10.16, where the conditional distribution of salaries for data analysts with no graduate degree ($Y|X = 0$) is $U(60, 100)$ and the conditional distribution of salaries for data analysts with a graduate degree ($Y|X = 1$) is $U(90, 210)$, let $\pi = P(X = 1)$ denote the probability that a data analyst has a graduate degree. Then the expected value of salary is related to the two conditional expectations of salaries as follows:*

$$\begin{aligned} E(Y) &= E(Y|X = 0) \cdot P(X = 0) + E(Y|X = 1) \cdot P(X = 1) \\ &= \tfrac{60+100}{2}(1 - \pi) + \tfrac{90+210}{2}\pi = 80 + 70\pi. \end{aligned}$$

For instance, if $\pi = 0.2$, or 20%, the expected value of salary is $E(Y) = 94$, or \$94,000.

10.6.3 Population Covariance and Population Correlation

For discrete random variables X and Y, section 8.4.3 introduced the population covariance

$$\sigma_{XY} = Cov(X, Y) = E\left[(X - \mu_X)(Y - \mu_Y)\right] = \sum_{(k,\ell)} (x_k^* - \mu_X)(y_\ell^* - \mu_Y)\, p_{XY}(x_k^*, y_\ell^*),$$

which is a weighted average of $(x_k^* - \mu_X)(y_\ell^* - \mu_Y)$ for all possible joint outcomes (x_k^*, y_ℓ^*), where the weights are the joint pmf values $p_{XY}(x_k^*, y_\ell^*)$. When X and Y are continuous random variables, an analogous definition involves a weighted average using weights given by the joint pdf values $f_{XY}(x, y)$, integrating over all (x, y).

Definition 10.18 *The **population covariance** of continuous random variables X and Y, denoted as σ_{XY} or $Cov(X, Y)$, is*

$$\sigma_{XY} = Cov(X, Y) = E\left[(X - \mu_X)(Y - \mu_Y)\right] = \int_{-\infty}^{\infty} \int_{-\infty}^{\infty} (x - \mu_X)(y - \mu_Y) f_{XY}(x, y)\, dx\, dy.$$

Definition 10.19 *The **population correlation** of continuous random variables X and Y, denoted as ρ_{XY} or $Corr(X, Y)$, is*

$$\rho_{XY} = Corr(X, Y) = \frac{\sigma_{XY}}{\sigma_X \sigma_Y}.$$

The properties for the population covariance and population correlation were provided earlier in proposition 8.4. The population correlation ρ_{XY} is unitless, with $0 \leq \rho_{XY} \leq 1$, and its sign is always the same as the sign of the population covariance ($sign(\rho_{XY}) = sign(\sigma_{XY})$).

Example 10.22 (Related uniform random variables) *We can determine the population covariance and population correlation for the random variables X and Y introduced in* *example 10.14, where the joint pdf was*

$$f_{XY}(x,y) = \begin{cases} 2 & \text{if } 0 \le x \le 0.5 \text{ and } 0 \le y \le 0.5 \\ 2 & \text{if } 0.5 < x \le 1 \text{ and } 0.5 < y \le 1 \\ 0 & \text{otherwise.} \end{cases}$$

The population covariance is

$$\begin{aligned} \sigma_{XY} &= \int_{-\infty}^{\infty} \int_{-\infty}^{\infty} (x-\mu_X)(y-\mu_Y) f_{XY}(x,y)\, dx\, dy \\ &= \int_0^{0.5} \int_0^{0.5} (x-0.5)(y-0.5)(2)\, dx\, dy + \int_{0.5}^{1} \int_{0.5}^{1} (x-0.5)(y-0.5)(2)\, dx\, dy \\ &= 2 \int_0^{0.5} \tfrac{-0.25}{2} (y-0.5)\, dy + 2 \int_{0.5}^{1} \tfrac{0.25}{2} (y-0.5)\, dy \\ &= (2)\left(\tfrac{-0.25}{2}\right)\left(\tfrac{-0.25}{2}\right) + (2)\left(\tfrac{0.25}{2}\right)\left(\tfrac{0.25}{2}\right) = 0.0625; \end{aligned}$$

and the population correlation is

$$\rho_{XY} = \frac{\sigma_{XY}}{\sigma_X \sigma_Y} = \frac{0.0625}{\left(\frac{1}{\sqrt{12}}\right)\left(\frac{1}{\sqrt{12}}\right)} = 0.75.$$

We have used $\mu_X = \mu_Y = 0.5$ and $\sigma_X = \sigma_Y = \frac{1}{\sqrt{12}}$ since X and Y are $U(0,1)$ random variables. The high positive correlation indicates a strong positive association, which is expected given the specification of the joint pdf.

It should be noted that the population covariance $\sigma_{XY} = Cov(X,Y) = E[(X-\mu_X)(Y-\mu_Y)]$ is a well-defined population concept that is more general than the specific cases considered in chapter 8 (X and Y both discrete) and this section (X and Y both continuous). Regardless of the form of X and Y, we can think of taking many draws of (x_i, y_i) from the population, and the population covariance σ_{XY} is the number to which the sample covariance $s_{xy} = \frac{1}{n-1}\sum_{i=1}^{n}(x_i - \bar{x})(y_i - \bar{y})$ eventually converges (i.e., for very large n).[3] One particular case of interest that has not been considered thus far is the covariance between a discrete X and a continuous Y. For simplicity, let's consider a binary (Bernoulli) $X \in \{0, 1\}$, with $X \sim Bernoulli(\pi)$, in which case[4]

$$\begin{aligned} \sigma_{XY} &= E[(X-\mu_X)(Y-\mu_Y)] \\ &= E[(X-\mu_X)(Y-\mu_Y)|X=0]\, P(X=0) \\ &\quad + E[(X-\mu_X)(Y-\mu_Y)|X=1]\, P(X=1) \\ &= E[(0-\pi)(Y-\mu_Y)|X=0]\,(1-\pi) + E[(1-\pi)(Y-\mu_Y)|X=1]\,\pi \\ &= \pi(1-\pi)\,(E[Y-\mu_Y|X=1] - E[Y-\mu_Y|X=0]) \\ &= \pi(1-\pi)\,(E[Y|X=1] - E[Y|X=0]). \end{aligned}$$

The second equality here follows from application of proposition 10.6. For the third equality, we plug in the value 0 for X when conditioning on $X=0$ and the value 1 when conditioning on $X=1$, and we use the fact that $P(X=1) = \mu_X = \pi$. For the fourth equality, we pull the constants ($-\pi$ and π, respectively) outside the conditional expectations and then simplify. Finally, for the fifth equality, the additive constants ($-\mu_Y$ for both) is pulled out of the two conditional expectations and cancel each other out. Since $\pi(1-\pi)$ is always positive, the population covariance σ_{XY}, and thus the population correlation ρ_{XY}, are positive when $E[Y|X=1] > E[Y|X=0]$, negative when $E[Y|X=1] < E[Y|X=0]$, and zero when $E[Y|X=1] = E[Y|X=0]$. This result is analogous to the discussion in section 7.2.3, where the sample correlation between a discrete x and a continuous y was considered.

Example 10.23 (Data analyst salaries) *Continuing example 10.16, where the conditional distribution of salaries for data analysts with no graduate degree ($Y|X=0$) is $U(60, 100)$ and the conditional distribution of salaries for data analysts with a graduate degree ($Y|X=1$) is $U(90, 210)$, the population covariance is*

$$\sigma_{XY} = \pi(1-\pi)\left(E[Y|X=1] - E[Y|X=0]\right) = \pi(1-\pi)(150-80) = 70\pi(1-\pi),$$

where $\pi = P(X=1)$ is the probability that a data analyst at the firm has a graduate degree. There is a positive covariance, and therefore a positive correlation, between the graduate-degree indicator X and the salary Y since the population conditional mean of salaries for graduate-degree data analysts is higher than the population conditional mean of nongraduate-degree data analysts. For instance, if $\pi = 0.20$, the population covariance is $\sigma_{XY} = (70)(0.2)(0.8) = 11.2$.

10.6.4 Independence of Random Variables

This section begins the discussion of independence by focusing on the case of two continuous random variables, but this concept of independence is then generalized to two random variables of any type (discrete, continuous, or a mixture of the two). The generalization allows several propositions related to independence to apply to two random variables of any type.

First, independence is defined for the case of two continuous random variables as in definition 10.20.

Definition 10.20 *The continuous random variables X and Y are **independent** if and only if*

$$f_{XY}(x, y) = f_X(x) f_Y(y) \text{ for every } x \text{ and } y$$

or, equivalently,

$$F_{XY}(x, y) = F_X(x) F_Y(y) \quad \text{for every possible outcome pair } (x, y),$$

*where $F_{XY}(x, y) = P(X \leq x, Y \leq y)$. If this equality fails for any (x, y), then the continuous random variables X and Y are **dependent**.*

Independence based upon the cdf (i.e., $F_{XY}(x, y) = F_X(x)F_Y(y)$ for all (x, y)) is the same condition for independence seen for discrete random variables in definition 8.17. In fact, since the cdf is a unifying concept for all random variables, including those that might be a mixture of both discrete and continuous outcomes, definition 10.21 presents a general description of independence that applies to any type of random variable.

Definition 10.21 *The random variables X and Y are **independent** if and only if*

$$F_{XY}(x, y) = F_X(x) F_Y(y) \quad \text{for every possible outcome pair } (x, y),$$

where $F_{XY}(x,y)=P(X\leq x, Y\leq y)$. If this equality fails for any (x,y), then the random variables X and Y are ***dependent****.*

This definition is general in the sense that X can be discrete, continuous, or some mixture of discrete and continuous, as can Y. For instance, X could be a binary discrete random variable, while Y is a uniform continuous random variable.

Proposition 10.7 *If the random variables X and Y are independent, the population covariance σ_{XY} and population correlation ρ_{XY} equal zero ($\sigma_{XY}=\rho_{XY}=0$). Equivalently, if the random variables X and Y have a nonzero population covariance or correlation, X and Y are dependent.*

As shown in example 8.24, it is possible for discrete random variables to be dependent and have a population covariance/correlation equal to zero. The same is true of continuous random variables. It is possible to have dependent continuous random variables with a population covariance/correlation equal to zero.[5]

A general characterization of independence can also be given in terms of conditional distributions as in proposition 10.8.

Proposition 10.8 *The random variables X and Y are* ***independent*** *if and only if:*

- *For any possible value y, the conditional distribution of X given $Y=y$ is the same as the marginal distribution of X. Mathematically, $F_{X|Y}(x|y)=F_X(x)$ for any x and any possible value y.*
- *For any possible value x, the conditional distribution of Y given $X=x$ is the same as the marginal distribution of Y. Mathematically, $F_{Y|X}(y|x)=F_Y(y)$ for any y and any possible value x.*

It is sufficient to show just one of the equivalences between the conditional pdfs and the marginal pdf. For instance, if we show that $F_{X|Y}(x|y)=F_X(x)$ for any x and any possible value y, it is unnecessary to also show $F_{Y|X}(y|x)=F_Y(y)$ for any y and any possible value x.

As with definition 10.21, proposition 10.8 does not restrict the type of random variables since the results are stated in terms of the conditional cdf. X can be discrete, continuous, or a mixture of the two, as can Y. If X and Y are both discrete, having $F_{X|Y}(x|y)=F_X(x)$ for any x and y is equivalent to having $p_{X|Y}(x|y)=p_X(x)$ for all possible outcomes (x,y). If X and Y are both continuous, having $F_{X|Y}(x|y)=F_X(x)$ for any x and y is equivalent to having $f_{X|Y}(x|y)=f_X(x)$ for all possible (x,y).

Example 10.24 (Unrelated uniform random variables) *In example 10.17, for the joint pdf*

$$f_{XY}(x,y)=\begin{cases}1 & \text{if } 0\leq x\leq 1 \text{ and } 0\leq y\leq 1\\ 0 & \text{otherwise,}\end{cases}$$

we showed that the conditional pdf of X given $Y=y$ (for $0\leq y\leq 1$) is the same as the marginal pdf of X. Also, we showed that the conditional pdf of Y given $X=x$ (for $0\leq x\leq 1$) is the same as the marginal pdf of Y. Then, from proposition 10.8, X and Y are independent. Since they are independent, we also know $\sigma_{XY}=\rho_{XY}=0$ without having to calculate $\sigma_{XY}=\int_{-\infty}^{\infty}\int_{-\infty}^{\infty}(x-\mu_X)(y-\mu_Y)f_{XY}(x,y)\,dx\,dy$.

If two random variables are independent, it becomes much simpler to determine joint probabilities. For two discrete random variables, the definition of independence itself (definition 8.17) states that the joint probability of an outcome (x, y) is equal to the product of the marginal probabilities of x and y. For continuous variables, the probability of a specific joint outcome (x, y) is zero, so instead we consider the joint probability of X being in some interval and Y being in some other interval. Proposition 10.9 says that this joint probability is equal to the product of the marginal probabilities of the two random variables being in their respective intervals.

Proposition 10.9 *If X and Y are independent random variables, then for $a \leq b$ and $c \leq d$,*

$$P(a \leq X \leq b,\ c \leq Y \leq d) = P(a \leq X \leq b) \cdot P(c \leq Y \leq d).$$

The statement of this proposition is not restricted to continuous X and Y and holds for any types of random variable.

Example 10.25 (Producing a movie) *The cost of producing a movie depends upon the number of days of filming (X) and the number of days of editing (Y), with filming costing* \$1 *million per day and editing costing* \$500,000, *or* \$0.5 *million, per day. Assume that X and Y are both uniformly distributed and independent, as follows:*

$$X \sim U(60, 80) \text{ and } Y \sim U(40, 50).$$

The assumption of uniform distributions allows for partial days. The marginal pdfs are

$$f_X(x) = \begin{cases} 0.05 & \text{if } 60 \leq x \leq 80 \\ 0 & \text{otherwise} \end{cases} \quad \text{and } f_Y(y) = \begin{cases} 0.1 & \text{if } 40 \leq y \leq 50 \\ 0 & \text{otherwise.} \end{cases}$$

Since X and Y are independent, the joint pdf $f_{XY}(x, y) = f_X(x) f_Y(y)$ for all (x, y), yielding

$$f_{XY}(x, y) = \begin{cases} 0.005 & \text{if } 60 \leq x \leq 80 \text{ and } 40 \leq y \leq 50 \\ 0 & \text{otherwise.} \end{cases}$$

For any y such that $40 \leq y \leq 50$, the conditional pdf of X is the same as the marginal pdf of X: $f_{X|Y}(x|y) = f_X(x)$, which is 0.05 if $60 \leq x \leq 80$ and 0 otherwise. Similarly, for any x such that $60 \leq x \leq 80$, the conditional pdf of Y is the same as the marginal pdf of Y: $f_{Y|X}(y|x) = f_Y(y)$, which is 0.1 if $40 \leq y \leq 50$ and 0 otherwise. The independence also simplifies the calculation of joint cdf probabilities. For instance, the probability that X (days of filming) is less than or equal to 70 and Y (days of editing) is less than or equal to 45 is

$$P(X \leq 70, Y \leq 45) = P(X \leq 70) P(Y \leq 45) = ((0.05)(10)) \times ((0.1)(5)) = (0.5)(0.5) = 0.25.$$

Proposition 10.9 generalizes to more than two random variables. First, we provide a definition of independence for multiple random variables in defiinition 10.22.

Definition 10.22 *The random variables $X_1, X_2, \ldots, X_m$, for $m \geq 2$, are* ***independent*** *if and only if*

$$F_{X_1 X_2 \cdots X_m}(x_1, x_2, \ldots, x_m) = F_{X_1}(x_1) F_{X_2}(x_2) \cdots F_{X_m}(x_m) = \prod_{j=1}^{m} F_{X_j}(x_j)$$

for any $(x_1, x_2, \ldots, x_m)$, *where* $F_{X_1X_2\cdots X_m}(x_1, x_2, \ldots, x_m) = P(X_1 \le x_1, X_2 \le x_2, \ldots, X_m \le x_m)$.

This definition is stated in terms of the joint cdf and marginal cdfs so it applies to any types of random variables. With independence, the joint probability of the random variables being in specified intervals is equal to the product of the marginal probabilities of each random variable being in its corresponding interval, as in proposition 10,10.

Proposition 10.10 *If* $X_1, X_2, \ldots, X_m$, *for* $m \ge 2$, *are independent random variables, then*

$$P(a_1 \le X_1 \le b_1, a_2 \le X_2 \le b_2, \ldots, a_m \le X_m \le b_m)$$
$$= P(a_1 \le X_1 \le b_1)P(a_2 \le X_2 \le b_2) \cdots P(a_m \le X_m \le b_m) = \prod_{j=1}^{m} P(a_j \le X_j \le b_j),$$

where $a_j \le b_j$ *for all* $j \in \{1, 2, \ldots, m\}$.

If $m = 3$ and X_1, X_2, and X_3 are independent random variables, then for $a_1 \le b_1$, $a_2 \le b_2$, and $a_3 \le b_3$,

$$P(a_1 \le X_1 \le b_1, a_2 \le X_2 \le b_2, a_3 \le X_3 \le b_3)$$
$$= P(a_1 \le X_1 \le b_1)P(a_2 \le X_2 \le b_2)P(a_3 \le X_3 \le b_3).$$

For example, if X_1, X_2, and X_3 are independent $U(0, 1)$ random variables,

$$P(0.2 \le X_1 \le 0.5, 0.7 \le X_2 \le 0.8, 0.1 \le X_3 \le 0.3)$$
$$= P(0.2 \le X_1 \le 0.5)P(0.7 \le X_2 \le 0.8)P(0.1 \le X_3 \le 0.3) = (0.3)(0.1)(0.2) = 0.006.$$

We conclude this section by generalizing the definition of i.i.d. random variables to cover continuous random variables. Recall from definition 9.2 that discrete random variables are i.i.d. if they are independent and have the same marginal pmfs.

Definition 10.23 *The random variables* $X_1, X_2, \ldots, X_n$ *are* ***independent and identically distributed (i.i.d.)*** *if (i)* $X_1, X_2, \ldots, X_n$ *are independent and (ii) each* X_j *has the same cdf* ($F_{X_j}(x) = F_X(x)$ *for all* $j \in \{1, 2, \ldots, n\}$ *and all* x).

This definition includes discrete random variables (definition 9.2) as a special case, but it also applies to continuous random variables and random variables that are a mixture of discrete and continuous outcomes.

10.7 Linear Transformations and Combinations of Multiple Random Variables

Section 8.5 and 8.6 considered linear transformations and linear combinations of discrete random variables. This section generalizes the results from those sections to all random variables.

10.7.1 Linear Transformations of Two Random Variables

For two linear transformations of different random variables, $V = a + bX$ and $W = c + dY$, proposition 10.11 generalizes the results from section 8.5.2 for the case of two discrete random variables.

Proposition 10.11 *If X and Y are random variables,* $V = a + bX$ *for known constants a and b, and* $W = c + dY$ *for known constants c and d, then the following is the case:*

i. *(population covariance)* $\sigma_{VW} = bd\sigma_{XY}$.

ii. *(population correlation)* $\rho_{VW} = \begin{cases} \rho_{XY} & \text{if } bd > 0 \\ -\rho_{XY} & \text{if } bd < 0. \end{cases}$

iii. *(transforming one random variable and not the other)* $\sigma_{VY} = b\sigma_{XY}$ *and* $\rho_{VY} = \begin{cases} \rho_{XY} & \text{if } b \geq 0 \\ -\rho_{XY} & \text{if } b < 0. \end{cases}$

Example 10.26 (Education and earnings) *Let X and Y represent the weekly earnings (in dollars) and educational attainment (in years) for an employed individual from the population, with population covariance* σ_{XY} *and population correlation* ρ_{XY}*. Then,* $W = 52Y$ *is the annualized earnings of an employed individual from the population. The population covariance* σ_{XW} *is 52 times the population covariance* σ_{XY}*, and the population correlation* ρ_{XW} *is the same as the population correlation* ρ_{XY}*. Focusing on the population correlation, the scale of the earnings random variable does not affect the correlation value. If the scale of education is changed as well, such as measuring education in months rather than years, the population correlation would remain unchanged.*

10.7.2 Linear Combination of Multiple Random Variables

For the linear combination of two random variables, $V = k + aX + bY$, proposition 10.12 generalizes the results from section 8.6 for the case of two discrete random variables.

Proposition 10.12 *If X and Y are random variables and* $V = k + aX + bY$ *for known constants k, a, and b, then the following is the case:*

i. *(population mean)* $\mu_V = k + a\mu_X + b\mu_Y$.

ii. *(population variance)* $\sigma_V^2 = a^2\sigma_X^2 + b^2\sigma_Y^2 + 2ab\sigma_{XY}$, *with the following special cases: (no correlation between X and Y) for* $\sigma_{XY} = 0$: $\sigma_V^2 = a^2\sigma_X^2 + b^2\sigma_Y^2$ *(sum of two random variables) for* $V = X + Y$: $\sigma_V^2 = \sigma_X^2 + \sigma_Y^2 + 2ab\sigma_{XY}$ *(difference of two random variables) for* $V = X - Y$: $\sigma_V^2 = \sigma_X^2 + \sigma_Y^2 - 2ab\sigma_{XY}$

iii. *(population standard deviation)* $\sigma_V = \sqrt{\sigma_V^2} = \sqrt{a^2\sigma_X^2 + b^2\sigma_Y^2 + 2ab\sigma_{XY}}$

Example 10.27 (Producing a movie) *Continuing example 10.25, where* $X \sim U(60, 80)$*, the number of days of filming, and* $Y \sim U(40, 50)$*, the number of days of editing, are independent random variables. The production cost, denoted as C, is \$1 million per day of filming and \$0.5 million per day of editing:*

$$C = X + 0.5Y \text{ (in millions of dollars)}.$$

The population mean of C, or the expected cost, is

$$\mu_C = \mu_X + 0.5\mu_Y = 70 + (0.5)(45) = 92.5,$$

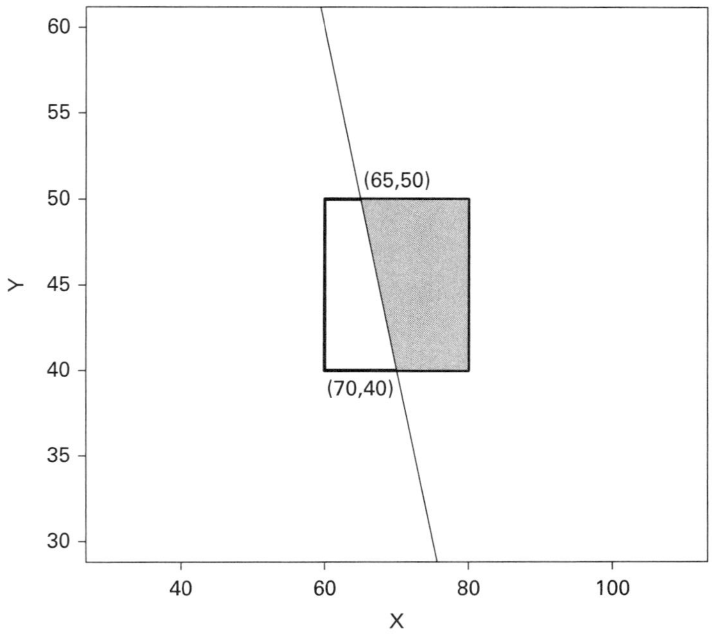

Figure 10.10
Over-budget region for the movie example

or \$92.5 *million. The population variance of C is*

$$\sigma_C^2 = \sigma_X^2 + (0.5)^2\sigma_Y^2 = \frac{(80-60)^2}{12} + (0.5)^2\frac{(50-40)^2}{12} = \frac{425}{12}$$

and the population standard deviation of C is

$$\sigma_C = \sqrt{\frac{425}{12}} \approx 5.95,$$

or approximately \$5.95 *million.*

Suppose that \$90 *million has been budgeted for the movie. What is the probability that the film goes over budget? The probability of going over budget is*

$$P(C > 90) = P(X + 0.5Y > 90),$$

so we need to find the region of possible (x, y) *values for which* $x + 0.5y > 90$*. Figure 10.10 helps to visualize the problem. The rectangle represents the range of the possible values for X and Y, with x between* 60 *and* 80 *and y between* 40 *and* 50*. The diagonal line is the* $y = 2(90 - x) = 180 - 2x$ *line, such that any* (x, y) *value above this line has* $y > 180 - 2x$ *or* $x + 0.5y > 90$*. Then,* $P(X + 0.5Y > 90)$ *is obtained by integrating the joint pdf* $f_{XY}(x, y) = 0.005$ *over this region. Since the pdf is constant within the region, the problem simplifies to determining the area of the gray region in figure 10.10 and multiplying it by* 0.005*, the height of the rectangular solid (given by the pdf). The area of the gray trapezoid is* $\frac{10+15}{2} \cdot 10 = 125$*, so that the probability of going over budget is* $P(C > 90) = (125)(0.005) = 0.625$*, or* 62.5%*.*

An alternative approach to calculate $P(C > 90) = P(X + 0.5Y > 90)$ is with computer simulation. Rather than analytically deriving the probability of the gray region in figure 10.10, the computer simulation approximates the probability by repeatedly making random draws of X and Y and seeing how often the linear combination $X + 0.5Y$ is larger than 90. The following R code approximates the over-budget probability with 100,000 simulated draws of X and Y:

```
set.seed(1234)

# initialize the number of simulations
num_simulations <- 100000

# create simulated draws
days_filming_draws <- runif(num_simulations,min=60,max=80)
days_editing_draws <- runif(num_simulations,min=40,max=50)

# calculate frequency corresponding to costs being over-budget
mean(days_filming_draws + 0.5*days_editing_draws > 90)

## [1] 0.6262
```

The approximated over-budget probability is 62.62%, which is very close to the true probability of 62.5%.

10.7.3 Linear Combination of Independent Random Variables

For a linear combination of multiple independent random variables, proposition 10.13 generalizes the results from section 8.6.1 for discrete random variables.

Proposition 10.13 *For independent random variables $X_1, X_2, \ldots, X_m$, the linear combination*

$$V = k + a_1X_1 + a_2X_2 + \cdots + a_mX_m = k + \sum_{j=1}^{m} a_jX_j$$

has the population mean

$$\mu_V = k + a_1\mu_{X_1} + a_2\mu_{X_2} + \cdots + a_m\mu_{X_m} = k + \sum_{j=1}^{m} a_j\mu_{X_j}$$

and the population variance

$$\sigma_V^2 = a_1^2\sigma_{X_1}^2 + a_2^2\sigma_{X_2}^2 + \cdots + a_m^2\sigma_{X_m}^2 = \sum_{j=1}^{m} a_j^2\sigma_{X_j}^2.$$

When $V = X_1 + X_2 + \cdots + X_m$ is the sum of independent random variables,

$$\mu_V = \mu_{X_1} + \mu_{X_2} + \cdots + \mu_{X_m} = \sum_{j=1}^{m} \mu_{X_j}$$

and

$$\sigma_V^2 = \sigma_{X_1}^2 + \sigma_{X_2}^2 + \cdots + \sigma_{X_m}^2 = \sum_{j=1}^{m} \sigma_{X_j}^2.$$

When $V = \frac{1}{m}(X_1 + X_2 + \cdots + X_m)$ *is the average of independent random variables,*

$$\mu_V = \frac{1}{m}\mu_{X_1} + \frac{1}{m}\mu_{X_2} + \cdots + \frac{1}{m}\mu_{X_m} = \frac{1}{m}\sum_{j=1}^{m}\mu_{X_j}$$

and

$$\sigma_V^2 = \frac{1}{m^2}\sigma_{X_1}^2 + \frac{1}{m^2}\sigma_{X_2}^2 + \cdots + \frac{1}{m^2}\sigma_{X_m}^2 = \frac{1}{m^2}\sum_{j=1}^{m}\sigma_{X_j}^2.$$

Example 10.28 (Sum of independent uniform random variables) *Suppose that* $X_1 \sim U(0,1)$ *and* $X_2 \sim U(0,1)$ *are independent uniform random variables. The sum* $V = X_1 + X_2$ *has the population mean*

$$\mu_V = \mu_{X_1} + \mu_{X_2} = 0.5 + 0.5 = 1,$$

the population variance

$$\sigma_V^2 = \sigma_{X_1}^2 + \sigma_{X_2}^2 = \frac{1}{12} + \frac{1}{12} = \frac{1}{6},$$

and the population standard deviation

$$\sigma_V = \sqrt{\frac{1}{6}} = \frac{1}{\sqrt{6}}.$$

Interestingly, $V = X_1 + X_2$ *is actually the triangular distribution from example 10.2, which the interested reader can verify (by finding* $F_V(\cdot)$ *and checking that it's the cdf of the triangular distribution). The population mean, variance, and standard deviation are the same as those derived for the triangular distribution in example 10.10. Knowing that the triangular distribution* V *is the sum of two independent* $U(0,1)$ *random variables greatly simplifies the calculation of these quantities, compared to the brute-force method using the population variance formula in example 10.10.*

How about the sum of three *independent* $U(0,1)$ *random variables? In this case,* $V = X_1 + X_2 + X_3$*, with the population mean*

$$\mu_V = \mu_{X_1} + \mu_{X_2} + \mu_{X_3} = (3)(0.5) = 1.5,$$

the population variance

$$\sigma_V^2 = \sigma_{X_1}^2 + \sigma_{X_2}^2 + \sigma_{X_3}^2 = (3)\left(\frac{1}{12}\right) = \frac{1}{4},$$

and the population standard deviation

$$\sigma_V = \sqrt{\frac{3}{12}} = \frac{1}{2}.$$

This idea extends to the sum of m independent $U(0,1)$ *random variables,* $V = X_1 + X_2 + \cdots + X_m$*, with the population mean*

$$\mu_V = \mu_{X_1} + \mu_{X_2} + \cdots + \mu_{X_m} = (m)(0.5) = 0.5m,$$

the population variance

$$\sigma_V^2 = \sigma_{X_1}^2 + \sigma_{X_2}^2 + \cdots + \sigma_{X_m}^2 = (m)\left(\frac{1}{12}\right) = \frac{m}{12},$$

and the population standard deviation

$$\sigma_V = \sqrt{\frac{m}{12}}.$$

For the sum of independent $U(0,1)$ random variables, the population mean is m times the common population mean of the X_js, and the population variance is m times the common population variance of the X_js.

If we are instead interested in the average of m independent $U(0,1)$ random variables, $V = \frac{1}{m}(X_1 + X_2 + \cdots + X_m)$, the population mean is

$$\mu_V = \frac{1}{m}\mu_{X_1} + \frac{1}{m}\mu_{X_2} + \cdots + \frac{1}{m}\mu_{X_m} = (m)\left(\frac{1}{m}\right)(0.5) = 0.5,$$

the population variance is

$$\sigma_V^2 = \frac{1}{m^2}\sigma_{X_1}^2 + \frac{1}{m^2}\sigma_{X_2}^2 + \cdots + \frac{1}{m^2}\sigma_{X_m}^2 = \left(\frac{1}{m^2}\right)(m)\left(\frac{1}{12}\right) = \frac{1}{12m},$$

and the population standard deviation is

$$\sigma_V = \frac{1}{\sqrt{12m}}.$$

For the average of independent $U(0,1)$ random variables, the population mean is the same as the common population mean of the X_js, the population variance is $\frac{1}{m}$ times the common population variance of the X_js, and the population standard deviation is $\frac{1}{\sqrt{m}}$ times the common population standard deviation of the X_j's. Figure 10.11 shows the distribution of a $U(0,1)$ random variable, along with the distribution of $V = \frac{1}{m}(X_1 + X_2 + \cdots + X_m)$ for three different values of m (2 in the upper-right graph, 3 in the lower-left graph, and 10 in the lower-right graph). For each distribution, the mean μ_V is 0.5, which is the mean of the $U(0,1)$ random variable. As m increases, the dispersion of the distribution of V decreases, with $\sigma_V^2 = \frac{1}{24}$ for $m = 2$, $\sigma_V^2 = \frac{1}{36}$ for $m = 3$, and $\sigma_V^2 = \frac{1}{120}$ for $m = 10$. We get a scaled version of the triangular distribution for $m = 2$. For $m = 3$, there is a much different bell-shaped distribution. For $m = 10$, there is also a bell-shaped distribution but more tightly distributed around the mean as compared to $m = 3$.

The results from example 10.28 for the sum and average of m independent $U(0,1)$ random variables can be generalized to *any* collection of i.i.d. random variables $X_1, X_2, \ldots, X_m$.

Proposition 10.14 *Suppose that $X_1, X_2, \ldots, X_m$ are i.i.d. random variables with common population mean μ_X and population variance σ_X^2.*

i. *For the sum of the i.i.d. random variables, $V = X_1 + X_2 + \cdots + X_m$,*

$$\mu_V = m\mu_X,\ \sigma_V^2 = m\sigma_X^2,\ \text{and}\ \sigma_V = \sqrt{m}\sigma_X.$$

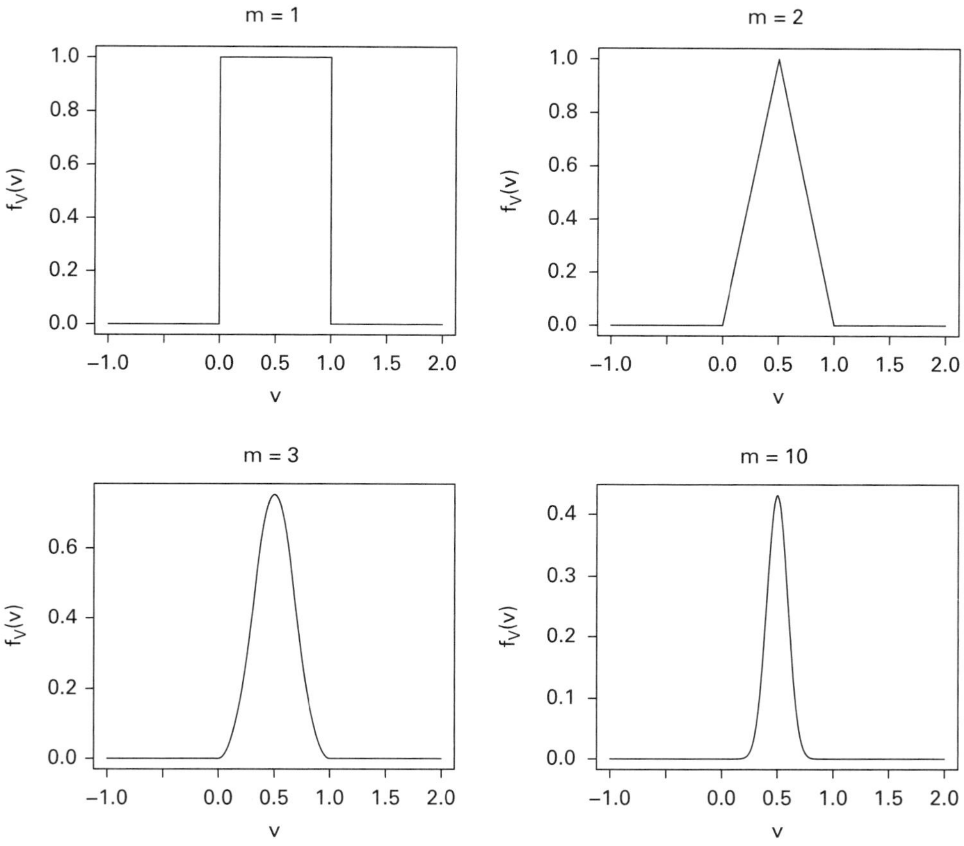

Figure 10.11
The pdfs for the average of m i.i.d. $U(0, 1)$ random variables

ii. *For the average of the i.i.d. random variables,* $V = \frac{1}{m}(X_1 + X_2 + \cdots + X_m)$,

$$\mu_V = \mu_X,\ \sigma_V^2 = \frac{\sigma_X^2}{m},\ \textit{and}\ \sigma_V = \frac{\sigma_X}{\sqrt{m}}.$$

In addition to the generalized results in proposition 10.14, the tendency for the sum $V = X_1 + X_2 + \cdots + X_m$ or the average $V = \frac{1}{m}(X_1 + X_2 + \cdots + X_m)$ to look like a bell-shaped distribution as m gets larger is also a general phenomenon that occurs for any i.i.d. random variables. This phenomenon was previously encountered in chapter 9 for binomial random variables since a $Binomial(n, \pi)$ random variable is the sum of n i.i.d. Bernoulli random variables. For example, figure 9.3 showed a $Binomial(100, 0.1)$ random variable with a distribution that appeared bell-shaped and approximately symmetric. Even though the pdf of a $Bernoulli(0.1)$ is certainly not bell-shaped, the sum of m $Bernoulli(0.1)$ random variables takes on the symmetric, bell-shaped distribution for the large $m = 100$ value.

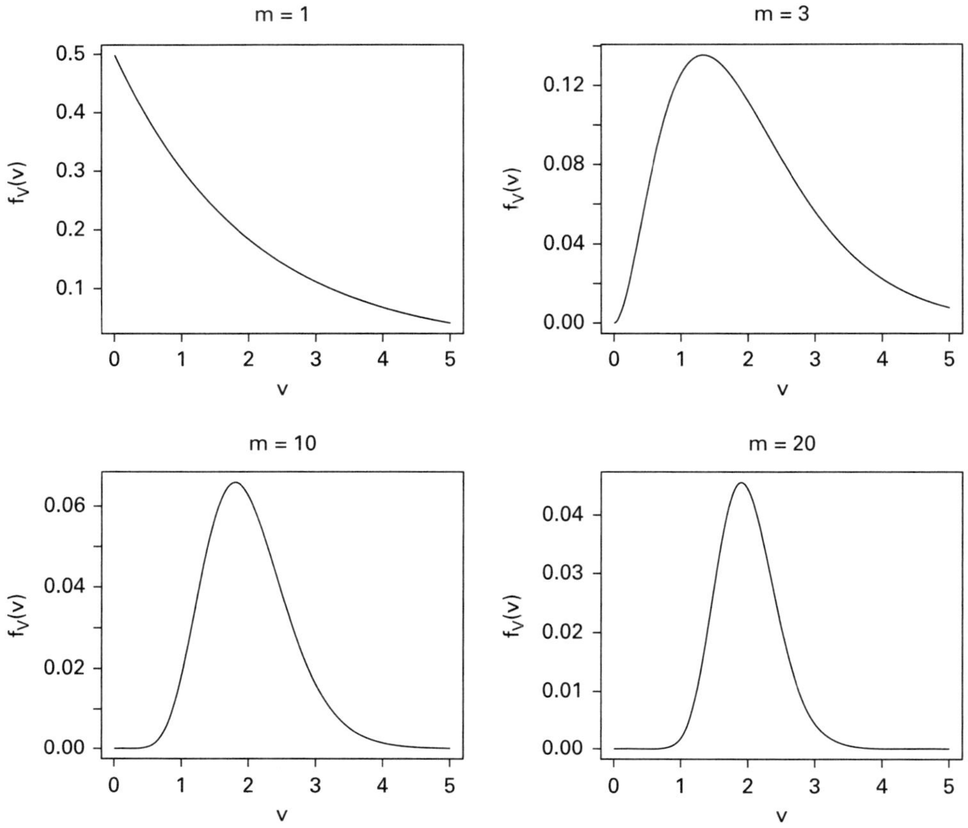

Figure 10.12
The pdfs for the average of m i.i.d. exponential random variables

Example 10.29 illustrates this same phenomenon for an asymmetric continuous distribution that is quite different from the uniform distribution.

Example 10.29 (Asymmetric distribution) *We consider a random variable with an asymmetric distribution, known as an* exponential random variable *(discussed in more detail in chapter* 11*). The top-left graph in figure* 10.12 *shows the pdf for a random variable X that is positive for positive values $x > 0$. In this example, the population mean is $\mu_X = 2$. Assume that $X_1, X_2, \ldots, X_m$ are i.i.d. random variables with this distribution. The remaining three graphs in figure* 10.12 *show the distribution of the average $V = \frac{1}{m}(X_1 + X_2 + \cdots + X_m)$ for three different values of m (*3 *in the top-right graph,* 10 *in the lower-left graph, and* 20 *in the lower-right graph). While each X_j is asymmetric, the distribution of V becomes more symmetrical as m increases. Even with $m = 3$, the distribution begins to look like a bell-shaped distribution, though it is more asymmetric than when m increases to* 10 *or* 20*. For the higher m values, the nearly symmetric distributions are approximately centered around $\mu_X = 2$ and the dispersion decreases as m is increased from* 10 *to* 20.

10.8 Expected Values of Functions of Continuous Random Variables

Section 8.7 provided formulas for the expected value of a function of a single discrete random variable $g(X)$ and a function of two discrete random variables $g(X, Y)$. In this section, we provide analogous formulas for continuous random variables, which involves replacing the summations in propositions 8.8 and 8.9 with integrals.

Proposition 10.15 considers a function of a single random variable.

Proposition 10.15 *For any function* $g(x)$ *and a continuous random variable* X *with pdf* $f_X(\cdot)$*, the population mean or expected value of* $g(X)$ *is*

$$\mu_{g(X)} = E(g(X)) = \int_{-\infty}^{\infty} g(x) f_X(x)\, dx.$$

A special case of proposition 10.15 is the population variance (definition 10.8), where $g(X) = (X - \mu_X)^2$.

Example 10.30 (Price experiment) *On a given day, a firm randomly sets the price* X *(in dollars) of its product according to a uniform random variable, with* $X \sim U(10, 20)$*. The quantity demanded of the product is* $100 - 3X$*, which ranges from a high of* 70 *(for* $X = 10$*) to a low of* 40 *(for* $X = 20$*). We are allowing the quantity demanded to be a noninteger value. The total revenue for the day, as a function of* X*, is*

$$g(X) = (100 - 3X)(X) = 100X - 3X^2.$$

The expected revenue is

$$\mu_{g(X)} = \int_{10}^{20} (100x - 3x^2)(0.1)\, dx = \int_{10}^{20} (10x - 0.3x^2)\, dx = 5x^2 \Big|_{10}^{20} - 0.1x^3 \Big|_{10}^{20} = 800.$$

Proposition 10.16 considers the expected value of a function of two continuous random variables.

Proposition 10.16 *For any function* $g(x, y)$ *and continuous random variables* X *and* Y *with joint pdf* $f_{XY}(\cdot, \cdot)$*, the expected value or population mean of* $g(X,Y)$ *is*

$$\mu_{g(X,Y)} = E(g(X, Y)) = \int_{-\infty}^{\infty} \int_{-\infty}^{\infty} g(x, y) f_{XY}(x, y)\, dx\, dy.$$

A special case of proposition 10.16 is the population covariance (definition 10.19), where $g(X, Y) = (X - \mu_X)(Y - \mu_Y)$.

Example 10.31 (Stock returns in consecutive years) *An investor invests* $\$1,000$ *in a certain stock. The asset return in the first year, denoted as* R_1*, is a uniform random variable with* $R_1 \sim U(-0.02, 0.08)$*. The asset return in the second year, denoted as* R_2*, is a uniform random variable with* $R_2 \sim U(-0.06, 0.14)$*. The returns* R_1 *and* R_2 *are independent. Then, the overall value of the investment after two years, in dollars, is*

$$g(R_1, R_2) = 1000(1 + R_1)(1 + R_2).$$

The expected value of the investment after two years, in dollars, is

$$\mu_{g(R_1,R_2)} = \int_{-0.06}^{0.14}\int_{-0.02}^{0.08} 1000(1+x)(1+y)\left(\frac{1}{0.1}\right)\left(\frac{1}{0.2}\right) dx\, dy$$
$$= \int_{-0.06}^{0.14}\int_{-0.02}^{0.08} 50000(1+x)(1+y)\, dx\, dy,$$

which, after evaluating the double integral, is equal to \$1,071.20.

10.9 Strictly Increasing Transformations of Random Variables

This section considers a situation where we are interested in the distribution of $g(X)$, where X is a random variable and $g(\cdot)$ is a *strictly increasing* transformation (function). When the transformation of a random variable is strictly increasing, we can often determine the cdf of $g(X)$ when the cdf of X is known. The main result of this section is proposition 10.17, which relates the cdf of $g(X)$ to the cdf of X.

Proposition 10.17 *If function $g(\cdot)$ is strictly increasing on the sample space associated with random variable X (i.e., $g(x_1) > g(x_2)$ for any possible outcomes x_1 and x_2 with $x_1 > x_2$), the cdf of the random variable $g(X)$ is*

$$F_{g(X)}(v) = P(g(X) \leq v) = P(X \leq g^{-1}(v)) = F_X(g^{-1}(v)),$$

where $g^{-1}(\cdot)$ is the inverse function of $g(\cdot)$ and v is any possible outcome in the sample space associated with $g(X)$.

The inverse function $g^{-1}(\cdot)$ basically undoes function $g(\cdot)$, so $g^{-1}(g(x)) = x$ for any x. If we have $g(x) = v$, then applying $g^{-1}(\cdot)$ to both sides yields[6]

$$g^{-1}(g(x)) = g^{-1}(v) \text{ or } x = g^{-1}(v).$$

For example, consider the exponential function $g(x) = e^x$. The function that undoes the exponentiation is the natural logarithm function $\ln(\cdot)$ since $\ln(e^x) = x$, meaning that the inverse function is $g^{-1}(v) = \ln(v)$.

Example 10.32 (Transformations of a uniform random variable) *If $X \sim U(0, 1)$, proposition 10.17 can be used directly to determine the distributions (cdfs) of random variables like $\sqrt{X}$ and X^2, both of which are strictly increasing functions of X on the range $[0, 1]$. We consider $\sqrt{X}$ first. The sample space associated with $\sqrt{X}$ is also $[0, 1]$ since $0 \leq X \leq 1$. Then, the cdf of $\sqrt{X}$ is*

$$F_{\sqrt{X}}(v) = P(\sqrt{X} \leq v) = P(X \leq v^2) = v^2 \text{ for any } v \in [0, 1].$$

The second equality is the key step for determining the cdf, as we square both sides of the inequality, which is the inverse of the square-root function.

For X^2, the sample space is also $[0, 1]$, and the cdf is

$$F_{X^2}(v) = P(X^2 \leq v) = P(X \leq \sqrt{v}) = \sqrt{v} \text{ for any } v \in [0, 1].$$

In this case, the second equality involves taking the square root of both sides of the inequality, which is the inverse of the square function.

Interestingly, a linear transformation of X with a positive coefficient on X is a special case of a strictly increasing transformation. Specifically, consider the linear transformation $g(X) = a + bX$ with $b > 0$. The sample space associated with $g(X)$ is $[a, a+b]$ since $0 \leq X \leq 1$, and the cdf of $g(X)$ is

$$F_{g(X)}(v) = P(a + bX \leq v) = P\left(X \leq \frac{v-a}{b}\right) = \frac{v-a}{b} \text{ for any } v \in [a, a+b].$$

Thus, $g(X)$ is also a uniform random variable, with $g(X) \sim U(a, a+b)$ and a pdf equal to $\frac{1}{b}$ for all values in the interval $[a, a+b]$.

Example 10.33 (Financial advisor investment fee) *A financial advisor at a certain investment company charges an investment fee based upon the annual portfolio return. Specifically, if a client's portfolio has an annual return represented by the random variable X, the financial advisor charges an investment fee equal to $g(X)$ times the overall portfolio value, where*

$$g(X) = 0.002 + \frac{0.013}{1 + e^{-40X}}.$$

Function $g(\cdot)$ is increasing, with an asymptote of 0.002 (or 0.2%) to the left and an asymptote of 0.015 (or 1.5%) to the right, as illustrated in figure 11.3. Assuming that the annual portfolio return for the customer is uniformly distributed, with $X \sim U(-0.10, 0.25)$, what is the cdf of the investment fee $g(X)$? We first determine the relevant sample space associated with $g(X)$. The minimum possible value and maximum possible value are

$$g(-0.10) = 0.002 + \frac{0.013}{1 + e^{-40(-0.10)}} \text{ and } g(0.25) = 0.002 + \frac{0.013}{1 + e^{-40(-0.25)}},$$

respectively. Then the cdf of $g(X)$ is

$$F_{g(X)}(v) = P\left(0.002 + \frac{0.013}{1 + e^{-40X}} \leq v\right) = P\left(\frac{1}{1 + e^{-40X}} \leq \frac{v - 0.002}{0.013}\right)$$
$$\text{for } g(-0.10) < v < g(0.25).$$

We require additional simplification to get an inequality in terms of X. For any v between $g(-0.10)$ and $g(0.25)$,

$$F_{g(X)}(v) = P\left(\frac{0.013}{v - 0.002} \leq 1 + e^{-40X}\right) = P\left(\frac{0.015 - v}{v - 0.002} \leq e^{-40X}\right)$$
$$= P\left(\ln\left(\frac{0.015 - v}{v - 0.002}\right) \leq -40X\right),$$

which yields

$$F_{g(X)}(v) = P\left(X \leq \frac{1}{40}\ln\left(\frac{v - 0.002}{0.015 - v}\right)\right) = \frac{1}{0.35}\left(\frac{1}{40}\ln\left(\frac{v - 0.002}{0.015 - v}\right) - (-0.10)\right).$$

The last equality follows from $X \sim U(-0.10, 0.25)$, which implies a pdf of $f_X(x) = \frac{1}{0.25-(-0.10)} = \frac{1}{0.35}$ for $x \in [-0.10, 0.25]$.

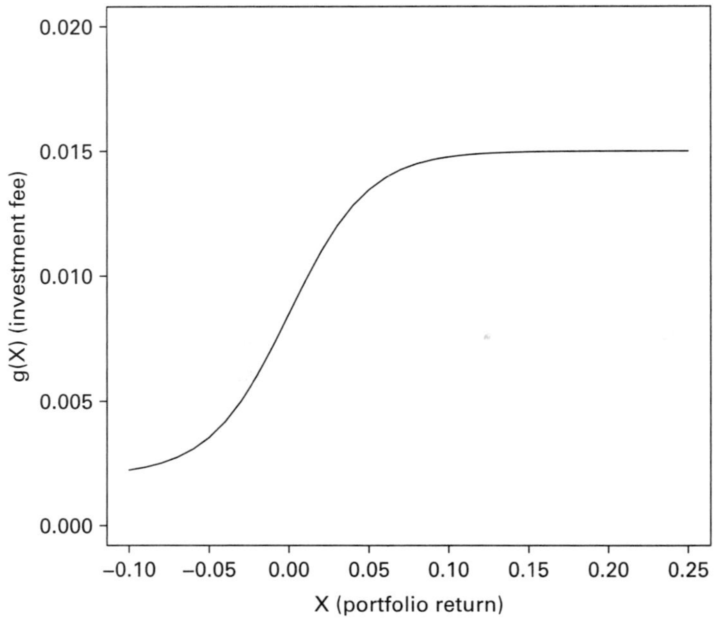

Figure 10.13
Financial advisor investment fee schedule

For a continuous random variable X, proposition 10.17 implies a particularly convenient relationship between the population quantiles of $g(X)$ and the population quantiles of X. The result is given in proposition 10.18:

Proposition 10.18 *If function $g(\cdot)$ is strictly increasing on the sample space associated with a continuous random variable X, the population qth quantile of the random variable $g(X)$ is*

$$\tau_{g(X),q} = g(\tau_{X,q}) \text{ for any } q \in (0,1).$$

To get the qth population quantile of $g(X)$, we apply the $g(\cdot)$ function to the qth population quantile of X. Why does this relationship hold? Recall that $\tau_{X,q}$ is the value for which $P(X \leq \tau_{X,q})$ is equal to q. Therefore, $P(g(X) \leq g(\tau_{X,q}))$ is also equal to q since $g(\cdot)$ is a strictly increasing function, meaning that the population qth quantile of $g(X)$ is $g(\tau_{X,q})$.

For example 10.32, where $X \sim U(0,1)$, the population qth quantile (for $q \in (0,1)$) of $\sqrt{X}$ is $\sqrt{q}$ since the population qth quantile of X is q; similarly, the population qth quantile of X^2 is q^2 for $q \in (0,1)$.

For example 10.33, what is the population median of the investment fee $g(X)$? According to proposition 10.18, since the population median of X is the middle of the $(-0.10, 0.25)$ interval, which is 0.075,

$$\tau_{g(X),0.5} = g(\tau_{0.5}) = g(0.075) = 0.002 + \frac{0.013}{1+e^{-40(0.075)}} \approx 0.01438.$$

10.10 Random Variables with Discrete and Continuous Outcomes

This chapter has already alluded to random variables that are a mixture or hybrid of discrete and continuous random variables, with the idea first introduced in section 10.1.1 and then again in section 10.3.2, when we discussed the generality of the cdf. Unfortunately, while the cdf applies generally to discrete random variables, continuous random variables, and mixtures of the two, our definitions of many population quantities have been quite different for discrete and continuous random variables. For example, the population mean and the population variance are both defined as summations for discrete random variables and as integrals for continuous random variables.

Although a completely unified framework that handles arbitrary mixtures of discrete and continuous random variables is beyond the scope of this book, we briefly discuss how to determine population quantities for such mixtures. The basic approach, which works in most cases of practical interest, is to apply summations for the discrete outcomes, to apply integrals for the continuous outcomes, and then sum the results. This idea is best illustrated through examples.

Example 10.34 (Uniform distribution with point masses) *As in example 10.5, suppose that the cdf of X is*

$$F_X(x) = \begin{cases} 0 & \text{if } x < 0 \\ 0.3 + 0.5x & \text{if } 0 \le x < 1 \\ 1 & \text{if } x \ge 1. \end{cases}$$

This cdf corresponds to probabilities $P(X=0) = 0.3$ and $P(X=1) = 0.2$ for the two possible discrete outcomes, which leaves a probability of 0.5 for the (uniform) continuous outcomes $x \in (0, 1)$. The population mean of X is

$$\mu_X = E(X) = (0)(0.3) + (1)(0.2) + \left(\int_0^1 x(1)dx\right)(0.5) = 0.45,$$

where the first two terms account for the two discrete outcomes and the integral accounts for the continuous outcomes. This calculation partitions the expected value $E(X)$ into a weighted average of conditional expectations, with

$$E(X) = E(X|X=0) \cdot P(X=0) + E(X|X=1) \cdot P(X=1) + E(X|0 < X < 1) \cdot P(0 < X < 1),$$

which simplifies to

$$E(X) = (0)(0.3) + (1)(0.2) + (0.5)(0.5) = 0.45.$$

Similarly, the population variance of X is

$$\sigma_X^2 = (0 - 0.45)^2(0.3) + (1 - 0.45)^2(0.2) + \left(\int_0^1 (x - 0.45)^2(1)dx\right)(0.5) \approx 0.164.$$

Example 10.35 (Average hourly wages, with zeros) *Suppose we are interested in the average hourly wage in a population of teenagers, but we want to include teenagers who are not working (i.e., zero hourly wage). If 40% of teenagers do not work and the remaining 60%*

have wages (in dollars) drawn from a $U(8,16)$ *random variable, the population mean of wages is*

$$(0.4)(0)+(0.6)\int_8^{16}\frac{x}{8}dx=(0.6)\left.\frac{x^2}{16}\right|_8^{16}=(0.6)\frac{256-64}{16}=7.2,$$

or \$7.20. *The population variance of wages is*

$$(0.4)(0-7.2)^2+(0.6)\int_8^{16}\frac{(x-7.2)^2}{8}dx=(0.4)(0-7.2)^2+(0.6)\left.\frac{(x-7.2)^3}{24}\right|_8^{16}=37.76,$$

meaning the population standard deviation of wages is $\sqrt{37.76}$ *or approximately* \$6.14.

Notes

1. Strictly speaking, the outcome $X=-1$ (which corresponds to the stock price dropping to zero) may not have zero probability, but we assume that it does for ease of exposition.
2. This result also holds for an infinite set of disjoint and exhaustive events, in which case the summation becomes an infinite summation.
3. In the terminology of chapter 14, the sample covariance is said to be a "consistent estimator" of the population covariance σ_{XY}.
4. For general discrete X, the same basic idea applies, except that there are more (possibly infinite) possible outcomes x_k^* for X, resulting in a summation of terms that are probabilities $p_X(x_k^*)$ times conditional expectations, given $X=x_k^*$.
5. In chapter 11, however, we will see that for X and Y with *normal distributions*, independence of X of Y holds if and only if the population covariance/correlation is equal to zero.
6. The inverse function $g^{-1}(\cdot)$ is well defined since $g(\cdot)$ is strictly increasing.

Exercises

1. Consider the pdf

$$f_X(x)=\begin{cases}3x^2 & \text{for } 0\le x\le 1\\ 0 & \text{otherwise.}\end{cases}$$

   a. Show that $f_X(\cdot)$ is a valid pdf.
   b. What is the population mean of X?
   c. What is the population variance of X?
   d. What is the value of the cdf of X evaluated at 0.4?
   e. What is the population median of X?
   f. Confirm your answers to (a), (b), and (d) using *R* by defining functions corresponding to $f_X(x)$ and $xf_X(x)$ and using the `integrate` function.

2. In the US, a FICO score is a measure of an individual's creditworthiness that companies (lenders, credit-card issuers, and other creditors) use to determine whether to extend credit to that person. An individual's FICO score is an integer between 300 and 850, with low scores indicating low creditworthiness and high scores indicating high creditworthiness. Assume that we can consider FICO scores to be an approximately continuous variable, even though they are integers. According to `fico.com`,

the population probabilities associated with different intervals in April 2023 were as follows:

Interval	300–499	500–549	550–599	600–649	650–699	700–749	750–799	800–850
Probability	0.028	0.057	0.068	0.086	0.120	0.164	0.236	0.241

Let X denote the random variable associated with an individual's FICO score.

a. What is $F_X(799)$?

b. What are the smallest and largest possible values of $F_X(520)$?

c. In which interval is the population median of X?

d. What are the smallest and largest possible values of the population mean of X?

3. Consider the following pdf for a random variable X defined on the interval [0, 1]:

$$f_X(x) = \begin{cases} 0.8 & \text{for } x \in [0, 0.5] \\ K & \text{for } x \in (0.5, 1] \\ 0 & \text{otherwise.} \end{cases}$$

a. What is K? Graph the pdf of X.

b. What is the cdf of X? Graph it.

c. What is the expected value of X?

d. What is the population variance of X?

e. What is the population 10% quantile of X?

f. What is the population 90% quantile of X?

g. Write an *R* function `rsplitunif` that takes a single argument `n` and returns a vector with `n` draws of the random variable X. (Hint: Think of X as being generated in the following way: with probability 0.4, X is drawn from a $U(0, 0.5)$ distribution, and with probability 0.6, X is drawn from a $U(0.5, 1)$ distribution.)

h. Using the function `rsplitunif`, conduct 10,000 simulations in *R* to confirm the answers to (c) and (d).

i. Suppose that $Y \sim U(0, 1)$ and X and Y are independent. Using the functions `runif` and `rsplitunif`, conduct 10,000 simulations in *R* to approximate $P(X > Y)$.

j. Suppose that $Y \sim U(0, 1)$ and X and Y are independent. Using the functions `runif` and `rsplitunif`, conduct 10,000 simulations in *R* to approximate $P(|X - Y| \leq 0.1)$, the probability that X and Y are within 0.1 of each other.

4. Let X be a random variable with pdf

$$f_X(x) = \begin{cases} 0.5x & \text{if } 0 \leq x \leq 1 \\ 1 - 0.5x & \text{if } 1 < x \leq 2 \\ 0.25 & \text{if } 2 < x \leq 4 \\ 0 & \text{otherwise.} \end{cases}$$

a. Show that $f_X(\cdot)$ is a valid pdf.

b. What is the cdf $F_X(\cdot)$?

c. For what value of x^* is the quantity $F_X(x^*+0.1) - F_X(x^*-0.1) = P(|X-x^*| \leq 0.1)$ maximized?

d. What is the population IQR of X?

e. What is $E(X)$?

5. Two office-supply companies, Office Plus and Paper King, constitute a duopoly in several cities. Suppose that the random variable X represents the market share of Office Plus in a given city, with X having the following pdf:

$$f_X(x) = \begin{cases} 12\left(x - \frac{1}{2}\right)^2 & \text{for } x \in [0, 1] \\ 0 & \text{otherwise.} \end{cases}$$

a. What is the maximum value of $f_X(\cdot)$, and at which x value(s) is it achieved?

b. What is the minimum value of $f_X(\cdot)$, and at which x value(s) is it achieved?

c. Graph $f_X(\cdot)$.

d. What is the cdf of X?

e. What is the probability that the market share of Office Plus in a given city is between 40% and 60%?

f. What is the population variance of X?

g. *The market share of Paper King in a given city is $Y = 1 - X$. Show that the pdf of Y is the same as the pdf of X.

h. *Let $M = \max(X, Y) = \max(X, 1-X)$ be the maximum of the two companies' market shares in a given city. What is the pdf of M?

i. *The following *R* function `rparabola` takes a single argument `n` and returns `n` random draws of the random variable X:

```
rparabola <- function(n) {
  q <- runif(n)
  sign_vec <- sign(2*q-1)
  return( sign_vec*( (sign_vec*(2*q-1)/8)^(1/3) ) + 0.5 )
}
```

i. Copy and paste the code into *R* to define the `rparabola` function. Using the `rparabola` function, draw a histogram of 100,000 simulated draws of X. How does the histogram compare to the pdf graph from (c)?

ii. Consider two cities, with associated market shares X_1 and X_2 for Office Plus, where X_1 and X_2 are i.i.d. random variables with pdf $f_X(\cdot)$. Using the `rparabola` function, draw a histogram of 100,000 simulated draws of $\frac{X_1+X_2}{2}$, the average market share for Office Plus in the two cities.

iii. Consider 10 cities, with associated market shares $X_1, \ldots, X_{10}$ for Office Plus, where $X_1, \ldots, X_{10}$ are i.i.d. random variables with pdf $f_X(\cdot)$. Using the `rparabola` function, draw a histogram of 100,000 simulated draws of $\frac{X_1+X_2+\cdots+X_{10}}{10}$, the average market share for Office Plus in the 10 cities.

6. *Consider a car insurance policy that has a deductible of \$500 and maximum coverage of \$15,000. If the policyholder gets into an accident and submits a claim, the policyholder must pay the first \$500 of the claim and the insurance company pays the rest of the claim up to the maximum of \$15,000. Therefore, the maximum amount that the insurance company pays on a claim is \$14,500, which happens when the claim is \$15,000 or more. Suppose that the pdf of the claim X is

$$f_X(x) = \begin{cases} \frac{6}{20{,}000^3}(20{,}000 - x)x & \text{for } 0 \leq x \leq 20{,}000 \\ 0 & \text{otherwise.} \end{cases}$$

Let Y be the amount that the insurance company pays on the claim. Y has discrete and continuous outcomes, with 0 and 14,500 being discrete outcomes with positive probabilities and $y \in (0, 14{,}500)$ being continuous outcomes.

a. What is the probability, $P(Y=0)$, that the insurance company pays nothing?

b. What is the probability, $P(Y=14{,}500)$, that the insurance company pays the maximum amount?

c. Determine the cdf of Y.

d. What is the probability that the insurance company pays between \$10,000 and \$12,000 on the claim?

7. The lifetime of a new restaurant, in years, is described by the random variable X with pdf

$$f_X(x) = \begin{cases} \frac{1}{(x+1)^2} & \text{for } x > 0 \\ 0 & \text{otherwise.} \end{cases}$$

For example, if $X=2$, the restaurant is in business for exactly two years and then closes.

a. Determine the cdf of X.

b. What is the population median of X?

c. What is the probability that the restaurant stays in business for at least two years?

d. If a company opens three restaurants, each of which has a lifetime that is an independent draw of X, what is the probability that none of the restaurants lasts two years?

8. *Let $X_1, X_2, \ldots, X_n$ be i.i.d. continuous random variables with population median $\tau_{X,0.5}$. Let $\min_X = \min(X_1, X_2, \ldots, X_n)$ and $\max_X = \max(X_1, X_2, \ldots, X_n)$.

a. Show that $P(\min_X \leq \tau_{X,0.5} \leq \max_X) = 1 - \left(\frac{1}{2}\right)^{n-1}$. (Hint: Consider the probability of the complement.)

b. For $n=4$, think about having many four-observation samples $\{x_1, x_2, x_3, x_4\}$, each of which is a realization of the $\{X_1, X_2, X_3, X_4\}$ i.i.d. draws. For what fraction of these samples is the population median $\tau_{X,0.5}$ between $\min(x_1, x_2, x_3, x_4)$ and $\max(x_1, x_2, x_3, x_4)$?

c. What is the smallest value of n for which the probability in (a) is at least 95%?

d. What is the smallest value of n for which the probability in (a) is at least 99%?

9. Each of the following *R* commands makes 10,000 random draws that are stored in vector `x`. For each command, (i) describe the distribution that the random draws are being drawn from and (ii) provide your best guess of what the value of `mean(x)` would be.

a. `x <- 10*runif(10000)-7`

b. `x <- 1*(runif(10000)<0.7)`

c. `x <- 2*(runif(10000)<0.4)-1`

d. `x <- runif(10000,1,3)+runif(10000,5,8)`

10. A store's sales depend on whether it is a weekday (Monday through Friday) or a weekend day (Saturday or Sunday). Specifically, sales (in thousands of dollars) are distributed as a uniform random variable $U(1,3)$ on a weekday and as a uniform random variable $U(2,5)$ on a weekend day. Let X denote the random variable associated with the store's sales on a random day of the week (i.e., the probability associated with each day is 1/7). We say that X is a *mixture of uniform random variables*.

a. Determine the cdf of X by applying the Law of Total Probability:

$$P(X \leq x) = P(X \leq x|weekday)P(weekday) + P(X \leq x|weekend)P(weekend).$$

b. Determine the pdf of X.

c. Conduct 10,000 simulations in *R*, making a random draw of X in each simulation, and plot a histogram of the draws to confirm your answer to (b).

d. What is the population mean of X? Either calculate the integral or use the fact that

$$E(X) = E(X|weekday)P(weekday) + E(X|weekend)P(weekend).$$

e. What is the population median of X?

11. An investor is considering two real estate properties, property A and property B. The annual return from property A (call it R_A), follows a uniform distribution between 4% and 9%, and the annual return from property B (call it R_B), follows a uniform distribution between 2% and 12%. Assume that R_A and R_B are independent.

a. What are the pdfs of R_A and R_B?

b. Which property has a higher probability of an annual return greater than 7%?

c. What is the probability that both R_A and R_B are greater than 7%?

d. What are the expected values and population standard deviations of R_A and R_B?

e. What is $P(R_B > R_A)$?

f. Conduct 10,000 simulations in *R*, making random draws of R_A and R_B for each simulation, to confirm your answer to (e).

12. Two buses arrive randomly and independently at a station between 8 a.m. and 8:20 a.m. The arrival time of each bus can be modeled as a uniform distribution.

a. If each bus stays at the station for two minutes after its arrival, what is the probability that the buses are simultaneously at the station at some point?

b. Conduct 10,000 simulations in R to confirm your answer to (a).

c. For this part, vary the amount of time t (in minutes) that the buses remain at the station. For each integer value $t \in \{1, 2, \cdots, 9, 10\}$, conduct 10,000 simulations in R to approximate the probability that the two buses are simultaneously at the station at some point. Plot the approximated probabilities versus t.

13. Derive the joint cdf in example 10.15.

14. An individual commutes to work five days a week. The length of the morning commute, in minutes, is described by a $U(20, 40)$ random variable, and the length of the return commute, in minutes, is described by a $U(25, 55)$ random variable. Assume that all commute times are independent of each other.

a. What is the expected value of the total commute time on a given day?

b. What is the standard deviation of the total commute time on a given day?

c. What is the expected value of the difference between the return commute time and the morning commute time on a given day?

d. What is the standard deviation of the difference between the return commute time and the morning commute time on a given day?

e. What is the expected value of the average commute time (morning and return combined) over five days?

f. What is the standard deviation of the average commute time (morning and return combined) over five days?

g. Conduct 10,000 simulations in R to approximate the probability that the return commute time is less than the morning commute time on a given day.

h. Conduct 10,000 simulations in R to approximate the probability that, over five days, the average return commute time is less than the average morning commute time.

15. Suppose that the random variables X and Y have the following joint pdf:

$$f_{XY}(x,y) = \begin{cases} x+y & \text{if } x \in [0,1] \text{ and } y \in [0,1] \\ 0 & \text{otherwise.} \end{cases}$$

a. What are the marginal pdfs $f_X(x)$ and $f_Y(y)$?

b. What is the conditional pdf $f_{X|Y}(x|y)$ for $0 \le y \le 1$?

c. Based upon the answers to (a) and (b), are X and Y independent?

d. What is $E(X)$?

e. What is $E(X+Y)$?

f. *What is $E(XY)$?

16. Suppose that the random variables X and Y have the following joint pdf:

$$f_{XY}(x,y) = \begin{cases} 4xy & \text{if } x \in [0,1] \text{ and } y \in [0,1] \\ 0 & \text{otherwise.} \end{cases}$$

a. What are the marginal pdfs $f_X(x)$ and $f_Y(y)$?

b. Show that X and Y are independent by showing that the joint pdf is equal to the product of the marginal pdfs for all possible (x, y) values.

c. Show that X and Y are independent by showing that the conditional pdf of X given $Y = y$ doesn't depend on y.

17. *A company has instituted a new recruiting strategy for executive hires, but it is uncertain how successful the strategy will be. Suppose that the probability of a successful hire is denoted as $P_H = P(H = 1)$, where "success" means that the company makes an offer that is accepted, and H is 1 for a successful hire and 0 for an unsuccessful hire. Given the uncertainty of the recruiting strategy, the recruiting manager initially thinks that P_H is equally likely to be any number between 0 and 1; that is, the "prior distribution" of P_H is $U(0, 1)$.

a. Suppose that the first attempted hire is successful ($H_1 = 1$). Using the following continuous version of Bayes' theorem,

$$f_{P_H}(p|H_1 = 1) = \frac{P(H_1 = 1|P_H = p)f_{P_H}(p)}{\int_0^1 P(H_1 = 1|P_H = p')f_{P_H}(p')\, dp'},$$

determine the posterior distribution of P_H conditional on $H_1 = 1$. That is, what is $f_{P_H}(p|H_1 = 1)$ as a function of p for $p \in [0, 1]$?

b. Suppose that the first two attempted hires are successful ($H_1 = 1$, $H_2 = 1$), where it is assumed that H_1 and H_2 are independent. Determine the posterior distribution of P_H conditional on $H_1 = 1$ *and* $H_2 = 1$. That is, what is $f_{P_H}(p|H_1 = 1, H_2 = 1)$ as a function of p for $p \in [0, 1]$? (Hint: Replace "$H_1 = 1$" by "$H_1 = 1, H_2 = 1$" in the formula from (a) and use the independence assumption.)

c. If the first hire is successful ($H_1 = 1$) and the second hire is unsuccessful ($H_2 = 0$), how would your answer to (b) change?

d. Rather than $P_H \sim U(0, 1)$, suppose that the prior distribution is instead described by the pdf

$$f_{P_H}(p) = \begin{cases} 6p(1-p) & \text{for } p \in [0, 1] \\ 0 & \text{otherwise.} \end{cases}$$

i. Confirm that $f_{P_H}(p)$ is a valid pdf.

ii. Without calculating integrals by hand, provide a graph in R that shows the prior distribution of P_H and the posterior distribution of P_H, conditional on the first hire being successful ($H_1 = 1$). (Hint: Use the formula from (a) and the `integrate` function to determine $f_{P_H}(p|H_1 = 1)$ for possible values of p.)

18. Conduct 10,000 simulations to approximate the expected value and the standard deviation of the investment after two years in example 10.31.

19. The winner of a raffle receives a payout of X^2 dollars, where X is drawn from a $U(20, 40)$ random variable.

a. What is the sample space associated with the payout X^2?

b. What is the cdf of the payout X^2?

c. What is the pdf of the payout X^2?

d. What is the expected value of the payout X^2?

e. *For this part, assume that the payout is $X^2 - 10X$ instead of X^2.

i. Confirm that the payout is an increasing function of X for the relevant values of X ($20 \leq X \leq 40$).

ii. What is the sample space associated with the payout $X^2 - 10X$?

iii. What is the cdf of the payout $X^2 - 10X$? (Hint: Use the quadratic formula to determine $P(X^2 - 10X \leq v)$.)

20. A cryptocurrency miner owns a computer server that mines cryptocurrency for 20 hours during the day (4 a.m. to midnight). Let $X \in [0, 20]$ denote a random variable that indicates, on a given day, how many hours the server runs without crashing. If $X < 20$, the server crashes before midnight; if $X = 20$, the server does not crash and runs until midnight. Suppose that the cdf of X is

$$F_X(x) = \begin{cases} 0 & \text{for } x \leq 0 \\ 0.005x + 0.00025x^2 & \text{for } 0 < x < 20 \\ 1 & \text{for } x \geq 20. \end{cases}$$

a. What is $P(X = 20)$? (Think about the jump in $F_X(x)$ that occurs at $x = 20$.)

b. If the cryptocurrency miner makes the equivalent of \$1,000 for each hour that the server runs, what is the probability that the miner makes more than \$16,000 on a given day?

c. What is the pdf associated with the conditional distribution of X, given $0 < X < 20$?

d. If the cryptocurrency miner makes the equivalent of \$1,000 for each hour that the server runs, what is the expected value of the amount of money made on a given day?

11

Models of Continuous Random Variables

11.1 Normal Random Variable

The **normal random variable** is the most important type of random variable in statistics. Not only does the normal distribution provide a good approximation to the distribution of many observed real-world variables, but it also plays a crucial role in statistical inference. A normal random random variable is completely described by two parameters: a mean parameter μ and a variance parameter σ^2.

Definition 11.1 *A **normal random variable** X with mean parameter μ and variance parameter σ^2, denoted as $X \sim N(\mu, \sigma^2)$, has the pdf*

$$f_X(x) = \frac{1}{\sigma\sqrt{2\pi}} e^{-\frac{1}{2}\left(\frac{x-\mu}{\sigma}\right)^2} \quad \text{for } -\infty < x < \infty.$$

Figure 11.1 shows the shape of the pdf $f_X(\cdot)$ for a normal random variable $X \sim N(\mu, \sigma^2)$. The pdf is bell-shaped and symmetric, with a peak at the center of the distribution where $x = \mu$. The pdf is positive for all values along the real line since e^v is always positive, even for negative v values. The pdf decreases as x moves either to the left or the right of μ.

Proposition 11.1 states some important properties of a normal random variable.

Proposition 11.1 *If $X \sim N(\mu, \sigma^2)$ is a normal random variable, then*

i. $\mu_X = E(X) = \mu$.

ii. $\sigma_X^2 = Var(X) = \sigma^2$.

iii. $\sigma_X = sd(X) = \sigma$.

iv. *X is symmetric around μ, with population median $\tau_{X,0.5} = \mu$.*

v. *The maximum value of the pdf $f_X(x)$ occurs at $x = \mu$, with the pdf strictly increasing to the left of μ (i.e., $f(x_1) < f(x_2)$ if $x_1 < x_2 < \mu$) and strictly decreasing to the right of μ (i.e., $f(x_1) > f(x_2)$ if $\mu > x_1 > x_2$).*

Properties (i) and (ii) can be verified by using the definitions of the population mean and the population variance, but the proofs are complicated due to the form of the

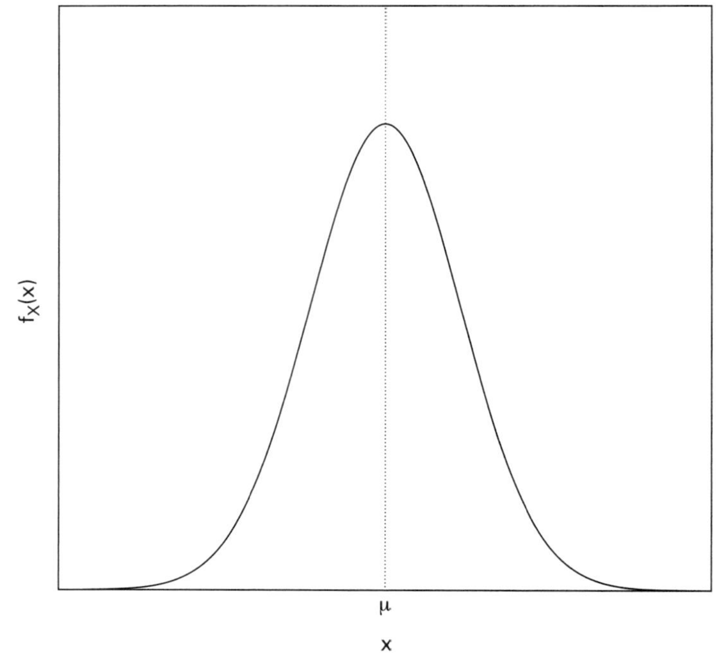

Figure 11.1
The pdf of a normal random variable

normal pdf.[1] Property (iv) holds since, for any $v > 0$,

$$f_X(\mu+v) = \frac{1}{\sigma\sqrt{2\pi}} e^{-\frac{1}{2}\left(\frac{v}{\sigma}\right)^2} = \frac{1}{\sigma\sqrt{2\pi}} e^{-\frac{1}{2}\left(\frac{-v}{\sigma}\right)^2} = f_X(\mu-v).$$

Property (v) holds since $f_X(x) = \frac{1}{\sigma\sqrt{2\pi}} e^{-\frac{1}{2}\left(\frac{x-\mu}{\sigma}\right)^2}$ is maximized when the exponent $-\frac{1}{2}\left(\frac{x-\mu}{\sigma}\right)^2$ equals zero, which happens at $x = \mu$.[2]

Since μ and σ^2 are the population mean and the population variance, respectively, the parameters of $X \sim N(\mu, \sigma^2)$ describe the location and dispersion of X. The top graph of figure 11.2 shows two normal random variables with the same population variance, where one distribution is centered at μ and the other (dotted) distribution is shifted to the right, with the same shape, and centered on a larger population mean. The bottom graph of figure 11.2 shows two normal random variables centered around the same population mean, where the dotted distribution has a larger population variance than the solid distribution. As the population variance increases, with the population mean fixed, more probability shifts to the left and right tails, increasing the spread of the distribution.

The cdf $F_X(x)$ of a normal random variable $X \sim N(\mu, \sigma^2)$ is

$$F_X(x_0) = P(X \leq x_0) = \int_{-\infty}^{x_0} \frac{1}{\sigma\sqrt{2\pi}} e^{-\frac{1}{2}\left(\frac{x-\mu}{\sigma}\right)^2} dx.$$

Unfortunately, there's no simple, closed-form formula for the integral needed to calculate $F_X(x_0)$. That said, as discussed next, *R* has built-in functions to calculate the pdf $f_X(x_0)$ and the cdf $F_X(x_0)$ for any x_0. Since the random variable $X \sim N(\mu, \sigma^2)$ is symmetric around μ,

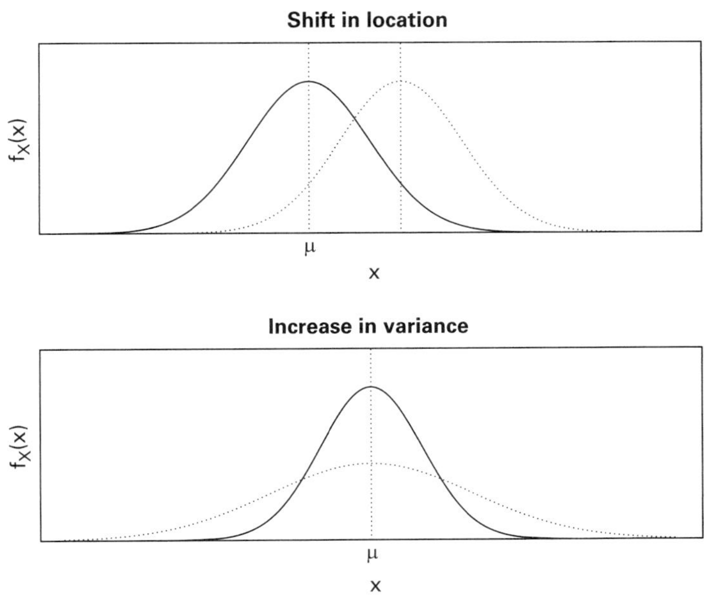

Figure 11.2
Location and variance of normal random variables

the following two properties of the normal cdf hold:

$$F_X(\mu) = 0.5$$

and

$$F_X(\mu - v) = 1 - F_X(\mu + v) \text{ for any } v > 0.$$

Figure 11.3 shows the normal cdf curve (bottom graph), along with the associated normal pdf curve (top graph). The cdf is an S-shaped curve. As the argument of $F_X(\cdot)$ gets more and more negative, the value of $F_X(\cdot)$ gets arbitrarily close to 0 but is always strictly greater than 0. Similarly, as the argument of $F_X(\cdot)$ gets more and more positive, the value of $F_X(\cdot)$ gets arbitrarily close to 1 but is always strictly less than 1.

The following *R* functions are useful for working with a normal random variable:

- `dnorm(x, mean=0, sd=1)`: Returns the pdf of a normal random variable, with the mean `mean` and standard deviation `sd`, evaluated at argument `x`, which may be a single number or a vector. The optional arguments `mean` and `sd` have the default values of `0` and `1`, respectively.
- `pnorm(x, mean=0, sd=1)`: Returns the cdf of a normal random variable, with the mean `mean` and standard deviation `sd`, evaluated at argument `x`, which may be a single number or a vector. The optional arguments `mean` and `sd` have the default values of `0` and `1`, respectively.
- `rnorm(n, mean=0, sd=1)`: Creates a vector of `n` i.i.d. random draws of a normal random variable with the mean `mean` and standard deviation `sd`. The optional arguments `mean` and `sd` have the default values of `0` and `1`, respectively.

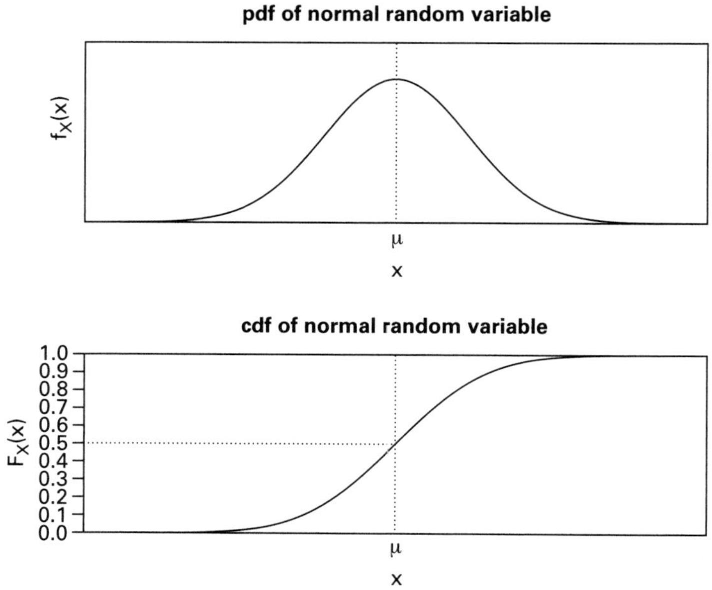

Figure 11.3
The cdf for a normal random variable

- `qnorm(p, mean=0, sd=1)`: Returns the population quantiles of a normal random variable, with the mean `mean` and standard deviation `sd`, specified by argument `p`, which may be a single number or a vector. The optional arguments `mean` and `sd` have default values of `0` and `1`, respectively.

The ability to easily calculate the pdf and cdf of a normal random variable is particularly appealing due to the complicated nature of the pdf and the lack of a closed-form expression for the cdf. The following *R* code illustrates the use of these functions:

```
set.seed(1234)

dnorm(0)

## [1] 0.3989423

dnorm(1)

## [1] 0.2419707

pnorm(1)

## [1] 0.8413447

rnorm(10)

##  [1] -1.2070657  0.2774292  1.0844412 -2.3456977  0.4291247  0.5060559
##  [7] -0.5747400 -0.5466319 -0.5644520 -0.8900378

dnorm(0,mean=0,sd=3)

## [1] 0.1329808
```

```
dnorm(1,mean=0,sd=3)

## [1] 0.1257944

pnorm(1,mean=0,sd=3)

## [1] 0.6305587

rnorm(10,mean=0,sd=3)

##  [1] -1.4315781 -2.9951593 -2.3287617  0.1933765  2.8784822 -0.3308565
##  [7] -1.5330285 -2.7335862 -2.5115150  7.2475055
```

After the random seed is set, the next four commands in the code use the default values `mean=0` and `sd=1`, corresponding to $X \sim N(0,1)$. Listed in order, they return the pdf values $f_X(0)$ and $f_X(1)$, the cdf value $F_X(1)$, and 10 i.i.d. draws for $X \sim N(0,1)$. The last four commands use `mean=0` and `sd=3`, returning (in order) the pdf values $f_X(0)$ and $f_X(1)$, the cdf value $F_X(1)$, and 10 i.i.d. draws for $X \sim N(0,9)$.

11.1.1 Rule-of-Thumb Probability Intervals

For a normal random variable $X \sim N(\mu, \sigma^2)$, it is possible to show the following:

- $P(\mu - \sigma \leq X \leq \mu + \sigma) \approx 0.6827$, meaning that the probability that X is within one standard deviation (σ) of its mean μ is approximately 68% or just over 2/3.
- $P(\mu - 2\sigma \leq X \leq \mu + 2\sigma) \approx 0.9545$, meaning that the probability that X is within two standard deviations (2σ) of its mean μ is approximately 95%.
- $P(\mu - 3\sigma \leq X \leq \mu + 3\sigma) \approx 0.9973$, meaning that the probability that X is within three standard deviations (3σ) of its mean μ is nearly 100%, as there is only a 0.27% probability that X is more than 3σ away from μ.

Figure 11.4 shows the pdf of a normal random variable $X \sim N(\mu, \sigma^2)$, with the gray region in the top graph indicating $P(\mu - \sigma \leq X \leq \mu + \sigma)$ and the gray region in the bottom graph indicating $P(\mu - 2\sigma \leq X \leq \mu + 2\sigma)$. The probability $P(\mu - \sigma \leq X \leq \mu + \sigma) \approx 0.6827$ is equal to the area under the pdf curve between $\mu - \sigma$ and $\mu + \sigma$, and the probability $P(\mu - 2\sigma \leq X \leq \mu + 2\sigma) \approx 0.9545$ is equal to the area under the pdf curve between $\mu - 2\sigma$ and $\mu + 2\sigma$.

A more exact 95% probability interval, commonly used for statistical inference, follows from the fact that

$$P(\mu - 1.96\sigma \leq X \leq \mu + 1.96\sigma) \approx 0.9500.$$

There is a 95% probability that X is within 1.96 standard deviations, rather than two standard deviations, of its mean, μ.

An exact 90% probability interval, also commonly used, follows from the fact that

$$P(\mu - 1.645 \leq X \leq \mu + 1.645) \approx 0.9000.$$

There is a 90% probability that X is within 1.645 standard deviations of its mean, μ.

Example 11.1 (Asset returns) *Suppose the annual return on an asset X is normally distributed, with $X \sim N(0.07, (0.08)^2)$. The annual return has population mean $\mu_X = 0.07$ (or 7%) and population standard deviation $\sigma_X = 0.08$ (or 8%). Then, there is approximately*

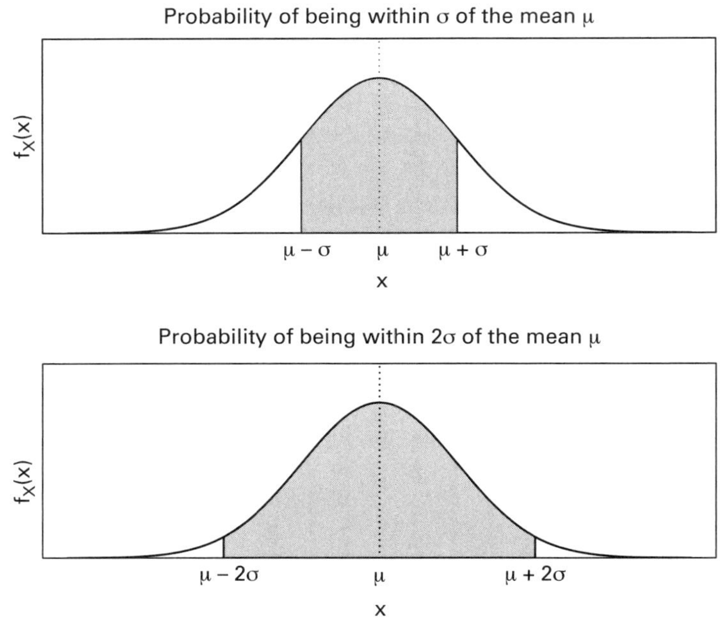

Figure 11.4
Probability intervals for a normal random variable

a 68% probability that the annual return is between $0.07-0.08=-0.01$ *and* $0.07+0.08=0.15$. *Also, there is a 95% probability that the annual return is between* $0.07-(1.96)(0.08)=-0.0868$ *and* $0.07+(1.96)(0.08)=0.2268$.

11.1.2 Standard Normal Random Variable and Probability Intervals

A normal random variable with mean zero ($\mu=0$) and a standard deviation of 1 ($\sigma^2=\sigma=1$) is a useful random variable known as the **standard normal random variable,** as described in definition 11.2.

Definition 11.2 *A* ***standard normal random variable****, often denoted as Z, is a normal random variable with* $\mu=0$, $\sigma^2=1$, *and* $\sigma=1$. *That is,* $Z\sim N(0,1)$.

Plugging $\mu=0$ and $\sigma=1$ into the pdf formula from definition 11.1, the standard normal $Z\sim N(0,1)$ has the pdf

$$f_Z(z)=\frac{1}{\sqrt{2\pi}}e^{-\frac{1}{2}z^2} \text{ for } -\infty<z<\infty.$$

The standard normal distribution $Z\sim N(0,1)$ is used so often that its pdf and cdf are often represented with special notation. The standard normal pdf is denoted as $\phi(\cdot)$, with

$$\phi(z)=f_Z(z),$$

and the standard normal cdf is denoted as $\Phi(\cdot)$, with

$$\Phi(z)=F_Z(z).$$

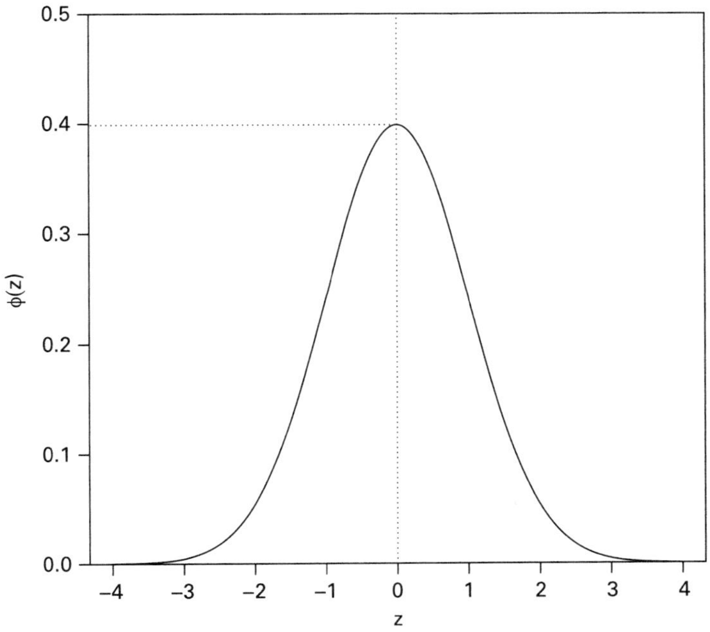

Figure 11.5
The pdf for a $N(0, 1)$ random variable

Figure 11.5 shows the pdf curve for a standard normal random variable $Z \sim N(0, 1)$. The distribution is symmetric around zero, with $\phi(-v) = \phi(v)$ and $\Phi(-v) = 1 - \Phi(v)$ for all $v > 0$. The peak of the standard normal distribution occurs at $z = 0$, with $\phi(0) \approx 0.3989$.

For the standard normal random variable, a 95% probability interval is

$$P(-1.96 \leq Z \leq 1.96) \approx 0.9500,$$

and a 90% probability interval is

$$P(-1.645 \leq Z \leq 1.645) \approx 0.9000.$$

Where do the values 1.96 and 1.645 come from? The `qnorm` function can be used to confirm that -1.96 and 1.96 are, respectively, the 2.5% and 97.5% quantiles of the $N(0, 1)$ distribution and that -1.645 and 1.645 are, respectively, the 5% and 95% quantiles of the $N(0, 1)$ distribution.

```
# return the 2.5%, 5%, 95%, and 97.5% quantiles of N(0,1)
qnorm(c(0.025,0.05,0.95,0.975))

## [1] -1.959964 -1.644854  1.644854  1.959964

# confirm the quantiles and probability intervals using the cdf
pnorm(-1.96)

## [1] 0.0249979

pnorm(1.96)
```

```
## [1] 0.9750021
pnorm(1.96)-pnorm(-1.96)
## [1] 0.9500042
pnorm(-1.645)
## [1] 0.04998491
pnorm(1.645)
## [1] 0.9500151
pnorm(1.645)-pnorm(-1.645)
## [1] 0.9000302
```

Due to the symmetry of the normal distribution, there is 2.5% probability that $Z < -1.96$ and a 2.5% probability that $Z > 1.96$. Therefore, the value 1.96 is the population 97.5% quantile of Z,

$$\Phi(1.96) = 0.975,$$

leaving 2.5% probability in the tail to the right of 1.96. Similarly, for the 90% probability interval $(-1.645, 1.645)$, there is a 5% probability that $Z < -1.645$ and a 5% probability that $Z > 1.645$. The value 1.645 is the population 95% quantile of Z,

$$\Phi(1.645) = 0.95,$$

leaving 5% probability in the tail to the right of 1.645.

This approach can be used for any symmetric probability interval of the standard normal random variable. For instance, for a 70% probability interval, we find the value c such that the $(-c, c)$ interval is associated with a 15% probability that $Z < -c$ and a 15% probability that $Z > c$. The appropriate value of c is the population 85% quantile of Z, which can be found in R:

```
qnorm(0.85)
## [1] 1.036433
```

The population 85% quantile is approximately 1.036, so $\Phi(1.036) \approx 0.85$, $\Phi(-1.036) \approx 0.15$, and

$$P(-1.036 \leq Z \leq 1.036) \approx 0.70.$$

Similarly, for a 80% probability interval, we find the population 90% quantile of Z, leaving 10% in the right tail:

```
qnorm(0.90)
## [1] 1.281552
```

The population 90% quantile is approximately 1.282, so that $\Phi(1.282) \approx 0.90$, $\Phi(-1.282) \approx 0.10$, and

$$P(-1.282 \leq Z \leq 1.282) \approx 0.80.$$

11.1.3 Standardizing a Normal Random Variable

A useful property of normal random variables is that taking a linear transformation of a normal random variable yields another normal random variable, as in proposition 11.2.

Proposition 11.2 *If $X \sim N(\mu, \sigma^2)$, the linear transformation $Y = a + bX$ is also a normal random variable, with $Y \sim N(a + b\mu, b^2\sigma^2)$. The population mean of Y is $\mu_Y = a + b\mu$, the population variance of Y is $\sigma_Y^2 = b^2\sigma^2$, and the population standard deviation of Y is $\sigma_Y = |b|\sigma$.*

Proposition 11.2 can be applied to standardize any normal random variable $X \sim N(\mu, \sigma^2)$ by de-meaning it (i.e., subtracting μ) and dividing the result by the standard deviation σ:

$$X \sim N(\mu, \sigma^2) \Longrightarrow \frac{X - \mu}{\sigma} \sim N(0, 1).$$

The standardized random variable $Z = \frac{X-\mu}{\sigma}$ is a linear transformation of X with the additive constant $a = -\frac{\mu}{\sigma}$ and scaling constant $b = \frac{1}{\sigma}$. Regardless of the original $X \sim N(\mu, \sigma^2)$ random variable, Z indicates the number of standard deviations σ that X is above or below μ. When Z is positive, X is above its mean. When Z is negative, X is below its mean.

We can determine interval probabilities for X in terms of interval probabilities for $Z \sim N(0, 1)$. For instance,

$$P(X \leq b) = P\left(\frac{X - \mu}{\sigma} \leq \frac{b - \mu}{\sigma}\right) = P\left(Z \leq \frac{b - \mu}{\sigma}\right) = \Phi\left(\frac{b - \mu}{\sigma}\right)$$

and

$$P(a \leq X \leq b) = P\left(\frac{a - \mu}{\sigma} \leq Z \leq \frac{b - \mu}{\sigma}\right) = \Phi\left(\frac{b - \mu}{\sigma}\right) - \Phi\left(\frac{a - \mu}{\sigma}\right).$$

Example 11.2 (Asset returns) *Example 11.1 considered annual asset returns given by the random variable $X \sim N(0.07, (0.08)^2)$. The normal random variable X can be standardized by de-meaning it and dividing the result by its standard deviation:*

$$Z = \frac{X - 0.07}{0.08} \sim N(0, 1).$$

If $Z = 1.5$, X is 1.5 standard deviations above its mean and $X = 0.19$. If $Z = -2.5$, X is 2.5 standard deviations below its mean and $X = -0.13$. The probability of a positive annual return is

$$P(X > 0) = P\left(\frac{X - 0.07}{0.08} > \frac{0 - 0.07}{0.08}\right) = P(Z > -0.875) = 1 - \Phi(-0.875).$$

For the first equality, we subtract 0.07 and then divide by 0.08 for both X (on the left of the $>$) and 0 (on the right of the $>$) so the probability remains the same. Then, we calculate $1 - \Phi(-0.875) \approx 0.8092$:

```
1-pnorm(-0.875)

## [1] 0.809213
```

Alternatively, since $1 - \Phi(-0.875) = \Phi(0.875)$ *by symmetry of* Z, `pnorm(0.875)` *would give the same answer.*

Using the results on probability intervals for $Z \sim N(0,1)$ in section 11.1.2, probability intervals for general $X \sim N(\mu, \sigma^2)$, including the rule-of-thumb intervals from section 11.1.1, can be constructed. These intervals are based on $X = \mu + \sigma Z$ being a linear transformation of Z. For example, for a 95% probability interval,

$$0.95 = P(-1.96 \le Z \le 1.96) = P(\mu - 1.96\sigma \le \mu + \sigma Z \le \mu + 1.96\sigma)$$
$$= P(\mu - 1.96\sigma \le X \le \mu + 1.96\sigma),$$

where the second equality is obtained by multiplying each of the three terms within the probability by σ and then adding μ to each term. Similarly, the 90% probability interval for X is

$$0.90 = P(-1.645 \le Z \le 1.645) = P(\mu - 1.645\sigma \le X \le \mu + 1.645\sigma).$$

For other symmetric probability intervals of X, centered around μ, the constant that provides the desired probability can be determined. As an example, suppose that we want an 85% probability interval for X, leaving 7.5% probability less than the lower end point of the interval and 7.5% probability greater than the upper end point of the interval. For the standard normal, the 92.5% quantile is equal to 1.440, so $\Phi(1.440) = 0.925$ and $\Phi(-1.440) = 0.075$.

```
qnorm(0.925)

## [1] 1.439531
```

Therefore,

$$P(-1.440 \le Z \le 1.440) = 0.85,$$

and, equivalently,

$$P(\mu - 1.440\sigma \le X \le \mu + 1.440\sigma) = 0.85.$$

Using the parameters from example 11.2 ($\mu = 0.07$, $\sigma = 0.08$), there is an 85% probability that X (annual asset return) is between $0.07 - (1.440)(0.08) = -0.0452$ and $0.07 + (1.440)(0.08) = 0.1852$.

11.1.4 Independence of Normal Random Variables

As discussed in chapter 10 (proposition 10.7), the independence of two random variables implies that their population covariance and population correlation both equal zero, as in proposition 11.3. While the reverse (zero covariance implying independence) is not always true in general, it does turn out to be true for normal random variables.

Proposition 11.3 *The normal random variables* X *and* Y *are independent if and only if* $\sigma_{XY} = \rho_{XY} = 0$.

If X and Y are normally distributed with zero correlation, then X and Y are independent, meaning that joint probabilities can be expressed in terms of marginal probabilities:

$$P(a < X < b,\ c < Y < d) = P(a < X < b)P(c < Y < d).$$

For instance, if X and Y are independent $N(0, 1)$ random variables, the probability that both X and Y are less than 1 is

$$P(X < 1, Y < 1) = P(X < 1)P(Y < 1) \approx (0.8413)^2 \approx 0.708.$$

11.1.5 Linear Combinations of Normal Random Variables

Another useful property of normal random variables is that any linear combination of two or more normal random variables is also a normal random variable. It is important to stress that this property is *not* a general result for other types of random variables. Recall, from example 10.28, that the sum of two independent $U(0, 1)$ random variables is not a uniform distribution but rather a triangular distribution. As another example, the sum of independent Bernoulli random variables is not a Bernoulli random variable, but rather a binomial random variable.

Proposition 11.4 *Suppose that $X \sim N(\mu_X, \sigma_X^2)$ and $Y \sim N(\mu_Y, \sigma_Y^2)$ are normal random variables with covariance σ_{XY}. Then, the linear combination*

$$V = k + aX + bY$$

is a normal random variable, with

$$V \sim N(k + a\mu_X + b\mu_Y,\ a^2\sigma_X^2 + b^2\sigma_Y^2 + 2ab\sigma_{XY}).$$

Proposition 10.12 already implies that the population mean of V is $k + a\mu_X + b\mu_Y$ and the population variance of V is $a^2\sigma_X^2 + b^2\sigma_Y^2 + 2ab\sigma_{XY}$. So the new result in proposition 11.4 is that the linear combination V is also normal, with these population mean and variance parameters characterizing the normal distribution.

Example 11.3 (Producing a movie) *Example 10.27 considered the probability that a movie's production costs were over budget based on an assumption of independent uniform distributions for the number of days of filming (X) and the number of days of editing (Y). Recall that the cost random variable is $C = X + 0.5Y$ (in millions) and the movie budget is \$90 million. In this example, let's assume independent normal distributions for X and Y rather than independent uniform distributions, as follows:*

$$X \sim N(70, 4^2) \text{ and } Y \sim N(45, 2^2), \text{ with } \sigma_{XY} = \rho_{XY} = 0.$$

The production costs have a normal distribution with population mean

$$\mu_C = \mu_X + 0.5\mu_Y = 70 + (0.5)(45) = 92.5,$$

population variance

$$\sigma_C^2 = \sigma_X^2 + (0.5)^2\sigma_Y^2 = 16 + (0.5)^2(4) = 17,$$

and population standard deviation $\sigma_C = \sqrt{\sigma_C^2} = \sqrt{17}$. The probability that production costs go over budget is

$$P(C > 90) = P\left(\frac{C - 92.5}{\sqrt{17}} > \frac{90 - 92.5}{\sqrt{17}}\right) = P\left(Z > -\frac{2.5}{\sqrt{17}}\right),$$

which is approximately 0.728, *or* 72.8%.

```
1-pnorm(-2.5/sqrt(17))

## [1] 0.7278552
```

Example 11.4 (Two-asset portfolio) *Consider two assets whose annual returns X and Y are described by the following two normal distributions:*

$$X \sim N(0.07, (0.08)^2) \text{ and } Y \sim N(0.03, (0.01)^2).$$

Think of asset X as a riskier asset, like a stock mutual fund, with a higher average return but more risk due to the higher variance. Think of asset Y as a less risky asset, like a bond mutual fund, with a lower average return but less risk. Suppose that a portfolio is constructed by investing half the money in asset X and half the money in asset Y, so the returns on the two-asset portfolio are

$$V = 0.5X + 0.5Y.$$

If the asset returns are uncorrelated ($\rho_{XY} = 0$), then

$$\mu_V = 0.5\mu_X + 0.5\mu_Y = (0.5)(0.07) + (0.5)(0.03) = 0.05,$$
$$\sigma_V^2 = (0.5)^2\sigma_X^2 + (0.5)^2\sigma_Y^2 = (0.5)^2(0.08)^2 + (0.5)^2(0.01)^2 = 0.001625,$$

and

$$\sigma_V = \sqrt{0.001625} \approx 0.0403.$$

Thus, $V \sim N(0.05, 0.001625)$ when $\rho_{XY} = 0$. The average return for the two-asset portfolio V is exactly at the midpoint of the averages of the two returns since the portfolio is equally weighted. The standard deviation and variance of V indicate that V is less risky than asset X and more risky than asset Y.

What if the asset returns X and Y are positively correlated instead, say with $\rho_{XY} = 0.2$? The population mean of $V = 0.5X + 0.5Y$ is unchanged, and the population variance of V becomes

$$\begin{aligned}\sigma_V^2 &= (0.5)^2\sigma_X^2 + (0.5)^2\sigma_Y^2 + 2(0.5)(0.5)\sigma_{XY}\\ &= (0.5)^2(0.08)^2 + (0.5)^2(0.01)^2 + (2)(0.5)(0.5)(0.2)(0.08)(0.01) = 0.001705,\end{aligned}$$

using $\sigma_{XY} = \rho_{XY}\sigma_X\sigma_Y$. The population standard deviation of V is $\sigma_V = \sqrt{0.001705} \approx$ 0.0413, which is 2.5% higher than the standard deviation of 0.403 for the case of no asset correlation. The positive correlation between X and Y leads to a tendency for the asset returns to move together, resulting in more variance than when X and Y are uncorrelated.

Even though the average return for X is greater than the average return for Y, it is possible that the realized Y value is larger than the realized X value due to the variances of the returns. The probability $P(Y > X)$ can be written as $P(Y > X) = P(Y - X > 0)$, which is a probability in terms of a linear combination. The difference $Y - X$ has the population mean

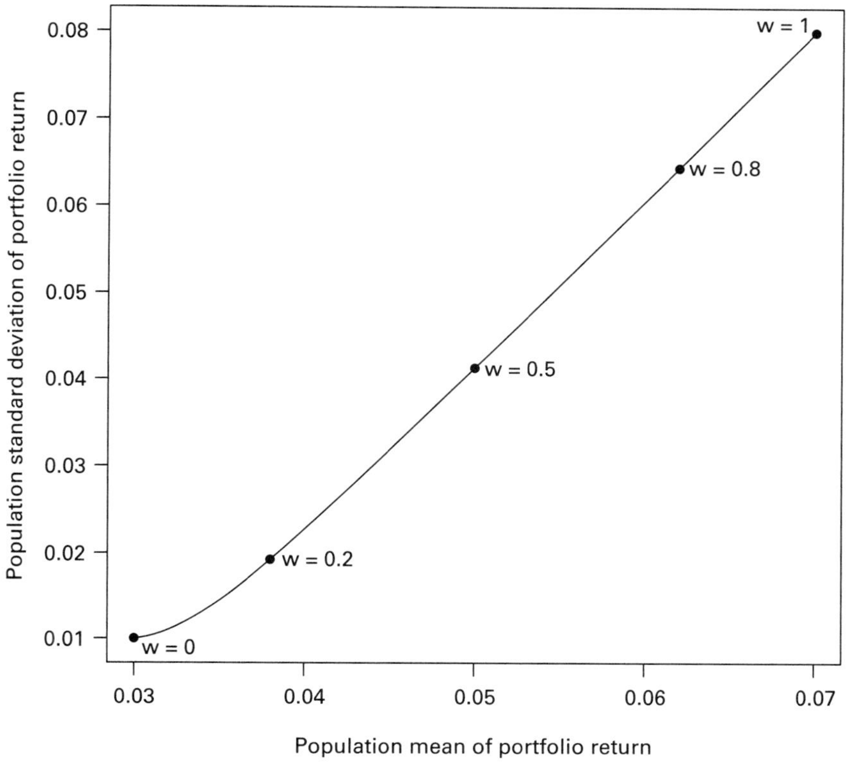

Figure 11.6
Population mean and standard deviation for weighted two-asset portfolios

$0.03 - 0.07 = -0.04$ *and population variance* $(0.01)^2 + (0.08)^2 - (2)(0.2)(0.08)(0.01) = 0.00618$, *so*

$$P(Y > X) = P(Y - X > 0) = P\left(\frac{Y - X - (-0.04)}{\sqrt{0.00618}} > \frac{0 - (-0.04)}{\sqrt{0.00618}}\right) = P\left(Z > \frac{0.04}{\sqrt{0.00618}}\right),$$

which is approximately 0.3054, *or* 30.54%.

```
1-pnorm(0.04/sqrt(0.00618))
## [1] 0.3054386
```

The population mean and population standard deviation can also be determined for a two-asset portfolio with unequal weights. Figure 11.6 summarizes the population mean and standard deviation for $V = wX + (1 - w)Y$, *with weights ranging from* $w = 0$ *to* $w = 1$. *There are black dots shown for portfolios with* $w = 0.2$, $w = 0.5$, *and* $w = 0.8$. *As* w *moves from* 0 *to* 1, *the mean and standard deviation move from the values associated with the* Y *asset to the values associated with the* X *asset.*

Proposition 11.4 generalizes to more than two normal random variables. Specifically, the linear combination of *any* number of normal random variables is also a normal random variable. For example, when $V = X_1 + X_2 + X_3$ for the normal random variables X_1, X_2, and X_3, proposition 11.4 implies that the linear combination $X_1 + X_2$ is normally distributed and, then, that the linear combination of $X_1 + X_2$ and X_3 is also normally distributed. This type of reasoning can be extended to additional random variables, as in proposition 11.5.

Proposition 11.5 *If $X_1, X_2, \ldots, X_m$ are normal random variables with population means $\mu_1, \mu_2, \cdots, \mu_m$ and population variances $\sigma_1^2, \sigma_2^2, \ldots, \sigma_m^2$, respectively, then*

$$V = k + a_1X_1 + a_2X_2 + \cdots + a_mX_m$$

is a normal random variable with population mean $\mu_V = k + a_1\mu_1 + a_2\mu_2 + \cdots + a_m\mu_m$.

The variance expression for the general linear combination is more complicated, as it depends upon the (possibly nonzero) covariances between the random variables (see proposition 8.7). For the case of independent normal random variables, the variance expression for the linear combination simplifies considerably, and proposition 11.6 provides a complete specification of the distribution of the linear combination.

Proposition 11.6 *If $X_1, X_2, \ldots, X_m$ are independent normal random variables with population means $\mu_1, \mu_2, \cdots, \mu_m$ and population variances $\sigma_1^2, \sigma_2^2, \ldots, \sigma_m^2$, respectively, then*

$$V = k + a_1X_1 + a_2X_2 + \cdots + a_mX_m$$

is a normal random variable, with

$$V \sim N(k + a_1\mu_1 + a_2\mu_2 + \cdots + a_m\mu_m,\ a_1^2\sigma_1^2 + a_2^2\sigma_2^2 + \cdots + a_m^2\sigma_m^2).$$

(i) For a sum of i.i.d. normal random variables, $V = X_1 + X_2 + \cdots + X_m$, with each $X_j \sim N(\mu, \sigma^2)$,

$$V \sim N(m\mu, m\sigma^2) \text{ and } \sigma_V = \sqrt{m}\sigma.$$

(ii) For an average of i.i.d. random variables, $V = \frac{1}{m}(X_1 + X_2 + \cdots + X_m)$, with each $X_j \sim N(\mu, \sigma^2)$,

$$V \sim N\left(\mu, \frac{\sigma^2}{m}\right) \text{ and } \sigma_V = \frac{\sigma}{\sqrt{m}}.$$

Example 11.5 (Monthly sales) *Assume that the monthly sales M at Hayden's Hardware, measured in thousands of dollars, are i.i.d. and normally distributed: $M \sim N(10, 4)$. The* total *sales in a given year are $Y = M_1 + M_2 + \cdots + M_{12}$, where each $M_j \sim M(10, 4)$ is i.i.d. The total sales in a given year is normally distributed, with*

$$Y \sim N(120, 48) \text{ and } \sigma_Y = \sqrt{48} \approx 6.928.$$

There is a 90% probability that Y is in the $120 \pm (1.645)(6.928)$ *interval, which is* $(108.6, 131.4)$. *There is a 95% probability that Y is in the* $120 \pm (1.96)(6.928)$ *interval, which is* $(106.4, 133.6)$.

The average *monthly sales in a given year, $A = \frac{1}{12}(M_1 + M_2 + \cdots + M_{12})$, is normally distributed with*

$$A \sim N\left(10, \frac{4}{12}\right) \text{ and } \sigma_A = \frac{1}{\sqrt{3}} \approx 0.577.$$

There is a 90% probability that A is in the $10 \pm (1.645)(0.577)$ *interval, which is* $(9.05, 10.95)$. *There is a 95% probability that A is in the* $10 \pm (1.96)(0.577)$ *interval, which is* $(8.87, 11.13)$. *The standard deviation of the average monthly sales in a given year (*0.577*) is considerably less than the standard deviation of monthly sales in any given month. Averaging over 12 months leads to a lower variance and less dispersion. If the average is taken over more months, the dispersion would continue to decrease. With an average taken over 24 months, the standard deviation is* $\frac{1}{\sqrt{6}} \approx 0.408$. *With an average taken over 36 months, the standard deviation is* $\frac{1}{3} \approx 0.333$.

11.2 Log-Normal Random Variable

The normal distribution is a good approximation for many variables encountered in economic applications, as well as in everyday life. But many variables in economics (wages, sales, firm size) are right-skewed, meaning that the symmetric normal distribution won't provide a good model or approximation of the underlying distribution in the population. In such cases, it can be useful to directly use a random variable with a distribution that is right-skewed. This section introduces one such random variable, known as the **log-normal random variable**, described in definition 11.3.

Definition 11.3 *A positive-valued random variable X is a* ***log-normal random variable****, with parameters* μ *and* σ^2, *if* $\ln(X) \sim N(\mu, \sigma^2)$.

Function $\ln(\cdot)$ is the *natural log* function, satisfying $e^{\ln(x)} = x$ for all $x > 0$. The possible outcomes for X are all positive real numbers since $e^{\ln(x)} > 0$ for any possible value of $\ln(x)$. Figure 11.7 shows an example of a pdf curve for a log-normal random variable. The pdf curve is always positive for $x > 0$ and is clearly asymmetric and right-skewed. (The pdf is equal to zero when $x \leq 0$.) The population median of X equals e^{μ} since the population median of $\ln(X)$ is μ. Since $\ln(\cdot)$ is an increasing function of its argument, this fact follows from

$$P(X < e^{\mu}) = P(\ln(X) < \mu) = 0.5.$$

Due to the right-skewness of the log-normal random variable, the population mean of X is *greater than* the population median e^{μ}.

Example 11.6 (Weekly earnings) *Consider the earnwk variable for employed individuals from the* ***cps*** *data set. Example 6.9 showed the right-skewed sample distribution of weekly earnings. The top graph in figure 11.8 is the histogram of weekly earnings, this time with a normal distribution (solid curve) drawn over the histogram. The graphed normal distribution has population mean equal to the sample mean of earnwk and population variance equal to the sample variance of earnwk, as that distribution would match the histogram pretty closely if the earnwk values were truly drawn from a normal distribution. It's evident that the normal distribution does not fit the histogram of earnwk well, as the histogram is too right-skewed to be matched by the normal distribution. The bottom graph in figure 11.8 shows the histogram of* $\ln(earnwk)$, *with a normal distribution (solid curve) drawn over the histogram. In this case, the population mean and variance are chosen to be the sample mean and variance of* $\ln(earnwk)$ *rather than earnwk. The histogram of* $\ln(earnwk)$ *looks*

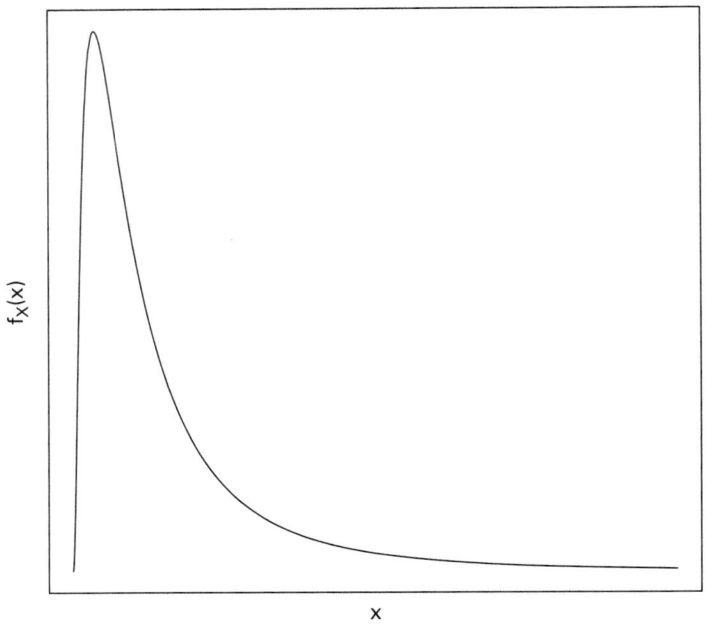

Figure 11.7
The pdf for a log-normal random variable

much more symmetric than the histogram of earnwk, with the right tail no longer evident. The normal distribution seems to provide a pretty good fit to the histogram—certainly much better than in the top graph. Taken together, these graphs suggest that the normal distribution might provide a good model for ln(*earnwk*) *but not earnwk, meaning a log-normal model for earnwk is more sensible than a normal model for earnwk.*

Here is the R code to create figure 11.8:

```
# create log-earnings variable
logwage <- log(cpsemployed$earnwk)

# graph-display format (two rows, one column)
par(mfrow=c(2,1))

# histogram of earnings, with normal distribution overlaid
hist(cpsemployed$earnwk, breaks=50, freq=FALSE,
     main="Distribution of weekly earnings", xlab="Weekly earnings($)")
lines(0:8000, dnorm(0:8000,mean(cpsemployed$earnwk),sd(cpsemployed$earnwk)), lwd=3)

# histogram of log-earnings, with normal distribution overlaid
hist(logwage, breaks=50, freq=FALSE, main="Distribution of ln(weekly earnings)",
      xlab="ln(weekly earnings)")
lines(seq(3,9,by=0.01), dnorm(seq(3,9,by=0.01),mean(logwage),sd(logwage)), lwd=2)
```

11.2.1 Probability Intervals for a Log-Normal Random Variable

We generally want to make statements about the random variable X (e.g., weekly earnings) rather than the random variable $\ln(X)$ (e.g., natural log of weekly earnings) since values of

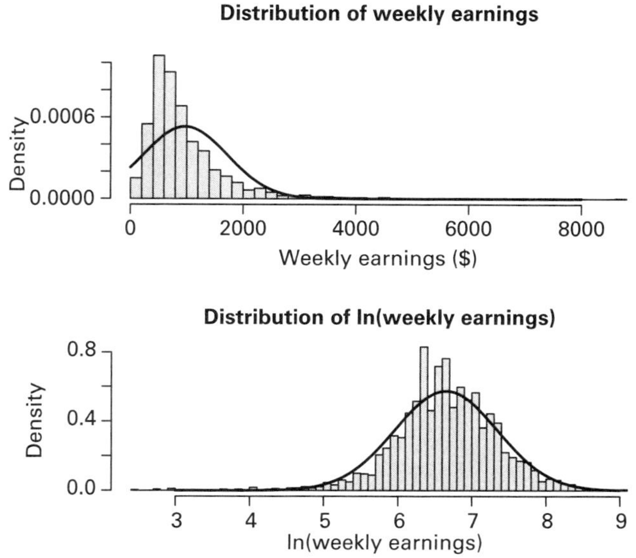

Figure 11.8
Distributions of weekly earnings and ln(weekly earnings)

X are easily interpretable, whereas values of $\ln(X)$ are not. If X is weekly earnings, X is in dollars, while $\ln(X)$ is in the less interpretable units of log-dollars.

To construct a probability interval for a log-normal random variable X, note that the cdf of X is directly related to the cdf of a normal random variable (see proposition 10.17). Specifically, for any $x_0 > 0$,

$$F_X(x_0) = P(X \le x_0) = P(\ln(X) \le \ln(x_0)) = F_{\ln(X)}(\ln(x_0)).$$

The second equality holds since $\ln(\cdot)$ is a strictly increasing function. Since $\ln(X) \sim N(\mu, \sigma^2)$ is a normal random variable, $F_{\ln(X)}(\ln(x_0))$ is the cdf of a $N(\mu, \sigma^2)$ random variable evaluated at $\ln(x_0)$.

Then the probability that X is in the interval $[a, b]$, for $0 < a < b$, can also be written in terms of the normal cdf:

$$P(a \le X \le b) = P(\ln(a) \le \ln(X) \le \ln(b)) = F_{\ln(X)}(\ln(b)) - F_{\ln(X)}(\ln(a)).$$

Using this relationship, 90% and 95% probability intervals for X can be constructed as follows:

$$P(\mu - 1.645\sigma \le \ln(X) \le \mu + 1.645\sigma) \approx 0.90 \implies P\left(e^{\mu-1.645\sigma} \le X \le e^{\mu+1.645\sigma}\right) \approx 0.90$$

$$P(\mu - 1.96\sigma \le \ln(X) \le \mu + 1.96\sigma) \approx 0.95 \implies P\left(e^{\mu-1.96\sigma} \le X \le e^{\mu+1.96\sigma}\right) \approx 0.95.$$

There is a 90% probability that X is between $e^{\mu-1.645\sigma}$ and $e^{\mu+1.645\sigma}$, with a 5% probability that X is below $e^{\mu-1.645\sigma}$ and a 5% probability that X is above $e^{\mu+1.645\sigma}$. There is a 95% probability that X is between $e^{\mu-1.96\sigma}$ and $e^{\mu+1.96\sigma}$, with a 2.5% probability that X is below $e^{\mu-1.96\sigma}$ and a 2.5% probability that X is above $e^{\mu+1.96\sigma}$. Unlike the probability intervals for a normal random variable, these probability intervals are asymmetric in the sense that the end points are not equidistant from the population median e^{μ}. For example, the difference $e^{\mu} - e^{\mu-1.96\sigma}$ is not the same as $e^{\mu+1.96\sigma} - e^{\mu}$.

This approach can be used for any probability interval. For instance, for a 70% probability interval, with 15% probabilities each in the left tail and the right tail, $\Phi(1.036) \approx 0.85$ implies

$$P(\mu - 1.036\sigma \leq \ln(X) \leq \mu + 1.036\sigma) \approx 0.70 \Longrightarrow P\left(e^{\mu-1.036\sigma} \leq X \leq e^{\mu+1.036\sigma}\right) \approx 0.70.$$

Example 11.7 (Weekly earnings) *Assume that weekly earnings X are log-normally distributed with* $\ln(X) \sim N(6.5, (0.7)^2)$. *These parameters roughly correspond to the log-normal distribution shown in figure 11.8. Then, a 95% probability interval for X is*

$$P(e^{6.5-1.96(0.7)} \leq X \leq e^{6.5+1.96(0.7)}) = P(168.68 \leq X \leq 2622.81) = 0.95.$$

There is a 95% probability that weekly earnings are between \$168.68 *and* \$2,622.81, *a 2.5% probability that weekly earnings are less than* \$168.68, *and a 2.5% probability that weekly earnings are greater than* \$2,622.81.

The following *R* functions are useful for working with a log-normal random variable:

- `dlnorm(x, meanlog=0, sdlog=1)`: Returns the pdf of a log-normal random variable evaluated at argument `x`, which may be a single number or a vector. The optional arguments `meanlog` and `sdlog` have the default values of `0` and `1`, respectively, and represent the mean and standard deviation of the natural log of the random variable.
- `plnorm(x, meanlog=0, sdlog=1)`: Returns the cdf of a log-normal random variable evaluated at argument `x`, which may be a single number or a vector. The optional arguments `meanlog` and `sdlog` have the default values of `0` and `1`, respectively, and represent the mean and standard deviation of the natural log of the random variable.
- `rlnorm(n, meanlog=0, sdlog=1)`: Creates a vector of `n` i.i.d. random draws of a log-normal random variable. The optional arguments `meanlog` and `sdlog` have the default values of `0` and `1`, respectively, and represent the mean and standard deviation of the natural log of the random variable.
- `qlnorm(p, meanlog=0, sdlog=1)`: Returns the population quantiles of a log-normal random variable specified by argument `p`, which may be a single number or a vector. The optional arguments `meanlog` and `sdlog` have the default values of `0` and `1`, respectively, and represent the mean and standard deviation of the natural log of the random variable.

For instance, although the probability intervals in example 11.7 were determined through the use of the normal distribution, they can also be calculated directly in *R* based upon the quantiles of the log-normal distribution:

```
qlnorm(c(0.025,0.975), meanlog=6.5, sdlog=0.7)

## [1]  168.6837 2622.7398
```

For the same distribution $\ln(X) \sim N(6.5, (0.7)^2)$, the following code calculates the probability that $X > 1000$ (weekly earnings over \$1,000) and makes 10 random draws from the distribution:

```
1-plnorm(1000, meanlog=6.5, sdlog=0.7)

## [1] 0.2801124

set.seed(1234)
rlnorm(10, meanlog=6.5, sdlog=0.7)

##  [1]  285.7319  807.7079 1420.9896  128.7679  898.1951  947.8906  444.8255
##  [8]  453.6644  448.0405  356.7280
```

11.2.2 Population Statistics for a Log-normal Random Variable

Proposition 11.7 provides the population mean, variance, and standard deviation of a log-normal random variable.

Proposition 11.7 *If X is a log-normal random variable with $\ln(X) \sim N(\mu, \sigma^2)$, the population mean of X is*

$$\mu_X = e^{\mu + \frac{1}{2}\sigma^2},$$

the population variance of X is

$$\sigma_X^2 = e^{2\mu+\sigma^2}\left(e^{\sigma^2} - 1\right),$$

and the population standard deviation of X is

$$\sigma_X = e^{\mu+\frac{1}{2}\sigma^2}\sqrt{e^{\sigma^2} - 1}.$$

It is worth noting that the population mean of X is *not* e^{μ} even though the population mean of $\ln(X)$ is μ. Instead, the population mean of X is a scaled-up version of e^{μ}, with a scaling factor of $e^{\frac{1}{2}\sigma^2} > 1$.[3] For the weekly earnings example (example 11.7), where $\ln(X) \sim N(6.5, (0.7)^2)$, the population mean of weekly earnings is $\mu_X = e^{6.5+\frac{1}{2}(0.7)^2} \approx \$849.80$ and the population standard deviation of weekly earnings is $\sigma_X = e^{6.5+\frac{1}{2}(0.7)^2}\sqrt{e^{(0.7)^2} - 1} \approx \$675.75$.

```
exp(6.5+0.5*(0.7)^2)

## [1] 849.7991

exp(6.5+0.5*(0.7)^2)*sqrt(exp(0.7^2)-1)

## [1] 675.7459
```

11.3 Chi-Square Random Variable

A family of distributions, known as **chi-square distributions**, or using the Greek letter, χ^2 **distributions**, turns out to be useful for statistical inference and is discussed in chapters 12 and 16. The simplest chi-square distribution is based on a standard normal random variable $Z \sim N(0, 1)$, with the chi-square random variable defined as the square of Z, as in definition 11.4.

Definition 11.4 *A random variable X is a* ***chi-square random variable****, denoted as $X \sim \chi^2$ or $X \sim \chi_1^2$, if $X = Z^2$ and $Z \sim N(0,1)$.*

All possible outcomes of $X \sim \chi_1^2$ are nonnegative since $X = Z^2$. This random variable is a special case of a more general chi-square random variable, which involves the sum of squared independent standard normals. Specifically, if $Z_1, Z_2, \ldots, Z_m$ are i.i.d. $N(0,1)$ random variables, $Z_1^2 + Z_2^2 + \cdots + Z_m^2$ is said to have a chi-square distribution with m degrees of freedom, as in definition 11.5.

Definition 11.5 *A random variable X is a* ***chi-square random variable with m degrees of freedom****, denoted as $X \sim \chi_m^2$, if $X = Z_1^2 + Z_2^2 + \cdots + Z_m^2$ and $Z_1, Z_2, \ldots, Z_m$ are i.i.d. $N(0,1)$.*

Figure 11.9 shows the distributions for four chi-squared random variables, corresponding to 4 degrees of freedom (χ_4^2), 6 degrees of freedom (χ_6^2), 8 degrees of freedom (χ_8^2), and 10 degrees of freedom (χ_{10}^2). The x-axis has been arbitrarily cut off at 20, but each of the four distributions has a right tail that extends forever. The four distributions are all right-skewed. And, as the value for the degrees of freedom increases, both the mean and the dispersion of the distributions increase. This last feature should not be surprising since every time we increase the degrees of freedom, we are adding Z_j^2 terms to the random variable. In fact, it turns out that, for $X \sim \chi_m^2$, the population mean and variance are $\mu_X = m$ and $\sigma_X^2 = 2m$, respectively.

The following *R* functions are useful for working with chi-square random variables:

- `dchisq(x, df)`: Returns the pdf of a chi-square random variable with `df` degrees of freedom evaluated at argument `x`, which may be a single number or a vector.
- `pchisq(x, df)`: Returns the cdf of a chi-square random variable with `df` degrees of freedom evaluated at argument `x`, which may be a single number or a vector.
- `rchisq(n, df)`: Creates a vector of `n` i.i.d. random draws of a chi-square random variable with `df` degrees of freedom.
- `qchisq(p, df)`: Returns the population quantiles of a chi-square random variable with `df` degrees of freedom specified by argument `p`, which may be a single number or a vector.

Here are some examples of these functions being used:

```
qchisq(0.90,4)

## [1] 7.77944

qchisq(0.90,6)

## [1] 10.64464

qchisq(0.90,8)

## [1] 13.36157

qchisq(0.90,10)

## [1] 15.98718

pchisq((1.96)^2,1)
```

```
## [1] 0.9500042
pchisq((1.645)^2,1)
## [1] 0.9000302
```

The `qchisq` commands calculate the population 90% quantiles associated with the four chi-square distributions shown in figure 11.9. As expected, the population 90% quantiles increase as the degrees of freedom increase, reflecting a higher likelihood to have values farther out in the right tail. The `pchisq` commands illustrate a connection between the cdf of a χ^2 random variable and the cdf of a $N(0,1)$ random variable. Whereas 1.96 is the population 97.5% quantile of $N(0,1)$, the `pchisq((1.96)^2,1)` command confirms that the value $(1.96)^2$ is the population 95% quantile of χ^2. This result arises since, if $X \sim \chi^2$ and $Z \sim N(0,1)$,

$$P\left(X < (1.96)^2\right) = P\left(Z^2 < (1.96)^2\right) = P(-1.96 < Z < 1.96) \approx 0.95,$$

where the second equality follows from the fact that $Z^2 < (1.96)^2$ can happen only when $|Z| < 1.96$. Similarly, with 1.645 being the population 95% quantile of $N(0,1)$, the

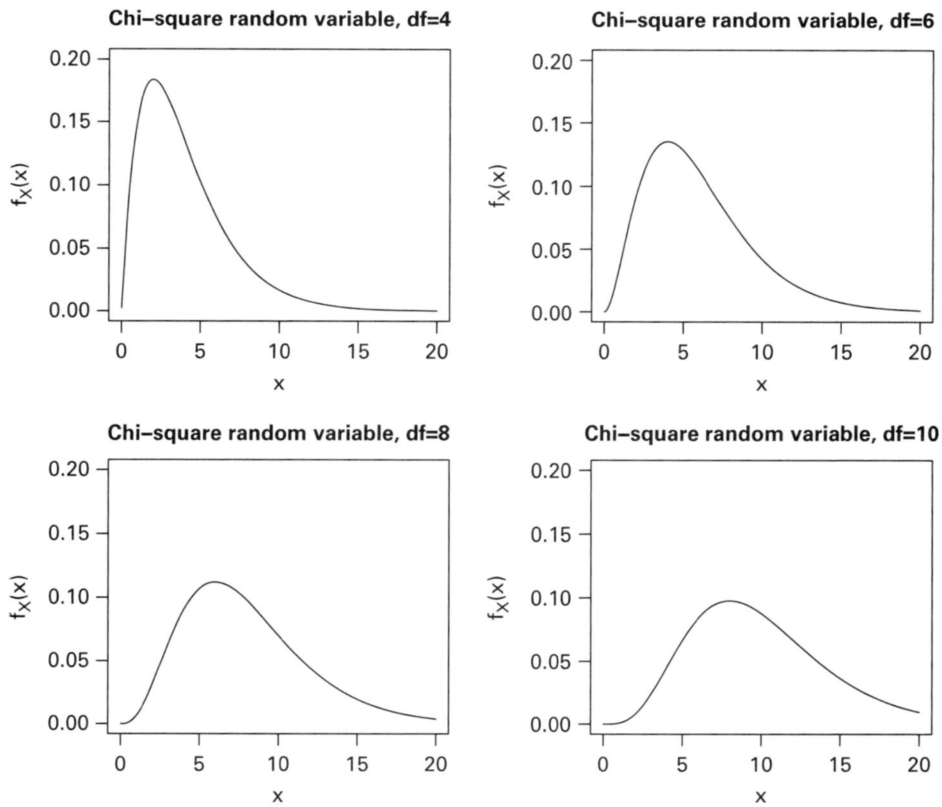

Figure 11.9
The pdfs for chi-square random variables

`pchisq((1.645)^2,1)` command confirms that the value $(1.645)^2$ is the population 90% quantile of χ^2, which follows from

$$P\left(X < (1.645)^2\right) = P\left(Z^2 < (1.645)^2\right) = P(-1.645 < Z < 1.645) \approx 0.90.$$

11.4 Exponential Random Variable

An **exponential random variable** is often used to model the amount of time that it takes for a certain event to occur. Some examples include the following:

- The time that it takes a user to leave a website
- The time that it takes for a worker strike to end
- The time that a call with a customer service agent lasts

Since the exponential model usually concerns time-related events, it is an example of a **duration model**. The exponential model can be thought of as modeling the duration of the website visit, the duration of the worker strike, or the duration of the customer service call. The formal definition of an exponential random variable is given in definition 11.6.

Definition 11.6 *An* ***exponential random variable*** *X with parameter $\theta > 0$, written as $X \sim Exp(\theta)$, is a positive-valued random variable with pdf*

$$f_X(x) = \theta e^{-\theta x} \text{ for } x > 0.$$

Parameter θ describes how quickly the underlying event is expected to occur. Larger values of θ correspond to events that are expected to occur more quickly (i.e., shorter durations), whereas smaller values of θ correspond to events that are not expected to occur quickly (i.e., longer durations). Since it turns out that the population mean of X is $\mu_X = E(X) = \frac{1}{\theta}$, the value $\frac{1}{\theta}$ can be thought of as the expected time until the event occurs. For $\theta = 1$, the expected time until the event occurs is 1 time unit, whereas for $\theta = 0.2$, the expected time until the event occurs is 5 time units.

The cdf $F_X(\cdot)$ for an exponential random variable X is obtained by integrating the pdf. For any $x_0 > 0$,

$$F_X(x_0) = \int_0^{x_0} \theta e^{-\theta x}\, dx = -e^{-\theta x}\Big|_0^{x_0} = 1 - e^{-\theta x_0}.$$

Figure 11.10 graphs three pdf curves corresponding to three values of the θ parameter, with $\theta = 1$ in the left graph, $\theta = 0.5$ in the middle graph, and $\theta = 0.2$ in the right graph. Each of these pdfs appears to be a strictly decreasing function of x, which is a property of any exponential random variable since the derivative $f'_X(x) = -\theta^2 e^{-\theta x} < 0$ for all $x > 0$. Regardless of the value of parameter θ, it is always more likely for an exponential random variable to have smaller values than larger values; for example, the probability that X is in $(0, 1)$ is larger than the probability that X is in $(1, 2)$. As the value of θ decreases, moving from left to right in figure 11.10, the height of the pdf near zero decreases and the right tail becomes thicker.

The population-descriptive statistics for an exponential random variable are given in the following proposition.[4]

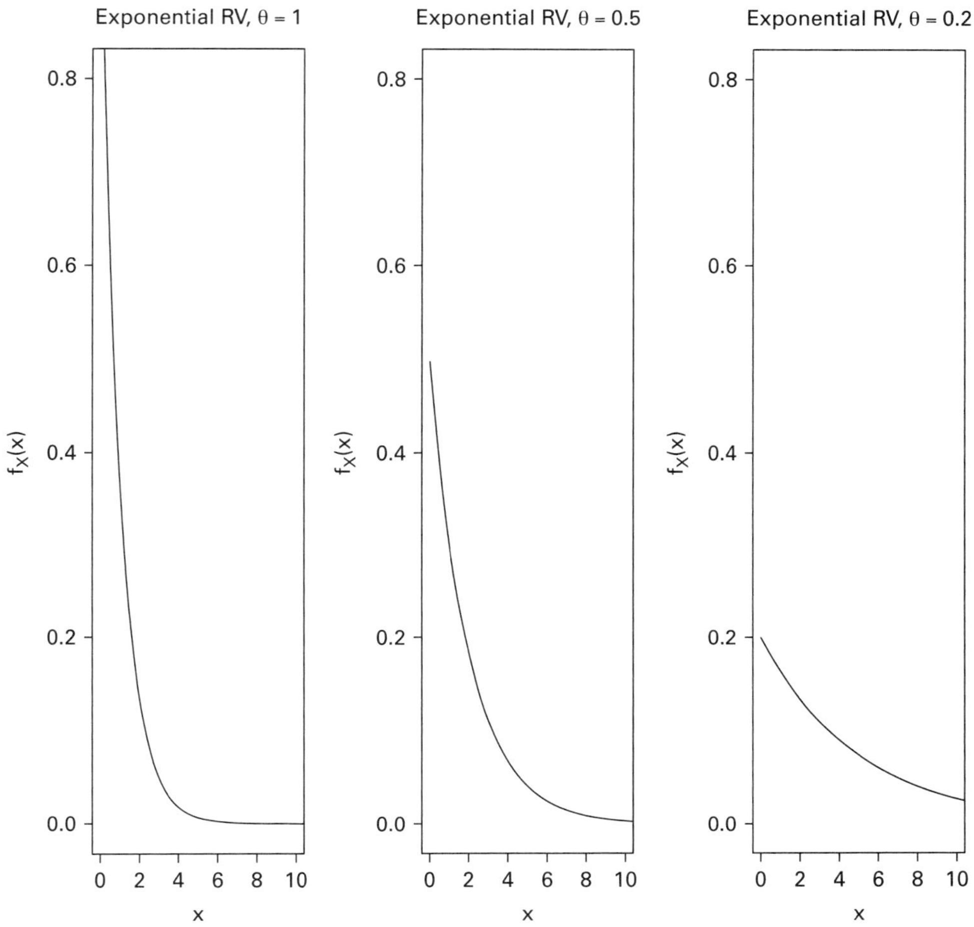

Figure 11.10
The pdfs for exponential random variables

Proposition 11.8 *If $X \sim Exp(\theta)$, the population mean of X is*

$$\mu_X = \frac{1}{\theta},$$

the population variance of X is

$$\sigma_X^2 = \frac{1}{\theta^2},$$

and the population standard deviation of X is

$$\sigma_X = \frac{1}{\theta}.$$

Example 11.8 (Duration of website visit) *For a particular website, assume that the number of minutes that any given visitor spends on the website before leaving is an exponential random variable $X \sim Exp(0.5)$. The expected duration of the website visit, μ_X, is $\frac{1}{0.5}$ or 2 minutes. The standard deviation of the duration of the website visit is also 2 minutes. The* *cdf, derived previously as $F_X(x) = 1 - e^{-\theta x}$, can be used to calculate the probabilities of intervals. For example, the probability that the duration of the website visit is between* 1 *and* 2 *minutes is*

$$F_X(2) - F_X(1) = \left(1 - e^{-(2)(0.5)}\right) - \left(1 - e^{-(1)(0.5)}\right) = e^{-0.5} - e^{-1} \approx 0.239.$$

The probability that the website-visit duration is between 2 *and* 3 *minutes is*

$$F_X(3) - F_X(2) = \left(1 - e^{-(3)(0.5)}\right) - \left(1 - e^{-(2)(0.5)}\right) = e^{-1} - e^{-1.5} \approx 0.145.$$

The largest probability for a 1-minute interval is for a duration between 0 *and* 1, *with probability* $e^{-0} - e^{-0.5} \approx 0.393$.

The following *R* functions are useful for working with exponential random variables:

- `dexp(x, rate=1)`: Returns the pdf of an exponential random variable with the rate θ equal to `rate` evaluated at argument `x`, which may be a single number or a vector.
- `pexp(x, rate=1)`: Returns the cdf of an exponential random variable with the rate θ equal to `rate` evaluated at argument `x`, which may be a single number or a vector.
- `rexp(n, rate=1)`: Creates a vector of `n` i.i.d. random draws of an exponential random variable with the rate θ equal to `rate`.
- `qexp(p, rate=1)`: Returns the population quantiles of an exponential random variable with the rate θ equal to `rate` specified by argument `p`, which may be a single number or a vector.

The following code confirms the calculations in example 11.8 and illustrates the use of the function `rexp`:

```
pexp(1,rate=0.5)-pexp(0,rate=0.5)

## [1] 0.3934693

pexp(2,rate=0.5)-pexp(1,rate=0.5)

## [1] 0.2386512

pexp(3,rate=0.5)-pexp(2,rate=0.5)

## [1] 0.1447493

set.seed(1234)
rexp(10,rate=0.5)

##  [1] 5.00351721 0.49351777 0.01316391 3.48549218 0.77436517 0.17989934
##  [7] 1.64816303 0.40523580 1.67608064 1.52086060

temp <- rexp(100000,rate=0.5)
mean(temp)

## [1] 1.998902

sd(temp)

## [1] 1.983554
```

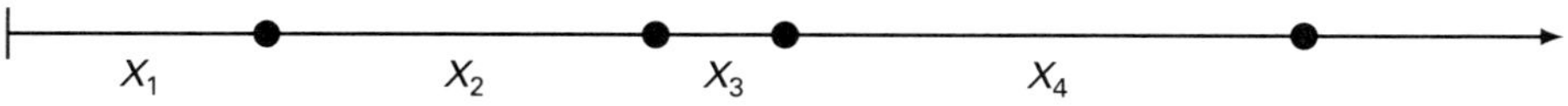

Figure 11.11
Sequence of exponential random variables

The first `rexp` command simulates 10 i.i.d. draws from an exponential random variable with $\theta = 0.5$. The second use of `rexp`, in the assignment of the variable `temp`, simulates 100,000 draws, and the values of `mean(temp)` and `sd(temp)` are both very close to the population mean and standard deviation of $\frac{1}{\theta} = 2$.

One potential drawback of the exponential model is that the exponential pdf is a strictly decreasing function of x, regardless of the parameter value θ. This feature means that the assumption of an exponential random variable implies that shorter durations are always more likely than longer durations. To relax this restriction, a more flexible duration model would be needed. An example of this is the *Weibull model*, a two-parameter model that generalizes the exponential model. Specifically, a Weibull random variable X has the pdf

$$f_X(x) = \alpha\theta(\theta x)^{\alpha-1}e^{-(\theta x)^{\alpha}} \text{ for } x > 0,$$

where the two parameters α and θ are both positive.[5] The exponential pdf is a special case, corresponding to $\alpha = 1$. Other values of α lead to pdf shapes different from those associated with the exponential model. For example, for certain θ and α values, the Weibull pdf can increase until reaching a peak and then decrease afterward, which is a more appropriate model if the most likely durations are not close to zero, but rather at some other value.

11.4.1 Sequence of Exponential Random Variables

The description of an exponential random variable relates it to the time that it takes a certain event to occur. What if we are interested in multiple occurrences of a particular event? For example, an exponential random variable can be used to model the time that it takes a new customer to arrive (since the last customer's arrival). Think of a "timer" that starts when the previous customer arrived and ends when a new customer arrives. Once the new customer arrives, another exponential random variable can be used to model the time that the next customer arrives, and so on. If these successive "times between events" are i.i.d. draws (denoted $X_1, X_2, X_3, \ldots$) of an exponential random variable $Exp(\theta)$, there is a direct connection to the Poisson random variable discussed in section 9.5, as given in proposition 11.9.

Proposition 11.9 *If the time scale starts at zero and the times between events ($X_1, X_2, X_3, \ldots$) are i.i.d. draws of an $Exp(\theta)$ random variable, then the number of events between time zero and time T is a $Poisson(T\theta)$ random variable.*

Figure 11.11 provides a visualization of the timeline, with the left end corresponding to time zero. X_1 is the time to the first event occurrence, X_2 is the time between the first and second occurrences, X_3 is the time between the second and third occurrences, and so on. The total number of events that occur depends on the time T that is specified. If T is less than X_1, the event count is 0; if T is between X_1 and $X_1 + X_2$, the event count is 1; if T is between $X_1 + X_2$ and $X_1 + X_2 + X_3$, the event count is 2; and so on.[6]

To illustrate the connection between exponential random variables and Poisson random variables, we revisit the example of customers visiting a coffee shop (example 9.6). In that example, the situation was modeled in terms of the number or count of customers, assumed to be a Poisson random variable. Here, we instead model the customer arrival times as exponential random variables and, using proposition 11.9, *infer* that the number of customers is a Poisson random variable. The advantage of this approach is that we can say something about the distributions of both the arrival times between customers (exponential random variables) and the count of customers (a Poisson random variable).

Example 11.9 (Coffee shop customers) *Suppose that the arrival time, in hours, of a new customer at a coffee shop (since the last customer arrived) is an* $Exp(20)$ *random variable, so that the average arrival time is* $\frac{1}{20} = 0.05$ *hours or* 3 *minutes. If it is assumed that the arrival time of each successive customer is also an i.i.d.* $Exp(20)$ *random variable, proposition 11.9 implies that the number of customers who arrive over the course of an hour is a* $Poisson(20)$ *random variable. Likewise, the number of customers that arrive over the course of 2 hours is a* $Poisson(40)$ *random variable.*

If we are interested in the arrival times themselves, rather than the count, the distribution associated with the $Exp(\theta)$ random variable can be used for any single arrival time (e.g., between customers in example 11.9). How about the time that it takes *two* customers to arrive? This time would be a draw from the random variable given by the sum of two i.i.d. $Exp(\theta)$ random variables. For example, the time for the first two customers to arrive is a draw from $X_1 + X_2$. Unfortunately, $X_1 + X_2$ is a more complicated distribution (*not* an exponential), but we can use computer simulations to approximate this distribution and its properties (mean, standard deviation). Similarly, the time for the first three customers to arrive is a draw from $X_1 + X_2 + X_3$, whose distribution can again be approximated via computer simulations. Perhaps not surprisingly, as seen in example 10.29 and figure 10.12 (for the average of exponential random variables), the shape of the distribution of the sum of exponential random variables begins to look bell-shaped as the number of random variables in the sum increases.[7]

11.5 Mixture of Normal Random Variables

One drawback of the normal random variable is that it is unimodal and, therefore, serves as a poor model or approximation for data that are drawn from a true distribution with two or more humps or modes. In this section, we describe an approach that mixes normal random variables together to create more flexible distributions that may be asymmetric and/or have multiple modes. While we focus here on the case of mixing two normal random variables, we start with a more general definition of a mixture distribution, given in definition 11.7.

Definition 11.7 *A* ***mixture distribution*** *is the probability distribution associated with a random variable* X *that is based on a collection of underlying random variables* $Y_1, Y_2, \ldots, Y_m$ *using the following two-step process: (i) a random variable is selected from the collection* $Y_1, Y_2, \ldots, Y_m$ *according to the probabilities* $\pi_1, \pi_2, \ldots, \pi_m$, *where* $\sum_{j=1}^{m} \pi_j = 1$; *and (ii) the realized value of* X *is the realized value of the selected random variable.*

To focus on mixtures of normal random variables, we provide definition 11.8 as a special case of definition 11.7.

Definition 11.8 *A* ***mixture of normal random variables*** *is a random variable with a mixture distribution based upon normal random variables* $Y_1, Y_2, \ldots, Y_m$.

In example 11.10, we revisit the data-analyst salary example (example 10.16), first noting that the original example involved a mixture of uniform random variables and then considering a different mixture distribution based upon normal random variables instead of uniform random variables.

Example 11.10 (Data analyst salaries) *Example 10.16 considered the salaries of data analysts at a large firm, where the salaries for nongraduate-degree data analysts and graduate-degree data analysts were modeled as different uniform random variables. Using the notation from definition 11.7, let* $Y_1 \sim U(60, 100)$ *and* $Y_2 \sim U(90, 210)$ *denote these two random variables. If the probability that a data analyst at the firm has a graduate degree is* 20%, *the random variable for data-analyst salaries is a mixture of* Y_1 *and* Y_2 *with probabilities* $\pi_1 = 0.8$ *and* $\pi_2 = 0.2$.

Now suppose that the two salary distributions are modeled as normal *random variables rather than uniform random variables, specifically with*

$$Y_1 \sim N(80, 10^2) \text{ and } Y_2 \sim N(150, 30^2),$$

and still with $\pi_1 = 0.8$ *and* $\pi_2 = 0.2$. *To visualize the mixture distribution* X, *based upon* Y_1 *and* Y_2 *with probabilities* 0.8 *and* 0.2, *respectively, the following R code simulates* 1,000,000 *draws of* X:

```
# create a temporary vector with 1,000,000 U(0,1) draws
set.seed(1234)
temp <- runif(1000000)

# salaries drawn from N(80,10^2) and N(150,30^2) distributions
y1 <- rnorm(1000000,mean = 80,sd = 10)
y2 <- rnorm(1000000,mean = 150,sd = 30)

# construct the mixture random variable, with probs 0.8 and 0.2
salary <- (temp<=0.8)*y1 + (temp>0.8)*y2

# draw smoothed density of salaries, with vertical lines at 80 and 150
plot(density(salary), main="", xlab="Data-analyst salary (in $1,000s)")
abline(v=80,lty=3)
abline(v=150,lty=3)
```

The `temp` *vector consists of* $U(0, 1)$ *draws that are used to determine whether* Y_1 *or* Y_2 *is the chosen random variable for a given draw of* X. Y_1 *is chosen with probability* 0.8, *or equivalently when the corresponding element of* `temp` *is less than or equal to* 0.8, *whereas* Y_2 *is chosen when the corresponding element of* `temp` *is greater than* 0.8. *The* `salary` *assignment command stores the full vector of simulated* X *draws. Figure 11.12 shows the smoothed density plot output by the R code, with dotted lines drawn at the values of* 80 *and* 150, *corresponding to the means of* Y_1 *and* Y_2, *respectively. The resulting distribution has two humps or modes at approximately* 80 *and* 150. *The hump near* 80 *is much higher than*

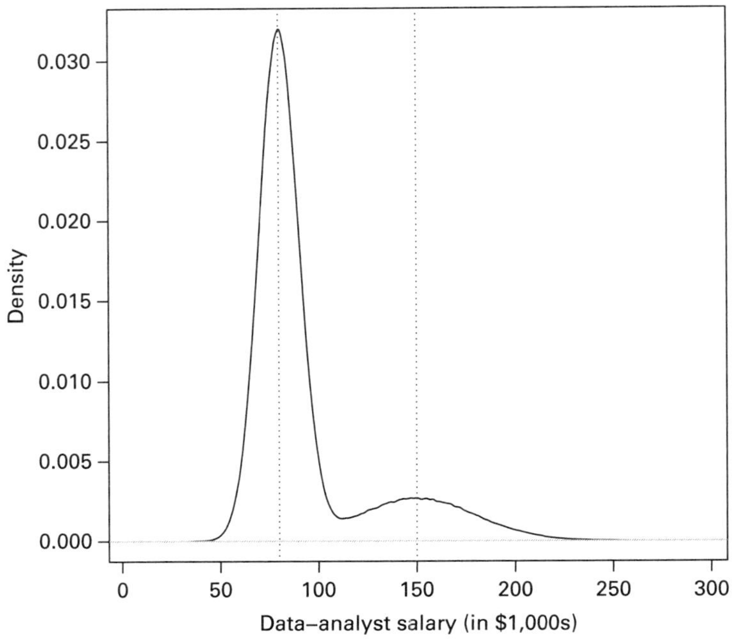

Figure 11.12
Mixture of two normal random variables

the one near 150 *since the probability (*80%*) that* X *is drawn from the* Y_1 *random variable is so much higher than the probability (*20%*) that* X *is drawn from the* Y_2 *random variable. To see the effect of changing the probabilities of the two underlying distributions, the interested reader can alter the code above by replacing the* π_1 *value (*`0.8`*) with other values.*

Notes

1. Alternatively, the population mean result ($\mu_X = \mu$) follows directly from property (iv).
2. The "strictly increasing" and "strictly decreasing" results can be shown by taking the derivative of $f_X(x)$ with respect to x and verifying that the derivative is positive for $x < \mu$ and negative for $x > \mu$.
3. That is, $e^{E(\ln(X))} > E\left(e^{\ln(X)}\right) = E(X)$, which is a special case of a result in statistics known as *Jensen's inequality*.
4. The interested reader can confirm these properties by working out the appropriate integrals. For example, $\mu_X = \frac{1}{\theta}$ can be shown by evaluating $\int_0^\infty \theta e^{-\theta x} x\, dx$.
5. The interested reader can look at the documentation for the Weibull-related *R* functions `dweibull`, `pweibull`, `rweibull`, and `qweibull`.
6. Proposition 11.9 allows an infinite number of possible events, so time T can be any value.
7. The exact distribution of a sum of i.i.d. exponential random variables is a special case of a distribution known as the *gamma distribution*.

Exercises

1. An airline knows that the duration of the flight from Austin to Nashville is uniformly distributed between 110 minutes and 130 minutes. The flight departs at 1 p.m.

a. If the airline wants the probability of a late arrival to be 20%, what time should it state as the arrival time?

b. When the plane lands, it may have to wait until an arrival gate is available for passengers to disembark. The time that the arrival gate becomes available is uniformly distributed between 2.50 p.m. and 3 p.m. If the flight time and the time that the arrival gate becomes available are independent, what is the probability that the plane will have to wait for its arrival gate when it lands?

c. Same as (b), but now assume that flight time and arrival-gate availability are independent normal random variables. Assume that the two random variables have the same means as the uniform random variables described previously, and the standard deviation of the flight time is 5 minutes and the standard deviation of the arrival-gate availability is 2.5 minutes.

2. A credit card company knows that the monthly balance X (in dollars) of a representative customer is normally distributed:

$$X \sim N(300, 2500).$$

Assume that the monthly balances of customers are independent draws from X.

a. If X_1 and X_2 are the monthly balances for two randomly chosen customers, what is the probability that $X_1 + X_2$ is greater than \$700?

b. If X_1 and X_2 are the monthly balances for two randomly chosen customers, let $Y = \frac{X_1}{X_2}$ denote the ratio of two customers' balances. Here, Y is a *nonlinear* combination of X_1 and X_2, and Y is *not* a normal random variable. Conduct 100,000 simulations in *R* to approximate the following quantities: (i) the population mean of Y, (ii) the population standard deviation of Y, (iii) the population median of Y, and (iv) the probability that $Y > 1.5$.

3. Maternal smoking during pregnancy has a negative association with birth weight. Suppose that the distribution of newborn child's birth weight (in grams) is $BW_S \sim N(3050, 590^2)$ if the mother smokes during pregnancy, while the distribution is $BW_{NS} \sim N(3260, 530^2)$ if the mother does not smoke during pregnancy.

a. Which of the following is larger: the pdf of BW_S evaluated at 3,050, or the pdf of BW_{NS}, evaluated at 3,260? Explain why.

b. Plot the two pdf's on the same graph, over the range between 2,000 grams and 4,500 grams.

c. A baby who weighs less than 2,500 grams is classified as "low birth weight." What is the probability of a low-birth-weight baby if the mother smokes during pregnancy? What is the probability of a low-birth-weight baby if the mother does not smoke during pregnancy?

d. What is the probability that the birth weight of a baby born to a smoking mother is greater than the birth weight of a baby born to the nonsmoking mother? (Treat the two births as independent events.)

e. Conduct 10,000 simulations in *R* to confirm your answer to (d).

f. Now consider two births associated with smoking mothers and two births associated with nonsmoking mothers, where the four birth weights are independent random variables. What is the probability that the average of the two birth weights for the smoking mothers is greater than the average of the two birth weights for the non-smoking mothers?

g. Conduct 10,000 simulations in *R* to confirm your answer to (f).

h. There are approximately 453.6 grams in a pound. What is the normal distribution associated with birth weight *in pounds* for a baby born to a mother who smokes during pregnancy?

4. The annual returns for two stocks, for the companies Widgetville and Planet Widget, are draws from the following normal distributions:

$$\text{Widgetville: } X \sim N(0.10, 0.0064) \text{ and Planet Widget: } Y \sim N(0.06, 0.0049),$$

with positive correlation $\rho_{XY} = 0.3$.

a. What is the probability that Widgetville's return is greater than Planet Widget's return in a given year?

b. If you buy \$100 of Widgetville stock and \$200 of Planet Widget stock, what is the distribution of the net gain/loss (in dollars) on your portfolio after one year?

c. For the portfolio in (b), what is the probability that the net gain is greater than \$30 after one year?

d. *Now suppose that you can invest only in Widgetville stock. Write a function `widgetgain(amt,yrs,numsim)` with three arguments: `amt` is the amount invested in Widgetville stock, `yrs` is the number of years that the money is invested, and `numsim` is the number of simulations. The function should return a vector of length `numsim`, where each element of the vector is a simulation of the net gain/loss over `yrs` years from an investment of `amt` in Widgetville stock. Assume that the annual return in each year is an independent draw from the random variable X, but make sure to allow compounding. For instance, for `yrs` = 2, if \$100 is invested with a 0.10 return in year 1 and a 0.05 return in year 2, you would have $\$100(1+0.10) = \$110$ after one year and $\$110(1+0.05) = \$115.50$ after two years, yielding \$15.50 net profit. Draw histograms of `widgetgain(100,10,10000)` and `widgetgain(100,20,10000)`, and calculate their respective simulated means and standard deviations.

5. A clothing store has to decide whether to spend money on advertising for the upcoming month. If it does not advertise, the distribution of monthly revenue (in thousands of dollars) is $NA \sim N(90, 36)$. If it does advertise, the distribution of monthly revenue (in thousands of dollars) is $A \sim N(95, 25)$. The cost of advertising is \$4,000.
The correlation between NA and A is $\rho_{NA,A}$. (This correlation is likely positive since there are factors that affect monthly revenue whether or not the store advertises.) The profitability X associated with advertising is a random variable, with $X = A - NA - 4$.

a. What is the expected value of X?

b. What is the population standard deviation of X, in terms of $\rho_{NA,A}$? Calculate σ_X for $\rho_{NA,A} = 0.6$ and $\rho_{NA,A} = 0.8$.

c. Using the fact that X is also normally distributed, plot the probability of positive profitability, $P(X > 0)$, versus $\rho_{NA,A}$ for $\rho_{NA,A} = \{0.1, 0.2, \ldots, 0.8, 0.9\}$.

6. Suppose that $X_1 \sim N(0,4)$, $X_2 \sim N(1,1)$, and $X_3 \sim N(2,9)$ are independent random variables.
   a. What is the expected value of the average of draws from X_1, X_2, and X_3?
   b. What is the population variance of the average of draws from X_1, X_2, and X_3?
   c. What is the probability that X_2 is larger than the average $\frac{X_1+X_3}{2}$?
   d. What is the probability that X_1 is larger than the average $\frac{X_2+X_3}{2}$?
   e. Conduct 10,000 simulations in R to approximate the probability that X_2 is closer to X_1 than it is to X_3, which is $P(|X_2 - X_1| < |X_2 - X_3|)$.
7. A worker has three projects (A, B, and C) that she needs to complete, but there is uncertainty about how long each project will take to complete. The completion times, in hours, are draws from normal random variables:

$$T_A \sim N(12,9), \; T_B \sim N(24,16), \text{ and } T_C \sim N(8,4).$$

   Assume that T_A, T_B, and T_C are independent.
   a. What is the distribution of the total completion time for the three projects?
   b. Suppose the worker does project A, and then B, and then C. Conduct 100,000 simulations in R to approximate the pmf of $X =$ # projects completed within 40 hours (a week of work)?
   c. Suppose the worker does project C, and then A, and then B. Conduct 100,000 simulations in R to approximate the pmf of $X =$ # projects completed within 40 hours (a week of work)?
   d. *For this part, drop the independence assumption and assume that the correlation between any two completion times is equal to 0.4 (i.e., $\rho_{T_AT_B} = \rho_{T_AT_C} = \rho_{T_BT_C} = 0.4$). What is the distribution of the total completion time for the three projects? How does this compare to your answer to (a)?
8. A *truncated normal random variable* is restricted to a range (a, b) and, within that range, has a pdf proportional to the pdf of a normal random variable. Specifically, the pdf of a truncated normal random variable based upon a $N(\mu, \sigma^2)$ random variable and restricted to the range (a, b) is

$$f_X(x) = \begin{cases} \frac{1}{\sigma} \cdot \frac{\phi\left(\frac{x-\mu}{\sigma}\right)}{\Phi\left(\frac{b-\mu}{\sigma}\right) - \Phi\left(\frac{a-\mu}{\sigma}\right)} & \text{if } a < x < b \\ 0 & \text{otherwise.} \end{cases}$$

   a. Explain why the variance of X is less than σ^2 (without actually determining σ_X^2).
   b. Write an R function `dtruncnorm(x,mean,sd,a,b)` that returns the pdf of a truncated normal, based on a normal with mean `mean` and standard deviation `sd` and restricted to the range between `a` and `b`, evaluated at each element of the vector `x`.

c. Write an *R* function `rtruncnorm(n,mean,sd,a,b)` that returns a vector of `n` i.i.d. random draws of a truncated normal, based on a normal with mean `mean` and standard deviation `sd` and restricted to the range between `a` and `b`. (Hint: Continually make i.i.d. draws from the normal distribution until there are `n` values between `a` and `b`.)

d. In a population of credit-card holders, each individual has a probability L of making a late payment in a given month. Suppose the distribution of these probabilities follows a truncated normal distribution on the range $(0, 1)$ based on a $N(0.1, 0.3^2)$ random variable.

   i. Using the `dtruncnorm` function, draw the pdf of L for values ranging between -0.5 and 1.5 on the x-axis.

   ii. Using the `rtruncnorm` function to create 100,000 simulated draws of L, what are the approximate values of $E(L)$, $sd(L)$, $\tau_{L,0.5}$, and $P(L < 0.1)$?

9. Suppose the wealth W (in dollars) of 70-year-olds in the US is described by a log-normal random variable:

$$\ln(W) \sim N(10, 1).$$

   a. Using only the normal distribution, provide a 90% probability interval for the wealth of a randomly chosen 70-year-old.

   b. Using only the normal distribution, what is the probability that a randomly chosen 70-year-old has wealth between \$10,000 and \$30,000?

   c. Conduct 100,000 simulations in *R*, using the `rlnorm` function, to confirm your answer to (b).

10. WebNet, a large technology company, owns many websites. The monthly traffic T for any given website (i.e., the number of unique visitors to the website) is a log-normal random variable with $\ln(T) \sim N(10, 4)$.

   a. What is the population median of T?

   b. What is the probability that a given website has monthly traffic greater than 20,000?

   c. Assume that WebNet earns two cents for every unique visitor to any of its websites. Fill in the blank in the following sentence: "There is a 90% probability that WebNet's monthly earnings for a given website is greater than ____________ dollars."

11. A homeowner has an outdoor light that is always kept on. After replacing the light bulb, suppose that the new bulb's life X (in years) is drawn from an exponential random variable with mean 0.8.

   a. What is the probability that the bulb lasts at least one year?

   b. What is the probability that the bulb lasts between one year and two years?

   c. *Suppose the homeowner immediately replaces a broken bulb with a new bulb, whose life is a new i.i.d. draw of X. Conduct 100,000 simulations in *R*, using exponential random variables, to approximate the pmf of the total number of bulbs B needed to keep the light illuminated for at least two years.

i. What are the approximate values of $P(B=1)$ and $P(B=2)$?

ii. What is the approximate value of $E(B)$?

iii. Based on proposition 11.9, how is B related to a Poisson random variable?

12. A company's customer service department takes calls throughout the day. After a given customer calls, the time (in minutes) before the next customer calls is an i.i.d. exponential random variable with mean 4. The length of any given call (in minutes) is also an exponential random variable, but it has mean 3 and is independent of the length of other calls and the arrival times of all calls. Suppose customer A calls at exactly 3 p.m. and the next two calls are by customer B and customer C (in that order).

   a. What is the probability that customer B calls before 3:05 p.m.?

   b. What is the probability that customer A's call ends after 3:05 p.m.?

   c. Conduct 100,000 simulations in R to approximate the following probabilities:

      i. The probability that customer B calls before customer A's call ends

      ii. The probability that both customer B and customer C call before customer A's call ends

      iii. The probability that customer B's call is still ongoing when *both* customer A's and customer C's calls end

13. *A consumer wants to purchase a product, and she knows that the product costs p_1 dollars at a major internet retailer. She has to decide whether to spend time searching the internet for a lower price. Suppose that a lower price $p_2 < p_1$ is available at another internet retailer, but it takes some time T (in minutes) to find that retailer and price. Assume that T is an exponential random variable, with $T \sim Exp(\theta)$, and the consumer's opportunity cost of time is c dollars per minute.

   a. What is the consumer's net gain associated with finding the lower price (in terms of p_1, p_2, T, and c)?

   b. The consumer wants to search for the lower price only if the *expected* net gain is positive. What must be true about c for the consumer to search? Your answer should be an inequality in terms of p_1, p_2, and θ.

   c. Now suppose the consumer never searches for more than m minutes (since she knows the exponential random variable has a long right tail). So, if she finds the lower price p_2 within m minutes, she faces that price; otherwise, she faces the original price p_1. What must be true about c for the consumer to search? Your answer should be an inequality in terms of p_1, p_2, θ, and m. (Hint: Use the fact that $\int_a^b x\theta e^{-\theta x}dx = \left(-xe^{-\theta x} + \frac{e^{-\theta x}}{\theta}\right)\Big|_a^b$.)

14. Suppose X is an exponential random variable, with $X \sim Exp(\theta)$.

   a. What is the population median of X?

   b. Provide a 95% probability interval (L, U) for X, where $P(X < L) = P(X > U) = 0.025$.

15. This question is a modified version of exercise 10.10, but now using normal random variables instead of uniform random variables. A store's sales depend on whether it is a

weekday (Monday through Friday) or a weekend day (Saturday or Sunday). Specifically, sales (in thousands of dollars) are distributed as a normal random variable $N(2, 0.5^2)$ on a weekday and as a normal random variable $N(3.5, 0.75^2)$ on a weekend day. Let X denote the random variable associated with the store's sales on a randomly chosen day of the week (i.e., the probability associated with each day is 1/7), so X is a mixture of normal random variables.

a. What is $E(X)$?

b. Determine $P(X \geq 3)$ analytically, in terms of the standard normal cdf $\Phi(\cdot)$. Use the `pnorm` function in *R* to calculate the probability.

c. Create a vector with 1,000,000 simulated draws of X in *R*. When simulating these draws, store the vector with the indicator of whether it is a weekday or not, as that data will be needed for (e).

   i. Confirm your answers to (a) and (b) based on the simulated draws.

   ii. What is the approximate population median of X based on the simulated draws?

   iii. Find the approximate population 2.5% and 97.5% quantiles to construct an approximate 95% probability interval for X.

   iv. Draw the smoothed density associated with the simulated draws. Does the density curve appear to be normal?

d. Suppose you know that a given day's sales are between \$2,000 and \$3,000. Use Bayes' theorem to determine the probability that it is a weekday. (Hint: Let A denote the event that it is a weekday, let B denote the event that $2 \leq X \leq 3$, and determine $P(A|B)$.)

e. Use the simulated draws from (c) to confirm your answer to (d).

16. Let $X \in \{0, 1\}$ be an indicator of whether an individual is female, with $X = 1$ for women and $X = 0$ for men. Height Y (in inches) in a certain population is distributed as a $N(64.5, 6.25)$ random variable for women and as a $N(70, 9)$ random variable for men. Assume $P(X = 1) = 0.5$.

a. What is $E(Y)$?

b. Create a vector with 1,000,000 simulated draws of Y in *R*.

   i. Confirm your answer to (a) based on the simulated draws.

   ii. What is the approximate population variance of Y?

   iii. What is the approximate population IQR of Y?

   iv. What is the approximate probability that Y is between 70 and 75? How does this probability compare to the conditional probabilities $P(70 \leq Y \leq 75|X = 0)$ and $P(70 \leq Y \leq 75|X = 1)$? (For each conditional probability, use the appropriate normal distribution to calculate the actual probability rather than using the simulations.)

c. *Find σ_Y^2 analytically, using the fact that

$$\sigma_Y^2 = E((Y - \mu_Y)^2) = E((Y - \mu_Y)^2|X = 0)P(X = 0) + E((Y - \mu_Y)^2|X = 1)P(X = 1).$$

12

Sampling Distributions: Exact

A central goal of statistics is to use an observed sample to say something about how the data are truly generated, sometimes called the **data-generating process (DGP)**. In other words, the observed sample characterizes the population from which the sample was drawn. To simplify matters, this chapter focuses on the case of a simple random sample. Recall from chapter 5 that a simple random sample is one for which each element of the population is equally likely to be sampled. To formalize matters, let's assume that the observed random sample $\{x_1, x_2, \ldots, x_n\}$ are the realizations of a collection of n independent and identically distributed (i.i.d.) random variables $\{X_1, X_2, \ldots, X_n\}$. That is, x_1 is the realized outcome of X_1, x_2 is the realized outcome of X_2, and so on. Since the random variables $\{X_1, X_2, \ldots, X_n\}$ are i.i.d., they share a common cdf, $F_X(\cdot)$. The goal is to use the sample $\{x_1, x_2, \ldots, x_n\}$ to characterize the distribution $F_X(\cdot)$ associated with the population.

Example 12.1 (Math SAT scores) *Let X be the random variable associated with math SAT score in a population of students. With an observed random sample, what can be said about the population mean μ_X of math SAT scores? Suppose that a random sample of 10 students is collected, with sample mean $\bar{x} = \frac{1}{10}\sum_{i=1}^{10} x_i = 618$. Intuitively, it seems like $\bar{x} = 618$ should be a good guess for the population mean μ_X, but how good is it? How close is the sample mean $\bar{x} = 618$ to μ_X? The precision of $\bar{x} = 618$ as a guess for μ_X depends on the variability of the sample mean itself. If a different set of 10 students had been collected from the population, would their sample mean also be close to 618? How about for yet another randomly chosen set of 10 students? After all, the observed $\{x_1, x_2, \ldots, x_{10}\}$ sample is just one possible realization of the random variables $\{X_1, X_2, \ldots, X_{10}\}$, and therefore the descriptive statistic $\bar{x} = \frac{1}{10}\sum_{i=1}^{10} x_i$ is just one possible realization of the sample mean $\frac{1}{10}\sum_{i=1}^{10} X_i$.*

Figure 12.1 provides a visual representation of how we can think about drawing our random sample of 10 students. There are many ways of drawing a random sample of 10 observations, and only one of those samples, represented by the gray shading, is observed. For the sample $\{x_1, x_2, \ldots, x_{10}\}$, descriptive statistics like the sample mean $\bar{x}$ and the sample variance s_x^2 can be calculated. Had one of the other samples been observed, a different sample mean and sample variance would have been obtained. We are interested in characterizing the distribution of the realizations of $\bar{x}$ (or s_x^2) over the possible random samples.

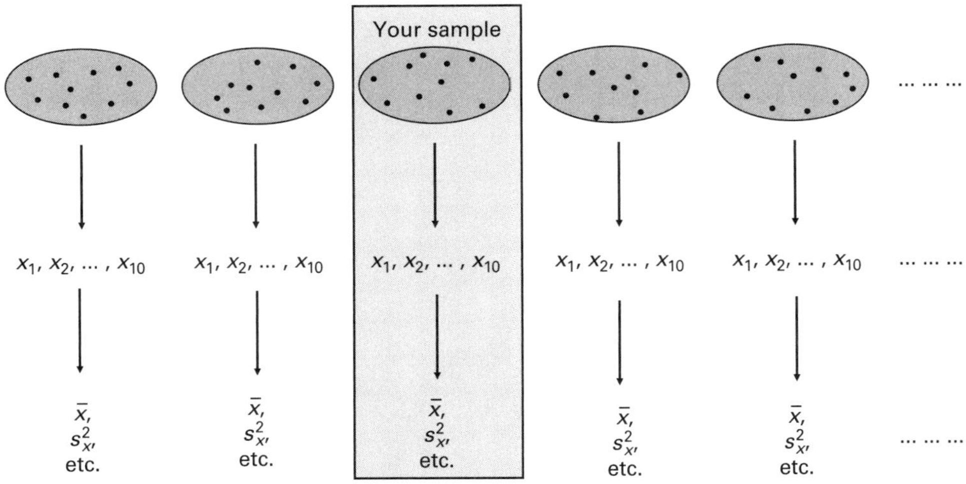

Figure 12.1
Random sampling and sampling distributions

In definition 6.1, a statistic $s(x_1, x_2, \ldots, x_n)$ was defined as a function of the observed sample data. The sample mean $\bar{x}$ and the sample variance s_x^2 are examples of statistics. For considering the distribution of a statistic over the possible random samples that can be drawn, the random variable $s(X_1, X_2, \ldots, X_n)$ is introduced. The random variable $s(X_1, X_2, \ldots, X_n)$ involves the same function $s(\cdot)$ applied to the random variables $X_1, X_2, \ldots, X_n$ rather than the observed variables $x_1, x_2, \ldots, x_n$. In the case of the sample mean,

$$s(x_1, x_2, \ldots, x_n) = \bar{x} = \frac{1}{n}\sum_{i=1}^{n} x_i$$

is the sample mean for a sample $\{x_1, x_2, \ldots, x_n\}$, and

$$s(X_1, X_2, \ldots, X_n) = \bar{X} = \frac{1}{n}\sum_{i=1}^{n} X_i$$

is the random variable associated with the sample mean. Before observing the data, $\bar{X}$ is itself a random variable since it depends on the random variables $X_1, X_2, \ldots, X_n$. After observing the data, x_1 is the realization of X_1, x_2 is the realization of X_2, and so on through x_n, and $\bar{x}$ is the realization of the random variable $\bar{X}$.

Similarly, in the case of the sample variance,

$$s(x_1, x_2, \ldots, x_n) = s_x^2 = \frac{1}{n-1}\sum_{i=1}^{n} (x_i - \bar{x})^2$$

and

$$s(X_1, X_2, \ldots, X_n) = s_X^2 = \frac{1}{n-1}\sum_{i=1}^{n} (X_i - \bar{X})^2.$$

Before observing the data, s_X^2 is itself a random variable. The subscript X indicates that s_X^2 is a random variable. After observing the data, the sample variance s_x^2 is the realization of the random variable s_X^2.

The distribution of a statistic over the possible random samples is known as the **sampling distribution** and is formally defined as in definition 12.1.

Definition 12.1 *The* ***sampling distribution*** *of a statistic $s(X_1, X_2, \ldots, X_n)$ is the probability distribution of the statistic over all possible random samples of size n from the population.*

For a given sample size n, the sampling distribution of a statistic is also sometimes called the **exact sampling distribution** or the **finite-sample distribution**. This chapter focuses on some examples where the exact sampling distribution of a statistic can be determined based on the specific form of the underlying random variables $X_1, X_2, \ldots, X_n$. For example, in the case of i.i.d. Bernoulli random variables or i.i.d. normal random variables, the exact sampling distribution of the sample mean for a random sample of size n can be determined. For i.i.d. normal random variables, the exact sampling distribution of the sample variance for a random sample of size n can also be determined. In more general cases, however, it can be difficult to characterize the exact sampling distribution of a statistic, even for a simple statistic like the sample mean. As it turns out, much more can be said when the sample size n is large, as more general results are available to provide an approximate sampling distribution, rather than an exact sampling distribution, for a wide range of statistics. This idea is the focus of chapter 13, which considers large-sample or asymptotic distributions for statistics.

12.1 Sampling Distribution of the Sample Mean

If the distribution of the i.i.d. random variables $X_1, X_2, \ldots, X_n$ is known, it is sometimes possible to explicitly determine the sampling distribution of $\bar{X} = \frac{1}{n}\sum_{i=1}^{n} X_i$. This section considers two such cases, Bernoulli random variables (section 12.1.1) and normal random variables (section 12.1.2), before discussing other cases that do not have a simple description of the sampling distribution of $\bar{X}$ (section 12.1.3).

12.1.1 Bernoulli Random Variables

Example 12.2 (Customer purchases) *Let a customer purchase be represented by the binary variable X (1 if purchase, 0 if not). A random sample of 10 customers from the population is observed, where the purchase behavior of each customer is an i.i.d. Bernoulli(π) random variable. In the notation introduced here, with $X_1, X_2, \ldots, X_{10}$ as i.i.d. Bernoulli(π) random variables, what is the distribution of $\bar{X} = \frac{1}{10}(X_1 + X_2 + \cdots + X_{10})$? First, note that the sum $X_1 + X_2 + \cdots + X_{10}$ is a sum of 10 i.i.d. Bernoulli(π) random variables, which is a Binomial($10, \pi$) random variable. Then, $\bar{X}$ is $\frac{1}{10}$ times a Binomial($10, \pi$) random variable. ($\bar{X}$ is the random variable representing the sample proportion, as detailed in section 9.2.1.) Using the pmf formula for binomial random variables, the pmf for $\bar{X}$ is*

p	$P(\bar{X} = p)$
0	$(1-\pi)^{10}$
1/10	$\binom{10}{1}\pi(1-\pi)^9$
2/10	$\binom{10}{2}\pi^2(1-\pi)^8$
3/10	$\binom{10}{3}\pi^3(1-\pi)^7$
4/10	$\binom{10}{4}\pi^4(1-\pi)^6$
5/10	$\binom{10}{5}\pi^5(1-\pi)^5$
6/10	$\binom{10}{6}\pi^6(1-\pi)^4$
7/10	$\binom{10}{7}\pi^7(1-\pi)^3$
8/10	$\binom{10}{8}\pi^8(1-\pi)^2$
9/10	$\binom{10}{9}\pi^9(1-\pi)$
1	π^{10}

For instance, $P(\bar{X} = 3/10)$ *is the probability that* 3 *of the* 10 *customers make a purchase, equal to* $\binom{10}{3}\pi^3(1-\pi)^7$. *Put another way, given that the true probability of purchase is* π, *the probability that a random sample of* 10 *customers has exactly* 3 *purchases is* $\binom{10}{3}\pi^3(1-\pi)^7$. *Since the population mean and variance of the Binomial*$(10, \pi)$ *are* 10π *and* $10\pi(1-\pi)$, *respectively, the population mean and variance of* $\bar{X}$ *are* π *and* $\frac{\pi(1-\pi)}{10}$, *respectively.*

The results of example 12.2 can be generalized to any sample size n, and proposition 12.1 summarizes the results for the sampling distribution of $\bar{X}$ for n i.i.d. Bernoulli random variables.

Proposition 12.1 *If* $X_1, X_2, \ldots, X_n$ *are i.i.d. Bernoulli*(π) *random variables, the sampling distribution of* $\bar{X}$ *is*

$$P\left(\bar{X} = \frac{j}{n}\right) = \binom{n}{j}\pi^j(1-\pi)^{n-j} \text{ for } j \in \{0, 1, 2, \ldots, n\}.$$

The population mean of $\bar{X}$ *is*

$$\mu_{\bar{X}} = E(\bar{X}) = \pi,$$

the population variance of $\bar{X}$ *is*

$$\sigma^2_{\bar{X}} = Var(\bar{X}) = \frac{\pi(1-\pi)}{n},$$

and the population standard deviation of $\bar{X}$ *is*

$$\sigma_{\bar{X}} = sd(\bar{X}) = \sqrt{\frac{\pi(1-\pi)}{n}}.$$

To illustrate the results of proposition 12.1, figure 12.2 shows the sampling distributions of $\bar{X}$ for $\pi = 0.2$ with four sample sizes ($n = 2$, $n = 4$, $n = 10$, and $n = 20$). The y-axis label is $p_{\bar{X}}(v)$, denoting the pmf of the random variable $\bar{X}$. For $n = 2$, the most likely outcome for $\bar{X}$ is 0, and of course, it's impossible to observe an outcome of 0.2 since the only possible outcomes are 0, 0.5, and 1. For $n = 4$, the most likely outcomes are 0 and 0.25, each with probability 0.4096. For $n = 10$, the most likely outcome is the true probability 0.2, with probability just over 0.3, and the bulk of the distribution is between 0 and 0.4. For $n = 20$, the most likely outcome is again the true probability 0.2, but now the next most likely outcomes are 0.15 and 0.25, neither of which was possible with the smaller $n = 10$. Overall, the dispersion of the sampling distribution decreases as the sample size increases. Comparing the bottom two distributions ($n = 10$ and $n = 20$), for instance, the larger sample size of $n = 20$ has a sampling distribution that is more tightly distributed around the true probability of $\pi = 0.2$. Put another way, the variability of the sample mean as a descriptive statistic decreases as the sample size increases. As will be seen later, this relationship implies that an observed sample mean is a more reliable guess or estimate of the population mean when the sample size is larger. In the context of the customer-purchase example (example 12.2), a more accurate guess for the purchase probability π would be expected if 20 customers are observed rather than 10 customers.

12.1.2 Normal Random Variables

Example 12.3 (Firm sales) *Suppose firm sales in a given year is a normal random variable X. Ten years of data on firm sales are observed, under the assumption that the sales in each year is an i.i.d. $N(\mu, \sigma^2)$ random variable. With the parameters μ and σ^2 unknown, can the observed data say something about their possible values? Before the sales are observed, the potential annual sales values can be thought of as random variables $X_1, X_2, \ldots, X_{10}$, which are i.i.d. $N(\mu, \sigma^2)$ random variables. Then, from proposition 11.6, the average sales over the 10 years, $\bar{X} = \frac{1}{10}(X_1 + X_2 + \cdots + X_{10})$, is also normally distributed with*

$$\bar{X} \sim N\left(\mu, \frac{\sigma^2}{10}\right) \text{ and } \sigma_{\bar{X}} = \frac{\sigma}{\sqrt{10}}.$$

Thus, the observed sample mean $\bar{x}$, the average of sales over the 10 years of data, is a single draw from a normal random variable with mean μ and standard deviation $\sigma/\sqrt{10}$. From the properties of normal random variables, the observed sample mean $\bar{x}$ is within $1.96\frac{\sigma}{\sqrt{10}}$ of the population mean μ with probability 95% and within $1.645\frac{\sigma}{\sqrt{10}}$ of the population mean μ with probability 90%.

The results of example 12.3 can be generalized to any sample size n, and the following proposition summarizes the results for the sampling distribution of $\bar{X}$ for n i.i.d. normal random variables:

Proposition 12.2 *If $X_1, X_2, \ldots, X_n$ are i.i.d. $N(\mu, \sigma^2)$ random variables, the sampling distribution of $\bar{X}$ is*

$$\bar{X} \sim N\left(\mu, \frac{\sigma^2}{n}\right).$$

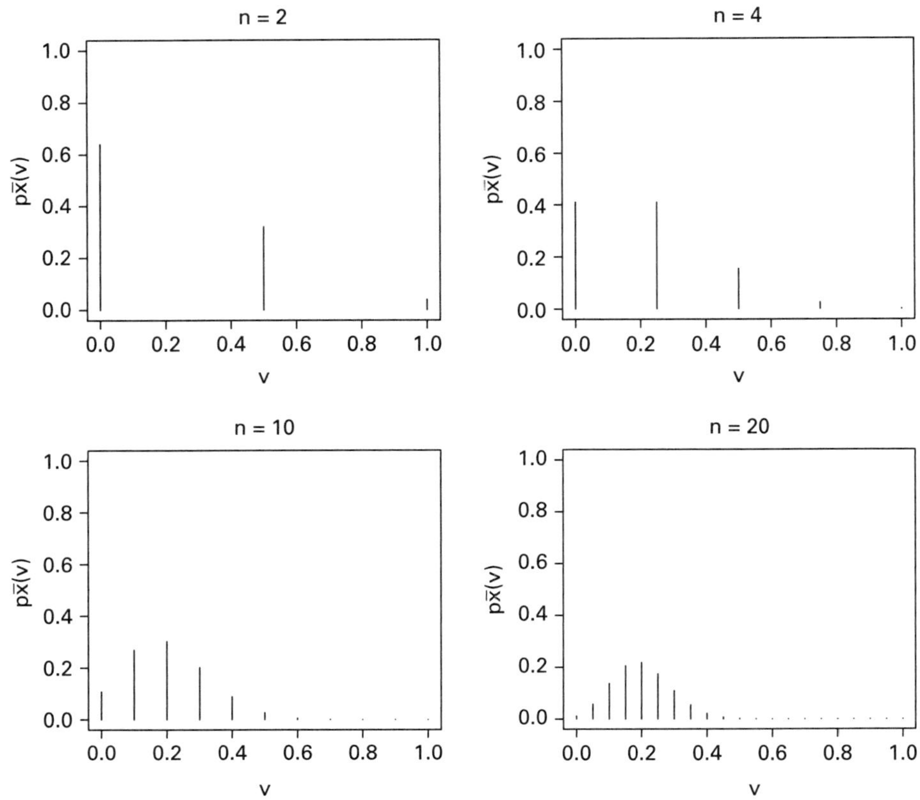

Figure 12.2
Sampling distributions of the sample mean of i.i.d. Bernoulli(0.2) random variables

The population mean of $\bar{X}$ *is*

$$\mu_{\bar{X}} = E(\bar{X}) = \mu,$$

the population variance of $\bar{X}$ *is*

$$\sigma^2_{\bar{X}} = Var(\bar{X}) = \frac{\sigma^2}{n},$$

and the population standard deviation of $\bar{X}$ *is*

$$\sigma_{\bar{X}} = sd(\bar{X}) = \frac{\sigma}{\sqrt{n}}.$$

The sampling distribution of $\bar{X}$, which is that of a normal random variable, is centered around the population mean μ. The variance, denoted as $\sigma^2_{\bar{X}}$, and the standard deviation, denoted as $\sigma_{\bar{X}}$, both depend on the sample size, with the variance proportional to $\frac{1}{n}$ and the standard deviation proportional to $\frac{1}{\sqrt{n}}$. As the sample size n increases, the dispersion of the sampling distribution of $\bar{X}$ decreases. For instance, comparing the sampling distribution for two sample sizes, say $n = 100$ and $n = 200$, the variances are $\frac{\sigma^2}{100}$ and $\frac{\sigma^2}{200}$, respectively, so the variance with $n = 200$ is half as large as the variance with $n = 100$. Similarly, for these sample sizes, the standard deviations are $\frac{\sigma}{\sqrt{100}}$ and $\frac{\sigma}{\sqrt{200}}$, respectively, so the standard deviation with $n=200$ is $\frac{1}{\sqrt{2}}$ times the standard deviation with $n=100$. More generally, doubling the sample size always results in a normal sampling distribution with the same mean μ, a variance that is half as large, and a standard deviation that is scaled by $\frac{1}{\sqrt{2}}$. Quadrupling the sample size results in a normal sampling distribution with the same mean, a variance that is $\frac{1}{4}$ as large, and a standard deviation that is $\frac{1}{2}$ as large. In fact, if the sample size n grows arbitrarily large ($n \to \infty$), the variance $\sigma^2_{\bar{X}}$ and standard deviation $\sigma_{\bar{X}}$ shrink toward zero.

12.1.3 Other Random Variables

Beyond the cases of Bernoulli and normal random variables, it may not be easy to provide a simple characterization of the sampling distribution of the sample mean $\bar{X}$. The Bernoulli and normal random variables are particularly convenient since the distribution of the average of i.i.d. draws is known, specifically a scaled binomial for the former and a normal for the latter. For other types of i.i.d. random variables, the distribution of the average of i.i.d. draws may be some other distribution not covered in this book. There are many distribution families that are used by researchers and practitioners. In some cases, it might be possible to analytically determine the exact form of the sampling distribution of $\bar{X}$, but in other cases, it might be infeasible to do so.

There are some results for the sampling distribution of $\bar{X}$ that hold regardless of the distribution of the underlying i.i.d. random variables $X_1, X_2, \ldots, X_n$. From proposition 10.13, the mean, variance, and standard deviation of $\bar{X}$ are

$$\mu_{\bar{X}} = \mu_X,\ \sigma^2_{\bar{X}} = \frac{\sigma^2_X}{n},\ \text{and } \sigma_{\bar{X}} = \frac{\sigma_X}{\sqrt{n}},$$

where μ_X, σ^2_X, and σ_X are the mean, variance, and standard deviation of X, respectively. These properties were shown previously for both the Bernoulli and the normal, but they hold for any distribution of X. Specifically, the mean of $\bar{X}$ is the mean of X, and the variance and standard deviation of $\bar{X}$ are scaled versions of the variance and standard deviation of X, with the variance scaled by $\frac{1}{n}$ and the standard deviation scaled by $\frac{1}{\sqrt{n}}$. Thus, the dispersion of the sampling distribution decreases with sample size in the same way as for the Bernoulli and normal cases. Unfortunately, knowing the mean and variance of $\bar{X}$ does not completely characterize the sampling distribution of $\bar{X}$, as the actual distribution (pmf, probability density function (pdf), and/or cumulative distribution function (cdf)) of X must be used to derive the distribution of $\bar{X}$.

Example 12.4 (Uniform random variables) *Suppose that $X_1, X_2, \ldots, X_n$ are i.i.d. $U(0,1)$ random variables. Recall that $X \sim U(0,1)$ has $\mu_X = \frac{1}{2}$, $\sigma^2_X = \frac{1}{12}$, and $\sigma_X = \frac{1}{\sqrt{12}}$, so*

$$\mu_{\bar{X}} = \mu_X = \frac{1}{2},\ \sigma^2_{\bar{X}} = \frac{\sigma^2_X}{n} = \frac{1}{12n},\ \textit{and } \sigma_{\bar{X}} = \frac{\sigma_X}{\sqrt{n}} = \frac{1}{\sqrt{12n}}.$$

The sampling distribution of $\bar{X}$ is not *uniform even though each X_i is uniform. For example, as seen in example 10.28 for $n=2$, the sum X_1+X_2 is a triangular distribution, meaning that $\frac{1}{2}(X_1+X_2)$ also has the shape of a triangular distribution. For larger n, the distribution*

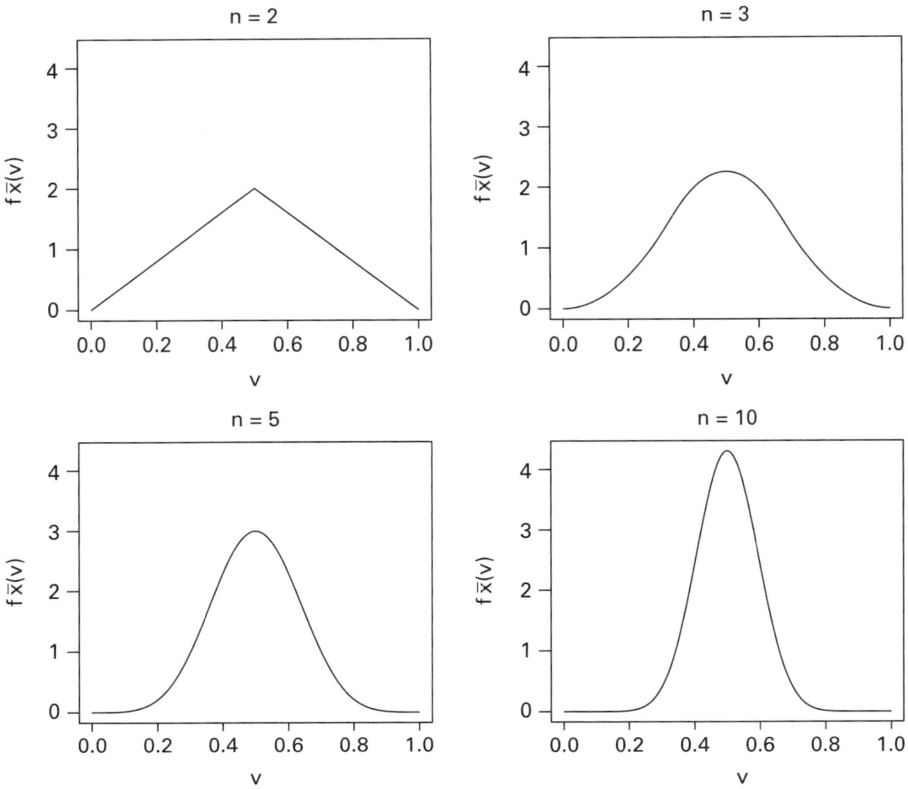

Figure 12.3
Sampling distributions of the sample mean of i.i.d. $U(0, 1)$ random variables

of $\bar{X}$ becomes smoother.[1] *Figure 12.3 shows the exact sampling distributions for $\bar{X}$ for four sample sizes ($n = 2$, $n = 3$, $n = 5$, and $n = 10$). The y-axis label is $f_{\bar{X}}(v)$, corresponding to the pdf of the random variable $\bar{X}$. The range of $\bar{X}$ is always between 0 and 1, regardless of n, since the average can never be less than the minimum value of X or greater than the maximum value of X. Each distribution has a mean at $\frac{1}{2}$, and the standard deviations are approximately* 0.204, 0.167, 0.129, *and* 0.091, *respectively.*

Example 12.5 (Log-normal random variables) *Suppose that $X_1, X_2, \ldots, X_n$ are i.i.d. log-normal random variables, with $\ln(X_i) \sim N(\mu, \sigma^2)$ for each $i \in \{1, 2, \ldots, n\}$. The sum of log-normal random variables does* not *have a log-normal distribution, and therefore the average of log-normal random variables does not have a log-normal distribution.*[2] *While there are no general results regarding the exact sampling distribution of $\bar{X}$ for i.i.d. log-normal random variables, computer simulation can approximate the sampling distribution for specific values of n, μ, and σ^2. For a log-normal distribution with $\ln(X) \sim N(0, 1)$ ($\mu = 0$, $\sigma^2 = \sigma = 1$), figure 12.4 shows the simulated sampling distributions for four sample sizes ($n = 2$, $n = 5$, $n = 10$, and $n = 20$). The assumed data-generating process is used to make simulated draws. For instance, for $n = 2$, X_1 and X_2 are drawn randomly and independently from a log-normal distribution with $\ln(X) \sim N(0, 1)$ and the average of the two draws is calculated,* *and this process is repeated. For the graphs shown in the figure,* 100,000 *simulations are used. The top-left graph shows the smoothed density associated with* 100,000 *draws of* $\bar{X} = \frac{1}{2}(X_1 + X_2)$ *for* $n = 2$*. A similar process is used for the other sample sizes. For* $n = 20$*,* $X_1, X_2, \ldots, X_{20}$ *are randomly and independently drawn from a log-normal distribution with* $\ln(X) \sim N(0, 1)$*, which gives simulated draws of* $\bar{X} = \frac{1}{20}(X_1 + X_2 + \cdots + X_{20})$*, and the lower-right graph shows the smoothed density associated with* 100,000 *draws of* $\bar{X}$*. Comparing across sample sizes, the dispersion in the distributions decreases as the sample size increases as expected. Also, the right skewness characteristic of the log-normal distribution is evident at the smaller sample sizes (*$n = 2$ *and* $n = 5$*), but it is less dramatic for* $n = 10$ *and has nearly disappeared for* $n = 20$*. In fact, the distribution of* $\bar{X}$ *for* $n = 20$ *looks more like a normal distribution than it does a log-normal distribution.*

The following R code approximates the sampling distributions for figure 12.4:

```
set.seed(1234)

# initialize the number of simulations
num_simulations <- 100000

# create a vector of sample averages for n=2
xmean_2 <- replicate(num_simulations, mean(rlnorm(2, meanlog=0, sdlog=1)))

# create a vector of sample averages for n=5
xmean_5 <- replicate(num_simulations, mean(rlnorm(5, meanlog=0, sdlog=1)))

# create a vector of sample averages for n=10
xmean_10 <- replicate(num_simulations, mean(rlnorm(10, meanlog=0, sdlog=1)))

# create a vector of sample averages for n=20
xmean_20 <- replicate(num_simulations, mean(rlnorm(20, meanlog=0, sdlog=1)))

# graph-display format (two rows, two columns)
par(mfrow=c(2,2))

# plot smoothed densities for all four sample sizes
plot(density(xmean_2), main=expression(n==2), xlim=c(0,8),
    xlab=expression(v), ylab=expression(f[bar(X)](v)))
plot(density(xmean_5), main=expression(n==5), xlim=c(0,8),
    xlab=expression(v), ylab=expression(f[bar(X)](v)))
plot(density(xmean_10), main=expression(n==10), xlim=c(0,8),
    xlab=expression(v), ylab=expression(f[bar(X)](v)))
plot(density(xmean_20), main=expression(n==20), xlim=c(0,8),
    xlab=expression(v), ylab=expression(f[bar(X)](v)))
```

The code uses the function `replicate`, which is useful for simulations involving random variables:

- `replicate(n, expr)`: Returns a vector of length `n` containing the results of evaluating the expression `expr` a total of `n` times.

The first use of `replicate` in the code creates the vector `xmean_2`, with 100,000 evaluations of the expression `mean(rlnorm(2, meanlog=0, sdlog=1))`, which calculates the sample mean over 2 draws of the specified log-normal distribution. The other three `replicate` expressions are similar, but with different sample sizes specified.

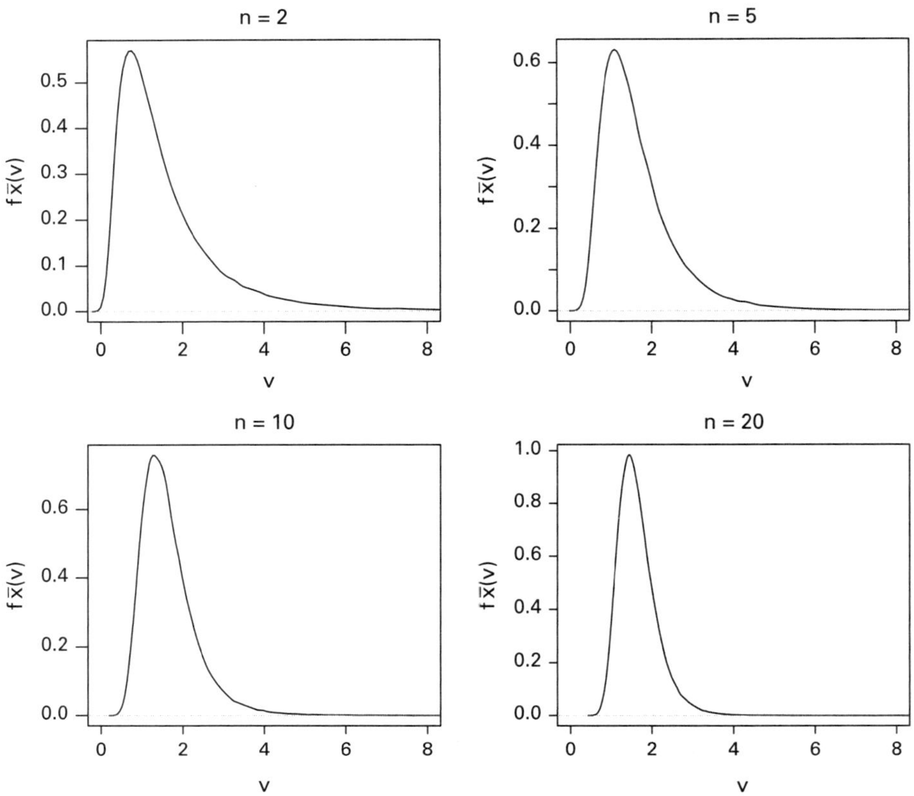

Figure 12.4
Sampling distributions of the sample mean of log-normal random variables

The simulation approach in example 12.5 is extremely general and can be used to approximate many different sampling distributions. As long as (1) the distribution of the underlying i.i.d. random variables is known and (2) a computer can simulate random draws from that distribution, simulation can always approximate the sampling distribution of $\bar{X}$ for a given sample size n. Moreover, since there is nothing special about the statistic $\bar{X}$, this approach can also be used to approximate the sampling distribution of other statistics. For example, to approximate the sampling distribution of the sample variance s_X^2, the only step that changes is how the statistic is calculated after the draws of $X_1, X_2, \ldots, X_n$. For $\bar{X}$, the value of the draw of $\bar{X} = \frac{1}{n}\sum_{i=1}^n X_i$ is used, whereas for s_X^2, the value of the draw of $s_X^2 = \frac{1}{n-1}\sum_{i=1}^n (X_i - \bar{X})^2$ is used. The approach can also be used for other statistics, like sample quantiles (including the sample median), sample interquartile range (IQR), and others.

12.2 Sampling Distribution of the Sample Variance

Like the sample average, the sample variance should be considered a random variable (s_X^2) before the sample is observed. After the sample is observed, from realizations of $X_1, X_2, \ldots, X_n$, the realization of the sample variance s_X^2 is the observed sample variance

s_X^2. Likewise, for the random variable s_X associated with the sample standard deviation, the realization of s_X is the sample standard deviation s_x of the observed sample. For the random variables $X_1, X_2, \ldots, X_n$, the random variables associated with the sample variance and the sample standard deviation are, respectively,

$$s_X^2 = \frac{1}{n-1}\sum_{i=1}^{n}(X_i - \bar{X})^2 \text{ and } s_X = \sqrt{s_X^2} = \sqrt{\frac{1}{n-1}\sum_{i=1}^{n}(X_i - \bar{X})^2}.$$

What can be said about the sampling distribution of s_X^2? If the mean and variance of the random variable X are known, even if the distribution of X is not fully known, proposition 12.3 provides results for the population statistics of the sample variance s_X^2.

Proposition 12.3 *If $X_1, X_2, \ldots, X_n$ are i.i.d. random variables with population mean μ_X and population variance σ_X^2, the population mean of s_X^2 is*

$$\mu_{s_X^2} = E(s_X^2) = \sigma_X^2,$$

the population variance of s_X^2 is

$$\sigma_{s_X^2}^2 = Var(s_X^2) = \frac{E\left[(X - \mu_X)^4\right]}{n} - \frac{\left(\sigma_X^2\right)^2 (n-3)}{n(n-1)},$$

and the population standard deviation of s_X^2 is

$$\sigma_{s_X^2} = sd(s_X^2) = \sqrt{\frac{E\left[(X - \mu_X)^4\right]}{n} - \frac{\left(\sigma_X^2\right)^2 (n-3)}{n(n-1)}}.$$

We make a few remarks about the results here. First, the mean of the random variable s_X^2 is the population variance σ_X^2. The $\frac{1}{n-1}$ scaling in the sample variance formula is crucial for this result to hold, as it would not hold if the scaling were $\frac{1}{n}$ instead of $\frac{1}{n-1}$. With this scaling, the result says that the possible realizations of s_X^2 are, on average, equal to σ_X^2. Second, although they take a somewhat complicated form, the population variance and standard deviation are both decreasing in the sample size n, meaning that the variability of the sample variance decreases for larger sample sizes. It may seem strange to think about the "variance of a variance," but that's exactly what we have here. The sample variance random variable s_X^2 is just an example of a statistic $s(X_1, X_2, \ldots, X_n)$, and its variance $\sigma_{s_X^2}^2 = Var(s_X^2)$ provides a measure of the dispersion of the sampling distribution of s_X^2. If the sample variance varies a lot depending on which sample happens to be observed, this would be associated with a large value of $\sigma_{s_X^2}^2 = Var(s_X^2)$, whereas there would be a lower value of $\sigma_{s_X^2}^2 = Var(s_X^2)$ if the sample variance does not vary a lot depending on the sample that happens to be observed. If the sample size n grows arbitrarily large ($n \to \infty$), the variance $\sigma_{s_X^2}^2$ and standard deviation $\sigma_{s_X^2}$ shrink toward zero.

12.2.1 Normal Random Variables

As with the sample average $\bar{X}$, much more can be said about the sampling distribution of the sample variance s_X^2 when i.i.d. random variables are normally distributed.

Proposition 12.4 *If $X_1, X_2, \ldots, X_n$ are i.i.d. $N(\mu, \sigma^2)$ random variables, the sampling distribution of s_X^2 is characterized by*

$$(n-1)\frac{s_X^2}{\sigma^2} \sim \chi_{n-1}^2,$$

where χ_{n-1}^2 is a chi-square distribution with $n-1$ degrees of freedom.

The chi-square random variable was introduced in section 11.3. For known sample size n and known variance σ^2, proposition 12.4 says that the sample variance s_X^2 is a scaled version of a χ_{n-1}^2 random variable, specifically a χ_{n-1}^2 random variable multiplied by $\frac{\sigma^2}{n-1}$. Thus, the exact sampling distribution for s_X^2 is completely characterized in the specific case of i.i.d. normal random variables.[3] Knowledge of the exact sampling distribution of s_X^2 allows probability intervals to be calculated for the sample variance or the sample standard deviation when the parameters of the underlying normal distribution are known.

Example 12.6 (Asset returns) *Suppose that the annual returns of an asset are i.i.d. normal random variables with mean 0.08 and standard deviation 0.06. Based upon an observed sample of 10 years of annual returns ($n = 10$), the sample variance s_x^2 and sample standard deviation s_x are calculated. Before observing the data or calculating those descriptive statistics, a probability interval can be determined for the random variable s_X^2 or the random variable s_X. Let's say that we are interested in 95% probability intervals, that is intervals for which there is a 95% chance that the calculated descriptive statistic falls in the specified interval. Starting with the sample variance,*

$$(n-1)\frac{s_X^2}{\sigma^2} \sim \chi_{n-1}^2 \implies 9\frac{s_X^2}{(0.06)^2} \sim \chi_9^2 \implies 2500 s_X^2 \sim \chi_9^2.$$

We calculate the population 2.5% and 97.5% quantiles of the χ_9^2 distribution with the `qchisq` *function in R :*

```
qchisq(c(0.025,0.975),9)

## [1]  2.700389 19.022768
```

The 2.5% and 97.5% quantiles of the χ_9^2 distribution are 2.700 and 19.023, respectively, so there is a 95% probability of a χ_9^2 random variable being in the interval $(2.700, 19.023)$. Since $2500 s_X^2 \sim \chi_9^2$, the probability of s_X^2 being in the interval $(2.700/2500, 19.023/2500) \approx (0.0011, 0.0076)$ is also equal to 95%. There is a 2.5% probability that s_X^2 is less than 0.0011 and a 2.5% probability that s_X^2 is greater than 0.0076. With a probability interval for the sample variance s_X^2, we can construct a probability interval for the sample standard deviation s_X since $s_X = \sqrt{s_X^2}$ and $\sqrt{\cdot}$ is an increasing function. Specifically, taking the square root of each end point, the 95% probability interval for s_X is $(\sqrt{0.0011}, \sqrt{0.0076}) \approx (0.033, 0.087)$.

We can also calculate the probability that the sample variance or the sample standard deviation is in some prespecified interval. For instance, what is the probability that the sample standard deviation is less than 0.05? When a sample standard deviation is calculated, it will either be less than 0.05 or greater than 0.05 (with certainty), but what is the probability before *the sample is observed? This probability is*

$$P(s_X < 0.05) = P(s_X^2 < (0.05)^2) = P(2500s_X^2 < 2500(0.05)^2) = P(2500s_X^2 < 6.25),$$

and since $2500s_X^2 \sim \chi_9^2$, *this probability is the cdf of the* χ_9^2 *distribution evaluated at* 6.25*:*

```
pchisq(6.25,9)

## [1] 0.2853401
```

Therefore, $P(s_X < 0.05) \approx 0.285$, *meaning there is a* 28.5% *chance that the observed sample standard deviation will be less than* 0.05.

12.2.2 Other Random Variables

Beyond i.i.d. normal random variables, it's generally not easy to provide a simple characterization of the sampling distribution of the sample variance s_X^2.[4] As illustrated in section 12.1.3 for the sample average $\bar{X}$, however, simulation methods can approximate the sampling distribution associated with a given sample size n and a specified underlying distribution of the i.i.d. random variables $X_1, X_2, \ldots, X_n$.

Example 12.7 (Uniform random variables) *Suppose that* $X_1, X_2, \ldots, X_n$ *are i.i.d.* $U(0,1)$ *random variables, with population statistics* $\mu_X = \frac{1}{2}$, $\sigma_X^2 = \frac{1}{12}$, *and* $\sigma_X = \frac{1}{\sqrt{12}}$. *Proposition 12.3 implies that the population mean of* s_X^2 *is* $\sigma_X^2 = \frac{1}{12}$ *(i.e.,* $E(s_X^2) = \frac{1}{12} \approx 0.0833$*) and also that the population variance and standard deviation decrease for larger sample sizes. To see the actual shapes of the distributions for particular sample sizes, computer simulations can approximate the sampling distributions of* s_X^2. *Figure 12.5 shows the simulated sampling distributions for four sample sizes (*$n=2$, $n=3$, $n=5$, *and* $n=10$*), in each case based on* 100,000 *simulated calculations of the sample variance. For instance, with* $n=5$, *the process of drawing* X_1, X_2, X_3, X_4, X_5 *as i.i.d.* $U(0,1)$ *random variables and calculating their sample variance is repeated. The lower-left graph (*$n=5$*) shows the smoothed density of the* 100,000 *simulated values of the sample variance. The four graphs are drawn with the same x-axes and y-axes for ease of comparison. The label on the y-axis is* $f_{s_X^2}(v)$, *corresponding to the pdf of the random variable* s_X^2. *The average associated with each of the four graphs should be* $E(s_X^2) = \frac{1}{12} \approx 0.0833$. *For the very small sample sizes (*$n=2$ *and* $n=3$*), the distributions for* s_X^2 *are extremely right-skewed. Even with* $n=5$, *the distribution becomes much more symmetric and bell-shaped. For* $n=10$, *the approximately symmetric and bell-shaped distribution becomes tighter (i.e., less dispersion) compared to* $n=5$ *and appears to be centered near the mean value of* 0.0833.

The following R code creates figure 12.5:

```
set.seed(1234)

# initialize the number of simulations
num_simulations <- 100000

# create a vector of sample variances for n=2, n=3, n=5, and n=10
varunif_2 <- replicate(num_simulations, var(runif(2)))
varunif_3 <- replicate(num_simulations, var(runif(3)))
```

```
varunif_5 <- replicate(num_simulations, var(runif(5)))
varunif_10 <- replicate(num_simulations, var(runif(10)))

# graph-display format (two rows, two columns)
par(mfrow=c(2,2))

# plot smoothed densities for the four sample sizes
plot(density(varunif_2), xlim=c(0,0.5), ylim=c(0,15), main=expression(n==2),
    xlab=expression(v), ylab=expression(f[s[X]^2](v)))
plot(density(varunif_3), xlim=c(0,0.5), ylim=c(0,15), main=expression(n==3),
    xlab=expression(v), ylab=expression(f[s[X]^2](v)))
plot(density(varunif_5), xlim=c(0,0.5), ylim=c(0,15), main=expression(n==5),
    xlab=expression(v), ylab=expression(f[s[X]^2](v)))
plot(density(varunif_10), xlim=c(0,0.5), ylim=c(0,15), main=expression(n==10),
    xlab=expression(v), ylab=expression(f[s[X]^2](v)))
```

The `replicate` *function conducts the* 100,000 *simulations with a single command for each of the sample sizes. Using an almost identical approach, we can simulate sampling distributions for the sample standard deviation* s_X. *Rather than calculating the sample variance in each of the* 100,000 *simulations, we calculate the sample standard deviation (replacing the* `var` *function with the* `sd` *function in the code shown here) and graph the smoothed*

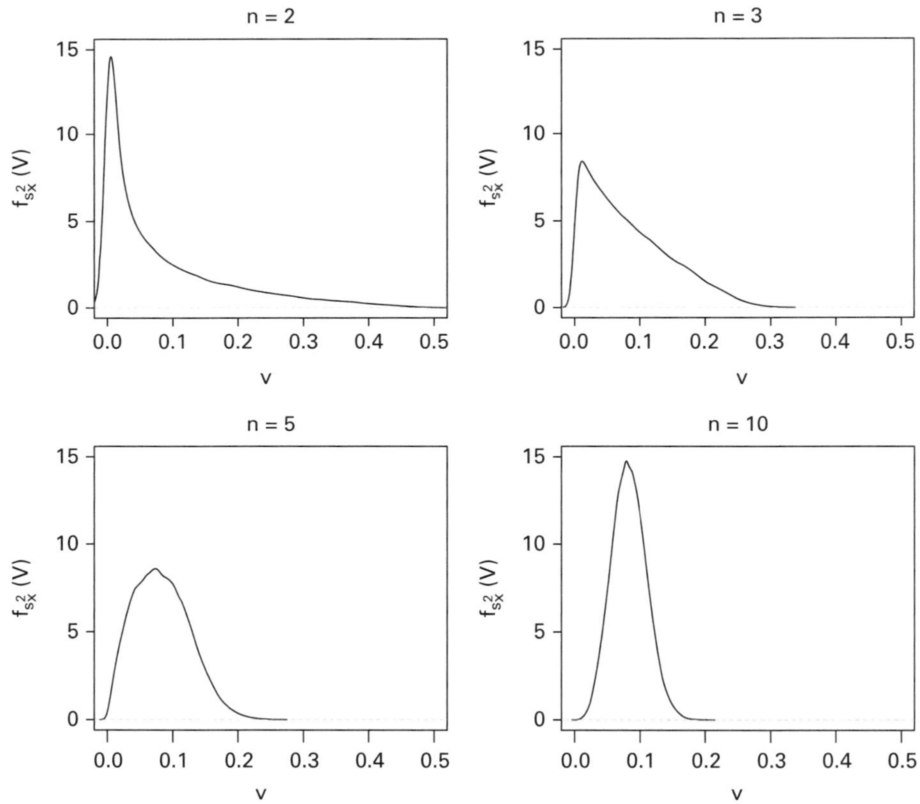

Figure 12.5
Sampling distributions of the sample variance for i.i.d. $U(0, 1)$ random variables

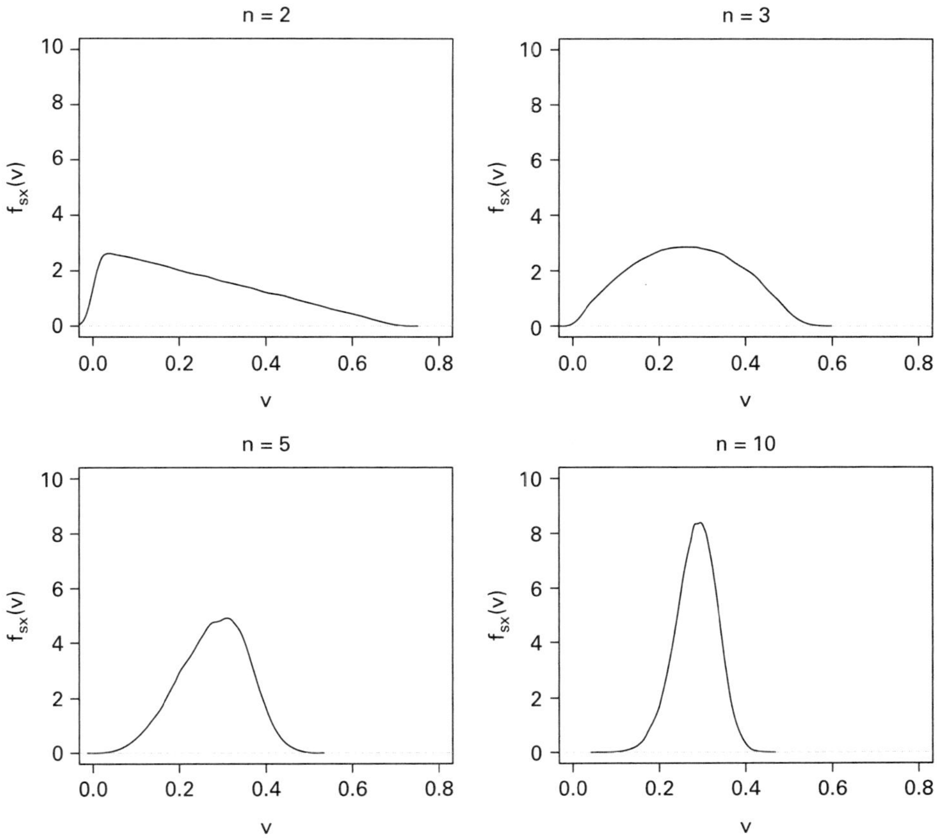

Figure 12.6
Sampling distributions of the sample standard deviation for i.i.d. $U(0, 1)$ random variables

densities over the 100,000 *simulated standard deviations for each sample size. Figure 12.6 shows these simulated distributions using the same four sample sizes as before. The y-axis label is $f_{s_X}(v)$, corresponding to the pdf of the random variable s_X. Again, the dispersion of the distributions decreases for the larger sample sizes, and symmetry of the sampling distribution approximately holds when $n = 10$.*

Example 12.8 (Log-normal random variables) *Suppose that $X_1, X_2, \ldots, X_n$ are i.i.d. log-normal random variables, with $\ln(X_i) \sim N(0, 1)$ for each $i \in \{1, 2, \ldots, n\}$. Figure 12.7 shows the simulated sampling distributions of the standard deviation s_X for $n = 10$ and $n = 30$, using* 100,000 *simulations for each sample size. Even with a sample size of $n = 30$ the sampling distribution of s_X is distinctly right-skewed. In fact, at first glance, the sampling distributions for $n = 10$ and $n = 30$ look quite similar. A closer look, especially for the lower values on the x-axis, reveals that the pdf for $n = 30$ starts increasing at slightly larger values than the pdf for $n = 10$ and the peak occurs at a slightly larger value for $n = 30$. It may not be obvious that there is reduced dispersion in the distribution when moving from $n = 10$ to $n = 30$, but we know from proposition 12.3 that the variance of the $n = 30$ distribution must be lower than the variance of the $n = 10$ distribution. Although neither sample size is large*

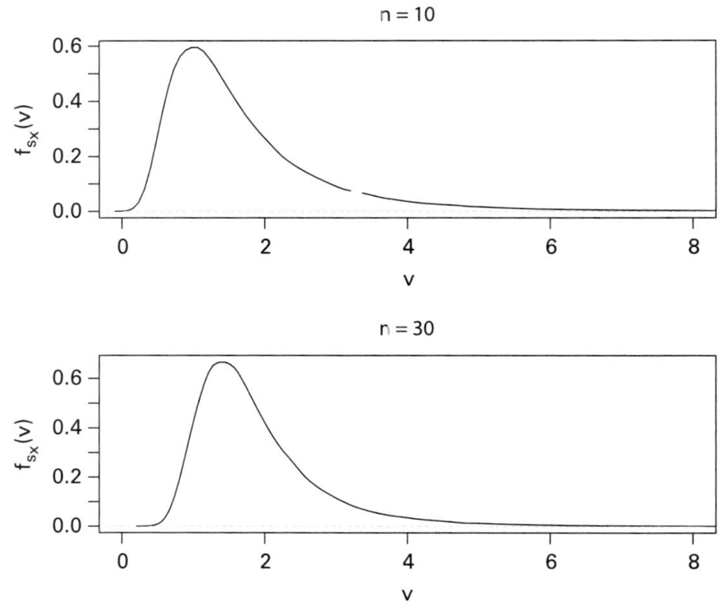

Figure 12.7
Sampling distributions of the sample standard deviation for i.i.d. log-normal random variables

enough to result in a symmetric and bell-shaped sampling distribution, we would get such a symmetric and bell-shaped sampling distribution eventually if we continue to increase the sample size n. This general idea is the focus of chapter 13, where the concept of asymptotic or large-sample sampling distributions is discussed.

Here is the R code to create figure 12.7, again using the `replicate` *function:*

```
set.seed(1234)

# initialize the number of simulations
num_simulations <- 100000

# create a vector of sample standard deviations for n=10
sdlogn_10 <- replicate(num_simulations, sd(rlnorm(10)))

# create a vector of sample standard deviations for n=30
sdlogn_30 <- replicate(num_simulations, sd(rlnorm(30)))

# graph-display format (two rows, one column)
par(mfrow=c(2,1))

# plot smoothed densities for the two sample sizes
plot(density(sdlogn_10), main=expression(n==10), xlim=c(0,8),
    xlab=expression(v), ylab=expression(f[s[X]](v)))
plot(density(sdlogn_30), main=expression(n==30), xlim=c(0,8),
    xlab=expression(v), ylab=expression(f[s[X]](v)))
```

12.3 Sampling Distribution of Other Statistics

This chapter has focused on the sampling distribution for three specific statistics: the sample mean $\bar{X}$, the sample variance s_X^2, and the sample standard deviation s_X. However, the basic idea of a sampling distribution extends to other statistics. This section provides a few examples.

Example 12.9 (Maximum of normal random variables) *Suppose that $X_1, X_2, \ldots, X_n$ are i.i.d. $N(\mu, \sigma^2)$ random variables, and we are interested in the sampling distribution of the* maximum *of the n random variables. Specifically, let* $\max_X = \max(X_1, X_2, \ldots, X_n)$*, where the value of the* $\max(\cdot)$ *function is the largest of its arguments, and the realized sample maximum is* $\max_x = \max(x_1, x_2, \ldots, x_n)$ *for the sample* $\{x_1, x_2, \ldots, x_n\}$*. For $n = 1$, the sampling distribution is just $N(\mu, \sigma^2)$. For $n = 2$, the cdf of* $\max_X$ *is determined as follows:*

$$
\begin{aligned}
P(max_X < v) &= P(\max(X_1, X_2) < v) \\
&= P(X_1 < v, X_2 < v) \\
&= P(X_1 < v)P(X_2 < v) \text{ since } X_1 \text{ and } X_2 \text{ are independent} \\
&= P\left(\frac{X_1 - \mu}{\sigma} < \frac{v - \mu}{\sigma}\right) P\left(\frac{X_2 - \mu}{\sigma} < \frac{v - \mu}{\sigma}\right) = \Phi\left(\frac{v - \mu}{\sigma}\right)^2.
\end{aligned}
$$

This reasoning extends to larger n, with the cdf given by

$$F_{max_X}(v) = P(max_X < v) = P(\max(X_1, X_2, \ldots, X_n)) = \Phi\left(\frac{v - \mu}{\sigma}\right)^n.$$

How about the pdf of max_X? Since the pdf is the derivative of the cdf (proposition 10.2), the pdf of max_X is

$$f_{max_X}(v) = F'_{max_X}(v) = \frac{n}{\sigma}\phi\left(\frac{v - \mu}{\sigma}\right)\Phi\left(\frac{v - \mu}{\sigma}\right)^{n-1}.$$

For the standard normal $N(0, 1)$, the cdf and pdf simplify to $F_{max_X}(v) = \Phi(v)^n$ and $f_{max_X}(v) = n\phi(v)\Phi(v)^{n-1}$, respectively. For this case, figure 12.8 shows graphs of the pdf of max_X for three sample sizes ($n = 5$, $n = 10$, and $n = 30$). As expected, the distribution shifts to the right as the sample size increases since the maximum is taken over a larger number of $N(0, 1)$ draws. Whereas the center of the distribution for $n = 5$ is located just above 1, the center of the distribution for $n = 30$ is closer to 2.

Interestingly, the approach for determining the sampling distribution of the maximum max_X can be extended to other continuous distributions rather easily. No special properties of the normal distribution were used to derive the cdf $F_{max_X}(v)$, so for some other distribution (like a uniform, a log-normal, or something else), the cdf and pdf of max_X are $F_{max_X}(v) = F_X(v)^n$ and $f_{max_X}(v) = nf_X(v)F_X(v)^{n-1}$, respectively. For instance, for i.i.d. $U(0, 1)$ random variables where $F_x(v) = v$ and $f_x(v) = 1$ for $v \in (0, 1)$, the distribution of $\max_X$ *has cdf and pdf $F_{max_X}(v) = v^n$ and $f_{max_X}(v) = nv^{n-1}$ for $v \in (0, 1)$, respectively.*

Example 12.10 (Sample median of normal random variables) *Suppose that $X_1, X_2, \ldots, X_n$ are i.i.d. $N(\mu, \sigma^2)$ random variables, and we are interested in the sampling distribution of the sample median. Since $\tilde{x}_{0.5}$ is the median of the sample $\{x_1, x_2, \ldots, x_n\}$, let $\tilde{X}_{0.5}$ denote the random variable associated with the sample median. Before observing the sample, the*

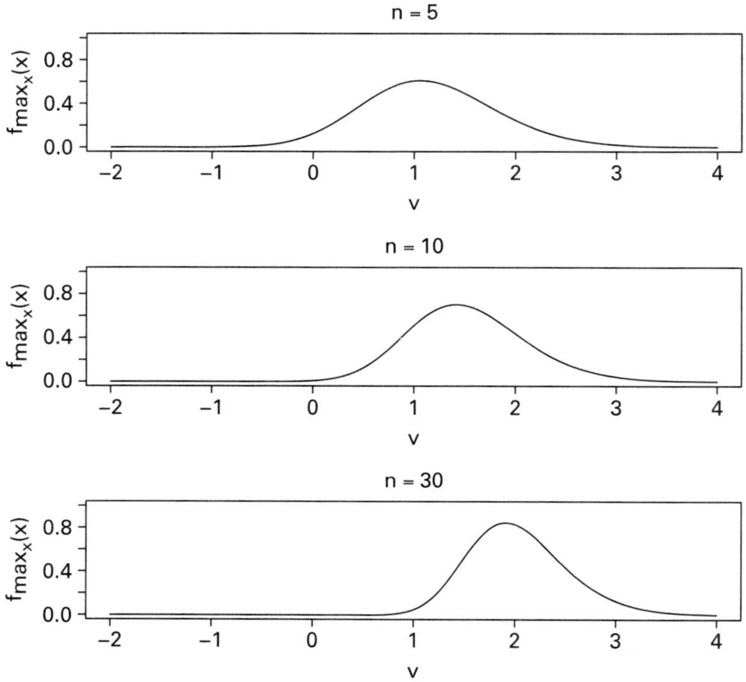

Figure 12.8
Sampling distributions of the sample maximum for i.i.d. $N(0, 1)$ random variables

sample median is the random variable $\tilde{X}_{0.5}$, *and the random sample leads to a realization* $\tilde{x}_{0.5}$ *of the random variable* $\tilde{X}_{0.5}$. *To fix ideas, consider a random sample of* $n = 3$ *observations. If the observations are sorted in order, the sample median is the middle value of the three observations. What would its sampling distribution be? Since the sample median is always the middle of the three observations, the sampling distribution should be less dispersed than the original* $N(\mu, \sigma^2)$ *distribution. It's less likely to have the middle observation in either of the tails since that would mean that there needs to be another observation even farther out in the tail. Using similar reasoning, the sampling distribution of the sample median should become tighter around the center* μ *of the original distribution as the sample size increases. Also, since the original random variables are symmetric, it would be surprising if the sampling distribution of the sample median was not also symmetric. While it's difficult to analytically determine the sampling distribution in this case, unlike the sample maximum in example 12.9, computer simulations can approximate the sampling distributions of the sample median for different sample sizes. Figure 12.9 shows the simulated sampling distributions of the sample median for three sample sizes (*$n = 3$, $n = 10$, *and* $n = 20$*), where the standard normal* $N(0, 1)$ *distribution is assumed. For each sample size, we use* 100,000 *simulations for the sampling distribution, where in each simulation, a random sample is drawn and the sample median is calculated. As predicted, the distributions become tighter around the center, at zero, as the sample size increases. The distributions all look symmetric. As a comparison, for the* $n = 20$ *graph at the bottom, the dotted line shows the sampling distribution of the sample mean* $\bar{X}$, *which is* $\bar{X} \sim N(0, 1/20)$ *since* $n = 20$. *So,*

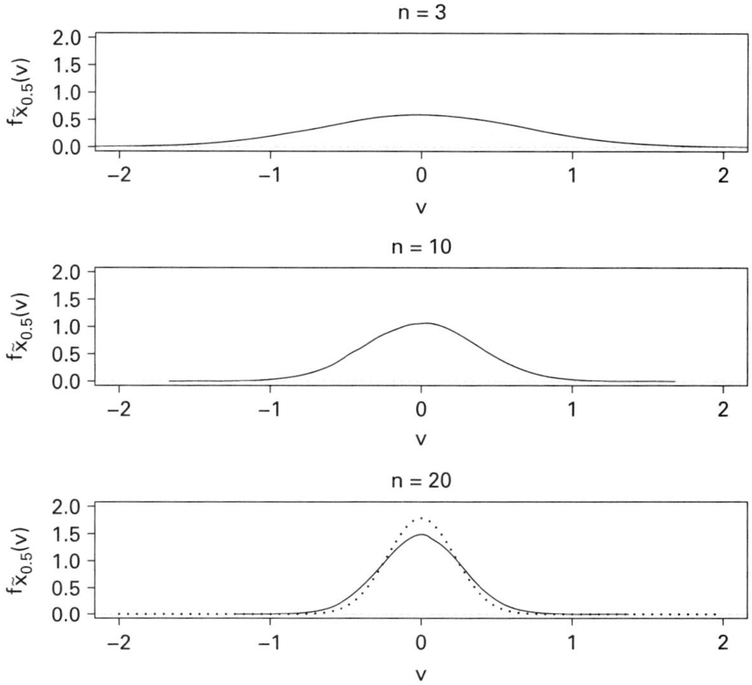

Figure 12.9
Sampling distributions of the sample median for i.i.d. $N(0, 1)$ random variables

the sampling distributions of $\tilde{X}_{0.5}$ and $\bar{X}$ are both centered around zero, but the sampling distribution of $\bar{X}$ exhibits less dispersion (lower variance) than the sampling distribution of $\tilde{X}_{0.5}$. Put another way, it is somewhat more likely that the sample mean will be closer to the population mean/median of zero than the sample median:

```
set.seed(1234)

# initialize the number of simulations
num_simulations <- 100000

# create a vector of sample medians for n=3, n=10, and n=20
mednorm_3 <- replicate(num_simulations, median(rnorm(3)))
mednorm_10 <- replicate(num_simulations, median(rnorm(10)))
mednorm_20 <- replicate(num_simulations, median(rnorm(20)))

# graph-display format (three rows, one column)
par(mfrow=c(3,1))

# plot smoothed densities for the three sample sizes
# for n=20, plot the sampling dist'n of the sample mean for comparison
plot(density(mednorm_3), ylim=c(0,2), xlim=c(-2,2), main=expression(n==3),
    xlab=expression(v), ylab=expression(f[widetilde(X)[0.5]](v)))
plot(density(mednorm_10), ylim=c(0,2), xlim=c(-2,2), main=expression(n==10),
    xlab=expression(v), ylab=expression(f[widetilde(X)[0.5]](v)))
plot(density(mednorm_20), ylim=c(0,2), xlim=c(-2,2), main=expression(n==20),
    xlab=expression(v), ylab=expression(f[widetilde(X)[0.5]](v)))
lines(seq(-2,2,0.01), dnorm(seq(-2,2,0.01),sd=1/sqrt(20)), lty=3, lwd=2)
```

Even when the sampling distribution is difficult or impossible to obtain analytically, the simulation approach illustrated in example 12.10 can simulate the sampling distribution for a statistic. Example 12.10 didn't require any special properties of the normal distribution, so the same approach can be used for any underlying distribution of the i.i.d. random variables in the case of the sample median. For other statistics, like other sample quantiles or the sample IQR, the only step in the simulation process that needs to be changed is calculating the appropriate statistic for each simulated random sample. This approach works even for bivariate statistics, like the sample covariance or the sample correlation, if the underlying joint distribution of the bivariate random variables is fully known.

Notes

1. The exact sampling distribution of $\bar{X}$ is not easy to derive, but it is known as the *Irwin-Hall distribution*.
2. In contrast, the *product* of log-normal random variables is log-normal. For instance, if $\ln(X_1)$ and $\ln(X_2)$ are i.i.d. $N(\mu, \sigma^2)$ random variables, then $\ln(X_1X_2) = \ln(X_1) + \ln(X_2)$, which has a $N(2\mu, 2\sigma^2)$ distribution, meaning that X_1X_2 is log-normal.
3. Although the proof of proposition 12.4 is complicated, some intuition can be developed for the result by noting that $(n-1)s_X^2 = \sum_{i=1}^n (X_i - \bar{X})^2$, which implies that $(n-1)\frac{s_X^2}{\sigma^2} = \sum_{i=1}^n \left(\frac{X_i - \bar{X}}{\sigma}\right)^2$. If this expression had the population mean μ rather than $\bar{X}$, it would be $\sum_{i=1}^n \left(\frac{X_i - \mu}{\sigma}\right)^2$, which is the sum of n squared i.i.d. $N(0, 1)$ random variables. That would imply that $\sum_{i=1}^n \left(\frac{X_i - \mu}{\sigma}\right)^2$ has a χ_n^2 distribution rather than the χ_{n-1}^2 distribution in the proposition. Thus, the fact that $\bar{X}$ appears in the expression for s_X^2, rather than the population mean μ, means that we lose a degree of freedom when using $\bar{X}$ in place of μ. The interested reader can try working this out explicitly for the case of $n = 2$, where the result simplifies to $\frac{s_X^2}{\sigma^2} \sim \chi_1^2$.
4. One exception is the case of i.i.d. Bernoulli random variables, but that's arguably not very interesting since the sample variance $s_X^2 = \bar{X}(1 - \bar{X})$, so the exact sampling distribution of $\bar{X}$ (a scaled binomial distribution) can be used directly to construct the sampling distribution for s_X^2.

Exercises

1. Let $X \in \{1, 2, 3\}$ be a discrete random variable with pmf

$$P(X = 1) = 0.2, \; P(X = 2) = 0.3, \; P(X = 3) = 0.5.$$

   a. Consider a random sample of two observations, where X_1 and X_2 are i.i.d. random variables with the pmf given here. What is the sampling distribution of $\bar{X}$?
   b. Consider a random sample of three observations, where X_1, X_2, and X_3 are i.i.d. random variables with the pmf given here. What is the sampling distribution of $\bar{X}$?
   c. Consider a random sample of three observations, where X_1, X_2, and X_3 are i.i.d. random variables with the pmf given here. What is the sampling distribution of the sample median?

2. Suppose the probability of a recession ($R = 1$) in the US in any given year is 10% and that the realizations of the Bernoulli random variable $R \sim \textit{Bernoulli}(0.1)$ in different years are independent. Consider a period of 10 consecutive years ($n = 10$) with realizations $R_1, R_2, \ldots, R_{10}$.

a. What is the sampling distribution of $\bar{R}$, the sample proportion of 10 years in which there is a recession? What are the mean and the standard deviation of the sampling distribution?

b. What is the sampling distribution of $T = \sum_{i=1}^{10} R_i$, the total number of recession years? What are the mean and the standard deviation of the sampling distribution? What is $P(T \geq 2)$?

3. Let $X \in \{1, 2, 3, 4, 5, 6\}$ be the random variable associated with the roll of a fair die.

a. What is the sampling distribution associated with the total of two independent rolls of a fair die?

b. What is the sampling distribution associated with the maximum of two independent rolls of a fair die?

c. Conduct 100,000 simulations in *R* and draw the histogram that approximates the sampling distribution associated with the total of five independent rolls of a fair die.

d. Conduct 100,000 simulations in *R* and draw the histogram that approximates the sampling distribution associated with the maximum of five independent rolls of a fair die.

e. What are the mean and variance of the random variable associated with the sum of 100 independent rolls of a fair die?

f. What are the mean and variance of the random variable associated with the average of 100 independent rolls of a fair die?

4. A credit card company knows that the monthly balance X (in dollars) of a representative customer is normally distributed:

$$X \sim N(300, 2500).$$

Assume that the monthly balances of customers are independent draws from X.

a. Let $X_1, X_2, \ldots, X_{100}$ denote the monthly balances for 100 randomly chosen customers.

i. What is the distribution of $\bar{X}$, the average monthly balance of the 100 customers?

ii. Determine a 99% probability interval for $\bar{X}$.

b. The credit card company considers a customer to be a "low-balance customer" if she or he has a monthly balance below \$200. Let L be an indicator variable equal to 1 for a low-balance customer and 0 otherwise.

i. What is the distribution of L?

ii. For 20 randomly chosen customers, what is the probability of at least one low-balance customer?

iii. If $L_1, L_2, \ldots, L_{100}$ denotes the low-balance random variables for 100 randomly chosen customers, what is the distribution of the sample proportion $\bar{L} = \frac{1}{100} \sum_{i=1}^{100} L_i$?

5. Three firms have annual profits (in millions of dollars), denoted as X_1, X_2, and X_3, that are log-normally distributed. Assume that

$$\ln X_1 \sim N(0,1), \;\; \ln X_2 \sim N(0.5, 0.25), \;\text{ and }\; \ln X_3 \sim N(0.75, 0.16)$$

are independent of each other. Based on 100,000 simulations in R, draw a histogram (with 100 bins) and a smoothed density corresponding to the distribution of $X_1 + X_2 + X_3$, the sum of annual profits. How does the average of the simulated draws of $X_1 + X_2 + X_3$ compare to the population mean of $X_1 + X_2 + X_3$?

6. Consider i.i.d. random variables $X_1, X_2, \ldots, X_n$ drawn from a $N(\mu, 4)$ distribution.
   a. For $n = 10$, what is $P(|\bar{X} - \mu| < 0.1)$ (i.e., the probability that $\bar{X}$ is within 0.1 of μ)?
   b. What is the smallest value of n that guarantees $P(|\bar{X} - \mu| < 0.1) > 0.95$?
   c. If you instead want $P(|\bar{X} - \mu| < 0.1) > 0.90$, would you require a larger or smaller n compared to (b)?
   d. If the random variables were instead drawn from a $N(\mu, 9)$ distribution and you want $P(|\bar{X} - \mu| < 0.1) > 0.95$, would you require a larger or smaller n compared to (b)?

7. In the population, IQ ("intelligence quotient") scores are normally distributed with a mean of 100 and a standard deviation of 15. You intend to obtain IQ scores for a random sample of 20 individuals, from which you will calculate the sample average and the sample standard deviation.
   a. For a single individual, what is the probability that the IQ score is greater than 105?
   b. For a sample of 20 individuals, what is the probability that the sample average is greater than 105?
   c. For a sample of 20 individuals, provide an 80% probability interval for the sample standard deviation. That is, what are the values a and b satisfying $P(s_X < a) = 0.1$ and $P(s_X > b) = 0.1$ so $P(a \le s_X \le b) = 0.8$?
   d. Conduct simulations in R using the assumed $N(100, 15^2)$ distribution and $n = 20$. Specifically, use 100,000 simulations, where for each simulation you draw an i.i.d. sample of size $n = 20$ from the assumed distribution.
      i. Draw a density histogram of the 100,000 sample averages.
      ii. What is the standard deviation of the 100,000 sample averages?
      iii. Draw a density histogram of the 100,000 sample standard deviations.
      iv. What is the standard deviation of the 100,000 sample standard deviations?
      v. What is the proportion of the 100,000 simulations for which the sample average is greater than 105? Does this proportion approximate the probability from (b)?

8. The triangular random variable X, introduced in chapter 10, has the pdf

$$f_X(x) = \begin{cases} x & \text{if } 0 \le x \le 1 \\ 2 - x & \text{if } 1 < x \le 2 \\ 0 & \text{otherwise.} \end{cases}$$

a. Using the fact that X is equivalent to the sum of two independent $U(0,1)$ random variables, write an R function `rtri` that takes a single argument `n` and returns a vector of `n` i.i.d. draws of the triangular random variable X.

b. For a sample size of $n = 10$, use the `rtri` function to conduct 100,000 simulations to approximate the sampling distributions of both the sample mean $\bar{X}$ and the sample standard deviation s_X. Draw the densities associated with each sampling distribution. What are the mean and standard deviation of each sampling distribution?

c. Repeat (b) for a sample size of $n = 40$.

9. For each of 10 pharmaceutical drug companies, assume that the number of new drug discoveries in a given year is an independent Poisson random variable with expected value equal to 0.5.

For this question, use the following fact about independent Poisson random variables: *If $X_1, X_2, \ldots, X_k$ are independent Poisson random variables with Poisson parameters $\lambda_1, \lambda_2, \ldots, \lambda_k$, respectively, the sum $X_1 + X_2 + \cdots + X_k$ is a Poisson$(\lambda_1 + \lambda_2 + \cdots + \lambda_k)$ random variable.*

a. What is the sampling distribution of the total number (over all 10 companies) of new drug discoveries in a given year?

b. What is the sampling distribution of the average number (per company) of new drug discoveries in a given year?

c. Conduct 10,000 simulations in R to approximate the sampling distribution of the sample median of the total number of new drug discoveries in a given year. Draw the histogram and calculate the mean and standard deviation of the sampling distribution.

10. Consider a sample of $n = 100$ observations, where the underlying random variables $X_1, X_2, \ldots, X_{100}$ are i.i.d. uniform $U(0,1)$ random variables.

a. Conduct 10,000 simulations in R to approximate the sampling distribution of the sample interquartile IQR_X. Draw the density and report the mean and standard deviation of the sampling distribution. Is the mean of the sampling distribution close to what you expected? (Think about the population IQR.)

b. *This part considers the *trimmed mean* of a sample, a descriptive statistic defined as the sample mean calculated on the sample *after dropping the most extreme observations*. For example, the 5% trimmed mean for $n = 100$ is the sample mean of the 90 observations that remain after dropping the 5 largest observations and the 5 smallest observations. Similarly, the 2% trimmed mean for $n = 100$ is the sample mean of the 96 observations that remain after dropping the 2 largest observations and the 2 smallest observations. Conduct 10,000 simulations in R to approximate the sampling distributions of the sample mean, the 5% trimmed mean, and the 2% trimmed mean. Draw their densities on the same graph to compare the distributions. Calculate the mean and standard deviation for each of the three sampling distributions, and comment on how the values compare to each other.

11. *In an *English auction*, the price of an object increases as bidders continue to bid on the object. A well-known theoretical prediction in economics is that the bidder with the highest valuation for the object wins the auction, and the highest bid is equal to the

second-highest valuation among the bidders. For instance, among a group of bidders, if the highest and second-highest valuations of an object are \$92 and \$88, respectively, the prediction is that the bidder with the highest valuation bids \$88, wins the auction, and realizes a surplus equal to the valuation minus the bid, or $\$92 - \$88 = \$4$.

For this question, assume that a seller holds an auction for an item that she values at \$85, and assume that the theoretical prediction described here holds in practice. There is a group of B bidders for the object, with valuations $V_1, V_2, \ldots, V_B$ for the object, where each valuation V_i is an i.i.d. draw of a $U(80, 100)$ random variable.

a. First, consider the case of two bidders ($B = 2$).
   i. What is the sampling distribution of the winning bid?
   ii. What is the expected value of the seller's surplus (equal to the winning bid minus \$85)?
   iii. What is the probability that the seller's surplus is negative?
   iv. What is the expected value of the winner's surplus (equal to the winner's valuation minus the winning bid)?
b. For $B = 3$, what is the sampling distribution of the winning bid?
c. For each B value in $\{5, 10, 15\}$, conduct 100,000 simulated auctions in R in which you determine the winning bid and the winner's valuation for each auction. Based on the simulated auctions, (i) graph the estimated distributions (pdfs) of both the seller's surplus and the winner's surplus and (ii) calculate the expected values of both the seller's surplus and the winner's surplus.

12. Let $\max_X = \max(X_1, X_2, \ldots, X_n)$ be the maximum of n random variables $X_1, X_2, \ldots, X_n$, as in section 12.3.
   a. If $X_1, X_2, \ldots, X_n$ are i.i.d. draws from a $U(0, 1)$ random variable, determine the probability that $\max_X$ is greater than 0.98 as a function of n. What is the smallest n for which this probability is at least 95%?
   b. If $X_1, X_2, \ldots, X_n$ are i.i.d. draws from a $N(0, 1)$ random variable, determine the probability that $\max_X$ is greater than 3 as a function of n. What is the smallest n for which this probability is at least 95%?
   c. If $X_1, X_2, \ldots, X_n$ are i.i.d. rolls of a fair die, with $X_i \in \{1, 2, 3, 4, 5, 6\}$, what are the cdf and pmf of $\max_X$?
   d. If $X_1, X_2, \ldots, X_n$ are i.i.d. draws from a $N(0, 1)$ random variable, what is the cdf of the *minimum* value $\min_X = \min(X_1, X_2, \ldots, X_n)$ as a function of n?
   e. If $X_1, X_2, \ldots, X_n$ are i.i.d. draws from a $U(0, 1)$ random variable, what is the cdf of the *second-largest* value as a function of n?

13

Sampling Distributions: Asymptotic

Chapter 12 explored sampling distributions of statistics when the sample size n is fixed and the complete distribution of the underlying independent and identically distributed (i.i.d.) random variables is known. Analytical characterizations were provided for the exact sampling distribution in some cases (e.g., the sample mean of Bernoulli random variables, the sample mean of normal random variables, and the sample variance of normal random variables), while simulations were used for others (e.g., the standard deviation of a uniform distribution and the sample median of a normal distribution). In this chapter, we shift the focus to the situation where the sample size n grows large, and then characterize the sampling distribution of a statistic in this context.

Two main motivations drive the consideration of large-sample or asymptotic sampling distributions. First, most real-world data sets used by economists and other practitioners tend to be large, which could mean hundreds or thousands (or even millions) of observations. For instance, the stock-return data set **sp500** used throughout this book has 364 observations, while the labor-force data set **cps** has several thousand. These sample sizes far exceed the small n examples in chapter 12. Second, with large n, remarkable statistical results enable the characterization of the sampling distribution of a statistic. These results typically indicate that the asymptotic sampling distribution is a normal distribution for most statistics discussed in this book. This contrasts with the exact sampling distributions in chapter 12, where normality holds only in very specific cases (e.g., the sample mean of i.i.d. normal random variables), but more generally, the sampling distribution's shape depends on the specific sample size, the specific statistic, and the specific underlying distribution of the random variables. As such, for large samples, we will apply properties of the normal distribution in chapter 14 to further analyze the variability of a given statistic and, when viewed as a guess or estimator of an underlying parameter, its precision.

13.1 Asymptotic Distribution of the Sample Mean

We begin our discussion of the asymptotic sampling distribution, or more concisely the **asymptotic distribution**, with the sample mean $\bar{X}$. As in chapter 12, the case of i.i.d. random variables $X_1, X_2, \ldots, X_n$ is considered throughout. The standard notation μ_X and σ_X^2 denotes the population mean and variance of X, but no specific distribution is assumed.

In this general case, from proposition 10.13, the population mean, variance, and standard deviation of $\bar{X}$ are

$$\mu_{\bar{X}} = \mu_X, \ \sigma^2_{\bar{X}} = \frac{\sigma^2_X}{n}, \text{ and } \sigma_{\bar{X}} = \frac{\sigma_X}{\sqrt{n}}.$$

To think about the sample size n growing large, we imagine what would happen if we continually drew new observations from the population, with each new observation being a realization of an i.i.d. random variable X_i. In this thought experiment, n grows arbitrarily large as more and more observations are drawn from the population, so $n \to \infty$ ("n goes to infinity"). Of course, in practice, our observed sample will not be infinite, but the thought experiment is useful since the theoretical results are based on the limit of what happens when $n \to \infty$. And these theoretical results can be applied to a given data set so long as n is large. A more complete discussion of "How large is large enough?" is provided next.

13.1.1 Law of Large Numbers

From the general results given here, the mean of the sampling distribution of $\bar{X}$ is equal to the mean μ_X of the underlying random variables. The variance and standard deviation of the sampling distribution of $\bar{X}$ are both decreasing in n, with $\sigma^2_{\bar{X}} = \frac{\sigma^2_X}{n} \to 0$ and $\sigma_{\bar{X}} = \frac{\sigma_X}{\sqrt{n}} \to 0$ as $n \to \infty$. As the sample size n gets arbitrarily large, the sampling distribution of $\bar{X}$ has a mean that stays at μ_X and a dispersion that shrinks to zero, meaning that the realizations of $\bar{X}$ eventually get arbitrarily close to μ_X. This intuition is formalized by the **Law of Large Numbers (LLN)**, given in proposition 13.1.[1]

Proposition 13.1 (Law of Large Numbers (LLN)) *If $X_1, X_2, \ldots, X_n$ are i.i.d. random variables with finite population mean μ_X, the sample mean $\bar{X} = \frac{1}{n}\sum_{i=1}^{n}$ gets arbitrarily close to μ_X as $n \to \infty$.*

For i.i.d. Bernoulli random variables, the LLN is closely related to the frequentist interpretation of probabilities discussed in chapter 2. If $X_1, X_2, \ldots, X_n$ are i.i.d. $Bernoulli(\pi)$ random variables, the sample mean $\bar{X}$ is the sample proportion of 1s (successes) among the n trials, and the LLN states that this sample proportion of successes gets arbitrarily close to π, the true probability of success, as n grows large. For a fair coin toss, the sample proportion of heads gets arbitrarily close to 0.5 as the number of tosses grows large. For a fair die roll, the sample proportion of 6s gets arbitrarily close to 1/6 as the number of rolls grows large. For a population of i.i.d. customers, each with purchase probability 0.2, the sample proportion of purchases gets arbitrarily close to 0.2 as the number of customers grows large.

The LLN holds regardless of the underlying distribution of the random variables. If a particular model has been assumed for the random variables and that model is correct, the LLN implies that $\bar{X}$ gets arbitrarily close to the model-based μ_X when n grows large. For i.i.d. normal random variables, with $X \sim N(\mu, \sigma^2)$, the sample mean gets arbitrarily close to μ as n grows large. For i.i.d. log-normal random variables, with $\ln(X) \sim N(\mu, \sigma^2)$, the sample mean gets arbitrarily close to $e^{\mu+\sigma^2/2}$ as n grows large. For i.i.d. uniform random variables, with $X \sim U(a,b)$, the sample mean gets arbitrarily close to $\frac{a+b}{2}$ as n grows large. For i.i.d. Poisson random variables, with $X \sim Poisson(\lambda)$, the sample mean gets arbitrarily close to λ as n grows large.

13.1.2 Central Limit Theorem

While the LLN is a powerful and general result about the sample mean $\bar{X}$ for large sample sizes, it doesn't say anything about the *shape* of the sampling distribution of $\bar{X}$. Another classic statistics result, the **Central Limit Theorem (CLT)**, fully characterizes the sampling distribution of $\bar{X}$ for large sample sizes, as given in proposition 13.2. Amazingly, the shape of this asymptotic distribution for $\bar{X}$ is always that of a normal distribution, regardless of the underlying distribution of the i.i.d. random variables.[2]

Proposition 13.2 (Central Limit Theorem (CLT)) *If $X_1, X_2, \ldots, X_n$ are i.i.d. random variables with finite population mean μ_X and finite population variance σ_X^2, then for sufficiently large n, the sample mean $\bar{X} = \frac{1}{n}\sum_{i=1}^{n} X_i$ is approximately normally distributed with mean μ_X and variance $\frac{\sigma_X^2}{n}$:*

$$\bar{X} \overset{a}{\sim} N\left(\mu_X, \frac{\sigma_X^2}{n}\right).$$

The notation $\overset{a}{\sim}$ in proposition 13.2 is read as "asymptotically distributed as." We say that "$\bar{X}$ is asymptotically distributed as a normal random variable with mean μ_X and variance $\frac{\sigma_X^2}{n}$," or alternatively, "$\bar{X}$ has an asymptotic normal distribution with mean μ_X and variance $\frac{\sigma_X^2}{n}$." The variance of the asymptotic distribution is called the **asymptotic variance**, and the standard deviation of the asymptotic distribution, which is $\frac{\sigma_X}{\sqrt{n}}$, is called the **asymptotic standard deviation**.

The asymptotic normal distribution of $\bar{X}$ from the CLT can be standardized to yield a random variable that has the standard normal as its asymptotic distribution. Subtracting the mean μ_X from $\bar{X}$ and then dividing the result by the asymptotic standard deviation $\frac{\sigma_X}{\sqrt{n}}$ yields

$$\frac{\bar{X} - \mu_X}{\sigma_X/\sqrt{n}} \overset{a}{\sim} N(0, 1).$$

While the mean and variance of $\bar{X}$ are already known to be μ_X and $\frac{\sigma_X^2}{n}$, respectively, the CLT provides the much stronger result that the asymptotic sampling distribution of $\bar{X}$ is a *normal distribution* with those mean and variance parameters. The CLT is a remarkable result since the underlying distribution of the random variables $X_1, X_2, \ldots, X_n$ can be anything. The theorem holds whether the underlying distribution is discrete or continuous (or some combination of the two) and whether the underlying distribution is symmetric or asymmetric.

Since a similar result for the specific case of i.i.d. normal random variables has been seen previously (section 12.1.2), it is important to understand the difference between that result and the general CLT result. For i.i.d. normal random variables, with $X \sim N(\mu, \sigma^2)$, the *exact* sampling distribution is $\bar{X} \sim N\left(\mu, \frac{\sigma^2}{n}\right)$. The use of $\sim$ rather than $\overset{a}{\sim}$ indicates an exact sampling distribution rather than an asymptotic distribution. The $\bar{X} \sim N\left(\mu, \frac{\sigma^2}{n}\right)$ distribution holds for any sample size, even for very small n, while the CLT result $\bar{X} \overset{a}{\sim} N\left(\mu_X, \frac{\sigma_X^2}{n}\right)$ is an approximate sampling distribution that requires large n.

The CLT requires that the sample size n is sufficiently large, but what does "sufficiently large" mean in practice? Many textbooks give simple rules of thumb, like saying that $n > 30$ (a sample with more than 30 observations) is sufficient for a large sample. Unfortunately, in reality, the number of observations required for the CLT to hold (i.e., for the normal approximation to be accurate) depends upon the distribution of the underlying random variables. For instance, a heavily right-skewed distribution of the random variables usually requires larger n than a nicely symmetric distribution of the random variables. In fact, in some of the distribution graphs seen in chapter 12, there were cases where the distribution of $\bar{X}$ looked normal for very small sample sizes (e.g., the uniform distribution) and other cases where it did not (e.g., the log-normal distribution). For the specific case of Bernoulli random variables, which arises frequently in practice and is discussed in more detail next, an oft-used and effective rule of thumb is that the normal approximation works well when both $n\pi > 10$ and $n(1-\pi) > 10$. This Bernoulli rule of thumb requires a larger sample size when the success probability π is closer to 0 or 1 (i.e., when the Bernoulli distribution is more asymmetric). When $\pi = 0.5$, the rule of thumb suggests having $n > 20$ is sufficient to use the CLT. In contrast, the rule of thumb suggests $n > 50$ for the CLT approximation when $\pi = 0.2$ or $\pi = 0.8$ and $n > 100$ when $\pi = 0.1$ or $\pi = 0.9$.

Example 13.1 (Restaurant franchises) *Suppose that a fast-food mogul owns 60 franchises of Burger Depot. The monthly revenue, in thousands of dollars, for each of the 60 franchises is an i.i.d. random variable drawn from a distribution with population mean $\mu_X = 20$ and population standard deviation $\sigma_X = 4$. In a given month, the approximate distribution of the sample average of monthly revenues at the 60 franchises, based on the CLT, is*

$$\bar{X} \overset{a}{\sim} N\left(20, \frac{4^2}{60}\right).$$

Then an approximate 95% probability interval for the sample average of monthly revenues at the 60 franchises is

$$\left(20 - 1.96\frac{4}{\sqrt{60}}, 20 + 1.96\frac{4}{\sqrt{60}}\right) \approx (18.99, 21.01),$$

and an approximate 90% probability interval for the sample average of monthly revenues at the 60 franchises is

$$\left(20 - 1.645\frac{4}{\sqrt{60}}, 20 + 1.645\frac{4}{\sqrt{60}}\right) \approx (19.15, 20.85).$$

If the average monthly revenue falls below \$18,500 in a given month, the fast-food mogul will need a bank loan. The probability of that happening in a given month is

$$P(\bar{X} < 18.5) = P\left(\frac{\bar{X} - 20}{4/\sqrt{60}} < \frac{18.5 - 20}{4/\sqrt{60}}\right) = \Phi\left(\frac{18.5 - 20}{4/\sqrt{60}}\right) \approx 0.00184, \textit{ or } 0.184\%.$$

```
pnorm((18.5-20)/(4/sqrt(60)))

## [1] 0.001837806
```

The CLT (proposition 13.2) provides the asymptotic distribution for the average $\bar{X}$ of i.i.d. random variables. Since the sum $X_1+X_2+\cdots+X_n$ is equal to $n\bar{X}$, an immediate corollary of the CLT is proposition 13.3, which describes the asymptotic distribution of the *sum* of i.i.d. random variables.

Proposition 13.3 (Asymptotic distribution of the sum of i.i.d. random variables) *If $X_1, X_2, \ldots, X_n$ are i.i.d. random variables with finite population mean μ_X and finite population variance σ_X^2, then for sufficiently large n, the sum $S=\sum_{i=1}^n X_i = n\bar{X}$ is approximately normally distributed, with*

$$S \overset{a}{\sim} N\left(n\mu_X, n\sigma_X^2\right).$$

$\bar{X}$ has an approximate normal distribution by the CLT, which means that $n\bar{X}$ also is approximately normal since it's a scaled version of $\bar{X}$. From proposition 10.14, the mean and variance of the normal distribution that approximates the sum $S=\sum_{i=1}^n X_i$ are n and n^2 times the mean and variance, respectively, of the underlying random variable X.

Example 13.2 (Restaurant franchises) *Continuing example 13.1, the approximate distribution for the total monthly revenues at the* 60 *franchises is*

$$S=60\bar{X} \overset{a}{\sim} N\left(60\cdot 20, 60\cdot 4^2\right) \text{ or } S \overset{a}{\sim} N(1200, 960).$$

Then an approximate 95% probability interval for the total monthly revenues at the 60 *franchises is*

$$\left(1200-1.96\sqrt{960}, 1200+1.96\sqrt{960}\right) \approx (1139, 1261).$$

This interval can also be obtained by multiplying the endpoints of the the interval $(18.99, 21.01)$ *for $\bar{X}$, calculated in example 13.1, by* 60.

13.1.3 Normal Approximation for the Sample Proportion or Binomial Random Variable

In section 12.1.1, the exact sampling distribution of $\bar{X}$ was determined when $X_1, X_2, \ldots, X_n$ are i.i.d. *Bernoulli*(π) random variables. Recall that $\bar{X}$, the sample mean or sample proportion of successes, has an exact sampling distribution equivalent to a *Binomial*(n,π) random variable scaled by $\frac{1}{n}$, which is true for any sample size n and success probability π. This result followed from the fact that $X_1+X_2+\cdots+X_n$ is, by definition, a *Binomial*(n,π) random variable, so $\bar{X}=\frac{1}{n}(X_1+X_2+\cdots+X_n)$ is a binomial random variable scaled by $\frac{1}{n}$. The population statistics for $\bar{X}$ are

$$\mu_{\bar{X}}=\pi,\ \sigma_{\bar{X}}^2=\frac{\pi(1-\pi)}{n},\ \text{and}\ \sigma_{\bar{X}}=\sqrt{\frac{\pi(1-\pi)}{n}}.$$

Applying the CLT, the asymptotic distribution of $\bar{X}$ is

$$\bar{X} \overset{a}{\sim} N\left(\pi, \frac{\pi(1-\pi)}{n}\right) \text{ or, equivalently, } \frac{\bar{X}-\pi}{\sqrt{\frac{\pi(1-\pi)}{n}}} \overset{a}{\sim} N(0,1).$$

For the binomial random variable $Y=X_1+X_2+\cdots+X_n$, where by definition, $Y\sim$ *Binomial*(n,π), the population mean is $\mu_Y=n\pi$ and the population variance is

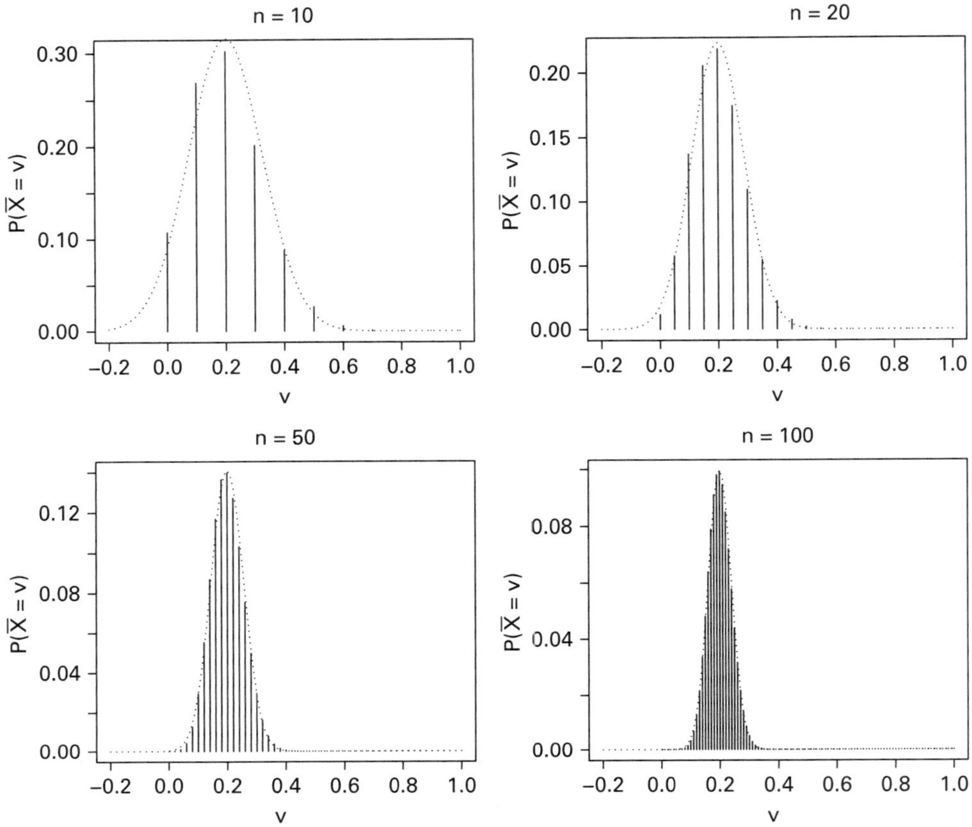

Figure 13.1
Sampling distributions of the sample mean for i.i.d. Bernoulli(0.2) random variables

$\sigma_Y^2 = n\pi(1-\pi)$. Since $Y = n\bar{X}$, proposition 13.3 implies

$$Y \overset{a}{\sim} N(n\pi, n\pi(1-\pi)) \text{ or, equivalently, } \frac{Y - n\pi}{\sqrt{n\pi(1-\pi)}} \overset{a}{\sim} N(0,1).$$

The rule of thumb discussed here stated that the CLT normal approximation works well when both $n\pi > 10$ and $n(1-\pi) > 10$. To illustrate this phenomenon, figure 13.1 considers four sample sizes ($n = 10$, $n = 20$, $n = 50$, and $n = 100$) when the success probability is $\pi = 0.2$. For $\pi = 0.2$, the rule of thumb suggests the asymptotic approximation works well when $n > 50$. For each graph in figure 13.1, the exact-distribution pmf for the sample proportion $\bar{X}$ is depicted with vertical bars, and the dotted curve associated with the asymptotic distribution $N\left(0.2, \frac{0.2(1-0.2)}{n}\right)$ is shown for comparison. For the smallest sample size ($n = 10$), the problem with the normal approximation appears in the left tail, as it has positive probabilities associated with negative values for $\bar{X}$. This problem largely disappears for $n = 20$, and even for that sample size (which is lower than the rule-of-thumb suggestion), the normal approximation looks pretty good. For the larger sample sizes ($n = 50$ and $n = 100$), the normal approximation matches the pmf almost exactly.

How do we calculate probabilities or probability intervals using the normal approximation? Let's start with a binomial random variable. As an example, let's say that $n = 50$ and $\pi = 0.2$, so $Y \sim Binomial(50, 0.2)$. The probability of having exactly 10 successes out of 50 trials is

$$P(Y = 10) = \binom{50}{10} 0.2^{10} 0.8^{40} \approx 0.1398.$$

```
dbinom(10,50,0.2)

## [1] 0.139819
```

Since $n\pi = 10$ and $n\pi(1-\pi) = 8$, the approximate sampling distribution of Y is $N(10, 8)$. The practical issue here is that only integer outcomes between 0 and 50 are possible, so none of the noninteger values are possible. To determine $P(Y = 10)$, then, we don't want to evaluate the probability density function (pdf) of $N(10, 8)$ at the value 10. Instead, we assume that the continuous interval (9.5, 10.5) corresponds to the discrete outcome 10, and it is similar for any other possible outcome; (10.5, 11.5) corresponds to the outcome 11, (35.5, 36.5) corresponds to the outcome 36, and so on. Then, if $W \sim N(10, 8)$ denotes the normal approximation, the probability of 10 successes based upon the normal approximation is

$$P(9.5 < W < 10.5) = \Phi\left(\frac{10.5 - 10}{\sqrt{8}}\right) - \Phi\left(\frac{9.5 - 10}{\sqrt{8}}\right) \approx 0.1403,$$

which is quite close to the true probability of 0.1398:

```
pnorm((10.5-10)/sqrt(8))-pnorm((9.5-10)/sqrt(8))

## [1] 0.1403162
```

This method of looking at the probability of a continuous bin to approximate the discrete-outcome probability is known as a **continuity correction**. To illustrate how the continuity correction can be used for different types of probability intervals, where the inequalities may be strict or weak, the following table considers additional examples based on the $Y \sim Binomial(50, 0.2)$ distribution:

Event probability	$P(Y=10)$	$P(6 < Y < 12)$	$P(6 < Y \leq 12)$	$P(6 \leq Y \leq 12)$
Probability based upon exact pmf	0.1398	0.6073	0.7105	0.7659
Normal approx. with continuity correction	$P(9.5 < W < 10.5)$	$P(6.5 < W < 11.5)$	$P(6.5 < W < 12.5)$	$P(5.5 < W < 12.5)$
Approx. probability based on normal	0.1403	0.5941	0.7037	0.7558

The continuity correction can be used in a similar way for calculating probabilities and probability intervals for the sample proportion $\bar{X}$ of i.i.d. Bernoulli random variables. The only difference is the scale of the outcomes associated with $\bar{X}$ compared to the binomial Y. For example, if $\bar{X}$ is the sample proportion of successes when $n = 50$ and $\pi = 0.2$, the possible outcomes for $\bar{X}$ are $\{0, 1/50, 2/50, \ldots, 49/50, 1\}$ or $\{0, 0.02, 0.04, \ldots, 0.98, 1\}$. For the outcome $\bar{X}=0.20$, the continuity correction uses the interval $(0.19, 0.21)$ since 0.19 is midway between 0.20 and the outcome below it (0.18) and 0.21 is midway between 0.20 and the outcome above it (0.22). For the probability $P(0.20 \leq \bar{X} \leq 0.26)$, the continuity correction uses the interval $(0.19, 0.27)$ to calculate an approximate probability based upon the asymptotic normal distribution $N\left(0.2, \frac{(0.2)(0.8)}{50}\right)$.

Example 13.3 (College-educated adults) *In the US, the probability that an adult aged 25 to 34 has at least a bachelor's degree is 40%. Suppose that a random sample of 100 adults aged 25 to 34 is drawn from the population. The sample proportion $\bar{X}$ with at least a bachelor's degree, among the $n=100$ individuals, has*

$$\mu_{\bar{X}}=0.40,\ \sigma_{\bar{X}}^2=\frac{(0.4)(0.6)}{100}=0.0024,\ \text{and}\ \sigma_{\bar{X}}=\sqrt{0.0024}\approx 0.0490.$$

The normal approximation can be used to calculate 90% and 95% probability intervals for $\bar{X}$. Without a continuity correction, an approximate 90% probability interval for $\bar{X}$ is

$$\begin{aligned}(\mu_{\bar{X}}-1.645\sigma_{\bar{X}}, \mu_{\bar{X}}+1.645\sigma_{\bar{X}}) &\approx (0.40-1.645(0.0490), 0.40+1.645(0.0490))\\ &\approx (0.3194, 0.4806),\end{aligned}$$

and an approximate 95% probability interval for $\bar{X}$ is

$$\begin{aligned}(\mu_{\bar{X}}-1.96\sigma_{\bar{X}}, \mu_{\bar{X}}+1.96\sigma_{\bar{X}}) &\approx (0.40-1.96(0.0490), 0.40+1.96(0.0490))\\ &\approx (0.3040, 0.4960).\end{aligned}$$

What is the probability that the observed proportion with at least a bachelor's degree is between 35% and 45% (inclusive)? The probability $P(0.35 \leq \bar{X} \leq 0.45)$, based on the exact distribution (the scaled binomial), is

$$\begin{aligned}P(0.35 \leq \bar{X} \leq 0.45) &= P(35 \leq Y \leq 45)\ \text{for}\ Y \sim Binomial(100, 0.4)\\ &= F_Y(45) - F_Y(34) \approx 0.7386.\end{aligned}$$

```
pbinom(45,100,0.40)-pbinom(34,100,0.40)

## [1] 0.738573
```

The approximate probability, based on the normal approximation and using a continuity correction, is

$$\begin{aligned}P(0.35 \leq \bar{X} \leq 0.45) &\approx P(0.345 < W < 0.455)\ \text{for}\ W \sim N(0.4, 0.0024)\\ &= \Phi\left(\frac{0.455-0.4}{0.0490}\right) - \Phi\left(\frac{0.345-0.4}{0.0490}\right) \approx 0.7383.\end{aligned}$$

```
pnorm((0.455-0.4)/0.0490)-pnorm((0.345-0.4)/0.0490)

## [1] 0.7383284
```

Now, consider a situation in which half of the sample (50 observations) is female and half (50 observations) is male. If the probability that a female adult aged 25 to 34 has at least a *bachelor's degree is 44% and the probability for a male adult aged 25 to 34 is 36%, what is the probability that the observed sample proportion of females with at least a bachelor's degree is greater than the observed sample proportion of males with at least a bachelor's degree? Letting $\bar{X}_f$ denote the sample proportion among 50 females and $\bar{X}_m$ denote the sample proportion among 50 males, this probability is*

$$P(\bar{X}_f > \bar{X}_m) = P(\bar{X}_f - \bar{X}_m > 0).$$

The random variable $\bar{X}_f - \bar{X}_m$ is a linear combination of two sample proportions—specifically, the difference between two sample proportions. While we don't have specific results about the distribution of the difference of scaled binomial random variables, we do have such results about the distribution of the difference of normal random variables. Therefore, a normal approximation can be used for both $\bar{X}_f$ and $\bar{X}_m$ to simplify the calculation of $P(\bar{X}_f - \bar{X}_m > 0)$.[3] Using the CLT, the asymptotic distributions of $\bar{X}_f$ and $\bar{X}_m$ are

$$\bar{X}_f \overset{a}{\sim} N\left(0.44, \frac{(0.44)(0.56)}{50}\right) \text{ and } \bar{X}_m \overset{a}{\sim} N\left(0.36, \frac{(0.36)(0.64)}{50}\right).$$

Moreover, $\bar{X}_f$ and $\bar{X}_m$ are independent random variables since they are each based on i.i.d. random variables from the population. Using results for linear combinations of normal random variables, $\bar{X}_f - \bar{X}_m$ is also approximately normal. To characterize the normal distribution, the population mean and variance of $\bar{X}_f - \bar{X}_m$ are determined as follows:

$$E(\bar{X}_f - \bar{X}_m) = E(\bar{X}_f) - E(\bar{X}_m) = 0.44 - 0.36 = 0.08$$

and

$$Var(\bar{X}_f - \bar{X}_m) = Var(\bar{X}_f) + Var(\bar{X}_m) = \frac{(0.44)(0.56) + (0.36)(0.64)}{50} = 0.009536.$$

Thus,

$$\bar{X}_f - \bar{X}_m \overset{a}{\sim} N(0.08, 0.009536),$$

which implies

$$P(\bar{X}_f - \bar{X}_m > 0) \approx 1 - \Phi\left(\frac{-0.08}{\sqrt{0.009536}}\right) \approx 0.7937.$$

```
1-pnorm(-0.08/sqrt(0.009536))

## [1] 0.7936729
```

Example 13.4 (Political polling) *Suppose that a political poll is conducted, where a random sample of voters is asked whether they intend to vote for candidate A or candidate B. Let π denote the true probability that a randomly chosen voter from the population intends to vote for candidate A. How many voters must be polled so the width of the (approximate) 95% probability interval for $\bar{X}$ is no greater than 6 percentage points wide? π is unknown here, which is why the poll is being conducted. The approximate 95% probability interval for $\bar{X}$ is*

$$\left(\pi - 1.96\sqrt{\frac{\pi(1-\pi)}{n}}, \pi + 1.96\sqrt{\frac{\pi(1-\pi)}{n}}\right),$$

so the width of the interval is

$$\pi + 1.96\sqrt{\frac{\pi(1-\pi)}{n}} - \left(\pi - 1.96\sqrt{\frac{\pi(1-\pi)}{n}}\right) = 3.92\sqrt{\frac{\pi(1-\pi)}{n}}.$$

To ensure that the width of the interval is less than or equal to 6 percentage points, we require

$$3.92\sqrt{\frac{\pi(1-\pi)}{n}} \le 0.06 \text{ or, equivalently, } n \ge \left(\frac{3.92}{0.06}\right)^2 \pi(1-\pi).$$

This inequality must hold for any possible value for π*. Since* $\pi(1-\pi)$ *is maximized when* $\pi = 0.5$*, we require*

$$n \ge \left(\frac{3.92}{0.06}\right)^2 (0.5)(1-0.5) \approx 1{,}067.1 \text{ or, equivalently, } n \ge 1{,}068.$$

Therefore, at least 1,068 *voters must be polled to get a* 95% *probability interval for* $\bar{X}$ *that is no greater than 6 percentage points wide.*

Example 13.5 (Simulation error) *In chapter 2, computer simulations illustrated that a probability is the limit to which a long-run frequency converges. For example, figure 2.1 showed the cumulative frequency of heads after* 10,000 *simulations. In that example, since the parameter of the underlying Bernoulli(*π*) random variable is known to be* $\pi = 0.5$*, the asymptotic sampling distribution provides information about the simulation error that would be expected with this many simulations. Specifically, the* 95% *probability interval is*

$$\left(0.5 - 1.96\sqrt{\frac{0.5(1-0.5)}{10000}}, 0.5 + 1.96\sqrt{\frac{0.5(1-0.5)}{10000}}\right) \approx (0.490, 0.510),$$

meaning there is a 95% *probability that the observed heads frequency for* 10,000 *coin tosses is between* 49.0% *and* 51.0%*.*

If the number of simulations is 100,000 *rather than* 10,000*, the* 95% *probability interval for the observed heads frequency is*

$$\left(0.5 - 1.96\sqrt{\frac{0.5(1-0.5)}{100000}}, 0.5 + 1.96\sqrt{\frac{0.5(1-0.5)}{100000}}\right) \approx (0.497, 0.503),$$

and, utilizing the fact that the 99.5% *quantile of a* $N(0,1)$ *random variable is* $\tau_{Z,0.995} \approx 2.576$*, a* 99% *probability interval for the observed heads frequency is*

$$\left(0.5 - 2.576\sqrt{\frac{0.5(1-0.5)}{100000}}, 0.5 + 2.576\sqrt{\frac{0.5(1-0.5)}{100000}}\right) \approx (0.496, 0.504).$$

This example illustrates how increasing the number of simulations leads to reduced simulation error. In fact, using the approach of example 13.4, we can determine the number of simulations needed to get a desired width of the probability interval. In the case of the 99% *confidence interval, to get an interval that has a width less than* 0.002 *(very narrow!), we need*

$$2 \times 2.576\sqrt{\frac{0.5(1-0.5)}{n}} < 0.002 \text{ or, equivalently, } n > \left(\frac{5.152}{0.002}\right)^2 (0.5)(1-0.5) = 1{,}658{,}944.$$

With this many simulated coin tosses, there is a 99% probability that the observed heads frequency is within 0.001, *or* 0.1%, *of the true heads probability of* 0.5, *or* 50%.

13.2 Asymptotic Distribution of the Sample Variance

In chapter 12, the exact sampling distribution for the sample variance was shown for the case of i.i.d. normal random variables. Recall that for i.i.d. $X_1, X_2, \ldots, X_n \sim N(\mu, \sigma^2)$ random variables, the sampling distribution for any sample size n is characterized by

$$(n-1)\frac{s_X^2}{\sigma^2} \sim \chi^2_{n-1}.$$

For other underlying random variables, proposition 12.3 provided the following population statistics for s_X^2, which hold in general for any i.i.d. $X_1, X_2, \ldots, X_n$:

$$\mu_{s_X^2} = E(s_X^2) = \sigma_X^2 \text{ and } \sigma^2_{s_X^2} = Var(s_X^2) = \frac{E\left((X-\mu_X)^4\right)}{n} - \frac{\left(\sigma_X^2\right)^2 (n-3)}{n(n-1)}.$$

Unfortunately, knowing the mean and variance of s_X^2 does not provide a complete characterization of the sampling distribution of s_X^2. The case of i.i.d. normal random variables is a special case where the exact sampling distribution is known, but in general, the exact sampling distribution is not known. In section 12.2.2, for instance, simulation methods were used as an alternative to approximate the sampling distribution of s_X^2 for i.i.d. uniform random variables and i.i.d. log-normal random variables.

Interestingly, while the sample variance s_X^2 is not exactly an sample average, it turns out to have similar large-sample properties to the ones presented in section 13.1 for the sample mean $\bar{X}$. First, the sample variance s_X^2 of i.i.d. random variables gets close to the population variance σ_X^2 for large samples, similar to the result for $\bar{X}$ from the LLN. Second, the sample variance s_X^2 of i.i.d. random variables has an approximate normal distribution for large samples, similar to the result for $\bar{X}$ from the CLT. These two properties are summarized in proposition 13.4.

Proposition 13.4 *If* $X_1, X_2, \ldots, X_n$ *are i.i.d. random variables with finite population mean* μ_X, *finite population variance* σ_X^2, *and* $E\left((X-\mu_X)^4\right) < \infty$, *then*

i. s_X^2 *gets arbitrarily close to* σ_X^2 *as* $n \to \infty$.
ii. *For sufficiently large n,* s_X^2 *is approximately normally distributed:*

$$s_X^2 \overset{a}{\sim} N\left(\sigma_X^2, \frac{E\left((X-\mu_X)^4\right) - \left(\sigma_X^2\right)^2}{n}\right).$$

The variance of the approximate normal distribution, $\frac{E\left((X-\mu_X)^4\right) - \left(\sigma_X^2\right)^2}{n}$, looks slightly different from the expression $\sigma^2_{s_X^2} = Var(s_X^2) = \frac{E\left[(X-\mu_X)^4\right]}{n} - \frac{\left(\sigma_X^2\right)^2 (n-3)}{n(n-1)}$. The reason is that

$\frac{n-3}{n-1} \rightarrow 1$ as $n \rightarrow \infty$, so $\frac{(\sigma_X^2)^2(n-3)}{n(n-1)}$ simplifies to $\frac{(\sigma_X^2)^2}{n}$. As n increases the variance of the approximate normal sampling distribution shrinks to zero since n is in the denominator. This shrinking variance for large n is consistent with the first property, that the sample variance s_X^2 gets arbitrarily close to σ_X^2, which is the mean and center of the asymptotic normal distribution. Again, the asymptotic normal sampling distribution given by this proposition is remarkable, as it holds regardless of the shape of the underlying distribution of the i.i.d. random variables.

Example 13.6 (Normal random variables) *When $X_1, X_2, \ldots, X_n$ are i.i.d. random variables, there is an exact sampling distribution for s_X^2, described by $(n-1)\frac{s_X^2}{\sigma^2} \sim \chi_{n-1}^2$, that holds for any sample size n, including very small sample sizes. For larger samples, there is also an asymptotic distribution given by proposition 13.4. Figure 13.2 shows both sampling distributions (the exact distribution and the asymptotic distribution) of s_X^2 when $X_1, X_2, \ldots, X_n$ are i.i.d. $N(0,1)$ random variables for four sample sizes ($n=10$, $n=20$, $n=50$, and $n=100$), with the exact distribution given by the solid curve and the asymptotic distribution given by the dotted curve. There is some difference between the exact distribution and asymptotic distribution for the smallest sample size ($n=10$), with the exact distribution peaking at a value less than 1 and exhibiting right skewness. At $n=20$, the two distributions are much closer, with the peak for the exact distribution just slightly lower than 1. The two distributions are extremely close to each other at $n=50$ and virtually identical at $n=100$, suggesting the asymptotic normal approximation works well for sample sizes of 50 and larger. That said, even at $n=20$, the normal approximation does pretty well in approximating the exact sampling distribution of s_X^2.*

Example 13.7 (Uniform random variables) *In example 12.7, simulation methods approximated the sampling distribution of s_X^2 for i.i.d. $U(0,1)$ random variables for some small sample sizes, and even with just 10 observations ($n=10$) the sampling distribution appeared symmetric and bell-shaped. For i.i.d. $U(0,1)$ random variables, the appropriate asymptotic distribution can be derived by determining the mean and variance of the normal distribution given in part (ii) of proposition 13.4. First, the mean of the distribution is σ_X^2, which is $\frac{1}{12}$ for $X \sim U(0,1)$. Second, the variance of the distribution is $\frac{E((X-\mu_X)^4)-(\sigma_X^2)^2}{n}$, which can be evaluated by determining $E\left((X-\mu_X)^4\right)$:*

$$E\left((X-\mu_X)^4\right) = \int_0^1 (x-0.5)^4 dx = \left.\frac{(x-0.5)^5}{5}\right|_0^1 = 0.00625 - (-0.00625) = 0.0125 = \frac{1}{80}.$$

Plugging into the variance expression yields

$$\frac{E\left((X-\mu_X)^4\right)-\left(\sigma_X^2\right)^2}{n} = \frac{1/80 - (1/12)^2}{n} = \frac{1}{180n},$$

so the asymptotic distribution is

$$s_X^2 \overset{a}{\sim} N\left(\frac{1}{12}, \frac{1}{180n}\right).$$

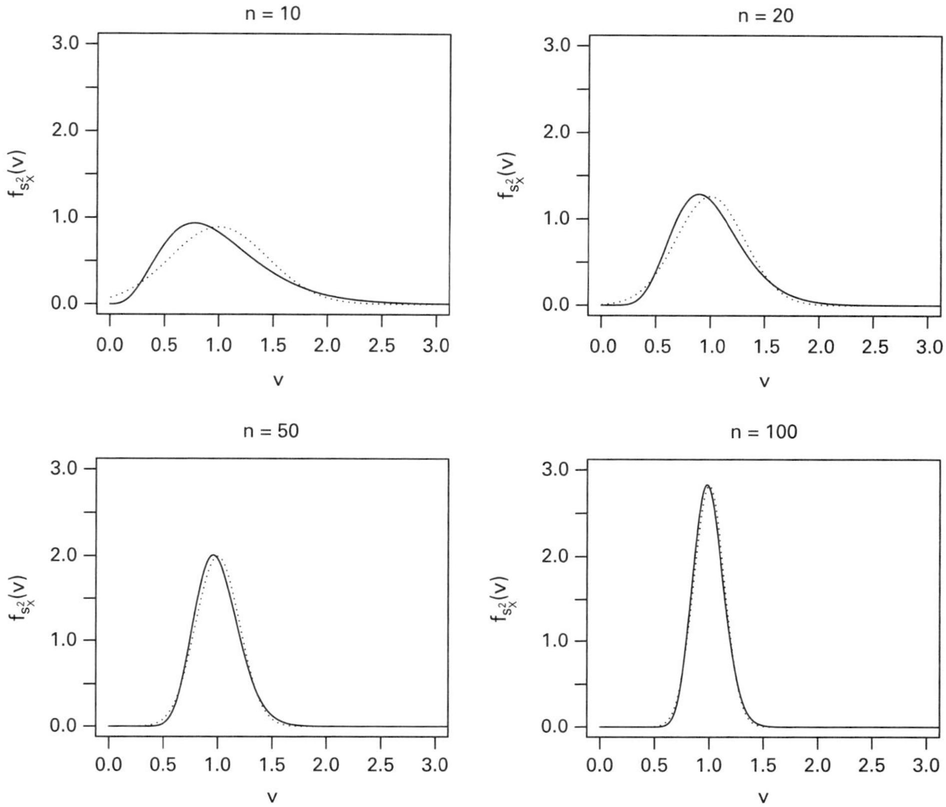

Figure 13.2
Sampling distributions of the sample variance for i.i.d. $N(0, 1)$ random variables

As an example, for $n = 50$*, the distribution is* $s_X^2 \overset{a}{\sim} N\left(\frac{1}{12}, \frac{1}{9000}\right)$*, so that an approximate 95% probability interval for* s_X^2 *is* $\left(\frac{1}{12} - 1.96\frac{1}{\sqrt{9000}}, \frac{1}{12} + 1.96\frac{1}{\sqrt{9000}}\right) \approx (0.0627, 0.1040)$*. There is approximately a 95% probability that the sample variance will be between* 0.0627 *and* 0.1040.

13.2.1 Asymptotic Distribution of the Sample Standard Deviation

We have now seen that the sample mean and the sample variance have asymptotic normal distributions that hold regardless of the distribution of the underlying i.i.d. random variables. The same is true of the sample standard deviation. The sample standard deviation s_X gets arbitrarily close to the population standard deviation σ_X for large samples, and its sampling distribution is approximated by a normal distribution in large samples, as in proposition 13.5.

Proposition 13.5 *If* $X_1, X_2, \ldots, X_n$ *are i.i.d. random variables with finite population mean* μ_X*, finite population variance* σ_X^2*, and* $E\left((X - \mu_X)^4\right) < \infty$*, then*

i. s_X *gets arbitrarily close to* σ_X *as* $n \to \infty$.
ii. *For sufficiently large n,* s_X *is approximately normally distributed:*

$$s_X \overset{a}{\sim} N\left(\sigma_X, \frac{E\left((X-\mu_X)^4\right)-\left(\sigma_X^2\right)^2}{4n\sigma_X^2}\right).$$

The asymptotic variance of s_X is equal to $\frac{1}{4\sigma_X^2}$ times the asymptotic variance of s_X^2 from proposition 13.4. (Given the asymptotic distribution of s_X^2, the asymptotic distribution of s_X follows from a result in statistics known as the *delta method*, discussed in section 14.5.) Thus, if the asymptotic variance of s_X^2 has already been determined, we can usually determine the asymptotic variance of s_X, as illustrated in example 13.8.

Example 13.8 (Uniform random variables) *Continuing example 12.7, the asymptotic distribution for s_X for larger sample sizes can be determined. Since*

$$\frac{E\left((X-\mu_X)^4\right)-\left(\sigma_X^2\right)^2}{n} = \frac{1}{180n},$$

the asymptotic variance of s_X is

$$\frac{E\left((X-\mu_X)^4\right)-\left(\sigma_X^2\right)^2}{4n\sigma_X^2} = \frac{1/180n}{4/12} = \frac{1}{60n},$$

which implies

$$s_X \overset{a}{\sim} N\left(\frac{1}{\sqrt{12}}, \frac{1}{60n}\right).$$

For instance, for $n=50$, the distribution is $s_X \overset{a}{\sim} N\left(\frac{1}{\sqrt{12}}, \frac{1}{3000}\right)$, meaning an approximate 95% probability interval for s_X is $\left(\frac{1}{\sqrt{12}} - 1.96\frac{1}{\sqrt{3000}}, \frac{1}{\sqrt{12}} + 1.96\frac{1}{\sqrt{3000}}\right) \approx$ $(0.2529, 0.3245)$. There is approximately a 95% probability that the sample standard deviation will be between 0.2529 and 0.3245.

13.3 Asymptotic Distribution of Other Statistics

In this chapter, we have seen that a normal distribution can approximate the large-sample sampling distribution of the sample mean $\bar{X}$, the sample variance s_X^2, and the sample standard deviation s_X. As it turns out, nearly all the statistics that have been considered in this book also have an asymptotic sampling distribution that can be approximated by a normal distribution. This phenomenon is extremely useful for the purposes of conducting statistical inference, as the normal distribution can be used across a wide range of statistics when the sample is large.

13.3.1 Asymptotic Distribution of Sample Quantiles and Interquartile Range

Like the sample mean, the sample median has a sampling distribution that is approximately normal for large sample sizes. Recall that $\tilde{X}_{0.5}$ is the random variable associated with the sample median, and $\tilde{x}_{0.5}$ is the sample median that is realized after a random sample is drawn from the population. The sample median gets arbitrarily close to the population median as $n \to \infty$, and the specific form of its asymptotic normal distribution is given in proposition 13.6.

Proposition 13.6 *If $X_1, X_2, \ldots, X_n$ are i.i.d. continuous random variables with pdf $f_X(\cdot)$ and population median $\tau_{X,0.5}$, then*

i. *$\tilde{X}_{0.5}$ gets arbitrarily close to $\tau_{X,0.5}$ as $n \to \infty$.*
ii. *For sufficiently large n, $\tilde{X}_{0.5}$ is approximately normally distributed:*

$$\tilde{X}_{0.5} \overset{a}{\sim} N\left(\tau_{X,0.5}, \frac{1}{4nf_X(\tau_{X,0.5})^2}\right).$$

The asymptotic variance has the familiar $1/n$ scaling seen for the asymptotic distributions of the sample mean, the sample variance, and the sample standard deviation. The asymptotic variance formula contains $f_X(\tau_{X,0.5})$, which is the pdf associated with X evaluated at the population median. With asymptotic distributions available for both the sample mean and the sample median, it can be interesting to compare their respective asymptotic variances, especially in the case of a symmetric distribution where both the population mean and population median are equal to the center of the distribution. Example 13.9 shows how the asymptotic variances of the sample mean and sample median can be compared in the case of the normal distribution.

Example 13.9 (Normal random variables) *Suppose that $X_1, X_2, \ldots, X_n$ are i.i.d. $N(0,1)$ random variables, so $\mu_X = \tau_{X,0.5} = 0$. For the sample mean $\bar{X}$, the asymptotic distribution is $\bar{X} \overset{a}{\sim} N(0, 1/n)$ from the CLT, and even the exact sampling distribution is $\bar{X} \sim N(0, 1/n)$. For the sample median $\tilde{X}_{0.5}$, proposition 13.6 implies the asymptotic distribution $\tilde{X}_{0.5} \overset{a}{\sim} N\left(0, \frac{1}{4n\phi(0)^2}\right)$, where $\phi(\cdot)$ is the pdf associated with the $N(0,1)$ distribution. How do these two asymptotic variances compare? Since $\phi(0) \approx 0.3989$, the asymptotic variance of $\tilde{X}_{0.5}$ is*

$$\frac{1}{4n\phi(0)^2} \approx \frac{1}{4n(0.3989)^2} \approx \frac{1.5711}{n},$$

which is greater than $1/n$, meaning that in large samples, the dispersion associated with the sample median's sampling distribution is larger than the dispersion associated with the sample mean's sampling distribution. To quantify this difference, probability intervals for the two statistics can be formed for a chosen sample size. With 100 observations ($n = 100$), the asymptotic standard deviation of $\bar{X}$ is $1/\sqrt{100} = 0.1$, so a 95% probability interval for $\bar{X}$ is $(-0.196, 0.196)$; the asymptotic standard deviation of $\tilde{X}_{0.5}$ is $\sqrt{1.5711/100} \approx 0.1253$, so a 95% probability interval for $\tilde{X}_{0.5}$ is $(-1.96(0.1253), 1.96(0.1253)) \approx (-0.246, 0.246)$. In the thought experiment where many different 100-observation samples are drawn from the population, this result says that the realizations of the sample mean (the $\bar{x}$ values) tend to be slightly closer to the true center of the distribution (at zero) than the realizations of the sample median (the $\tilde{x}_{0.5}$ values). Over many 100-observations samples, the sample mean is between -0.196 and 0.196 for 95% of the samples, whereas the sample median is between -0.246 and 0.246 for 95% of the samples. So, in the case of the $N(0,1)$ distribution, the sample mean provides a more precise measure of the center of the distribution. On the other hand, recall that the sample median is a more robust measure of the center of the distribution since it is less affected by outliers than the sample mean. There is a precision-robustness trade-off here, with the sample mean being more precise and less robust and the sample median being less precise and more robust. The interested reader can show that *this idea generalizes to other normal random variables (i.e., $X_1, X_2, \ldots, X_n$ i.i.d. $N(\mu, \sigma^2)$ random variables), with the asymptotic variance of the sample mean being larger than the asymptotic variance of the sample median.*

Since other quantiles of a distribution may be of interest, we generalize proposition 13.6 to other sample quantiles. For i.i.d. random variables, any sample quantile gets arbitrarily close to its corresponding population quantile in large samples and has an asymptotic distribution that is normal.[4]

Proposition 13.7 *If $X_1, X_2, \ldots, X_n$ are i.i.d. continuous random variables with pdf $f_X(\cdot)$ and population quantiles $\tau_{X,q}$, then for any $q \in (0,1)$,*

i. *$\tilde{X}_q$ gets arbitrarily close to $\tau_{X,q}$ as $n \to \infty$.*

ii. *For sufficiently large n, $\tilde{X}_q$ is approximately normally distributed:*

$$\tilde{X}_q \overset{a}{\sim} N\left(\tau_{X,q}, \frac{q(1-q)}{nf_X(\tau_{X,q})^2}\right).$$

The sample median ($q = 0.5$) is a special case of proposition 13.7, with $q(1-q) = \frac{1}{4}$ leading to the asymptotic variance in proposition 13.6. Beyond the sample size n, which enters in the usual $1/n$ form, both q and $f_X(\tau_{X,q})$ affect the value of the asymptotic variance, so the asymptotic variance generally varies as q varies. Example 13.10 illustrates how the asymptotic distributions and probability intervals can be determined at different quantiles.

Example 13.10 (Normal random variables) *Continuing example 13.9, suppose that $X_1, X_2, \ldots, X_n$ are i.i.d. $N(0,1)$ random variables, and again consider a sample of 100 observations ($n = 100$). Using the asymptotic variance formula from proposition 13.7, the asymptotic standard deviation for any quantile q is*

$$\frac{\sqrt{q(1-q)}}{\sqrt{100}\phi(\tau_{X,q})}.$$

The 95% probability interval for $\tilde{X}_q$ is constructed as the true quantile $\tau_{X,q}$ plus or minus 1.96 times the asymptotic standard deviation. The following table shows the asymptotic standard deviations and 95% probability intervals for five quantiles ($q = 0.1$, $q = 0.25$, $q = 0.5$, $q = 0.75$, and $q = 0.9$):

q	$\tau_{X,q}$	$\phi(\tau_{X,q})$	$\frac{\sqrt{q(1-q)}}{\sqrt{100}\phi(\tau_{X,q})}$	95% *Interval for* $\tilde{X}_q$
0.1	−1.2816	0.1755	0.1709	(−1.617, −0.947)
0.25	−0.6745	0.3178	0.1363	(−0.942, −0.407)
0.5	0	0.3989	0.1253	(−0.246, 0.246)
0.75	0.6745	0.3178	0.1363	(0.407, 0.942)
0.9	1.2816	0.1755	0.1709	(0.947, 1.617)

For $q = 0.5$, the values in the table correspond to those in example 13.9. While $q(1-q)$ is largest at $q = 0.5$, the asymptotic standard deviation for $q = 0.5$ is actually the smallest among those shown, which occurs since the value of the pdf $\phi(0)$ is much larger than the *pdf $\phi(\cdot)$ evaluated at the other quantiles. The asymptotic standard deviations are largest at the extreme quantiles ($q=0.1$ and $q=0.9$), so the probability intervals for $\tilde{X}_{0.1}$ and $\tilde{X}_{0.9}$ are also the widest. Due to the symmetry of the $N(0,1)$ distribution, the asymptotic standard deviation for $q=0.1$ is the same as that for $q=0.9$, and similarly the asymptotic standard deviation for $q=0.25$ is the same as that for $q=0.75$. As a result, the widths of the probability intervals for $\tilde{X}_{0.1}$ and $\tilde{X}_{0.9}$ are the same, as are the widths of the probability intervals for $\tilde{X}_{0.25}$ and $\tilde{X}_{0.75}$.*

How about the interquartile range (IQR)? Recall that the sample statistic IQR_x is the difference between the 75% sample quantile $\tilde{x}_{0.75}$ and the 25% sample quantile $\tilde{x}_{0.25}$. The associated random variable is $\tilde{X}_{0.75}-\tilde{X}_{0.25}$ for underlying i.i.d. random variables $X_1, X_2, \ldots, X_n$, with the IQR_x statistic being a realization of $\tilde{X}_{0.75}-\tilde{X}_{0.25}$ for the sample that happens to be observed. Then, $\tilde{X}_{0.75}-\tilde{X}_{0.25}$ has an asymptotic normal distribution since, from proposition 13.7, both $\tilde{X}_{0.25}$ and $\tilde{X}_{0.75}$ have asymptotic normal distributions. Proposition 13.8 gives the specific asymptotic normal distribution associated with $\tilde{X}_{0.75}-\tilde{X}_{0.25}$.[5]

Proposition 13.8 *If $X_1, X_2, \ldots, X_n$ are i.i.d. continuous random variables with pdf $f_X(\cdot)$ and population quantiles $\tau_{X,q}$, then*

i. *$\tilde{X}_{0.75}-\tilde{X}_{0.25}$ gets arbitrarily close to $\tau_{X,0.75}-\tau_{X,0.25}$ as $n \to \infty$.*
ii. *For sufficiently large n, $\tilde{X}_{0.75}-\tilde{X}_{0.25}$ is approximately normally distributed:*

$$\tilde{X}_{0.75}-\tilde{X}_{0.25} \overset{a}{\sim} N\left(\tau_{X,0.75}-\tau_{X,0.25},\right.$$
$$\left.\frac{1}{16n}\left(\frac{3}{f_X(\tau_{X,0.25})^2}+\frac{3}{f_X(\tau_{X,0.75})^2}-\frac{2}{f_X(\tau_{X,0.25})f_X(\tau_{X,0.75})}\right)\right).$$

Example 13.11 (Uniform random variables) *Suppose that $X_1, X_2, \ldots, X_n$ are i.i.d. $U(0,1)$ random variables, so $\tau_{X,0.25}=0.25$ and $\tau_{X,0.75}=0.75$. Then, since $f_X(v)=1$ for all $v \in (0,1)$, the asymptotic distribution of $\tilde{X}_{0.75}-\tilde{X}_{0.25}$ is*

$$\tilde{X}_{0.75}-\tilde{X}_{0.25} \overset{a}{\sim} N\left(0.5, \frac{1}{4n}\right).$$

For a sample of 100 observations ($n=100$), the asymptotic variance is $1/400$ and the asymptotic standard deviation is $\sqrt{1/400}=0.05$, so a 95% probability interval for $\tilde{X}_{0.75}-\tilde{X}_{0.25}$ is

$$(0.5-1.96(0.05), 0.5+1.96(0.05)) \approx (0.402, 0.598).$$

Over the possible 100-observation samples that can be drawn from the population, there is a 95% probability that the realized IQR_x is between 0.402 and 0.598.

13.3.2 Asymptotic Distribution of the Sample Correlation

Thus far, the focus of chapter 12 and this chapter has been the sampling distributions of statistics for univariate data. This section considers the large-sample sampling distribution for the sample correlation, a statistic based on bivariate data that measures the linear association between two variables. The focus here is on the sample correlation rather than the sample covariance since, as previously discussed, the sample correlation is far more useful in practice, as it is not affected by the units or scale of the underlying variables.

We first discuss the population and i.i.d. sampling in the context of bivariate data. Let X and Y denote two random variables, which have a population correlation ρ_{XY}. A draw from the population is a realization of (x, y) that comes from the joint distribution of X and Y. When i.i.d. draws are taken from the population to get a sample $\{(x_1, y_1), (x_2, y_2), \ldots, (x_n, y_n)\}$, each draw of an observation pair (x_i, y_i) should provide no information about any other draw of an observation pair (x_j, y_j). While x and y may be related, as would certainly be the case for a nonzero population correlation, the "independent" part of i.i.d. refers to independence across different bivariate draws, not within a bivariate pair. As an example, if data are collected on education and earnings from a random sample of employed individuals, there is likely a relationship between education and earnings, but the random draw of an individual (and their education and earnings values) is independent of the draws of any other individuals in the sample.

For sample size n, let $\{(X_1, Y_1), (X_2, Y_2), \ldots, (X_n, Y_n)\}$ denote the random variables associated with the n draws of bivariate data from the population. For any realized sample, the sample correlation r_{xy} can be calculated. Denote the associated random variable as r_{XY}, where r_{xy} is the realization of r_{XY} that arises from the particular sample that happens to be observed. Proposition 13.9 formally states that the sample correlation r_{XY} gets arbitrarily close to the population correlation ρ_{XY} in large samples and has an asymptotic normal distribution.[6]

Proposition 13.9 *If $(X_1, Y_1), (X_2, Y_2), \ldots, (X_n, Y_n)$ are i.i.d. bivariate random variables with population correlation ρ_{XY}, then*

i. *r_{XY} gets arbitrarily close to ρ_{XY} as $n \to \infty$.*

ii. *For sufficiently large n, r_{XY} is approximately normally distributed:*

$$r_{XY} \overset{a}{\sim} N\left(\rho_{XY}, \frac{(1-\rho_{XY}^2)^2}{n}\right).$$

Interestingly, the asymptotic distribution of the sample correlation depends only on the sample size n and the population correlation ρ_{XY}, not on any other feature of the joint distribution of X and Y. When the population correlation is zero ($\rho_{XY} = 0$), the asymptotic distribution of the sample correlation simplifies to $r_{XY} \overset{a}{\sim} N(0, 1/n)$, in which case a 95% probability interval for r_{XY} is $(-1.96/\sqrt{n}, 1.96/\sqrt{n})$. Thus, even though the population correlation is zero, the observed sample correlation will not be exactly zero, except by some rare coincidence, and a probability interval for r_{XY} can be quantified. For $n = 100$, the 95% probability interval for r_{XY} is $(-0.196, 0.196)$ when $\rho_{XY} = 0$; for $n = 400$, the 95% probability interval for r_{XY} is $(-0.098, 0.098)$ when $\rho_{XY} = 0$; and so on.

Example 13.12 (Bivariate normal random variables) *This example considers two random variables, X and Y, that have a* ***bivariate normal distribution****, which is the case when $aX + bY$ is normal for any values a and b. For bivariate normal random variables X and Y, the marginal distributions of X and Y are both normal (plugging in $a = 1, b = 0$*

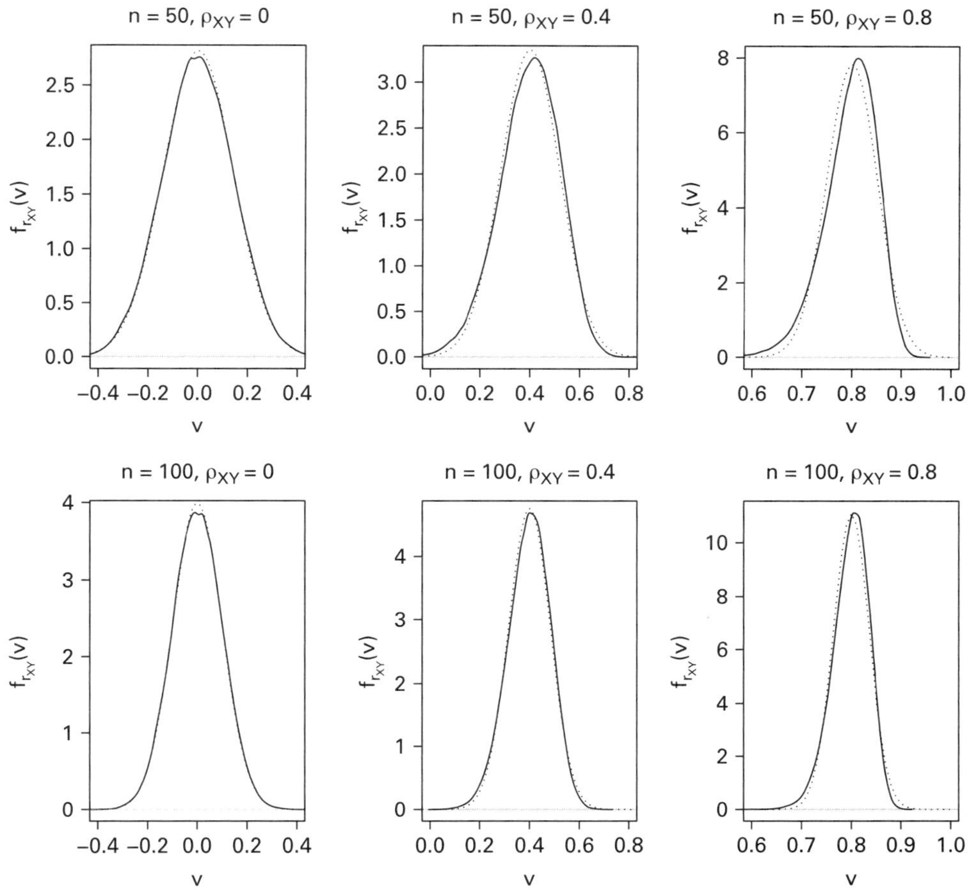

Figure 13.3
Sampling distributions of the sample correlation for bivariate normal random variables

for the former and $a = 0, b = 1$ for the latter). Denoting the marginal distributions as $X \sim N(\mu_X, \sigma_X^2)$ and $Y \sim N(\mu_Y, \sigma_Y^2)$ and the population correlation between X and Y as ρ_{XY}, computer simulations can be used to determine the exact sampling distribution of r_{XY} since most statistical packages provide the ability to draw (x, y) randomly from bivariate normal random variables X and Y.[7] For simplicity, consider the case where the marginal distributions of X and Y are both standard normal, so that $\mu_X = \mu_Y = 0$ and $\sigma_X = \sigma_Y = 1$. Figure 13.3 compares the simulation-based exact sampling distribution of r_{XY} with the asymptotic sampling distribution of r_{XY} for two sample sizes ($n = 50$ and $n = 100$) and three population correlation values ($\rho_{XY} = 0$, $\rho_{XY} = 0.4$, and $\rho_{XY} = 0.8$). For the simulation-based exact sampling distributions, 100,000 simulations are used, and the solid black lines in the graphs show the density plot of the realized r_{xy} values.[8] As a comparison, the dotted curves show the asymptotic normal distribution given in proposition 13.9, which is $r_{XY} \overset{a}{\sim} N\left(0, \frac{1}{n}\right)$ for $\rho_{XY} = 0$, $r_{XY} \overset{a}{\sim} N\left(0.4, \frac{(0.84)^2}{n}\right)$ for $\rho_{XY} = 0.4$, and $r_{XY} \overset{a}{\sim} N\left(0.8, \frac{(0.36)^2}{n}\right)$ for $\rho_{XY} = 0.8$. Even with a sample size of $n = 50$, the asymptotic distributions appear to be quite close to the exact sampling distributions, although there are some small differences *for* $\rho_{XY} = 0.4$ *and slightly bigger differences for* $\rho_{XY} = 0.8$*. Statisticians have previously documented that it takes larger samples for the asymptotic distribution to provide a good approximation when the population correlation* ρ_{XY} *is very large, which is consistent with the evidence from the* $n = 50$ *graphs in the top row. At the larger sample size of* $n = 100$*, there is still a slight discrepancy between the exact sampling distribution and the asymptotic distribution for* $\rho_{XY} = 0.8$*.*

For this example, knowing the specific type of distribution (i.e., the bivariate normal) is only important for being able to simulate the exact sampling distributions of r_{XY}*. Even without knowing the form of the joint distribution of* X *and* Y*, proposition 13.9 provides the large-sample sampling distribution from just the population correlation* ρ_{XY}*. For example, in the case of a sample with* 100 *observations (*$n = 100$*) and* $\rho_{XY} = 0.4$*, the asymptotic distribution of* r_{XY} *is* $r_{XY} \overset{a}{\sim} N\left(0, \frac{(0.84)^2}{100}\right)$*, meaning the asymptotic standard deviation is* $\sqrt{\frac{(0.84)^2}{100}} = 0.084$ *and a 95% probability interval for* r_{XY} *is* $(0.4 - (1.96)(0.084), 0.4 + (1.96)(0.084)) \approx (0.235, 0.565)$*. Thus, over the possible* 100*-observation i.i.d. samples that can be drawn from the population (with* $\rho_{XY} = 0.4$*), there is a 95% probability that the realized sample correlation* r_{xy} *is between* 0.235 *and* 0.565*. For* $n = 100$ *and* $\rho_{XY} = 0.8$*, the asymptotic distribution of* r_{XY} *is* $r_{XY} \overset{a}{\sim} N\left(0, \frac{(0.36)^2}{100}\right)$*, meaning the asymptotic standard deviation is* $\sqrt{\frac{(0.36)^2}{100}} = 0.036$ *and a 95% probability interval for* r_{XY} *is* $(0.8 - (1.96)(0.036), 0.8 + (1.96)(0.036)) \approx (0.729, 0.871)$*.*

13.3.3 Not All Statistics Are Asymptotically Normal

This chapter has shown that many of the statistics considered in this book are associated with random variables that have a normal asymptotic distribution. Such statistics are called *asymptotically normal*, and this property allows easy calculation of probability intervals based on the normal distribution. And, while nearly all the statistics used by practitioners are asymptotically normal, there are instances in which a statistic has a nonnormal sampling distribution even in large samples. While we won't spend much time on statistics that are not asymptotically normal, we illustrate this point by revisiting an example from chapter 12.

Example 13.13 (Maximum of normal random variables) *Example 12.9 considered the exact sampling distribution of* $\max_X = \max(X_1, X_2, \ldots, X_n)$ *when* $X_1, X_2, \ldots, X_n$ *are i.i.d. normal random variables. Formulas for the exact sampling distribution were provided for a given sample size* n *and distributional parameters* μ *and* σ*. For the* $N(0,1)$ *distribution, these formulas were used to graph the sampling distributions for three small sample sizes (*$n = 5$*,* $n = 10$*, and* $n = 30$*). But what happens as the sample size* n *gets large? Does the asymptotic distribution of* $\max_X$ *look normal? Figure 13.4 shows the exact sampling distributions, based upon the formulas from example 12.9, for the much larger sample sizes of* $n = 5{,}000$ *and* $n = 20{,}000$*. Even with these very large sample sizes, the sampling distributions of* $\max_X$ *are asymmetric and right-skewed, suggesting that the asymptotic sampling distribution is nonnormal. In fact, it is known that* $\max_X$ *has an asymptotic distribution known as a* Gumbel distribution *rather than a normal distribution.*

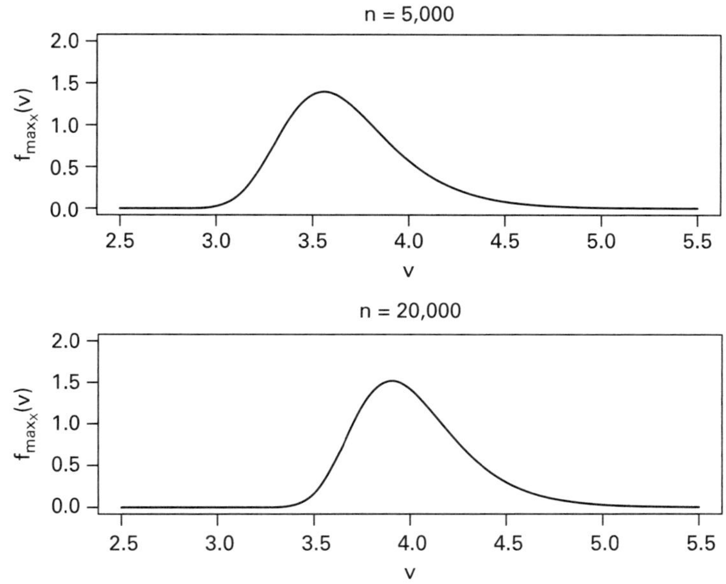

Figure 13.4
Sampling distributions of the sample maximum for i.i.d. $N(0, 1)$ random variables

Notes

1. A more formal statement of the LLN requires a concept known as *convergence in probability*. The mathematical condition corresponding to the phrase "gets arbitrarily close to μ_X" is that, for any $\epsilon > 0$,

$$\lim_{n\to\infty} P\left(|\bar{X} - \mu_X| < \epsilon\right) = 1.$$

No matter how small ϵ is, there is always a large enough sample such that there is a probability arbitrarily close to 1 that the distance between $\bar{X}$ and μ_X is less than ϵ.

2. As with the LLN, a formal statement of the CLT requires more advanced statistical concepts. Specifically, a commonly used version of the CLT uses the concept of *convergence in distribution* and states that the random variable $\sqrt{n}(\bar{X} - \mu_X)$ converges in distribution to the normal distribution $N(0, \sigma_X^2)$.
3. Alternatively, to use the exact distributions of $\bar{X}_f$ and $\bar{X}_m$ (i.e., the scaled binomial random variables), simulation methods can be used to approximate $P(\bar{X}_f > \bar{X}_m)$.
4. This result was first shown in Cramér (1946) .
5. The mean of the asymptotic distribution is $E(\tilde{X}_{0.75}) - E(\tilde{X}_{0.25}) = \tau_{X,0.75} - \tau_{X,0.25}$. The random variables $\tilde{X}_{0.25}$ and $\tilde{X}_{0.75}$ are *not* independent. The asymptotic variance is the sum of the asymptotic variances of $\tilde{X}_{0.25}$ and $\tilde{X}_{0.75}$, each of which is obtained from proposition 13.7, minus two times the asymptotic covariance. The asymptotic covariance of $\tilde{X}_{0.25}$ and $\tilde{X}_{0.75}$ is $\frac{0.25^2}{nf_X(\tau_{X,0.25})f_X(\tau_{X,0.75})}$. This covariance result was first shown in Babu and Rao (1988) .
6. The standard deviation for the sample correlation was first provided by Bowley (1928).
7. For bivariate normal random variables X and Y, where the marginal distributions are $X \sim N(\mu_X, \sigma_X^2)$ and $Y \sim N(\mu_Y, \sigma_Y^2)$, the joint pdf is

$$f_{XY}(x, y) = \frac{1}{2\pi\sigma_X\sigma_Y\sqrt{1-\rho_{XY}^2}} e^{-\frac{1}{2(1-\rho_{XY}^2)}\left[\left(\frac{x-\mu_X}{\sigma_X}\right)^2 + \left(\frac{y-\mu_Y}{\sigma_Y}\right)^2 - 2\rho_{XY}\frac{(x-\mu_X)(y-\mu_Y)}{\sigma_X\sigma_Y}\right]},$$

where ρ_{XY} is the population correlation between X and Y.

8. A straightforward approach is used to sample from bivariate normal random variables. Using $\sigma_X = \sigma_Y = 1$, x is drawn from $X \sim N(0, 1)$ and y is drawn from $N(\rho_{XY}x, 1 - \rho^2_{XY})$. The interested reader can confirm that $Y = \rho_{XY}X + \sqrt{1 - \rho^2_{XY}}Z$, where $Z \sim N(0, 1)$ and Z is independent of X, has a standard normal distribution $N(0, 1)$.

Exercises

1. 50 individuals are randomly selected from a population in which the probability of owning a dog is 25%.
   a. Based on the binomial distribution, what is the probability that the sample proportion of dog owners is strictly greater than 30%?
   b. Based on the binomial distribution, what is the probability that the sample proportion of dog owners is between 25% and 35%?
   c. Based on the normal-distribution approximation to the binomial distribution, what is the probability that the sample proportion of dog owners is between 25% and 35%? Use the continuity correction.
2. Major League Baseball (MLB) teams play 162 games during the regular season. A team's final winning percentage is equal to its total number of wins divided by 162. Assume that any given team has a true win probability and that their wins/losses are 162 i.i.d. draws from a Bernoulli random variable with this win probability.
   a. If team A has the win probability $\pi_A = 0.60$, what is the asymptotic distribution of team A's winning percentage, W_A?
   b. If team A has the win probability $\pi_A = 0.60$ and team B has the win probability $\pi_B = 0.55$, what is the asymptotic distribution of the *difference* in winning percentages, $W_A - W_B$, for the two teams? (Assume that W_A and W_B are independent of each other.)
   c. Based upon your answer to (b), what is $P(W_B > W_A)$?
   d. If you look at winning percentages halfway through the season (i.e., after 81 games) rather than at the end of the season, what is the probability that team B's winning percentage is higher than team A's winning percentage?
   e. Conduct 100,000 simulations in *R* to approximate the probabilities in (c) and (d) using the exact sampling distributions (based on the binomial) of the winning percentage. Use a strict inequality in the simulations so equal winning percentages are not counted.
   f. Returning to asymptotic distributions, how many games in a season would be required to ensure that $P(W_A > W_B)$ is at least 99%?
3. Consider a two-candidate election between candidates A and B. The probability that an older registered voter (aged 65 or over) favors candidate A is 70%. The probability that a younger registered voter (under age 65) favors candidate B is 56%. In the election, 200 older voters and 1,000 younger voters turn out. Assume that the voters who turn out are randomly drawn from the subpopulations of registered voters.

a. What is the asymptotic distribution associated with the sample proportion of votes for candidate A?

b. What is the approximate 95% probability interval for the sample proportion of votes for candidate A?

c. What is the approximate probability that candidate A wins the election?

4. A worker at a data-entry company enters numbers into a spreadsheet and makes errors at a 0.1% rate; that is, the probability that the worker makes an error for any given spreadsheet cell is $\pi = 0.1\%$, or 0.001. Consider the number of errors, given by the random variable X, that the worker makes when entering 50,000 cells of data. Assume that the underlying $Bernoulli(\pi)$ trials are i.i.d.

   a. What is the exact sampling distribution of X?

   b. Using the exact sampling distribution, what is $P(45 \leq X \leq 55)$?

   c. What is the asymptotic distribution of X?

   d. Using the asymptotic distribution and the continuity correction, what is $P(45 \leq X \leq 55)$?

   e. Using the asymptotic distribution, provide a 95% probability interval for X.

   f. Another worker at the company makes errors at a 0.11% rate, slightly higher than the worker described before. Again, assume that the underlying Bernoulli trials are i.i.d.

      i. Using the asymptotic distributions, what is the probability that this worker makes more errors than the other worker if both enter 50,000 cells of data?

      ii. Using the asymptotic distributions, what is a 95% probability interval for the total number of errors for the two workers?

5. Frank's Factory produces computer chips on two separate production lines. The output of each production line is 10,000 computer chips per day, with a 2% probability that any given chip is defective. Assume that the quality/defectiveness of each computer chip is independent.

   a. What is the asymptotic distribution of the sample proportion of defects among the 20,000 chips produced on a given day?

   b. What is the asymptotic distribution of the difference between the total number of defects on one production line and the total number of defects on the other production line?

   c. What is the approximate probability that the magnitude of the difference in (b) is greater than 10? (Do not worry about a continuity correction.)

   d. How would your answer to (c) change if one production line has a defect probability of 2.1% instead of 2%?

6. For a random sample of 400 unemployed workers drawn from the population, the duration of unemployment (in weeks) for each worker is an i.i.d. draw of a random variable X. Given the large sample size, you can assume that the CLT implies that $\bar{X}$ has an approximately normal distribution.

a. If $P(\bar{X} > 21) = 0.5$, what is $E(X)$?

b. If $P(\bar{X} > 21) = 0.4$, what can be said about $E(X)$?

c. If $E(X) = 20.7$ and $\sigma_X = 11.2$, what is $P(\bar{X} > 21)$?

d. How would the answer to (c) change if $n = 1600$ rather than $n = 400$?

7. The Air Quality Index (AQI) is used by the U.S. Environmental Protection Agency (EPA) as an overall measure of air quality. The AQI has a scale of 0 to 500, with lower values for better air quality. For instance, the range 0 to 50 is considered "good," the range 51 to 100 is considered "moderate" (acceptable air quality, minimal risk), the range 101 to 150 is considered "unhealthy for sensitive groups," and higher values indicate even more unhealthy air quality.

   Suppose that 52 weekly AQI measures are taken during the year in both Augusta (in Maine) and Los Angeles (in California). Assume that all AQI measures are independent of each other, with Augusta's measures i.i.d. draws from the Q_A random variable with expected value 25 and standard deviation 15 and Los Angeles's measures i.i.d. draws from the Q_L random variable with expected value 45 and standard deviation 25.

   a. What is the asymptotic distribution of $\bar{Q}_A$, the sample average of AQI for Augusta over 52 weeks?

   b. What is the asymptotic distribution of $\bar{Q}_L$, the sample average of AQI for Los Angeles over 52 weeks?

   c. What is the asymptotic distribution of $\bar{Q}_L - \bar{Q}_A$?

   d. Are you able to say anything about the probability $P(Q_A > Q_L)$ in a given week? Explain why or why not.

8. Cindy's Cereals sells cereal in 20-ounce boxes. Its manufacturing process leads to actual weights (in ounces) that are i.i.d. draws from the random variable $X \sim N\left(20, \frac{1}{900}\right)$. Suppose that it repackages any box weighing less than 19.9 ounces.

   a. What is the probability that any given box is repackaged?

   b. For a manufacturing run of 20,000 boxes, what is the approximate normal distribution of the number of boxes that are repackaged?

   c. The profits per box are P_1 for boxes weighing at least 19.9 ounces and P_2 for boxes weighing less than 19.9 ounces, where $P_1 > P_2$ due to the repackaging required for the latter. For a manufacturing run of 20,000 boxes, what is the approximate normal distribution of total profits in terms of P_1 and P_2?

9. *Allison's Apparel, a women's clothing store, is open eight hours each day. On any given day, it is known that the arrival time for the next customer has an expected value of 5 minutes and a population standard deviation of 3 minutes. (Arrival time is measured since the store opening for the first customer and since the last customer's arrival for each subsequent customer.) Assume that all arrival times are independent of each other. Let T denote the random variable associated with the total number of customers that shop at Allison's Apparel on a given day.

   a. Ignore for a moment that the store eventually closes. What is the asymptotic distribution of the average arrival time (in minutes) for n customers? What is the asymptotic distribution of the total amount of time (in minutes) that it takes n customers to arrive?

b. What is the approximate probability that the total time that it takes 100 customers to arrive is less than 480 minutes?

c. Explain why the probability in (b) is equal to $P(T \geq 100)$.

d. Using the same reasoning as in (b) and (c), approximate $P(T \geq 101)$ and $P(T = 100) = P(T \geq 100) - P(T \geq 101)$.

e. The pmf of T, evaluated at 100, was determined in (d). Using the same reasoning, plot the pmf of T over the range of values $\{80, 81, \ldots, 119, 120\}$ in *R*.

10. The number of visitors X to a popular website during the one minute between 10 a.m. and 10:01 a.m. is a Poisson random variable with $\lambda = 120$.

 a. Thinking of X as the sum of 60 i.i.d. random variables $Y_1, Y_2, \ldots, Y_{60} \sim Poisson(2)$, where each Y_i is the number of visitors in a given second, what is the asymptotic normal distribution associated with X?

 b. Using the asymptotic normal distribution, provide an approximate 90% probability interval for X.

 c. Calculate the exact probability, based on the $Poisson(120)$ distribution, that X is within the interval from (b).

11. For example 13.8, conduct 100,000 simulations with the `replicate` function in *R* to approximate the 95% probability interval for the sample standard deviation and confirm that it is close to $(0.2529, 0.3245)$.

12. Use propositions 13.6 and 13.7 for this question.

 a. What is the asymptotic distribution of the sample median $\tilde{X}_{0.5}$ if $X \sim U(0, 1)$ and $n = 400$?

 b. What is the asymptotic distribution of the sample median $\tilde{X}_{0.5}$ if $X \sim U(a, b)$ and $n = 400$?

 c. What is the asymptotic distribution of the sample 75% quantile $\tilde{X}_{0.75}$ if $X \sim U(0, 1)$ and $n = 400$? Provide an approximate 90% probability interval for $\tilde{X}_{0.75}$.

13. Use proposition 13.8 to determine a 95% probability interval for $\tilde{X}_{0.75} - \tilde{X}_{0.25}$ when $X \sim N(0, 1)$ and $n = 100$.

14. *As mentioned in section 13.3.2, there are some concerns with the approximation provided by the asymptotic distribution of the sample correlation,

$$r_{XY} \overset{a}{\sim} N\left(\rho_{XY}, \frac{(1-\rho_{XY}^2)^2}{n}\right),$$

especially when the magnitude of the population correlation, $|\rho_{XY}|$, is large. In particular, when n is not large, the actual sampling distribution of r_{XY} may be quite asymmetric for large $|\rho_{XY}|$. An alternative approach, proposed by statistician R. A. Fisher (Fisher, 1915) and known as the *Fisher transformation*, can provide more accurate confidence intervals for r_{XY}. The idea is to consider the asymptotic distribution of the following increasing function of r_{XY},

$$\frac{1}{2}\ln\left(\frac{1+r_{XY}}{1-r_{XY}}\right),$$

rather than r_{XY} itself. The asymptotic distribution of the Fisher transformation of r_{XY} is

$$\frac{1}{2}\ln\left(\frac{1+r_{XY}}{1-r_{XY}}\right) \overset{a}{\sim} N\left(\frac{1}{2}\ln\left(\frac{1+\rho_{XY}}{1-\rho_{XY}}\right), \frac{1}{n}\right).$$

a. Provide a 95% probability interval for $\frac{1}{2}\ln\left(\frac{1+r_{XY}}{1-r_{XY}}\right)$ in terms of n if $\rho_{XY}=0$.

b. Suppose that a probability interval for $\frac{1}{2}\ln\left(\frac{1+r_{XY}}{1-r_{XY}}\right)$ has been calculated to be (L, U). That is,

$$P\left(L \leq \frac{1}{2}\ln\left(\frac{1+r_{XY}}{1-r_{XY}}\right) \leq U\right) = p$$

for a probability p. Using this probability expression, construct a probability interval (r_L, r_U) for r_{XY}, with $P(r_L \leq r_{XY} \leq r_U) = p$. (Hint: Exponentiate the three quantities within the parentheses.)

c. How does the 95% probability interval based on the Fisher transformation compare to the 95% probability interval based on the original r_{XY} distribution when $\rho_{XY}=0$ and $n=100$?

d. How does the 95% probability interval based on the Fisher transformation compare to the 95% probability interval based on the original r_{XY} distribution when $\rho_{XY}=0.85$ and $n=40$?

14

Estimation and Confidence Intervals

Building on the concept of sampling distributions from chapters 12 and 13, this chapter introduces estimation, which involves the use of a statistic as a guess, or *estimate*, of an underlying quantity of interest. As an example, the sample mean might be used to estimate the true population mean in the usual situation where it is unknown. The sample mean has been introduced as a statistic based on an observed sample, and it can be viewed as playing two different roles—first as a descriptive statistic for the observed sample and second as an estimate of the unknown population mean. This idea is formalized below and generalized to other statistics that may be used for estimation. In addition, to quantify the precision associated with a given estimate of an underlying quantity of interest, this chapter also introduces confidence intervals, which provide a range of plausible values for the quantity of interest based on the estimation procedure and some prespecified confidence level. In the case of the sample mean, for instance, methods to construct confidence intervals for the unknown population mean, based on the realization of the sample mean, are introduced.

14.1 Estimation and Properties of Estimators

14.1.1 Terminology and Notation

Some new terminology is required for estimation, as introduced in definition 14.1.

Definition 14.1 *An* ***estimator*** *is the random variable associated with a statistic whose purpose is to estimate a quantity of interest, known as the* ***estimand*** *or* ***parameter****. The realization of the estimator, which is the calculated statistic based upon an observed sample of data, is known as the* ***estimate****.*

We introduce some general notation associated with definition 14.1. If $s(\cdot)$ is the statistic for estimation, let the estimand or parameter be denoted as θ, the estimator be denoted as

$$\hat{\theta}_X = s(X_1, X_2, \ldots, X_n),$$

and the estimate be denoted as

$$\hat{\theta}_x = s(x_1, x_2, \ldots, x_n).$$

The estimator $\hat{\theta}_X$ is a random variable before the sample is observed, and the estimate $\hat{\theta}_x$ is the realization of the random variable $\hat{\theta}_X$ that depends on the actual observed sample.

For instance, if the sample mean is used as an estimator of the population mean, the estimand or parameter is

$$\theta = \mu_X,$$

and the estimator and estimate are, respectively,

$$\hat{\theta}_X = \bar{X} = \frac{1}{n}\sum_{i=1}^{n} X_i \text{ and } \hat{\theta}_x = \bar{x} = \frac{1}{n}\sum_{i=1}^{n} x_i.$$

Similarly, if the sample variance is used as an estimator of the population variance, the estimand or parameter is

$$\theta = \sigma_X^2,$$

and the estimator and estimate are, respectively,

$$\hat{\theta}_X = s_X^2 = \frac{1}{n-1}\sum_{i=1}^{n}(X_i - \bar{X})^2 \text{ and } \hat{\theta}_x = s_x^2 = \frac{1}{n-1}\sum_{i=1}^{n}(x_i - \bar{x})^2.$$

In the interest of brevity, and without loss of generality, the notation here specifies a statistic that is a function of univariate data. For these definitions and everything that follow in this chapter, we can generalize to the case of bivariate data, or even data with more than two variables. For instance, the estimator with bivariate data is denoted as

$$\hat{\theta}_{XY} = s((X_1, Y_1), (X_2, Y_2), \ldots, (X_n, Y_n)),$$

and the estimate is denoted as

$$\hat{\theta}_{xy} = s((x_1, y_1), (x_2, y_2), \ldots, (x_n, y_n)).$$

The following table provides a more complete list of the descriptive statistics considered thus far. For each descriptive statistic, the estimand is the underlying population quantity estimated by the descriptive statistic. For instance, the sample median estimates the population median $\tau_{X,0.5}$, with the estimator $\tilde{X}_{0.5}$ representing the random variable associated with the sample median *before* the sample is observed and the estimate $\tilde{x}_{0.5}$ representing the realized sample median *after* the sample is observed. Similarly, the sample correlation estimates the population correlation ρ_{XY}, with the estimator r_{XY} representing the random sample correlation before the sample is observed and the estimate r_{xy} representing the realized sample correlation after the sample is observed.

	Estimand θ	Estimator $\hat{\theta}_X$	Estimate $\hat{\theta}_x$
Mean	μ_X	$\bar{X}$	$\bar{x}$
Variance	σ_X^2	s_X^2	s_x^2
Standard deviation	σ_X	s_X	s_x
Median	$\tau_{X,0.5}$	$\tilde{X}_{0.5}$	$\tilde{x}_{0.5}$
Quantile	$\tau_{X,q}$	$\tilde{X}_q$	$\tilde{x}_q$
IQR	$\tau_{X,0.75} - \tau_{X,0.25}$	$\tilde{X}_{0.75} - \tilde{X}_{0.25}$	$\tilde{x}_{0.75} - \tilde{x}_{0.25}$
Correlation	ρ_{XY}	r_{XY}	r_{xy}

14.1.2 Properties of Estimators

There are various properties that describe how an estimator should be expected to perform. The first property that we introduce is **unbiasedness**, which says that, on average, an estimator provides the right answer. The formal definition of an **unbiased estimator** is provided in definition 14.2.

Definition 14.2 *An estimator $\hat{\theta}_X = s(X_1, X_2, \ldots, X_n)$ is an **unbiased estimator** of θ if*

$$E\left(\hat{\theta}_X\right) = E(s(X_1, X_2, \ldots, X_n)) = \theta$$

*for any sample size n. If $E\left(\hat{\theta}_X\right) \neq \theta$, then $\hat{\theta}_X$ is a **biased estimator**.*

Unbiasedness is known as a *finite-sample property* since the $E\left(\hat{\theta}_X\right) = \theta$ condition must hold regardless of the sample size n for the estimator $\hat{\theta}_X$ to be unbiased. The sample mean $\bar{X}$, as an estimator of the population mean μ_X, and the sample variance s_X^2, as an estimator of the population variance σ_X^2, are two examples of unbiased estimators.

Proposition 14.1 *(i) The sample mean $\bar{X} = \frac{1}{n}\sum_{i=1}^{n} X_i$ is an unbiased estimator of the population mean $\mu_X = E(X)$. (ii) The sample variance $s_X^2 = \frac{1}{n-1}\sum_{i=1}^{n}(X_i - \bar{X})^2$ is an unbiased estimator of the population variance $\sigma_X^2 = Var(X) = E\left((X - \mu_X)^2\right)$.*

In fact, the unbiasedness of the sample mean and the sample variance was already discussed in chapter 12, and proposition 14.1 gives a restatement of results from that chapter. The unbiasedness property means that, on average, the estimator gives the right answer (the estimand). More precisely, over all the possible samples of size n that can be drawn from the population, the expected value of the sample mean $\bar{X}$ equals the population mean μ_X. Similarly, over all the possible samples of size n that can be drawn from the population, the expected value of the sample variance s_X^2 equals the population variance σ_X^2. The $\frac{1}{n-1}$ scaling for s_X^2 is required to make it an unbiased estimator, as it would be biased if the scaling were $\frac{1}{n}$ instead.

The unbiasedness of the sample mean in part (i) of proposition 14.1 has several interesting applications to random variable models discussed previously in this book:

- For independent and identically distributed (i.i.d.) *Bernoulli*(π) random variables, the sample mean $\bar{X}$ is an unbiased estimator of the success probability π, which also means that $n\bar{X}$ is an unbiased estimator of the population mean $n\pi$ of a *Binomial*(n, π) random variable.
- For i.i.d. *Poisson*(λ) random variables, the sample mean $\bar{X}$ is an unbiased estimator of λ since $\mu_X = \lambda$.
- For i.i.d. $N(\mu, \sigma^2)$ random variables, the sample mean $\bar{X}$ is an unbiased estimator of μ.
- For i.i.d. *Exp*(θ) random variables, the sample mean $\bar{X}$ is an unbiased estimator of $\frac{1}{\theta}$ since $\mu_X = \frac{1}{\theta}$.

While the sample mean and sample variance are unbiased, many other estimators are not. For instance, the sample standard deviation and the sample correlation are generally biased estimators, with $E(s_X) \neq \sigma_X$ and $E(r_{XY}) \neq \rho_{XY}$. In fact, among the estimators listed in the table in section 14.1.1, only the sample mean and sample variance are guaranteed to be unbiased estimators. While having a biased estimator might seem problematic, the bias of estimators like s_X or r_{XY} is generally an issue only in very small samples. Rather than concerning ourselves with what happens in small samples, a much more important property of an estimator $\hat{\theta}_X$ is that it gets close to the "right answer" (i.e., the estimand θ) for large sample sizes. This property is known as the **consistency** of an estimator, and the formal definition of a **consistent estimator** is provided in definition 14.3.

Definition 14.3 *An estimator $\hat{\theta}_X = s(X_1, X_2, \ldots, X_n)$ is a **consistent estimator** of θ if $\hat{\theta}_X$ gets arbitrarily close to θ as $n \to \infty$.*

Consistency is generally considered the minimal requirement for a statistical estimator to be useful in practice. The consistency of many estimators, including all those listed in the table in section 14.1.1, has already been stated in the propositions in chapter 13, when the large-sample or asymptotic sampling distributions of various descriptive statistics were discussed. For example, the Law of Large Numbers (LLN; described in proposition 13.1) states that the sample mean $\bar{X}$, when viewed as an estimator, is a consistent estimator of the population mean μ_X since $\bar{X} = \frac{1}{n}\sum_{i=1}^{n}$ gets arbitrarily close to μ_X as $n \to \infty$. As another example, for the sample correlation, part (i) of proposition 13.9 states that r_{XY} is a consistent estimator of the population correlation ρ_{XY}. Thus, even though r_{XY} may be a biased estimator of ρ_{XY}, r_{XY} still gets arbitrarily close to ρ_{XY} as $n \to \infty$. Thus, for a large sample, the sample correlation is an appropriate estimator for the population correlation.

In addition to the consistency properties provided in chapter 13 for several descriptive statistics, as estimators of their associated population quantities, the propositions in that chapter also stated that each of these statistics has an asymptotic sampling distribution that is normally distributed. When an estimator has an asymptotic sampling distribution that is normally distributed, it is said to be an **asymptotically normal estimator**, as in definition 14.4.

Definition 14.4 *An estimator $\hat{\theta}_X = s(X_1, X_2, \ldots, X_n)$ is said to be a **$\sqrt{n}$-consistent and asymptotically normal estimator** (or, more concisely, an **asymptotically normal estimator**) if*

$$\hat{\theta}_X \overset{a}{\sim} N\left(\theta, \frac{V}{n}\right)$$

for some V that does not depend on n, or equivalently,

$$\sqrt{n}\left(\hat{\theta}_X - \theta\right) \overset{a}{\sim} N(0, V).$$

For example, from the Central Limit Theorem (CLT; described in proposition 13.2), the sample mean $\bar{X}$ is an asymptotically normal estimator of the population mean μ_X, with

$$\bar{X} \overset{a}{\sim} N\left(\mu_X, \frac{\sigma_X^2}{n}\right) \text{ or, equivalently, } \sqrt{n}\left(\bar{X} - \mu_X\right) \overset{a}{\sim} N\left(0, \sigma_X^2\right).$$

The "$\sqrt{n}$-consistent" phrase in definition 14.4 refers to the rate at which the estimator $\bar{X}$ approaches the estimand μ_X. To see why that is the case, note that the asymptotic standard deviation of $\bar{X}$ is $\frac{\sigma_X}{\sqrt{n}}$, so the width of any probability interval for $\bar{X}$ is proportional to $\frac{1}{\sqrt{n}}$.

The proportionality factor $\frac{1}{\sqrt{n}}$ is not unique to the sample mean; rather, it is a general feature of any $\sqrt{n}$-consistent and asymptotically normal estimator. From definition 14.4, if the asymptotic variance of an estimator $\hat{\theta}_X$ is $\frac{V}{n}$, where V does not depend on n, the asymptotic standard deviation of $\hat{\theta}_X$ is $\frac{\sqrt{V}}{\sqrt{n}}$, which is proportional to $\frac{1}{\sqrt{n}}$.

As another example, part (ii) of proposition 13.9 provides the asymptotic distribution of the sample correlation r_{XY}:

$$r_{XY} \overset{a}{\sim} N\left(\rho_{XY}, \frac{(1-\rho_{XY}^2)^2}{n}\right) \text{ or, equivalently, } \sqrt{n}(r_{XY} - \rho_{XY}) \overset{a}{\sim} N\left(0, (1-\rho_{XY}^2)^2\right).$$

The asymptotic standard deviation of the estimator r_{XY} is $\frac{1-\rho_{XY}^2}{\sqrt{n}}$. As the sample size n increases, the standard deviation of r_{XY} shrinks toward zero, corresponding to the consistency of the sample correlation r_{XY}, which says that r_{XY} gets arbitrarily close to ρ_{XY} as $n \to \infty$.

14.1.3 Asymptotic Efficiency

What if there are two or more estimators that can be used to estimate some underlying parameter θ? Is it possible to say that one estimator is "better" than another? To be more precise, consider a case in which there are two alternative estimators $\hat{\theta}_X^a$ and $\hat{\theta}_X^b$, both of which are consistent and asymptotically normal estimators of θ.[1] As an example, for the population mean μ_X, we could specify $\hat{\theta}_X^a$ as the usual sample mean $\bar{X}$ and $\hat{\theta}_X^b$ as the sample mean based on only the first half of the observations. Assuming that n is even, so $n/2$ is an integer, these estimators are

$$\hat{\theta}_X^a = \frac{1}{n}\sum_{i=1}^{n} X_i \text{ and } \hat{\theta}_X^b = \frac{1}{n/2}\sum_{i=1}^{n/2} X_i.$$

Both estimators are consistent and asymptotically normal estimators of μ_X, but it seems as though $\hat{\theta}_X^a$ should be preferred since it is based on more information, using all n observations, as opposed to the $n/2$ observations used for $\hat{\theta}_X^b$. The way to quantify that $\hat{\theta}_X^a$ is a more precise estimator than $\hat{\theta}_X^b$ is to compare the asymptotic variances or the asymptotic standard deviations of the two estimators. The asymptotic variances of $\hat{\theta}_X^a$ and $\hat{\theta}_X^b$ are, respectively,

$$\frac{\sigma_X^2}{n} \text{ and } \frac{\sigma_X^2}{n/2} = \frac{2\sigma_X^2}{n};$$

and the asymptotic standard deviations of $\hat{\theta}_X^a$ and $\hat{\theta}_X^b$ are, respectively,

$$\frac{\sigma_X}{\sqrt{n}} \text{ and } \frac{\sqrt{2}\sigma_X}{\sqrt{n}}.$$

Therefore, the asymptotic standard deviation of the half-sample mean $\hat{\theta}_X^b$ is $\sqrt{2}$ times larger than the asymptotic standard deviation of the full-sample mean $\hat{\theta}_X^a$. On this basis, the full-sample mean $\hat{\theta}_X^a$ should be the preferred estimator. In statistical terminology, it is said that the estimator $\hat{\theta}_X^a$ is *more efficient* than the estimator $\hat{\theta}_X^b$, an idea that is stated more generally in definition 14.5.

Definition 14.5 *If there are two asymptotically normal estimators $\hat{\theta}_X^a$ and $\hat{\theta}_X^b$ of parameter θ, the estimator $\hat{\theta}_X^a$ is (asymptotically)* ***more efficient*** *than estimator $\hat{\theta}_X^b$ if the asymptotic variance of $\hat{\theta}_X^a$ is less than the asymptotic variance of estimator $\hat{\theta}_X^b$. Estimator $\hat{\theta}_X$ is* ***asymptotically efficient*** *among all asymptotically normal estimators if it is more efficient than any other asymptotically normal estimator.*

For the previous example, the full-sample mean $\hat{\theta}_X^a$ is more efficient than the half-sample mean $\hat{\theta}_X^b$, but that doesn't necessarily mean that there isn't some other estimator that is more efficient than $\hat{\theta}_X^a$. That is, we have not shown that the sample mean $\bar{X}$ is the asymptotically efficient estimator since that would require showing that it has a lower asymptotic variance than *any* other asymptotically normal estimator of μ_X.

Example 14.1 (Normal random variables, mean versus median) *For a normal random variable $N(\mu, \sigma^2)$, the population mean and population median both equal μ. The sample mean and sample median are two possible estimators of parameter μ. Example 13.6 considered the standard normal distribution, where $X_1, X_2, \ldots, X_n$ are i.i.d. $N(0,1)$ random variables and $\mu = 0$. In that example, it was shown that the asymptotic variance of the sample mean $\bar{X}$ was smaller than the asymptotic variance for the sample median $\tilde{X}_{0.5}$ since*

$$\bar{X} \overset{a}{\sim} N(0, 1/n)$$

and

$$\tilde{X}_{0.5} \overset{a}{\sim} N\left(0, \frac{1}{4n\phi(0)^2}\right),$$

with $\frac{1}{4n\phi(0)^2} \approx \frac{1}{4n(0.3989)^2} \approx \frac{1.5711}{n} > \frac{1}{n}$. Therefore, when the underlying random variables have a standard normal distribution, the sample mean $\bar{X}$ is a more efficient estimator of $\mu = 0$ than the sample median $\tilde{X}_{0.5}$. The interested reader can show that this result holds for the general case of i.i.d. $X_1, X_2, \ldots, X_n \sim N(\mu, \sigma^2)$ random variables.

14.2 Finite-Sample Confidence Intervals: Population Mean of i.i.d. Normal Random Variables

This chapter considers how confidence intervals for parameters (estimands) can be constructed for a wide range of estimators. For unknown parameter θ, the goal is to provide an interval of plausible values for θ based on estimator $\hat{\theta}_X$. For example, a 95% confidence interval for θ is an interval for which, before observing the sample, there is a 95% probability that parameter θ is in the interval created by the estimation procedure. The first case considered in this section is a confidence interval for the population mean μ of normally distributed i.i.d. random variables $X_1, X_2, \ldots, X_n \sim N(\mu, \sigma^2)$. For this case, the exact sampling distribution results for the sample mean $\bar{X}$ from section 12.1.2 are used to derive a confidence interval for μ. The resulting confidence interval is an example of a finite-sample or exact confidence interval, as it is valid for any sample size n, even very small n. While this specific case is interesting and sometimes useful, the resulting confidence interval does not generalize to other settings. Therefore, to provide a more general method of constructing confidence intervals, subsequent sections of this chapter consider the appropriate confidence interval based on any asymptotically normal estimator.

The resulting confidence interval is an asymptotic confidence interval valid in large samples.

This section considers an observed sample $\{x_1, x_2, \ldots, x_n\}$, where the underlying i.i.d. random variables $X_1, X_2, \ldots, X_n$ have a normal distribution $N(\mu, \sigma^2)$. The goal is to construct a confidence interval for parameter μ based on the sample mean estimator $\bar{X}$. From section 12.1.2, the exact sampling distribution for the sample mean $\bar{X}$ is

$$\bar{X} \sim N\left(\mu, \frac{\sigma^2}{n}\right) \text{ or, equivalently, } \frac{\bar{X}-\mu}{\sigma/\sqrt{n}} \sim N(0,1).$$

Based on this exact sampling distribution, we can construct a probability interval for $\bar{X}$ when μ and σ^2 are *assumed to be known*. For instance, a 95% probability interval for $\bar{X}$ is

$$\left(\mu - 1.96\frac{\sigma}{\sqrt{n}}, \mu + 1.96\frac{\sigma}{\sqrt{n}}\right),$$

meaning that, over all possible samples of size n that can be drawn from the population, the probability that $\bar{X}$ is in the $\left(\mu - 1.96\frac{\sigma}{\sqrt{n}}, \mu + 1.96\frac{\sigma}{\sqrt{n}}\right)$ interval is equal to 0.95:

$$P\left(\bar{X} \in \left(\mu - 1.96\frac{\sigma}{\sqrt{n}}, \mu + 1.96\frac{\sigma}{\sqrt{n}}\right)\right) = 0.95.$$

In most cases of interest, however, the parameters μ and σ^2 of the normal distribution are not known to researchers. As a result, we would like to flip things around here, forming a probability interval for parameter μ based on the estimator rather than a probability for the estimator based on parameters μ and σ^2. To do so, the first step is to rewrite the exact sampling distribution in terms of the difference $\bar{X} - \mu$, as follows:

$$\bar{X} - \mu \sim N\left(0, \frac{\sigma^2}{n}\right) \text{ or, equivalently, } \mu - \bar{X} \sim N\left(0, \frac{\sigma^2}{n}\right).$$

Both $\bar{X} - \mu$ and $\mu - \bar{X}$ have the same $N\left(0, \frac{\sigma^2}{n}\right)$ distribution due to the symmetry of the normal distribution. The latter distribution says that $\mu - \bar{X}$ is normally distributed, with mean zero and standard deviation $\frac{\sigma}{\sqrt{n}}$. Therefore, over all possible samples of size n from the population, there is a 95% probability that $\mu - \bar{X}$ is between $-1.96\frac{\sigma}{\sqrt{n}}$ and $1.96\frac{\sigma}{\sqrt{n}}$:

$$P\left(\mu - \bar{X} \in \left(-1.96\frac{\sigma}{\sqrt{n}}, 1.96\frac{\sigma}{\sqrt{n}}\right)\right) = 0.95$$

or, equivalently, by adding $\bar{X}$ to the $\mu - \bar{X}$ term and to the interval end points:

$$P\left(\mu \in \left(\bar{X} - 1.96\frac{\sigma}{\sqrt{n}}, \bar{X} + 1.96\frac{\sigma}{\sqrt{n}}\right)\right) = 0.95.$$

If σ were known, this probability interval would be useful as a confidence interval for μ. Unfortunately, since σ is unknown, this probability interval is not directly applicable since the end points can't be calculated.

Rather than using the unknown population standard deviation σ, a sensible alternative is to use the sample standard deviation s_X, which is an estimator of σ. The complication introduced by using the estimator s_X in place of the parameter σ is that, while $\frac{\bar{X}-\mu}{\sigma/\sqrt{n}} \sim N(0,1)$, the same is not true for the exact sampling distribution of $\frac{\bar{X}-\mu}{s_X/\sqrt{n}}$. In fact, $\frac{\bar{X}-\mu}{s_X/\sqrt{n}}$ does not have a normal distribution, but rather a distribution known as a ***t*-distribution**, as stated in proposition 14.2.

Proposition 14.2 *If $X_1, X_2, \ldots, X_n$ are i.i.d. $N(\mu, \sigma^2)$ random variables,*

$$\frac{\bar{X}-\mu}{s_X/\sqrt{n}} \sim t_{n-1},$$

*where $\bar{X}$ is the sample mean, s_X is the sample standard deviation, and t_{n-1} is a **t-distribution with** $n-1$ **degrees of freedom**.*

The exact form of the *t*-distribution, as given by the pdf or cdf, is somewhat complicated, and therefore not explicitly shown here. The following *R* functions are useful for working with random variables that follow a *t*-distribution:

- `dt(x, df)`: Returns the pdf of a *t*-distributed random variable with `df` degrees of freedom evaluated at argument `x`, which may be a single number or a vector.
- `pt(x, df)`: Returns the cdf of a *t*-distributed random variable with `df` degrees of freedom evaluated at argument `x`, which may be a single number or a vector.
- `rt(n, df)`: Creates a vector of `n` i.i.d. random draws of a *t*-distributed random variable with `df` degrees of freedom.
- `qt(p, df)`: Returns the population quantiles of a *t*-distributed random variable with `df` degrees of freedom specified by argument `p`, which may be a single number or a vector.

The *t*-distribution has the following properties:

- t_{n-1} is symmetric and bell-shaped.
- t_{n-1} has thicker tails than the standard normal $N(0,1)$ distribution.
- The tails of t_{n-1} become less thick as n increases, and the limit (as $n \to \infty$) of the t_{n-1} distribution is the standard normal $N(0,1)$ distribution.

To illustrate these properties of *t*-distributions, figure 14.1 shows the distributions for four values (3, 5, 10, and 30) of the degrees of freedom. For each of the graphs, the pdf of the *t*-distribution is indicated by the black curve, and the standard normal pdf is shown as a dotted curve for comparison. The thicker tails, also associated with a lower peak at zero, are quite evident in the top two graphs, for degrees of freedom equal to 3 and 5. For degrees of freedom equal to 10, the *t*-distribution is closer to the $N(0,1)$ distribution, with less thick tails than the graphs in the top row. For degrees of freedom equal to 30, the *t*-distribution is nearly indistinguishable from the $N(0,1)$ distribution.

The intuition is that using the estimator s_X in place of the true σ introduces uncertainty into the ratio $\frac{\bar{X}-\mu}{s_X/\sqrt{n}}$, and the thicker tails of the t_{n-1} distribution, as compared to the $N(0,1)$ distribution, account for this additional uncertainty. The uncertainty is larger for smaller sample sizes since s_X is a less precise estimator of σ when n is small, reflected by the thicker tails associated with lower degrees of freedom in the *t*-distribution. For larger sample sizes,

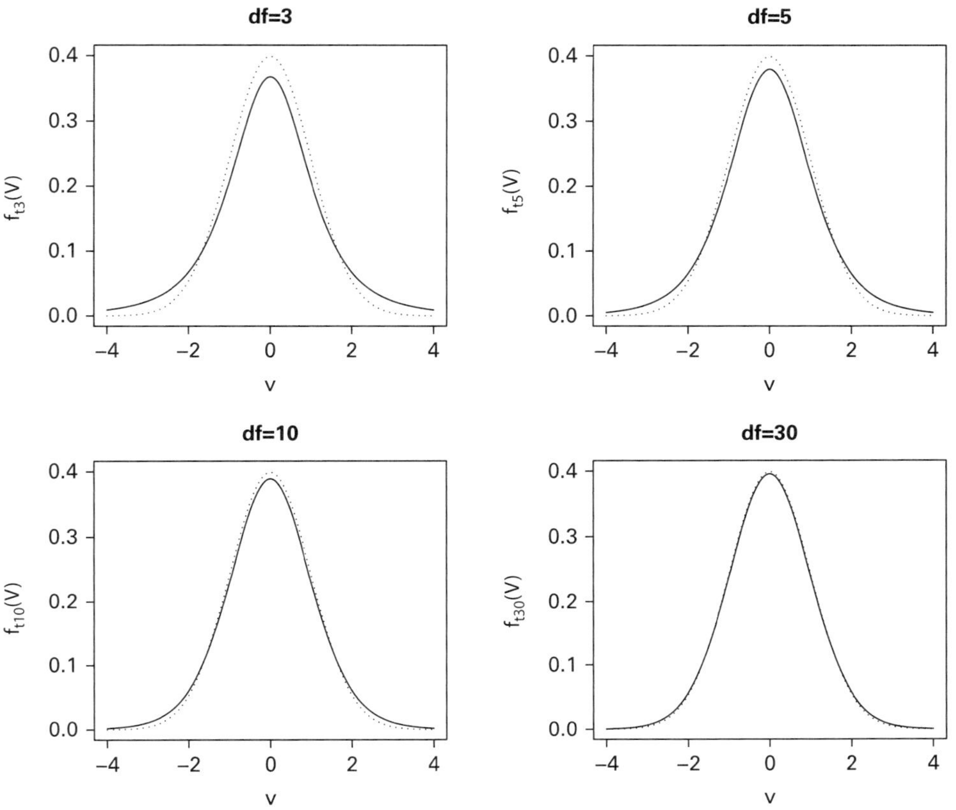

Figure 14.1
t-distributions for various degrees of freedom

s_X becomes more precise as an estimator of σ, corresponding to thinner tails for the t_{n-1} distribution. Since s_X gets arbitrarily close to σ as n increases, it should not be surprising that the limit of the t_{n-1} distribution is the $N(0, 1)$ distribution, as $\frac{\bar{X}-\mu}{s_X/\sqrt{n}} \approx \frac{\bar{X}-\mu}{\sigma/\sqrt{n}}$ for large n.

To quantify the difference in probability intervals associated with a t-distribution rather than a $N(0, 1)$ distribution, the following table provides the 97.5% quantile (used to construct a 95% probability interval) and the 95% quantile (used to construct a 90% probability interval) for several values of the degrees of freedom:

Sample size n	Distribution (t_{n-1})	97.5% quantile ($t_{n-1,0.025}$)	95% quantile ($t_{n-1,0.05}$)
4	t_3	3.182	2.353
6	t_5	2.571	2.015
11	t_{10}	2.228	1.812
31	t_{30}	2.042	1.697
101	t_{100}	1.984	1.660
501	t_{500}	1.965	1.648
“large”	$N(0, 1)$	1.960	1.645

The table introduces new notation corresponding to the quantiles of the t_{n-1} distribution, specifically the $t_{n-1,0.025}$ notation in the 97.5% quantile column and the $t_{n-1,0.05}$ notation in the 95% quantile column. This notation is formally defined as in definition 14.6.

Definition 14.6 *The **critical value** $t_{n-1,q}$ denotes the $(1-q)$ quantile of the t_{n-1} distribution. For example, $t_{n-1,0.025}$ is the 97.5% quantile of the t_{n-1} distribution, and $t_{n-1,0.05}$ is the 95% quantile of the t_{n-1} distribution.*

For the t_{30} distribution, corresponding to $n=31$, the 97.5% quantile $t_{30,0.025}$ is 2.042, which is approximately 4% larger than the 97.5% quantile (1.960) of the $N(0, 1)$ distribution, meaning that a 95% probability interval based on the t_{30} distribution has a width that is approximately 4% larger than the interval based on the $N(0, 1)$ distribution. As n increases, the quantile values for the t_{n-1} distribution approach those of the $N(0, 1)$ distribution. For $n=101$ and the t_{100} distribution, the quantile values in the table are approximately 1% larger than the quantile values for the $N(0, 1)$ distribution. The difference becomes almost entirely negligible for $n=501$ and the t_{500} distribution, with the quantiles only about 0.2% larger than the $N(0, 1)$ quantiles.

The following *R* code shows how the critical values in this table are calculated with the `qt` function:

```
qt(c(0.975,0.95),3)

## [1] 3.182446 2.353363

qt(c(0.975,0.95),5)

## [1] 2.570582 2.015048

qt(c(0.975,0.95),10)

## [1] 2.228139 1.812461

qt(c(0.975,0.95),30)

## [1] 2.042272 1.697261

qt(c(0.975,0.95),100)

## [1] 1.983972 1.660234

qt(c(0.975,0.95),500)

## [1] 1.964720 1.647907
```

Figure 14.2 provides some graphical examples of critical values. The top two graphs show the 2.5% critical values for the t_5 and t_{30} distributions. In the top-left graph, the gray area to the right of the $t_{5,0.025} \approx 2.571$ critical value has a probability of 2.5%, and due to symmetry, the gray area to the left of $-t_{5,0.025}$ has a probability of 2.5%. Therefore, the probability for the interval between $-t_{5,0.025}$ and $t_{5,0.025}$ is equal to 95%. The top-right graph, for the t_{30} distribution, is similar, except that the critical value $t_{30,0.025} \approx 2.042$ is not as large as the $t_{5,0.025} \approx 2.571$ critical value. Since the t_{30} distribution has less thick tails than the t_5 distribution, the critical value must be farther to the left to still have the probability to the right of $t_{30,0.025}$, again represented by the gray area, equal to 2.5%. The bottom two

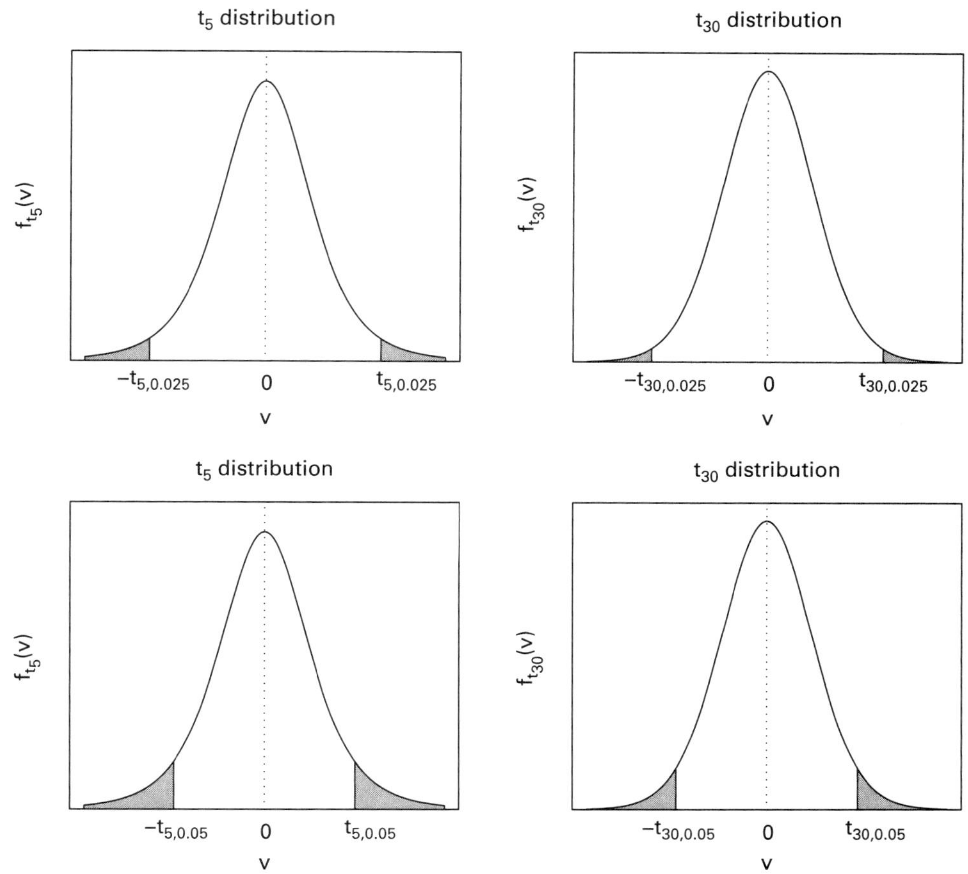

Figure 14.2
Critical values for a t-distribution

graphs show the 5% critical values for the t_5 and t_{30} distributions. For each of these two graphs, the gray area in the right tail has 5% probability, as does the gray area in the left tail, meaning that there is a 90% probability of being between $-t_{5,0.05}$ and $t_{5,0.05}$ for the t_5 distribution and between $-t_{30,0.05}$ and $t_{30,0.05}$ for the t_{30} distribution.

The sampling distribution from proposition 14.2 is

$$\frac{\bar{X}-\mu}{s_X/\sqrt{n}} \sim t_{n-1}$$

or, equivalently, since t_{n-1} is a symmetric distribution,

$$\frac{\mu-\bar{X}}{s_X/\sqrt{n}} \sim t_{n-1}.$$

Thus, there is a 95% probability that the quantity $\frac{\mu-\bar{X}}{s_X/\sqrt{n}}$ is between $-t_{n-1,0.025}$ and $t_{n-1,0.025}$. This probability can be written as

$$P\left(\frac{\mu-\bar{X}}{s_X/\sqrt{n}} \in (-t_{n-1,0.025}, t_{n-1,0.025})\right) = 0.95.$$

After multiplying by $s_X/\sqrt{n}$ and adding $\bar{X}$, for both the ratio and the end points, an equivalent probability is

$$P\left(\mu \in \left(\bar{X} - t_{n-1,0.025}\frac{s_X}{\sqrt{n}}, \bar{X} + t_{n-1,0.025}\frac{s_X}{\sqrt{n}}\right)\right) = 0.95.$$

Therefore, over all possible samples of size n from the population, there is a 95% probability that μ is in the interval

$$\left(\bar{X} - t_{n-1,0.025}\frac{s_X}{\sqrt{n}}, \bar{X} + t_{n-1,0.025}\frac{s_X}{\sqrt{n}}\right).$$

Other than the fact that it's commonly used by researchers, there is nothing special about the use of a 95% probability here. For instance, if we are interested in a 90% probability interval, the only change is to use the $t_{n-1,0.05}$ critical value rather than the $t_{n-1,0.025}$ critical value so there is a total of 10% probability in the tails rather than a total of 5% probability in the tails. This probability can be written

$$P\left(\mu \in \left(\bar{X} - t_{n-1,0.05}\frac{s_X}{\sqrt{n}}, \bar{X} + t_{n-1,0.05}\frac{s_X}{\sqrt{n}}\right)\right) = 0.90.$$

This idea can be generalized to other probability levels. For example, for a sample size of $n = 20$ and an 80% confidence interval, the critical value $t_{19,0.10} \approx 1.328$ is used.

```
qt(0.90,19)

## [1] 1.327728
```

To make these probability intervals useful in practice, the only remaining step is to use the realized estimates $\bar{x}$ and s_x, associated with the observed sample, in place of the estimators $\bar{X}$ and s_X. The standard deviation of the sampling distribution of $\bar{X}$ is equal to $\frac{s_X}{\sqrt{n}}$, which is replaced by $\frac{s_x}{\sqrt{n}}$. This latter quantity is known as the **standard error** of the sample mean estimator, where definition 14.7 introduces the general concept of a standard error.

Definition 14.7 *The **standard error** is the estimated standard deviation of the sampling distribution of estimator* $\hat{\theta}_X$.

Then the general result is that, for a value α, the $(1-\alpha)$ **confidence interval** for μ is

$$\left(\bar{x} - t_{n-1,\alpha/2}\frac{s_x}{\sqrt{n}}, \bar{x} + t_{n-1,\alpha/2}\frac{s_x}{\sqrt{n}}\right).$$

When $\alpha = 0.05$, the 95% confidence interval for μ is

$$\left(\bar{x} - t_{n-1,0.025}\frac{s_x}{\sqrt{n}}, \bar{x} + t_{n-1,0.025}\frac{s_x}{\sqrt{n}}\right).$$

When $\alpha = 0.10$, the 90% confidence interval for μ is

$$\left(\bar{x} - t_{n-1,0.05}\frac{s_x}{\sqrt{n}}, \bar{x} + t_{n-1,0.05}\frac{s_x}{\sqrt{n}}\right).$$

This type of confidence interval is called a **two-sided confidence interval** since there are end points on both sides. Moreover, this type of confidence interval is symmetric around estimate $\bar{x}$ since the two end points are equidistant from $\bar{x}$, as described in proposition 14.3.

Proposition 14.3 (Two-sided confidence interval for the sample mean of i.i.d. normal random variables) *If $X_1, X_2, \ldots, X_n$ are i.i.d. $N(\mu, \sigma^2)$ random variables, the probability that μ is in the interval*

$$\left(\bar{X} - t_{n-1,\alpha/2}\frac{s_X}{\sqrt{n}}, \bar{X} + t_{n-1,\alpha/2}\frac{s_X}{\sqrt{n}}\right)$$

equals $1-\alpha$. The associated $1-\alpha$ confidence interval for μ is

$$\left(\bar{x} - t_{n-1,\alpha/2}\frac{s_x}{\sqrt{n}}, \bar{x} + t_{n-1,\alpha/2}\frac{s_x}{\sqrt{n}}\right),$$

where $\bar{x}$ is the sample mean and $\frac{s_x}{\sqrt{n}}$ is the standard error based on the sample standard deviation s_x.

Example 14.2 (Monte Carlo simulations) *This example considers a set of Monte Carlo simulations to illustrate the concept of a 95% confidence interval. Specifically, we consider a sample size of $n=8$, where the underlying i.i.d. random variables $X_1, X_2, \ldots, X_8 \sim N(10,4)$. For each simulation, associated with an eight-observation sample drawn from the population, a 95% confidence interval for μ is constructed as in proposition 14.3. Figure 14.3 provides a graphical representation of the confidence intervals for 100 simulations, with each bar representing the confidence interval from a single simulation. The gray confidence intervals are those for which the true μ value of 10 is within the confidence interval, and the black confidence intervals are those for which $\mu=10$ is not within the confidence interval. For 95% confidence intervals, μ should be within the confidence interval roughly 95% of the time and outside the confidence interval roughly 5% of the time. In figure 14.3, there are four (black) confidence intervals that do not contain $\mu=10$, representing 4% of the simulations.*[2]

Example 14.3 (Food truck) *The weekly profits X of a food truck are normally distributed with unknown mean μ and variance σ^2. Suppose that weekly profits are recorded for a total of six weeks, with the sample average of weekly profits equal to \$1,200 and sample standard deviation equal to \$200. If each week's profits can be considered an i.i.d. normal random variable, what is a 95% confidence interval for μ, the population average of weekly profits? The confidence interval is centered at $\bar{x}=1{,}200$, and the standard error is $\frac{s_x}{\sqrt{n}} = \frac{200}{\sqrt{6}} \approx 81.65$, so the 95% confidence interval for μ is*

$$\left(\bar{x} - t_{6-1,0.025}\frac{s_x}{\sqrt{n}}, \bar{x} + t_{6-1,0.025}\frac{s_x}{\sqrt{n}}\right)$$

$$\approx (1200 - (2.571)(81.65), 1200 + (2.571)(81.65)) \approx (990, 1410).$$

With 95% confidence, the true average μ of weekly sales is between \$990 and \$1,410.

How about a 90% confidence interval for μ? The only quantity in the confidence interval formula that changes is the critical value, which is now $t_{5,0.05}$ rather than $t_{5,0.025}$. The 90%

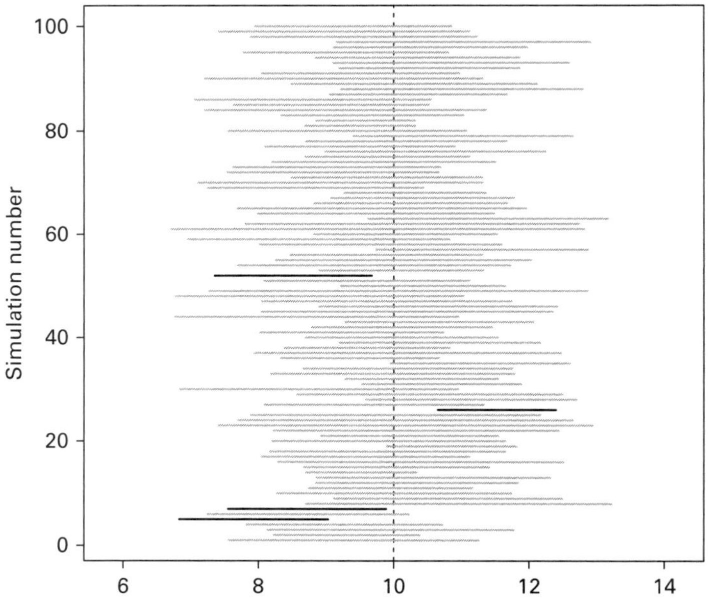

Figure 14.3
Monte Carlo simulations of confidence intervals

confidence interval for μ is

$$\left(\bar{x} - t_{6-1,0.05}\frac{s_x}{\sqrt{n}}, \bar{x} + t_{6-1,0.05}\frac{s_x}{\sqrt{n}}\right)$$

$$\approx (1200 - (2.015)(81.65), 1200 + (2.015)(81.65)) \approx (1035, 1365).$$

This confidence interval is more narrow than the 95% confidence interval. The more narrow interval is associated with less confidence, 90% probability, as compared to larger 95% confidence associated with the wider interval.

14.2.1 One-Sided Confidence Intervals

What if we are interested in a one-sided confidence interval for μ rather than a two-sided confidence interval? Again using the fact that

$$\frac{\mu - \bar{X}}{s_X/\sqrt{n}} \sim t_{n-1},$$

there is a 95% probability that the quantity $\frac{\mu-\bar{X}}{s_X/\sqrt{n}}$ is above $-t_{n-1,0.05}$. The critical value here uses the 5% probability, rather than the 2.5% probability used for a two-sided confidence interval, since the full 5% probability must be to the left of $-t_{n-1,0.05}$. This probability can be written as

$$P\left(\frac{\mu - \bar{X}}{s_X/\sqrt{n}} > -t_{n-1,0.05}\right) = 0.95$$

or, equivalently, as

$$P\left(\mu > \bar{X} - t_{n-1,0.05}\frac{s_X}{\sqrt{n}}\right) = 0.95.$$

This probability can be also be written in terms of a one-sided interval, as

$$P\left(\mu \in \left(\bar{X} - t_{n-1,0.05}\frac{s_X}{\sqrt{n}}, \infty\right)\right) = 0.95.$$

Using similar reasoning, since there is a 95% probability that the quantity $\frac{\mu-\bar{X}}{s_X/\sqrt{n}}$ is below $t_{n-1,0.05}$, it follows that

$$P\left(\frac{\mu - \bar{X}}{s_X/\sqrt{n}} < t_{n-1,0.05}\right) = 0.95,$$

which implies

$$P\left(\mu < \bar{X} + t_{n-1,0.05}\frac{s_X}{\sqrt{n}}\right) = 0.95$$

or

$$P\left(\mu \in \left(-\infty, \bar{X} + t_{n-1,0.05}\frac{s_X}{\sqrt{n}}\right)\right) = 0.95.$$

As with a two-sided confidence interval, a one-sided confidence interval is implemented by replacing $\bar{X}$ with the sample mean $\bar{x}$ and s_X with the sample standard deviation s_x. These one-sided confidence intervals can also be generalized to other probabilities, as stated in proposition 14.4.

Proposition 14.4 (One-sided confidence intervals for the sample mean of i.i.d. normal random variables) *If $X_1, X_2, \ldots, X_n$ are i.i.d. $N(\mu, \sigma^2)$ random variables, the probability that*

$$\mu > \bar{X} - t_{n-1,\alpha}\frac{s_X}{\sqrt{n}}$$

is equal to $1-\alpha$, and the probability that

$$\mu < \bar{X} + t_{n-1,\alpha}\frac{s_X}{\sqrt{n}}$$

is equal to $1-\alpha$. The associated $1-\alpha$ confidence intervals for μ are

$$\left(\bar{x} - t_{n-1,\alpha}\frac{s_x}{\sqrt{n}}, \infty\right)$$

and

$$\left(-\infty, \bar{x} + t_{n-1,\alpha}\frac{s_x}{\sqrt{n}}\right),$$

where $\bar{x}$ is the sample mean and $\frac{s_x}{\sqrt{n}}$ is the standard error based on the sample standard deviation s_x.

Example 14.4 (Food truck) *Continuing example 14.3, one-sided confidence intervals can be constructed based on proposition 14.4. For instance, the owner of the food truck may be most concerned about the downside risk. As such, the owner would be more interested in having 95% confidence, or some other confidence level, that weekly profits are above a* *certain value. A one-sided 95% confidence interval for* μ *is*

$$\left(\bar{x} - t_{6-1,0.05}\frac{s_X}{\sqrt{n}}, \infty\right) \approx (1200 - (2.015)(81.65), \infty) \approx (1035, \infty).$$

The owner can be 95% confident that the true average of weekly sales is greater than \$1,035. Note that \$1,035 is the same value as the lower end of the two-sided 90% confidence interval found in example 14.3, as the same critical value $t_{5,0.05}$ *is used for both calculations.*

What if the owner would like even more certainty, such as 99% confidence rather than 95% confidence? The critical value $t_{6-1,0.05} \approx 2.015$ *is replaced with* $t_{6-1,0.01} \approx 3.365$*, yielding the one-sided 99% confidence interval*

$$\left(\bar{x} - t_{6-1,0.01}\frac{s_X}{\sqrt{n}}, \infty\right) \approx (1200 - (3.365)(81.65), \infty) \approx (925, \infty).$$

Since the owner wants to be more certain here (99% probability), the lower end of the one-sided confidence interval (\$925) is considerably lower than the lower end of the 95% one-sided confidence interval (\$1,035).

14.3 Asymptotic Confidence Intervals: Population Mean of i.i.d. Random Variables

This section considers the case of a large sample, in which case the asymptotic sampling distribution results from chapter 13 apply. To make a clear comparison with the finite-sample confidence intervals introduced in section 14.2, this section focuses on the sample mean estimator. The generalization to other asymptotically normal estimators is discussed in section 14.4.

For a large sample, we can relax the assumption of normality that was crucial for the finite-sample confidence intervals in section 14.2. The observed sample is $\{x_1, x_2, \ldots, x_n\}$, for large n, where the underlying random variables $X_1, X_2, \ldots, X_n$ are assumed to be i.i.d. but not necessarily normal. By the CLT, the asymptotic sampling distribution of the sample mean estimator $\bar{X}$ is

$$\bar{X} \overset{a}{\sim} N\left(\mu_X, \frac{\sigma_X^2}{n}\right) \quad \text{or, equivalently,} \quad \frac{\bar{X} - \mu_X}{\sigma_X/\sqrt{n}} \overset{a}{\sim} N(0, 1).$$

The important difference from the finite-sample case is that $\frac{s_X}{\sqrt{n}}$ gets arbitrarily close to the asymptotic standard deviation $\frac{\sigma_X}{\sqrt{n}}$ as the sample size n gets large. As such, there is no need to adjust for the uncertainty in estimating σ_X, as we did with the t-distribution for the finite-sample case. Since s_X is a consistent estimator of σ_X, the asymptotic distribution for $\bar{X}$ can also be written as

$$\frac{\bar{X} - \mu_X}{s_X/\sqrt{n}} \overset{a}{\sim} N(0, 1)$$

by plugging in s_X for σ_X. As a result, asymptotic confidence intervals are based on critical values from the standard normal distribution $N(0, 1)$ rather than critical values from a t-distribution. Definition 14.8 introduces the notation for critical values based on the $N(0, 1)$ distribution.

Definition 14.8 *The* ***critical value*** z_q *denotes the* $(1-q)$ *quantile of the* $N(0,1)$ *distribution. For example,* $z_{0.025} \approx 1.96$ *is the 97.5% quantile of the* $N(0,1)$ *distribution, and* $z_{0.05} \approx 1.645$ *is the 95% quantile of the* $N(0,1)$ *distribution.*

Following the same reasoning from section 14.2, but now using z critical values rather than t critical values,

$$P\left(\frac{\mu_X - \bar{X}}{s_X/\sqrt{n}} \in (-z_{0.025}, z_{0.025})\right) = 0.95$$

and

$$P\left(\mu_X \in \left(\bar{X} - z_{0.025}\frac{s_X}{\sqrt{n}}, \bar{X} + z_{0.025}\frac{s_X}{\sqrt{n}}\right)\right) = 0.95.$$

For i.i.d. normal random variables, this 95% probability interval is the natural large-sample version of the finite-sample interval $P\left(\mu_X \in \left(\bar{X} - t_{n-1,0.025}\frac{s_X}{\sqrt{n}}, \bar{X} + t_{n-1,0.025}\frac{s_X}{\sqrt{n}}\right)\right) = 0.95$ since the limit of the t_{n-1} distribution as n increases is the $N(0,1)$ distribution, meaning that the limit of the $t_{n-1,0.025}$ critical value as n increases, is the $z_{0.025}$ critical value. Importantly, however, the asymptotic 95% probability interval also applies to nonnormal i.i.d. random variables.

The 95% probability interval can be generalized to other probability intervals, with the $(1-\alpha)$ probability interval

$$P\left(\mu_X \in \left(\bar{X} - z_{\alpha/2}\frac{s_X}{\sqrt{n}}, \bar{X} + z_{\alpha/2}\frac{s_X}{\sqrt{n}}\right)\right) = 1 - \alpha.$$

The critical value $z_{\alpha/2}$ is the value for which there is a probability of $\alpha/2$ that a $N(0,1)$ random variable is larger than $z_{\alpha/2}$, and by symmetry, $-z_{\alpha/2}$ is the value for which there is a probability of $\alpha/2$ that a $N(0,1)$ random variable is less than $-z_{\alpha/2}$. Proposition 14.5 formally states the results for a two-sided confidence interval based upon this general probability interval.

Proposition 14.5 (Two-sided confidence interval for the sample mean of i.i.d. random variables) *If* $X_1, X_2, \ldots, X_n$ *are i.i.d. random variables with population mean* μ_X *and large* n*, the probability that* μ_X *is in the interval*

$$\left(\bar{X} - z_{\alpha/2}\frac{s_X}{\sqrt{n}}, \bar{X} + z_{\alpha/2}\frac{s_X}{\sqrt{n}}\right)$$

is approximately equal to $1-\alpha$*. The associated asymptotic* $1-\alpha$ *confidence interval for* μ_X *is*

$$\left(\bar{x} - z_{\alpha/2}\frac{s_x}{\sqrt{n}}, \bar{x} + z_{\alpha/2}\frac{s_x}{\sqrt{n}}\right),$$

where $\bar{x}$ *is the sample mean and* $\frac{s_x}{\sqrt{n}}$ *is the standard error based on the sample standard deviation* s_x*.*

The standard error $\frac{s_x}{\sqrt{n}}$ is sometimes called an **asymptotic standard error** since it is based on the asymptotic sampling distribution, but we often simply call it a "standard error" so the same terminology can be used when referring to a finite-sample standard error or an asymptotic standard error. The following *R* code defines the function se_meanx, which calculates the standard error of $\bar{x}$, given by $\frac{s_x}{\sqrt{n}}$, for any vector of data:

```
# function to calculate the se of the sample mean for a vector
se_meanx <- function (x, na.rm = FALSE) {
    if (na.rm) {
        x <- x[!is.na(x)]
    }
    return(sd(x)/sqrt(length(x)))
}
```

The function se_meanx has an optional argument na.rm so vectors with missing (NA) values can be handled in a, similar way as the optional argument na.rm available for built-in functions like mean and sd.

For i.i.d. *Bernoulli*(π) random variables, the sample mean $\bar{X}$ is an estimator of the population mean or success probability π. Since the population variance is $\sigma_X = \pi(1-\pi)$, the asymptotic standard deviation of $\bar{X}$ is

$$\frac{\sigma_X}{\sqrt{n}} = \sqrt{\frac{\pi(1-\pi)}{n}},$$

which suggests two possible approaches to calculating the standard error: (1) plug in the sample standard deviation s_x for σ_X, so the standard error is $\frac{s_x}{\sqrt{n}}$, or (2) plug in the sample mean $\bar{x}$ for π, so the standard error is $\frac{\bar{x}(1-\bar{x})}{\sqrt{n}}$. Approach (2) is valid since the sample mean is a consistent estimator of π, so $\sqrt{\bar{x}(1-\bar{x})}$ will be arbitrarily close to $\sqrt{\pi(1-\pi)}$. Since $\bar{x}$ is just the sample proportion of successes, or the fraction of 1s observed, this approach is appealing since it doesn't require the extra step of calculating s_x. Approach (1) and approach (2) give nearly identical standard errors, as it can be shown that $s_x = \sqrt{\frac{n}{n-1}}\sqrt{\bar{x}(1-\bar{x})}$ and the scaling factor $\sqrt{\frac{n}{n-1}} \approx 1$ for large n.[3]

Example 14.5 (Widget website) *In example 2.1, Widgets.com conducted an e-mail campaign experiment in which* 300 *users received e-mail A,* 300 *users received e-mail B, and* 2,400 *users received no e-mail. The resulting purchase probabilities were* 20% *(*60 *out of* 300*) for e-mail A recipients,* 22% *(*66 *out of* 300*) for e-mail B recipients, and* 15% *(*360 *out of* 2,400*) for nonrecipients. Using proposition 14.5, confidence intervals for the purchase probabilities of the three groups can be calculated. Let* π_A *denote the purchase probability of an e-mail A recipient,* π_B *denote the purchase probability of an e-mail B recipient, and* π_C *denote the purchase probability of a nonrecipient. Starting with the sample of e-mail A recipients, assume that the random variables* $X_1, X_2, \ldots, X_{300}$ *are i.i.d. Bernoulli(*π_A*), with a* 1 *(success) associated with a purchase. The observed proportion of successes (purchases) is* $p_A = 0.20$*, so the sample mean is*

$$\bar{x} = p_A = 0.20,$$

and the standard error (of the sample mean) is

$$\frac{s_x}{\sqrt{n}} = \sqrt{\frac{p_A(1-p_A)}{n}} = \sqrt{\frac{(0.2)(0.8)}{300}} \approx 0.0231.$$

Then, the asymptotic 95% *confidence interval for* π_A *is*

$$\left(\bar{x} - z_{0.025}\frac{s_X}{\sqrt{n}}, \bar{x} + z_{0.025}\frac{s_X}{\sqrt{n}}\right)$$
$$\approx (0.20 - (1.96)(0.0231), 0.20 + (1.96)(0.0231)) \approx (0.155, 0.245).$$

Similar calculations yield an asymptotic 95% *confidence interval for* π_B*,*

$$\left(\bar{x} - z_{0.025}\frac{s_X}{\sqrt{n}}, \bar{x} + z_{0.025}\frac{s_X}{\sqrt{n}}\right)$$
$$\approx \left(0.22 - (1.96)\sqrt{\frac{(0.22)(0.78)}{300}}, 0.22 + (1.96)\sqrt{\frac{(0.22)(0.78)}{300}}\right) \approx (0.173, 0.267),$$

and an asymptotic 95% *confidence interval for* π_C*,*

$$\left(\bar{x} - z_{0.025}\frac{s_X}{\sqrt{n}}, \bar{x} + z_{0.025}\frac{s_X}{\sqrt{n}}\right)$$
$$\approx \left(0.15 - (1.96)\sqrt{\frac{(0.15)(0.85)}{2400}}, 0.15 + (1.96)\sqrt{\frac{(0.15)(0.85)}{2400}}\right) \approx (0.136, 0.164).$$

The standard errors for the three confidence intervals are 0.023, 0.024, *and* 0.007, *respectively. The first two standard errors are similar since the sample size* $n = 300$ *is the same and the sample standard deviations differ only slightly* ($\sqrt{(0.20)(0.80)}$ *versus* $\sqrt{(0.22)(0.78)}$). *The third standard error is much lower due to the larger sample size* $n = 2{,}400$.

Figure 14.4 provides a graphical representation of the 95% *confidence intervals for the purchase probabilities of the three groups. The confidence intervals for* π_A *and* π_B *have similar widths since they have similar standard errors. The confidence interval for* π_C *(no e-mail) is much narrower since the sample size* $n = 2{,}400$ *is much larger, corresponding to the fact that* $p_C = 0.15$ *is a more precise estimate of* π_C *than the estimates for the smaller samples. Comparing the confidence intervals for* π_A *and* π_B*, it seems that there is no strong evidence that the purchase probability of e-mail B recipients,* π_B*, is greater than the purchase probability of e-mail A recipients,* π_A*. Even though the observed purchase probability is larger for e-mail B recipients, there is a lot of overlap between the confidence intervals of the two purchase probabilities* π_A *and* π_B*. (Later in this chapter, in section 14.4.5, a more formal method for directly looking at the difference* $\pi_A - \pi_B$ *is considered.) In contrast, the narrow confidence interval for* π_C *has no overlap with the* π_B *confidence interval and only small overlap with the* π_A *confidence interval, offering some evidence that the purchase probability* π_C *is lower than either* π_A *or* π_B*. (Again, a more formal examination of the differences* $\pi_A - \pi_C$ *and* $\pi_B - \pi_C$ *is provided later.)*

Example 14.6 (Labor force data) *The confidence interval for a sample mean estimator can be applied to variables from a cross-sectional data set, at least in cases where the observations can plausibly be considered to be realizations of i.i.d. random variables. This idea is illustrated by calculating standard errors and confidence intervals for several variables from the* **cps** *data set. The following table shows the sample mean* $\bar{x}$*, the standard error* $\frac{s_X}{\sqrt{n}}$*, the asymptotic* 95% *confidence interval for* μ_X*, and the asymptotic* 90% *confidence interval*

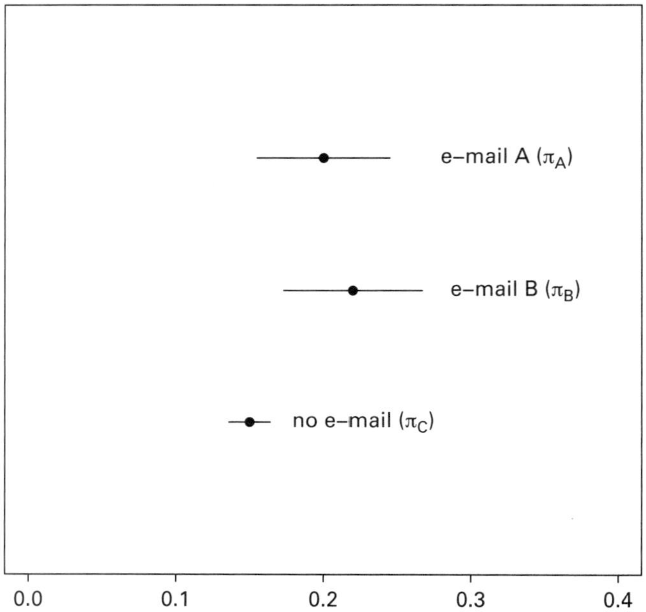

Figure 14.4
Asymptotic 95% confidence intervals for true widget purchase probabilities

*for μ_X for five variables: age (age in years), educ (education in years), ownchild (number of children in household), earnwk (weekly earnings), and union (*1 *if union member,* 0 *if not). For the first three variables, the quantities are calculated based on the full sample size $n = 4{,}013$. For the last two variables, the quantities are calculated based on the employed sample size $n = 2{,}809$.*

Variable	n	$\bar{x}$	$se = \frac{s_X}{\sqrt{n}}$	*95% CI for* μ_X	*90% CI for* μ_X
age	4,013	45.02	0.142	(44.74, 45.30)	(44.78, 45.25)
educ	4,013	12.57	0.039	(12.50, 12.65)	(12.51, 12.64)
ownchild	4,013	0.748	0.0176	(0.713, 782)	(0.719, 0.777)
earnwk	2,809	971.18	14.16	(943.42, 998.93)	(947.89, 994.47)
union	2,809	0.098	0.0056	(0.087, 0.109)	(0.089, 0.107)

The abbreviation "se" is used for the standard error $\frac{s_X}{\sqrt{n}}$. For the education variable educ, the asymptotic 95% *confidence interval for the population mean is* (12.50, 12.65). *Since this confidence interval is quite narrow, the sample mean* 12.57 *is providing a precise estimate of the population mean, which occurs since the sample size $n = 4{,}013$ is so large here. For the weekly earnings variable earnwk, the sample mean of* \$971.18 *is the estimate of the population mean of weekly earnings for the* population of employed individuals, *with the associated asymptotic* 95% *confidence interval being between* \$943.42 *and* \$998.93. *The union variable is an indicator variable, so its observations are realizations of Bernoulli random variables. The estimate of the probability that an employed individual from the population is in a union is* 9.8%, *with an asymptotic* 95% *confidence interval for the probability of union membership being between* 8.7% *and* 10.9%.

Here is the R code used to calculate the n, $\bar{x}$, *and* $se = \frac{s_x}{\sqrt{n}}$ *columns in the table:*

```
# create union indicator variable from unionstatus factor variable
cps$union <- 1*(cps$unionstatus=="Union")

# loop through variables of interest
for (varname in c("age","educ","ownchild","earnwk","union")) {
    # sample size is number of non-missing observations
    nobs <- sum(!is.na(cps[,varname]))

    # calculate sample mean of variable
    mean_var <- mean(cps[,varname], na.rm=TRUE)

    # calculate standard error of sample mean
    se_mean <- se_meanx(cps[,varname], na.rm=TRUE)

    # output results
    print(paste(varname,":","n",nobs,"Mean",signif(mean_var,digits=5),"SE(mean)",
    signif(se_mean,digits=5)))
}

## [1] "age : n 4013 Mean 45.017 SE(mean) 0.14205"
## [1] "educ : n 4013 Mean 12.573 SE(mean) 0.038863"
## [1] "ownchild : n 4013 Mean 0.74782 SE(mean) 0.017606"
## [1] "earnwk : n 2809 Mean 971.18 SE(mean) 14.16"
## [1] "union : n 2809 Mean 0.098256 SE(mean) 0.0056172"
```

The code loops through the five variable names. For each variable given by `varname`, *the expression* `sum(!is.na(cps[,varname]))` *returns the number of nonmissing observations, which is the effective sample size n. To limit the number of significant digits reported by R, the function* `signif` *is used so that five significant digits, as specified by the second argument* `digits=5`, *are reported for the values of* `mean_var` *and* `se_mean`.

Example 14.7 (Simulation error: the likelihood of streaks) *In example 4.16, computer simulations estimated the probability that a streak of at least five heads occurs during a sequence of* 100 *coin tosses. Over* 100,000 *simulations, the calculated frequency of streaks was* 0.81156, *corresponding to* 81,156 *of the* 100,000 *simulations. The associated standard error is*

$$\sqrt{\frac{0.81156(1-0.81156)}{100000}} \approx 0.001237,$$

so an asymptotic 99% *confidence interval for the probability (of observing a sequence of at least five heads in* 100 *coin tosses) is*

$$(0.81156 - z_{0.005}(0.001237), 0.81156 + z_{0.005}(0.001237)) \approx (0.8084, 0.8147),$$

using $z_{0.005} \approx 2.576$. *Thus, it is very likely that the probability is close to* 81% *with this many simulations.*

14.3.1 One-Sided Confidence Intervals

In section 14.2.1, one-sided confidence intervals for the population mean of i.i.d. normal random variables were constructed using finite-sample distribution results and the t-distribution. Similarly, one-sided asymptotic confidence intervals can be constructed for the population mean of i.i.d. random variables, which may be nonnormal, by using the asymptotic distribution results and the $N(0,1)$ distribution. These confidence intervals are described in proposition 14.6.

Proposition 14.6 (One-sided confidence intervals for the sample mean of i.i.d. random variables) *If $X_1, X_2, \ldots, X_n$ are i.i.d. random variables with population mean μ_X and large n, the probability that*

$$\mu_X > \bar{X} - z_\alpha \frac{s_X}{\sqrt{n}}$$

is approximately equal to $1-\alpha$, and the probability that

$$\mu_X < \bar{X} + z_\alpha \frac{s_X}{\sqrt{n}}$$

is approximately equal to $1-\alpha$. The associated $1-\alpha$ asymptotic confidence intervals for μ_X are

$$\left(\bar{x} - z_\alpha \frac{s_x}{\sqrt{n}}, \infty\right)$$

and

$$\left(-\infty, \bar{x} + z_\alpha \frac{s_x}{\sqrt{n}}\right),$$

where $\bar{x}$ is the sample mean and $\frac{s_x}{\sqrt{n}}$ is the standard error based on the sample standard deviation s_x.

Example 14.8 (Worksite accidents) *At a certain factory, assume that the occurrence of a worker-involved accident on any given day is an i.i.d. draw of a Bernoulli(π) random variable. Over the course of 1,000 work days, there are 20 days on which a worker-involved accident occurs. The factory owner is interested in the one-sided confidence interval for π, in which only the upper end point is specified. A 95% one-sided confidence interval for π is*

$$\left(-\infty, \bar{x} + z_{0.05} \frac{s_x}{\sqrt{n}}\right) = \left(-\infty, 0.02 + 1.645\sqrt{\frac{(0.02)(0.98)}{1000}}\right) \approx (-\infty, 0.0273),$$

so the owner can have 95% confidence that the daily accident probability π is less than 2.73%. Similarly, a 99% one-sided confidence interval for π is

$$\left(-\infty, \bar{x} + z_{0.01} \frac{s_x}{\sqrt{n}}\right) = \left(-\infty, 0.02 + 2.326\sqrt{\frac{(0.02)(0.98)}{1000}}\right) \approx (-\infty, 0.0303),$$

so the owner can have 99% confidence that the daily accident probability π is less than 3.03%.

14.4 Asymptotic Confidence Intervals: Parameters with Asymptotically Normal Estimators

Section 14.3 introduced an asymptotic confidence interval for a population mean based on an i.i.d. sample drawn from the population. This section generalizes the concept of an asymptotic confidence interval to *any* parameter estimated by an asymptotically normal estimator. Chapter 13 provided the asymptotic sampling distributions associated with several statistics of interest, including the sample mean, the sample variance, the sample standard deviation, sample quantiles, the sample interquartile range (IQR), and the sample correlation. Each of these statistics, when viewed as an estimator of its population analog (the parameter or estimand of interest), has an asymptotic sampling distribution that is normally distributed. For example, proposition 13.6 provided the asymptotic distribution for the sample median,

$$\tilde{X}_{0.5} \overset{a}{\sim} N\left(\tau_{X,0.5}, \frac{1}{4nf_X(\tau_{X,0.5})^2}\right),$$

and proposition 13.9 provided the asymptotic distribution for the sample correlation,

$$r_{XY} \overset{a}{\sim} N\left(\rho_{XY}, \frac{(1-\rho_{XY}^2)^2}{n}\right).$$

The sample median $\tilde{X}_{0.5}$, as an estimator of the population median $\tau_{X,0.5}$, and the sample correlation r_{XY}, as an estimator of the population correlation ρ_{XY}, are just two examples of $\sqrt{n}$-consistent and asymptotically normal estimators. As defined in definition 14.4, a $\sqrt{n}$-consistent and asymptotically normal estimator has an asymptotic sampling distribution of the form

$$\hat{\theta}_X \overset{a}{\sim} N\left(\theta, \frac{V}{n}\right),$$

where θ is the parameter (estimand) of interest, $\hat{\theta}_X$ is the estimator of θ, and the associated asymptotic standard deviation is $\sqrt{\frac{V}{n}}$. Standardizing the random variable $\hat{\theta}_X$ gives the equivalent asymptotic distribution

$$\frac{\hat{\theta}_X - \theta}{\sqrt{V/n}} \overset{a}{\sim} N(0, 1).$$

For this general case, an asymptotic 95% confidence interval is based on

$$P\left(-1.96 < \frac{\hat{\theta}_X - \theta}{\sqrt{V/n}} < 1.96\right) = 0.95$$

or, equivalently,

$$P\left(\hat{\theta}_X - 1.96\sqrt{\frac{V}{n}} < \theta < \hat{\theta}_X + 1.96\sqrt{\frac{V}{n}}\right) = 0.95.$$

To calculate an asymptotic 95% confidence interval for θ, the realized estimate $\hat{\theta}_x$ is used in place of the estimator $\hat{\theta}_X$ and the standard error associated with $\hat{\theta}_x$ is used in place of $\sqrt{V/n}$. For the standard error, a consistent estimate of V is required (call it $\hat{V}$), so the standard error is $\sqrt{\hat{V}/n}$. For example, in the case of the sample mean $\bar{X}$ as an estimator of the population mean μ_X, the results from section 14.4 use $\hat{V} = s_x^2$, so the standard error is $\sqrt{\hat{V}/n} = \frac{s_x}{\sqrt{n}}$. As another example, in the case of the sample correlation r_{XY} as an estimator of the population correlation ρ_{XY}, the asymptotic variance shown here has $V = (1-\rho_{XY}^2)^2$, which can be consistently estimated by plugging in r_{xy} for ρ_{XY}, so $\hat{V} = (1-r_{xy}^2)^2$ and the standard error is $\sqrt{\hat{V}/n} = \frac{1-r_{xy}^2}{\sqrt{n}}$.

The 95% probability statement here can be generalized to other probability levels by changing the $z_{0.025} = 1.96$ critical value to the appropriate critical value needed for the chosen level of probability. To get a probability of $1-\alpha$, the appropriate critical value is $z_{\alpha/2}$:

$$P\left(\hat{\theta}_X - z_{\alpha/2}\sqrt{\frac{V}{n}} < \theta < \hat{\theta}_X + z_{\alpha/2}\sqrt{\frac{V}{n}}\right) = 1-\alpha.$$

The general form of proposition 14.5, which holds for any asymptotically normal estimator $\hat{\theta}_X$, is given by proposition 14.7.

Proposition 14.7 (Two-sided confidence intervals based on an asymptotically normal estimator) *If $\hat{\theta}_X$ is an asymptotically normal estimator of θ with $\hat{\theta}_X \overset{a}{\sim} N\left(\theta, \frac{V}{n}\right)$, the probability that θ is in the interval*

$$\left(\hat{\theta}_X - z_{\alpha/2}\sqrt{\frac{V}{n}}, \hat{\theta}_X + z_{\alpha/2}\sqrt{\frac{V}{n}}\right)$$

is approximately equal to $1-\alpha$ for large n. The associated asymptotic $1-\alpha$ confidence interval for θ is

$$\left(\hat{\theta}_x - z_{\alpha/2}\sqrt{\frac{\hat{V}}{n}}, \hat{\theta}_x + z_{\alpha/2}\sqrt{\frac{\hat{V}}{n}}\right),$$

where $\hat{\theta}_x$ is the realized estimate of θ and $\sqrt{\frac{\hat{V}}{n}}$ is the standard error based on a consistent estimate $\hat{V}$ of V.

A convenient notation is $se(\hat{\theta}_x)$, denoting the standard error for estimate $\hat{\theta}_x$, with

$$se(\hat{\theta}_x) = \sqrt{\frac{\hat{V}}{n}},$$

and the asymptotic $1-\alpha$ confidence interval for θ, from proposition 14.7, is

$$(\hat{\theta}_x - z_{\alpha/2}se(\hat{\theta}_x), \hat{\theta}_x + z_{\alpha/2}se(\hat{\theta}_x)).$$

Proposition 14.7 provides a very powerful result, as it covers all the asymptotically normal estimators introduced in this book. More generally, beyond the estimators that we've covered, if there is an asymptotically normal estimator $\hat{\theta}_X$ of θ for which a computer is able to produce both estimate $\hat{\theta}_x$ and standard error $se(\hat{\theta}_x)$, the asymptotic $1-\alpha$ confidence interval is always $(\hat{\theta}_x - z_{\alpha/2}se(\hat{\theta}_x), \hat{\theta}_x + z_{\alpha/2}se(\hat{\theta}_x))$.

In section 14.3, the asymptotic confidence interval was given for the population mean, based upon the sample mean of an i.i.d. sample, with

$$\text{asymptotic standard deviation } \sqrt{\frac{V}{n}} = \frac{s_X}{\sqrt{n}} \text{ and standard error } se\,(\bar{x}) = \sqrt{\frac{\hat{V}}{n}} = \frac{s_x}{\sqrt{n}}.$$

For the specific case of underlying i.i.d. *Bernoulli*(π) random variables, where the sample mean is an estimator of the probability $\pi = \mu_X$,

$$\text{asymptotic standard deviation } \sqrt{\frac{V}{n}} = \sqrt{\frac{\pi(1-\pi)}{n}}$$

$$\text{and standard error } se\,(\bar{x}) = \sqrt{\frac{\hat{V}}{n}} = \sqrt{\frac{\bar{x}(1-\bar{x})}{n}}.$$

In the remainder of this section, several other examples of asymptotic confidence intervals are considered. Before those examples, however, we state the general result for one-sided confidence intervals in proposition 14.8.

Proposition 14.8 (One-sided confidence intervals based on an asymptotically normal estimator) *If $\hat{\theta}_X$ is an asymptotically normal estimator of θ with $\hat{\theta}_X \overset{a}{\sim} N\left(\theta, \frac{V}{n}\right)$, the probability that*

$$\theta > \hat{\theta}_X - z_\alpha \sqrt{\frac{V}{n}}$$

is approximately equal to $1-\alpha$ for large n, and the probability that

$$\theta < \hat{\theta}_X + z_\alpha \sqrt{\frac{V}{n}}$$

is approximately equal to $1-\alpha$ for large n. The associated asymptotic $1-\alpha$ confidence intervals for θ are

$$\left(\hat{\theta}_x - z_\alpha \sqrt{\frac{\hat{V}}{n}}, \infty\right) \text{ and } \left(-\infty, \hat{\theta}_x + z_\alpha \sqrt{\frac{\hat{V}}{n}}\right),$$

where $\hat{\theta}_x$ is the realized estimate of θ and $\sqrt{\frac{\hat{V}}{n}}$ is the standard error based on a consistent estimate $\hat{V}$ of V.

From proposition 14.8, the one-sided asymptotic $(1-\alpha)$ confidence intervals for θ are

$$(\hat{\theta}_x - z_\alpha se(\hat{\theta}_x), \infty) \text{ and } (-\infty, \hat{\theta}_x + z_\alpha se(\hat{\theta}_x)),$$

where $\hat{\theta}_x$ is the estimate of θ and $se(\hat{\theta}_x)$ is its standard error.

14.4.1 Confidence Intervals for the Population Standard Deviation

Proposition 13.5 provided the asymptotic sampling distribution of the sample standard deviation s_X for i.i.d. random variables $X_1, X_2, \ldots, X_n$:

$$s_X \overset{a}{\sim} N\left(\sigma_X, \frac{E\left((X-\mu_X)^4\right) - \left(\sigma_X^2\right)^2}{4n\sigma_X^2}\right).$$

The expression for V in this case, then, is

$$V = \frac{E\left((X-\mu_X)^4\right) - \left(\sigma_X^2\right)^2}{4\sigma_X^2}.$$

The asymptotic standard deviation is

$$\sqrt{\frac{V}{n}} = \frac{\sqrt{E\left((X-\mu_X)^4\right) - \left(\sigma_X^2\right)^2}}{2\sqrt{n}\sigma_X}.$$

To consistently estimate V, which is necessary to derive a formula for the standard error $se(s_x)$, the realized sample standard deviation s_x is plugged in for σ_X and the summation

$$\frac{1}{n}\sum_{i=1}^{n}(x_i - \bar{x})^4$$

is plugged in for $E\left((X-\mu_X)^4\right)$, leading to

$$\hat{V} = \frac{\frac{1}{n}\sum_{i=1}^{n}(x_i-\bar{x})^4 - \left(s_x^2\right)^2}{4s_x^2} \text{ and } se(s_x) = \sqrt{\frac{\hat{V}}{n}} = \frac{\sqrt{\frac{1}{n}\sum_{i=1}^{n}(x_i-\bar{x})^4 - \left(s_x^2\right)^2}}{2\sqrt{n}s_x}.$$

With this formula for $se(s_x)$, the asymptotic $1-\alpha$ confidence interval for the population standard deviation is

$$\left(s_x - z_{\alpha/2}se\left(s_x\right), s_x + z_{\alpha/2}se\left(s_x\right)\right).$$

The following *R* code defines the function `se_sx`, which calculates the standard error $se(s_x)$ for any vector of data:

```
# function to calculate the se of the sample stdev for a vector
se_sx <- function (x, na.rm = FALSE) {
    if (na.rm) {
        x <- x[!is.na(x)]
    }
    nobs <- length(x)
    return(sqrt(mean((x-mean(x))^4) - sd(x)^4)/(2 * sqrt(nobs) * sd(x)))
}
```

Example 14.9 (Standard deviation of monthly stock returns) *Using the monthly stock return data from* ***sp500****, confidence intervals can be formed for the population standard deviation of monthly returns for any given company. The following table shows the sample standard deviation* s_x*, its standard error, and the associated* 95% *confidence interval (in the column labeled "*95% *CI for* σ_X*") for six stocks (HD, LOW, BAC, WFC, MRO, COP).*

Company	s_x	$se(s_x)$	95% *CI for* σ_X
HD	0.0737	0.00312	(0.0676, 0.0798)
LOW	0.0916	0.00361	(0.0845, 0.0987)
BAC	0.1053	0.00882	(0.0880, 0.1226)
WFC	0.0816	0.00538	(0.0710, 0.0921)
MRO	0.1210	0.01064	(0.1001, 0.1418)
COP	0.0817	0.00502	(0.0719, 0.0916)

For example, the sample standard deviation s_x *of Home Depot (HD) monthly returns is* 0.0737*, and we can say with* 95% *confidence that the population standard deviation* σ_X *is between* 0.0676 *and* 0.0798*. The standard errors in the "*$se(s_x)$*" column vary a lot, with the largest value of* 0.01064 *for MRO, indicating that its standard deviation estimate is the least precise or, equivalently, that the associated confidence interval is the widest of the six stocks.*

The following R code calculates the s_x *and* $se(s_x)$ *values for this table:*

```
# loop through the stocks of interest
for (stock in c("HD","LOW","BAC","WFC","MRO","COP")) {
    returns <- sp500[,stock]

    # calculate the stdev of returns
    sx <- sd(returns)
```

```
    # calculate the se of the stdev of returns
    stderror_sx <- se_sx(returns)

    # output the stock name, along with stdev and se(stdev)
        print(paste(stock,":","sx",round(sx,4),"se_sx",round(stderror_sx,5)))
}

## [1] "HD : sx 0.0737 se_sx 0.00312"
## [1] "LOW : sx 0.0916 se_sx 0.00361"
## [1] "BAC : sx 0.1053 se_sx 0.00882"
## [1] "WFC : sx 0.0816 se_sx 0.00538"
## [1] "MRO : sx 0.121 se_sx 0.01064"
## [1] "COP : sx 0.0817 se_sx 0.00502"
```

14.4.2 Confidence Intervals for the Population Correlation

Proposition 13.9 provided the asymptotic sampling distribution of the sample standard deviation r_{XY} for i.i.d. bivariate random variables $(X_1, Y_1), (X_2, Y_2), \ldots, (X_n, Y_n)$:

$$r_{XY} \overset{a}{\sim} N\left(\rho_{XY}, \frac{(1-\rho_{XY}^2)^2}{n}\right).$$

In this case, $V = (1-\rho_{XY}^2)^2$, so the asymptotic standard deviation is

$$\sqrt{\frac{V}{n}} = \frac{1-\rho_{XY}^2}{\sqrt{n}}.$$

Since the sample correlation is a consistent estimator of the population correlation ρ_{XY}, the standard error $se(r_{xy})$ is obtained by plugging in r_{xy} for ρ_{XY} in the expression for V, yielding

$$se(r_{xy}) = \sqrt{\frac{\hat{V}}{n}} = \frac{1-r_{xy}^2}{\sqrt{n}}$$

and an asymptotic $1-\alpha$ confidence interval

$$\left(r_{xy} - z_{\alpha/2} se\left(r_{xy}\right), r_{xy} + z_{\alpha/2} se\left(r_{xy}\right)\right).$$

The following *R* code defines the function `se_rxy`, which calculates the standard error $se(r_{xy})$ for any two vectors of data:

```
# function to calculate the se of the sample stdev for a vector
se_rxy <- function (x, y, na.rm = FALSE) {
    # if na.rm is TRUE, keep only observations where both are non-missing
    if (na.rm) {
       nonmissing <- (!is.na(x) & !is.na(y))
       x <- x[nonmissing]
       y <- y[nonmissing]
    }
    nobs <- length(x)
    return((1-cor(x,y)^2)/sqrt(nobs))
}
```

Example 14.10 (Education and earnings) *For the sample of $n = 2{,}809$ employed individuals from the* **cps** *data set, the sample correlation between education ($x = educ$) and weekly* *earnings* ($y = earnwk$) *is* $r_{xy} \approx 0.325$. *Its standard error is*

$$se(r_{xy}) = \frac{1 - r_{xy}^2}{\sqrt{n}} \approx \frac{1 - (0.325)^2}{\sqrt{2809}} \approx 0.0169.$$

```
se_rxy(cps$educ, cps$earnwk, na.rm = TRUE)

## [1] 0.01687314
```

The asymptotic 95% confidence interval for the population correlation ρ_{XY} *between education and earnings is*

$$\left(r_{xy} - z_{0.025} se\left(r_{xy}\right), r_{xy} + z_{0.025} se\left(r_{xy}\right)\right)$$
$$\approx (0.325 - (1.96)(0.0169), 0.325 + (1.96)(0.0169)) \approx (0.292, 0.358),$$

and the asymptotic 99% confidence interval for the population correlation ρ_{XY} *is*

$$\left(r_{xy} - z_{0.005} se\left(r_{xy}\right), r_{xy} + z_{0.005} se\left(r_{xy}\right)\right)$$
$$\approx (0.325 - (2.576)(0.0169), 0.325 + (2.576)(0.0169)) \approx (0.281, 0.369).$$

It can be said, with 99% confidence, that the population correlation ρ_{XY} *is between* 0.281 *and* 0.369. *This 99% confidence interval provides strong evidence that the population correlation between education and earnings is positive since a correlation value of zero is not in the interval and, in fact, is far below the lower end point. Since the standard error is* 0.0169, *the number of standard errors that zero is below the lower end point* 0.281 *is* $\frac{0.281}{0.0169} \approx 16.6$.

Example 14.11 (Monthly stock returns) *Example 7.13 provided a sample correlation matrix for six stocks (HD, LOW, BAC, WFC, MRO, and COP) from the* ***sp500*** *data set. Applying the* $se(r_{xy}) = \frac{1-r_{xy}^2}{\sqrt{n}}$ *formula, the matrix can be augmented to provide standard errors, reported in parentheses, alongside the sample correlation values:*

	HD	*LOW*	*BAC*	*WFC*	*MRO*	*COP*
HD	1.000	0.648 (0.030)	0.331 (0.047)	0.280 (0.048)	0.189 (0.051)	0.215 (0.050)
LOW		1.000	0.357 (0.046)	0.262 (0.049)	0.181 (0.051)	0.256 (0.049)
BAC			1.000	0.692 (0.027)	0.331 (0.047)	0.339 (0.046)
WFC				1.000	0.379 (0.045)	0.396 (0.044)
MRO					1.000	0.771 (0.021)
COP						1.000

Since each stock pair has the same sample size $n = 364$, *the standard error* $se(r_{xy}) = \frac{1-r_{xy}^2}{\sqrt{n}}$ *is a decreasing function of* r_{xy}, *so larger correlation values are associated with lower standard errors. The largest sample correlation of* 0.771, *between MRO and COP, has* $se(r_{xy})$ *equal to* 0.021, *whereas the lowest sample correlation of* 0.181, *between LOW and MRO, has* $se(r_{xy})$ *equal to* 0.051.

The confidence intervals for different correlations can be compared to each other. For instance, the asymptotic 95% confidence interval for $\rho_{HD,LOW}$ *is* (0.589, 0.707), *whereas the asymptotic 95% confidence interval for* $\rho_{HD,BAC}$ *is* (0.239, 0.423). *The sample correlation*

$r_{HD,LOW} = 0.648$ is considerably higher than the sample correlation $r_{HD,BAC} = 0.331$, and these two confidence intervals provide strong evidence that $r_{HD,LOW} > r_{HD,BAC}$ is not happening by chance. The two 95% confidence intervals for $\rho_{HD,LOW}$ and $\rho_{HD,BAC}$ have no overlap at all and, in terms of the standard error magnitudes, are separated by a large distance.

In general, it is very useful to report the standard error of an estimate alongside the estimate itself, as the reader can then form any asymptotic confidence interval based on those two numbers. For instance, for the sample correlation $r_{xy} = 0.648$ and the standard error $se(r_{HD,LOW}) = 0.030$ for the HD and LOW monthly stock returns, a reader could "ballpark" an asymptotic 95% confidence interval by using a 0.648 plus-or-minus two-standard-error (0.030) interval. Alternatively, to more formally calculate a confidence interval, the appropriate critical value can be used, for example an asymptotic 90% confidence interval for ρ_{XY} of $(0.648 - (1.645)(0.030), 0.648 + (1.645)(0.030))$ using $z_{0.05} \approx 1.645$.

Here is the R code to create this table:

```
# determine the sample size
nobs <- nrow(sp500)

# initialize variables with stock names and number of stocks
stocks <- c("HD","LOW","BAC","WFC","MRO","COP")
num_stocks <- length(stocks)

# double loop to loop through possible pairs of stocks
for (i in 1:(num_stocks-1)) {
    for (j in (i+1):num_stocks) {
        # calculate sample correlation
        rxy <- cor(sp500[,stocks[i]],sp500[,stocks[j]])

        # calculate standard error of sample correlation
        stderr_rxy <- se_rxy(sp500[,stocks[i]],sp500[,stocks[j]])

        # output stock-pair names, along with corr and se(corr)
        print(paste("Pair",stocks[i],stocks[j],": rxy",round(rxy,3),
        "se(rxy)",round (stderr_rxy,3)))
    }
}

## [1] "Pair HD LOW : rxy 0.648 se(rxy) 0.03"
## [1] "Pair HD BAC : rxy 0.331 se(rxy) 0.047"
## [1] "Pair HD WFC : rxy 0.28 se(rxy) 0.048"
## [1] "Pair HD MRO : rxy 0.189 se(rxy) 0.051"
## [1] "Pair HD COP : rxy 0.215 se(rxy) 0.05"
## [1] "Pair LOW BAC : rxy 0.357 se(rxy) 0.046"
## [1] "Pair LOW WFC : rxy 0.262 se(rxy) 0.049"
## [1] "Pair LOW MRO : rxy 0.181 se(rxy) 0.051"
## [1] "Pair LOW COP : rxy 0.256 se(rxy) 0.049"
## [1] "Pair BAC WFC : rxy 0.692 se(rxy) 0.027"
## [1] "Pair BAC MRO : rxy 0.331 se(rxy) 0.047"
## [1] "Pair BAC COP : rxy 0.339 se(rxy) 0.046"
## [1] "Pair WFC MRO : rxy 0.379 se(rxy) 0.045"
## [1] "Pair WFC COP : rxy 0.396 se(rxy) 0.044"
## [1] "Pair MRO COP : rxy 0.771 se(rxy) 0.021"
```

14.4.3 Confidence Intervals for Population Quantiles

Forming an asymptotic confidence interval for a population quantile turns out to be more difficult than the examples considered thus far. Consider the asymptotic distribution of the sample median as an example:

$$\tilde{X}_{0.5} \overset{a}{\sim} N\left(\tau_{X,0.5}, \frac{1}{4nf_X(\tau_{X,0.5})^2}\right).$$

The problem is that, to consistently estimate the asymptotic variance here, a consistent estimator of $f_X(\tau_{X,0.5})$ is required. While the realized sample median $\tilde{x}_{0.5}$ can be plugged in for the population median $\tau_{X,0.5}$, the pdf $f_X(\cdot)$ is an unknown function when the distribution of X has not been completely specified. Since it is generally not desirable to fully assume the distribution of X when estimating a population quantile, there are two alternative approaches implemented by statistical packages:

1. Estimate $f_X(\tau_{X,0.5})$ by plugging the realized sample median $\tilde{x}_{0.5}$ in for the population median $\tau_{X,0.5}$ and then estimating $f_X(\tilde{x}_{0.5})$ using a *nonparametric density estimator*. Denoting the resulting estimate as $\hat{f}_X(\tilde{x}_{0.5})$, the standard error for the sample median is

$$se(\tilde{x}_{0.5}) = \frac{1}{2\sqrt{n}\hat{f}_X(\tilde{x}_{0.5})}.$$

   Discussing nonparametric density estimation is beyond the scope of this book, but it is worth noting that this estimation approach requires a choice by the researcher, similar to the choice of the bin width for a histogram, that can affect the density estimate. The sensitivity of the standard-error estimate to this choice is a drawback of this approach.

2. Estimate the standard error $se(\tilde{x}_{0.5})$ using a simulation-based approach known as *the bootstrap*, which is discussed in detail in chapter 15.

Moving to the general case, proposition 13.7 provided the asymptotic distribution for the sample quantile $\tilde{X}_q$, which is an estimator of the population quantile $\tau_{X,q}$:

$$\tilde{X}_q \overset{a}{\sim} N\left(\tau_{X,q}, \frac{q(1-q)}{nf_X(\tau_{X,q})^2}\right).$$

The estimate is the realized sample quantile $\tilde{x}_q$. The asymptotic standard deviation is

$$\sqrt{\frac{V}{n}} = \frac{\sqrt{q(1-q)}}{\sqrt{n}f_X(\tau_{X,q})},$$

which again contains the problematic density $f_X(\tau_{X,q})$. At this point, given the difficulty associated with estimating this density, let's assume that the sample quantile $\tilde{x}_q$ and its associated standard error $se(\tilde{x}_q)$ can be calculated by a statistical package like *R*. Then, an asymptotic $1-\alpha$ confidence interval for the population quantile $\tau_{X,q}$ is calculated in the usual way:

$$(\tilde{x}_q - z_{\alpha/2}se(\tilde{x}_q), \tilde{x}_q + z_{\alpha/2}se(\tilde{x}_q)).$$

Example 14.12 (Labor force data) *In example 6.11, various sample quantiles, including all sample deciles and sample quartiles, were calculated for the weekly earnings (earnwk) variable from the* ***cps*** *data set. The following table augments the table from example 6.11 by also including standard errors and asymptotic 95% confidence intervals. Where do the standard errors $se(\tilde{x}_q)$ come from? They are calculated using the bootstrap, covered in chapter 15, and the R code/results are specifically provided in example 15.7:*

q	$\tilde{x}_q$	$se(\tilde{x}_q)$	95% *CI for* $\tau_{X,q}$
0.1	355	7.0	(341, 368)
0.2	480	7.4	(465, 495)
0.25	520	11.7	(497, 543)
0.3	576	7.4	(562, 590)
0.4	670	10.4	(650, 690)
0.5	770	11.1	(748, 792)
0.6	900	16.2	(868, 932)
0.7	1,080	21.9	(1037, 1123)
0.75	1,194	23.7	(1147, 1240)
0.8	1,346	23.9	(1299, 1393)
0.9	1,750	52.0	(1648, 1852)

For the sample median, the estimate is \$770 *(per week), with a standard error of* \$11.1 *(per week) and an asymptotic* 95% *confidence interval* (748, 792) *for the population median* $\tau_{X,0.5}$ *of weekly earnings. From the asymptotic standard deviation formula,* $\frac{\sqrt{q(1-q)}}{\sqrt{n}f_X(\tau_{X,q})}$*, there are two factors that affect the size of the standard error. The term* $q(1-q)$ *is maximized at* $q=0.5$*, so this term leads to smaller standard errors for* q *values closer to* 0 *or* 1*. On the other hand, the term* $f_X(\tau_{X,q})$ *leads to smaller standard errors at quantiles that have higher pdf* $f_X(\cdot)$ *values. Looking at figure 6.7, the density of weekly earnings peaks just above the* 0.25 *quantile of the distribution. In the right tail, the sparsity of the wage data, associated with very low pdf values, leads to much higher standard errors even though* $q(1-q)$ *is relatively lower. As a result, the estimate of the population* 90% *quantile is considerably less precise than the estimates of the other quantiles, whereas the most precise estimates are at the* 10%*,* 20%*, and* 30% *quantiles.*

14.4.4 Confidence Intervals for One-Sample Population Mean Differences

This section considers a situation in which a multivariate, cross-sectional data set is available and interest lies in a confidence interval for the difference in population means, $\mu_X - \mu_Y$, where X and Y are random variables associated with two variables in the data set. The key feature that will be exploited is that values of both variables x and y are observed for each cross-sectional unit. Here are a few examples where such a confidence interval would be of interest:

- *Exam score data*: An instructor gives two exams, where X is the random variable associated with scores on the first exam and Y is the random variable associated with scores on the second exam. If both exam scores are available for a sample of students, a confidence interval for $\mu_X - \mu_Y$ provides information about the difference in the true average scores for the two exams.
- *Asset return data*: X and Y are random variables associated with the returns on two different assets. If both asset returns are available for observations over the same time period, a confidence interval for $\mu_X - \mu_Y$ provides information about the difference in the true average returns of the two assets.

- *Website user activity*: X and Y are binary variables associated with two different actions by a website user, where 1 indicates the action is taken and 0 indicates the action is not taken. For example, X and Y could indicate whether the user clicks on two different links on the website homepage. If both binary variables are observed for a sample of website visitors, a confidence interval for $\mu_X - \mu_Y$ provides information about the difference in the probabilities of the two actions.

Assume that a sample $\{(x_1,y_1),(x_2,y_2),\ldots,(x_n,y_n)\}$ is observed, with underlying i.i.d. bivariate random variables $\{(X_1,Y_1),(X_2,Y_2),\ldots,(X_n,Y_n)\}$. The random variables may be discrete or continuous, and we make no assumptions about their distributions. In practice, as indicated by the examples given here, X and Y would be expected to have the same units, so the difference $\mu_X - \mu_Y$ is meaningful and generally would have the same type of distribution (e.g., both binary in the website example, and both continuous in the asset return example, and both approximately continuous in the exam score example).

The logical estimator for $\mu_X - \mu_Y$ is the difference in sample means $\bar{X} - \bar{Y}$. Since both $\bar{X}$ and $\bar{Y}$ have asymptotically normal sampling distributions, the linear combination $\bar{X} - \bar{Y}$ has one as well. Let $W = X - Y$ denote the linear combination (difference) of the random variables X and Y, and let $w_i = x_i - y_i$ denote the corresponding linear combination (difference) of the observed variables. Applying the result for the asymptotic sampling distribution of the sample mean to the random variable W yields

$$\bar{W} \overset{a}{\sim} N\left(\mu_W, \frac{\sigma_W^2}{n}\right)$$

or, equivalently, since $\bar{W} = \bar{X} - \bar{Y}$ and $\mu_W = \mu_{X-Y} = \mu_X - \mu_Y$,

$$\bar{X} - \bar{Y} \overset{a}{\sim} N\left(\mu_X - \mu_Y, \frac{\sigma_W^2}{n}\right).$$

The asymptotic variance σ_W^2 has not been simplified since it depends on the covariance between X and Y. The appropriate standard error is $\frac{s_W}{\sqrt{n}}$, as

$$s_W = \sqrt{\frac{1}{n-1}\sum_{i=1}^{n}(w_i - \bar{w})^2} = \sqrt{\frac{1}{n-1}\sum_{i=1}^{n}(x_i - y_i - (\bar{x} - \bar{y}))^2}$$

is a consistent estimate of σ_W.

Then the two-sided asymptotic $1-\alpha$ confidence interval for $\mu_X - \mu_Y$, the difference in population means, is

$$\left((\bar{x}-\bar{y}) - z_{\alpha/2}\frac{s_W}{\sqrt{n}}, (\bar{x}-\bar{y}) + z_{\alpha/2}\frac{s_W}{\sqrt{n}}\right) = \left((\bar{x}-\bar{y}) - z_{\alpha/2}\frac{s_{X-Y}}{\sqrt{n}}, (\bar{x}-\bar{y}) + z_{\alpha/2}\frac{s_{X-Y}}{\sqrt{n}}\right).$$

One-sided asymptotic confidence intervals can also be constructed based on the estimate $\bar{x} - \bar{y}$, the standard error $\frac{s_W}{\sqrt{n}}$, and appropriate critical values.

Example 14.13 (Exam score data) *The data set **exams** contains the scores on two exams, out of 100 points, for a sample of 77 students. There are two variables, exam1 and exam2, indicating the scores on the first exam and the second exam, respectively. The sample means are*

$$\overline{exam1} = 77.31 \text{ and } \overline{exam2} = 70.90,$$

so

$$\overline{exam1} - \overline{exam2} = 6.42.$$

The standard deviation of $w = exam1 - exam2$ *is*

$$s_w = s_{exam1-exam2} = \sqrt{\frac{1}{n-1}\sum_{i=1}^{n}(exam1_i - exam2_i - (\overline{exam1} - \overline{exam2}))^2} \approx 10.81,$$

giving a standard error of $\frac{s_w}{\sqrt{n}} = \frac{10.81}{\sqrt{77}} \approx 1.23$. *Here is the R code used to perform these calculations:*

```
# sample mean of exam-score difference
mean(exams$exam1-exams$exam2)

## [1] 6.415584

# sample stdev of exam-score difference
sd(exams$exam1-exams$exam2)

## [1] 10.80734

# std error of sample mean of exam-score difference
se_meanx(exams$exam1-exams$exam2)

## [1] 1.231611
```

The asymptotic 95% confidence interval for $\mu_{exam1} - \mu_{exam2}$ *is*

$$(6.42 - (1.96)(1.23), 6.42 + (1.96)(1.23)) \approx (4.01, 8.83).$$

With 95% confidence, it can be said that the population average of exam1 *scores is between* 4.01 *and* 8.83 *points higher than the population average of* exam2 *scores. The observed difference of* 6.42 *in the sample means is thus associated with statistical evidence that the second exam was truly more difficult, at least as reflected by a lower population mean. If the exams were equally difficult, with* $\mu_{exam1} = \mu_{exam2}$, *it would be expected that the value of zero would either be in the 95% confidence interval or at least close to the interval.*

Example 14.14 (Website user activity) *A company is interested in gauging the effectiveness of two alternative links on the home page of its website. A visitor to the website has the opportunity to click on link A, an event denoted by* $linkA = 1$ *if clicked and* 0 *if not, and/or link B, an event denoted by* $linkB = 1$ *if clicked and* 0 *if not. The following table summarizes the click behavior for a sample of* 200 *visitors:*

		linkB	
		0	1
linkA	0	110	30
	1	44	16

Let π_A *and* π_B *be the probabilities that a website user from the population clicks on link A and link B, respectively. Since linkA and linkB are indicator variables, generated by a Bernoulli random variable,* $\mu_{linkA} = \pi_A$, $\mu_{linkB} = \pi_B$, *and* $\mu_{linkA} - \mu_{linkB} = \pi_A - \pi_B$. *The sample average of linkA, equal to the observed frequency of 1s (clicks), is* $\overline{linkA} = \frac{60}{200} =$ 0.30, *and the sample average of linkB is* $\overline{linkB} = \frac{46}{200} = 0.23$. *The sample standard deviation of linkA − linkB can be determined in two different, but equivalent, ways. The first is to calculate the sample standard deviation based on the sample proportions of linkA − linkB. The second is to calculate the sample standard deviation based on the full sample of observations. Although we don't have a data set corresponding to this table, it is easy enough to generate a data frame in R with the sample. Based on the table, there are* 110 *rows with* $linkA = linkB = 0$, 30 *rows with* $linkA = 0$ *and* $linkB = 1$, 44 *rows with* $linkA = 1$ *and* $linkB = 0$, *and* 16 *rows with* $linkA = linkB = 1$. *The following R code shows how the standard deviation* $s_{linkA-linkB}$ *can be calculated by the two methods:*

```
# calculate means based upon table of counts
meanA <- (44+16)/200
meanB <- (30+16)/200
meandiff <- meanA-meanB

# calculate stdev of linkA-linkB using sample proportion method
sd_1 <- sqrt((200/199)*((110/200)*((0-0)-meandiff)^2+(30/200)*((0-1)-meandiff)^2
                         +(44/200)*((1-0)-meandiff)^2+(16/200)*((1-1)-meandiff)^2))

print(paste("SD of linkA-linkB from sample proportion method:",round(sd_1,5)))

## [1] "SD of linkA-linkB from sample proportion method: 0.60575"

# create a data frame based upon the table of counts
links <- data.frame(linkA = c(rep(0,110),rep(0,30),rep(1,44),rep(1,16)),
                    linkB = c(rep(0,110),rep(1,30),rep(0,44),rep(1,16)))

# calculate stdev of linkA-linkB using the sample
sd_2 <- sd(links$linkA - links$linkB)

print(paste("SD of linkA-linkB from sample stdev method:",round(sd_2,5)))

## [1] "SD of linkA-linkB from sample stdev method: 0.60575"
```

The standard deviation $s_{linkA-linkB} \approx 0.60575$ *yields the standard error*

$$se(\overline{linkA} - \overline{linkB}) = \frac{s_{linkA-linkB}}{\sqrt{200}} \approx 0.0428.$$

Therefore, an asymptotic 95% *confidence interval for the difference in click probabilities* $\pi_A - \pi_B$ *is*

$$((0.30 - 0.23) - (1.96)(0.0428), (0.30 - 0.23) + (1.96)(0.0428)) \approx (-0.0140, 0.1540),$$

so it can be said with 95% *confidence that* $\pi_A - \pi_B$ *is between* −1.40% *and* 15.40%.

14.4.5 Confidence Intervals for Two-Sample Parameter Differences

This section considers a problem similar to that in section 14.4.4, except that the two variables of interest come from *different* samples rather than the same sample. This situation is called a two-sample problem, as one estimated parameter is based on one random sample and another estimated parameter is based on another random sample. Here are examples where the difference in parameters for a two-sample setting would be of interest:

- *Widget purchases*: Example 2.1 considered the purchase probabilities associated with three samples: 300 e-mail A recipients, 300 e-mail B recipients, and 2,400 non-recipients. How do we construct a confidence interval for $\pi_A - \pi_B$, where π_A is the purchase probability for e-mail A recipients and π_B is the purchase probability for e-mail B recipients? How about the difference between e-mail A recipients and nonrecipients?
- *Union versus nonunion workers*: If a sample of employed individuals is split into two samples based on union status, with one subsample being union employees and the other subsample being nonunion employees, there may be interest in the difference between the population mean of weekly earnings for the two subpopulations. Similarly, the difference in population variance or population standard deviation of weekly earnings could be examined, or the difference in the population correlation between education and earnings for the two subpopulations could be examined.

To formalize the two-sample setting, assume that one sample is labeled sample A and the other sample is labeled sample B, where θ_A and θ_B are the underlying parameters of interest for the two corresponding populations. The sizes of the two samples are n_A and n_B, respectively. Assume that $\hat{\theta}_{X_A}$ and $\hat{\theta}_{X_B}$ are asymptotically normal estimators of θ_A and θ_B, with

$$\hat{\theta}_{X_A} \overset{a}{\sim} N\left(\theta_A, \frac{V_A}{n_A}\right) \text{ and } \hat{\theta}_{X_B} \overset{a}{\sim} N\left(\theta_B, \frac{V_B}{n_B}\right),$$

where the realized parameter estimates are denoted as $\hat{\theta}_{x_A}$ and $\hat{\theta}_{x_B}$. The associated standard errors are $se(\hat{\theta}_{x_A}) = \sqrt{\frac{\hat{V}_A}{n_A}}$ and $se(\hat{\theta}_{x_B}) = \sqrt{\frac{\hat{V}_B}{n_B}}$, where $\hat{V}_A$ and $\hat{V}_B$ are consistent estimates of V_A and V_B.

A key property of the two-sample setting is that the two estimators $\hat{\theta}_{X_A}$ and $\hat{\theta}_{X_B}$ are independent of each other since they are based on different random samples. Therefore, when considering the asymptotic variance of the difference $\hat{\theta}_{X_A} - \hat{\theta}_{X_B}$, it is not necessary to consider the covariance between the two estimators since it equals zero. The difference $\hat{\theta}_{X_A} - \hat{\theta}_{X_B}$, as a linear combination of $\hat{\theta}_{X_A}$ and $\hat{\theta}_{X_B}$, is asymptotically normal. The mean of the asymptotic sampling distribution is $\theta_A - \theta_B$, and the variance of the asymptotic sampling distribution is $\frac{V_A}{n_A} + \frac{V_B}{n_B}$, so

$$\hat{\theta}_{X_A} - \hat{\theta}_{X_B} \overset{a}{\sim} N\left(\theta_A - \theta_B, \frac{V_A}{n_A} + \frac{V_B}{n_B}\right).$$

To find the standard error for this estimator, the consistent estimates $\hat{V}_A$ and $\hat{V}_B$ are plugged in for V_A and V_B and a square root is taken, leading to

$$se(\hat{\theta}_{x_A} - \hat{\theta}_{x_B}) = \sqrt{\frac{\hat{V}_A}{n_A} + \frac{\hat{V}_B}{n_B}} = \sqrt{se(\hat{\theta}_{x_A})^2 + se(\hat{\theta}_{x_B})^2}.$$

Thus, the two-sided asymptotic $1-\alpha$ confidence interval for the parameter difference $\theta_A - \theta_B$ is

$$\left((\hat{\theta}_{x_A} - \hat{\theta}_{x_B}) - z_{\alpha/2}\sqrt{se(\hat{\theta}_{x_A})^2 + se(\hat{\theta}_{x_B})^2}, (\hat{\theta}_{x_A} - \hat{\theta}_{x_B}) + z_{\alpha/2}\sqrt{se(\hat{\theta}_{x_A})^2 + se(\hat{\theta}_{x_B})^2}\right).$$

Conveniently, only the estimate $\hat{\theta}_{x_A}$ and standard error $se(\hat{\theta}_{x_A})$ from sample A and the estimate $\hat{\theta}_{x_B}$ and standard error $se(\hat{\theta}_{x_B})$ from sample B are needed to calculate this confidence interval.

Example 14.15 (Widget website) *For the Widgets.com e-mail experiment, example 14.5 provided confidence intervals for the purchase probabilities π_A, π_B, and π_C associated with the subpopulations of e-mail A recipients, e-mail B recipients, and nonrecipients. The estimates for these three parameters are the observed purchase frequencies*

$$p_A = \frac{60}{300} = 0.20,\ p_B = \frac{66}{300} = 0.22,\ \textit{and}\ p_C = \frac{360}{2400} = 0.15,$$

with standard errors

$$se(p_A) = \sqrt{\frac{p_A(1-p_A)}{300}} = \sqrt{\frac{(0.20)(0.80)}{300}} \approx 0.0231,$$

$$se(p_B) = \sqrt{\frac{p_B(1-p_B)}{300}} = \sqrt{\frac{(0.22)(0.78)}{300}} \approx 0.0239,$$

and

$$se(p_C) = \sqrt{\frac{p_C(1-p_C)}{2400}} = \sqrt{\frac{(0.15)(0.85)}{2400}} \approx 0.0068.$$

In this example, confidence intervals are calculated for the difference in purchase probabilities for two of the subpopulations: $\pi_A - \pi_B$, $\pi_A - \pi_C$, and $\pi_B - \pi_C$. (It's unnecessary to separately consider the differences $\pi_B - \pi_A$ and $\pi_C - \pi_B$ since their confidence intervals can be inferred directly from the confidence intervals for $\pi_A - \pi_B$ and $\pi_B - \pi_C$, respectively.) The standard error of $p_A - p_B$, as an estimator for $\pi_A - \pi_B$, is

$$se(p_A - p_B) = \sqrt{\frac{(0.20)(0.80)}{300} + \frac{(0.22)(0.78)}{300}} \approx 0.0332.$$

The asymptotic 95% confidence interval for $\pi_A - \pi_B$ is

$$((0.20-0.22)-(1.96)(0.0332), (0.20-0.22)+(1.96)(0.0332)) \approx (-0.085, 0.045).$$

This interval is quite wide, with 95% confidence that the difference in purchase probabilities $\pi_A - \pi_B$ is between −8.5% and 4.5%. The value of zero, corresponding to no difference ($\pi_A = \pi_B$), is within this interval and therefore is plausible.

The asymptotic 95% confidence intervals for $\pi_A - \pi_C$ and $\pi_B - \pi_C$ can be constructed similarly. The associated standard errors are

$$se(p_A - p_C) = \sqrt{\frac{(0.20)(0.80)}{300} + \frac{(0.15)(0.85)}{2400}} \approx 0.0242.$$

and

$$se(p_B - p_C) = \sqrt{\frac{(0.22)(0.78)}{300} + \frac{(0.15)(0.85)}{2400}} \approx 0.0250.$$

The asymptotic 95% confidence interval for $\pi_A - \pi_C$ is

$$((0.20-0.15)-(1.96)(0.0242), (0.20-0.15)+(1.96)(0.0242)) \approx (0.026, 0.074),$$

and the 95% asymptotic confidence interval for $\pi_B - \pi_C$ is

$$((0.22 - 0.15) - (1.96)(0.0250), (0.22 - 0.15) + (1.96)(0.0250)) \approx (0.045, 0.095).$$

Unlike the confidence interval for $\pi_A - \pi_B$, these two confidence intervals provide statistical evidence of significant differences in the purchase probabilities, with the first indicating 95% confidence that the difference $\pi_A - \pi_C$ is between 2.6% and 7.4% and the second indicating 95% confidence that the difference $\pi_B - \pi_C$ is between 4.5% and 9.5%. The value of zero, corresponding to no difference, is not contained within either of the two confidence intervals.

Example 14.16 (Union workers versus nonunion workers) *Example 6.20 provided descriptive statistics for weekly earnings of union and nonunion workers, based on the* ***cps*** *data set. Part of the table from example 6.20 is reproduced here:*

Sample	n	$\bar{x}$	s_x
Union workers	276	1197.7	720.0
Nonunion workers	2533	946.5	749.7

Using U and NU subscripts to distinguish union and nonunion workers, the standard errors of the sample means $\bar{x}_U = 1{,}197.7$ and $\bar{x}_{NU} = 946.5$ are

$$se(\bar{x}_U) = \frac{720.0}{\sqrt{276}} \text{ and } se(\bar{x}_{NU}) = \frac{749.7}{\sqrt{2533}},$$

so the standard error of $\bar{x}_U - \bar{x}_{NU}$, as an estimator of $\mu_{X,U} - \mu_{X,NU}$, is

$$se(\bar{x}_U - \bar{x}_{NU}) = \sqrt{\frac{720.0^2}{276} + \frac{749.7^2}{2533}} \approx 45.83.$$

The following R code uses the function `se_meanx`*, defined in section 14.3, to calculate these quantities:*

```
# calculate std errors of the sample averages of union and non-union wages
se_union <- se_meanx(cps[cps$unionstatus=="Union","earnwk"], na.rm=TRUE)
se_nonunion <- se_meanx(cps[cps$unionstatus=="Non-union","earnwk"], na.rm=TRUE)

se_union

## [1] 43.33802

se_nonunion

## [1] 14.89694

# calculate std error of the difference in sample averages
sqrt(se_union^2 + se_nonunion^2)

## [1] 45.82688
```

The asymptotic 95% confidence interval for the difference in the population means of weekly earnings, $\mu_{X,U} - \mu_{X,NU}$, is

$$((1197.7 - 946.5) - (1.96)(45.83), (1197.7 - 946.5) + (1.96)(45.83)) \approx (161, 341).$$

It can be said with 95% confidence that the difference between the population means of weekly earnings between union and nonunion workers is between \$161 *and* \$341. *This confidence interval provides fairly strong evidence that the difference is indeed positive, although the size of the interval indicates that the estimated difference of* \$251 *is not very precise.*

How about the difference in the population standard deviations of weekly earnings for union versus nonunion workers? That is, what can be said about the difference in the variation of wage distributions for the two subpopulations? The sample standard deviations $s_{x,U} = 720.0$ *and* $s_{x,NU} = 749.7$ *are estimates of the population standard deviations* $\sigma_{X,U}$ *and* $\sigma_{X,NU}$. *Based on the formula from section 14.4.1 for the standard error of the sample standard deviation, the associated standard errors are*

$$se(s_{x,U}) = 55.18 \text{ and } se(s_{x,NU}) = 34.56.$$

Then the standard error of $s_{x,U} - s_{x,NU}$, *as an estimator of* $\sigma_{X,U} - \sigma_{X,NU}$, *is*

$$se(s_{x,U} - s_{x,NU}) = \sqrt{se(s_{x,U})^2 + se(s_{x,NU})^2} \approx 65.1.$$

The following R code uses the function `se_sx`, *defined in section 14.4.1, to calculate these quantities:*

```
# calculate std errors of the stdevs of earnings
se_sx_union <- se_sx(cps[cps$unionstatus=="Union","earnwk"], na.rm=TRUE)
se_sx_nonunion <- se_sx(cps[cps$unionstatus=="Non-union","earnwk"], na.rm=TRUE)

se_sx_union

## [1] 55.18488

se_sx_nonunion

## [1] 34.56397

# calculate std error of difference in stdevs of earnings
sqrt(se_sx_union^2 + se_sx_nonunion^2)

## [1] 65.11558
```

The asymptotic 95% confidence interval for the difference in population standard deviations of weekly earnings, $\sigma_{X,U} - \sigma_{X,NU}$, *is*

$$((720.0 - 749.7) - (1.96)(130.2), (720.0 - 749.7) + (1.96)(130.2) \approx (-157, 98),$$

indicating no statistical evidence of a difference between the two population standard deviations.

To illustrate the generality of this two-sample approach, we consider looking at a difference in correlations for the two subpopulations. Specifically, what can be said about the difference in the population correlations between weekly earnings and education for union versus nonunion workers? Does the confidence interval for the difference provide any statistical evidence that the earnings-education relationship is different for union and nonunion workers? Let $r_{xy,U}$ *and* $r_{xy,NU}$ *denote the sample correlations between earnwk and educ for union and nonunion workers, respectively, and let* $\rho_{XY,U}$ *and* $\rho_{XY,NU}$ *denote the corresponding population correlations. The sample correlations between earnwk and educ for the two subsamples are*

$$r_{xy,U} \approx 0.253 \text{ and } r_{xy,NU} \approx 0.329.$$

```
cor(cps[cps$unionstatus=="Union","earnwk"],cps[cps$unionstatus=="Union","educ"],
use="complete.obs")

## [1] 0.2529239

cor(cps[cps$unionstatus=="Non-union","earnwk"],cps[cps$unionstatus=="Non-union",
"educ"],use="complete.obs")

## [1] 0.3287519
```

The optional argument `use="complete.obs"` *tells the* `cor` *function to use only those observations for which all variables have nonmissing (non*`NA`*) values. This argument is similar to the* `na.rm=TRUE` *optional argument for functions like* `mean` *and* `sd`*.*

Using the formula for the standard error of a sample correlation (section 14.4.2), the standard errors are

$$se(r_{xy,U}) = \frac{1-0.253^2}{\sqrt{276}} \approx 0.056 \text{ and } se(r_{xy,NU}) = \frac{1-0.329^2}{\sqrt{2533}} \approx 0.018.$$

Then, the standard error of $r_{xy,U} - r_{xy,NU}$, as an estimator of $\rho_{XY,U} - \rho_{XY,NU}$, is

$$se(r_{xy,U} - r_{xy,NU}) = \sqrt{se(r_{xy,U})^2 + se(r_{xy,NU})^2} \approx 0.059.$$

The following R code uses the function `se_rxy`*, defined in section 14.4.2, to calculate these quantities:*

```
se_rxy_union <- se_rxy(cps[cps$unionstatus=="Union","earnwk"],
                       cps[cps$unionstatus=="Union","educ"], na.rm=TRUE)
se_rxy_nonunion <- se_rxy(cps[cps$unionstatus=="Non-union","earnwk"],
                       cps[cps$unionstatus=="Non-union","educ"], na.rm=TRUE)

se_rxy_union

## [1] 0.05634235

se_rxy_nonunion

## [1] 0.01772186

sqrt(se_rxy_union^2 + se_rxy_nonunion^2)

## [1] 0.05906374
```

Therefore, the asymptotic 95% confidence interval for $r_{xy,U} - r_{xy,NU}$ is

$$((0.253-0.329)-(1.96)(0.059), (0.253-0.329)+(1.96)(0.059)) \approx (-0.19, 0.04).$$

While there is some evidence that $r_{xy,U} - r_{xy,NU} < 0$, this confidence interval contains the value zero that corresponds to $r_{xy,U} = r_{xy,NU}$.

14.5 Functions of Consistent Estimators

When we have one or more parameters of interest, there may be situations in which we are interested in estimating a function of the parameters. Due to an important result, known as the **continuous mapping theorem**, it is usually straightforward to estimate such a function of one or more parameters. We first state in proposition 14.9 the result for the case of a single parameter θ, for which we have a consistent estimator θ_X.

Proposition 14.9 (Continuous mapping theorem) *If θ_X is a consistent estimator of θ and $f(\cdot)$ is a continuous function, then $f(\theta_X)$ is a consistent estimator of $f(\theta)$.*

As an example we've already seen, when $\theta = \sigma_X^2$ is the population variance of X, the sample variance $\theta_X = s_X^2$ is a consistent estimator of θ. For $f(\cdot) = \sqrt{\cdot}$, the continuous mapping theorem says that $\sqrt{\theta_X} = \sqrt{s_X^2} = s_X$ is a consistent estimator of the population standard deviation σ_X.

Another example is a quantity known as **odds**, which is based on the Bernoulli parameter π. Specifically, odds are defined as the ratio

$$f(\pi) = \frac{\pi}{1-\pi}.$$

For instance, when $\pi = 0.5$, the odds are $f(0.5) = 1$, or "even odds." When $\pi = \frac{2}{3}$, the odds are $f(\frac{2}{3}) = 2$, meaning that the probability of success is twice the probability of failure. Since $\bar{X}$ is a consistent estimator of π when $X_1, X_2, \ldots, X_n$ are i.i.d. *Bernoulli*(π) random variables, the continuous mapping theorem says that

$$f(\bar{X}) = \frac{\bar{X}}{1-\bar{X}}$$

is a consistent estimator of the odds.

The continuous mapping theorem extends more generally to a case with multiple consistent estimators. To keep things simple, we state the result for the case of two consistent estimators in proposition 14.10, but it can be generalized to more estimators.

Proposition 14.10 (Continuous mapping theorem) *If θ_X^a and θ_X^b are consistent estimators of θ^a and θ^b, respectively, and $f(\cdot,\cdot)$ is a continuous function, then $f(\theta_X^a, \theta_X^b)$ is a consistent estimator of $f(\theta^a, \theta^b)$. Special cases include the following:*

i. $\theta_X^a + \theta_X^b$ *is a consistent estimator of* $\theta^a + \theta^b$.

ii. $\theta_X^a \theta_X^b$ *is a consistent estimator of* $\theta^a \theta^b$.

iii. $\frac{\theta_X^a}{\theta_X^b}$ *is a consistent estimator of* $\frac{\theta^a}{\theta^b}$ *(if $\theta^b \neq 0$).*

As an example, suppose that we are interested in the quantity $\frac{IQR_X}{\sigma_X}$, which is the population IQR of random variable X in terms of standard deviations (e.g., a value of 4 indicates that the IQR is 4 standard deviations wide). For i.i.d. random variables $X_1, X_2, \ldots, X_n$, consistent estimators of IQR_X and σ_X are $\tilde{X}_{0.75} - \tilde{X}_{0.25}$ and s_X, respectively, and proposition 14.10 implies that

$$\frac{\tilde{X}_{0.75} - \tilde{X}_{0.25}}{s_X}$$

is a consistent estimator of $\frac{IQR_X}{\sigma_X}$.

As another example, suppose that we want to compare two population correlations $\rho_{X_1X_2}$ and $\rho_{X_3X_4}$ based on data from a single data set (like the **sp500** data set). Consistent estimators are the sample correlations $r_{X_1X_2}$ and $r_{X_3X_4}$, respectively. Two alternative ways to compare correlations are to look at differences and to look at ratios. Proposition 14.10 implies that the difference $r_{X_1X_2} - r_{X_3X_4}$ is a consistent estimator of the difference $\rho_{X_1X_2} - \rho_{X_3X_4}$ and also that the ratio $\frac{r_{X_1X_2}}{r_{X_3X_4}}$ is a consistent estimator of the ratio $\frac{\rho_{X_1X_2}}{\rho_{X_3X_4}}$ (if $\rho_{X_3X_4} \neq 0$).

How about asymptotic normality? The good news is that, due to a result known as the **delta method**, functions of estimators will generally be asymptotically normal if the underlying estimators are themselves asymptotically normal. For the single-parameter case, the only additional assumption needed is that function $f(\cdot)$ is differentiable at the true parameter θ. Proposition 14.11 provides the asymptotic-variance formula for the single-parameter case.

Proposition 14.11 (Delta method) *If θ_X is an asymptotically normal estimator of θ, with $\sqrt{n}(\hat{\theta}_X - \theta) \overset{a}{\sim} N(0, V)$, and $f(\cdot)$ is a continuous function that is differentiable at θ, then $f(\theta_X)$ is an asymptotically normal estimator of $f(\theta)$, with*

$$\sqrt{n}(f(\hat{\theta}_X) - f(\theta)) \overset{a}{\sim} N(0, f'(\theta)^2 V)$$

or, equivalently,

$$f(\hat{\theta}_X) \overset{a}{\sim} N\left(f(\theta), \frac{f'(\theta)^2 V}{n}\right).$$

In the one-parameter case with estimator $f(\theta_X)$, the mean of the asymptotic distribution is $f(\theta)$ since $f(\theta_X)$ is a consistent estimator of $f(\theta)$; the variance of the asymptotic distribution is $f'(\theta)^2 \frac{V}{n}$, where $\frac{V}{n}$ is the asymptotic variance of the original estimator. For example, for the odds estimator $f(\bar{X}) = \frac{\bar{X}}{1-\bar{X}}$, $f'(\pi) = \frac{1}{(1-\pi)^2}$ is obtained by taking the derivative of $\frac{\pi}{1-\pi}$ with respect to π. Therefore, the asymptotic variance of $f(\bar{X}) = \frac{\bar{X}}{1-\bar{X}}$, as an estimator of $f(\pi) = \frac{\pi}{1-\pi}$, is

$$\left(\frac{1}{(1-\pi)^2}\right)^2 \frac{\pi(1-\pi)}{n} = \frac{\pi}{n(1-\pi)^3}$$

since $\frac{\pi(1-\pi)}{n}$ is the asymptotic variance of $\bar{X}$ as an estimator of π. The asymptotic standard deviation is $\sqrt{\frac{\pi}{n(1-\pi)^3}}$, leading to the following standard error for the odds estimator:

$$se\left(\frac{\bar{x}}{1-\bar{x}}\right) = \sqrt{\frac{\bar{x}}{n(1-\bar{x})^3}}.$$

In the two-parameter case with an estimator $f(\theta_X^a, \theta_X^b)$, the mean of the asymptotic normal distribution is $f(\theta^a, \theta^b)$; the variance of the asymptotic distribution is more complicated than the one-parameter case since it involves partial derivatives and the covariance between the two estimators.

14.6 Asymptotic Predictive Intervals for Continuous Random Variables

For a continuous random variable X, suppose that we would like to provide an interval for which the probability that X lies in the interval is equal to a certain value, like 95% or 99%.

In general, a two-sided $1-\alpha$ probability interval for X is

$$(\tau_{X,\alpha/2}, \tau_{X,1-\alpha/2}),$$

where the lower end point is the population $\alpha/2$th quantile and the upper end point is the population $(1-\alpha/2)$th quantile. Since these population quantiles are unknown in most cases of practical interest, we must estimate them. The ability to consistently estimate the two quantiles at the end points leads to an interval, which we call an **asymptotic predictive interval**, that has end points that become arbitrarily close to the true end points as the sample size increases.

Definition 14.9 *An* ***asymptotic*** $(1-\alpha)$ ***predictive interval*** *for X is $(\hat{\theta}_{\alpha/2}, \hat{\theta}_{1-\alpha/2})$, where $\hat{\theta}_{\alpha/2}$ is a consistent estimate of the population $\alpha/2$th quantile $\tau_{X,\alpha/2}$ and $\hat{\theta}_{1-\alpha/2}$ is a consistent estimate of the population $(1-\alpha/2)$th quantile $\tau_{X,1-\alpha/2}$.*

There are two types of asymptotic predictive intervals, *model-free intervals* and *model-based intervals*. A model-free interval uses sample quantiles as consistent estimates of population quantiles, so the model-free asymptotic $(1-\alpha)$ predictive interval is

$$(\hat{\theta}_{\alpha/2}, \hat{\theta}_{1-\alpha/2}) = (\tilde{x}_{\alpha/2}, \tilde{x}_{1-\alpha/2}).$$

An advantage of this interval is that it doesn't require any knowledge or assumptions about the distribution of X. For the case of $\alpha = 0.05$, the asymptotic 95% predictive interval has the lower end point $\tilde{x}_{0.025}$ and the upper end point $\tilde{x}_{0.975}$. For a large sample, there is approximately a 95% probability that a new draw of X from the population is between $\tilde{x}_{0.025}$ and $\tilde{x}_{0.975}$. Since $\tilde{x}_{0.025}$ and $\tilde{x}_{0.975}$ are only estimates of the population quantiles $\tau_{X,0.025}$ and $\tau_{X,0.975}$, it is a good idea to calculate the standard errors for both $\tilde{x}_{0.025}$ and $\tilde{x}_{0.975}$ to assess how close the estimated end points are likely to be to the true end points.

A model-based interval uses an assumed model and estimates of the model's parameters rather than the sample quantiles. As an example, for a normal random variable $X \sim N(\mu, \sigma^2)$, section 11.1 provided the $(1-\alpha)$ probability interval

$$(\tau_{X,\alpha/2}, \tau_{X,1-\alpha/2}) = (\mu - z_{\alpha/2}\sigma, \mu + z_{\alpha/2}\sigma),$$

which suggests the model-based asymptotic $(1-\alpha)$ predictive interval

$$(\hat{\theta}_{\alpha/2}, \hat{\theta}_{1-\alpha/2}) = (\bar{x} - z_{\alpha/2}s_x, \bar{x} + z_{\alpha/2}s_x).$$

Since $\bar{X}$ is a consistent estimator of μ and s_X is a consistent estimator of σ, $\bar{X} - z_{\alpha/2}s_X$ and $\bar{X} + z_{\alpha/2}s_X$ are consistent estimators of $\mu - z_{\alpha/2}\sigma$ and $\mu + z_{\alpha/2}\sigma$, respectively. The realized end points $\bar{x} - z_{\alpha/2}s_x$ and $\bar{x} + z_{\alpha/2}s_x$ become arbitrarily close to the true end points as the sample size gets larger. The model-based interval for a normal random variable has its end points equidistant from $\bar{x}$, whereas a model-free interval would not necessarily have this property. Also, if the model is true, it's likely that a model-based interval provides more precise estimates of the end points than a model-free interval does. The intuition is that the model-based interval uses additional information, specifically the model assumed for X, as opposed to the model-free interval. That said, for very large samples, the model-free and model-based intervals should look quite similar since both are based on consistent estimates of the end points.

Notes

1. If one estimator is inconsistent, then the consistent estimator is generally preferred.
2. The number of confidence intervals, out of 100, for which μ is outside the confidence interval is a $Binomial(100, 0.05)$ random variable. If the number of Monte Carlo simulations grows large, the percentage of times that μ falls outside the 95% confidence interval gets arbitrarily close to 5%.
3. The sample variance s_x^2 of a Bernoulli random variable is $\frac{1}{n-1}\sum_{i=1}^{n}(x_i-\bar{x})^2$, which simplifies to

$$\frac{1}{n-1}\sum_{i=1}^{n}(x_i^2-2x_i\bar{x}+\bar{x}^2)=\frac{1}{n-1}\left(\sum_{i=1}^{n}x_i-2\bar{x}\sum_{i=1}^{n}x_i+n\bar{x}^2\right)=\frac{1}{n-1}\left(n\bar{x}-2n\bar{x}+n\bar{x}^2\right)$$

$$=\frac{1}{n-1}n\bar{x}(1-\bar{x}),$$

applying the fact that $x_i^2=x_i$ for an indicator variable x_i. Thus, $s_x=\sqrt{\frac{n}{n-1}}\sqrt{\bar{x}(1-\bar{x})}$.

Exercises

1. A professor wants to estimate the probability of cannabis use (in the last year) in the population of students who take economic statistics. Concerned that students might not honestly answer a direct question about cannabis use, she uses a method known as *randomized response* to elicit honest responses. There are 300 students in her economic statistics class. Before class, she creates 300 pieces of paper, numbered 1 through 300. As students come into class, they randomly pick a piece of paper (without the professor knowing the numbers). Once everyone is seated, she gives the following instructions: "Please answer the question 'Is the last digit of your phone number even?' if your piece of paper has a number less than or equal to 150. Please answer the question 'Have you used cannabis in the last year?' if your piece of paper has a number greater than 150." Using a phone-response system, she observes the total number of "Yes" responses and nothing else. Let Y denote the random variable associated with the total number of "Yes" responses from the 300-student class. Let π denote the probability of cannabis use in the last year. Assume that (i) students honestly respond to their question and (ii) the probability of an even last-digit is 0.5.
   a. What is $E(Y)$ in terms of π?
   b. Propose an estimator of π that is a function of Y, and show that the proposed estimator is unbiased. What is the value of the estimate when the realization of Y is 125?
   c. *The estimator in (b) estimates the *unconditional* probability of cannabis use in the last year. Using Bayes' theorem, determine both the conditional probability of cannabis use given a "Yes" response and the conditional probability of cannabis use given a "No" response in terms of π. Plug in the estimate of π from (b) (when $Y=125$) to yield estimates of these conditional probabilities.
   d. Suppose that an alternative randomization technique is used. Rather than the paper system, each student is asked to flip a coin and then to answer the first question if the flip is heads and the second question if the flip is tails. In this case, while the expected number of students answering the first question is 150, the actual number

may be different from 150. Is the estimator from (b) still unbiased? How would you expect the variance of the coin flip–based estimator to compare to the variance of the paper-based estimator?

e. Returning to the paper randomization, suppose that the professor alters the instructions to have the students answer the first question if their number is less than or equal to 75 and the second question if their number is greater than 75. What is an appropriate estimator of π (in terms of Y) in this case?

f. In this part, computer simulations will be used to compare the performance of the estimators for the three randomized-response alternatives (alternative 1: original question, alternative 2: part (d), alternative 3: part (e)). Assume that the probability of cannabis use in the last year is known to be 30%. For each of the three alternatives, conduct 10,000 simulations in *R* of the 300-student class responses and calculate the associate estimates. What are the averages of the estimates for each of the three alternatives? What are the standard deviations of the estimates for each of the three alternatives? Do the relative sizes of the standard deviations make sense?

2. *Consider a negative binomial random variable $X \sim NegBin(r, \pi)$.

a. For $r \geq 2$, show that $\hat{\pi}_X = \frac{r-1}{X+r-1}$ is an unbiased estimator of π.

b. An employee at an investment management company has been asked to call the company's clients to see which ones are interested in receiving information about a new mutual fund. The employee has decided that she will take a break after the fourth successful call, where "success" means that the client is interested in receiving information about the new mutual fund. Based on (a), what is the estimate of the success probability if the successes occur on the 5'th, 12'th, 14'th, and 21'st calls?

c. For $r \geq 1$, show that $\hat{\theta}_X = \frac{X+r}{r}$ is an unbiased estimator of $\theta = \frac{1}{\pi}$.

d. The result in (c) suggests an alternative estimator of π, given by $\tilde{\pi}_X = \frac{r}{X+r}$. Using this estimator, what is the estimate of the success probability for the scenario described in (b)?

e. It turns out that both $\hat{\pi}_X$ and $\tilde{\pi}_X$ are consistent estimators of π (when we think of r growing very large). In this part, computer simulations will be used to compare the performance of the alternative estimators. Specifically, we consider the scenario from (b), but we assume that we know that the true success probability is 17% ($\pi = 0.17$). For $r = 4$, conduct 100,000 simulated i.i.d. draws of $X \sim NegBin(r, \pi)$ in *R*; and, for each draw, calculate the two estimates based on $\hat{\pi}_X$ and $\tilde{\pi}_X$.

i. What are the average values of the two estimators (over the simulations)?

ii. One way to compare estimators is to calculate the *mean absolute error*, defined as the average of |estimate – true value| over the simulations. What are the mean absolute errors of the two estimators?

iii. Another way to compare estimators is to calculate the *mean squared error*, defined as the average of (estimate – true value)^2 over the simulations. What are the mean squared errors of the two estimators?

iv. Repeat the simulations and parts (i)–(iii) for $r = 100$ instead of $r = 4$.

3. Referring to proposition 13.3, explain why the sum of i.i.d. random variables, $S = \sum_{i=1}^{n} X_i$, is an unbiased estimator of $n\mu_X$ but not a consistent estimator of $n\mu_X$.
4. Consider an i.i.d. random sample $x_1, x_2, \ldots, x_n$ drawn from a $N(10, 4)$ distribution, where n is very large. In thinking about the box plot with whiskers and outliers (section 6.5.1), provide your best guesses for the values of the following quantities:
   a. Sample median (line within the box)
   b. Top and bottom of the box
   c. Upper and lower whiskers
   d. The percentage of points ("outliers") above the upper whisker
5. Assume that the monthly returns on a certain asset are i.i.d. draws from a normally distributed random variable X, with unknown mean and standard deviation. You observe the monthly returns for one year ($n = 12$), and calculate the sample average to be 0.01 and the sample standard deviation to be 0.08.
   a. Using the appropriate t-distribution, provide 95% and 99% confidence intervals for the population mean μ_X.
   b. How do the confidence intervals in (a) compare to the confidence intervals based on a $N(0, 1)$ distribution, rather than the t-distribution, for calculating the critical values?
   c. Suppose instead that $n = 120$, and you calculate the same sample average (0.01) and sample standard deviation (0.08). Using the normal approximation, provide 95% and 99% confidence intervals for μ_X.
6. Assume that IQ scores are normally distributed, but the population mean and standard deviation of the normal distribution are unknown. You collect a random sample of 20 individuals for which you calculate $\bar{x} = 98$ and $s_x = 12$.
   a. Using the appropriate t-distribution, provide 90% and 95% confidence intervals for the population mean μ_X.
   b. How do the confidence intervals in (a) compare to the confidence intervals based upon a $N(0, 1)$ distribution, rather than the t-distribution, for calculating the critical values?
   c. If you instead had $\bar{x} = 102$ and $s_x = 12$, how would the width of the 95% confidence interval for μ_X compare to the one in (a)?
   d. If you instead had $\bar{x} = 98$ and $s_x = 16$, how would the width of the 95% confidence interval for μ_X compare to the one in (a)?
7. A company's weekly profits are i.i.d. draws of a normal random variable $X \sim N(\mu, \sigma^2)$. After n weeks of profits $(x_1, x_2, \ldots, x_n)$ are observed, the following two confidence intervals for μ, based on t-distribution inference, are constructed:

   $$95\% \text{ confidence interval for } \mu\text{: } (7.901, 12.099)$$

   and

   $$90\% \text{ confidence interval for } \mu\text{: } (8.355, 11.645).$$

a. What is the sample average of the weekly profits?

b. What is n? (Hint: Think about what the ratio of confidence-interval widths says about the ratio between the critical values. Then use R to determine the value of n consistent with that ratio.)

c. What is the sample standard deviation of weekly profits?

d. What are the one-sided 95% confidence intervals for μ? There are two such intervals, one of the form (L, ∞) and one of the form $(-\infty, U)$.

8. A car factory implements a new production process and, over the course of the first 7 days, produces 198, 208, 206, 225, 234, 210, and 187 cars. Assume that the daily production numbers are i.i.d. draws from a normal distribution.

a. Calculate the sample mean and sample standard deviation of daily production.

b. Determine the one-sided 95% confidence interval, of the form (L, ∞), for the population average of daily production.

9. You are a fast-food mogul and own 50 franchises of Longhorn Burgers and 40 franchises of Perfect Pitas. Suppose the monthly revenue (in thousands of dollars) for every franchise is an i.i.d. random variable, with Longhorn Burgers revenues drawn from a distribution with population mean 20 and population standard deviation 4 and Perfect Pitas revenues drawn from a distribution with population mean 15 and population standard deviation 3.

a. In a given month, what is the approximate distribution of the sample average of the monthly revenues at the 40 franchises of Perfect Pitas?

b. In a given month, what is the approximate distribution of the sample average of monthly revenues at all 90 franchises?

10. The two-sided asymptotic 95% confidence interval for a certain parameter is $(10, 20)$.

a. What is the two-sided asymptotic 80% confidence interval for the same parameter?

b. For the one-sided asymptotic 95% confidence interval (L, ∞), what is the value of L?

11. A university's information technology department has a large inventory of old computer monitors, some of which are no longer working. To estimate the proportion of nonworking monitors, the staff tests 60 of them (at random) and finds that 15 are not working. Find the asymptotic 95% confidence interval for the population proportion of monitors that are not working.

12. A landscaping company offers a promotion in a suburban neighborhood, whereby they mow a house's lawn for no charge the first time and then ask the homeowners if they would like to continue having regular (paid) service thereafter. Suppose n homeowners allow the company to mow their lawn for free. The company's owner is interested in the probability π that such homeowners will continue with the paid service. After observing the proportion of n homeowners who continue with the paid service, the company's owner finds that the asymptotic 95% confidence interval for π is $(0.1503, 0.2997)$.

a. What is the sample proportion of homeowners who continued with the paid service?

b. What is n?

13. An advertising company wishes to estimate the population mean of the distribution of hours of television watched per household per day. Suppose the population standard deviation of hours watched per household per day is known to be 2.8 hours. The company decides that it wants the asymptotic 99% confidence interval for the population mean to be no wider than 0.5 hours. What is the minimum sample size that results in a small-enough confidence interval?

14. Use the **exams** data set, which contains data for 77 students on two exams (`exam1` and `exam2`). Suppose `exam1` scores are i.i.d. draws from a random variable with mean μ_1 and standard deviation σ_1 and `exam2` scores are i.i.d. draws from a random variable with mean μ_2 and standard deviation σ_2.

    a. Which is larger: the middle of the asymptotic 95% confidence interval for μ_1 or the middle of the asymptotic 95% confidence interval for μ_2?

    b. Which is larger: the width of the asymptotic 95% confidence interval for μ_1 or the width of the asymptotic 95% confidence interval for μ_2?

    c. What is the asymptotic 90% confidence interval for μ_1?

    d. What is the asymptotic 90% confidence interval for $\mu_1 + \mu_2$, the sum of the exam population means?

    e. What is the asymptotic 90% confidence interval for the population correlation $\rho_{exam1,exam2}$?

    f. The standard errors of the sample standard deviations of `exam1` and `exam2` are 2.1122 and 2.8632, respectively. Which is larger: the upper end point of the asymptotic 95% confidence interval of σ_{exam1} or the upper end point of the asymptotic 95% confidence interval of σ_{exam2}?

15. A researcher surveys 500 individuals on their level of happiness x (on a scale from 1 to 10) and their annual income y (in thousands of dollars). Assume that the observed data are i.i.d. draws from the joint distribution of the underlying random variables (X, Y).

    a. Before the researcher observes the data, what is the largest possible width of the asymptotic 95% confidence interval for the population correlation ρ_{XY} (i.e., over all possible realizations of r_{xy})?

    b. The researcher calculates $r_{xy} = 0.23$ based on the observed sample. What is the asymptotic 95% confidence interval for ρ_{XY}?

    c. Using $r_{xy} = 0.23$ again, what is the one-sided asymptotic 95% confidence interval, of the form $(L, 1)$, for ρ_{XY}? (The upper end is 1 and not ∞ since $\rho_{XY} \leq 1$.)

16. In a random survey of 250 undergraduates, 150 respondents indicate that they consume caffeine daily.

    a. Provide an asymptotic 95% confidence interval for the probability π_C that an undergraduate from the population consumes caffeine daily.

    b. Another survey of a different sample of 250 undergraduates indicates that 140 out of the 250 respondents sleep at least 7 hours on a daily basis. Let π_S denote the probability that an undergraduate sleeps at least 7 hours on a daily basis. Provide an asymptotic 95% confidence interval for the difference $\pi_C - \pi_S$.

17. In the **sp500** data, 236 of the 360 monthly returns for the S&P 500 Index (`idx`) are positive. Assume that each monthly return is an i.i.d. draw from some underlying random variable.

    a. Provide an asymptotic 90% confidence interval for the probability that the S&P 500 Index monthly return is positive.

    b. A risk-averse investor is worried about large negative returns. In the **sp500** data, 33 of the 360 monthly returns for the S&P 500 Index are less than -0.05. Provide a one-sided asymptotic 95% confidence interval, of the form $(0, U)$, for the probability that the S&P 500 Index monthly return is less than -0.05. (The lower end point is zero, and not negative infinity, since the parameter is a probability.)

18. Use the **cps** data set for this question.

    a. Form a table of `gender` and `union` to determine how many male workers are union versus nonunion and how many female workers are union versus nonunion.

    b. What is the asymptotic 95% confidence interval for the probability that a male worker is in a union?

    c. What is the asymptotic 95% confidence interval for the probability that a female worker is in a union?

    d. What is the asymptotic 95% confidence interval for the difference between the probability that a male worker is in a union and the probability that a female worker is in a union?

19. A survey of 150 students from college A finds that the average SAT math score is 610, with a sample standard deviation of 50. A similar survey of 200 students from college B finds that the average and standard deviation are 580 and 45, respectively.

    a. What is the asymptotic 90% confidence interval for the population average of SAT math scores at college A?

    b. What is the asymptotic 90% confidence interval for the population average of SAT math scores at college B?

    c. What is the asymptotic 90% confidence interval for the difference between the population average at college A and the population average at college B?

    d. Based on the confidence interval from (c), do you think that it is likely that the population average for college A students is larger than the population average for college B students?

20. *An economist wants to study the gender wage gap in a particular industry by collecting salary data from male and female workers. In this industry, 80% of workers are male. Let $\mu_m = E(X_m)$ and $\mu_f = E(X_f)$ denote the population means of salaries for male and female workers, respectively. The economist is interested in forming confidence intervals for the difference $\mu_m - \mu_f$, and the size of the total sample collected (male and female combined) is n. Let γ be the proportion of male workers in the sample, so there are γn male workers and $(1-\gamma)n$ female workers.

    a. If $\sigma_m^2 = Var(X_m)$ and $\sigma_f^2 = Var(X_f)$, what is the asymptotic variance of $\bar{X}_m - \bar{X}_f$ in terms of n, γ, σ_m^2, and σ_f^2?

b. What value of γ (in terms of σ_m, σ_f, and/or n) minimizes the asymptotic variance of $\bar{X}_m - \bar{X}_f$?

c. If the variances of male and female wages are the same ($\sigma_m^2 = \sigma_f^2$), what value of γ minimizes the asymptotic variance of $\bar{X}_m - \bar{X}_f$?

d. If the economist collects the data with a simple random sample, the proportion of male workers in the sample will be approximately 80%. For the case of equal wage variances ($\sigma_m^2 = \sigma_f^2$), how would the width of the asymptotic confidence interval based on $\gamma = 0.8$ (simple random sample) compare to the width of the asymptotic confidence interval using the optimal γ found in (c)? Does this argue for oversampling or undersampling of female workers?

21. The number of customers that enter a coffee shop during a given minute of the day (say, between 2 p.m. and 2.01 p.m.) is distributed as a $X \sim Poisson(\lambda)$ random variable, where λ is an unknown parameter. Suppose the coffee shop gathers data over the course of 100 days, each day recording the number of customers that enter between 2 p.m. and 2:01 p.m., and finds that the sample average is 2.12 and the sample standard deviation is 1.53. You may assume that the number of customers on each day is independent from other days.

a. The population mean of a Poisson random variable X is $\mu_X = \lambda$. Provide an asymptotic 95% confidence interval for λ.

b. Thinking of the sample average $\bar{x}$ as an estimate of $\mu_X = \lambda$, what is the estimated probability of having exactly two customers enter between 2 p.m. and 2:01 p.m. on a given day? (Plug $\bar{x}$ in for λ in the Poisson pmf formula.)

c. Repeat (b), but now use the lower and upper end points of the confidence interval from (a) to calculate the estimated probability of having exactly two customers enter between 2 p.m. and 2:01 p.m. on a given day. (Since there is uncertainty in our estimate of λ, as reflected by the confidence interval, this part shows how that uncertainty translates to estimation of the probability value.)

22. Use the **strikes** data set for this question. This data set contains information on worker contract strikes within US manufacturing for the period 1968–1976. There are 566 observations on the variable *duration* (strike duration, in weeks).

a. What are the sample mean, sample median, and sample standard deviation of *duration*?

b. Draw a (density) histogram of *duration* with 10 bins.

c. Given the right-skewed nature of *duration*, consider whether a log-normal distribution might be a good description of *duration*. Generate a new variable *lndur* equal to the natural logarithm of *duration*.

d. If $lndur \sim N(\mu, \sigma^2)$ is true, the sample mean of *lndur* is a consistent estimator of μ. Provide an asymptotic 95% confidence interval for μ.

e. If $lndur \sim N(\mu, \sigma^2)$ is true, the expected value of *duration* is $e^{\mu + \frac{1}{2}\sigma^2}$. Using the sample mean and sample standard deviation of *lndur* as estimates of μ and σ, respectively, plug into the expected-value formula to get an estimated expected value of *duration*. How does this estimate compare to the sample mean of *duration*?

f. Draw a density histogram of *lndur* with 10 bins. What do you conclude about the log-normal distribution being a good model for *duration*?

23. The number of workplace injuries at a certain factory is tracked over 200 weeks. The average of the weekly number of injuries is 0.4. Assume that each weekly observation is an i.i.d. draw from a $Poisson(\lambda)$ random variable.

    a. Provide an asymptotic 90% confidence interval for λ.

    b. Show that $P(X > 0) = 1 - P(X = 0)$ is an increasing function of λ.

    c. Based on (b), the end points of an asymptotic 90% confidence interval for $P(X > 0)$ can be determined from the the following two probabilities: the probability that there are any workplace injuries in a given week based on the $Poisson(\underline{\lambda})$ distribution, and the probability that there are any workplace injuries in a given week based on the $Poisson(\overline{\lambda})$ distribution. Calculate these end points.

24. A convenience store sells lottery tickets. On the day of a large drawing, the time (in minutes) between lottery-ticket purchases can be considered i.i.d. draws of an exponential random variable $X \sim Exp(\theta)$. Suppose that the average of 100 observed times-between-purchases is 1.8 minutes.

    a. What is the asymptotic 90% confidence interval for $\frac{1}{\theta}$?

    b. Use the end points of the interval from (a) to form an asymptotic 90% confidence interval for θ. The resulting interval need not be symmetric. (Hint: $P(L \leq 1/\theta \leq U) = P(1/U \leq \theta \leq 1/L)$.)

    c. Based on the continuous mapping theorem (proposition 14.9), how would you consistently estimate θ given the consistent estimator of $\frac{1}{\theta}$?

    d. *Based on the delta method, what is the asymptotic standard error associated with the estimate from (c)?

    e. Provide an asymptotic 90% confidence interval for θ using the estimate from (c) and the standard error from (d). How does this interval compare to the one found in (b)?

25. A prolific inventor submits 150 patent applications to the US Patent and Trademark Office, and 110 are successful (i.e., resulting in a patent being issued). Assume that the success of each patent application is an i.i.d. $Bernoulli(\pi)$ random variable.

    a. What are the estimated odds of a patent application being successful? (Recall that the odds are defined as $\frac{\pi}{1-\pi}$.)

    b. Provide an asymptotic 95% confidence interval for the odds of a patent application being successful.

26. For i.i.d. $X_1, X_2, \ldots, X_n \sim Poisson(\lambda)$ random variables, a consistent estimator of $\mu_X = \lambda$ is the sample average.

    a. Based on the continuous mapping theorem (proposition 14.9), how would you consistently estimate the population standard deviation ($\sigma_X = \sqrt{\lambda}$) of the underlying Poisson random variable?

    b. *Based on the delta method, provide a formula for the asymptotic standard error associated with the estimate from (a). If $\bar{x} = 1.2$ and $n = 200$, what is the asymptotic 95% confidence interval for $sd(X) = \sqrt{\lambda}$?

15

The Bootstrap

Chapter 14 considered the estimation of standard errors and confidence intervals based on two types of statistical inference: finite-sample inference and asymptotic inference. For finite-sample inference, the finite-sample (exact) sampling distribution results from chapter 12 were used as the basis for confidence intervals for the population mean of i.i.d. normal random variables. For asymptotic inference, the more general asymptotic (large-sample) sampling distribution results from chapter 13 were used as the basis for confidence intervals associated with any asymptotic normal estimator.

This chapter introduces another type of statistical inference, known as **bootstrap inference** or, more concisely, the **bootstrap**. The bootstrap is a resampling method that can be used as an alternative to the inference approaches in chapter 14. Why might the bootstrap be needed as an alternative method for statistical inference? Here are three reasons:

1. Even with a very large sample, when the asymptotic distribution provides a good approximation, the asymptotic variance (or standard deviation) formula might not provide a simple approach for the estimation of an estimator's standard error. For example, as seen in chapter 13, when using a sample quantile to estimate a population quantile, the formula for the asymptotic standard deviation involves an unknown pdf $f_X(\cdot)$ value that needs to be estimated for a standard error to be calculated. The bootstrap provides an alternative approach that avoids estimation of the pdf altogether.
2. In certain situations, we may want to estimate a parameter for which the asymptotic variance is not readily available and/or difficult to determine. For example, suppose that the two random variables X and Y are associated with two variables in an observed sample and we are interested in the difference between the population standard deviations, $\sigma_X - \sigma_Y$. Section 14.4.4 considered a similar situation, in which an asymptotic confidence interval for $\mu_X - \mu_Y$ was proposed, but the reasoning used there does not obviously extend to $\sigma_X - \sigma_Y$. While $s_X - s_Y$ is an appropriate estimator of $\sigma_X - \sigma_Y$, determining the asymptotic standard deviation of $s_X - s_Y$, as an estimator of $\sigma_X - \sigma_Y$, is difficult since s_X and s_Y are not necessarily independent of each other. The bootstrap provides an alternative approach for this situation that does not require the researcher to analytically determine the asymptotic variance. There are many other similar situations, in which the asymptotic variance is difficult to obtain, that the bootstrap is an

appealing alternative. For example, the bootstrap can be used to construct a confidence interval for the difference between the population mean μ_X and population median $\tau_{X,0.5}$, based on a random sample associated with a random variable X. As another example, the bootstrap can be used to construct a confidence interval for the difference between two population correlations, based on a random sample of multivariate data. For the labor force data, for instance, if we are interested in the relative magnitudes of $\rho_{earnwk,educ}$ and $\rho_{earnwk,age}$, a confidence interval for $\rho_{earnwk,educ} - \rho_{earnwk,age}$ would be useful.

3. The observed sample may not be large enough for the asymptotic distribution to provide a good approximation of an estimator's true sampling distribution. While a finite-sample distribution can be used in some cases, like estimation of the population mean of i.i.d. normal random variables, it is more likely that a finite-sample distribution will not be available. For instance, for the estimation of the population mean μ_X of i.i.d. random variables with an unknown distribution, there is no finite-sample distribution that can be used, but the asymptotic sampling distribution may not provide a good approximation if the sample is very small. As another example, the asymptotic distribution associated with estimation of the population correlation ρ_{XY} between two random variables X and Y is known to provide a poor approximation of the true sampling distribution when the sample is small and the true correlation ρ_{XY} is close to 1 or -1. For these examples and others where we might be concerned that the sample is not large enough, the bootstrap provides an alternative approach for constructing confidence intervals.

15.1 Bootstrap Sampling

Whereas finite-sample and asymptotic sampling distributions are based on formulas derived from statistical results, the bootstrap instead uses a resampling approach that relies entirely on computer simulations. The fundamental idea underlying the bootstrap is that the distribution of the variables in the observed sample can be used as an approximation of the distribution of the random variables from the underlying population. While it is not possible to make additional draws from the population itself, we will create new samples, known as **bootstrap samples**, by instead making draws from the observed sample itself.

Definition 15.1 *For a given sample of size n, a* ***bootstrap sample*** *of size n is constructed by treating the original sample as the population of interest and sampling* with replacement *by making n i.i.d. draws from the sample.*

This resampling is done *many* times, and the number of bootstrap replications is denoted as B. Some additional notation is required to distinguish bootstrap sample observations from the original sample observations. For a univariate sample $\{x_1, x_2, \ldots, x_n\}$, a given bootstrap sample is denoted as

$$\{x_1^b, x_2^b, \ldots, x_n^b\} \text{ for } b \in \{1, 2, \ldots, B\}.$$

For a bivariate sample $\{(x_1, y_1), (x_2, y_2), \ldots, (x_n, y_n)\}$, a given bootstrap sample is denoted as

$$\{(x_1^b, y_1^b), (x_2^b, y_2^b), \ldots, (x_n^b, y_n^b)\} \text{ for } b \in \{1, 2, \ldots, B\}.$$

If there are more than two variables, this notation can be generalized to include additional variables.

Example 15.1 *Consider the following bivariate data with seven observations ($n = 7$):*

$$\{(x_i, y_i)\}_{i=1}^{7} = \{(4,8),(3,6),(8,10),(12,1),(0,15),(10,3),(5,6)\}.$$

These data can be shown in a table, similar to how they might appear in a spreadsheet:

i	x_i	y_i
1	4	8
2	3	6
3	8	10
4	12	1
5	0	15
6	10	3
7	5	6

To create the first bootstrap sample, associated with $b = 1$, the computer draws a sample of 7 observations, by sampling with replacement from the original sample of 7 observations. We make 7 i.i.d. draws from the set of row numbers $\{1,2,3,4,5,6,7\}$, where each row number is equally likely to be drawn, each with probability 1/7. Suppose that the 7 draws are

$$7, 5, 2, 5, 6, 7, 3,$$

meaning that the 7th row is drawn first, the 5th row is drawn second, the 2nd row is drawn third, and so on, yielding the first bootstrap sample:

i	x_i^1	y_i^1
1	5	6
2	0	15
3	3	6
4	0	15
5	10	3
6	5	6
7	8	10

There are a couple of important things to note. First, the entire row *is drawn each time that a bootstrap observation is created. That is, the x and y variables are jointly drawn for each bootstrap observation, which is important so the relationship between the x and y variables from the original sample is preserved in the bootstrap sample. Second, due to the sampling being done* with replacement, *repeat observations are to be expected in the boostrap sample. In this first bootstrap sample, the 5th and 7th observations from the original sample both appear twice.*

Suppose that the 7 draws for the second bootstrap sample ($b = 2$) are

$$5, 2, 2, 2, 1, 4, 6,$$

the 7 draws for the third bootstrap sample ($b = 3$) are

$$2, 4, 6, 5, 7, 2, 1,$$

and the 7 draws for the fourth bootstrap sample ($b = 4$) are

$$4, 3, 3, 5, 4, 1, 4.$$

The following table shows the original sample alongside the four bootstrap samples. For the original sample and the bootstrap samples, the table also shows seven different descriptive statistics: means of x and y, medians of x and y, standard deviations of x and y, and the correlation between x and y:

	Sample		Bootstrap samples							
i	x_i	y_i	x_i^1	y_i^1	x_i^2	y_i^2	x_i^3	y_i^3	x_i^4	y_i^4
1	4	8	5	6	0	15	3	6	12	1
2	3	6	0	15	3	6	12	1	8	10
3	8	10	3	6	3	6	10	3	8	10
4	12	1	0	15	3	6	0	15	0	15
5	0	15	10	3	4	8	5	6	12	1
6	10	3	5	6	12	1	3	6	4	8
7	5	6	8	10	10	3	4	8	12	1

$\bar{x}$	6	4.43	5	5.29	8
$\tilde{x}_{0.5}$	5	5	3	4	8
s_x	4.20	3.78	4.32	4.23	4.62
$\bar{y}$	7	8.71	6.43	6.43	6.57
$\tilde{y}_{0.5}$	6	6	6	6	8
s_y	4.62	4.75	4.43	4.43	5.62
r_{xy}	−0.790	−0.762	−0.845	−0.870	−0.898

Focusing first on the mean of the x variable, the original sample has sample mean $\bar{x} = 6$. Due to the randomness associated with the construction of the bootstrap samples, the sample means for the bootstrap samples should not be expected to equal 6, and the sample means of x are 4.43, 5, 5.29, and 8 for the four bootstrap samples. Similarly, for the other descriptive statistics, the statistic associated with any bootstrap sample can differ from the statistic associated with the original sample.

The following R code constructs a single bootstrap sample:

```
set.seed(1234)

# create a data frame with the original data
df <- data.frame(x = c(4,3,8,12,0,10,5), y = c(8,6,10,1,15,3,6))

# determine the number of observations
nobs <- nrow(df)

# create a vector of index values by sampling with replacement
bs_index <- sample(1:nobs, nobs, replace = TRUE)
bs_index

## [1] 4 2 6 5 4 7 1
```

```
# create a data frame corresponding for the bootstrap sample
bs_df <- df[bs_index,]

bs_df

##       x  y
## 4    12  1
## 2     3  6
## 6    10  3
## 5     0 15
## 4.1  12  1
## 7     5  6
## 1     4  8

print(paste("Means for bootstrap sample: x", round(mean(bs_df$x),2), ", y",
      round(mean(bs_df$y),2)))

## [1] "Means for bootstrap sample: x 6.57 , y 5.71"

print(paste("Medians for bootstrap sample: x", median(bs_df$x), ", y",
      median(bs_df$y)))

## [1] "Medians for bootstrap sample: x 5 , y 6"

print(paste("Stdevs for bootstrap sample: x", round(sd(bs_df$x),2), ", y",
      round(sd(bs_df$y),2)))

## [1] "Stdevs for bootstrap sample: x 4.76 , y 4.89"

print(paste("Correlation for bootstrap sample:", round(cor(bs_df$x,bs_df$y),3)))

## [1] "Correlation for bootstrap sample: -0.924"
```

The data frame `df` *contains the original sample of $n = 7$ observations for x and y. The* `sample` *command randomly draws 7 index values from the set $\{1, 2, \ldots, 7\}$ with replacement (*`replace = TRUE` *argument). From the value of* `bs_index`*, we see that the index value 4 is drawn twice and the index value 3 is not drawn at all. The data frame* `bs_df` *is assigned to be the bootstrap sample associated with the index values* `bs_index`*. The* `print` *commands summarize the descriptive statistics for this bootstrap sample.*

Example 15.1 shows four bootstrap samples associated with an original sample. In practice, however, many more bootstrap samples are used for inference (i.e., a large value of B). Ideally, we would want to construct all possible bootstrap samples from the original sample, but that's not feasible. After all, by the multiplication rule, the total number of possible distinct bootstrap samples equals n^n. Even with $n = 7$, the total number of distinct bootstrap samples is 823,543.

The next section considers bootstrap sampling when the number of bootstrap samples B is chosen to be large. With the statistic or estimator of interest calculated for each bootstrap sample, we can consider the distribution of the statistic or estimator over the large number of bootstrap samples.

15.2 Bootstrap Sampling Distribution

This section considers the **bootstrap sampling distribution** of a statistic or estimator, defined as in definition 15.2.

Definition 15.2 *The* ***bootstrap sampling distribution*** *of a statistic or estimator $s(\cdot)$ is the distribution of the values of the statistic or estimator applied to each of the B bootstrap samples.*

For a univariate statistic $s(x_1, x_2, \ldots, x_n)$, the statistic for the bth bootstrap sample is $s(x_1^b, x_2^b, \ldots, x_n^b)$. For example, the sample mean $\bar{x}$ is the statistic or estimate associated with the estimator $\bar{X}$, and its bootstrap sampling distribution is the distribution of the sample means for each of the B bootstrap samples, $\{x_1^b, x_2^b, \ldots, x_n^b\}$ for $b \in \{1, 2, \ldots, B\}$. If the bootstrap sample means are denoted as $\bar{x}^b$, then

$$\bar{x}^b = \frac{1}{n}\sum_{i=1}^{n} x_i^b \text{ for each } b \in \{1, 2, \ldots, B\}.$$

The collection of the B values, $\{\bar{x}^1, \bar{x}^2, \ldots, \bar{x}^B\}$, is the bootstrap sampling distribution of the sample mean.

As another example, the sample correlation r_{xy} is the estimate associated with the estimator r_{XY}, and the bootstrap sample correlations are

$$r_{xy}^b = \frac{\frac{1}{n-1}\sum_{i=1}^{n}\left(x_i^b - \bar{x}^b\right)\left(y_i^b - \bar{y}^b\right)}{\sqrt{\frac{1}{n-1}\sum_{i=1}^{n}\left(x_i^b - \bar{x}^b\right)^2}\sqrt{\frac{1}{n-1}\sum_{i=1}^{n}\left(y_i^b - \bar{y}^b\right)^2}} \text{ for each } b \in \{1, 2, \ldots, B\},$$

where $\bar{x}^b = \frac{1}{n}\sum_{i=1}^{n} x_i^b$ and $\bar{y}^b = \frac{1}{n}\sum_{i=1}^{n} y_i^b$. The collection of the B values, $\{r_{xy}^1, r_{xy}^2, \ldots, r_{xy}^B\}$, is the bootstrap sampling distribution of the sample correlation between x and y.

Example 15.2 *Continuing example 15.1, consider the bootstrap sampling distributions of four statistics: the sample mean $\bar{x}$, the sample mean $\bar{y}$, the sample standard deviation s_x, and the sample correlation r_{xy}. Increasing B to 1,000, figure 15.1 shows histograms of the statistics calculated for each of the 1,000 bootstrap samples. For each histogram, a corresponding density curve is drawn, and the value of the original sample statistic is indicated by a vertical dashed line. For the sample mean $\bar{x}$ (top-left graph), the bootstrap sampling distribution looks fairly symmetric around $\bar{x} = 6$, with nearly the entire distribution between 3 and 9. For the sample mean $\bar{y}$ (top-right graph), the bootstrap sampling distribution also looks fairly symmetric, this time around $\bar{y} = 7$. In contrast, the bootstrap sampling distributions for the sample standard deviation s_x (bottom-left graph) and the sample correlation r_{xy} (bottom-right graph) appear asymmetric. For the sample correlation, the sample statistic $r_{xy} = -0.790$ is clearly to the right of the peak of the distribution, which occurs below -0.9. The bootstrap sampling distribution for r_{xy} has a long right tail, while there is no left tail since the sample correlation is never below -1.*

Here is the R code to create the $B = 1{,}000$ bootstrap samples, calculate their descriptive statistics, and draw the graphs in figure 15.1:

```
set.seed(1234)

# create a data frame with the original data
df <- data.frame(x = c(4,3,8,12,0,10,5), y = c(8,6,10,1,15,3,6))

# determine the number of observations
nobs <- nrow(df)
```

```
# initialize the number of bootstrap replications
B <- 1000

# initialize vectors to hold bootstrap statistics
bs_meanx <- rep(0,B)
bs_meany <- rep(0,B)
bs_sx <- rep(0,B)
bs_rxy <- rep(0,B)

# loop B times to create bootstrap samples and calculate bootstrap statistics
for (i in 1:B) {
    # create a vector of index values by sampling with replacement
    bs_index <- sample(1:nobs, nobs, replace = TRUE)

    # create a data frame corresponding for the bootstrap sample
    bs_df <- df[bs_index,]

    # calculate the bootstrap statistics
    bs_meanx[i] <- mean(bs_df$x)
    bs_meany[i] <- mean(bs_df$y)
    bs_sx[i] <- sd(bs_df$x)
    bs_rxy[i] <- cor(bs_df$x,bs_df$y)
}

# graph-display format (two rows, two columns)
par(mfrow=c(2,2))

hist(bs_meanx, freq=FALSE, xlab="", ylab="",
     main=expression(paste ("Sample mean of ",x)),
     xlim=c(2,10), yaxt='n', breaks=20)
lines(density(bs_meanx), lwd=2)
abline(v=mean(df$x), lwd=3, lty=2)

hist(bs_meany, freq=FALSE, xlab="", ylab="",
     main=expression(paste("Sample mean of ",y)),
     xlim=c(2,12), yaxt='n', breaks=20)
lines(density(bs_meany), lwd=2)
abline(v=mean(df$y), lwd=3, lty=2)

hist(bs_sx, freq=FALSE, xlab="", ylab="",
     main=expression(paste ("Sample standard deviation of ",x)),
     xlim=c(1,6), yaxt='n', breaks=40)
lines(density(bs_sx), lwd=2)
abline(v=sd(df$x), lwd=3, lty=2)

hist(bs_rxy, freq=FALSE, xlab="", ylab="",
     main=expression(paste("Sample correlation of ",x," and ",y)),
     xlim=c(-1,0), yaxt='n', breaks=40)
lines(density(bs_rxy), lwd=2)
abline(v=cor(df$x,df$y), lwd=3, lty=2)
```

15.3 Bootstrap Standard Errors and Bootstrap Confidence Intervals

This section shows how the bootstrap sampling distribution from section 15.2 can be used for statistical inference, specifically to calculate the standard error of an estimate and construct confidence intervals for an underlying parameter. Denote the statistic or estimator of interest by $s()$, where the arguments have been omitted to allow for the generality of handling univariate data, bivariate data, or multivariate data with more than two variables. For univariate data, $s()$ is short for $s(x_1, x_2, \ldots, x_n)$. For a large number B of bootstrap samples,

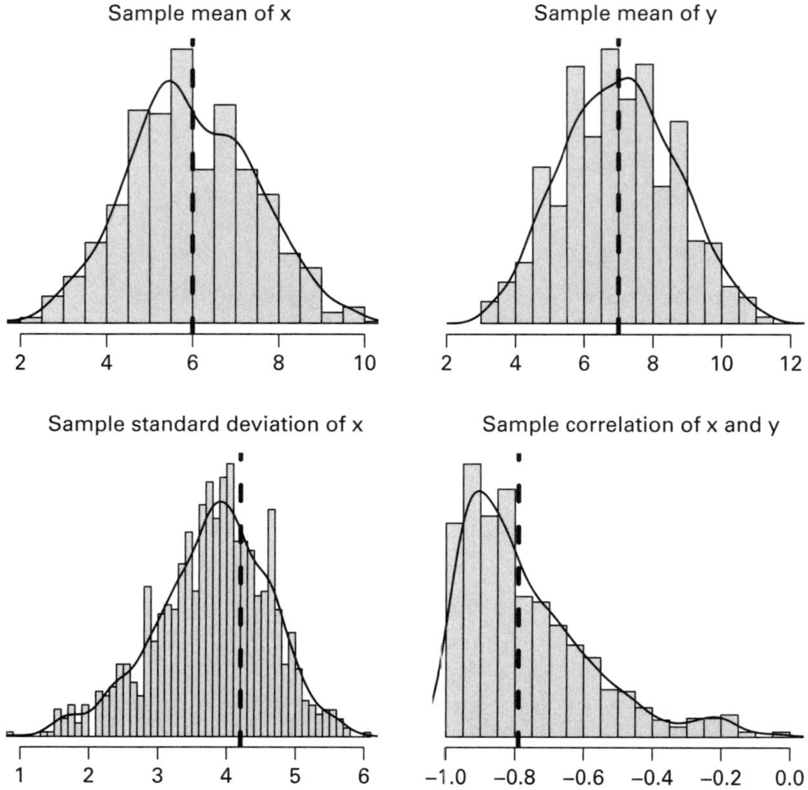

Figure 15.1
Bootstrap sampling distributions

this statistic is calculated for each bootstrap sample. In definition 15.3, the notation s^b is used as shorthand for the statistic calculated on the bth bootstrap sample. With this notation, s^1 is the statistic for the first bootstrap sample, s^2 is the statistic for the second bootstrap sample, and so on. For univariate data, s^b is shorthand for $s(x_1^b, x_2^b, \ldots, x_n^b)$. Using this notation, the **bootstrap standard error** is defined as in definition 15.3.

Definition 15.3 *The **bootstrap standard error** of a statistic s(), denoted as se_B, is the standard deviation of the s() statistic over the B bootstrap samples,*

$$se_B = \sqrt{\frac{1}{B-1}\sum_{b=1}^{B}\left(s^b - \bar{s}_B\right)},$$

where s^b is the statistic for the bth bootstrap sample and $\bar{s}_B$ is the average of the statistic over the B bootstrap samples.

For large samples, the bootstrap standard error can be used as an alternative to the asymptotic standard error. Even in cases where the asymptotic distribution provides a good approximation of an estimator's sampling distribution, the bootstrap standard error can be useful if the asymptotic standard error is difficult to calculate (e.g., sample quantiles) or if the formula for the asymptotic standard deviation may be unknown (as in the examples at the beginning of the chapter).

When does the bootstrap standard error provide an appropriate alternative to the asymptotic standard error? Generally, so long as an estimator $\hat{\theta}_X$ is $\sqrt{n}$-consistent and asymptotically normal, the bootstrap standard error se_B gets arbitrarily close to the asymptotic standard deviation of $\hat{\theta}_X$ as $n \to \infty$ and $B \to \infty$. The reasons for the two $\to \infty$ conditions are quite different. The $n \to \infty$ requirement is needed so the sample is large enough for the asymptotic normal sampling distribution to be arbitrarily close to the estimator's true sampling distribution. The $B \to \infty$ requirement, on the other hand, is needed to eliminate the sampling error associated with bootstrap sampling, as there is an inherent randomness due to the resampling process. In practice, B should be chosen to be as large as possible, subject to the constraints of the computer, to minimize sampling error. In most applied work, B is chosen to be in the thousands, and a larger choice is always preferred if possible.

The bootstrap standard error can be used for any of the $\sqrt{n}$-consistent and asymptotically normal estimators already discussed in this book, as well as the regression estimators discussed in chapters 17 and 18. For estimators that are not $\sqrt{n}$-consistent and asymptotically normal, the bootstrap is not guaranteed to provide valid inference. An example is the sample maximum, $\max_X = \max(X_1, X_2, \dots, X_n)$, for which the bootstrap should not be used.

Example 15.3 (Mean and median of a log-normal random variable) *Suppose that the sample $x_1, x_2, \dots, x_{100}$ consists of i.i.d. draws from a log-normal distribution, with $\ln(X) \sim N(0,1)$. The sample size is $n = 100$. The following R code first draws the sample $x_1, x_2, \dots, x_{100}$ from the population, calculates the sample mean $\bar{x}$ and sample median $\tilde{x}_{0.5}$, and calculates the asymptotic standard error $se(\bar{x}) = \frac{s_x}{\sqrt{n}}$. For the sample median, the asymptotic standard error is difficult to calculate, as previously discussed. Using $B = 10{,}000$ bootstrap iterations, the code calculates bootstrap standard errors for the sample mean $\bar{x}$ and the sample median $\tilde{x}_{0.5}$:*

```
set.seed(1234)

nobs <- 100
x <- rlnorm(nobs)
mean(x)

## [1] 1.523783

median(x)

## [1] 0.6808154

# asymptotic standard error of sample mean
sd(x)/sqrt(nobs)

## [1] 0.2170919

# use bootstrap sampling to calculate bootstrap standard errors
B <- 10000
bs_meanx <- rep(0,B)
bs_medianx <- rep(0,B)

for (i in 1:B) {
    bs_index <- sample(1:nobs, nobs, replace = TRUE)
```

```
    bs_x <- x[bs_index]
    bs_meanx[i] <- mean(bs_x)
    bs_medianx[i] <- median(bs_x)
}

print(paste("Bootstrap standard error of the sample mean:", round(sd(bs_meanx),4)))

## [1] "Bootstrap standard error of the sample mean: 0.2165"

print(paste("Bootstrap standard error of the sample median:",
      round(sd(bs_medianx),4)))

## [1] "Bootstrap standard error of the sample median: 0.0877"
```

The original sample has sample mean $\bar{x} = 1.524$, sample median $\tilde{x}_{0.5} = 0.681$, and sample standard deviation $s_x = 2.171$. For the sample mean, the asymptotic standard error is $se(\bar{x}) = \frac{s_x}{\sqrt{n}} = 0.2171$. Using $B = 10{,}000$ bootstrap samples, the bootstrap standard error for the sample mean is $se_B(\bar{x}) = 0.2165$, and the bootstrap standard error for the sample median is $se_B(\tilde{x}_{0.5}) = 0.0877$. The bootstrap standard error $se_B(\bar{x}) = 0.2165$ is very close to the asymptotic standard error $se(\bar{x}) = 0.2171$.

How about statistical inference for the difference between the population mean μ_X and the population median $\tau_{X,0.5}$? The difference between the sample mean and the sample median, $\bar{x} - \tilde{x}_{0.5} = 1.5239 - 0.6808 = 0.8431$, is an estimate of $\mu_X - \tau_{X,0.5}$. (The difference $\mu_X - \tau_{X,0.5}$ is nonzero here since the log-normal is a right-skewed distribution, with population mean greater than population median.) After bootstrap sampling, we can calculate the bootstrap standard error $se_B(\bar{x} - \tilde{x}_{0.5})$ by calculating the difference $\bar{x} - \tilde{x}_{0.5}$ for each bootstrap sample and taking the standard deviation of the resulting B values. This process, based on the $B = 10{,}000$ bootstrap samples, yields the bootstrap standard error $se_B(\bar{x} - \tilde{x}_{0.5}) = 0.1874$.

```
print(paste("Bootstrap standard error of the difference:",
      round(sd(bs_meanx-bs_medianx),4)))

## [1] "Bootstrap standard error of the difference: 0.1874"
```

Example 15.4 (Monthly stock returns: correlation differences) *Example 14.11 provided a sample correlation matrix, with asymptotic standard errors, for six stocks from the* ***sp500*** *data set. Each sample correlation in that matrix is an estimate of the underlying population correlation. For example, the sample correlation $r_{HD,LOW}$ is an estimate of the population correlation $\rho_{HD,LOW}$, and its asymptotic standard error can be used for statistical inference. What if we are instead interested in the* difference *between two population correlations? For instance, the difference $\rho_{HD,LOW} - \rho_{HD,BAC}$ is of interest if we want to know if the correlation between HD and LOW returns is larger or smaller than the correlation between HD and BAC returns. While the difference $r_{HD,LOW} - r_{HD,BAC}$ provides a logical estimator of $\rho_{HD,LOW} - \rho_{HD,BAC}$, it is difficult to calculate an asymptotic standard error for $r_{HD,LOW} - r_{HD,BAC}$. This situation is different from the two-sample setting in section 14.4.5, where the asymptotic standard error of the correlation difference was easy to determine since the two correlations for a two-sample problem are independent of each other (see example 14.16). Here, since they are based on the same sample, the sample correlations $r_{HD,LOW}$ and $r_{HD,BAC}$ are not independent, meaning that the asymptotic standard* *error is complicated and depends on their covariance/correlation. Rather than attempting to derive this more complicated asymptotic standard error, the bootstrap provides a simpler alternative. We repeatedly create bootstrap samples, and for each bootstrap sample, the bootstrap statistic is the difference between the HD-LOW correlation and the HD-BAC correlation.*

The following R code calculates the bootstrap standard error for $r_{HD,LOW} - r_{HD,BAC}$ *using* $B = 1{,}000$:

```
set.seed(1234)

# calculate difference r(HD,LOW)-r(HD,BAC)
cor(sp500$HD,sp500$LOW)-cor(sp500$HD,sp500$BAC)

## [1] 0.317298

# initialize variables
nobs <- nrow(sp500)
B <- 1000
bs_rdiff <- rep(0,B)

# loop to create bootstrap samples and calculate the bootstrap statistic
for (i in 1:B) {
    bs_index <- sample(1:nobs, nobs, replace = TRUE)
    bs_df <- sp500[bs_index,]
    bs_rdiff[i] <- cor(bs_df$HD,bs_df$LOW)-cor(bs_df$HD,bs_df$BAC)
}

# output the bootstrap standard error for r(HD,LOW)-r(HD,BAC)
sd(bs_rdiff)

## [1] 0.06108325
```

The estimate of the difference $\rho_{HD,LOW} - \rho_{HD,BAC}$ *is* $r_{HD,LOW} - r_{HD,BAC} = 0.317$, *and the bootstrap standard error is* $se_B(r_{HD,LOW} - r_{HD,BAC}) = 0.061$. *There is nothing special about the pairs HD-LOW and HD-BAC in this example, and the same approach could be used to calculate a bootstrap standard error for the difference between any two correlations (e.g.,* $r_{BAC,WFC} - r_{MRO,COP}$*).*

15.3.1 Normal-Based Bootstrap Confidence Interval

For an estimator that is $\sqrt{n}$-consistent and asymptotically normal, section 15.3 discussed that the bootstrap standard error is a consistent estimator of the asymptotic standard deviation. Similar to how asymptotic standard errors were used to construct confidence intervals in chapter 14, bootstrap standard errors can also be used to construct confidence intervals. Since the normality of the estimator's sampling distribution is still used, this confidence interval is known as a **normal-based bootstrap confidence interval**.

Specifically, for an estimate $\hat{\theta}$ of an unknown parameter θ, a two-sided normal-based bootstrap $1 - \alpha$ confidence interval is

$$(\hat{\theta} - z_{\alpha/2} se_B(\hat{\theta}), \hat{\theta} + z_{\alpha/2} se_B(\hat{\theta})),$$

where $se_B(\hat{\theta})$ is the bootstrap standard error of estimate $\hat{\theta}$. Similarly, one-sided normal-based bootstrap $1 - \alpha$ confidence intervals are

$$(\hat{\theta} - z_{\alpha} se_B(\hat{\theta}), \infty) \text{ and } (-\infty, \hat{\theta} + z_{\alpha} se_B(\hat{\theta})).$$

Example 15.5 (Mean and median of a log-normal random variable) *In example 15.3, the sample created by* 100 *i.i.d. draws from a log-normal random variable had the following:*

$$\bar{x} = 1.5238, \text{ with } se_B(\bar{x}) = 0.2165,$$

$$\tilde{x}_{0.5} = 0.6808, \text{ with } se_B(\tilde{x}_{0.5}) = 0.0877,$$

and

$$\bar{x} - \tilde{x}_{0.5} = 0.8430, \text{ with } se_B(\bar{x} - \tilde{x}_{0.5}) = 0.1874.$$

Based on these estimates and bootstrap standard errors, two-sided normal-based bootstrap confidence intervals can be constructed. The 95% *confidence interval for the population mean* μ_X *is*

$$(\bar{x} - 1.96se_B(\bar{x}), \bar{x} + 1.96se_B(\bar{x})) = (1.5238 - (1.96)(0.2165), 1.5238 + (1.96)(0.2165))$$
$$\approx (1.099, 1.948).$$

The 95% *confidence interval for the population median* $\tau_{X,0.5}$ *is*

$$(\tilde{x}_{0.5} - 1.96se_B(\tilde{x}_{0.5}), \tilde{x}_{0.5} + 1.96se_B(\tilde{x}_{0.5})) \approx (0.509, 0.853).$$

The 95% *confidence interval for the difference* $\mu_X - \tau_{X,0.5}$ *is*

$$((\bar{x} - \tilde{x}_{0.5}) - 1.96se_B(\bar{x} - \tilde{x}_{0.5}), (\bar{x} - \tilde{x}_{0.5}) + 1.96se_B(\bar{x} - \tilde{x}_{0.5})) \approx (0.476, 1.210).$$

For this last interval, it can be said with 95% *confidence that the difference between the population mean and population median,* $\mu_X - \tau_{X,0.5}$*, is between* 0.476 *and* 1.210*. Therefore, this confidence interval provides strong statistical evidence that the population mean* μ_X *is greater than the population median* $\tau_{X,0.5}$*, as zero is not in the interval and is quite far (approximately* $\frac{0.476}{0.1874} \approx 2.5$ *bootstrap standard errors) from the lower end point of the interval.*

Example 15.6 (Monthly stock returns: correlation differences) *In example 15.4, the bootstrap was used to calculate a bootstrap standard error of the estimate* ($r_{HD,LOW} - r_{HD,BAC} = 0.317$) *of the difference in correlations* $\rho_{HD,LOW} - \rho_{HD,BAC}$*. The calculated bootstrap standard error was* $se_B(r_{HD,LOW} - r_{HD,BAC}) = 0.061$*. Therefore, a* 95% *confidence interval for* $\rho_{HD,LOW} - \rho_{HD,BAC}$ *is*

$$(r_{HD,LOW} - r_{HD,BAC}) \pm 1.96se_B(r_{HD,LOW} - r_{HD,BAC}) \approx (0.20, 0.44),$$

and a 99% *confidence interval for* $\rho_{HD,LOW} - \rho_{HD,BAC}$ *is*

$$(r_{HD,LOW} - r_{HD,BAC}) \pm z_{0.005}se_B(r_{HD,LOW} - r_{HD,BAC}) \approx (0.16, 0.47).$$

The 99% *confidence interval provides strong statistical evidence that* $\rho_{HD,LOW} - \rho_{HD,BAC}$ *is positive or, equivalently, that* $\rho_{HD,LOW}$ *is greater than* $\rho_{HD,BAC}$*.*

Example 15.7 (Labor force data) *Example 14.12 reported the bootstrap standard errors for various sample quantiles, including the sample deciles and sample quartiles, of the weekly earnings variable. Here is the R code used to create the table in example 14.12, with* $B = 5{,}000$ *bootstrap iterations used to calculate the bootstrap standard errors and the normal-based confidence intervals:*

```
set.seed(1234)

# construct a vector with the quantiles of interest (deciles and quartiles)
quantiles <- sort(c(0.1,0.2,0.25,0.3,0.4,0.5,0.6,0.7,0.75,0.8,0.9))

# initialize variables
nobs <- nrow(cpsemployed)
B <- 5000

# loop over the quantiles of interest
for (q in quantiles) {
    bs_quantile <- rep(0,B)

    # loop to create bootstrap samples and calculate the bootstrap statistics
    for (i in 1:B) {
        bs_index <- sample(1:nobs, nobs, replace = TRUE)
        bs_quantile[i] <- quantile(cpsemployed[bs_index,"earnwk"],q)
    }

    # output the bootstrap se and normal-based bootstrap CI for this quantile
    quantile_estimate <- quantile(cpsemployed$earnwk,q)
    print(paste("Quantile ", q, ": sample quantile ", round(quantile_estimate),
                ", bootstrap se ", round(sd(bs_quantile),1),
                ", 95% CI (", round(quantile_estimate-1.96*sd(bs_quantile),1),
                ",", round(quantile_estimate+1.96*sd(bs_quantile),1), ")", sep=""))
}

## [1] "Quantile 0.1: sample quantile 355, bootstrap se 7, 95% CI (340.8,368.3)"
## [1] "Quantile 0.2: sample quantile 480, bootstrap se 7.4, 95% CI (465.4,494.6)"
## [1] "Quantile 0.25: sample quantile 520, bootstrap se 11.7, 95% CI (497,543)"
## [1] "Quantile 0.3: sample quantile 576, bootstrap se 7.4, 95% CI (561.6,590.4)"
## [1] "Quantile 0.4: sample quantile 670, bootstrap se 10.4, 95% CI (649.6,690.4)"
## [1] "Quantile 0.5: sample quantile 770, bootstrap se 11.1, 95% CI (748.1,791.9)"
## [1] "Quantile 0.6: sample quantile 900, bootstrap se 16.2, 95% CI (868.2,931.8)"
## [1] "Quantile 0.7: sample quantile 1080, bootstrap se 21.9, 95% CI (1037.1,
        1122.9)"
## [1] "Quantile 0.75: sample quantile 1194, bootstrap se 23.7, 95% CI (1147.1,
        1240.1)"
## [1] "Quantile 0.8: sample quantile 1346, bootstrap se 23.9, 95% CI (1299.1,
        1392.9)"
## [1] "Quantile 0.9: sample quantile 1750, bootstrap se 52, 95% CI (1648,1852)"
```

How about calculating a bootstrap standard error and normal-based bootstrap interval for the interquartile range $\tau_{X,0.75} - \tau_{X,0.25}$? The following R code calculates these quantities, again using $B = 5{,}000$ bootstrap iterations:

```
set.seed(1234)

# initialize variables
nobs <- nrow(cpsemployed)
B <- 5000
bs_iqr <- rep(0,B)

# loop to create bootstrap samples and calculate the bootstrap statistics
for (i in 1:B) {
    bs_index <- sample(1:nobs, nobs, replace = TRUE)
    bs_iqr[i] <- IQR(cpsemployed[bs_index,"earnwk"])
}

# output the bootstrap standard error and normal-based CI for the IQR
iqr_earnwk <- IQR(cpsemployed$earnwk)
```

```
print(paste("IQR: ", round(iqr_earnwk,1),
            ", bootstrap se ", round(sd(bs_iqr),1),
            ", 95% CI (", round(iqr_earnwk-1.96*sd(bs_iqr),1),
            ",", round(iqr_earnwk+1.96*sd(bs_iqr),1), ")", sep=""))

## [1] "IQR: 673.6, bootstrap se 23.3, 95% CI (628,719.2)"
```

The estimated IQR_{earnwk} *is* 673.6, *with a bootstrap standard error* $se_B(IQR_{earnwk}) = 23.3$ *and a normal-based bootstrap* 95% *confidence interval of* $(628.0, 719.2)$. *With* 95% *confidence, it can be said that the population interquartile range (IQR) is between* 628.0 *and* 719.2.

15.3.2 Bootstrap Percentile Interval

When there is concern that the asymptotic normal distribution does not accurately approximate the true sampling distribution of an estimator, an alternative to the normal-based confidence interval might be used. The asymptotic normal distribution may not work well because the sample size n is not large enough, which might occur because n is truly small or because the estimator under consideration, like a sample correlation (especially when close to 1 or -1) or a sample standard deviation, requires larger n for asymptotic normality to hold.

An alternative bootstrap confidence interval that allows nonnormality and asymmetry is known as a **bootstrap percentile interval**. Once the bootstrap sampling distribution for a given statistic is obtained, the bootstrap percentile interval is easy to construct. Here are the steps required:

1. For the statistic or estimate $s()$ of interest, calculate the statistic or estimate for B bootstrap samples to yield the bootstrap statistics $s^1, s^2, \ldots, s^B$. These B statistics constitute the bootstrap sampling distribution.
2. For a two-sided bootstrap percentile interval, let $c_{\alpha/2}$ be the sample $\frac{\alpha}{2}$th quantile of the $\{s^1, s^2, \ldots, s^B\}$ values and $c_{1-\alpha/2}$ be the sample $(1-\frac{\alpha}{2})$th quantile of the $\{s^1, s^2, \ldots, s^B\}$ values. The two-sided bootstrap $1-\alpha$ percentile interval is
$$\left(c_{\alpha/2}, c_{1-\alpha/2}\right).$$
For example, the two-sided bootstrap 95% percentile interval is $(c_{0.025}, c_{0.975})$, where $c_{0.025}$ and $c_{0.975}$ are the sample 2.5% and 97.5% quantiles of the $\{s^1, s^2, \ldots, s^B\}$ values, respectively.
3. For a one-sided bootstrap percentile interval, let c_α be the sample αth quantile of the $\{s^1, s^2, \ldots, s^B\}$ values and $c_{1-\alpha}$ be the sample $(1-\alpha)$-th quantile of the $\{s^1, s^2, \ldots, s^B\}$ values. The one-sided bootstrap $1-\alpha$ percentile intervals are
$$(c_\alpha, \infty) \text{ and } (-\infty, c_{1-\alpha}).$$
For example, the one-sided bootstrap 95% percentile intervals are $(c_{0.05}, \infty)$ and $(-\infty, c_{0.95})$, where $c_{0.05}$ and $c_{0.95}$ are the sample 5% and 95% quantiles of the $\{s^1, s^2, \ldots, s^B\}$ values, respectively.

There is nothing in the construction of the two-sided percentile interval that causes estimate $s()$ to be at the midpoint of the interval. Therefore, unlike a two-sided confidence interval based on the normal distribution, the bootstrap percentile interval may be asymmetric, with the distance between the lower end point and the estimate potentially being different from the distance between the estimate and the upper end point.[1]

While the importance of choosing a large value for B has already been discussed, it is particularly important in the context of bootstrap percentile intervals. These intervals are based on the extremes of the bootstrap sampling distribution (e.g., the 2.5% and 97.5% quantiles in the case of a two-sided 95% interval), and estimators of extreme quantiles are less precise than estimators of other quantities like the standard deviation. Therefore, all else equal, we want a larger choice of B for calculating percentile intervals than would be needed for the calculation of a bootstrap standard error.

Example 15.8 (Mean and median of a log-normal random variable) *Example 15.3 provided R code to calculate the bootstrap distribution ($B = 10{,}000$ bootstrap statistics) associated with the sample mean $\bar{x}$, the sample median $\tilde{x}_{0.5}$, and the difference $\bar{x} - \tilde{x}_{0.5}$. The bootstrap statistics for the sample mean and the sample median were stored in the vectors* `bs_meanx` *and* `bs_medianx`*, respectively. With these two vectors available, we can calculate bootstrap 95% percentile intervals with the following R code:*

```
# output the 2.5% and 97.5% quantiles of the bootstrap distributions
quantile(bs_meanx, c(0.025,0.975))

##     2.5%    97.5%
## 1.136476 1.976367

quantile(bs_medianx, c(0.025,0.975))

##      2.5%     97.5%
## 0.6041670 0.8766832

quantile(bs_meanx-bs_medianx, c(0.025,0.975))

##      2.5%     97.5%
## 0.4870959 1.2108018
```

The results are as follows: a bootstrap 95% percentile interval of $(1.136, 1.976)$ *for μ_X (as compared to the normal-based bootstrap 95% confidence interval* $(1.099, 1.948)$*), a bootstrap 95% percentile interval of* $(0.604, 0.877)$ *for $\tau_{X,0.5}$ (as compared to the normal-based bootstrap 95% confidence interval* $(0.509, 0.853)$*), and a bootstrap 95% percentile interval of* $(0.487, 1.211)$ *for $\mu_X - \tau_{X,0.5}$ (as compared to the normal-based bootstrap 95% confidence interval* $(0.476, 1.210)$*). The only meaningful difference seems to arise for the $\tau_{X,0.5}$ interval, which may suggest that the sample size ($n = 100$) is not large enough for asymptotic normality of the sample median estimator with the underlying log-normal random variable.*[2]

Changing the confidence level of the percentile interval is straightforward. The following R code instead calculates 90% bootstrap percentile intervals by changing the quantiles from 2.5% and 97.5% to 5% and 95%, respectively:

```
# output the 5% and 95% quantiles of the bootstrap distributions
quantile(bs_meanx, c(0.05,0.95))

##       5%      95%
## 1.185190 1.900511
```

```
quantile(bs_medianx, c(0.05,0.95))

##         5%       95%
## 0.6088298 0.8469239

quantile(bs_meanx-bs_medianx, c(0.05,0.95))

##         5%       95%
## 0.5295099 1.1399688
```

The resulting bootstrap 90% *percentile intervals are* $(1.185, 1.901)$, $(0.609, 0.847)$, *and* $(0.530, 1.140)$ *for* μ_X, $\tau_{X,0.5}$, *and* $\mu_X - \tau_{X,0.5}$, *respectively.*

Example 15.9 (Monthly stock returns: correlation differences) *Example 15.4 provided R code to calculate the bootstrap distribution (*$B = 1{,}000$ *bootstrap statistics) associated with* $r_{HD,LOW} - r_{HD,BAC}$, *with the results in the vector* `bs_rdiff`. *With this vector available, we can calculate a bootstrap* 95% *percentile interval by calculating the sample* 2.5% *and* 97.5% *quantiles:*

```
# output the 2.5% and 97.5% quantiles of the bootstrap distribution
quantile(bs_rdiff, c(0.025,0.975))

##      2.5%     97.5%
## 0.2019175 0.4382012
```

The bootstrap 95% *percentile interval is* $(0.20, 0.44)$, *which is the same (to two decimal places) as the normal-based bootstrap* 95% *confidence interval calculated in example 15.6.*

Example 15.10 (Labor market data) *Example 15.7 provided R code to calculate the bootstrap distribution (*$B = 5{,}000$*) for the IQR of the weekly earnings variable (earnwk) from the* ***cps*** *data set, with the results in the vector* `bs_iqr`. *With this vector available, we can calculate a bootstrap* 95% *percentile interval:*

```
# output the 2.5% and 97.5% quantiles of the bootstrap distribution
quantile(bs_iqr, c(0.025,0.975))

##    2.5%   97.5%
## 617.663 701.000
```

The bootstrap 95% *percentile interval for the population IQR is* $(617.7, 701.0)$, *as compared to the normal-based bootstrap* 95% *confidence interval* $(628.0, 719.2)$.

Notes

1. While the ability of the bootstrap percentile interval to handle nonnormality and asymmetry is appealing, there are no general theoretical results showing that the bootstrap percentile interval actually performs better than a normal-based interval (either asymptotic or bootstrap). In fact, the main theoretical results supporting the use of the bootstrap involve large samples ($n \to \infty$), in which case the asymptotic normal confidence interval should work well. For large samples, another bootstrap method known as the *studentized bootstrap* or *bootstrap-t method* has been shown to perform better than asymptotic normal confidence intervals; this method is beyond the scope of this book, but the interested reader can refer to a more comprehensive introduction to the bootstrap, such as Efron and Tibshirani (1993) or Davison and Hinkley (1997).

2. Indeed, the histogram of the bootstrap median estimates confirms that the bootstrap distribution does not have a nice bell shape like that of the distribution of the bootstrap mean estimates. The interested reader can confirm with the commands `hist(bs_meanx,breaks=50)` and `hist(bs_medianx,breaks=50)`.

Exercises

1. Suppose that a bootstrap sample of size n is created by sampling with replacement from the i.i.d. sample $\{x_1, x_2, \ldots, x_n\}$.
   a. What is the probability that a specific observation x_j, for some $j \in \{1, 2, \ldots, n\}$, is in the bootstrap sample at least once?
   b. What is the probability that a specific observation x_j, for some $j \in \{1, 2, \ldots, n\}$, is in the bootstrap sample at least twice?
   c. Evaluate the probabilities from (a) and (b) for $n = 100$.
2. Exercise 14.25 considered a prolific inventor who submits 150 patent applications to the US Patent and Trademark Office, with 110 being successful (resulting in a patent being issued). Assume that the success of each patent application can be considered an i.i.d. *Bernoulli*(π) random variable.
   a. Rather than creating a data set with 110 successes (1s) and 40 failures (0s), a shortcut to create bootstrap samples for i.i.d. Bernoulli data is to repeatedly draw from the appropriate binomial distribution, which in this case is *Binomial* $\left(150, \frac{110}{150}\right)$. For instance, the first draw from this binomial might be 105, corresponding to a bootstrap sample with 105 successes (1s) and 45 failures (0s). Use this approach with 5,000 iterations to calculate a bootstrap standard error and construct a two-sided normal-based bootstrap 95% confidence interval for π.
   b. Using the same approach, conduct 5,000 bootstrap iterations to calculate a bootstrap standard error and construct a two-sided normal-based bootstrap 95% confidence interval for the odds $\frac{\pi}{1-\pi}$.
   c. Using the same approach, conduct 5,000 bootstrap iterations to construct a two-sided bootstrap 95% percentile interval for the odds $\frac{\pi}{1-\pi}$.
3. Use the **strikes** data set for this question. This data set contains information on worker contract strikes within US manufacturing for the period 1968–1976. There are 566 observations on the variable *duration* (strike duration, in weeks). Let X denote the random variable associated with *duration*.
   a. What are the sample 75% quantile and sample 90% quantile of *duration*? What is the sample IQR of *duration*?
   b. Use the bootstrap with 5,000 iterations to construct two-sided normal-based bootstrap 95% confidence intervals for $\tau_{X,0.75}$, $\tau_{X,0.90}$, and IQR_X.
   c. Let $\pi_{long} = P(X > 52)$ be the probability that a strike lasts longer than a year. Use the bootstrap with 5,000 iterations to construct a two-sided normal-based bootstrap 95% confidence interval for π_{long}. How does this interval compare to the asymptotic 95% confidence interval for π_{long}?

4. Due to concerns about a dangerous intersection, a town gathers data on the weekly number of car accidents in the intersection. The data for 50 consecutive weeks are summarized in the following table:

# accidents	0	1	2	3	4	5	6
# weeks	23	14	5	2	5	0	1

Assume that the number of accidents each week is an i.i.d. draw from a random variable X.

a. Create a data frame or a vector of 50 observations in *R* based on the table.

b. Calculate the sample mean and the sample variance.

c. Use the bootstrap with 5,000 iterations to construct a two-sided normal-based bootstrap 95% confidence interval for μ_X and a two-sided normal-based bootstrap 95% confidence interval for σ_X^2.

d. A town official studied statistics in college and wonders whether X is a Poisson random variable. She recalls that a feature of a *Poisson*(λ) random variable is that the population mean and population variance both equal λ and therefore equal each other. Let $\theta = \mu_X - \sigma_X^2$ be the difference between the population mean and population variance, which would be $\theta = 0$ if X were truly Poisson. Use the bootstrap with 5,000 iterations to construct a two-sided normal-based bootstrap 95% confidence interval for $\theta = \mu_X - \sigma_X^2$. What does the confidence interval say about X being Poisson?

e. Same as (d), but construct a two-sided bootstrap 95% percentile interval for $\theta = \mu_X - \sigma_X^2$.

5. There are 106 unemployed individuals in the **cps** data set, for whom the variable *lfstatus* (labor-force status) is equal to "Unemployed" and the variable *unempwks* (weeks unemployed) is reported. For this question, focus on this subsample of 106 unemployed individuals.

a. Draw a histogram of *unempwks* for the unemployed individuals.

b. Given the right-skewed distribution of *unempwks*, a classmate suggests taking a log transformation of the variable to get a distribution that is more symmetric. Create a new variable *lnunempwks* that does so, and calculate its sample mean and sample median.

c. If the distribution of *lnunempwks* is symmetric, it should be the case that the population mean μ_X equals the population median $\tau_{X,0.5}$ for the underlying random variable X. Use the bootstrap with 5,000 iterations to construct a two-sided normal-based bootstrap 95% confidence interval for $\theta = \mu_X - \tau_{X,0.5}$. What does the confidence interval say about the population mean of X being equal to the population median of X?

d. Same as (c), but do so for the ratio parameter $\theta = \frac{\mu_X}{\tau_{X,0.5}}$.

6. There are 2,809 employed individuals in the **cps** data set, which is the sample of interest for this question. Suppose the probability of union membership, in the population, is π_m for male workers and π_f for female workers.

a. What is the estimate of the ratio $\frac{\pi_m}{\pi_f}$, given by $\frac{\hat{\pi}_m}{\hat{\pi}_f}$, where $\hat{\pi}_m$ is the observed sample proportion of union members among male workers and $\hat{\pi}_f$ is the observed sample proportion of union members among female workers?

b. Use the bootstrap with 5,000 iterations to construct a two-sided normal-based bootstrap 95% confidence interval for the ratio $\frac{\pi_m}{\pi_f}$. (Hint: Create bootstrap samples separately for the male-worker subsample and the female-worker subsample.)

c. The **odds ratio (OR)** is a measure used by statisticians to compare the likelihood of a certain outcome occurring in two different groups. In the context of this union-gender example, the OR is

$$\frac{\pi_m/(1-\pi_m)}{\pi_f/(1-\pi_f)},$$

which is the ratio between the odds of a male worker being in a union and the odds of a female worker being in a union. Plugging $\hat{\pi}_m$ and $\hat{\pi}_f$ in for π_m and π_f, what is the estimated OR? Use the bootstrap with 5,000 iterations to construct a two-sided normal-based bootstrap 95% confidence interval for the OR.

16

Hypothesis Testing

This chapter introduces the concept of **hypothesis testing**. Sections 16.1–16.3 focus on testing a *hypothesis* about the value of a single unknown parameter, and section 16.4 extends the framework to consider tests of multiple hypotheses. For the case of a single unknown parameter, we fix ideas by denoting the unknown parameter of interest by θ. The following examples motivate the usefulness of hypothesis tests for a single parameter.

Example 16.1 (Widget website) *Examples 2.1, 14.5, and 14.15 considered the purchase probabilities for three groups of Widgets.com users: recipients of e-mail A, recipients of e-mail B, and nonrecipients. The parameters π_A, π_B, and π_C denote the purchase probabilities of these three groups, respectively. To compare the effectiveness of e-mail A versus e-mail B, the difference $\pi_A - \pi_B$ is the quantity of interest. If this difference is positive, then e-mail A is more effective than e-mail B; if this difference is negative, then e-mail B is more effective than e-mail A; and, if this difference is zero, then e-mail A and e-mail B are equally effective. If $\theta = \pi_A - \pi_B$ is the parameter equal to the difference between the purchase probabilities, we want to know whether $\theta = 0$ (no difference in e-mail effectiveness) or $\theta \neq 0$ (difference in e-mail effectiveness). Example 14.15 constructed a 95% confidence interval $(-0.085, 0.045)$ for $\theta = \pi_A - \pi_B$ based on the estimate $p_A = \frac{60}{300} = 0.20$ of π_A, $p_B = \frac{66}{300} = 0.22$ of π_B, and the standard errors $se(p_A)$ and $se(p_B)$. From this confidence interval, it appears that zero is a plausible value for θ since it falls within the confidence interval. Therefore, it would be expected that a formal statistical test would not be able to rule out $\theta = 0$ with a high level of confidence. A test of the hypothesis $\theta = 0$ is known as a* two-sided test *since statistical evidence of either $\theta < 0$ or $\theta > 0$ would call into question the hypothesis $\theta = 0$.*

Example 16.2 (Investment opportunity) *You are interested in the possibility of buying a business that produces and sells a certain product. By your calculations, the true average of weekly sales would need to be at least \$10,000 for the investment to be worthwhile. As part of due diligence, you obtain weekly sales figures from the business for 10 randomly chosen weeks. For those 10 weeks, the sample mean of weekly sales is \$11,200, and the sample standard deviation of weekly sales is \$3,400. If θ denotes the population mean of weekly sales, measured in thousands of dollars, you are interested in knowing whether $\theta \leq 10$ (the business is not a worthwhile investment) or $\theta > 10$ (the business is a worthwhile investment). The observed sample is $\{x_1, x_2, \ldots, x_{10}\}$, where x_i is weekly sales in thousands of* *dollars for a particular week. If these observations can be considered i.i.d. draws from some distribution, the sample mean $\bar{x} = 11.2$ serves as an estimate of θ and, therefore, provides some evidence against $\theta \leq 10$. But how strong is this evidence? A more formal test needs to take into account the fact that the sample mean is a random variable. As seen in the following discussion, the estimate's standard error accounts for the potential imprecision of the estimator, similar to what was done for confidence intervals in chapter 14. A test of the hypothesis $\theta \leq 10$ is known as a* one-sided test *since only statistical evidence of $\theta > 10$ would call into question the hypothesis $\theta \leq 10$.*

As done for confidence intervals in chapter 14, this chapter first considers hypothesis testing for the unknown population mean of i.i.d. normal random variables and then considers hypothesis testing for the more general case of an unknown parameter for which an asymptotic normal estimator is available. For the former case, covered in section 16.1, testing is based on the exact sampling distribution of the sample mean estimator. For the latter case, covered in section 16.2, testing is based on the asymptotic distribution of the estimator.

Before getting into the details of the tests, some additional notation and terminology is needed, described in definition 16.1.

Definition 16.1 *The **null hypothesis** is the hypothesis to be tested and is denoted H_0. The **alternative hypothesis** is the opposite of the null hypothesis and is denoted as H_1 (although some other sources use the notation H_a).*

For a **two-sided test** of an unknown parameter θ, the null hypothesis is

$$H_0 : \theta = c,$$

for some known constant c specified by the researcher. The alternative hypothesis, which is the opposite of the null hypothesis, is

$$H_1 : \theta \neq c.$$

The alternative hypothesis is true whenever the null hypothesis is false, and vice versa.[1] The hypothesis test of H_0 determines whether there is statistical evidence to *reject* $H_0 : \theta = c$. In example 16.1, the null hypothesis is $H_0 : \theta = 0$. Since θ is unknown, an estimate of θ that is far from the hypothesized value c should provide statistical evidence against $H_0 : \theta = c$, but what does "far" mean? Due to the randomness and noise inherent in the estimate of θ, the estimate's standard error will help to quantify how far the estimate of θ is from c.

For a **one-sided test** of an unknown parameter θ, the null hypothesis is either

$$H_0 : \theta \geq c$$

or

$$H_0 : \theta \leq c,$$

for some known constant c specified by the researcher. The direction of the inequality in the null hypothesis H_0 depends on the situation. In example 16.2, for example, the null hypothesis of interest is $H_0 : \theta \leq 10$.

For the null hypothesis $H_0 : \theta \geq c$, the alternative hypothesis is

$$H_1 : \theta < c,$$

and the hypothesis test of H_0 determines whether there is statistical evidence to *reject* $H_0 : \theta \geq c$. Statistical evidence against $H_0 : \theta \geq c$ comes from an estimate of θ that is far below c.

For the null hypothesis $H_0 : \theta \leq c$, the alternative hypothesis is

$$H_1 : \theta > c,$$

and the hypothesis test of H_0 determines whether there is statistical evidence to *reject* $H_0 : \theta \leq c$. Statistical evidence against $H_0 : \theta \leq c$ comes from an estimate of θ that is far above c. Again, the notions of "far below" and "far above" will be formalized in terms of an estimate's standard error.

16.1 Finite-Sample Hypothesis Testing: Population Mean of i.i.d. Normal Random Variables

This section considers hypothesis tests for the population mean μ of normally distributed i.i.d. random variables $X_1, X_2, \ldots, X_n \sim N(\mu, \sigma^2)$. Section 14.2 covered this case in detail and used the exact sampling distribution results for the sample mean estimator $\bar{X}$ from section 12.1.2. Proposition 14.2 provided the key result to construct confidence intervals for μ:

$$\frac{\bar{X} - \mu}{s_X / \sqrt{n}} \sim t_{n-1},$$

where $\bar{X}$ is the sample mean estimator, s_X is the sample standard deviation estimator, and t_{n-1} is the t-distribution with $n-1$ degrees of freedom. The standard deviation of the estimator $\bar{X}$ is $s_X/\sqrt{n}$. The ratio $\frac{\bar{X}-\mu}{s_X/\sqrt{n}}$ is known as the ***t*-ratio** and indicates the number of standard deviations that the estimator $\bar{X}$ is away from the parameter μ. The t-ratio is positive when $\bar{X}$ is greater than μ and negative when $\bar{X}$ is less than μ. Unfortunately, even after observing the realized sample mean $\bar{x}$ and standard deviation s_x, the t-ratio is unknown since the parameter μ is unknown.

Consider the two-sided test of the null hypothesis,

$$H_0 : \mu = c$$

versus the alternative hypothesis,

$$H_1 : \mu \neq c,$$

where c is a known constant. Let's conduct the thought experiment where we *assume that the null hypothesis* $H_0 : \mu = c$ *is true*. Plugging c into the t-ratio for μ implies

$$\frac{\bar{X} - c}{s_X / \sqrt{n}} \sim t_{n-1} \text{ when } H_0 \text{ is true.}$$

The ratio $\frac{\bar{X}-c}{s_X/\sqrt{n}}$ is called the ***t*-statistic** and indicates the number of standard deviations that the estimator $\bar{X}$ is away from c. Unlike the t-ratio, the t-statistic can be calculated after the realized estimates $\bar{x}$ and s_x are observed. A positive t-statistic corresponds to an estimate above c, and a negative t-statistic corresponds to an estimate below c. For example, a t-statistic of 1.4 indicates that the sample mean is 1.4 standard deviations above c, whereas a t-statistic of -2.1 indicates that the sample mean is 2.1 standard deviations below c.

And, to be clear, when the term "standard deviation" is used here, it refers to the standard deviation of the estimator, which equals $s_X/\sqrt{n}$, rather than the standard deviation of X.

The t-statistic $\frac{\bar{X}-c}{s_X/\sqrt{n}}$ is a random variable *before* the sample is observed, and the result given here says that this t-statistic is distributed as a t_{n-1} distribution if H_0 is true. Therefore, intuitively, if H_0 is true, the realized t-statistic $\frac{\bar{x}-c}{s_x/\sqrt{n}}$ should look like it was drawn from a t_{n-1} distribution. This intuition can be formalized with probability intervals. Before providing a more general result, let's first consider the case of a two-sided 95% probability interval for the t-statistic $\frac{\bar{X}-c}{s_X/\sqrt{n}}$ when H_0 is true:

$$P\left(-t_{n-1,0.025} < \frac{\bar{X}-c}{s_X/\sqrt{n}} < t_{n-1,0.025}\right) = 0.95 \text{ when } H_0 : \mu = c \text{ is true}$$

or, equivalently,

$$P\left(\left|\frac{\bar{X}-c}{s_X/\sqrt{n}}\right| < t_{n-1,0.025}\right) = 0.95 \text{ when } H_0 : \mu = c \text{ is true.}$$

Importantly, these probability statements are *not* true under the alternative hypothesis $H_1 : \mu \neq c$ since the t-statistic $\frac{\bar{X}-c}{s_X/\sqrt{n}}$ does not have a t_{n-1} distribution when $\mu \neq c$. The second probability statement says that there is a 95% probability that the t-statistic has a magnitude or absolute value less than the critical value $t_{n-1,0.025}$ when H_0 is true. After observing the data, having the magnitude of the t-statistic, $\left|\frac{\bar{x}-c}{s_x/\sqrt{n}}\right|$, larger than $t_{n-1,0.025}$ is considered evidence against $H_0 : \mu_X = c$. A large magnitude for the t-statistic indicates that the estimate of μ, which is the sample mean $\bar{x}$, is far from the hypothesized value c, as measured by the number of standard errors $s_x/\sqrt{n}$.

The test based on the t-statistic, known as a t-**test**, uses the following rejection rule, suggested by the second probability statement:

Rejection rule based on the t-statistic (test at the 5% level):

- Reject $H_0 : \mu = c$ at the 5% level if |t-statistic| $= \left|\frac{\bar{x}-c}{s_x/\sqrt{n}}\right| \geq t_{n-1,0.025}$.
- Do not reject $H_0 : \mu = c$ at the 5% level if |t-statistic| $= \left|\frac{\bar{x}-c}{s_x/\sqrt{n}}\right| < t_{n-1,0.025}$.

There are two possible conclusions to be drawn from this t-test rejection rule. Either we "reject" the null hypothesis H_0, which occurs when the magnitude of the t-statistic is above the critical value, or we "do not reject" the hypothesis, which occurs when the magnitude of the t-statistic is below the critical value. A t-statistic with large magnitude (greater than the critical value) provides evidence against $H_0 : \mu = c$ and, therefore, it is said that $H_0 : \mu = c$ is rejected. On the other hand, a t-statistic with a small magnitude (less than the critical value) does not provide evidence against $H_0 : \mu = c$; such a t-statistic is not surprising given the underlying t_{n-1} distribution that holds when H_0 is true. For this case, it is said that $H_0 : \mu = c$ is not rejected. While some discussions of hypothesis testing use the term "accept H_0" rather than "do not reject H_0," the use of the term "accept H_0" is not advisable. After all, it is never possible to have strong evidence that $H_0 : \mu = c$ since there is always some uncertainty about the estimation of μ. The most that can be said, in the case that $\left|\frac{\bar{x}-c}{s_x/\sqrt{n}}\right| < t_{n-1,0.025}$, is that there is insufficient statistical evidence against the null hypothesis H_0. As such, using the phrase "do not reject H_0" is appropriate. Even in cases where there is a strong prior belief that H_0 is false, there might be a failure to reject H_0 just because a small sample size leads to a large standard error, $s_x/\sqrt{n}$.

The magnitude of the t-statistic, $\left|\frac{\bar{x}-c}{s_x/\sqrt{n}}\right|$, can be thought of as a "statistical distance" from estimate $\bar{x}$ to the hypothesized value c. While $|\bar{x}-c|$ gives the actual distance from estimate $\bar{x}$ to the hypothesized value c, there is no way to know if the actual distance $|\bar{x}-c|$ is small or large due to the uncertainty associated with estimate $\bar{x}$. Dividing by the standard error $s_x/\sqrt{n}$ accounts for this uncertainty, so the statistical distance $\left|\frac{\bar{x}-c}{s_x/\sqrt{n}}\right|$ is a distance in terms of the number of standard errors that $\bar{x}$ is from c. For this statistical distance, unlike the actual distance, statistical theory tells us what types of values should be expected if the null hypothesis is true; specifically, as discussed previously, the statistical distance given by the magnitude t-statistic $\left|\frac{\bar{x}-c}{s_x/\sqrt{n}}\right|$ should be the absolute value of a realized draw from the t_{n-1} distribution if the null hypothesis $H_0: \mu = c$ is true.

The **level** of the test, a term introduced in the rejection rule given here, is formally defined as in definition 16.2.

Definition 16.2 *The* ***level*** *or* ***significance level*** *of a hypothesis test, denoted as α, is the probability that the null hypothesis H_0 is rejected when the null hypothesis H_0 is true. The level of a test is also called the* ***type I error*** *of the test.*

For the rejection rule given here, the level of the hypothesis test is $\alpha = 5\%$ or $\alpha = 0.05$. Figure 16.1 provides a graphical view of the rejection regions on the t_{n-1} distribution. If the magnitude of the t-statistic is larger than $t_{n-1,0.025}$, its value falls into either the gray region for the left tail (if the t-statistic is negative) or the gray region for the right tail (if the t-statistic is positive). If H_0 is true, there is a 95% probability that the t-statistic falls in the middle region between the two values $-t_{n-1,0.025}$ and $t_{n-1,0.025}$. But, even if H_0 is true, there is still a 5% probability that the previous rejection rule says to "reject H_0" due to the t-statistic falling in the left or right tail; for such cases, the test is wrong about the rejection, and the level of the test ($\alpha = 5\%$ here) indicates the probability that the test rejects when H_0 is true.

We can generalize the t-test and the rejection rule to levels other than $\alpha = 0.05$. Letting α denote the level of the test, the relevant probability statement is that

$$P\left(\left|\frac{\bar{X}-c}{s_X/\sqrt{n}}\right| < t_{n-1,\alpha/2}\right) = 1-\alpha \text{ when } H_0: \mu = c \text{ is true.}$$

This probability statement leads to the following rejection rule for a t-test at the α level:

Rejection rule based on the t-statistic (test at the α level):

- Reject $H_0: \mu = c$ at the α level if $|\text{t-statistic}| = \left|\frac{\bar{x}-c}{s_x/\sqrt{n}}\right| \geq t_{n-1,\alpha/2}$.
- Do not reject $H_0: \mu = c$ at the α level if $|\text{t-statistic}| = \left|\frac{\bar{x}-c}{s_x/\sqrt{n}}\right| < t_{n-1,\alpha/2}$.

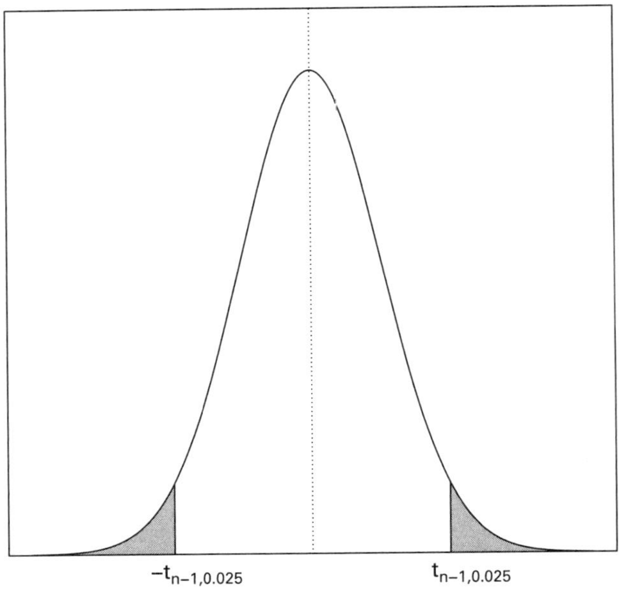

Figure 16.1
Rejection areas for the t-test at a 5% level

There is a trade-off involved in choosing the level of the test. By definition 16.2, the level indicates how likely it is to reject H_0 when H_0 is actually true. Therefore, decreasing the level from 10% to 5% lowers the probability that an incorrect rejection of H_0 occurs. On the other hand, since the t-statistic is the same regardless of the level of the test, decreasing the level from 10% to 5% leads to a lower chance that H_0 is rejected (even if H_0 is false). This lower rejection rate occurs since the critical value increases from $t_{n-1,0.05}$ to $t_{n-1,0.025}$ when the level is changed from 10% to 5%: if the magnitude of the t-statistic is less than $t_{n-1,0.05}$, the test would not reject H_0 at either level; if the magnitude of the t-statistic is greater than $t_{n-1,0.025}$, the test would reject H_0 at either level; and, if the magnitude of the t-statistic is between $t_{n-1,0.05}$ and $t_{n-1,0.025}$, the test would reject H_0 at the 10% level but not the 5% level.

Example 16.3 (Food truck) *In example 14.3, the weekly profits X of a food truck were assumed to be normally distributed, with unknown mean μ and variance σ^2, with each week's profits being an i.i.d. draw. Weekly profits were recorded for a total of six weeks, with the sample average of weekly profits equal to* \$1,200 *and the sample standard deviation equal to* \$200. *Suppose the food truck owner,* prior to seeing the data, *believed that the true average μ of weekly profits was* \$1,000. *This belief corresponds to the null hypothesis $H_0: \mu = 1{,}000$. For a test at the 5% level ($\alpha = 0.05$), the critical value is $t_{n-1,0.025} = t_{5,0.025} \approx 2.571$. Therefore, the t-test does not reject $H_0: \mu = 1{,}000$ at a 5% level since $|2.449| < 2.571$. The t-statistic and critical value can be calculated in R, as follows:*

```
# calculate t-statistic
tstat <- (1200-1000)/(200/sqrt(6))
tstat

## [1] 2.44949

# critical value for t-test at 5% level
qt(0.975,5)

## [1] 2.570582
```

For a test at the 10% level ($\alpha = 0.10$), the critical value is $t_{n-1,0.05} = t_{5,0.05} \approx 2.015$. Therefore, the t-test does reject $H_0 : \mu = 1{,}000$ at a 10% level since $|2.449| \geq 2.015$. For this null hypothesis, then, there is rejection at the 10% level but not the 5% level. The critical value can be calculated in R, as follows:

```
# critical value for t-test at 10% level
qt(0.95,5)

## [1] 2.015048
```

Example 14.3 calculated confidence intervals for μ based on the sample information provided here. The two-sided 95% confidence interval for μ is

$$\left(\bar{x} - t_{6-1,0.025}\frac{s_x}{\sqrt{n}}, \bar{x} + t_{6-1,0.025}\frac{s_x}{\sqrt{n}}\right) \approx (990, 1410),$$

and the two-sided 90% confidence interval for μ is

$$\left(\bar{x} - t_{6-1,0.05}\frac{s_x}{\sqrt{n}}, \bar{x} + t_{6-1,0.05}\frac{s_x}{\sqrt{n}}\right) \approx (1035, 1365).$$

The hypothesized value of $c = 1{,}000$ lies within the 95% confidence interval for μ but not within the 90% confidence interval for μ. With respect to the 95% confidence interval, the value of 1,000 is plausible since it lies within that interval. Not coincidentally, this fact corresponds with $H_0 : c = 1{,}000$ not being rejected at the 5% level. On the other hand, with respect to the 90% confidence interval, the value of 1,000 does not appear likely since it lies outside the interval. Again, not coincidentally, this fact corresponds with $H_0 : c = 1{,}000$ being rejected at the 10% level.

The last part of example 16.3 highlights the connection between two-sided $1-\alpha$ confidence intervals for μ and the rejection decision for tests at the α level. Proposition 16.1 formally states this relationship.

Proposition 16.1 *For a t-test of the null hypothesis $H_0 : \mu = c$ at the α-level, the null hypothesis H_0 is rejected if c lies outside the two-sided $1-\alpha$ confidence interval for μ and is not rejected if c lies inside the two-sided $1-\alpha$ confidence interval for μ.*

To show this result, note that the two-sided $1-\alpha$ confidence interval for μ is

$$\left(\bar{x} - t_{n-1,\alpha/2}\frac{s_x}{\sqrt{n}}, \bar{x} + t_{n-1,\alpha/2}\frac{s_x}{\sqrt{n}}\right),$$

so c is inside this interval if and only if

$$\bar{x} - t_{n-1,\alpha/2}\frac{s_x}{\sqrt{n}} < c < \bar{x} + t_{n-1,\alpha/2}\frac{s_x}{\sqrt{n}}.$$

The first inequality is equivalent to

$$\frac{\bar{x} - c}{s_x/\sqrt{n}} < t_{n-1,\alpha/2},$$

and the second inequality is equivalent to

$$\frac{\bar{x} - c}{s_x/\sqrt{n}} > -t_{n-1,\alpha/2}.$$

Putting these two inequalities together yields

$$-t_{n-1,\alpha/2} < \frac{\bar{x} - c}{s_x/\sqrt{n}} < t_{n-1,\alpha/2} \text{ or, equivalently, } \left|\frac{\bar{x} - c}{s_x/\sqrt{n}}\right| < t_{n-1,\alpha/2},$$

which corresponds to the "do not reject" rule for testing $H_0 : \mu = c$ at the α level. Therefore, c is within the two-sided $1 - \alpha$ confidence interval if and only if $H_0 : \mu = c$ is not rejected at the α level.

Therefore, an alternative and equivalent rejection rule for the t-test is the following:

Rejection rule based on the confidence interval (test at the α level):

- Reject $H_0 : \mu = c$ at the α level if c is not within the two-sided $1 - \alpha$ confidence interval for μ.
- Do not reject $H_0 : \mu = c$ at the α level if c is within the two-sided $1 - \alpha$ confidence interval for μ.

16.1.1 p-Values for the *t*-Test

A drawback of the rejection-rule approach to hypothesis testing is that the rejection rule says only whether H_0 is rejected based on the α-level and the associated critical value. For example, with a sample size of $n = 11$ and a level of $\alpha = 0.05$, the critical value is $t_{10,0.025} = 2.228$. A t-statistic of 2.2 would fail to reject H_0, whereas a slightly higher t-statistic of 2.3 would reject H_0. In addition, t-statistics of 2.3 and 6.3 would both lead to rejection of H_0, but the rejection rule by itself fails to indicate that the statistical evidence against H_0 is much stronger for a t-statistic of 6.3 than a t-statistic of 2.3.

To address this drawback, the concept of a **p-value** is introduced. We first provide the general definition of a p-value in definition 16.3, and then consider how the p-value is determined in the specific case of a two-sided t-test.

Definition 16.3 *The **p-value** of a test of the null hypothesis H_0 is the smallest level α^* such that the test rejects H_0 at the level α^*.*

In the case of the two-sided t-test of $H_0 : \mu = c$, the p-value is

$$\text{p-value} = P(|T| > |\text{t-statistic}|) = P\left(|T| > \left|\frac{\bar{x} - c}{s_x/\sqrt{n}}\right|\right) \text{ when } H_0 \text{ is true, where } T \sim t_{n-1}.$$

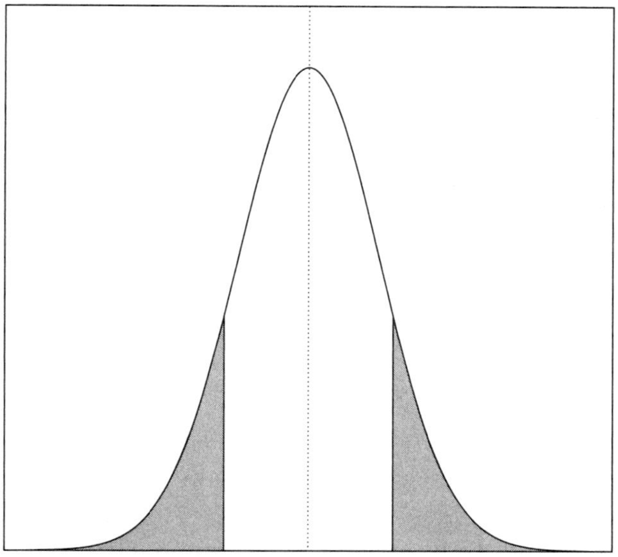

Figure 16.2
The p-value for a t-test

If the null hypothesis H_0 is true, the p-value is the probability of observing a t-statistic *at least as large in magnitude* as the one actually observed.

Figure 16.2 provides a graphical depiction of the p-value. The value of the realized t-statistic is denoted as *t-stat* with |*t-stat*| on the right side and −|*t-stat*| on the left side. The p-value equals the *total* area of the two tails, which includes the area to the left of −|*t-stat*| and the area to the right of |*t-stat*| and is represented by the gray shading.

As an example, if the t-statistic for $H_0 : \mu = c$ is calculated to be -1.2 for a sample size of $n = 15$, *R* can be used to find that $P(T > 1.2) = P(T < -1.2) \approx 0.125$ for $T \sim t_{14}$. Then the associated p-value is $P(|T| > 1.2) \approx (2)(0.125) = 0.250$, meaning that there is a 25% chance of seeing a t-statistic at least as large as 1.2 in magnitude if the null hypothesis H_0 is true. Due to the symmetry of the t-distribution, the p-value can be calculated as either two times the area of the left tail or two times the area of the right tail. The following *R* code illustrates the calculation of this p-value:

```
# calculate p-value as area in both tails
## calculation based upon two times left-tail area
2*pt(-1.2,14)

## [1] 0.2500528

## calculation based upon two times right-tail area
2*(1-pt(1.2,14))

## [1] 0.2500528
```

For a t-test at the 5% level, the critical value $t_{n-1,0.025}$ is the value that gives 5% total area in the tail to the right of $t_{n-1,0.025}$ and the tail to the left of $-t_{n-1,0.025}$. If the p-value for this t-test is less than 0.05 or 5%, it must be the case that the associated t-statistic

t^* is larger in magnitude than $t_{n-1,0.025}$ so the total area of the tails to the right of the $|t^*|$ and to the left of $-|t^*|$ is less than 0.05. Similarly, if the p-value for this t-test is greater than 0.05 or 5%, it must be the case that the associated t-statistic is smaller in magnitude than $t_{n-1,0.025}$ so the total area of the tails to the right of the $|t^*|$ and to the left of $-|t^*|$ is greater than 0.05. This same idea holds for any level of the test, suggesting the following rejection rule for the t-test based on the p-value:

Rejection rule based on the p-value (test at the α level):

- Reject $H_0 : \mu = c$ at the α level if the p-value $< \alpha$.
- Do not reject $H_0 : \mu = c$ at the α level if the p-value $> \alpha$.

The advantage of reporting the p-value associated with the t-test is that it immediately indicates whether the null hypothesis H_0 would be rejected at *any* specified level. For example, if the p-value is 0.08, the null hypothesis H_0 would be rejected for any level α greater than 0.08, and the null hypothesis H_0 would not be rejected for any level α less than 0.08. Larger t-statistic magnitudes are associated with lower p-values, meaning that H_0 is more likely to be rejected, and smaller t-statistic magnitudes are associated with higher p-values, meaning that H_0 is less likely to be rejected.

Example 16.4 (Food truck) *In example 16.3, a t-statistic of* 2.449 *was calculated for the null hypothesis* $H_0 : \mu = 1{,}000$. *Since the sample size is* $n = 6$, *the distribution of interest is the* t_5 *distribution. For a random variable* $T \sim t_5$, *R calculates that the p-value is* 0.058*:*

```
# calculate t-statistic
tstat <- (1200-1000)/(200/sqrt(6))

# calculate the p-value
2*pt(-abs(tstat),5)

## [1] 0.05797277

# confirm with the left and right tail area
## left-tail area
pt(-abs(tstat),5)

## [1] 0.02898639

## right-tail area
1-pt(abs(tstat),5)

## [1] 0.02898639
```

The left tail, corresponding to $P(T < -2.449)$, *has an area of* 0.029, *and the right tail, corresponding to* $P(T > 2.449)$, *also has an area of* 0.029, *yielding the p-value of* 0.058. *Therefore, the null hypothesis* $H_0 : \mu = 1{,}000$ *is rejected for any level above* 0.058 *and not rejected for any level below* 0.058. *This finding is consistent with example 16.3, where* H_0 *was rejected at the* 10% *level but not at the* 5% *level.*

16.1.2 Power of the t-Test

The level α of a test indicates how likely it is for the null hypothesis H_0 to be incorrectly rejected when H_0 is actually true. But what if H_0 is *not* true? An effective test should have a high probability of rejecting H_0 when H_0 is not true. This probability, known as the **power** of the test, depends on the specific alternative hypothesis that is true. In the case of the null hypothesis $H_0 : \mu = c$, a *specific alternative hypothesis* would be $\mu = d$ for some value $d \neq c$.

Definition 16.4 *For a specific alternative hypothesis, the* **power** *of a test is the probability that the test correctly rejects the null hypothesis H_0. For a specific alternative hypothesis, the probability that the test does not reject the null hypothesis, which is equal to 1 minus the power, is called the* **type II error** *of the test.*

If the specific alternative hypothesis $\mu = d$ (for $d \neq c$) is true, then

$$\frac{\bar{X} - d}{s_X/\sqrt{n}} \sim t_{n-1}.$$

Since H_0 is not true, the t-statistic $\frac{\bar{X}-c}{s_X/\sqrt{n}}$ is not distributed as a t_{n-1} random variable. Instead, the t-statistic can be written as

$$\frac{\bar{X} - c}{s_X/\sqrt{n}} = \frac{\bar{X} - d}{s_X/\sqrt{n}} + \frac{d - c}{s_X/\sqrt{n}},$$

such that the t-statistic is a t_{n-1} random variable plus the term $\frac{d-c}{s_X/\sqrt{n}}$. If d is much larger than c, this term is a large positive number so the observed t-statistic should look like a random draw from t_{n-1} plus a large positive number, which makes it more likely to be above the right-tail critical value for rejection. Similarly, if d is much smaller than c, this term is a large negative number so the observed t-statistic should look like a random draw from t_{n-1} minus a large number, which makes it more likely to be below the left-tail critical value for rejection.

To help visualize the power of the two-sided t-test, figure 16.3 shows the power of the test under two specific alternative hypotheses. In both graphs, the thin curve corresponds to the t_{n-1} distribution for the t-statistic that would hold under the null hypothesis H_0, along with the critical values $-t_{n-1,0.025}$ and $t_{n-1,0.025}$. The bold curves represent the actual distributions for the t-statistic under two specific alternative hypotheses, with d being 2.5 standard deviations above c in the top graph and 4 standard deviations above c in the bottom graph. The gray areas indicate the rejection probabilities using the critical-value cutoff rejection rule. For both graphs, the rejection probability is clearly much larger than 5%, which is the rejection rate under the null hypothesis. Since d is farther from c in the bottom graph, the power of the test is larger in the bottom graph, as indicated by the larger gray area. The type II errors are given by the white area under the bold curves, and since the type II error is just 1 minus the power, the type II error is smaller in the bottom graph.

For any specific alternative $\mu = d$, with $d \neq c$, the power of the t-test is an increasing function of $\left|\frac{d-c}{s_X/\sqrt{n}}\right|$, so the power depends on the sample size and the value of d. For a given sample size n, the power of the test increases when the specific alternative hypothesis has d farther from c. Intuitively, this relationship makes sense since it is easier to find evidence against $H_0 : \mu = c$ when d is not close to c. For any specific alternative hypothesis, the power of the test is increasing in the sample size n. Larger samples make it easier to reject the null hypothesis when it is false. In fact, if the sample size n gets arbitrarily large ($n \to \infty$), the power of the test becomes arbitrarily close to 100% if the null hypothesis H_0 is false.

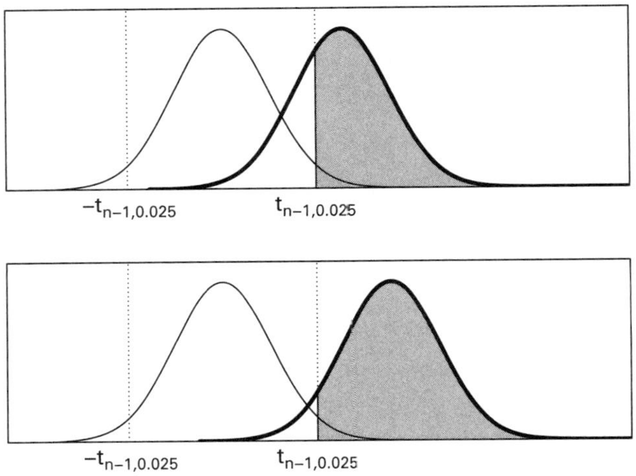

Figure 16.3
Power of a t-test

16.1.3 One-Sided *t*-Test

This section considers testing a one-sided null hypothesis. Suppose that there is interest in testing the null hypothesis,

$$H_0 : \mu \geq c,$$

versus the alternative hypothesis,

$$H_1 : \mu < c.$$

Thinking about the *t*-statistic $\frac{\bar{x}-c}{s_x/\sqrt{n}}$, it no longer makes sense to reject H_0 when the *t*-statistic is in the right tail of the t_{n-1} distribution. A positive *t*-statistic corresponds to an estimate of μ that is greater than c, and therefore consistent with the null hypothesis H_0 being true. The null hypothesis should be rejected only when a *t*-statistic is unusually negative. To have a test at the α level, the area under the t_{n-1} distribution to the left of the critical value must be equal to α. The corresponding rejection rule is as follows:

Rejection rule based on the *t*-statistic (test at the α level):

- Reject $H_0 : \mu \geq c$ at the α level if t-statistic $= \frac{\bar{x}-c}{s_x/\sqrt{n}} \leq -t_{n-1,\alpha}$.
- Do not reject $H_0 : \mu \geq c$ at the α level if t-statistic $= \frac{\bar{x}-c}{s_x/\sqrt{n}} > -t_{n-1,\alpha}$.

Similarly, for testing the null hypothesis,

$$H_0 : \mu \leq c,$$

versus the alternative hypothesis,

$$H_1 : \mu > c,$$

the null hypothesis should be rejected only when the *t*-statistic is in the right tail of the t_{n-1} distribution. For this case, negative *t*-statistics are associated with estimates of μ that are less than c, and therefore consistent with the null hypothesis. Only unusually large positive

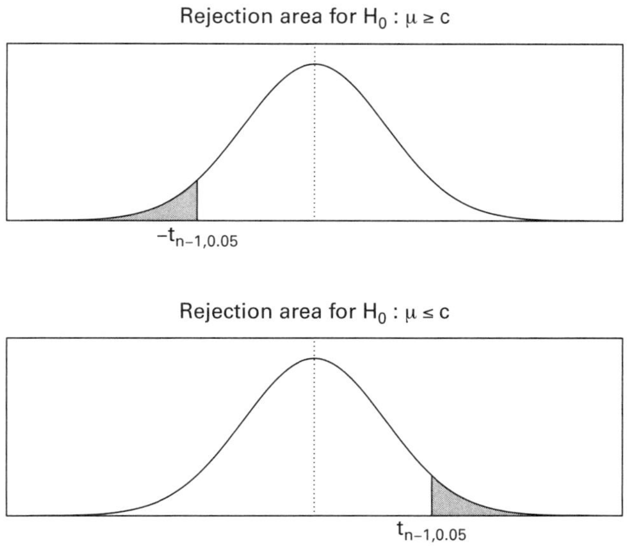

Figure 16.4
Rejection areas for one-sided t-tests at a 5% level

t-statistics provide evidence against the null hypothesis. Therefore, the corresponding rejection rule is as follows:

> **Rejection rule based on the t-statistic (test at the α level):**
>
> - Reject $H_0 : \mu \leq c$ at the α level if t-statistic $= \frac{\bar{x}-c}{s_x/\sqrt{n}} \geq t_{n-1,\alpha}$.
> - Do not reject $H_0 : \mu \leq c$ at the α level if t-statistic $= \frac{\bar{x}-c}{s_x/\sqrt{n}} < t_{n-1,\alpha}$.

Figure 16.4 shows the rejection areas for both types of one-sided tests for a 5% level. The top graph shows the rejection area for testing $H_0 : \mu \geq c$, with the gray region corresponding to values less than $-t_{n-1,0.05}$ and having an area of 5%. The bottom graph shows the rejection area for testing $H_0 : \mu \leq c$, with the gray region corresponding to values greater than $t_{n-1,0.05}$ and having an area of 5%.

Example 16.5 (Investment opportunity) *Continuing example 16.2, let's make the additional assumption that the* 10 *observed weekly sales figures are i.i.d. draws from a normal distribution. Knowing that the sample mean of weekly sales is* \$11,200 ($\bar{x} = 11.2$) *and the sample standard deviation of weekly sales is* \$3,400 ($s_x = 3.4$), *would the null hypothesis* $H_0 : \theta \leq 10$, *which corresponds to the business not being a worthwhile investment, be rejected at the 5% level? The t-statistic is*

$$\frac{\bar{x}-c}{s_x/\sqrt{n}} = \frac{11.2-10}{3.4/\sqrt{10}} \approx 1.116.$$

Since $t_{n-1,0.05} = t_{9,0.05} \approx 1.833$, *the null hypothesis* $H_0 : \theta \leq 10$ *is not rejected since the t-statistic* 1.116 *is less than the critical value* 1.833. *How about a one-sided test at the* 10% *level? The critical value is* $t_{9,0.10} \approx 1.383$, *so the null hypothesis is still not rejected at the* 10% *level since* $1.116 < 1.383$. *Here is the R code for the necessary calculations:*

```
# calculate t-statistic
tstat <- (11.2-10)/(3.4/sqrt(10))
tstat

## [1] 1.116098

# critical value for test at 5% level
qt(0.95,9)

## [1] 1.833113

# critical value for test at 10% level
qt(0.90,9)

## [1] 1.383029
```

The concept of p-values can be extended to one-sided hypothesis testing. Again, the difference from two-sided testing is that only one of the tails is considered, with the left tail used for $H_0 : \mu \geq c$ and the right tail used for $H_0 : \mu \leq c$. For the one-sided t-test of $H_0 : \mu \geq c$, the p-value is

$$\text{p-value} = P(T < \text{t-statistic}) = P\left(T < \frac{\bar{x} - c}{s_x/\sqrt{n}}\right) \text{ when } H_0 \text{ is true, where } T \sim t_{n-1}.$$

For the one-sided t-test of $H_0 : \mu \leq c$, the p-value is

$$\text{p-value} = P(T > \text{t-statistic}) = P\left(T > \frac{\bar{x} - c}{s_x/\sqrt{n}}\right) \text{ when } H_0 \text{ is true, where } T \sim t_{n-1}.$$

As with a two-sided test, the p-value can be used to determine whether a test of a one-sided null hypothesis should be rejected. The null hypothesis H_0 is rejected if the p-value is less than the level α of the test and not rejected if the p-value is greater than the level α of the test.

Example 16.6 (Investment opportunity) *Continuing example 16.5, the p-value associated with the t-statistic* 1.116 *is the area to the right of* 1.116 *under the* t_9 *distribution, which can be calculated in R as follows:*

```
# calculate p-value for one-sided t-test
1-pt(1.116,9)

## [1] 0.1466655
```

Since $P(T > 1.116) \approx 0.147$ *for a random variable* $T \sim t_9$, *the p-value for the one-sided test of* $H_0 : \mu \leq 10$ *is* 0.147. *Thus, a one-sided t-test does not reject* $H_0 : \mu \leq 10$ *for any level below* 14.7%. *Even if* $\mu = 10$, *this p-value tells us that there would be a* 14.7% *probability of seeing a t-statistic at least as large as the one observed (*1.116*).*

Finally, the connection between one-sided $1 - \alpha$ confidence intervals and one-sided tests at the α level can be established. Recall that the one-sided $1 - \alpha$ confidence intervals for μ are

$$\left(\bar{x} - t_{n-1,\alpha}\frac{s_x}{\sqrt{n}}, \infty\right) \text{ and } \left(-\infty, \bar{x} + t_{n-1,\alpha}\frac{s_x}{\sqrt{n}}\right).$$

Proposition 16.2 *For a t-test of the null hypothesis $H_0 : \mu \geq c$ at the α level, the null hypothesis H_0 is rejected if c lies outside the one-sided $1-\alpha$ confidence interval $\left(-\infty, \bar{x} + t_{n-1,\alpha}\frac{s_x}{\sqrt{n}}\right)$ for μ and is not rejected if c lies inside the one-sided $1-\alpha$ confidence interval $\left(-\infty, \bar{x} + t_{n-1,\alpha}\frac{s_x}{\sqrt{n}}\right)$ for μ. For a t-test of the null hypothesis $H_0 : \mu \leq c$ at the α-level, the null hypothesis H_0 is rejected if c lies outside the one-sided $1-\alpha$ confidence interval $\left(\bar{x} - t_{n-1,\alpha}\frac{s_x}{\sqrt{n}}, \infty\right)$ for μ and is not rejected if c lies inside the one-sided $1-\alpha$ confidence interval $\left(\bar{x} - t_{n-1,\alpha}\frac{s_x}{\sqrt{n}}, \infty\right)$ for μ.*

Example 16.7 (Investment opportunity) *Re-visiting example 16.5, the same conclusions for the one-sided tests at the 5% and 10% levels can be obtained by using one-sided confidence intervals and proposition 16.2. The one-sided 95% confidence interval for μ is*

$$\left(\bar{x} - t_{n-1,\alpha}\frac{s_x}{\sqrt{n}}, \infty\right) = \left(11.2 - t_{9,0.05}\frac{3.4}{\sqrt{10}}, \infty\right) \approx (9.23, \infty),$$

meaning that $H_0 : \mu \leq 10$ is not rejected at a 5% level since the value 10 is within this confidence interval. Similarly, the one-sided 90% confidence interval for μ is

$$\left(\bar{x} - t_{n-1,\alpha}\frac{s_x}{\sqrt{n}}, \infty\right) = \left(11.2 - t_{9,0.10}\frac{3.4}{\sqrt{10}}, \infty\right) \approx (9.71, \infty),$$

meaning $H_0 : \mu \leq 10$ is not rejected at a 10% level since the value 10 is within this confidence interval. But for the one-sided 85% confidence interval for μ, which is

$$\left(\bar{x} - t_{n-1,\alpha}\frac{s_x}{\sqrt{n}}, \infty\right) = \left(11.2 - t_{9,0.15}\frac{3.4}{\sqrt{10}}, \infty\right) = \left(11.2 - 1.0997\frac{3.4}{\sqrt{10}}, \infty\right)$$
$$\approx (10.01, \infty),$$

the value 10 is not within the interval. Thus, $H_0 : \mu \leq 10$ is rejected at a 15% level, which agrees with the p-value of 0.147 found in example 16.6.

16.2 Asymptotic Hypothesis Testing: Parameters with Asymptotically Normal Estimators

The hypothesis tests of section 16.1 only work for the specific case of an unknown population mean μ of underlying i.i.d. normal random variables, as the exact sampling distribution of the sample mean estimator is related to the t-distribution. This section considers the more general problem of testing a null hypothesis that involves an unknown parameter θ that can be estimated by an asymptotically normal estimator. In the same way that section 14.4 provided asymptotic confidence intervals for this general case, this section considers hypothesis tests related to θ for a large sample.

We fix ideas by considering an estimator $\hat{\theta}_X$ and estimate $\hat{\theta}_x$ based on a univariate random variable and data, respectively, but the results generalize to other estimators and estimates, including those based on multivariate data and multiple samples. Following the treatment in section 14.4, consider an asymptotically normal estimator $\hat{\theta}_X$ with asymptotic distribution

$$\hat{\theta}_X \overset{a}{\sim} N\left(\theta, \frac{V}{n}\right) \text{ or, equivalently, } \frac{\hat{\theta}_X - \theta}{\sqrt{V/n}} \overset{a}{\sim} N(0, 1),$$

where θ is the parameter (estimand) of interest and the associated asymptotic standard deviation is $\sqrt{\frac{V}{n}}$. For the estimate $\hat{\theta}_x$, based on the observed sample, the z-**ratio** is defined as

$$z\text{-ratio} = \frac{\hat{\theta}_x - \theta}{se(\hat{\theta}_x)},$$

where $se(\hat{\theta}_x)$ is the standard error associated with estimate $\hat{\theta}_x$. For a large sample, the z-ratio is a realized draw from the standard normal distribution $N(0, 1)$ since $se(\hat{\theta}_x)$ is a consistent estimator of the asymptotic standard deviation $\sqrt{\frac{V}{n}}$. However, since θ is an unknown parameter, the z-ratio cannot be calculated.

Following the same approach as section 16.1, consider the thought experiment in which the null hypothesis $H_0 : \theta = c$ is assumed to be true for a constant c. Define the z-**statistic** as

$$z\text{-statistic} = \frac{\hat{\theta}_x - c}{se(\hat{\theta}_x)},$$

which can be calculated since c is known. The z-statistic is the number of standard errors that estimate $\hat{\theta}_x$ is from c, with the z-statistic being positive if estimate $\hat{\theta}_x$ is above c and negative if estimate $\hat{\theta}_x$ is below c. If the null hypothesis $H_0 : \theta = c$ is true, the z-statistic should be a realized draw from the standard normal distribution $N(0, 1)$. Before observing the data, there is approximately a 95% probability that the realized z-statistic is between -1.96 and 1.96 if $H_0 : \theta = c$ is true or, more generally, a $1 - \alpha$ probability that the realized z-statistic is between $-z_{\alpha/2}$ and $z_{\alpha/2}$ if $H_0 : \theta = c$ is true. Thus, the rejection rule for the two-sided test of $H_0 : \theta = c$ based on the z-statistic is as follows:

Rejection rule based on the z-statistic (test at the α level):

- Reject $H_0 : \theta = c$ at the α level if $|\text{z-statistic}| = \left|\frac{\hat{\theta}_x - c}{se(\hat{\theta}_x)}\right| \geq z_{\alpha/2}$.
- Do not reject $H_0 : \theta = c$ at the α level if $|\text{z-statistic}| = \left|\frac{\hat{\theta}_x - c}{se(\hat{\theta}_x)}\right| < z_{\alpha/2}$.

The resulting test is known as a z-**test**, as it uses critical values from the normal distribution, in contrast to the t-test of section 16.1 that uses critical values from the t-distribution. Unlike the t-test, the critical values for the z-test do not depend on the sample size n. The magnitude of the z-statistic, $\left|\frac{\hat{\theta}_x - c}{se(\hat{\theta}_x)}\right|$, can be viewed as a statistical distance, measuring the number of standard errors that estimate $\hat{\theta}_x$ is from the hypothesized value c. By dividing by the standard error $se(\hat{\theta}_x)$, this statistical distance takes into account the uncertainty associated with estimate $\hat{\theta}_x$ and, based on statistical theory, should be the absolute value of a realized draw from the $N(0, 1)$ distribution if the null hypothesis $H_0 : \theta = c$ is true.

Example 16.8 (Widget website) *For the e-mail experiment summarized in example 16.1, suppose that the e-mail marketing director at Widgets.com believes that the true purchase probability for e-mail A recipients is 25%. The null hypothesis for this preexperiment* *belief is*

$$H_0 : \pi_A = 0.25.$$

Given that 60 *out of* 300 *e-mail A recipients made a purchase, what would a z-test of this null hypothesis conclude? The unknown parameter here is* $\theta = \pi_A$, *and the hypothesized value is* $c = 0.25$. *The estimate of* π_A *is* $\hat{\theta}_x = \bar{x} = p_A = \frac{60}{300} = 0.20$. *Then, since the standard error of* $\bar{x}$ *is* 0.0231 *as calculated in example 14.5, the z-statistic is*

$$\frac{\hat{\theta}_x - c}{se(\hat{\theta}_x)} = \frac{0.20 - 0.25}{0.0231} \approx -2.16.$$

This z-statistic says that the estimated probability of 0.20 *is* 2.16 *standard errors below the hypothesized probability of* 0.25. *For a test at the* 5% *level, the null hypothesis* $H_0 : \pi_A = 0.25$ *is rejected since* $|z\text{-statistic}| = |2.16| \geq 1.96 = z_{0.025}$. *For a test at the* 1% *level, the null hypothesis* H_0 *is not rejected since* $|z\text{-statistic}| = |2.16| < 2.576 = z_{0.005}$. *The following R code calculates the z-statistic and critical values:*

```
# calculate pa and its standard error
pa <- 60/300
se_pa <- sqrt(pa*(1-pa)/300)

# calculate z-statistic
zstat <- (pa-0.25)/se_pa
zstat

## [1] -2.165064

# critical value for test at 5% level
qnorm(0.975)

## [1] 1.959964

# critical value for test at 1% level
qnorm(0.995)

## [1] 2.575829
```

How about testing whether e-mail campaign A and e-mail campaign B are equally effective? As discussed in example 16.1, the null hypothesis of interest is

$$H_0 : \pi_A = \pi_B \text{ or, equivalently, } H_0 : \pi_A - \pi_B = 0.$$

This hypothesis fits into the framework developed here for the unknown parameter $\theta = \pi_A - \pi_B$, *and the alternative hypothesis is*

$$H_1 : \pi_A \neq \pi_B \text{ or, equivalently, } H_1 : \pi_A - \pi_B \neq 0.$$

The estimate of $\theta = \pi_A - \pi_B$ *is* $\hat{\theta}_x = p_A - p_B = \frac{60}{300} - \frac{66}{300} = 0.20 - 0.22 = -0.02$. *From example 14.15, the standard error of this estimate is* $se(p_A - p_B) = 0.0332$. *Then, the z-statistic for testing* $H_0 : \pi_A - \pi_B = 0$ *is*

$$\frac{(p_A - p_B) - 0}{se(p_A - p_B)} = \frac{-0.02 - 0}{0.0332} \approx -0.60.$$

Therefore, the null hypothesis $H_0 : \pi_A - \pi_B = 0$ *is not rejected at the* 5% *level since* $|-0.60| < 1.96$. *The null hypothesis is also not rejected at the* 10% *level since*

$|-0.60| < 1.645$. *As a result, there does not appear to be strong evidence that, based on the sample of e-mail A recipients and e-mail B recipients observed, there is a statistically significant difference between* π_A *and* π_B.

Using the same approach, we can test whether there are differences between (i) the purchase probability of e-mail A recipients and nonrecipients and (ii) the purchase probability of e-mail B recipients and nonrecipients. For (i), the null hypothesis is

$$H_0 : \pi_A - \pi_C = 0,$$

which has z-statistic

$$\frac{(p_A - p_C) - 0}{se(p_A - p_C)} = \frac{(0.20 - 0.15) - 0}{0.0242} \approx 2.07.$$

The null hypothesis $H_0 : \pi_A - \pi_C = 0$ *is rejected at the 5% level since* $|2.07| \geq 1.96$, *providing evidence of a statistically significant difference between* π_A *and* π_C. *For (ii), the null hypothesis is*

$$H_0 : \pi_B - \pi_C = 0,$$

which has z-statistic

$$\frac{(p_B - p_C) - 0}{se(p_B - p_C)} = \frac{(0.22 - 0.15) - 0}{0.0250} \approx 2.80.$$

The null hypothesis $H_0 : \pi_B - \pi_C = 0$ *is rejected at the 5% level since* $|2.80| \geq 1.96$, *providing evidence of a statistically significant difference between* π_B *and* π_C. *Taking these hypothesis tests together, there appears to be evidence supporting the hypotheses that each of the e-mail recipient groups has a greater purchase probability than that of the nonrecipient group; on the other hand, there is no strong evidence of a difference between the purchase probability of e-mail A recipients and the purchase probability of e-mail B recipients.*

Example 16.9 (Education and earnings) *For the* **cps** *data set, the correlation between weekly earnings (*$y = earnwk$*) and education (*$x = educ$*) among the* $n = 2{,}809$ *employed individuals is* $r_{xy} = 0.325$. *From example 14.10, the standard error of* r_{xy}, *as an estimate of the population correlation* ρ_{XY}, *equals* 0.0169. *To test whether there is* any *correlation between weekly earnings and education in the population, the null hypothesis is*

$$H_0 : \rho_{XY} = 0.$$

The z-statistic for testing $H_0 : \rho_{XY} = 0$ *is*

$$\frac{\hat{\theta}_x - c}{se(\hat{\theta}_x)} = \frac{r_{xy} - c}{se(r_{xy})} = \frac{0.325 - 0}{0.0169} \approx 19.23,$$

indicating that the sample correlation is 19.23 *standard errors above* 0, *which is a lot! Thus, the null hypothesis* $H_0 : \rho_{XY} = 0$ *is rejected at the 5% level since* $|19.23| \geq 1.96$ *and the 1% level since* $|19.23| \geq 2.576$. *These rejections are certainly not borderline since the z-statistic is so large in magnitude. Testing whether the population correlation equals zero is often of interest in other settings, as it provides an easy way to statistically support the idea that two variables are related by rejecting that they are unrelated. To perform such a test using asymptotic theory and a z-statistic, we only need the sample correlation and its standard error.*[2]

Example 16.10 (Exam score data) *Example 14.13 considered the data set* **exams**, *which contains scores on two* 100*-point exams for* 77 *students. The two exam scores are given by the variables* $exam1$ *and* $exam2$*. To test whether the exams are equally difficult, a null hypothesis of interest is whether their population means are the same:*

$$H_0 : \mu_{exam1} = \mu_{exam2} \text{ or, equivalently, } H_0 : \mu_{exam1} - \mu_{exam2} = 0.$$

The unknown parameter is $\theta = \mu_{exam1} - \mu_{exam2}$*. From example 14.13, the estimated score difference is*

$$\overline{exam1} - \overline{exam2} = 6.42,$$

with a standard error of

$$se(\overline{exam1} - \overline{exam2}) = 1.23.$$

The z-statistic associated with $H_0 : \mu_{exam1} - \mu_{exam2} = 0$ *is*

$$\frac{6.42 - 0}{1.23} \approx 5.22,$$

which provides evidence of a statistically significant difference between the population average of $exam1$ *and the population average of* $exam2$*. The null hypothesis is rejected at a 5% level since* $|5.22| \geq 1.96$ *and also at a 1% level since* $|5.22| \geq 2.576$.

Similar to the t-test of section 16.1, the z-test can also be conducted with confidence intervals. The rejection rule for the z-test based on the two-sided confidence interval is as follows:

Rejection rule based on the confidence interval (test at the α level):

- Reject $H_0 : \theta = c$ at the α level if c is not within the two-sided, $1 - \alpha$ confidence interval for θ.
- Do not reject $H_0 : \theta = c$ at the α level if c is within the two-sided, $1 - \alpha$ confidence interval for θ.

Example 16.11 (Widget website) *For the z-test of the null hypothesis* $H_0 : \pi_A = 0.25$ *in example 16.8, the estimated purchase probability is* $\hat{\theta}_x = \bar{x} = p_A = \frac{60}{300} = 0.20$*, with a standard error of* $\bar{x}$ *of* 0.0231*. As shown in example 14.5, a 95% confidence interval for* π_A *is*

$$(0.20 - (1.96)(0.0231), 0.20 + (1.96)(0.0231)) \approx (0.155, 0.245).$$

Since 0.25 *does not fall within this 95% confidence interval,* $H_0 : \pi_A = 0.25$ *is rejected at the 5% level, the same conclusion as reached in example 16.8. On the other hand, a 99% confidence interval for* π_A *is*

$$(0.20 - (2.576)(0.0231), 0.20 + (2.576)(0.0231)) \approx (0.140, 0.260),$$

meaning $H_0 : \pi_A = 0.25$ *is not rejected at the 1% level since* 0.25 *is within the interval.*

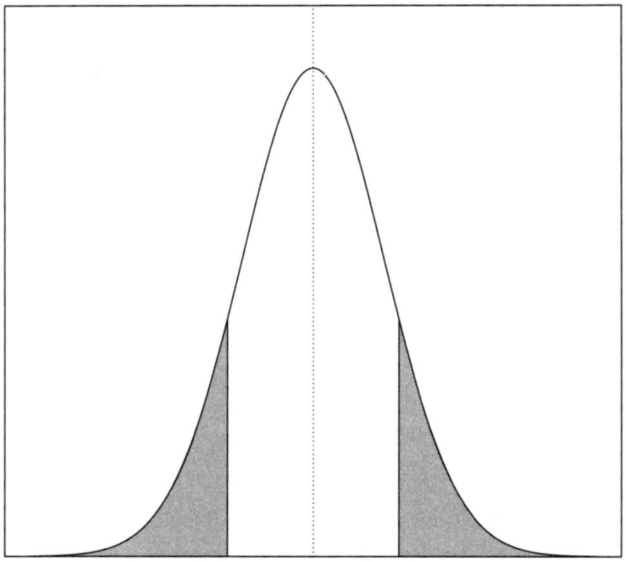

Figure 16.5
The p-value for a z-test

In the case of the two-sided z-test of $H_0: \theta = c$, the p-value is

$$\text{p-value} = P(|Z| > |\text{z-statistic}|) = P\left(|Z| > \left|\frac{\hat{\theta}_x - c}{se(\hat{\theta}_x)}\right|\right) \text{ when } H_0 \text{ is true, where } Z \sim N(0,1).$$

If the null hypothesis H_0 is true, the p-value is the probability of observing a z-statistic *at least as large in magnitude* as the one actually observed. Graphically, as shown in figure 16.5, the p-value is equal to the total area of the two tails, adding the area of the tail to the left of $-|\text{z-stat}|$ and the area of the tail to the right of $|\text{z-stat}|$. This graph is identical to figure 16.2, except that the distribution in figure 16.5 is the $N(0, 1)$ distribution, while the distribution in figure 16.2 is the t_{n-1} distribution.

For the z-test, like the t-test of section 16.1, knowing the p-value tells us whether the null hypothesis $H_0: \theta = c$ is rejected at *any* level α. When the p-value is less than α, the z-statistic must be in either the left tail (less than $-z_{\alpha/2}$) or in the right tail (greater than $z_{\alpha/2}$), indicating rejection at the level α. On the other hand, when the p-value is greater than α, the z-statistic is between $-z_{\alpha/2}$ and $z_{\alpha/2}$, indicating a lack of rejection at the level α.

Rejection rule based on the p-value (test at the α level):

- Reject $H_0: \theta = c$ at the α level if p-value $< \alpha$.
- Do not reject $H_0: \theta = c$ at the α level if p-value $> \alpha$.

Example 16.12 (Widget website) *For the z-test of the null hypothesis $H_0: \pi_A = 0.25$ in example 16.8, the calculated z-statistic is*

$$\frac{\hat{\theta}_x - c}{se(\hat{\theta}_x)} = \frac{0.20 - 0.25}{0.0231} \approx -2.16.$$

The associated p-value, for $Z \sim N(0, 1)$, is

$$p\text{-}value = P(|Z| > |z\text{-}statistic|) = P(|Z| > 2.16) \approx 0.030.$$

The following R code calculates the z-statistic and p-value:

```
# calculate pa and its standard error
pa <- 60/300
se_pa <- sqrt(pa*(1-pa)/300)

# calculate z-statistic
zstat <- (pa-0.25)/se_pa

# calculate the p-value for the z-test
2*(1-pnorm(abs(zstat)))

## [1] 0.03038282
```

The null hypothesis $H_0 : \pi_A = 0.25$ is rejected at any level above 3.0% and not rejected at any level below 3.0%, which agrees with the conclusion in examples 16.8 and 16.11 that H_0 is rejected at a 5% level, but not a 1% level.

Similarly, we can calculate p-values associated with the z-tests of the three null hypotheses $H_0 : \pi_A - \pi_B = 0$, $H_0 : \pi_A - \pi_C = 0$, and $H_0 : \pi_B - \pi_C = 0$ considered in example 16.8. The following R code calculates these p-values:

```
# calculate purchase probabilities and standard errors
pa <- 60/300
se_pa <- sqrt(pa*(1-pa)/300)
pb <- 66/300
se_pb <- sqrt(pb*(1-pb)/300)
pc <- 360/2400
se_pc <- sqrt(pc*(1-pc)/2400)

# calculate z-statistics
zstat_abdiff <- (pa-pb)/sqrt(se_pa^2 + se_pb^2)
zstat_acdiff <- (pa-pc)/sqrt(se_pa^2 + se_pc^2)
zstat_bcdiff <- (pb-pc)/sqrt(se_pb^2 + se_pc^2)
zstat_abdiff

## [1] -0.6015661

zstat_acdiff

## [1] 2.064674

zstat_bcdiff

## [1] 2.79972

# calculate the p-values for the z-tests
2*(1-pnorm(abs(zstat_abdiff)))

## [1] 0.547463

2*(1-pnorm(abs(zstat_acdiff)))

## [1] 0.03895389
```

```
2*(1-pnorm(abs(zstat_bcdiff)))

## [1] 0.005114694
```

The following table summarizes the z-statistics and p-values for the three null hypotheses:

Null hypothesis	*z-statistic*	*p-value*
$H_0: \pi_A - \pi_B = 0$	-0.60	0.547
$H_0: \pi_A - \pi_C = 0$	2.06	0.039
$H_0: \pi_B - \pi_C = 0$	2.80	0.005

Having the p-values makes it easy to see that the second and third null hypotheses are rejected at a 5% level and the first hypothesis is not. The strongest evidence of a difference is between π_B and π_C, with an associated p-value of 0.005, for which $H_0: \pi_B - \pi_C = 0$ is rejected at any level above 0.5%. The weakest evidence of a difference is between π_A and π_B, with a large p-value of 0.547 indicating that $H_0: \pi_A - \pi_B = 0$ is not rejected at any level below 54.7%.

16.2.1 One-Sided *z*-Test

Similar to the one-sided *t*-tests in section 16.1.3, this section considers one-sided versions of the *z*-test. Suppose that there is interest in testing the null hypothesis,

$$H_0: \theta \geq c,$$

versus the alternative hypothesis,

$$H_1: \theta < c.$$

Rejection of H_0 should occur only when the *z*-statistic is negative and large in magnitude since a positive *z*-statistic is consistent with the null hypothesis that θ is greater than c. For a one-sided test of level α, the area under the $N(0, 1)$ distribution to the left of the critical value must equal α. The appropriate critical value is $-z_\alpha$, and the corresponding rejection rule is as follows:

Rejection rule based on the *z*-statistic (test at the α level):

- Reject $H_0: \theta \geq c$ at the α level if the z-statistic $= \frac{\hat{\theta}_x - c}{se(\hat{\theta}_x)} \leq -z_\alpha$.
- Do not reject $H_0: \theta \geq c$ at the α level if the z-statistic $= \frac{\hat{\theta}_x - c}{se(\hat{\theta}_x)} > -z_\alpha$.

Similarly, for testing the null hypothesis,

$$H_0: \theta \leq c,$$

versus the alternative hypothesis

$$H_1: \theta > c,$$

the null hypothesis should be rejected only when the *z*-statistic is large and positive, corresponding to a *z*-statistic in the right tail of the $N(0, 1)$ distribution. For this case, negative

z-statistics are associated with estimates of θ that are less than c, and therefore consistent with the null hypothesis. The corresponding rejection rule is as follows:

Rejection rule based on the z-statistic (test at the α level):

- Reject $H_0 : \theta \leq c$ at the α level if z-statistic $= \frac{\hat{\theta}_x - c}{se(\hat{\theta}_x)} \geq z_\alpha$.
- Do not reject $H_0 : \theta \leq c$ at the α level if z-statistic $= \frac{\hat{\theta}_x - c}{se(\hat{\theta}_x)} < z_\alpha$.

The rejection areas for one-sided z-tests are similar to those depicted in figure 16.4 for one-sided t-tests. For instance, for a test at the 5% level, the 5% rejection area for testing $H_0 : \theta \geq c$ corresponds to all z-statistic values less than $-z_{0.05}$, whereas the 5% rejection area for testing $H_0 : \theta \leq c$ corresponds to all z-statistic values greater than $z_{0.05}$.

Example 16.13 (Betting strategy) *An experienced sports gambler is convinced that she has a strategy that is profitable at a certain casino. She has tested the strategy betting on* 120 *games, with a* 55% *success rate (*66 *winning bets on the* 120 *total games). Due to fees that the casino charges, the gambler needs a success rate* π *of at least* 52% *to be profitable over the long run. The gambler is therefore interested in testing, and hopes to be able to reject, the one-sided null hypothesis*

$$H_0 : \pi \leq 0.52,$$

which corresponds to unprofitable π *values. The alternative hypothesis* $H_1 : \pi > 0.52$ *corresponds to profitable* π *values. The observed success rate is consistent with the alternative hypothesis* H_1 *being true, but the gambler is concerned that the high realized success rate may have arisen due to chance. The estimate of* π *is* 0.55, *and the only additional information needed for the z-test is the standard error of this estimate. Under the assumption that the success of each bet is an i.i.d. Bernoulli(*π*) draw, the standard error is* $\sqrt{\frac{(0.55)(0.45)}{120}} \approx 0.0454$. *The z-statistic is* $\frac{0.55-0.52}{0.0454} \approx 0.661$, *meaning the null hypothesis* $H_0 : \pi \leq 0.52$ *is not rejected at a* 5% *level since* $0.661 < 1.645 = z_{0.05}$. *The gambler is right to be concerned. What if the* 55% *success rate had occurred for a much larger set of games (say* 360 *games instead of* 120*)? In that case, the standard error would be* $\sqrt{\frac{(0.55)(0.45)}{360}} \approx 0.0262$, *yielding a z-statistic of* $\frac{0.55-0.52}{0.0262} \approx 1.144$. *Even with this larger sample,* $H_0 : \pi \leq 0.52$ *is not rejected at a* 5% *level since* $1.144 < 1.645 = z_{0.05}$.

Similar to one-sided t-tests, the concept of p-values can be extended to one-sided z-tests. For the one-sided z-test of $H_0 : \theta \geq c$, the p-value is

$$\text{p-value} = P(Z < \text{z-statistic}) = P\left(Z < \frac{\hat{\theta}_x - c}{se(\hat{\theta}_x)}\right) \text{ when } H_0 \text{ is true, where } Z \sim N(0,1).$$

For the one-sided z-test of $H_0 : \theta \leq c$, the p-value is

$$\text{p-value} = P(Z > \text{z-statistic}) = P\left(Z > \frac{\hat{\theta}_x - c}{se(\hat{\theta}_x)}\right) \text{ when } H_0 \text{ is true, where } Z \sim N(0,1).$$

The null hypothesis H_0 is rejected if the p-value is less than the level α of the test and not rejected if the p-value is greater than the level α of the test.

Example 16.14 (Betting strategy) *Continuing example 16.13, the p-value associated with testing $H_0 : \pi \leq 0.52$, based on the 55% success rate on 120 games, is*

$$\text{p-value} = P(Z > \text{z-statistic}) = P(Z > 0.661) \approx 0.254 \text{ for } Z \sim N(0, 1).$$

```
# calculate z-statistic
zstat <- (0.55-0.52)/sqrt(0.55*(1-0.55)/120)
zstat

## [1] 0.6605783

# calculate one-sided p-value
1-pnorm(zstat)

## [1] 0.2544414
```

The null hypothesis $H_0 : \pi \leq 0.52$ is not rejected at any level below 25.4%. Example 16.13 also considered how things would change with a larger set of games (360) and the same success rate (55%), in which case the z-statistic increases to 1.144 due to the lower standard error. The p-value is $P(Z > 1.14) \approx 0.126$, considerably lower than 0.254 but still implying that $H_0 : \pi \leq 0.52$ is not rejected at any level below 12.6%. The following R code calculates the z-statistic and p-value:

```
# calculate z-statistic
zstat <- (0.55-0.52)/sqrt(0.55*(1-0.55)/360)
zstat

## [1] 1.144155

# calculate one-sided p-value
1-pnorm(zstat)

## [1] 0.1262797
```

16.3 Statistical Significance Versus Practical Significance

This section briefly discusses the important difference between **statistical significance** and **practical significance** in the context of the estimation and testing of parameter estimates. Statistical significance arises when there is statistical evidence that a parameter estimate is not equal to zero; that is, seeing a realized nonzero estimate of the parameter is not likely due to chance. On the other hand, practical significance relates to the magnitude of the estimated parameter and whether the magnitude makes the estimate practically important.

In examples 16.8 and 16.12, for instance, the estimated difference in purchase probabilities between e-mail B recipients and e-mail nonrecipients $(0.22 - 0.15 = 0.07)$ was statistically significant at any level above 0.5%, with a z-statistic of 2.80 and a p-value of 0.005. The estimated difference of 0.07 is almost certainly practically significant as well, since e-mail B recipients are much more likely (approximately 46.7% more likely) to make a purchase than e-mail nonrecipients.

In example 14.13, the difference between average scores on two exams (6.42) was statistically significant at any level, with a z-statistic of 5.22 and a p-value of 0.000. Is the magnitude of this estimated difference also practically significant? A simple way of answering that question is to compare the estimated difference to the standard deviations of the exam scores, which are $s_{exam1} = 14.19$ and $s_{exam2} = 14.15$. The estimated difference of the average scores is approximately 45% of the standard deviation on either exam, which is a practically important magnitude.

It is especially important to think about practical significance when sample sizes are large. A t-statistic or z-statistic has the standard error in its denominator, and since the standard error is proportional to $\frac{1}{\sqrt{n}}$, the test statistic becomes arbitrarily large in magnitude as n increases for any fixed value of its numerator. As a result, even if the estimate in the numerator is very small in magnitude, with little or no practical significance, it is possible to find that the estimate is statistically significant with a very low p-value. In the exam example, suppose the estimated difference in averages is 0.52 rather than 6.42 and the sample standard deviations are unchanged. With an extremely large sample size, the standard error would eventually be low enough that the p-value for testing equality of the exam score averages would be rejected at a 5% level, leading to a statistically significant estimated difference. The estimated difference would not, however, be very practically significant since it represents only about 3.7% of the standard deviation on either exam.

The idea here is that with a very large sample, we can precisely estimate parameters whose true values are close to zero from a practical point of view. The precision of the estimate can lead to a low p-value (when testing against zero) and thus statistical significance, even though the magnitude of the estimate is not practically significant. Example 16.15 illustrates this issue.

Example 16.15 (Fundraising campaign) *A nonprofit organization would like to increase the average level of giving among its* 100,000 *past donors, and a consultant has suggested a physical mail campaign to augment its usual e-mail outreach efforts. The nonprofit randomly selects* 50,000 *of its past donors as the treatment group to receive the physical mailing, and the other* 50,000 *past donors serve as the control group that receives only the e-mail outreach. The average donation received from the treatment donors is* \$32.20, *with a sample standard deviation of* \$28.90, *and the average donation received from the control donors is* \$31.80, *with a sample standard deviation of* \$29.10. *If* μ_T *and* μ_C *denote the population means of donations for the treatment and control subpopulations, respectively, a test of the null hypothesis*

$$H_0 : \mu_T = \mu_C \text{ or } H_0 : \mu_T - \mu_C = 0$$

has a z-statistic equal to

$$\frac{32.20 - 31.80}{\sqrt{\frac{28.90^2}{50000} + \frac{29.10^2}{50000}}} \approx 2.18$$

and a p-value of approximately 0.029. *Therefore, the estimated difference in average donations (*\$0.40, *or* 40 *cents) is statistically significant at a* 5% *level. But the* 40*-cent differential is likely not practically significant for the organization, as* 40 *cents is just over* 1% *of* *the average donation level and might also be offset or negated by the costs of the mail campaign.*

16.4 Hypothesis Testing for Multiple Hypotheses: The Wald Test

The previous sections of this chapter have considered testing a single null hypothesis involving an unknown parameter. In some situations, however, we may be interested in testing *multiple* hypotheses about unknown parameters. This section discusses a test, known as a **Wald test**, that tests multiple hypotheses simultaneously.[3] While many of the technical details underlying the Wald test are relegated to the appendix of this chapter, this section introduces a null hypothesis containing multiple linear restrictions, describes the intuition behind the Wald test, and discusses the use of the test's p-value to draw conclusions about the null hypothesis.

Example 16.16 (Widget website) *Using data on purchases by e-mail A recipients, e-mail B recipients, and nonrecipients, example 16.8 showed how z-tests could test whether the purchase probabilities of any two of the groups were equal to each other. Specifically, a z-test was used to separately test three null hypotheses against their alternatives:*

$$H_0 : \pi_A = \pi_B \text{ versus } H_1 : \pi_A \neq \pi_B,$$

$$H_0 : \pi_A = \pi_C \text{ versus } H_1 : \pi_A \neq \pi_C,$$

and

$$H_0 : \pi_B = \pi_C \text{ versus } H_1 : \pi_B \neq \pi_C.$$

The p-values for these three z-tests were 0.547, 0.039, *and* 0.005, *respectively. What if instead we wanted to simultaneously test the equality of all three purchase probabilities, $\pi_A = \pi_B = \pi_C$? In order for $\pi_A = \pi_B = \pi_C$ to be true, it must be the case that* both $\pi_A = \pi_B$ and $\pi_B = \pi_C$ *are true. Therefore, the null hypothesis can be written as*

$$H_0 : \pi_A = \pi_B,\ \pi_B = \pi_C,$$

where the convention is to read the comma in H_0 (between the individual hypotheses) as "and." The null hypothesis H_0 is false when either $\pi_A \neq \pi_B$ or $\pi_B \neq \pi_C$, *giving the alternative hypothesis*

$$H_1 : \pi_A \neq \pi_B \text{ or } \pi_B \neq \pi_C.$$

For H_0 to be false and H_1 to be true, it is enough to have either $\pi_A \neq \pi_B$ or $\pi_B \neq \pi_C$. For example, if $\pi_A \neq \pi_B$ and $\pi_B = \pi_C$, the null hypothesis is false. A few remarks about the formulation of the null hypothesis ($H_0 : \pi_A = \pi_B,\ \pi_B = \pi_C$) are necessary. First, it is unnecessary to also include $\pi_A = \pi_C$ in the statement of H_0 since that equality is implied by the other two. Inclusion of $\pi_A = \pi_C$ is redundant. Second, the choice of the two equalities in the statement of H_0 doesn't matter, so long as all three purchase probabilities are involved. That is, it is equally appropriate to specify the null hypothesis as $H_0 : \pi_A = \pi_C,\ \pi_B = \pi_C$ or $H_0 : \pi_A = \pi_B,\ \pi_A = \pi_C$. For any of these equivalent statements of the null hypothesis, the conclusion of the Wald test described next will be the same.[4]

To formalize the test of multiple hypotheses, suppose Q hypotheses are to be tested simultaneously. The notation $\theta_1, \theta_2, \ldots, \theta_Q$ denotes the unknown parameters to be tested against hypothesized values $c_1, c_2, \ldots, c_Q$, respectively. To keep things general, each of the θ_j parameters may itself be a linear function of multiple unknown parameters, an idea illustrated in the examples considered here. The null hypothesis of interest is

$$H_0 : \theta_1 = c_1,\ \theta_2 = c_2,\ \ldots,\ \theta_Q = c_Q.$$

This null hypothesis is said to consist of Q **linear restrictions** since each θ_j may be a linear function of parameters. The associated test of the null hypothesis H_0 tests whether *all* the Q linear restrictions are true. The alternative hypothesis, which is true when one or more of the q linear restrictions are false, is

$$H_1 : \theta_1 \neq c_1 \text{ or } \theta_2 \neq c_2 \text{ or } \ldots \text{ or } \theta_Q \neq c_Q.$$

Example 16.17 (Widget website) *For the null hypothesis*

$$H_0 : \pi_A = \pi_B,\ \pi_B = \pi_C$$

discussed in example 16.16, there are two linear restrictions ($Q = 2$). Using the notation developed here, this null hypothesis can be written as

$$H_0 : \theta_1 = 0,\ \theta_2 = 0,$$

where $\theta_1 = \pi_A - \pi_B$, $\theta_2 = \pi_B - \pi_C$, and $c_1 = c_2 = 0$. The alternative hypothesis

$$H_1 : \theta_1 \neq 0 \textit{ or } \theta_2 \neq 0$$

is true when $\pi_A \neq \pi_B$ or $\pi_B \neq \pi_C$.

Example 16.18 (Asset correlation) *Suppose an investor's portfolio currently consists of three assets: A, B, and C. The investor is considering whether to add asset D to the portfolio but wants to do so only if there is no evidence that asset D's daily returns have a correlation with the daily returns of the three other assets. Let ρ_{AD} denote the true correlation of the daily returns of asset A and asset D, and similar for ρ_{BD} and ρ_{CD}. The null hypothesis is*

$$H_0 : \rho_{AD} = 0,\ \rho_{BD} = 0,\ \rho_{CD} = 0,$$

which has three linear restrictions ($Q = 3$), with $\theta_1 = \rho_{AD}$, $\theta_2 = \rho_{BD}$, $\theta_3 = \rho_{CD}$, and $c_1 = c_2 = c_3 = 0$. The alternative hypothesis is

$$H_1 : \rho_{AD} \neq 0 \textit{ or } \rho_{BD} \neq 0 \textit{ or } \rho_{CD} \neq 0.$$

If the investor uses historical data to estimate the three correlations ($\rho_{AD}, \rho_{BD}, \rho_{CD}$) and test the null hypothesis H_0, a rejection of H_0 would provide statistical evidence that at least one of the correlations is nonzero, in which case the investor would not want to add asset D to the portfolio.

Example 16.19 (Earnings and marital status) *The labor-force data in* **cps** *contain a categorical variable (marstatus) indicating marital status, with four possible values: "Married," "Divorced," "Widowed," and "Never married." Suppose that we want to test whether* *there is any relationship between average weekly earnings (earnwk) and marital status. Put another way, are average weekly earnings (earnwk) the same or different for the four groups of workers as delineated by marital status? For notation, use M for married workers, D for divorced workers, W for widowed workers, and N for never-married workers, and let* $\mu_M, \mu_D, \mu_W, \mu_N$ *denote the population mean of weekly earnings for the four subpopulations of employed individuals. Since we want to test* $\mu_M = \mu_D = \mu_W = \mu_N$*, the null hypothesis can be written as*

$$H_0 : \mu_M = \mu_D,\ \mu_D = \mu_W,\ \mu_W = \mu_N,$$

which has $Q = 3$ *linear restrictions, with* $\theta_1 = \mu_M - \mu_D$*,* $\theta_2 = \mu_D - \mu_W$*,* $\theta_3 = \mu_W - \mu_N$*, and* $c_1 = c_2 = c_3 = 0$*. Again, as in example 16.16, there are equivalent ways of writing the null hypothesis (e.g.,* $H_0 : \mu_M = \mu_D,\ \mu_M = \mu_W,\ \mu_M = \mu_N$*). The alternative hypothesis is*

$$H_1 : \mu_M \neq \mu_D \text{ or } \mu_D \neq \mu_W \text{ or } \mu_W \neq \mu_N.$$

If the null hypothesis H_0 *is tested and rejected, there is evidence of a statistically significant difference in average weekly earnings between at least two of the four subpopulations of employed individuals.*

To conduct a Wald test of a null hypothesis with multiple linear restrictions, we require $\sqrt{n}$-consistent and asymptotically normal estimators of the Q parameters $\theta_1, \theta_2, \ldots, \theta_Q$. Let the realized estimates of those parameters be denoted as $\hat{\theta}_1, \hat{\theta}_2, \ldots, \hat{\theta}_Q$. Intuitively, when these estimates are close to the hypothesized values $c_1, c_2, \ldots, c_Q$, respectively, we are in a situation that is consistent with the null hypothesis H_0 being true. On the other hand, when one or more of the estimates is far from the hypothesized values $c_1, c_2, \ldots, c_Q$, respectively, we are in a situation that is consistent with the alternative hypothesis H_1 being true. The z-test discussed in section 16.2 can be used to test whether any individual $\hat{\theta}_j$ is close to an individual c_j, but it is only useful for testing a single linear restriction rather than multiple linear restrictions simultaneously.

To provide more intuition, consider the simplest case of $Q = 2$, where the null hypothesis is

$$H_0 : \theta_1 = c_1,\ \theta_2 = c_2,$$

and the alternative hypothesis is

$$H_1 : \theta_1 \neq c_1 \text{ or } \theta_2 \neq c_2.$$

The estimates of θ_1 and θ_2 are $\hat{\theta}_1$ and $\hat{\theta}_2$, respectively. The differences $\hat{\theta}_1 - c_1$ and $\hat{\theta}_2 - c_2$ form the basis of the Wald test, and the conclusion of the test is based on how far each of these differences is from zero. If $\hat{\theta}_1 - c_1$ and $\hat{\theta}_2 - c_2$ are both very close to zero, in statistical terms, that situation would be consistent with the null hypothesis H_0 being true. If the null hypothesis H_0 is true, meaning that $\theta_1 = c_1$ and $\theta_2 = c_2$, then both $\hat{\theta}_1 - c_1$ and $\hat{\theta}_2 - c_2$ should be realizations of draws from normal distributions that are centered at zero. For the z-test, recall that the magnitude of the z-statistic, $\left|\frac{\hat{\theta}_x - c}{se(\hat{\theta}_x)}\right|$, is a statistical distance between $\hat{\theta}_x$ and c, obtained by dividing the actual distance $|\hat{\theta}_x - c|$ by the standard error $se(\hat{\theta}_x)$. Unlike this z-test, there are now two distances, $|\hat{\theta}_1 - c_1|$ and $|\hat{\theta}_2 - c_2|$, to consider with $Q = 2$ hypotheses being tested.

The **Wald statistic** generalizes the notion of a statistical distance to handle both higher dimensions (two for the $Q = 2$ case) and possible correlation between the two estimators of θ_1 and θ_2. As the formula for the Wald statistic requires more advanced mathematics for the general case, a more complete description of the Wald statistic is left for the appendix. Instead, for the remaining discussion in this section, we assume that *R* is able to calculate the Wald statistic. In the case of $Q = 2$, the realized Wald statistic is a draw from a χ^2_2 distribution (a chi-square distribution with 2 degrees of freedom) if the null hypothesis H_0 is true. The Wald statistic is always nonnegative, which is intuitive since it is a statistical distance measure. Values of the Wald statistic close to zero are consistent with the null hypothesis H_0 since such values arise when $\hat{\theta}_1$ is close to c_1 and $\hat{\theta}_2$ is close to c_2. On the other hand, large positive values of the Wald statistic are consistent with the alternative hypothesis since such values arise when $\hat{\theta}_1$ is far from c_1 and/or $\hat{\theta}_2$ is far from c_2. As a result, the resulting **Wald test** is a one-sided test, where for $Q = 2$, rejection occurs only in the right tail of the χ^2_2 distribution. For instance, for a test at a 5% level, we reject H_0 if the Wald statistic is greater than the 95% quantile of the χ^2_2 distribution, which is approximately 5.991. For a test at a 10% level, we reject H_0 if the Wald statistic is greater than the 90% quantile of the χ^2_2 distribution, which is approximately 4.605. These critical values are calculated in *R* with the `qchisq` function:

```
qchisq(0.95,2)

## [1] 5.991465

qchisq(0.90,2)

## [1] 4.60517
```

The Wald statistic generalizes to additional linear restrictions. For higher Q, there are more estimators to be considered, and therefore more distances (between realized estimates and hypothesized values) to be taken into account by the Wald statistic. The biggest difference for higher Q is the sampling distribution of the Wald statistic when the null hypothesis is true. Specifically, the realized Wald statistic is a draw from a χ^2_Q distribution (a chi-square distribution with Q degrees of freedom) when the null hypothesis $H_0: \theta_1 = c_1, \theta_2 = c_2, \ldots, \theta_Q = c_Q$ is true. Intuitively, as more distances are added to the Wald statistic, the overall statistical distance measure is expected to increase, which corresponds to the thicker right tails of the chi-square distribution as the degrees of freedom increase.

If a statistical package can conduct a Wald test, it usually provides the value of the Wald statistic and/or the p-value for the test. As with *t*-tests and *z*-tests, the p-value is most useful, as it immediately tells us whether the null hypothesis H_0 would be rejected at any level. In the interest of completeness, however, we also describe how to conduct the test based on the Wald statistic itself. First, notation for the critical value of a chi-square distribution, as in definition 16.5, is required.

Definition 16.5 *The* ***critical value*** $w_{Q,q}$ *denotes the* $(1 - q)$ *quantile of the* χ^2_Q *distribution. For example,* $w_{2,0.05}$ *is the* 95% *quantile of the* χ^2_2 *distribution, and* $w_{2,0.10}$ *is the* 90% *quantile of the* χ^2_2 *distribution.*

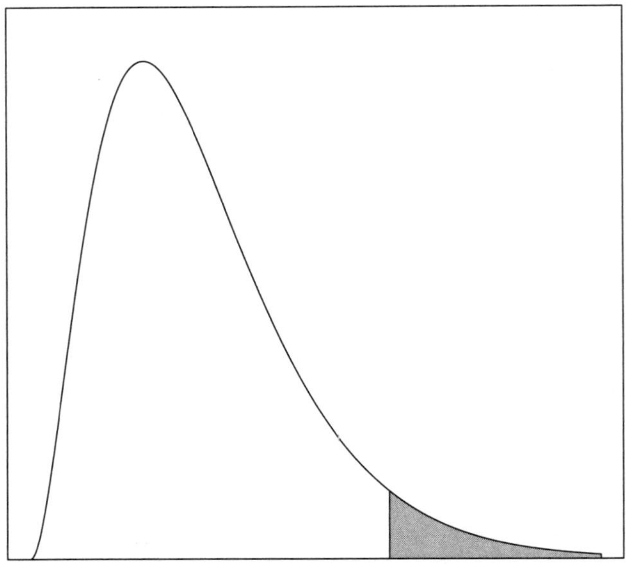

Figure 16.6
Rejection area for the Wald test at a 5% level

Proposition 16.3 states the sampling distribution of the Wald statistic when the null hypothesis is true.

Proposition 16.3 *The Wald statistic associated with testing the null hypothesis,*

$$H_0 : \theta_1 = c_1,\ \theta_2 = c_2,\ \ldots,\ \theta_Q = c_Q,$$

is distributed as a χ^2_Q *random variable if* H_0 *is true. The probability that the Wald statistic is greater than the critical value* $w_{Q,\alpha}$ *equals* α *if* H_0 *is true.*

For Q hypotheses being tested, given the χ^2_Q sampling distribution of the Wald statistic when H_0 is true, the following rejection rule based on critical values can be used:

Rejection rule for the Wald test based on the Wald statistic (test at the α level):

- Reject $H_0 : \theta_1 = c_1,\ \theta_2 = c_2,\ \ldots,\ \theta_Q = c_Q$ at the α level if the Wald statistic $\geq w_{Q,\alpha}$.
- Do not reject $H_0 : \theta_1 = c_1,\ \theta_2 = c_2,\ \ldots,\ \theta_Q = c_Q$ at the α level if the Wald statistic $< w_{Q,\alpha}$.

Figure 16.6 provides a graphical view of the rejection area for a test at the 5% level, where the critical value is $w_{Q,0.05}$. The gray region, corresponding to any Wald statistic above the critical value $w_{Q,0.05}$, indicates the area where H_0 is rejected. Therefore, the probability of rejecting H_0 when H_0 is true is equal to 5%, which is the level of the test.

Alternatively, if the p-value for the Wald test is available, this p-value can be directly used to test the null hypothesis. For the Wald test, the p-value is the probability that a χ^2_Q

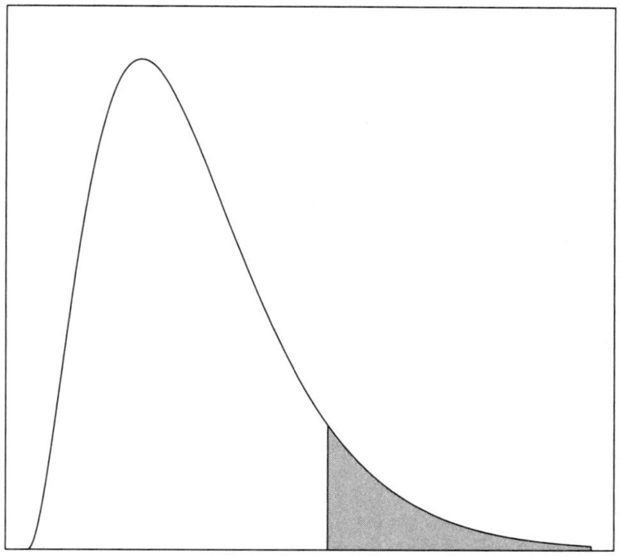

Figure 16.7
The p-value for a Wald test

random variable is greater than the Wald statistic. If the null hypothesis H_0 is true, the p-value provides the probability that a realized Wald statistic is at least as large as the one calculated for the observed sample. With the p-value, the rejection rule should look familiar:

Rejection rule for the Wald test based on the p-value (test at the α level):

- Reject $H_0 : \theta_1 = c_1,\ \theta_2 = c_2,\ \ldots,\ \theta_Q = c_Q$ at the α level if the p-value $< \alpha$.
- Do not reject $H_0 : \theta_1 = c_1,\ \theta_2 = c_2,\ \ldots,\ \theta_Q = c_Q$ at the α level if the p-value $> \alpha$.

Figure 16.7 shows how the p-value relates to the Wald statistic and the χ^2_Q distribution. For a given Wald statistic, labeled "Wald stat" in the graph, the p-value is the area of the gray region to the right of the Wald statistic under the χ^2_Q distribution.

We now revisit two of the examples from the beginning of this section.

Example 16.20 (Widget website) *Continuing example 16.17, the null hypothesis*

$$H_0 : \pi_A = \pi_B,\ \pi_B = \pi_C$$

or, equivalently,

$$H_0 : \theta_1 = 0,\ \theta_2 = 0,$$

where $\theta_1 = \pi_A - \pi_B$ and $\theta_2 = \pi_B - \pi_C$, corresponds to the purchase probabilities being the same for the three groups (e-mail A recipients, e-mail B recipients, and nonrecipients). There are $Q = 2$ hypotheses being tested. The Wald statistic associated with H_0 turns out to be 11.17. *The p-value is calculated in R with the* `pchisq` *function:*

```
1-pchisq(11.17,2)

## [1] 0.00375375
```

The p-value is approximately 0.004*, meaning that the null hypothesis is rejected at any level above* 0.4% *and providing strong statistical evidence that the three purchase probabilities are not all equal to each other. By itself, the Wald test does not tell which of the tested equalities is causing the rejection, although the previous evidence from the z-tests in this example suggests that the rejection is being driven by the fact that the estimate of* π_C *(*$p_C = 0.15$*) is much lower than the estimates of* π_A *(*$p_A = 0.20$*) and* π_B *(*$p_B = 0.22$*).*

Example 16.21 (Earnings and marital status) *Continuing example 16.19, where* μ_M, μ_D, μ_W, μ_N *denoted the population mean of weekly earnings for the four subpopulations based on marital status ("Married," "Divorced," "Widowed," and "Never married," respectively), the null hypothesis that these population means are equal to each other is*

$$H_0 : \mu_M = \mu_D,\ \mu_D = \mu_W,\ \mu_W = \mu_N.$$

$Q = 3$ *hypotheses are being tested. The sample means of weekly earnings for the four groups are*

$$\bar{x}_M = 1047,\ \bar{x}_D = 902,\ \bar{x}_W = 661,\ \textit{and}\ \bar{x}_N = 820.$$

The Wald statistic associated with H_0 *turns out to be* 80.90*, which has a p-value of* 0.000*. This p-value is calculated in R, as follows:*

```
1-pchisq(81.59,3)

## [1] 0
```

Thus, the null hypothesis is rejected at any level, providing strong statistical evidence against the population means being the same for the four groups. We might wonder if this result is being driven by the much lower sample average of weekly earnings observed for the "Widowed" group ($\bar{x}_W = 661$*). To ignore the "Widowed" group, we could test the null hypothesis,*

$$H_0 : \mu_M = \mu_D,\ \mu_D = \mu_N,$$

which has $Q = 2$ *hypotheses. For this null hypothesis, the Wald statistic is* 49.31*, still with a p-value of* 0.000*, indicating again that there is strong statistical evidence against the population means being the same for the three remaining groups.*

What happens for the Wald test when there is only a single restriction ($Q = 1$)? In this case, the null hypothesis is $H_0 : \theta_1 = c_1$, which can be tested with a z-test. Would we get a different answer using a Wald test? Thankfully, the answer is no. Whether the test of H_0 is conducted using a z-test or a Wald test, the p-value for the test will be numerically identical, meaning that the rejection conclusion is also the same for the two tests. This equivalence arises since, in the $Q = 1$ case, the Wald statistic is exactly equal to the square of the z-statistic, and the critical value $w_{1,\alpha}$ is exactly equal to the square of the critical value $z_{\alpha/2}$. The latter fact follows from a χ_1^2 random variable being equal to the square of a $Z \sim N(0, 1)$ random variable.

Appendix: Details of the Wald Test

This appendix provides more technical details of the Wald test, including the calculation of the Wald statistic and the associated p-value. Some additional notation is required. We let L denote the number of original parameters being estimated and Q denote the number of linear restrictions being tested. For example, for testing whether three Bernoulli parameters (π_A, π_B, π_C) are equal ($H_0: \pi_A = \pi_B = \pi_C$), we have $L = 3$ parameters and $Q = 2$ linear restrictions ($\pi_A = \pi_B$ and $\pi_B = \pi_C$).

We collect the L original parameters into a *vector* denoted as $\boldsymbol{\gamma}$, with elements $\gamma_1, \gamma_2, \ldots, \gamma_L$, and similarly, we collect the L estimates of these parameters into a vector denoted as $\hat{\boldsymbol{\gamma}}$, with elements $\hat{\gamma}_1, \hat{\gamma}_2, \ldots, \hat{\gamma}_L$:

$$\boldsymbol{\gamma} = \begin{bmatrix} \gamma_1 \\ \gamma_2 \\ \vdots \\ \vdots \\ \gamma_L \end{bmatrix} \text{ and } \hat{\boldsymbol{\gamma}} = \begin{bmatrix} \hat{\gamma}_1 \\ \hat{\gamma}_2 \\ \vdots \\ \vdots \\ \hat{\gamma}_L \end{bmatrix}.$$

We assume that each component of $\hat{\boldsymbol{\gamma}}$ is an asymptotically normal estimate of the corresponding component in $\boldsymbol{\gamma}$. That is, $\hat{\gamma}_1$ is an asymptotically normal estimate of γ_1, $\hat{\gamma}_2$ is an asymptotically normal estimate of γ_2, and so on.

These estimates will each have an estimated asymptotic variance, but it is also possible that there will be nonzero covariances between two or more of the estimates (e.g., if the estimates are based on the same sample). To allow for this possibility, we introduce the estimated asymptotic variance *matrix* with L rows and L columns, denoted as $\hat{V}$:

$$\hat{V} = \begin{bmatrix} \hat{v}_{11} & \hat{v}_{12} & \cdots & \cdots & \hat{v}_{1L} \\ \hat{v}_{21} & \hat{v}_{22} & \cdots & \cdots & \hat{v}_{2L} \\ \vdots & \vdots & \ddots & & \vdots \\ \vdots & \vdots & & \ddots & \vdots \\ \hat{v}_{L1} & \hat{v}_{L2} & \cdots & \cdots & \hat{v}_{LL} \end{bmatrix}.$$

The diagonal elements are the estimated asymptotic variances of the estimates $\hat{\gamma}_1, \hat{\gamma}_2, \ldots, \hat{\gamma}_L$. For example, $\hat{v}_{22}$ is the estimated asymptotic variance of $\hat{\gamma}_2$. The off-diagonal elements are the estimated asymptotic covariances between two estimates. For example, $\hat{v}_{12}$ is the estimated asymptotic covariance between $\hat{\gamma}_1$ and $\hat{\gamma}_2$. (The true covariance will be zero if the two underlying estimators are independent.)

To represent the Q linear restrictions being tested, we use a matrix denoted as $\mathbf{R}$ (with Q rows and L columns) and a vector denoted as $\mathbf{c}$ (with Q rows):

$$\mathbf{R} = \begin{bmatrix} R_{11} & R_{12} & \cdots & \cdots & R_{1L} \\ R_{21} & R_{22} & \cdots & \cdots & R_{2L} \\ \vdots & \vdots & & & \vdots \\ \vdots & \vdots & & & \vdots \\ R_{Q1} & R_{Q2} & \cdots & \cdots & R_{QL} \end{bmatrix} \text{ and } \mathbf{c} = \begin{bmatrix} c_1 \\ c_2 \\ \vdots \\ \vdots \\ c_Q \end{bmatrix}.$$

The elements of **R** and **c** are all constants associated with the null hypothesis being tested. Each row of **R** and **c** corresponds to one of the Q linear restrictions being tested. For example, the first row represents the restriction

$$R_{11}\gamma_1 + R_{12}\gamma_2 + \cdots + R_{1L}\gamma_L = c_1,$$

the second row represents the restriction

$$R_{21}\gamma_1 + R_{22}\gamma_2 + \cdots + R_{2L}\gamma_L = c_2,$$

and so on. The notation in section 16.4 had $\theta_1, \theta_2, \ldots, \theta_Q$ representing the linear functions of underlying parameters, such that the restrictions in the null hypothesis H_0 are

$$\theta_1 = R_{11}\gamma_1 + R_{12}\gamma_2 + \cdots + R_{1L}\gamma_L = c_1$$
$$\theta_2 = R_{21}\gamma_1 + R_{22}\gamma_2 + \cdots + R_{2L}\gamma_L = c_2$$
$$\vdots$$
$$\theta_Q = R_{Q1}\gamma_1 + R_{Q2}\gamma_2 + \cdots + R_{QL}\gamma_L = c_Q.$$

To develop a better understanding of how matrix **R** and vector **c** are specified in practice, let's consider some examples:

- **Testing the equality of three Bernoulli parameters**: For the Bernoulli parameters (π_A, π_B, π_C), we let $\gamma_1 = \pi_A$, $\gamma_2 = \pi_B$, and $\gamma_3 = \pi_C$ using the notation from above. The null hypothesis $H_0 : \gamma_1 = \gamma_2 = \gamma_3$ has two linear restrictions, $\gamma_1 = \gamma_2$ and $\gamma_2 = \gamma_3$, which can be rewritten as $\gamma_1 - \gamma_2 = 0$ and $\gamma_2 - \gamma_3 = 0$. In this case, the number of parameters is $L = 3$ and the number of restrictions is $Q = 2$. The appropriate **R** and **c** are

$$\mathbf{R} = \begin{bmatrix} 1 & -1 & 0 \\ 0 & 1 & -1 \end{bmatrix} \text{ and } \mathbf{c} = \begin{bmatrix} 0 \\ 0 \end{bmatrix},$$

  with the first row corresponding to $1 \cdot \gamma_1 + (-1) \cdot \gamma_2 + 0 \cdot \gamma_3 = 0$, which is $\gamma_1 - \gamma_2 = 0$, and the second row corresponding to $0 \cdot \gamma_1 + 1 \cdot \gamma_2 + (-1) \cdot \gamma_3 = 0$, which is $\gamma_2 - \gamma_3 = 0$.

- **Testing the equality of four population averages**: For the population averages (μ_A, μ_B, μ_C, μ_D), we let $\gamma_1 = \mu_A$, $\gamma_2 = \mu_B$, $\gamma_3 = \mu_C$, and $\gamma_4 = \mu_D$. The null hypothesis $H_0 : \mu_A = \mu_B = \mu_C = \mu_D$ has three linear restrictions, given by $\gamma_1 = \gamma_2$, $\gamma_2 = \gamma_3$, and $\gamma_3 = \gamma_4$. We have $L = 4$ parameters and $Q = 3$ restrictions. Then, **R** and **c** are

$$\mathbf{R} = \begin{bmatrix} 1 & -1 & 0 & 0 \\ 0 & 1 & -1 & 0 \\ 0 & 0 & 1 & -1 \end{bmatrix} \text{ and } \mathbf{c} = \begin{bmatrix} 0 \\ 0 \\ 0 \end{bmatrix}.$$

- **Testing whether all parameters equal zero**: In some situations, it may be of interest to test whether a set of parameters all equal zero. For instance, this test is often used in the context of multiple regression models (chapter 18). If we have L parameters $\gamma_1, \gamma_2, \ldots, \gamma_L$ and want to test

$$H_0 : \gamma_1 = \gamma_2 = \cdots = \gamma_L = 0,$$

there are $Q = L$ linear restrictions ($\gamma_1 = 0, \gamma_2 = 0, \ldots, \gamma_L = 0$). Then, $\mathbf{R}$ and $\mathbf{c}$ are

$$\mathbf{R} = \begin{bmatrix} 1 & 0 & \cdots & \cdots & 0 \\ 0 & 1 & \cdots & \cdots & 0 \\ \vdots & \vdots & \ddots & & \vdots \\ \vdots & \vdots & & \ddots & \vdots \\ 0 & 0 & \cdots & \cdots & 1 \end{bmatrix} \text{ and } \mathbf{c} = \begin{bmatrix} 0 \\ 0 \\ \vdots \\ \vdots \\ 0 \end{bmatrix}.$$

In this case, $\mathbf{R}$ is a matrix with 1s along the diagonal and 0s everywhere else.

- **Other tests**: The framework using $\mathbf{R}$ and $\mathbf{c}$ is quite general. Suppose we have $L = 5$ parameters and want to jointly test the following three restrictions:

$$H_0 : \gamma_1 + \gamma_2 = 4, \gamma_3 = 2\gamma_4, \gamma_5 = 10.$$

This null hypothesis says that the first two parameters sum to 4, the third parameter is two times the size of the fourth parameter, and the fifth parameter equals 10. The corresponding $\mathbf{R}$ and $\mathbf{c}$ are

$$\mathbf{R} = \begin{bmatrix} 1 & 1 & 0 & 0 & 0 \\ 0 & 0 & 1 & -2 & 0 \\ 0 & 0 & 0 & 0 & 1 \end{bmatrix} \text{ and } \mathbf{c} = \begin{bmatrix} 4 \\ 0 \\ 10 \end{bmatrix}.$$

The first row has the restriction $\gamma_1 + \gamma_2 = 4$, the second row has the restriction $\gamma_3 - 2\gamma_4 = 0$, and the third row has the restriction $\gamma_5 = 10$.

Since *R* does not have a built-in function for general Wald tests, we introduce a user-defined function `wald_test` here. The function `wald_test` takes four arguments: the estimate vector `gamma_hat` ($\hat{\boldsymbol{\gamma}}$), the asymptotic variance matrix `var_gamma_hat` ($\hat{\mathbf{V}}$), the restriction matrix `R` ($\mathbf{R}$), and the constant vector `c` ($\mathbf{c}$). The restriction matrix `R` and the constant vector `c` are optional arguments, with default values equal to the identity matrix (1s along the diagonal and 0s for all nondiagonal elements) and the zero vector, respectively; these default values correspond to a null hypothesis whose restrictions are that each element of $\boldsymbol{\gamma}$ equals zero. The function `wald_test` returns a list containing two elements, the Wald statistic (`W`) and its associated p-value (`p_value`). The p-value is calculated by determining the area to the right of the Wald statistic for the χ^2_Q distribution.

Here is the R code that defines the `wald_test` function:

```
wald_test <- function(gamma_hat, var_gamma_hat, R=diag(length(gamma_hat)),
            c=rep(0,length(gamma_hat))) {
    # gamma_hat: L x 1 vector of parameter estimates
    # var_theta_hat: L x L variance-covariance matrix of the parameter estimates
    # R: Q x L matrix of linear constraints to be tested
      ## default value of R is the identity matrix (for testing parameters equal
      ## to zero)
    # c: Q x 1 vector of values of the linear constraints to be tested
      ## default value of c is a vector of zeros (for testing parameters equal
      ## to zero)

    # when R has one row (one restriction), make sure R has matrix type
    if (!is.matrix(R)) {
```

```
        R <- t(as.matrix(R))
    }

    # calculate the Wald statistic using linear algebra
    W <- t(R %*% gamma_hat - c) %*%
    solve(R %*% var_gamma_hat %*% t(R)) %*% (R %*% gamma_hat - c)
    W <- as.numeric(W)

    # calculate the p-value of the Wald test
    p_value <- 1 - pchisq(W, nrow(R))

    # return the Wald test statistic and p-value
    return(list(W = W, p_value = p_value))
}
```

We define additional *R* functions to handle the estimation of the asymptotic variance matrix $\hat{V}$ for some of the more common examples of the Wald test.

Sample proportions of independent samples: The function `var_prop_indep` estimates the asymptotic variance matrix for a vector of sample proportions estimated on different independent samples. Its arguments are `pi_hat`, the vector of sample proportions, and `nobs`, the vector of underlying sample sizes for each sample:

```
var_prop_indep <- function(pi_hat, nobs) {
    # pi_hat: vector of sample proportions
    # nobs: vector of sample sizes

    # calculate the variance-covariance matrix
    var_pi_hat <- diag(pi_hat*(1-pi_hat)/nobs)

    return(var_pi_hat)
}
```

Example 16.22 (Widget website) *We revisit example 16.20 to detail how the Wald statistic and p-value are calculated for the null hypothesis $H_0 : \pi_A = \pi_B = \pi_C$. Here is the R code that uses the functions* `var_prop_indep` *and* `wald_test`*:*

```
# create a vector of the pi-hat estimates
gamma_hat <- c(0.20,0.22,0.15)

# calculate the estimated asymptotic variance matrix
var_hat <- var_prop_indep(gamma_hat, c(300,300,2400))

# set up the linear restrictions with R and c
R <- rbind(c(1,-1,0),c(0,1,-1))
c <- c(0,0)

# conduct the Wald test
wald_test(gamma_hat, var_hat, R, c)

## $W
## [1] 11.17296
##
## $p_value
## [1] 0.003748198
```

First, we store the estimated purchase probabilities ($\hat{\gamma}_1 = \hat{\pi}_A = 0.20$, $\hat{\gamma}_2 = \hat{\pi}_B = 0.22$, $\hat{\gamma}_3 = \hat{\pi}_C = 0.15$) *in the vector* `gamma_hat`*. Then, we estimate the asymptotic covariance matrix using* `var_prop_indep` *with the arguments* `gamma_hat` *(the Bernoulli parameter estimates) and the vector of sample sizes,* `c(300,300,2400)`*. Matrix* **R** *and vector* **c** *are defined to correspond to the linear restrictions* $\gamma_1 - \gamma_2 = 0$ *and* $\gamma_2 - \gamma_3 = 0$*. We use the* `rbind` *function to construct the* **R** *matrix. The* `rbind` *function takes vectors as arguments, where each vector corresponds to one row, and stacks the vectors as rows in a matrix. The command* `R <- rbind(c(1,-1,0),c(0,1,-1))` *stores the matrix*

$$\begin{bmatrix} 1 & -1 & 0 \\ 0 & 1 & -1 \end{bmatrix}$$

in variable R*. Finally, the function* `wald_test` *is called and outputs both the Wald statistic and the p-value, as seen previously in example 16.20.*

Sample means of independent samples: The function `var_mean_indep` estimates the asymptotic variance matrix for sample means estimated on different independent samples. Its argument is `x_vectors`, a *list* of the sample vectors (i.e., the actual observations in each sample). Each of the sample vectors within the list `x_vectors` may have different lengths:

```
var_mean_indep <- function(x_vectors) {
    # x_vectors: list of sample vectors

    # initialize variables
    num_means <- length(x_vectors)
    tempvec <- rep(0, num_means)

    # calculate the variance-covariance matrix
    for (i in 1:num_means) {
        tempvec[i] <- var(x_vectors[[i]])/length(x_vectors[[i]])
    }
    var_mean <- diag(tempvec)

    return(var_mean)
}
```

What is a list? A list in *R* is a collection of data objects, which may be of different data types. The list `x_vectors` is a collection of the sample vectors for each of the independent samples. For example, in the case of two samples, given by `sample1` and `sample2`, the function call would be `var_mean_indep(list(sample1,sample2))`, where function `list` combines its arguments into a list object.

Example 16.23 (Earnings and marital status) *We revisit example 16.21 to detail how the Wald statistic and p-value are caclulated for the null hypothesis* $H_0 : \mu_M = \mu_D = \mu_W = \mu_N$*, corresponding to the population mean of weekly earnings being the same for the four subpopulations based on marital status ("Married," "Divorced," "Widowed," and "Never Married," respectively). In this case, the number of parameters is* $L = 4$*, and the number of linear restrictions in* H_0 *is* $Q = 3$*. Here is the R code to conduct the Wald test:*

```
# construct the four subsamples based on marital status
sample_m <- cpsemployed[cpsemployed$marstatus=="Married",]$earnwk
sample_d <- cpsemployed[cpsemployed$marstatus=="Divorced",]$earnwk
sample_w <- cpsemployed[cpsemployed$marstatus=="Widowed",]$earnwk
sample_n <- cpsemployed[cpsemployed$marstatus=="Never married",]$earnwk

# create a vector of the sample mean estimates
gamma_hat <- c(mean(sample_m),mean(sample_d),mean(sample_w),mean(sample_n))

# calculate the estimated asymptotic variance matrix
var_hat <- var_mean_indep(list(sample_m,sample_d,sample_w,sample_n))

# set up the linear restrictions with R and c
R <- rbind(c(1,-1,0,0),c(0,1,-1,0),c(0,0,1,-1))
c <- c(0,0,0)

# conduct the Wald test
wald_test(gamma_hat, var_hat, R, c)

## $W
## [1] 80.90417
##
## $p_value
## [1] 0
```

Example 16.21 also considered dropping the "Widowed" subsample and testing H_0 : $\mu_M = \mu_D = \mu_N$. In this case, the number of parameters is $L=3$, and the number of linear restrictions in H_0 is $Q=2$. Here is the R code to conduct the Wald test:

```
# create a vector of the sample mean estimates
gamma_hat <- c(mean(sample_m),mean(sample_d),mean(sample_n))

# calculate the estimated asymptotic variance matrix
var_hat <- var_mean_indep(list(sample_m,sample_d,sample_n))

# set up the linear restrictions with R and c
R <- rbind(c(1,-1,0),c(0,1,-1))
c <- c(0,0)

# conduct the Wald test
wald_test(gamma_hat, var_hat, R, c)

## $W
## [1] 49.30529
##
## $p_value
## [1] 0.00000000001965583
```

Sample means from the same sample: Function `var_mean_onesample` estimates the asymptotic variance matrix for the sample means of several variables from the same sample. Its arguments are `df`, the data frame containing the sample, and `vars`, a vector of either variable names or indices that identifies which elements of the data frame `df` to consider:

```
var_mean_onesample <- function(df, vars) {
    # df: data frame containing the sample
    # vars: vector of variable names or indices

    # calculate the variance-covariance matrix
```

```
    var_mean <- cov(df[,vars])/nrow(df)

    return(var_mean)
}
```

Example 16.24 (Exam score data) *Example 16.10 implemented a z-test to test the equality of μ_{exam1} and μ_{exam2}, the population means of the variables exam1 and exam2 in the* ***exams*** *data set. There are $L=2$ parameters, and the z-test is a special case of the Wald test with $Q=1$ linear restriction. The null hypothesis is $H_0 : \mu_{exam1} - \mu_{exam2} = 0$. Here is the R code to conduct the Wald test:*

```
# create a vector of the sample mean estimates
gamma_hat <- c(mean(exams$exam1),mean(exams$exam2))

# calculate the estimated asymptotic variance matrix
var_hat <- var_mean_onesample(exams,c("exam1","exam2"))

# set up the linear restrictions with R and c
R <- c(1,-1)
c <- 0

# conduct the Wald test
wald_test(gamma_hat, var_hat, R, c)

## $W
## [1] 27.13471
##
## $p_value
## [1] 0.0000001897602
```

The p-value from this Wald test is the same as the p-value from the z-test in example 16.10, and the Wald statistic (approximately 27.13) is equal to the z-statistic (approximately 5.22) squared. If there were another exam within the data set, such as exam3, we could test the equality of all three population means, $H_0 : \mu_{exam1} = \mu_{exam2} = \mu_{exam3}$, by modifying the R code to account for the extra parameter ($L=3$) and the extra restriction ($Q=2$).

Notes

1. The use of the word "null" in "null hypothesis" stems from the fact that many tests of interest consider $c=0$, so $H_0 : \theta = 0$. If θ is a parameter that measures some sort of effect, which is common in regression models, the null hypothesis H_0 corresponds to no effect, or a "null" effect.
2. An alternative approach is to use a simple linear regression of one variable on the other variable, which is covered in chapter 17. For the simple linear regression, the z-test uses the slope estimate and its standard error.
3. The test is named after Abraham Wald, who first introduced the idea (Wald 1943).
4. More formally, the p-values associated with the tests of these three null hypotheses are numerically identical.

Exercises

1. You are considering investing in a company called Tech Trove. The company has been listed on a stock exchange for 10 years, with the following annual returns:

$$0.04, -0.10, 0.17, 0.02, -0.19, -0.08, -0.01, 0.09, 0.22, \text{ and } 0.06.$$

Assume that the annual returns are i.i.d. draws from a normal distribution with unknown mean μ and unknown variance σ^2.

a. What is the 95% confidence interval for μ?

b. What is the t-statistic for the test of the null hypothesis $H_0 : \mu = 0$? Do you reject H_0 at a 5% level?

c. What is the p-value for the test of the null hypothesis $H_0 : \mu = 0$?

d. Without doing any additional calculations, how would the t-statistic for testing $H_0 : \mu = -0.01$ compare to the t-statistic found in (b)?

e. Without doing any additional calculations, how would the p-value for testing $H_0 : \mu = -0.01$ compare to the p-value found in (c)?

2. A random sample of 12 undergraduates enrolled in an introductory economics course are asked to predict the annual income for their first job after graduation. The sample average of the responses is \$49,000, and the sample standard deviation of the responses is \$8,000. Assume that the individual responses are drawn independently from a normal distribution.

a. If μ_X denotes the population average of the predicted annual income, what is the t-statistic associated with testing $H_0 : \mu_X = 50{,}000$?

b. Do you reject $H_0 : \mu_X = 50{,}000$ at a 5% level?

c. What is the p-value for the t-test of $H_0 : \mu_X = 50{,}000$?

d. If the same sample average and sample standard deviation were observed for a sample size of $n = 1{,}200$ (instead of $n = 12$), would you reject $H_0 : \mu_X = 50{,}000$ at a 5% level?

e. For (d), is it necessary to assume that the individual responses are drawn from a normal distribution?

3. The average grade among fourth graders on a math aptitude test in a certain state is 62.5 out of 100 points. (Assume that 62.5 is the population average.) An educational company has developed an app for elementary-school math and would like to test whether the app improves average performance on the aptitude test among fourth graders. Let μ_X denote the population mean of test scores among students who use the app.

a. To provide statistical evidence that the app *increases* the true average of test scores, what are the appropriate one-sided null hypothesis and alternative hypothesis?

b. The company provides the app to 15 randomly selected students. The sample mean and sample standard deviation of test scores are 68.3 and 13.5, respectively. Under the assumption that the test scores are i.i.d. and normally distributed, would the t-test of the null hypothesis in (a) be rejected at the 5% level? At the 10% level?

c. What is the p-value associated with the t-test in (b)?

4. Suppose 200 students are chosen at random to do a blind taste test of green M&Ms versus red M&Ms. Of the 200 students, 112 students prefer green M&Ms and 88 prefer red M&Ms. Let π denote the population probability that a randomly selected student prefers green M&Ms.

a. What is the asymptotic 95% confidence interval for π?

b. State the null hypothesis that corresponds to green M&Ms and red M&Ms being equally preferred in the population.

c. What is the p-value associated with the test of the null hypothesis in (b)?

d. Suppose another 200 students are chosen to do the blind taste test, and once again(!) 112 prefer green M&Ms and 88 prefer red M&Ms. How do the following quantities for the 400-student sample compare to the associated quantities for the initial 200-student sample?

   i. Middle of the asymptotic 95% confidence interval for π

   ii. Width of the asymptotic 95% confidence interval for π

   iii. The p-value for testing the null hypothesis that green M&Ms and red M&Ms are equally preferred in the population

5. Use the **metricsgrades** data set for this question. These data are from a graduate econometrics course with 68 students.

   a. Provide an asymptotic 95% confidence interval for the difference in the population means of *total* (the composite course score) for the subpopulations of domestic ($domestic = 1$) and international ($domestic = 0$) students.

   b. What is the p-value for the z-test of the null hypothesis that the population means of *total* (composite course score) for the subpopulations of domestic ($domestic = 1$) and international ($domestic = 0$) students are the same?

   c. Define an indicator variable *hiscore* that equals 1 if $total > 80$ and 0 otherwise. What is the p-value for the z-test of the null hypothesis that $P(hiscore = 1|domestic = 1) = P(hiscore = 1|domestic = 0)$?

6. *A classmate gives you a coin to toss and claims that it is fair, but you have your suspicions. Let π be the probability of heads with this coin.

   a. If you toss the coin 100 times and get 58 heads, what is the p-value for testing $H_0 : \pi = 0.50$?

   b. Suppose the true heads probability is $\pi = 0.51$, so

$$\frac{\bar{X} - 0.51}{\sqrt{\frac{(0.51)(0.49)}{n}}} \sim N(0, 1).$$

   The z-statistic for testing $H_0 : \pi = 0.50$, before observing outcomes of the coin tosses, can be written as

$$\frac{\bar{X} - 0.50}{\sqrt{\frac{(0.51)(0.49)}{n}}} = \frac{\bar{X} - 0.51}{\sqrt{\frac{(0.51)(0.49)}{n}}} + \frac{0.51 - 0.50}{\sqrt{\frac{(0.51)(0.49)}{n}}},$$

   where the first term is a $N(0, 1)$ random variable and the second term is a constant (depending on n). Given this expression, how many tosses n^* would be required for there to be a 99% probability that the z-statistic is greater than 1.96?

c. For the n^* found in (b), and still assuming $\pi = 0.51$, conduct 100,000 simulations in R to determine the probability that the z-statistic is greater than 1.96. For this part, use the usual $se(\bar{x}) = \sqrt{\frac{\bar{x}(1-\bar{x})}{n}}$ formula rather than the true $\sqrt{\frac{(0.51)(0.49)}{n}}$ formula.

d. Repeat (b) for different values of the true heads probability, with $\pi = \{0.51, 0.52, \ldots, 0.59, 0.60\}$ being the possibilities. Plot a graph of required sample size n^* against probability π.

7. You have an estimate $\hat{\theta}_x = 1.2$ of an unknown parameter θ based on a sample size of $n = 100$. You calculate that the p-value associated with a z-test of $H_0 : \theta = 1$ is p^*. How do the following quantities compare with p^*?

   a. The p-value for a z-test of $H_0 : \theta = 1.1$

   b. The p-value for a one-sided z-test of $H_0 : \theta \leq 1$

8. Suppose that the smoking rate among adults aged 25–44 in the US is 11.2%. A public-health researcher has an informational video that she is convinced will lower the prevalence of smoking in this age group. She randomly selects a sample (including smokers and nonsmokers) from this age group and then follows up in six months to ask whether they are a smoker. Let π denote the probability of smoking for an adult aged 25–44 who sees the video.

   a. To provide statistical evidence that the informational video *decreases* the smoking rate, what are the appropriate one-sided null hypothesis and the alternative hypothesis?

   b. If the researcher finds that 10% of participants are smokers in her follow-up survey, how large would her sample need to be to reject the null hypothesis in (a) at a 5% level?

9. Use the **sp500** data set for this question.

   a. What are the z-statistic and p-value for testing that the population mean of Home Depot (*HD*) monthly returns is equal to the population mean of Lowe's (*LOW*) monthly returns?

   b. The sample mean of the market-index (*IDX*) monthly returns is 0.0078. If you are interested in testing whether the population mean of Home Depot monthly returns is the same as the population mean of the market-index monthly returns, would it be appropriate to test $H_0 : \mu_{HD} = 0.0078$? Explain.

   c. *Use the bootstrap for this part to calculate the standard error required for the test. What are the z-statistic and p-value for testing that the population median of Home Depot monthly returns equals the population median of Lowe's monthly returns?

   d. *Use the bootstrap for this part to calculate the standard error required for the test. What are the z-statistic and p-value for testing that the population standard deviation of Home Depot monthly returns equals the population standard deviation of Lowe's monthly returns?

e. *Use the bootstrap for this part to calculate the standard error required for the test. What are the z-statistic and p-value for testing that $\rho_{HD,IDX}$ (the population correlation between Home Depot returns and the market-index returns) is equal to $\rho_{LOW,IDX}$ (the population correlation between Lowe's returns and the market-index returns)?

10. Use the **cps** data set for this question. Focus on the sample of 2,809 employed individuals.

a. Provide the sample correlation matrix for the variables *hrslastwk*, *age*, *educ*, and *ownchild*.

b. For each sample correlation in the correlation matrix from (a), calculate the p-value for the z-test of the null hypothesis that the population correlation is equal to zero. Which of the correlations are statistically significant at a 5% level?

c. Define an indicator variable *collgrad* that equals 1 if $educ \geq 16$ and 0 otherwise. Calculate the sample standard deviation of *hrslastwk* for the two subsamples of college graduates ($collgrad = 1$) and noncollege graduates ($collgrad = 0$). What are the z-statistic and p-value for the test of the null hypothesis that the population standard deviations for the two subpopulations (college graduates and noncollege graduates) are the same? (Hint: Use the `se_sx` function defined in section 14.4.)

11. *A podcast provider has 1,000 customers who have signed up for a free three-month trial. It would like to do an A/B test of two alternative plans to get customers to subscribe after the free trial, where plan A guarantees a $2.99 monthly fee forever if the customer pays 12 months up front, and plan B guarantees a $3.99 monthly fee forever but allows cancellation at any time. Let π_A and π_B be the true probabilities of subscription for plans A and B, respectively. The podcast provider chooses the number of customers offered plan A, denoted as n^*, with the $1{,}000 - n^*$ other customers offered plan B. Let X_A and X_B be the Bernoulli random variables for the two plans, with success indicating a subscriber.

a. If $n^* = 300$, what is the asymptotic distribution of $\bar{X}_A - \bar{X}_B$ in terms of π_A and π_B?

b. What value of n^* minimizes the asymptotic variance of $\bar{X}_A - \bar{X}_B$ (in terms of π_A and π_B)?

c. If the null hypothesis $H_0 : \pi_A = \pi_B$ is true, what value of n^* minimizes the asymptotic variance of $\bar{X}_A - \bar{X}_B$?

d. Suppose that plan A is *slightly* more effective than plan B, with $\pi_A - \pi_B = \epsilon$, for some small number $\epsilon > 0$. Then, the z-statistic for testing $H_0 : \pi_A = \pi_B$, before observing outcomes, can be written as

$$\frac{\bar{X}_A - \bar{X}_B}{sd(\bar{X}_A - \bar{X}_B)} = \frac{\bar{X}_A - \bar{X}_B - \epsilon}{sd(\bar{X}_A - \bar{X}_B)} + \frac{\epsilon}{sd(\bar{X}_A - \bar{X}_B)}.$$

Since ϵ is small, you may assume that $sd(\bar{X}_A - \bar{X}_B)$ is approximately the same as it would be for $\pi_A = \pi_B$. Explain why the choice of n^* found in (c) is the best choice for testing $H_0 : \pi_A = \pi_B$, in the sense that it is the choice that makes rejection the most likely when $\pi_A - \pi_B = \epsilon$.

12. Use the **brands** data set for this question. The data set consists of 14,560 observations on customers who purchased a candy bar in their last visit to a certain market. There are five brands, numbered 1 through 5, and the *last_brand* variable indicates the brand that was purchased on the last visit. The *purchase* variable is 1 if the customer purchases a candy bar during the current visit and 0 otherwise. If a purchase is made ($purchase = 1$), the variable *brand* indicates the brand purchased on the current visit; if no purchase is made ($purchase = 0$), the variable *brand* has a value of 0.

    a. For each of the five conditional purchase probabilities (purchase given brand 1 on last visit, purchase given brand 2 on last visit, and so on), provide the estimated probability and an asymptotic 95% confidence interval.

    b. For the 10 possible pairs of different brands (b_1 and b_2), provide the p-value for the test of

    $$H_0 : P(purchase = 1|last\_brand = b_1) = P(purchase = 1|last\_brand = b_2).$$

    How many pairs indicate a statistically significant difference at a 5% level?

    c. Create a new variable *same_brand* equal to 1 if $brand = last\_brand$ and 0 otherwise. For each brand $b \in \{1, 2, 3, 4, 5\}$, provide the estimate of the conditional probability

    $$P(same\_brand = 1|last\_brand = b, purchase = 1),$$

    along with an asymptotic 95% confidence interval.

    d. For the 10 possible pairs of different brands (b_1 and b_2), provide the p-value for the test of

    $$H_0 : P(same\_brand = 1|last\_brand = b_1, purchase = 1)$$
    $$= P(same\_brand = 1|last\_brand = b_2, purchase = 1).$$

    How many pairs indicate a statistically significant difference at a 5% level?

    e. *Returning to the conditional purchase probabilities in (a), use the `wald_test` function to test the null hypothesis that all five probabilities are equal. What is the p-value? Do you reject the null hypothesis at the 5% level?

17

Simple Linear Regression

The linear regression model describes the relationship between an outcome variable and one or more explanatory variables. The linear regression estimator, also known as the *least-squares estimator*, is one of the most commonly used techniques for data analysis. This estimator is used for two main purposes: (1) making predictions about the outcome variable based on values of the explanatory variables and (2) estimating the causal effect of one or more explanatory variables on the outcome variable.

This chapter focuses on the simplest form of this model, *the simple linear regression (SLR) model*, with a single explanatory variable. Chapter 18 considers a more general model, the *multiple linear regression model*, which allows for multiple explanatory variables. The linear regression model is called "linear" since it assumes a linear relationship between the outcome variable and explanatory variables, but it allows for additional random noise in the outcome variable. As discussed in chapter 18, the linear nature is not as restrictive as the terminology suggests since nonlinearity can be incorporated with additional explanatory variables.

We focus on cross-sectional data in both this chapter and chapter 18, with the sample assumed to consist of independent and identically distributed (i.i.d.) draws from the underlying population. While linear regression models and estimators certainly apply to other types of data, including time-series data and panel data, the treatment of these cases is beyond the scope of this book.

17.1 The Simple Linear Regression Model

The **simple linear regression (SLR) model** assumes a linear association between two random variables, an **outcome variable** Y and an **explanatory variable** X. Y is sometimes called the *dependent variable*, and X is sometimes called the *independent variable* or the *covariate*. We use the terms "outcome variable" and "explanatory variable" since they are more descriptive of the variables' roles in the model. For now, attention is restricted to a continuous outcome variable Y. The case of a binary outcome variable Y is considered in section 18.9. The explanatory variable X may be continuous or discrete.

The population of interest is characterized by the bivariate joint distribution of (Y, X). The SLR model specifies that the outcome Y is a linear function of X plus a random "noise" term, represented by the random variable U:

$$Y = \alpha + \beta X + U \text{ with } E(U|X) = 0.$$

Y and X are both **observables**, meaning they are observed in the data as realizations (y, x) of the random variables (Y, X). In contrast, U is an **unobservable** since its realization (call it u) is not observed in the data. The parameter α is the **intercept** of the SLR model, and the parameter β is the **slope** of the SLR model. α and β are population parameters, part of a model that describes the relationship between Y and X in the population.

Two assumptions are embodied in the SLR model:

- *Linearity assumption*: The association between Y and X is assumed to be linear, captured by $\alpha + \beta X$ in the model. This assumption implicitly rules out a nonlinear association, like $Y = \alpha + \beta_1 X + \beta_2 X^2 + U$ or $Y = \alpha + \beta \ln(X) + U$. That said, a nonlinear association between Y and X can be easily accommodated: for a quadratic relationship, another explanatory variable X^2 can be added to the model (section 18.6.2); for a linear relationship between Y and $\ln(X)$, the model can be specified with $\ln(X)$ as the explanatory variable rather than X.
- *Exogeneity assumption*: The assumption $E(U|X) = 0$, known as the **exogeneity** assumption, states that the explanatory variable X provides no information about the expected value of U. Regardless of the value of X, the conditional expectation of U is always equal to zero.[1]

The exogeneity assumption $E(U|X) = 0$ requires that the random variables X and U are uncorrelated ($\sigma_{XU} = \rho_{XU} = 0$), but exogeneity is a stronger assumption than $\sigma_{XU} = \rho_{XU} = 0$.

Proposition 17.1 *If the exogeneity assumption $E(U|X) = 0$ holds, then $\sigma_{XU} = \rho_{XU} = 0$. Moreover, for any function $g(\cdot)$, if the exogeneity assumption holds, then $g(X)$ and U are uncorrelated: $\sigma_{g(X)U} = \rho_{g(X)U} = 0$.*

While the exogeneity assumption implies that X and U are uncorrelated, the reverse is not necessarily true. Even if X and U are uncorrelated, $E(U|X = x)$ may be nonzero for some x if U is correlated with some nonlinear function of X.

Under the exogeneity assumption, the conditional expectation $E(Y|X)$ is a linear function of X:

$$E(Y|X) = E(\alpha + \beta X + U|X) = E(\alpha|X) + E(\beta X|X) + E(U|X) = \alpha + \beta X.$$

The first equality comes from plugging in $\alpha + \beta X + U$ for Y based on the SLR model. The second equality follows from the fact that the expected value of the sum of random variables is the sum of the expected values of the random variables. The third equality uses several facts: α is a constant, so $E(\alpha|X) = \alpha$; after conditioning on X, βX is also a constant, so $E(\beta X|X) = \beta X$; and, $E(U|X) = 0$ by the exogeneity assumption.

Therefore, the SLR model can be thought of as a linear model for $E(Y|X)$. The meaning of the two parameters, the intercept α and the slope β, follows directly from the equation $E(Y|X) = \alpha + \beta X$:

- *Meaning of the intercept α*: When $X = 0$, the conditional expectation is $E(Y|X = 0) = \alpha + \beta \cdot 0 = \alpha$, implying

$$\alpha = E(Y|X = 0).$$

α is the population mean of Y conditional on $X=0$. Graphically, α is the y-intercept of the line $E(Y|X)=\alpha+\beta X$. The units of α are the same as the units of the random variable Y. Whether the value of α has a meaningful interpretation depends on the particular application. Generally, if zero is a sensible value for the X-variable, then the value of α has a meaningful interpretation. For instance, if Y measures an individual's earnings and X measures years of work experience, α is the population average of earnings conditional on no work experience (i.e., individuals newly in the workforce). On the other hand, if Y measures an individual's earnings and X measures an individual's age in years, α would not be meaningful by itself since it corresponds to the population earnings conditional on an individual being zero years of age, which is nonsensical.

- *Meaning of the slope* β: Parameter β is the slope of the $E(Y|X)=\alpha+\beta X$ line, indicating how much $E(Y|X)$ changes when X changes by 1 unit. Mathematically, for any value x^*,

$$E(Y|X=x^*+1)-E(Y|X=x^*)=\alpha+\beta(x^*+1)-(\alpha+\beta x^*)$$
$$=\alpha+\beta x^*+\beta-(\alpha+\beta x^*)=\beta,$$

which yields

$$\beta=E(Y|X=x^*+1)-E(Y|X=x^*).$$

Since the SLR model specifies a line extending endlessly with slope β, parameter β also describes the difference in conditional expectations for any two distinct values x^* and x^{**}:

$$E(Y|X=x^{**})-E(Y|X=x^*)=\alpha+\beta x^{**}-(\alpha+\beta x^*)=\beta(x^{**}-x^*)$$

or, equivalently,

$$\beta=\frac{E(Y|X=x^{**})-E(Y|X=x^*)}{x^{**}-x^*}.$$

The units of β are the units of Y divided by the units of X.

The SLR model provides a framework to think about the data-generating process for the population associated with (Y,X). First, consider value x^*, drawn from the unconditional distribution of the random variable X. Then, for the random variable Y, the draw from the population has expected value $E(Y|X=x^*)=\alpha+\beta x^*$. But this conditional draw is not necessarily equal to $\alpha+\beta x^*$ since $Y=\alpha+\beta x^*+U$ contains the unobservable noise U. If u^* is a draw from the conditional distribution of U given $X=x^*$, the realized draw of Y is $y^*=\alpha+\beta x^*+u^*$. Figure 17.1 provides a graphical representation of this data-generating process. In the graph, the conditional expectation $E(Y|X)=\alpha+\beta X$ is drawn with a positive intercept α and a positive slope β. The point $(0,\alpha)$ is where the $E(Y|X)=\alpha+\beta X$ line crosses the y-axis. The graph shows a single realized data point (x^*,y^*), arising from the proceess just described. For the x^*-value drawn from the unconditional distribution of X, the conditional expectation $E(Y|X=x)=\alpha+\beta x^*$ is read off the line. For the point in the graph, the draw of u^* from the conditional distribution of U given $X=x^*$ is a positive value, leading to a $y^*=\alpha+\beta x^*+u^*$ value that is above the line ($y^*>\alpha+\beta x^*$). If the draw of u^* had been negative, the observed data point would have been below the line.

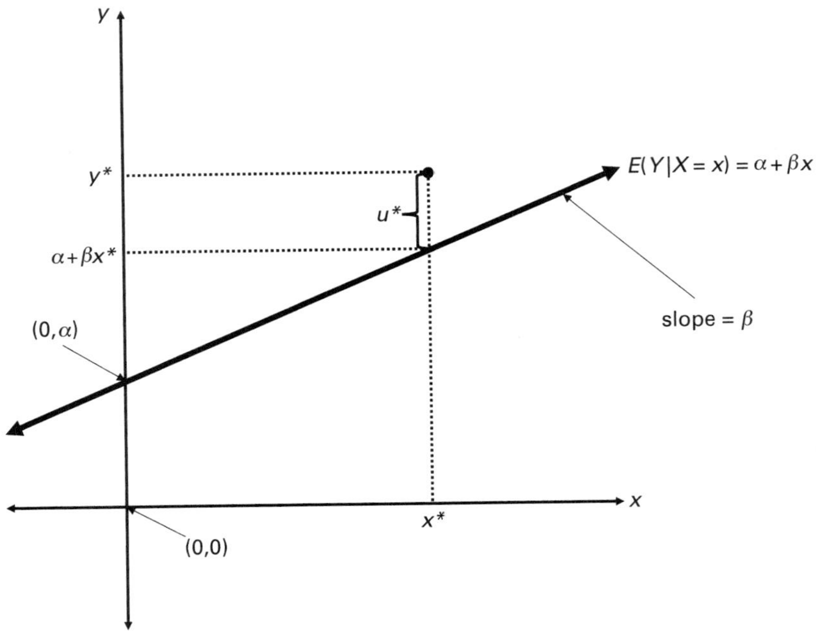

Figure 17.1
Data-generating process for the SLR model

Example 17.1 (Advertising and sales) *Suppose that a retail company has stores in* 80 *cities. During a particular week, the company randomly chooses* 20 *cities for on-line targeted advertising on a social media website. The targeted advertising is costly, which is why the company advertises to a subset of the 80 cities. If SALES is the random variable for weekly sales at a given location and AD is the random variable indicating whether the location receives target advertising* ($AD = 1$) *or not* ($AD = 0$)*, the SLR model relating SALES and AD is*

$$SALES = \alpha + \beta AD + U \text{ with } E(U|AD) = 0.$$

The exogeneity assumption $E(U|AD) = 0$ *is likely to hold here. Since AD is assigned randomly by the company to be* 0 *or* 1*, there is no reason to expect that there is a relationship between AD and the unobservable U. This situation is an example of an A/B experiment where the SLR model allows the estimation of the causal effect of targeted advertising (AD) on weekly sales (SALES). With the exogeneity assumption* $E(U|AD) = 0$ *holding, the SLR model implies that*

$$E(SALES|AD = 0) = \alpha \text{ and } E(SALES|AD = 1) = \alpha + \beta.$$

Figure 17.2 shows the SLR model for this example. While the line $E(SALES|AD) = \alpha + \beta AD$ *is defined for all values of AD, the only relevant values are* $AD = 0$ *(no advertising) and* $AD = 1$ *(advertising), and the conditional expectations for these two values are indicated by the solid points. The graph is drawn assuming a positive slope* $\beta > 0$*; of course, since* β *is unknown, it is possible that* $\beta = 0$ *(advertising has no effect) or even* $\beta < 0$ *(advertising has a negative effect).*

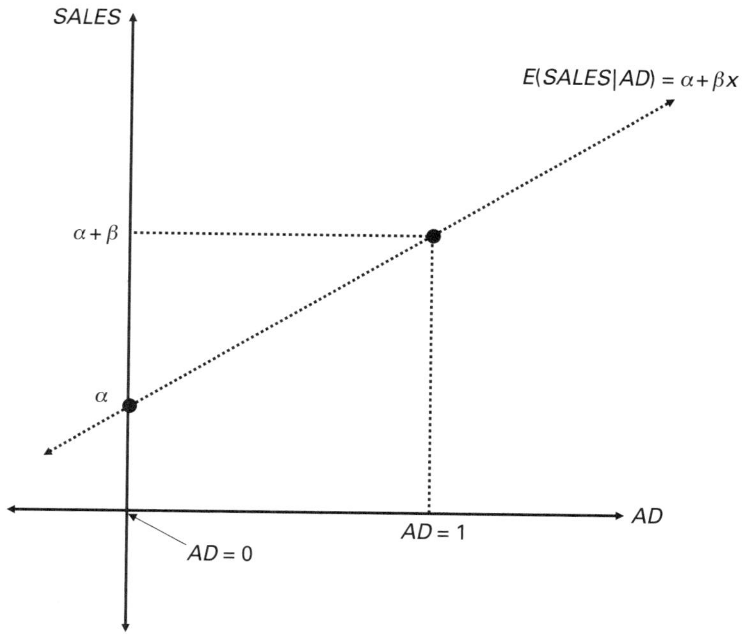

Figure 17.2
SLR model for sales and advertising

Parameter α is the population mean of sales for cities not receiving targeted advertising, and $\alpha + \beta$ is the population mean of sales for cities receiving targeted advertising. Subtracting the $E(SALES|AD=0)$ equation from the $E(SALES|AD=1)$ equation yields

$$\beta = E(SALES|AD=1) - E(SALES|AD=0),$$

*which is the **causal effect** of targeted advertising, equal to the difference in the population mean of sales between targeted cities and nontargeted cities. As such, since we will discuss how to estimate β next, the SLR model provides a natural framework for estimating the difference in average outcomes for "A" and "B" groups in an A/B experiment.*

Example 17.2 (Earnings and union status) *Suppose we are interested in modeling the relationship between weekly earnings, measured by the random variable EARNWK, and union status, measured by the random variable UNION, among employed individuals in the population. EARNWK is a continuous random variable, and UNION is an indicator variable equal to 1 for a union member and 0 for a nonmember. The SLR model relating EARNWK and UNION is*

$$EARNWK = \alpha + \beta UNION + U \text{ with } E(U|UNION) = 0.$$

This example is similar to example 17.1 since the explanatory variable is an indicator variable. α is the conditional population mean of weekly earnings for nonmembers,

$$\alpha = E(EARNWK|UNION=0),$$

and β is the difference in the population mean of weekly earnings between union members and nonmembers,

$$\beta = E(EARNWK|UNION = 1) - E(EARNWK|UNION = 0).$$

Unlike example 17.1, however, the explanatory variable UNION is not randomly assigned. Should we expect that the exogeneity assumption $E(U|UNION) = 0$ holds here? There is the possibility that union status may be related to the unobservable U. For instance, if union members tend to have higher skill levels than nonmembers and, thus, higher productivity than nonmembers, it could be the case that UNION and U are positively correlated. Such a correlation between UNION and U would imply that the exogeneity assumption $E(U|UNION) = 0$ does not hold. As discussed next, failure of the exogeneity assumption makes causal inference difficult (e.g., testing whether union membership causes *higher earnings) but does not preclude estimating the association between earnings and union membership (e.g., estimating the difference between average earnings for union members and average earnings for nonmembers).*

Example 17.3 (Earnings and education) *Rather than looking at the relationship between weekly earnings and union membership (example 17.2), suppose we are interested in the relationship between weekly earnings (EARNWK) and years of educational attainment (EDUC) among employed individuals in the population. The SLR model relating EARNWK and EDUC is*

$$EARNWK = \alpha + \beta EDUC + U \text{ with } E(U|EDUC) = 0.$$

EDUC is a discrete explanatory variable that can take several different values, in contrast to the binary explanatory variable UNION. As a result, the linearity assumption is considerably stronger here since it requires that the linear relationship, given by intercept α and slope β, holds over the entire range of the EDUC variable, as opposed to holding only for the two possible values of UNION.[2] Slope β is the change in the conditional expectation of weekly earnings associated with a one-year change in education. Intercept α is not really meaningful by itself since $EDUC = 0$ is far below the minimum education level traditionally observed in labor-force data sets in the US. As in example 17.2, there is a concern that the exogeneity assumption may not hold here. Specifically, it is generally believed by labor economists that EDUC is positively correlated with U since education likely has a positive association with unobserved productivity, unobserved family wealth, etc. This positive association would imply that $E(U|EDUC) = 0$ does not hold over the range of possible EDUC values.

Example 17.4 (Cigarette sales and cigarette taxes) *As discussed in example 5.3, there is a lot of variation in state-level cigarette taxes in the US. Standard economic theory predicts a negative relationship between cigarette sales (the "demand for cigarettes") and cigarette taxes since a higher tax is associated with a higher price. We use per-capita sales to make the data comparable across states. Let CIGSALES denote the number of packs per capita sold in a year in a given state, and let CIGTAX denote the state tax (in dollars) on a pack of cigarettes. Both CIGSALES and CIGTAX are continuous variables, and the SLR model relating the two variables is*

$$CIGSALES = \alpha + \beta CIGTAX + U \text{ with } E(U|CIGTAX) = 0.$$

Since cigarette taxes are the result of a political and legislative process in each state, there might be a concern that the exogeneity assumption $E(U|CIGTAX) = 0$ *does not hold. For instance, if there are very few smokers in a given state, there may be little resistance to a law that increases cigarette taxes. On the other hand, if there are many smokers in a given state, there might be more resistance to such a law. If true, these arguments imply a negative correlation between* $CIGTAX$ *and the unobservable* U *since* U *is higher in states where people have a greater inclination to smoke.*

Example 17.5 (Monthly stock returns and the overall market) *One of the simplest and most commonly used models for stock returns is an SLR model that relates the return of an individual stock to the return of the overall stock market. The standard way to measure the return of the overall stock market is to use an index, like the S&P 500 or the Russell 2000. Generally, the individual stock would be part of the index being used. To fix ideas, let* R_{STOCK} *denote the monthly return of a specific stock and* R_{IDX} *denote the monthly return of the S&P 500 index. Then, the SLR model relating the individual stock return to the overall market return is*[3]

$$R_{STOCK} = \alpha + \beta R_{IDX} + U \text{ with } E(U|R_{IDX}) = 0.$$

The intercept

$$\alpha = E(R_{STOCK}|R_{IDX} = 0)$$

is the population mean of the individual stock return R_{STOCK} *conditional on* $R_{IDX} = 0$*, which corresponds to the S&P 500 index being unchanged (zero market return) during a month. Slope* β *is the change in the conditional expectation of the individual stock's return* R_{STOCK}*, associated with a 1-unit change in the index return* R_{IDX}*. In terms of returns, a 1-unit change is huge since it corresponds to a* 100 *percentage-point change. Instead,* β *can be interpreted in terms of a smaller change in* R_{IDX}*, like a 1 percentage-point change. For a 1 percentage-point (*0.01*) change in* R_{IDX}*, the associated change in the conditional expectation of* R_{STOCK} *is* 0.01β*. The nature of the individual stock returns' relationship to the overall market returns depends on* β *as follows:*

- $\beta = 0$*: The expected stock return is not related to the market return.*
- $\beta = 1$*: The expected stock return moves exactly in tandem with the market return. For instance, if the market return is* 0.02 *higher, the expected stock return is also* 0.02 *higher.*
- $0 < \beta < 1$*: The expected stock return moves in the same direction as the market return, but with a smaller magnitude. For instance, if the market return is* 0.02 *higher, the expected stock return is higher by an amount less than* 0.02*.*
- $\beta > 1$*: The expected stock return moves in the same direction as the market return, but with a larger magnitude. For instance, if the market return is* 0.02 *higher, the expected stock return is higher by an amount greater than* 0.02*.*
- $\beta < 0$*: The expected stock return moves in the opposite direction as the market return. The expected stock return decreases when the market return increases, and the expected stock return increases when the market return decreases. Due to the behavior of the expected stock return in this case, such a stock is often called* countercyclical.

17.2 The Least-Squares Estimator

For the population described by the joint distribution of (Y,X), assume that the observed sample consists of i.i.d. draws

$$(y_1,x_1), (y_2,x_2), \dots, (y_n,x_n),$$

where n is the sample size. If the SLR model $Y=\alpha+\beta X+U$ holds for (Y,X), then for each observation $i=1,\dots,n$,

$$y_i=\alpha+\beta x_i+u_i,$$

where u_i is a realized draw of U conditional on $X=x_i$ and is known as the **residual** for observation i. Residual u_i is not observable, but it can be rewritten in terms of the observables y_i and x_i as

$$u_i=y_i-\alpha-\beta x_i.$$

Since $E(Y|X=x_i)=\alpha+\beta x_i$, the residual u_i indicates how much the observed outcome y_i differs from the conditional expectation of Y given $X=x_i$. A positive residual means that outcome y_i is higher than expected (greater than $\alpha+\beta x_i$), and a negative residual means that outcome y_i is lower than expected (less than $\alpha+\beta x_i$). As α and β are unknown parameters, residual u_i is also unknown. But if α and β can be estimated, as we do next, residual u_i can also be estimated.

Figure 17.3 shows an example of a data set, with $n=12$ observations, generated by the SLR model $Y=\alpha+\beta X+U$ with $E(U|X)=0$. The line $E(Y|X=x)=\alpha+\beta x$ is shown with a positive intercept α and a positive slope β in the graph. None of the observed data points fall *exactly* on the line, although a few are close. The data points above the line have positive residual values ($u_i>0$). For instance, the data point (x^*,y^*)-associated with the x^*-value lies above the line, with $y^*>\alpha+\beta x^*$ implying $u^*=y^*-\alpha-\beta x^*>0$. In contrast, the data points below the line have negative residual values ($u_i<0$), and the data point (x^{**},y^{**}) is an example with $y^{**}<\alpha+\beta x^{**}$ and $u^{**}=y^{**}-\alpha-\beta x^{**}<0$. For each data point (x_i,y_i) in the graph, the value of the residual $u_i=y_i-\alpha-\beta x_i$ is equal to the vertical distance from the data point to the $\alpha+\beta x$ line, with points above the line having positive residual values and points below the line having negative residual values.

To develop estimators of the SLR model parameters, we use proposition 17.2, which states how parameters α and β are related to the population-descriptive statistics of the random variables (Y,X). The key idea is that, once the parameters are expressed in terms of population-descriptive statistics, estimators can be developed by using *sample* descriptive statistics in place of population-descriptive statistics.

Proposition 17.2 *If the SLR model holds and $\sigma_X^2>0$, then slope β and intercept α are related to the population-descriptive statistics as follows:*

$$\beta=\frac{\sigma_{XY}}{\sigma_X^2} \text{ or, equivalently, } \beta=\rho_{XY}\frac{\sigma_Y}{\sigma_X}$$

and

$$\alpha=\mu_Y-\beta\mu_X=\mu_Y-\frac{\sigma_{XY}}{\sigma_X^2}\mu_X.$$

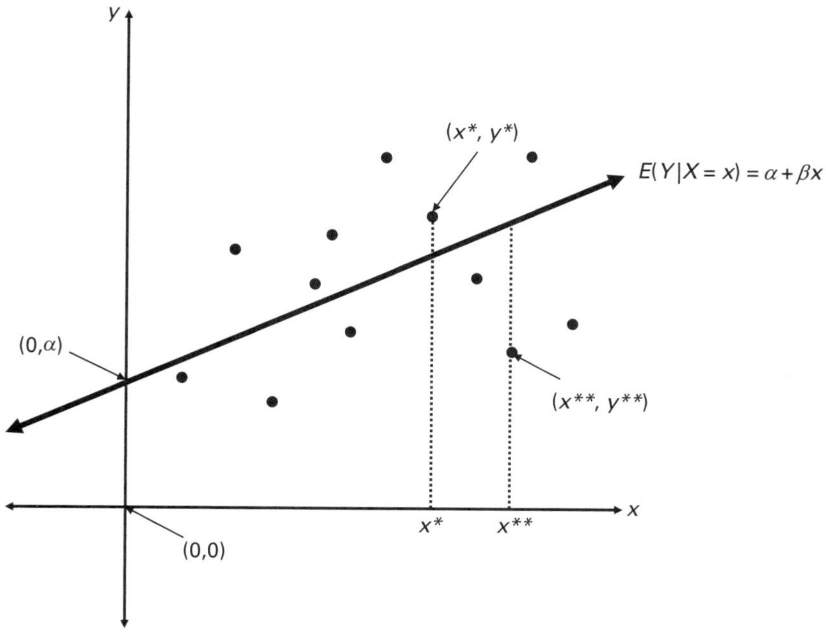

Figure 17.3
SLR model and observed data

The assumption that $\sigma_X^2 > 0$ means that X is not constant and, since σ_X^2 is in the denominator of the β expression, ensures that β is a well-defined parameter. To show the first result for β, note that

$$\begin{aligned} Cov(X, Y) &= Cov(X, \alpha + \beta X + U) \\ &= Cov(X, \alpha) + Cov(X, \beta X) + Cov(X, U) \\ &= \beta Var(X). \end{aligned}$$

The last equality follows from $Cov(X, \alpha) = 0$, since α is constant, and $Cov(X, U) = 0$, which is implied by the exogeneity assumption $E(U|X) = 0$. Dividing both sides by $Var(X)$, which is possible since $Var(X)$ is assumed to be positive, yields $\beta = \frac{Cov(X,Y)}{Var(X)} = \frac{\sigma_{XY}}{\sigma_X^2}$. The second result for β follows from the relationship between a population covariance and a population correlation, $\sigma_{XY} = \sigma_X \sigma_Y \rho_{XY}$, which implies

$$\beta = \frac{\sigma_{XY}}{\sigma_X^2} = \frac{\sigma_X \sigma_Y \rho_{XY}}{\sigma_X^2} = \rho_{XY} \frac{\sigma_Y}{\sigma_X}.$$

To show the result for α, taking expectations of both sides of the SLR model yields

$$E(Y) = E(\alpha + \beta X + U) = \alpha + \beta E(X),$$

from which it follows that $\alpha = E(Y) - \beta E(X) = \mu_Y - \beta \mu_X$.

An immediate implication of proposition 17.2 is that the sign of β is the same as the sign of the population covariance or population correlation between Y and X:

$$\text{sign}(\beta) = \text{sign}(\sigma_{XY}) = \text{sign}(\rho_{XY}).$$

For positively correlated random variables Y and X, slope β in the SLR model is positive. For negatively correlated Y and X, slope β in the SLR model is negative. Another implication of proposition 17.2 is that

$$\mu_Y = \alpha + \beta\mu_X = E(Y|X = \mu_X),$$

meaning that the point of population means (μ_X, μ_Y) lies on the model's conditional-expectation line $E(Y|X) = \alpha + \beta X$.

Knowing $\beta = \frac{\sigma_{XY}}{\sigma_X^2}$ suggests that an estimator of β can be obtained by using s_{XY} as an estimator of σ_{XY} and s_X^2 as an estimator of σ_X^2. In large samples, since both are consistent estimators, s_{XY} gets arbitrarily close to σ_{XY} and s_X^2 gets arbitrarily close to σ_X^2. Therefore, the fraction $\frac{s_{XY}}{s_X^2}$ should get arbitrarily close to $\beta = \frac{\sigma_{XY}}{\sigma_X^2}$ and, by the continuous mapping theorem (proposition 14.10), it should be a consistent estimator of β. This estimator is known as the **least-squares slope estimator** for the SLR model. The reason for the "least-squares" terminology will become clear shortly. Using the notation $\hat{\beta}_{XY}$ to denote the least-squares slope estimator,

$$\hat{\beta}_{XY} = \frac{s_{XY}}{s_X^2}.$$

The estimator $\hat{\beta}_{XY}$ is a random variable, as its value depends on the particular sample that is observed. After the sample $(y_1, x_1), (y_2, x_2), \ldots, (y_n, x_n)$ is observed, there is a realization of $\hat{\beta}_{XY}$. This realization is the **least-squares slope estimate**, denoted as $\hat{\beta}$, and is given by

$$\hat{\beta} = \frac{s_{xy}}{s_x^2} \text{ or, equivalently, } \hat{\beta} = r_{xy}\frac{s_y}{s_x}.$$

The notation $\hat{\beta}$ has been used, rather than $\hat{\beta}_{xy}$, in the interest of brevity. Stated in words, the least-squares slope estimate $\hat{\beta}$ is the sample covariance between x and y divided by the sample variance of x. For $\hat{\beta}$ to be well defined, we need $s_x^2 > 0$, which is true so long as the x_i values are not all equal. The formulas for $\hat{\beta}$ imply that

$$\text{sign}(\hat{\beta}) = \text{sign}(s_{xy}) = \text{sign}(r_{xy}).$$

The least-squares slope estimate has the same sign as the sample covariance and sample correlation between x and y. Based on the $\hat{\beta} = r_{xy}\frac{s_y}{s_x}$ formula, one way to think about the least-squares slope estimate is that it is a scaled version of the sample correlation r_{xy}, where the scaling is the ratio between the standard deviation of y and the standard deviation of x. This formula also implies that the units of estimate $\hat{\beta}$ are the units of y divided by the units of x, as is the case for parameter β itself.

For the intercept parameter α, the relationship $\alpha = \mu_Y - \beta\mu_X$ suggests that an estimator can be obtained by plugging in the estimator $\hat{\beta}_{XY}$ for β and using $\bar{Y}$ as an estimator of μ_Y and $\bar{X}$ as an estimator of μ_X and $\hat{\beta}$. This estimator is known as the **least-squares intercept estimator** for the SLR model. Using the notation $\hat{\alpha}_{XY}$ to denote this estimator,

$$\hat{\alpha}_{XY} = \bar{Y} - \hat{\beta}_{XY}\bar{X}.$$

After observing the sample, the realization is the **least-squares intercept estimate**, denoted as $\hat{\alpha}$, and is given by

$$\hat{\alpha} = \bar{y} - \hat{\beta}\bar{x}.$$

Based on the least-squares estimates $\hat{\alpha}$ and $\hat{\beta}$, the **estimated regression line** is

$$\hat{E}(Y|X=x)=\hat{\alpha}+\hat{\beta}x,$$

where $\hat{E}(Y|X=x)$ denotes the estimate of the true conditional expectation $E(Y|X=x)$. Since the formula for $\hat{\alpha}$ can be rewritten as

$$\bar{y}=\hat{\alpha}+\hat{\beta}\bar{x},$$

it follows that the point of sample means $(\bar{x},\bar{y})$ falls on the estimated regression line. That is, $\bar{y}$ is the estimated conditional expectation of Y given that X is equal to the sample mean $\bar{x}$.

Proposition 17.3 *If the SLR model holds and $s_X^2>0$, the least-squares slope estimator*

$$\hat{\beta}_{XY}=\frac{s_{XY}}{s_X^2}=r_{XY}\frac{s_Y}{s_X}$$

and the least-squares intercept estimator

$$\hat{\alpha}_{XY}=\bar{Y}-\hat{\beta}_{XY}\bar{X}$$

are consistent estimators of β and α, respectively.

Consistency follows from the continuous mapping theorem (proposition 14.10). $\hat{\beta}_{XY}=\frac{s_{XY}}{s_X^2}$ consistently estimates $\beta=\frac{\sigma_{XY}}{\sigma_X^2}$ since s_{XY} and s_X^2 are consistent estimators of σ_{XY} and σ_X^2, respectively. Also, $\hat{\alpha}_{XY}=\bar{Y}-\hat{\beta}_{XY}\bar{X}$ consistently estimates $\alpha=\mu_Y-\beta\mu_X$ since $\bar{Y}$ and $\bar{X}$ are consistent estimators of μ_Y and μ_X, respectively. It turns out that the least-squares estimators $\hat{\beta}_{XY}$ and $\hat{\alpha}_{XY}$ are also unbiased estimators of β and α, respectively. But since we will focus only on asymptotic statistical inference for the least-squares estimators, the consistency property is the important one. The estimators are also asymptotically normal, as discussed in detail next.

The `lm` function, where "`lm`" is short for "linear model," is the primary function used for least-squares estimation in *R*. Although `lm` has many different arguments available, the following provides the basic syntax that handles most linear-regression problems of interest:

- `lm(formula, data, subset)`: Returns the results from the least-squares estimation of the model specified by `formula`. The optional argument `data` specifies the data frame to be used and can greatly simplify how `formula` is written. The optional argument `subset` is a logical vector that specifies the subset of data to be used for estimation.

The `lm` function automatically ignores observations for which the variables used in the `formula` argument have missing (`NA`) values. Therefore, it is not necessary to remove rows of the vector or data frame with `NA` values.

Here are some simple examples that illustrate the usage of the `lm` function:

- `lm(df$y~df$x)`: This command returns the results from the least-squares estimation of an SLR model with `df$y` as the outcome variable and `df$x` as the explanatory variable. The syntax for the `formula` argument puts the outcome variable before the tilde (~) and the explanatory variable after the tilde. Here, `df` is a data frame that contains the two variables.

- `lm(y~x, data=df)`: This command is identical to `lm(df$y~df$x)`, with the `data=df` argument indicating that the variables in the `formula` argument are in the `df` data frame.
- `lm(y~x, data=df, subset=(x>10))`: This command does the least-squares estimation, with `df$y` as the outcome variable and `df$x` as the explanatory variable, on the subsample for which `(x>10)` is true.

Example 17.6 (Earnings and union status) *Example 17.2 considered an SLR model for the relationship between weekly earnings (EARNWK) and union membership (UNION):*

$$EARNWK = \alpha + \beta UNION + U.$$

To estimate this model, consider the sample of employed individuals ($n = 2{,}809$) from the ***cps*** *data set. The outcome variable is earnwk, and the explanatory variable is an indicator variable that indicates union membership. Although unionstatus is a categorical variable (with categories "Nonunion" and "Union") in the data set, for the purposes of estimating the SLR model, we define a binary variable with union = 1 for the "Union" category and union = 0 for the "Nonunion" category. The least-squares estimates of α and β can be calculated in R using the* `lm` *function:*

```
cps$union <- 1*(cps$unionstatus=="Union")

lm(earnwk~union, data=cps)

##
## Call:
## lm(formula = earnwk ~ union, data = cps)
##
## Coefficients:
## (Intercept)        union
##         947          251
```

The least-squares estimates of α and β are

$$\hat{\alpha} = 947 \text{ and } \hat{\beta} = 251.$$

The intercept estimate $\hat{\alpha}$ is an estimate of $\alpha = E(EARNWK|UNION = 0)$, the population mean of weekly earnings for nonunion members:

$$\hat{\alpha} = \hat{E}(EARNWK|UNION = 0) = 947.$$

This estimate turns out to be exactly *equal to the sample mean of weekly earnings for the 2,533 nonunion workers in the sample. The slope estimate $\hat{\beta}$ is an estimate of the difference in the population mean of weekly earnings between union members and nonmembers. The estimate $\hat{\beta} = 251$ turns out to be* exactly *equal to the difference between the sample mean of weekly earnings of the 276 union workers (\$1,198) and the sample mean of weekly earnings of the 2,533 nonunion workers (\$947). Thus, the estimated union premium in expected weekly earnings is \$251.*

Interestingly, R can do the least-squares estimation here without formally defining a new union indicator variable:

```
lm(earnwk~unionstatus, data=cps)

##
## Call:
## lm(formula = earnwk ~ unionstatus, data = cps)
##
## Coefficients:
##      (Intercept)  unionstatusUnion
##              947               251
```

The `lm` *function recognizes* `unionstatus` *as a categorical variable and automatically creates an indicator variable, which appears as* `unionstatusUnion` *in the ouput and is equal to* 1 *when* `unionstatus` *is* `Union` *and* 0 *when* `unionstatus` *is* `Nonunion`. *This capability extends to categorical variables with more than two categories. For example, the command* `lm(earnwk~race, data=cps)` *would automatically create two indicator variables based on the categorical variable* `race` *since it has three categories. Whether we explicitly create the indicator variables based on a categorical variable or let* `lm` *do it for us is largely a matter of preference, although creation of the indicator variables allows us to explicitly indicate which is the omitted category in the model.*

The relationship between the least-squares estimates and the different subsample averages of y, seen in example 17.6, is a general property for least-squares estimates when x is a binary variable, as described in proposition 17.4.

Proposition 17.4 *For an SLR model with a binary explanatory variable X, the least-squares estimates $\hat{\alpha}$ and $\hat{\beta}$ satisfy*

$$\hat{\alpha} = \bar{y}_0$$

and

$$\hat{\beta} = \bar{y}_1 - \bar{y}_0, \textit{ respectively,}$$

where $\bar{y}_0$ is the average of y-values in the $x = 0$ subsample and $\bar{y}_1$ is the average of y values in the $x = 1$ subsample.

The intercept estimate $\hat{\alpha}$ corresponds to the sample average of y-values for the $x = 0$ subsample, whereas the slope estimate $\hat{\beta}$ is the difference in the sample average of y-values for the $x = 1$ subsample and the sample average of y-values for the $x = 0$ subsample. Therefore, proposition 17.4 implies that the estimated line $\hat{\alpha} + \hat{\beta}x$ passes through the two points corresponding to the sample averages of y-values for the $x = 0$ and $x = 1$ subsamples. Specifically, the points $(0, \bar{y}_0)$ and $(1, \bar{y}_1)$ lie on the estimated line, corresponding to the slope estimate $\hat{\beta} = \frac{\bar{y}_1 - \bar{y}_0}{1-0} = \bar{y}_1 - \bar{y}_0$. It is important to stress that this property does not generalize to an x-variable that has more than two possible values.

Example 17.7 (Cigarette sales and cigarette taxes) *Example 17.4 considered an SLR model for the relationship between state-level annual cigarette sales (number of packs per capita), given by the random variable CIGSALES, and state-level cigarette taxes (dollars per pack), given by the random variable CIGTAX:*

$$CIGSALES = \alpha + \beta CIGTAX + U.$$

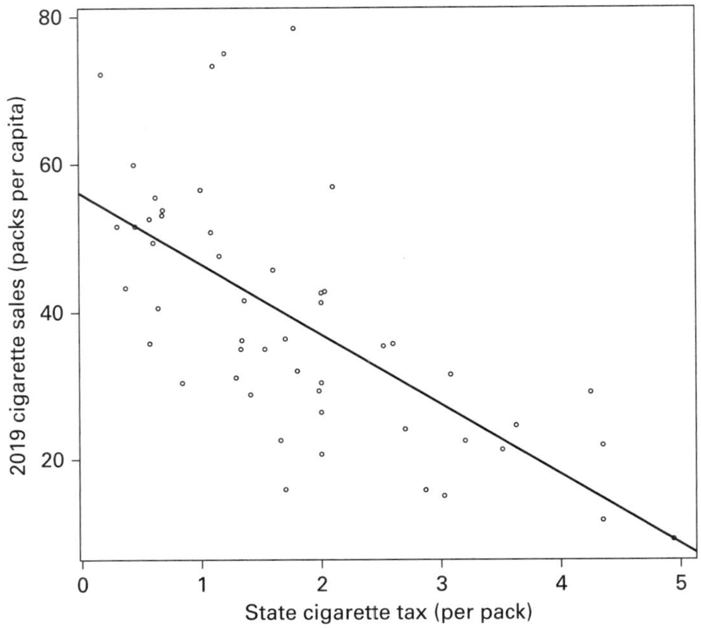

Figure 17.4
Least-squares estimated line for cigarette data

*The data set **cigdata** contains data for 2019 that can be used to estimate this SLR model. Specifically, data for the* 50 *individual states plus the District of Columbia yield a sample with* $n = 51$ *observations. The realized outcome variable is cigsales, and the realized explanatory variable is cigtax.*

```
lm(cigsales~cigtax, data=cigdata)

##
## Call:
## lm(formula = cigsales ~ cigtax, data = cigdata)
##
## Coefficients:
## (Intercept)       cigtax
##       55.95        -9.49
```

The least-squares estimates of α *and* β *are*

$$\hat{\alpha} = 55.95 \text{ and } \hat{\beta} = -9.49.$$

Figure 17.4 shows the scatter plot of cigsales versus cigtax, with the estimated SLR line,

$$\hat{\alpha} + \hat{\beta} cigtax = 55.95 - 9.49 cigtax,$$

drawn as a solid line. The intercept estimate $\hat{\alpha}$ *indicates that the estimated conditional expectation of CIGSALES, given* $CIGTAX = 0$*, is*

$$\hat{E}(CIGSALES|CIGTAX = 0) = 55.95,$$

corresponding to an estimated average cigarette sales of 55.95 *packs per capita among the subpopulation of (hypothetical) states with no state-level cigarette tax. In the graph,* $\hat{\alpha} = 55.95$ *is the value read off the least-squares estimated line at cigtax* $= 0$. *The estimated slope* $\hat{\beta}$ *is negative, indicating a negative association between cigsales and cigtax. The sign of* $\hat{\beta}$ *is the same as the sign of the sample correlation between cigtax and cigsales, which is equal to* -0.668 *for this sample. The meaning of the slope estimate of* $\hat{\beta} = -9.49$ *is that a* \$1 *in the state-level per-pack cigarette tax (CIGTAX) is associated with an estimated decrease of* 9.49 *packs per capita in expected state-level annual cigarette sales. As explained in section 17.1, slope parameter* β *also describes the difference between expected outcomes at different values of the explanatory variable. For instance, the difference between the expected CIGSALES at CIGTAX* $= 3$ *(\$3 per-pack tax) and the expected CIGSALES at CIGTAX* $= 1$ *(\$1 per-pack tax) is*

$$E(CIGSALES|CIGTAX = 3) - E(CIGSALES|CIGTAX = 1) = (3-1)\beta = 2\beta.$$

An estimate of this difference is $2\hat{\beta} = (2)(-9.49) = -18.98$, *or expected state-level sales that are estimated to be* 18.98 *packs per-capita lower among the population of states with a* \$3 *per-pack tax compared to the population of states with a* \$1 *per-pack tax. This estimate says something about* expected *state-level sales. It is possible that the realized state-level sales in a specific* \$3*-tax state may be higher than the realized state-level sales in a specific* \$1*-tax state. But averaging over many such states with* \$3 *taxes and* \$1 *taxes, the negative slope estimate says that expected state-level sales are estimated to be lower in states with* \$3 *taxes than in states with* \$1 *taxes.*

Here is the R code used to draw the scatter plot and least-squares regression line in figure 17.4:

```
par(mfrow = c(1,1))

plot(cigdata$cigtax,cigdata$cigsales,cex=0.5,xlab="State cigarette tax (per pack)",
    ylab="2019 cigarette sales (packs per capita)",xaxt="n",yaxt="n")
abline(lm(cigsales~cigtax,data=cigdata),lwd=2)
axis(1, at=seq(0,5,by=1), las=1)
axis(2, at=seq(0,80,by=20), las=1)
```

After the scatter plot is drawn with the `plot` *function, the least-squares regression line is drawn by using the* `lm` *regression itself as the argument of the* `abline` *function. The* `abline` *function determines the intercept and slope of the regression line from the* `lm` *regression results.*

Example 17.8 (Monthly stock returns and the overall market) *Example 17.5 introduced an SLR model of the relationship between the returns of an individual stock and the returns of a market index:*

$$R_{STOCK} = \alpha + \beta R_{IDX} + U.$$

We estimate the parameters of this model using a sample from the ***sp500*** *data set, which has* $n = 364$ *monthly observations. The variable IDX contains the monthly returns for the S&P 500 index, which we use as the realized values of the random variable* R_{IDX} *from the model here. Since the model can be used for any individual stock, let's first examine Home Depot (HD). We estimate the SLR model in R, as follows:*

```
lm(HD~IDX, data=sp500)

##
## Call:
## lm(formula = HD ~ IDX, data = sp500)
##
## Coefficients:
## (Intercept)          IDX
##     0.00873      1.02045
```

The least-squares estimates, using $y = HD$ *and* $x = IDX$*, are*

$$\hat{\alpha} = 0.0087 \text{ and } \hat{\beta} = 1.020.$$

Figure 17.5 shows the scatter plot of Home Depot monthly returns versus the S&P 500 index monthly returns. The solid line is the least-squares estimate line, with estimated intercept $\hat{\alpha} = 0.0087$ *and estimated slope* $\hat{\beta} = 1.020$*. The vertical dotted line is drawn through* $IDX = 0$*, corresponding to a monthly market index return of zero (no change in the market index). The horizontal dotted line is drawn through the intercept estimate, which is where the least-squares line intersects the vertical* $IDX = 0$ *line. The intercept estimate indicates that the estimated expected monthly return for Home Depot is* 0.0087*, or* 0.87%*, when the monthly return of the market index is zero. This estimate only tells us about the* expected *Home Depot return when the market index return is zero, as it is evident from the scatter plot that there is considerable variation to Home Depot monthly returns for observations that are close to the vertical* $IDX = 0$ *line, with values that range from about* -0.10 *(negative* 10% *monthly return) to* 0.15 *(positive* 15% *monthly return). While we often interpret slope estimates in terms of 1-unit changes in the explanatory variable, a 1-unit change is too large to be meaningful for this example since the full range of the IDX variable is between* -0.20 *and* 0.15*. Instead, think about a change of* 0.01*, or 1 percentage point, in the value of IDX. Then the slope estimate* $\hat{\beta} = 1.020$ *indicates that an increase of* 0.01 *in the market index monthly return is associated with an increase of* $(0.01)(1.020) = 0.0102$ *in the expected Home Depot monthly return. This estimated increase is very close to one-to-one since the estimated slope is numerically close to 1.*

The following table reports the results from least-squares estimation of the model for six choices of the outcome variable. Each row provides the least-squares estimates for an SLR model relating the individual stock monthly returns to the S&P 500 market index (IDX) monthly returns. The first row contains the results for Home Depot (HD), and the remaining rows contain estimation results for five other stocks:

	$\hat{\alpha}$	$\hat{\beta}$
Home Depot (HD)	0.0087	1.020
Lowe's (LOW)	0.0117	1.107
Bank of America (BAC)	0.0014	1.489
Wells Fargo (WFC)	0.0054	1.048
Marathon Oil (MRO)	−0.0006	1.449
ConocoPhillips (COP)	0.0028	1.042

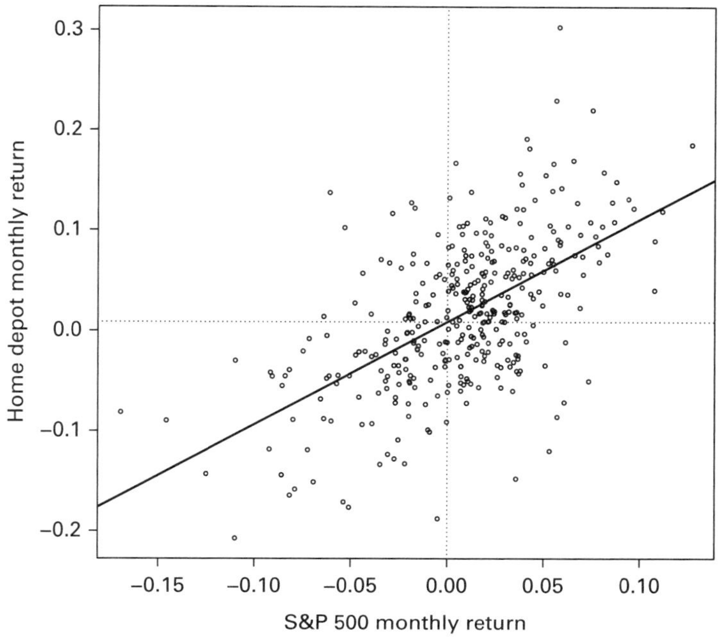

Figure 17.5
Least-squares estimated line for Home Depot versus S&P 500 monthly returns

The interpretation of the estimates is similar to that for the Home Depot results shown here. The stock with the highest estimated slope is Bank of America, with $\hat{\beta} = 1.489$, *which indicates that an increase of* 0.01 *in the market index monthly return is associated with an increase of* $(0.01)(1.489) = 0.0149$ *in the expected Bank of America monthly return. For Bank of America, the intercept estimate of* $\hat{\alpha} = 0.0014$ *indicates that the estimated expected monthly return of Bank of America is* 0.0014, *or* 0.14%, *when the market index monthly return is zero.*

17.3 Fitted Values, Estimated Residuals, and Regression Fit

To gain a better understanding of how well least-squares estimation fits the data, we now introduce some additional terminology and notation based on the observed sample and the least-squares estimates $\hat{\alpha}$ and $\hat{\beta}$.

Definition 17.1 *The **fitted value** for observation i, denoted as* $\hat{y}_i$, *is*

$$\hat{y}_i = \hat{\alpha} + \hat{\beta} x_i,$$

where $\hat{\alpha}$ *and* $\hat{\beta}$ *are the least-squares estimates of* α *and* β, *respectively.*

The fitted value is an estimate of the conditional expectation of the outcome variable, given that the explanatory variable equals x_i. The conditional expectation is $E(Y|X = x_i) = \alpha + \beta x_i$, so its estimate is obtained by plugging in $\hat{\alpha}$ for α and $\hat{\beta}$ for $\beta : \hat{E}(Y|X = x_i)$

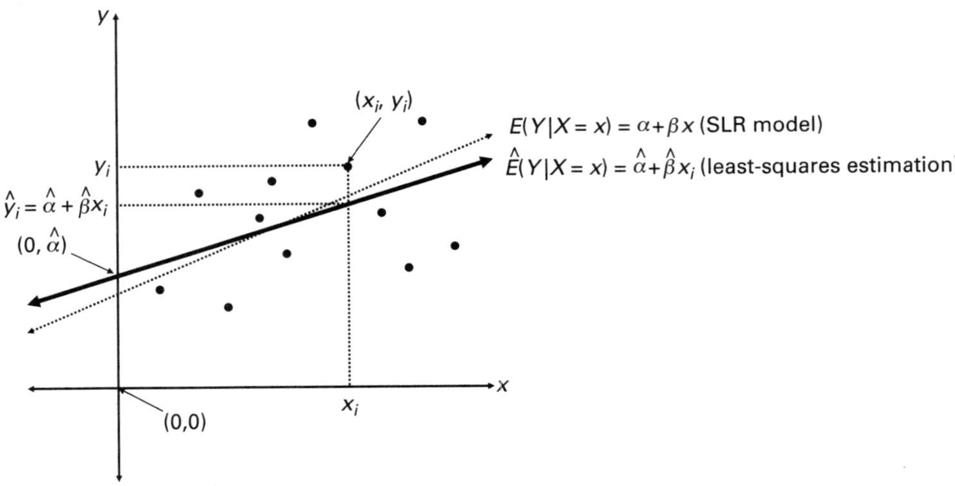

Figure 17.6
Least-squares estimated line

$=\hat{\alpha}+\hat{\beta}x_i$. Graphically, for a value of x_i, the fitted value $\hat{y}_i$ is read off the estimated regression line at $x=x_i$. For this reason, the least-squares regression line is also sometimes called the *fitted line*.

Figure 17.6 shows the same data set as in figure 17.3, but now with the least-squares estimated line $\hat{E}(Y|X=x)=\hat{\alpha}+\hat{\beta}x$ drawn. After estimation, the values of the estimates $\hat{\alpha}$ and $\hat{\beta}$ are known, so the least-squares estimated line is also known. In contrast, the true conditional-expectation line from the SLR model, $E(Y|X=x)=\alpha+\beta x$, is unknown since α and β are unknown parameters. It has been drawn in this graph, with a thin dotted line to make clear that this line, although unknown, is the one that we are trying to estimate with the least-squares estimated line. As shown in the graph, the estimated intercept $\hat{\alpha}$ for this sample turns out to be greater than the true α, and the estimated slope $\hat{\beta}$ turns out to be less (i.e., a flatter slope) than the true β. Looking at the specific point (x_i, y_i) highlighted on the graph, the fitted value $\hat{y}_i=\hat{\alpha}+\hat{\beta}x_i$ is read off the estimated line, with the realized outcome y_i greater than the fitted value $\hat{y}_i$ and the point being above the estimated line.

Definition 17.2 *The* ***estimated residual*** *for observation i, denoted as $\hat{u}_i$, is*

$$\hat{u}_i=y_i-\hat{y}_i=y_i-\hat{\alpha}-\hat{\beta}x_i,$$

where $\hat{\alpha}$ and $\hat{\beta}$ are the least-squares estimates of α and β, respectively.

The estimated residual is the difference between the realized outcome y_i and its fitted value $\hat{y}_i$. Graphically, observations with y_i above the estimated regression line have positive estimated residuals, and observations with y_i below the estimated regression line have negative estimated residuals. Looking at figure 17.6, the highlighted point (x_i, y_i) is above the estimated line and has a positive estimated residual since the realized outcome y_i is greater than the fitted value $\hat{y}_i$. Among the 12 data points shown in the graph, 6 lie above the estimated line and have positive estimated residuals, while 6 lie below the estimated line and have negative estimated residuals.

Proposition 17.5 states that the fitted values and estimated residuals provide consistent estimates of their population counterparts.[4]

Proposition 17.5 (Consistency of fitted values and estimated residuals) *Assume that the SLR model holds and $s_x^2 > 0$. Then, for any $i \in \{1, 2, \ldots, n\}$, the fitted value $\hat{y}_i$ is a consistent estimate of $E(Y|X = x_i) = \alpha + \beta x_i$, and the estimated residual $\hat{u}_i$ is a consistent estimate of the population residual $u_i = y_i - \alpha - \beta x_i$.*

This result follows directly from the fact that the least-squares estimators of α and β are consistent.

Example 17.9 (Cigarette sales and cigarette taxes) *Example 17.7 provided least-squares estimates of an SLR model with cigarette sales (cigsales) as the outcome variable y, measured in packs per capita in 2019, and cigarette tax (cigtax) as the explanatory variable x, measured in dollars per pack. Using the intercept estimate $\hat{\alpha} = 55.95$ and the slope estimate $\hat{\beta} = -9.49$, the fitted values and estimated residuals for the 51 observations can be calculated, respectively, as*

$$\hat{y}_i = \hat{\alpha} + \hat{\beta}x_i = 55.95 - 9.49x_i$$

and

$$u_i = y_i - \hat{y}_i = y_i - (55.95 - 9.49x_i).$$

These fitted values and estimated residuals are calculated by the `lm` *function in R. Specifically, if the results from* `lm` *estimation are stored in a variable called* `results`, *the fitted values and estimated residuals are contained in* `results$residuals` *and* `results$fitted.values`, *respectively:*

```
# store results of least-squares estimation
results <- lm(cigsales~cigtax, data=cigdata)

# access estimated residuals and fitted values
uhat <- results$residuals
yhat <- results$fitted.values

# output first seven estimated residuals and fitted values
uhat[1:7]

##        1        2        3        4        5        6        7
##  -6.5746   3.5549   2.5613 -16.2746 -12.9208 -17.5797   7.2201

yhat[1:7]

##      1      2      3      4      5      6      7
## 36.975 49.545 45.039 36.975 28.721 47.980 14.680
```

The following table shows fitted values $\hat{y}_i$ and estimated residuals $\hat{u}_i$ for the first seven states in the data:

State	x_i *(cigtax$_i$)*	y_i *(cigsales$_i$)*	$\hat{y}_i$	$\hat{u}_i$
Alaska	2.00	30.4	37.0	−6.6
Alabama	0.675	53.1	49.5	3.6
Arkansas	1.15	47.6	45.0	2.6
Arizona	2.00	20.7	37.0	−16.3
California	2.87	15.8	28.7	−12.9
Colorado	0.84	30.4	48.0	−17.6
Connecticut	4.35	21.9	14.7	7.2

The negative slope estimate is reflected by the fact that the fitted values are larger in states with lower tax rates. For example, Alabama has a low tax rate of \$0.675 *per pack and a high fitted value of* 49.5 *packs per-capita, whereas Connecticut has a high tax rate of* \$4.35 *per pack and a low fitted value of* 14.7 *packs per-capita. The fitted value* $\hat{y}_i$ *can be thought of as an* ***in-sample prediction*** *since it's an estimate of the conditional expectation of the outcome given that the explanatory variable is equal to* x_i. *For Alabama, the fitted value of* 49.5 *is what the estimates say we should expect, on average, for a state having a tax rate of* \$0.675. *The actual observed outcome for Alabama is* $y_i = 53.1$ *packs per-capita, which is* 3.6 *higher than the fitted value or in-sample prediction. The* 3.6 *value is the estimated residual* $\hat{u}_i$ *for Alabama, which can thought of as an* ***in-sample prediction error***. *For Connecticut, the observed outcome is* 21.9 *packs per-capita, which is* 7.2 *higher than the fitted value of* 14.7. *That is, the fitted value* $\hat{y}_i = 14.7$ *packs per-capita is what would be expected, on average, for a state with a tax rate* $x_i = 4.35$, *and the estimated residual* $\hat{u}_i = 7.2$ *indicates that the observed outcome is* 7.2 *packs per capita higher than the in-sample prediction.*

Estimation of the SLR model is often called *least-squares* estimation. The "least-squares" descriptor is used since an important property of estimates $\hat{\alpha}$ and $\hat{\beta}$ is that they minimize a summation that involves squared estimated residuals. This least-squares property is stated in the proposition 17.6.

Proposition 17.6 *If* $s_x^2 > 0$, *the least-squares estimates* $\hat{\alpha} = \bar{y} - \hat{\beta}\bar{x}$ *and* $\hat{\beta} = \frac{s_{xy}}{s_x^2}$ *are the values of a and b that minimize the function*

$$S(a, b) = \sum_{i=1}^{n} (y_i - a - bx_i)^2.$$

The values a and b can be thought of as guesses for intercept α and slope β, respectively. Then, the term $y_i - a - bx_i$ is an estimated residual associated with those guesses. The function $S(a, b)$ is a summation of the squared values of these estimated residuals over all the sample observations. Why do we want to minimize the function $S(a, b)$? The intuition is that good guesses for a and b are associated with an estimated regression line that goes through the data points in such a way that the vertical distances from the outcomes (y_i) to the estimated line ($a + bx_i$) are relatively small. The proof of proposition 17.6 uses partial derivatives.[5]

Evaluated at the least-squares estimates, the function $S(a, b)$ from proposition 17.6 is

$$S(\hat{\alpha}, \hat{\beta}) = \sum_{i=1}^{n} (y_i - \hat{\alpha} - \hat{\beta}x_i)^2 = \sum_{i=1}^{n} \hat{u}_i^2,$$

which is the sum of squared estimated residuals. This summation provides a measure of the overall noise in the y_i values that is not explained by the estimated regression line, but it is not normalized by the number of observations and therefore can be large simply because of a large sample size. A more useful measure is one that normalizes by sample size n. Definition 17.3 introduces the **residual variance estimate**, which is based on the sum of squared estimated residuals but also normalizes by the sample size.

Definition 17.3 *The* ***residual variance estimate****, denoted as* $\hat{\sigma}_U^2$*, is*

$$\hat{\sigma}_U^2 = \frac{1}{n-2}\sum_{i=1}^{n} \hat{u}_i^2.$$

Definition 17.4 *The* ***residual standard deviation estimate****, denoted as* $\hat{\sigma}_U$*, is*

$$\hat{\sigma}_U = \sqrt{\hat{\sigma}_U^2} = \sqrt{\frac{1}{n-2}\sum_{i=1}^{n} \hat{u}_i^2}.$$

The division by $n-2$, rather than n or $n-1$, used in these two definitions is discussed next.

Proposition 17.7 summarizes some important properties of the estimated residuals and the residual variance estimate.

Proposition 17.7 (Properties of estimated residuals) *The estimated residuals* $\hat{u}_i = y_i - \hat{\alpha} - \hat{\beta}x_i$*, based on the least-squares estimates* $\hat{\alpha}$ *and* $\hat{\beta}$*, have the following properties:*

i. *The sample average of the estimated residuals is zero:*

$$\frac{1}{n}\sum_{i=1}^{n} \hat{u}_i = 0.$$

ii. *The sample correlation between the values of the explanatory variable and the estimated residuals is zero:*

$$r_{x\hat{u}} = 0.$$

iii. *The sample correlation between the fitted values and the estimated residuals is zero:*

$$r_{\hat{y}\hat{u}} = 0.$$

iv. *The residual variance estimate* $\hat{\sigma}_U^2 = \frac{1}{n-2}\sum_{i=1}^{n} \hat{u}_i^2$ *is a consistent (and unbiased) estimate of* σ_U^2*, the unconditional variance of the random variable* U*.*

v. *The residual standard deviation estimate* $\hat{\sigma}_U = \sqrt{\frac{1}{n-2}\sum_{i=1}^{n} \hat{u}_i^2}$ *is a consistent estimate of* σ_U*, the unconditional standard deviation of the random variable* U*.*

Properties (i) and (ii) follow directly from the minimization problem in proposition 17.6 (see note 17.5). Property (i) says that, on average, the difference between the realized outcome y_i and its fitted value $\hat{y}_i$ equals zero. Equivalently, property (i) says that, on average, the fitted values are equal to the sample mean $\bar{y}$:

$$\frac{1}{n}\sum_{i=1}^{n} \hat{y}_i = \frac{1}{n}\sum_{i=1}^{n} y_i + \frac{1}{n}\sum_{i=1}^{n} \hat{u}_i = \frac{1}{n}\sum_{i=1}^{n} y_i = \bar{y}.$$

Thinking of the fitted value $\hat{y}_i$ as an in-sample prediction of the outcome Y associated with $X = x_i$, residual $\hat{u}_i$ can be thought of as a prediction error. This property, which says that the average prediction error equals zero, is desirable since a nonzero average prediction error would indicate that the estimation is systematically overestimating or underestimating the expected outcomes. Property (ii) says that the explanatory variable x_i and the estimated residual $\hat{u}_i$ are uncorrelated. Knowing whether x_i is above or below sample mean $\bar{x}$ does not provide information about whether $\hat{u}_i$ tends to be above or below its sample mean (which is zero). This property is desirable since a nonzero correlation between x_i and $\hat{u}_i$ would indicate that there is information in the x_i values that is not being utilized. For example, if the correlation between x_i and $\hat{u}_i$ were positive, it would imply that $\hat{u}_i$ tends to be positive for higher values of x_i ($x_i > \bar{x}$) and negative for lower values of x_i ($x_i < \bar{x}$), meaning that a line with a larger slope would provide a better fit than the estimated regression line.

Property (ii) is a sample counterpart to the population property that X and U are uncorrelated, which is implied by the exogeneity assumption $E(U|X) = 0$. That said, the fact that x_i and $\hat{u}_i$ are uncorrelated is a property of the least-squares estimates that holds *regardless of whether the exogeneity assumption is true*. It might be tempting to try to test whether X and U are uncorrelated by looking at the sample correlation between x_i and $\hat{u}_i$. Unfortunately, this approach is not useful since that sample correlation is always zero, even if X and U are correlated.

Property (iii) follows directly from property (ii) since $\hat{y}_i$ is a perfect linear function of x_i. Specifically, the covariance between fitted values and estimated residuals is

$$s_{\hat{y}\hat{u}} = Cov(\hat{\alpha} + \hat{\beta}x, \hat{u}) = \hat{\beta}Cov(x, \hat{u}) = 0,$$

which implies $r_{\hat{y}\hat{u}} = 0$.

Property (iv) provides a way to estimate the overall noise or variation contained in the unobservable U. The unconditional variance of U is σ_U^2, and the estimate $\hat{\sigma}_U^2 = \frac{1}{n-2}\sum_{i=1}^n \hat{u}_i^2$ is consistent, getting arbitrarily close to σ_U^2 as the sample size increases. The $\frac{1}{n-2}$ scaling is usually used in practice since it also leads to $\hat{\sigma}_U^2$ being unbiased. Compared to the $\frac{1}{n-1}$ scaling used for a sample variance estimate, the 2 in the $\frac{1}{n-2}$ scaling accounts for the two estimates $\hat{\alpha}$ and $\hat{\beta}$ needed to calculate the $\hat{u}_i$ values. Since we focus on the asymptotic properties of least-squares estimation, the use of the $\frac{1}{n-2}$ scaling, rather than, say, a $\frac{1}{n}$ scaling, becomes inconsequential for a large sample size n. As a summary measure, the estimated residual variance $\hat{\sigma}_U^2$ may not be ideal since its units are expressed in terms of y squared. On the other hand, the estimated residual standard deviation $\hat{\sigma}_U$ is a measure expressed in the units of y and, therefore, more easily interpretable. Property (v) states that $\hat{\sigma}_U$ is consistent, so it gets arbitrarily close to σ_U as the sample size increases.

The properties of estimated residuals are illustrated using the cigarette sales-tax regression example.

Example 17.10 (Cigarette sales and cigarette taxes) *Example 17.9 calculated the estimated residuals and fitted values from least-squares estimation in R, storing them as the variables* `uhat` *and* `yhat`*. The following R output confirms that the sample average of estimated residuals is equal to zero (property (i)) and, equivalently, that the sample average of fitted values is equal to the sample average of the outcome variable:*

```
mean(uhat)

## [1] 0.000000000000000019355

mean(cigdata$cigsales)

## [1] 38.82

mean(yhat)

## [1] 38.82
```

The following R output confirms that the correlation between the estimated residuals and the explanatory variable is equal to zero (property (ii)), as is the correlation between the estimated residuals and the fitted values (property (iii)):

```
cor(uhat,cigdata$cigtax)

## [1] -0.000000000000000023861

cor(uhat,yhat)

## [1] 0.000000000000000066505
```

Finally, the following R code calculates the residual standard deviation estimate:

```
sighatsq_u <- sum(uhat^2)/(nrow(cigdata)-2)
sighat_u <- sqrt(sighatsq_u)
sighat_u

## [1] 12.521

summary(results)$sigma

## [1] 12.521
```

The residual standard deviation estimate is $\hat{\sigma}_U \approx 12.52$ packs per capita. To get a sense of how large this residual standard deviation is, it can be compared to the standard deviation of the outcome variable, which is $s_y \approx 16.65$ packs per capita. The estimate $\hat{\sigma}_U$ is determined in two ways by the code shown here. The first way is to use the formulas for $\hat{\sigma}_U^2$ and $\hat{\sigma}_U$ directly, based on the estimated residual $\hat{u}_i$ values. The second way is to access the stored results from `lm` *estimation using the* `summary(results)$sigma` *command. The expression* `summary(results)` *provides a nice summary view of the* `lm` *estimation results, and some of its components (like* `$sigma` *for the residual standard deviation estimate) can be easily accessed. Here is the output from* `summary(results)` *for this example:*

```
summary(results)

##
## Call:
## lm(formula = cigsales ~ cigtax, data = cigdata)
##
## Residuals:
##    Min     1Q Median     3Q    Max
## -23.92  -8.10  -0.86   5.01  39.34
##
```

```
## Coefficients:
##              Estimate Std. Error t value           Pr(>|t|)
## (Intercept)    55.95       3.24   17.25 < 0.0000000000000002 ***
## cigtax         -9.49       1.51   -6.28         0.000000088 ***
## -
## Signif. codes:  0 '***' 0.001 '**' 0.01 '*' 0.05 '.' 0.1 ' ' 1
##
## Residual standard error: 12.5 on 49 degrees of freedom
## Multiple R-squared:  0.446, Adjusted R-squared:  0.434
## F-statistic: 39.4 on 1 and 49 DF,  p-value: 0.0000000875
```

This output contains many elements that have not yet been discussed, but the $\hat{\sigma}_U \approx 12.52$ *estimate can be seen near the end, with the line beginning* `Residual standard error`.

Working directly from the definition for estimated residuals, $\hat{u}_i = y_i - \hat{y}_i$, the outcome variable y_i can be written as

$$y_i = \hat{y}_i + \hat{u}_i.$$

This equation shows that y_i can be decomposed into two parts: the fitted value $\hat{y}_i = \hat{\alpha} + \hat{\beta}x_i$, which is the part of the outcome that is directly related to the explanatory variable x_i; and the estimated residual $\hat{u}_i$, which is the part of the outcome that is uncorrelated with the explanatory variable x_i. This decomposition can be used to determine how well least-squares estimation, based on the explanatory variable x_i, explains the variation in the y_i outcomes. From property (iii) of proposition 17.7, we know that the fitted values $\hat{y}_i$ and estimated residuals $\hat{u}_i$ are uncorrelated with each other ($s_{\hat{y}\hat{u}} = 0$), meaning that the sample variance of y is

$$s_y^2 = s_{\hat{y}}^2 + s_{\hat{u}}^2.$$

Thus, the variation in the outcome variable y_i is also decomposed into two parts: the variation explained by the explanatory variable x_i, given by $s_{\hat{y}}^2 = \frac{1}{n-1}\sum_{i=1}^{n}(\hat{y}_i - \bar{y})^2$; and the variation left unexplained by the explanatory variable, given by $s_{\hat{u}}^2 = \frac{1}{n-1}\sum_{i=1}^{n}\hat{u}_i^2$. ($s_{\hat{u}}^2$ and $\hat{\sigma}_U^2$ are slightly different since the former has a $\frac{1}{n-1}$ scaling and the latter has a $\frac{1}{n-2}$ scaling.) Therefore, the fraction of the variation in the outcome variable that is explained by the explanatory variable equals

$$\frac{s_{\hat{y}}^2}{s_y^2},$$

which can also be written as

$$1 - \frac{s_{\hat{u}}^2}{s_y^2}$$

since $s_{\hat{y}}^2 = s_y^2 - s_{\hat{u}}^2$ implies $\frac{s_{\hat{y}}^2}{s_y^2} = \frac{s_y^2 - s_{\hat{u}}^2}{s_y^2} = 1 - \frac{s_{\hat{u}}^2}{s_y^2}$. This measure of overall regression fit is known as the **R-squared** value, described in definition 17.5.

Definition 17.5 *The* ***R-squared*** *value associated with least-squares estimation of the SLR model, denoted as* R^2, *is*

$$R^2 = \frac{s_{\hat{y}}^2}{s_y^2} = 1 - \frac{s_{\hat{u}}^2}{s_y^2}.$$

For instance, if $R^2 = 0.24$, we say that "the explanatory variable explains 24% of the variation in the outcome variable." It is important to note that the sample size n does not have a direct impact on R^2. While having a very large sample improves the precision of the least-squares estimates, there is no reason to expect that R^2 increases for larger n. A large sample still has the same inherent residual noise as a smaller sample. In fact, looking at the expression for R^2, we see that $R^2 = 1 - \frac{s_{\hat{u}}^2}{s_y^2}$ should get arbitrarily close to $1 - \frac{\sigma_U^2}{\sigma_Y^2}$ as $n \to \infty$.

The terminology "R-squared" comes from the fact that R^2 equals the square of the sample correlation between the outcomes y_i and the fitted values $\hat{y}_i$, as stated in proposition 17.8.

Proposition 17.8 *The R-squared value is equal to the square of the correlation between outcomes y_i and fitted values $\hat{y}_i$:*

$$R^2 = r_{y\hat{y}}^2.$$

For simple linear regression, it is also the case that $R^2 = r_{yx}^2$.

This result can be shown as follows:

$$r_{y\hat{y}}^2 = \frac{(s_{y\hat{y}})^2}{s_y^2 s_{\hat{y}}^2} = \frac{\left(s_{\hat{y}}^2\right)^2}{s_y^2 s_{\hat{y}}^2} = \frac{s_{\hat{y}}^2}{s_y^2} = R^2,$$

where the first equality follows from the definition of correlation and the second equality follows from the fact that $s_{y\hat{y}} = s_{\hat{y}\hat{y}} + s_{\hat{u}\hat{y}} = s_{\hat{y}}^2 + 0 = s_{\hat{y}}^2$. Since $\hat{y}$ is a linear transformation of x, it also follows that $r_{y\hat{y}}^2 = r_{yx}^2$ and, therefore, $R^2 = r_{yx}^2$.

Since the correlation $r_{y\hat{y}}$ is between -1 and 1 (inclusive), it follows from proposition 17.8 that

$$0 \leq R^2 \leq 1.$$

At one extreme, $R^2 = 0$ corresponds to a case where there is no correlation between y_i and $\hat{y}_i$, which can happen only if $\hat{\beta} = 0$. Graphically, this case has a flat estimated regression line, with $\hat{\alpha} = \bar{y}$ and $\hat{\beta} = 0$. For $R^2 = 0$, the explanatory variable x_i explains none of the variation in outcome y_i. At the other extreme, $R^2 = 1$ corresponds to a case where there is perfect correlation, either positive ($r_{y\hat{y}} = 1$) or negative ($r_{y\hat{y}} = -1$). Graphically, the estimated regression line has $\hat{\beta} \neq 0$ and passes exactly through *all* the sample observation points (y_i, x_i). For $R^2 = 1$, the explanatory variable x_i explains 100% of the variation in outcome y_i.

For values of R^2 strictly between 0 and 1, there is a correlation between y_i and $\hat{y}_i$ but not perfect correlation. A regression with a larger magnitude of $r_{y\hat{y}}$ has a larger R^2. The sign of the correlation $r_{y\hat{y}}$ does not affect R^2, so for instance, the correlations $r_{y\hat{y}} = 0.8$ and $r_{y\hat{y}} = -0.8$ both have $R^2 = 0.8^2 = 0.64$.

Example 17.11 (Cigarette sales and cigarette taxes) *Continuing example 17.7, the decomposition of the outcome variance*

$$s_y^2 = s_{\hat{y}}^2 + s_{\hat{u}}^2$$

involves the following sample variances:

$$s_y^2 = 277.155, s_{\hat{y}}^2 = 123.522, \text{ and } s_{\hat{u}}^2 = 153.632.$$

The R-squared value is

$$R^2 = \frac{123.522}{277.155} \approx 0.446 \text{ or } 44.6\%,$$

which indicates that state-level cigarette taxes explain 44.6% of the variation in state-level cigarette sales. Therefore, 55.4% of the variation in state-level cigarette sales is left unexplained by state-level cigarette taxes. Another way to determine the R-squared value, which doesn't require the least-squares estimates at all, is to square the sample correlation between cigsales and cigtax. This sample correlation is $r_{cigsales,cigtax} = -0.6676$ for the observed sample, meaning that $R^2 = (-0.6676)^2 \approx 0.446$. The following R code illustrates the various equivalent methods of calculating R^2 for this example:

```
y <- cigdata$cigsales
x <- cigdata$cigtax

var(yhat)/var(y)

## [1] 0.44568

1-var(uhat)/var(y)

## [1] 0.44568

cor(y,yhat)^2

## [1] 0.44568

cor(y,x)^2

## [1] 0.44568

summary(results)$r.squared

## [1] 0.44568
```

The first two calculations of R^2 are based on definition 17.5. The second two calculations are based on proposition 17.8. The last expression accesses the R^2 value directly from the `lm` *regression results.*

Example 17.12 (Monthly stock returns and the overall market) *Continuing example 17.8, we consider the overall fit of the least-squares estimation using an individual stock return as outcome y and the market index return (IDX) as the explanatory variable x. For the six stocks considered in example 17.8, the following table reports the R-squared value and the estimated residual standard deviation associated with the least-squares estimates:*

	R^2	$\hat{\sigma}_U$
Home Depot (HD)	0.336	0.0602
Lowe's (LOW)	0.256	0.0791
Bank of America (BAC)	0.350	0.0850
Wells Fargo (WFC)	0.289	0.0689
Marathon Oil (MRO)	0.251	0.1048
ConocoPhillips (COP)	0.285	0.0692

The R-squared values are fairly similar across the six individual stocks. The lowest R-squared is for Marathon Oil, indicating that 25.1% of the variation in Marathon Oil's monthly returns are explained by S&P 500 monthly returns, and the highest R-squared is for Bank of America, indicating that 35.0% of the variation in Bank of America's monthly returns are explained by S&P 500 monthly returns. For these two stocks, since R^2 is equal to the square of the sample correlation between x and y, the corresponding sample correlations are $r_{MRO,IDX} = \sqrt{0.251} \approx 0.501$ and $r_{BAC,IDX} = \sqrt{0.350} \approx 0.592$. (These sample correlations must be positive, rather than negative, because the least-squares slope estimates reported in example 17.8 were both positive.)

Here is the R code to produce this table of R^2 and $\hat{\sigma}_U$ values:

```
# loop through the six stocks of interest
for (varname in c("HD","LOW","BAC","WFC","MRO","COP")) {
    # run the SLR regression of company returns vs index (IDX) returns
    results <- lm(sp500[,varname]~sp500$IDX)

    # access statistics from the regression results
    rsq <- summary(results)$r.squared
    sighat_u <- summary(results)$sigma

    # output the statistics
    print(paste(varname,"versus index:", "R-sq", round(rsq,3), "sighat_u",
          round(sighat_u,4)))
}

## [1] "HD versus index: R-sq 0.336 sighat_u 0.0602"
## [1] "LOW versus index: R-sq 0.256 sighat_u 0.0791"
## [1] "BAC versus index: R-sq 0.35 sighat_u 0.085"
## [1] "WFC versus index: R-sq 0.289 sighat_u 0.0689"
## [1] "MRO versus index: R-sq 0.251 sighat_u 0.1048"
## [1] "COP versus index: R-sq 0.285 sighat_u 0.0692"
```

17.4 Asymptotic Normality and Statistical Inference

For the purposes of statistical inference, we assume that sample size n is large enough for asymptotic inference. In this case, the least-squares estimators $\hat{\alpha}_{XY}$ and $\hat{\beta}_{XY}$ are each asymptotically normal estimators of their estimands α and β.

Proposition 17.9 *If the SLR model holds and $\sigma_X^2 > 0$, the least-squares estimators $\hat{\alpha}_{XY}$ and $\hat{\beta}_{XY}$ are $\sqrt{n}$-consistent and asymptotically normal estimators of α and β, respectively, with*

$$\sqrt{n}(\hat{\alpha}_{XY} - \alpha) \overset{a}{\sim} N(0, V_\alpha)$$

and

$$\sqrt{n}(\hat{\beta}_{XY} - \beta) \overset{a}{\sim} N\left(0, V_\beta\right),$$

so the asymptotic variance of $\hat{\alpha}_{XY}$ is $\frac{V_\alpha}{n}$ and the asymptotic variance of $\hat{\beta}_{XY}$ is $\frac{V_\beta}{n}$.

The proof of asymptotic normality is beyond the scope of this book, but it relies primarily on versions of the LLN and CLT introduced in chapter 13.[6]

To say more about the asymptotic variances $\frac{V_\alpha}{n}$ and $\frac{V_\beta}{n}$ and the associated standard errors, it is useful to introduce terminology related to the conditional variance of the SLR model's residuals.

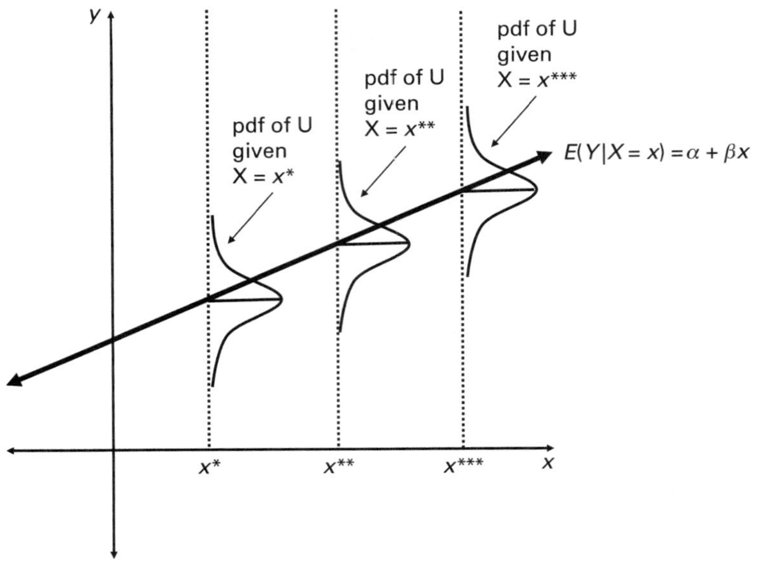

Figure 17.7
SLR model with homoskedastic residuals

Definition 17.6 *The residuals of an SLR model are* ***homoskedastic*** *if the conditional variance* $Var(U|X=x)$ *is constant and does not depend on x. Equivalently, since* σ_U^2 *is the unconditional variance of U, the residuals are homoskedastic if* $Var(U|X=x)=\sigma_U^2$ *for all x. The residuals are said to exhibit* ***homoskedasticity****.*

Figure 17.7 provides a graphical representation of the data-generating process for an SLR model with homoskedastic residuals. The graph shows three possible values (x^*, x^{**}, x^{***}) for the random variable X. At each of these three values, the conditional expectation $E(Y|X=x)$ is the value for which the associated vertical line passes through the $E(Y|X=x)=\alpha+\beta x$ line. The realized y outcome depends on this conditional expectation and the realized residual. The realized residual is a draw of U, conditional on the value of X. For $X=x^*$, this conditional distribution is shown as the rotated pdf curve for U conditional on $X=x^*$. This pdf curve is centered on the SLR model line, and its variance describes the variance of residual draws associated with $X=x^*$. Similarly, for x^{**} and x^{***}, the rotated pdf curves shown in the graph represent the distributions of U conditional on $X=x^{**}$ and $X=x^{***}$, respectively. In figure 17.7, the shape of the conditional distribution of U is exactly the same for $X=x^*$, $X=x^{**}$, and $X=x^{***}$, meaning that the conditional variance $Var(U|X=x)$ is the same for these possible values of X. Although the constant conditional residual variance has been shown for only three values in the figure, the same conditional residual variance would arise for any other value of X if the residuals are homoskedastic.

Definition 17.7 *The residuals of an SLR model are* ***heteroskedastic*** *if the conditional variance* $Var(U|X=x)$ *is nonconstant and depends on x. The residuals are said to exhibit* ***heteroskedasticity****.*

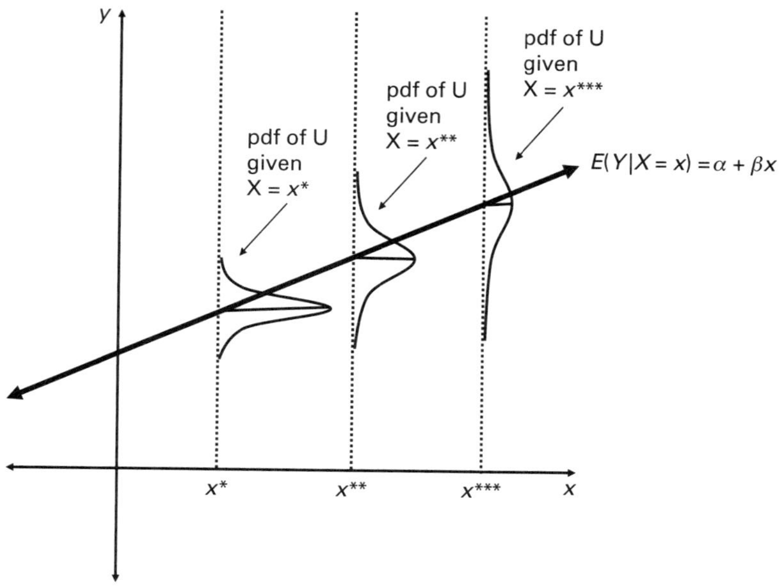

Figure 17.8
The SLR model with heteroskedastic residuals

Figure 17.8 shows how the data-generating process differs when the SLR model has heteroskedastic residuals. The graph shows a situation in which the variance of the residual U depends on the value of X, with the residual variance increasing for larger values of X. The distribution of U conditional on $X = x^*$ has a lower variance, with its pdf being more tightly distributed around the SLR line. The variance of the distribution of U conditional on $X = x^{**}$ is larger, as indicated by the increased dispersion of the pdf curve, and the variance of the distribution of U conditional on $X = x^{***}$ is even larger with a more dispersed pdf.

To see how the cases of homoskedastic errors and heteroskedastic errors affect the realized sample, we artificially create two samples based on the SLR model: one exhibiting homoskedasticity and one exhibiting heteroskedasticity. Specifically, we assume that the SLR model

$$Y = 1 + 8X + U$$

describes the conditional expectation $E(Y|X) = 1 + 8X$ for both samples, but the conditional distribution of U differs for the two samples, as follows:

- Sample 1: Homoskedastic errors, where the distribution of U given $X = x$ is $N(0, 1)$
- Sample 2: Heteroskedastic errors, where the distribution of U given $X = x$ is $N(0, 4x^2)$

Figure 17.9 shows the two samples generated by the SLR model under the two assumptions on the residuals. For each sample, the size is $n = 500$, and the x-values are draws from the $U(0, 1)$ distribution. The y-values are generated as $y = 1 + 8x + u$, where u is a draw from the distribution of U given $X = x$, as specified for the two samples here. The scatter plot on the top corresponds to sample 1, with homoskedastic errors, and the scatter plot on

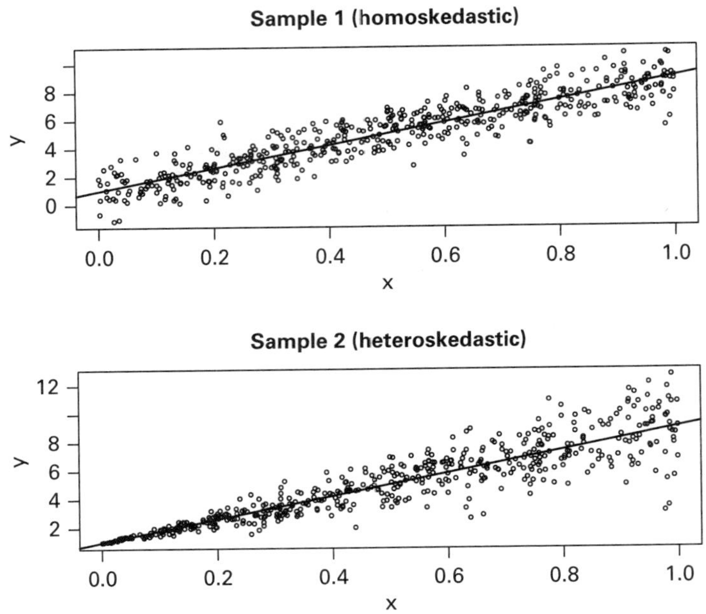

Figure 17.9
Homoskedasticity versus heteroskedasticity

the bottom corresponds to sample 2, with heteroskedastic errors. A solid line corresponding to the SLR line $E(Y|X) = 1 + 8X$ is drawn for reference in both plots. For the sample with homoskedastic errors, the vertical spread of data points around the SLR line stays roughly the same throughout the full range of possible x-values. On the other hand, for the sample with heteroskedastic errors, the vertical spread of data points around the SLR line changes a lot over the range of x-values, with very low conditional variance of the residuals for x-values near zero and a steadily increasing conditional variance of the residuals as x increases.

Practitioners care about the homoskedasticity/heteroskedasticity of residuals for two main reasons. First, the appropriate way to calculate standard errors associated with the least-squares estimates turns out to depend on whether the residuals are homoskedastic or heteroskedastic. This issue is discussed in more detail next. Second, if we are interested in determining a predictive interval for Y given $X = x$, this interval clearly depends on the conditional variance of U given $X = x$. For instance, looking at sample 2 in figure 17.9, a predictive interval for Y associated with a low value of x should be much narrower than a predictive interval for Y associated with a high value of x since the conditional variance of the residuals is much higher for higher values of x. We return to this idea in section 18.8, where we discuss predictive intervals based on least-squares estimation.

The formulas for the asymptotic variances $\frac{V_\alpha}{n}$ and $\frac{V_\beta}{n}$ in the general case of heteroskedasticity are quite complicated, and therefore are omitted from our discussion. That said, statistical packages like R are able to calculate standard errors based on these formulas, and these standard errors are known as **heteroskedasticity-robust standard errors** or sometimes, more concisely, as **robust standard errors**. For least-squares estimates $\hat{\alpha}$ and $\hat{\beta}$, which are the realizations of the estimators $\hat{\alpha}_{XY}$ and $\hat{\beta}_{XY}$, the standard errors are

$$se(\hat{\alpha}) = \sqrt{\frac{\widehat{V_\alpha}}{n}}, \text{ where } \widehat{V_\alpha} \text{ is a consistent estimator of } V_\alpha,$$

and

$$se(\hat{\beta}) = \sqrt{\frac{\widehat{V_\beta}}{n}}, \text{ where } \widehat{V_\beta} \text{ is a consistent estimator of } V_\beta.$$

Even if we suspect that residuals are homoskedastic, there is no harm in calculating heteroskedasticity-robust standard errors since they are valid regardless of whether homoskedasticity holds. Therefore, it is an accepted practice in economics to always use heteroskedasticity-robust standard errors for i.i.d. cross-sectional data.

Even though heteroskedasticity-robust standard errors are used in practice, it is useful to briefly consider the case of homoskedastic residuals for pedagogical reasons since the formulas for the asymptotic variances $\frac{V_\alpha}{n}$ and $\frac{V_\beta}{n}$ simplify greatly. Proposition 17.10 provides these formulas.

Proposition 17.10 *If the SLR model holds, $\sigma_X^2 > 0$, and the residuals are homoskedastic, the asymptotic variance of $\hat{\alpha}_{XY}$ is*

$$\frac{V_\alpha}{n} = \frac{\sigma_U^2}{n}\left(1 + \frac{\mu_X^2}{\sigma_X^2}\right),$$

and the asymptotic variance of $\hat{\beta}_{XY}$ is

$$\frac{V_\beta}{n} = \frac{\sigma_U^2}{n\sigma_X^2}.$$

The asymptotic-variance formulas in proposition 17.10 highlight the features of the population and sample that affect the precision of the least-squares estimators. To obtain standard errors for the case of homoskedasticity, sample descriptive statistics can be plugged in for population statistics, so

$$se(\hat{\alpha}) = \sqrt{\frac{\widehat{V_\alpha}}{n}} = \sqrt{\frac{s_{\hat{u}}^2}{n}\left(1 + \frac{\bar{x}^2}{s_x^2}\right)} = \frac{s_{\hat{u}}}{\sqrt{n}}\sqrt{1 + \frac{\bar{x}^2}{s_x^2}}$$

and

$$se(\hat{\beta}) = \sqrt{\frac{\widehat{V_\beta}}{n}} = \sqrt{\frac{s_{\hat{u}}^2}{ns_x^2}} = \frac{s_{\hat{u}}}{\sqrt{n}\,s_x}.$$

Let's focus on the slope estimate $\hat{\beta}$ first, as its standard error expression is a bit simpler. There are three factors that affect the standard error of $\hat{\beta}$:

- *Sample size:* Larger n leads to a smaller standard error $se(\hat{\beta})$, so having more observations improves precision. The standard error has the usual $\frac{1}{\sqrt{n}}$ scaling for $\sqrt{n}$-consistent and asymptotically normal estimators. For example, quadrupling the sample size leads to a standard error that is roughly half as large.

- *Residual noise:* A smaller residual variance σ_U^2, as estimated by $s_{\hat{u}}^2$, leads to a smaller standard error $se(\hat{\beta})$. Although it is not within our control, less noise in the residuals is better for precision.
- *Variation of the explanatory variable:* A larger variance σ_X^2 of the explanatory variable, as estimated by s_x^2, leads to a smaller standard error $se(\hat{\beta})$. Since slope β represents the change in $E(Y|X=x)$ associated with changes in x, the intuition is that it is better to have observations that exhibit a wide range of x-values. In cases where the distribution of X may be within our control, as might be the case when choosing a survey population (e.g., if age is the explanatory variable, choosing to interview people between 25 and 45 rather than between 30 and 40), a choice with higher variance σ_X^2 leads to a lower standard error $se(\hat{\beta})$, holding n and σ_U^2 fixed.

Now, looking at the formula for the standard error of intercept estimate $\hat{\alpha}$, these three factors affect $se(\hat{\alpha})$ in the same way. Larger n, smaller residual variance σ_U^2, and larger explanatory variable variance σ_X^2 are all associated with lower $se(\hat{\alpha})$ and more precise intercept estimates. There is a fourth factor for $se(\hat{\alpha})$, which is that a larger value of μ_X^2 or $\bar{x}^2$ leads to a larger standard error $se(\hat{\alpha})$. Recall that $\hat{\alpha}$ is an estimate of $E(Y|X=0)$. When the average X-value is far from zero (i.e., μ_X very negative or μ_X very positive), as indicated by μ_X^2 being large, this relationship says that it becomes more difficult to precisely estimate intercept α. The best-case scenario, for $se(\hat{\alpha})$, is when $\bar{x}=0$, in which case the standard-error formula simplies to $se(\hat{\alpha})=\frac{s_{\hat{u}}}{\sqrt{n}}$, which is similar to the formula for the standard error of a sample mean.

Unfortunately, the built-in *R* function `lm` does not calculate heteroskedasticity-robust standard errors and instead reports standard errors based on the restrictive assumption of homoskedastic residuals. Thankfully, there are several *R* packages with functions to calculate robust standard errors. We will use a package called `estimatr`, whose least-squares regression function is `lm_robust` and will be illustrated in the examples throughout this chapter and chapter 18. For now, here is the code that installs the `estimatr` package:

```
install.packages("estimatr")
library(estimatr)
```

Example 17.13 *(Cigarette sales and cigarette taxes) Example 17.7 reported the least-squares estimates for the SLR model using cigarette sales as the outcome variable and cigarette tax as the explanatory variable:*

$$\hat{\alpha}=55.95 \text{ and } \hat{\beta}=-9.49.$$

Using the `lm_robust` *function, the heteroskedasticity-robust standard errors are*

$$se(\hat{\alpha})=2.92 \text{ and } se(\hat{\beta})=1.06.$$

In R, the `lm_robust` *function is used like the* `lm` *function, with the specification of the* `formula` *argument (for the model to be estimated) and the optional* `data` *and* `subset` *arguments:*[7]

```
results <- lm_robust(cigsales~cigtax, data=cigdata)
results

##              Estimate Std. Error t value   Pr(>|t|) CI Lower CI Upper DF
## (Intercept)  55.9489     2.9236 19.1367 2.2024e-24   50.074  61.8242 49
## cigtax       -9.4871     1.0635 -8.9206 7.7435e-12  -11.624  -7.3499 49

summary(results)

##
## Call:
## lm_robust(formula = cigsales ~ cigtax, data = cigdata)
##
## Standard error type:  HC2
##
## Coefficients:
##             Estimate Std. Error t value Pr(>|t|) CI Lower CI Upper DF
## (Intercept)    55.95       2.92   19.14 2.20e-24     50.1    61.82 49
## cigtax         -9.49       1.06   -8.92 7.74e-12    -11.6    -7.35 49
##
## Multiple R-squared:  0.446 , Adjusted R-squared:  0.434
## F-statistic: 79.6 on 1 and 49 DF,  p-value: 7.74e-12
```

The commands `results` *and* `summary(results)` *provide alternative ways of displaying the* `lm_robust` *estimation results, with the latter providing additional information like the R-squared value. In addition to the parameter estimates and standard errors, the output includes additional columns for hypothesis testing and confidence intervals, which will be discussed in section 17.4.1. In the* `summary(results)` *output, the* `HC2` *specified as the "Standard error type" is the method that* `lm_robust` *uses to calculate the robust standard errors.*

When `results` *is assigned to the* `lm_robust` *regression function call, lots of useful information about the regression is stored in* `results`*, including the following:*

- `results$res_var`*: residual variance estimate* $\hat{\sigma}_U^2$
- `results$r.squared`*: R-squared value*
- `results$fitted.values`*: a vector (of length n) with the fitted values*
- `results$coefficients`*: a vector with the estimates, in the same order as the output*
- `results$std.error`*: a vector with the standard errors, in the same order as the output*

Here are some examples of how these quantities can be accessed after the regression given here:

```
# R-squared value
results$r.squared

## [1] 0.4456806

# residual standard deviation estimate
sqrt(results$res_var)

## [1] 12.52068
```

```
# slope estimate and its standard error
results$coefficients[2]

##    cigtax
## -9.487131

results$std.error[2]

##   cigtax
## 1.063502
```

Example 17.14 (Monthly stock returns and the overall market) *Example 17.8 reported the least-squares estimates for SLR models that related monthly returns of six individual stocks to monthly returns of the S&P 500 index. The following table augments those estimates with their heteroskedasticity-robust standard errors, reported in parentheses:*

	$\hat{\alpha}$ *(se)*	$\hat{\beta}$ *(se)*
Home Depot (HD)	0.0087 (0.0032)	1.020 (0.082)
Lowe's (LOW)	0.0117 (0.0042)	1.107 (0.104)
Bank of America (BAC)	0.0014 (0.0046)	1.489 (0.146)
Wells Fargo (WFC)	0.0054 (0.0039)	1.048 (0.131)
Marathon Oil (MRO)	−0.0006 (0.0053)	1.449 (0.203)
ConocoPhillips (COP)	0.0028 (0.0037)	1.042 (0.116)

The results from this table are obtained using the `lm_robust` *function for each of the six SLR models. For example, the first row corresponds to the SLR model with HD as the outcome variable and IDX as the explanatory variable, with results given by the following code:*

```
results <- lm_robust(HD~IDX, data=sp500)
results

##              Estimate Std. Error t value  Pr(>|t|)  CI Lower CI Upper  DF
## (Intercept) 0.0087316  0.0032119  2.7185 6.8739e-03 0.0024152 0.015048 362
## IDX         1.0204494  0.0816816 12.4930 5.0256e-30 0.8598195 1.181079 362
```

17.4.1 Confidence Intervals and Hypothesis Testing

Since least-squares estimators are $\sqrt{n}$-consistent and asymptotically normal, statistical inference can be conducted by using the general theory developed for confidence intervals in section 14.4, and for z-tests in section 16.2. While we focus on two-sided confidence intervals and two-sided tests, we can also construct one-sided confidence intervals and conduct one-sided tests as in those previous sections.

Calculation of confidence intervals requires only the least-squares estimates and their standard errors, with the $1-\alpha$ confidence intervals for α and β given by

$$(\hat{\alpha} - z_{\alpha/2}se(\hat{\alpha}), \hat{\alpha} + z_{\alpha/2}se(\hat{\alpha}))$$

and

$$(\hat{\beta} - z_{\alpha/2}se(\hat{\beta}), \hat{\beta} + z_{\alpha/2}se(\hat{\beta})),$$

respectively. For example, the 95% confidence interval for β is $(\hat{\beta} - 1.96se(\hat{\beta}), \hat{\beta} + 1.96se(\hat{\beta}))$, and the 90% confidence interval for β is $(\hat{\beta} - 1.645se(\hat{\beta}), \hat{\beta} + 1.645se(\hat{\beta}))$.

Example 17.15 *(Cigarette sales and cigarette taxes) Using the standard errors of the least-squares estimates from example 17.13, we can construct confidence intervals for α and β. The 95% confidence interval for α is*

$$(\hat{\alpha} - z_{0.025}se(\hat{\alpha}), \hat{\alpha} + z_{0.025}se(\hat{\alpha})) = (55.95 - (1.96)(2.924), 55.95 + (1.96)(2.924))$$
$$\approx (50.07, 61.82).$$

It can be said with 95% confidence that intercept parameter α, or equivalently the conditional expectation $E[CIGSALES|CIGTAX = 0]$, is between 50.07 and 61.82 packs per capita. The 95% confidence interval for β is

$$(\hat{\beta} - z_{0.025}se(\hat{\beta}), \hat{\beta} + z_{0.025}se(\hat{\beta})) = (-9.49 - (1.96)(1.064), -9.49 + (1.96)(1.064))$$
$$\approx (-11.62, -7.35).$$

It can be said with 95% confidence that slope parameter β, or equivalently the change in expected cigarette sales associated with a \$1 increase in cigarette taxes, is between -11.62 and -7.35 packs per capita.

The `lm_robust` *function automatically calculates these confidence intervals in R. The default is 95% confidence intervals, but the optional argument* `alpha` *can be set at other values (different from the default* `alpha=0.05`*) to get other confidence intervals. Here is the code for calculating 95% confidence intervals and 90% confidence intervals:*

```
# output results with 95% confidence intervals (default alpha=0.05)
lm_robust(cigsales~cigtax, data=cigdata)

##             Estimate Std. Error t value   Pr(>|t|) CI Lower CI Upper DF
## (Intercept)  55.9489     2.9236 19.1367 2.2024e-24   50.074  61.8242 49
## cigtax       -9.4871     1.0635 -8.9206 7.7435e-12  -11.624  -7.3499 49

# output results with 90% confidence intervals (alpha=0.10)
lm_robust(cigsales~cigtax, data=cigdata, alpha=0.10)

##             Estimate Std. Error t value   Pr(>|t|) CI Lower CI Upper DF
## (Intercept)  55.9489     2.9236 19.1367 2.2024e-24   51.047  60.8505 49
## cigtax       -9.4871     1.0635 -8.9206 7.7435e-12  -11.270  -7.7041 49
```

Moving to hypothesis testing, suppose that we want to test whether the slope parameter β is equal to a constant c, so the null hypothesis is

$$H_0 : \beta = c.$$

The z-statistic for testing this null hypothesis is

$$\text{z-statistic} = \frac{\hat{\beta} - c}{se(\hat{\beta})},$$

and it indicates the number of standard errors that $\hat{\beta}$ is above c (positive z-statistic) or the number of standard errors that $\hat{\beta}$ is below c (negative z-statistic). Applying the z-test approach from section 16.2, the rejection rule for testing $H_0 : \beta = c$ at the α level is:

- Reject $H_0 : \beta = c$ at the α level if the |z-statistic| $= \left|\frac{\hat{\beta}-c}{se(\hat{\beta})}\right| \geq z_{\alpha/2}$.
- Do not reject $H_0 : \beta = c$ at the α level if the |z-statistic| $= \left|\frac{\hat{\beta}-c}{se(\hat{\beta})}\right| < z_{\alpha/2}$.

As with other asymptotically normal estimators, we can calculate a p-value based on the z-statistic, and this p-value can be used to test the null hypothesis at *any* level α. For the null hypothesis $H_0 : \beta = c$, the p-value is

$$\text{p-value} = P\left(|Z| > |\text{z-statistic}|\right) = P\left(|Z| > \left|\frac{\hat{\beta} - c}{se(\hat{\beta})}\right|\right), \quad \text{where } Z \sim N(0, 1),$$

and the rejection rule for a test at the α level is:

- Reject $H_0 : \beta = c$ at the α level if the p-value $< \alpha$.
- Do not reject $H_0 : \beta = c$ at the α level if the p-value $> \alpha$.

A particularly important hypothesis test for the slope parameter uses $c = 0$, which corresponds to testing

$$H_0 : \beta = 0.$$

If $H_0 : \beta = 0$ is true, the SLR model doesn't contain the explanatory variable at all since $Y = \alpha + 0X + U = \alpha + U$. Therefore, a test of $H_0 : \beta = 0$ is essentially a test of whether X belongs in the model. If $H_0 : \beta = 0$ is rejected, that provides evidence that X belongs in the model or, in other words, that X is a significant factor in explaining Y. Also, since $\beta = \rho_{YX}\frac{\sigma_Y}{\sigma_X}$ is a scaled version of ρ_{YX}, the null hypothesis $H_0 : \beta = 0$ can be true only if $\rho_{YX} = 0$. Thus, the null hypothesis $H_0 : \beta = 0$ is equivalent to the null hypothesis that there is no correlation between Y and X, and least-squares estimation provides a simple way to establish a statistically significant correlation between the outcome variable and the explanatory variable (by rejecting $H_0 : \beta = 0$).

For the null hypothesis $H_0 : \beta = 0$, the z-statistic simplifies to

$$\text{z-statistic} = \frac{\hat{\beta}}{se(\hat{\beta})},$$

which is the ratio between estimate $\hat{\beta}$ and its standard error and indicates how many standard errors estimate $\hat{\beta}$ is from zero. The rejection rule is a special case of the rejection rule given previously:

- Reject $H_0 : \beta = 0$ at the α level if the |z-statistic| $= \left|\frac{\hat{\beta}}{se(\hat{\beta})}\right| \geq z_{\alpha/2}$.
- Do not reject $H_0 : \beta = 0$ at the α level if the |z-statistic| $= \left|\frac{\hat{\beta}}{se(\hat{\beta})}\right| < z_{\alpha/2}$.

Testing this null hypothesis is so commonly done by practitioners that it is standard for statistical packages to automatically report the z-statistic and p-value for the test of $H_0 : \beta = 0$, in addition to estimate $\hat{\beta}$ and its standard error $se(\hat{\beta})$. The output provided by the `lm_robust` function in the examples given here provides the z-statistic and p-value for $H_0: \beta = 0$ in the columns labeled `t value` and `Pr(>|t|)`. (More appropriate column labels would be "z-stat" and "p-value," respectively.)

While the discussion here has focused on z-tests for slope parameter β, the same ideas also apply to intercept parameter α. For a test of $H_0: \alpha = c$, the z-statistic is

$$\text{z-statistic} = \frac{\hat{\alpha} - c}{se(\hat{\alpha})},$$

and the z-test proceeds in the usual way using this z-statistic or its associated p-value. The z-statistic and p-value for testing $H_0: \alpha = 0$, which may or may not be of interest in a particular application, are also provided by the `lm_robust` results.

Example 17.16 (Cigarette sales and cigarette taxes) *Here is the code and output from example 17.13:*

```
results <- lm_robust(cigsales~cigtax, data=cigdata)
results

##              Estimate Std. Error t value   Pr(>|t|) CI Lower CI Upper DF
## (Intercept)  55.9489     2.9236 19.1367 2.2024e-24   50.074  61.8242 49
## cigtax       -9.4871     1.0635 -8.9206 7.7435e-12  -11.624  -7.3499 49
```

The first and second columns provide the estimates ($\hat{\alpha}$ in the first row, $\hat{\beta}$ in the second row) and the standard errors of the estimates ($se(\hat{\alpha})$ in the first row, $se(\hat{\beta})$ in the second row). The third column is the z-statistic for testing $H_0: \alpha = 0$ (first row) and for testing $H_0: \beta = 0$ (second row). The test of $H_0: \alpha = 0$ has a z-statistic equal to $\frac{\hat{\alpha}}{se(\hat{\alpha})}$, and the test of $H_0: \beta = 0$ has a z-statistic equal to $\frac{\hat{\beta}}{se(\hat{\beta})}$. Therefore, the z-statistics in the third column are equal to the estimates in the first column divided by the standard errors in the second column. Finally, the fourth column reports the p-value for the z-tests, which is the p-value for testing $H_0: \alpha = 0$ in the first row and the p-value for testing $H_0: \beta = 0$ in the second row.

Seeing such a low p-value for the α parameter is to be expected since it would be surprising if we could not reject the fact that the expected cigarette sales equals zero when there is no cigarette tax. In fact, from its z-statistic, estimate $\hat{\alpha}$ is 19.14 standard errors above zero! The p-value associated with $H_0: \beta = 0$ is more interesting. This p-value is again very small (zero to 11 decimal places), meaning the null hypothesis $H_0: \beta = 0$ is rejected at any level. In other words, the slope estimate is statistically significant at any level, meaning there is strong statistical support for the existence of a negative relationship between state-level cigarette sales and state-level cigarette taxes. The magnitude of this relationship can be provided by using a confidence interval, like the 95% confidence interval for β provided in example 17.15.

We've already seen that `results` contains information about the regression, and several quantities associated with the confidence intervals and testing are accessible after `results` is assigned to the `lm_robust` function call:

- `results$statistic`: A vector with the z-statistics of the estimates, in the same order as the output

- `results$p.value`: A vector with the p-values for the two-sided test against 0, in the same order as the output
- `results$conf.low` and `results$conf.high`: Vectors with the upper and lower end points of the confidence interval, respectively, in the same order as the output

For instance, `results$statistic[2]` is the slope parameter's z-statistic, and `results$conf.high[2]` is the upper end point of the confidence interval for the slope parameter.

Example 17.17 (Monthly stock returns and the overall market) *Using the estimates and standard errors reported in example 17.14, we can conduct z-tests on the SLR model parameters for any of the six stocks considered. For example, here are the least-squares estimation results previously seen for the Home Depot (HD) monthly stock returns, with the z-statistics and p-values reported for the tests of* $H_0 : \alpha = 0$ *and* $H_0 : \beta = 0$*:*

```
results <- lm_robust(HD~IDX, data=sp500)
results

##              Estimate Std. Error t value    Pr(>|t|)  CI Lower CI Upper  DF
## (Intercept) 0.0087316  0.0032119  2.7185 6.8739e-03 0.0024152 0.015048 362
## IDX         1.0204494  0.0816816 12.4930 5.0256e-30 0.8598195 1.181079 362
```

A test of the null hypothesis $H_0 : \alpha = 0$ *is interesting here since it is a test of whether the expected Home Depot monthly return equals zero when the S&P 500 monthly return equals zero. Based on the p-value of* 0.0069*, this null hypothesis is rejected at any level above* 0.69%*. The null hypothesis* $H_0 : \beta = 0$ *is less interesting here since it would be surprising to find no relationship (i.e., no correlation) between Home Depot returns and S&P 500 returns. Indeed, the p-value is equal to zero (to 29 decimal places), with the estimated slope* 12.55 *standard errors above zero. A more interesting null hypothesis to test is* $H_0 : \beta = 1$*. When* $\beta = 1$*, a* 0.01 *change in the S&P 500 monthly return is associated with an expected change of* 0.01 *in the Home Depot monthly return. The z-statistic associated with* $H_0 : \beta = 1$ *is*

$$z\text{-}statistic = \frac{\hat{\beta} - 1}{se(\hat{\beta})} = \frac{1.020 - 1}{0.0817} \approx 0.245.$$

The associated p-value is $2(1 - \Phi(0.245)) \approx 0.806$*. Therefore, using either the z-statistic rejection rule or the p-value rejection rule, the null hypothesis* $H_0 : \beta = 1$ *would not be rejected at a 5% level. Based on the p-value,* $H_0 : \beta = 1$ *would not be rejected at any reasonable level.*

Example 17.18 (Earnings and union status) *Example 17.6 provided the least-squares estimates for the SLR model with weekly earnings as the outcome variable and union membership as the explanatory variable. The more complete results are reported here:*

```
results <- lm_robust(earnwk~union, data=cps)
results

##             Estimate Std. Error t value   Pr(>|t|) CI Lower CI Upper   DF
## (Intercept)   946.50     14.897 63.5366 0.0000e+00   917.29   975.71 2807
## union         251.15     45.827  5.4805 4.6184e-08   161.30   341.01 2807
```

The test of $H_0 : \alpha = 0$ is not interesting here since the average earnings of nonunion workers is certainly nonzero. Instead, we focus on the test of $H_0 : \beta = 0$. Recall that

$$\beta = E(EARNWK|UNION = 1) - E(EARNWK|UNION = 0),$$

meaning that the null hypothesis $H_0 : \beta = 0$ is true if there is no difference between the expected weekly earnings of union workers and the expected weekly earnings of nonunion workers. The p-value of zero (to 7 decimal places) indicates that $H_0 : \beta = 0$ is rejected at any level, providing evidence that the association of weekly earnings with union status is statistically significant. The 95% confidence interval for β is

$$(\hat{\beta} - z_{0.025}se(\hat{\beta}), \hat{\beta} + z_{0.025}se(\hat{\beta})) = (251.2 - (1.96)(45.83), 251.2 + (1.96)(45.83))$$
$$\approx (161.3, 341.0),$$

meaning that we can say with 95% confidence that the true difference between the expected weekly earnings of union workers and the expected weekly earnings of nonunion workers is between \$161.30 *and* \$341.00.

Example 17.19 (A/B testing) *Example 17.18 showed that the SLR model and least-squares estimation can be used to directly test whether the expected value of an outcome variable Y differs over two subpopulations, where the binary variable X indicates which of the two subpopulations an observation is in. In example 17.18, the random variable Y is weekly earnings and the random variable X is union membership ($X = 1$ for union member, $X = 0$ for nonmember). The case of an A/B test, where the outcome of interest is a continuous outcome Y, also fits in to this framework. If there are two possible treatments (A or B) and X is a binary variable indicating the treatment (say, $X = 1$ for treatment B, $X = 0$ for treatment A), then*

$$\alpha = E(Y|X = 0) = \text{expected value of } Y \text{ for treatment } A$$

and

$$\alpha + \beta = E(Y|X = 1) = \text{expected value of } Y \text{ for treatment } B.$$

This setup is a generalization of the advertising experiment considered in example 17.1, where the Y random variable was SALES and the X random variable was AD (equal to 1 *for cities receiving targeted advertising and* 0 *for cities not receiving targeted advertising). To test whether the expected value of Y differs for the two treatments, the null hypothesis of interest is $H_0 : \beta = 0$. The z-test involves using either the z-statistic $\frac{\hat{\beta}}{se(\hat{\beta})}$ or its corresponding p-value. A confidence interval for the estimated difference between the expected outcome Y for treatment A and the expected outcome Y for treatment B is just the standard confidence interval for β.*

17.5 Causality and Prediction

In section 17.1, the SLR model $Y = \alpha + \beta X + U$, with the exogeneity assumption $E(U|X) = 0$, was introduced, and the several examples discussed in that section considered whether the exogeneity assumption was likely to hold. In this section, we provide a more detailed discussion of the exogeneity assumption and, in particular, its importance for establishing a causal effect of the explanatory variable X on the outcome variable Y. Even though least-squares estimation can be used more generally for the purposes of prediction (i.e., predicting Y given X), estimating the causal effect of X on Y requires the strong assumption of exogeneity.

First, let's consider the relationship between random variables Y and X *without specifying an exogeneity assumption*. To do so, proposition 17.11 provides an important result concerning the decomposition of Y into two parts, one that is a linear function of X and another that is a random variable uncorrelated with X.

Proposition 17.11 *If X and Y are random variables with $\sigma_X^2 > 0$, Y can be decomposed into a linear function of X and another random variable V such that*

$$Y = \alpha^* + \beta^* X + V \text{ with } Cov(X, V) = 0 \text{ and } E(V) = 0.$$

For this decomposition, parameter β^* is

$$\beta^* = \frac{Cov(X, Y)}{Var(X)} = \frac{\sigma_{XY}}{\sigma_X^2},$$

and parameter α^* is

$$\alpha^* = \mu_Y - \beta^* \mu_X = \mu_Y - \frac{\sigma_{XY}}{\sigma_X^2}\mu_X.$$

While the decomposition equation in proposition 17.11 looks very similar to the specification of the SLR model, it is important to stress that the relationship described by the proposition is *not* a model, in the sense that no assumptions have been made to state the relationship between Y and X. The observant reader will notice that β^* is the population analog of the least-squares slope estimator $\hat{\beta}_{XY} = \frac{s_{XY}}{s_X^2}$, and likewise α^* is the population analog of the least-squares intercept estimator $\hat{\alpha}_{XY} = \bar{Y} - \hat{\beta}_{XY}\bar{X}$.

Proposition 17.12 *For the decomposition in proposition 17.11, the least-squares slope estimators $\hat{\alpha}_{XY}$ and $\hat{\beta}_{XY}$ are consistent estimators of parameters α^* and β^*, respectively, if $s_X^2 > 0$. Moreover, parameters α^* and β^* provide the* ***best linear prediction*** *of Y given X in the population, in the sense that these parameters minimize $E\left[(Y - a - bX)^2\right]$ over all possible a and b values.*

Proposition 17.12 formally states that the least-squares estimators consistently estimate parameters α^* and β^* in the decomposition in proposition 17.11. An immediate implication is that the realized least-squares estimates $\hat{\alpha}$ and $\hat{\beta}$ get arbitrarily close to α^* and β^*, respectively, when the sample size grows large. The second part of proposition 17.12 states that parameters α^* and β^* provide the best linear prediction of Y given X in the population. A general linear predictor for Y is $a + bX$ for some numbers a and b, in which case the prediction error is $Y - a - bX$. The proposition says that choosing $a = \alpha^*$ and $b = \beta^*$ does the best job at minimizing the expected squared prediction error, which is $E\left[(Y - a - bX)^2\right]$. In fact, we already know from proposition 17.6 that the least-squares estimates $\hat{\alpha}$ and $\hat{\beta}$ minimize the sample version of this average, which is $\frac{1}{n}\sum_{i=1}^{n}(y_i - a - bx_i)^2$. Therefore, the least-squares estimates can be said to provide the best linear prediction of y given x for the realized sample, in the sense of minimizing the average squared in-sample prediction error. Importantly, this property of least-squares estimation holds even without assuming that the SLR model holds.

Now, turning to the idea of causality, let's consider the linear equation specified in the SLR model, but *without the exogeneity assumption*:

$$Y = \alpha + \beta X + U \text{ with } E(U) = 0.$$

We have stated $E(U) = 0$, which is just a normalization (relative to parameter α), rather than $E(U|X) = 0$. Suppose that we are able to shift the explanatory variable X by 1 unit while *holding the unobservable U fixed*. In this case, the outcome variable Y would change by β units. Thus, β measures the causal effect of a 1-unit change of X on Y, holding U fixed, and is the object of interest when a practitioner would like to establish and estimate a causal effect of X on Y. The practical problem arises when the explanatory variable X and the unobservable U are related to each other, in which case it doesn't make sense to hold U fixed while changing X. For instance, if X and U are positively correlated, it is more likely than not that U increases when X increases by 1 unit.

In the absence of the exogeneity assumption $E(U|X) = 0$, least-squares estimation does not generally provide a consistent estimate of the causal (slope) parameter β. To see the issue clearly, it is useful to work out the expression for $Cov(X, Y)$ without the exogeneity assumption imposed:

$$Cov(X, Y) = Cov(X, \alpha + \beta X + U) = \beta Var(X) + Cov(X, U),$$

which implies

$$\beta = \frac{\sigma_{XY} - \sigma_{XU}}{\sigma_X^2} = \beta^* - \frac{\sigma_{XU}}{\sigma_X^2}.$$

From the previous discussion, we know that the least-squares slope estimator always provides a consistent estimator of parameter β^*, but the causal-effect (slope) parameter β is equal to β^* only when X and U are uncorrelated ($\sigma_{XU} = 0$).

Proposition 17.13 *Suppose that α^* and β^* are the parameters from the decomposition in proposition 17.11. If $s_X^2 > 0$ and the SLR model holds without the exogeneity assumption imposed, which is*

$$Y = \alpha + \beta X + U \text{ with } E(U) = 0,$$

then the least-squares slope estimator $\hat{\beta}_{XY}$ is a consistent estimator of

$$\beta^* = \beta + \frac{\sigma_{XU}}{\sigma_X^2}.$$

When X and U are uncorrelated ($\sigma_{XU} = 0$), $\hat{\beta}_{XY}$ is a consistent estimator of β.

Proposition 17.13 contains two important results. First, in the case that X and U are uncorrelated, the model becomes

$$Y = \alpha + \beta X + U \text{ with } Cov(X, U) = 0 \text{ and } E(U) = 0,$$

which coincides exactly with the decomposition in proposition 17.11, with $\alpha^* = \alpha$, $\beta^* = \beta$, and $V = U$. For this case, the least-squares slope estimator provides a consistent estimator of the causal effect of X on Y. The SLR model's exogeneity assumption $E(U|X) = 0$ implies

$Cov(X, U) = 0$, and therefore implies consistent estimation of the causal-effect parameter β. Moreover, with the exogeneity assumption $E(U|X) = 0$, we know from section 17.1 that β has a causal intepretation in terms of conditional expectations since

$$\beta = E(Y|X = x^* + 1) - E(Y|X = x^*).$$

Second, in cases where X and U are correlated, knowing that the least-squares slope estimator consistently estimates $\beta + \frac{\sigma_{XU}}{\sigma_X^2}$ may tell us something about whether $\hat{\beta}$ is providing an overestimate or an underestimate of the causal-effect parameter β. If there is a positive correlation between X and U, the term $\frac{\sigma_{XU}}{\sigma_X^2}$ is positive, meaning that $\hat{\beta}$ is estimating a population quantity larger than β; that is, $\hat{\beta}$ provides an overestimate of β in this case. On the other hand, if there is a negative correlation between X and U, the term $\frac{\sigma_{XU}}{\sigma_X^2}$ is negative, meaning that $\hat{\beta}$ provides an underestimate of β.

How can we know whether the correlation between X and U is positive or negative? After all, U is never observed, so it is not possible to calculate the sample correlation between x and u. And while it is true that we can never definitively know the sign of the correlation between X and U, it may be possible to make an educated guess about its sign for particular examples of X and Y random variables.

Example 17.20 (Cigarette sales and cigarette taxes) *In example 17.4, where cigarette sales (Y) were modeled as a function of cigarette taxes (X), we argued that a negative correlation between X and the unobservable U might exist since higher (lower) cigarette taxes are more likely to be legislated in states with fewer (more) smokers. If $\sigma_{XU} < 0$, then the least-squares estimate $\hat{\beta} = -9.49$ (from example 17.7) is likely to be an underestimate of β (i.e., too negative). Unfortunately, we can't say much more than that. While it would be desirable to know* how much *we are underestimating β, that would require direct knowledge of the magnitude of σ_{XU} or ρ_{XU}, which we do not have.*

Example 17.21 (Earnings and education) *In example 17.3, where weekly earnings (Y) were modeled as a function of educational attainment (X), we argued that a positive correlation between X and the unobservable U might exist since unobserved productivity is likely to be higher (lower) for people with more (less) education. The slope parameter β is a measure of the causal effect of a 1-year change in education on expected weekly earnings. With the positive correlation between X and U ($\sigma_{XU} > 0$), the least-squares slope estimate $\hat{\beta}$ is an overestimate of this causal-effect parameter β. Therefore, even if a positive slope estimate $\hat{\beta}$ is obtained, it's still possible that there is no causal effect of education on expected earnings ($\beta = 0$) since $\hat{\beta}$ is providing an over-estimate.*

Notes

1. It is not restrictive to assume that $E(U|X)$ is equal to zero, rather than some other constant, since parameter α could always be changed to yield $E(U|X) = 0$. For instance, if we had $Y = \alpha' + \beta X + U'$ with $E(U'|X) = c$ for a constant c, the model could be rewritten as $Y = \alpha + \beta X + U$ with $\alpha = \alpha' + c$ and $U = U' - c$.
2. In fact, linearity is not really an assumption at all for the case of a binary explanatory variable since a line can always be drawn exactly through the two points: $E(Y|X = 0)$, at $X = 0$; and $E(Y|X = 1)$, at $X = 1$. Once a third value is possible for the explanatory variable, the linearity assumption becomes important since it requires that all three conditional expectations lie along a line.

3. A slightly more complicated version of this model, which incorporates the "risk-free rate," is the *capital asset pricing model* (*CAPM*). In the CAPM, the outcome variable is the stock return minus the risk-free rate (e.g., the interest rate on a US Treasury bond), and the explanatory variable is the market index return minus the risk-free rate.
4. We are being a little loose with the terminology here. Consistency is a property of an estimator, so the term "consistent estimate" refers to the realization of a consistent estimator.
5. Take partial derivatives of $S(a,b)$ with respect to a and b and set them both equal to zero:

$$\frac{\partial S(a,b)}{\partial a} = -2\sum_{i=1}^{n}(y_i - a - bx_i) = 0$$

and

$$\frac{\partial S(a,b)}{\partial b} = -2\sum_{i=1}^{n}(y_i - a - bx_i)x_i = 0.$$

The first equation implies $n\bar{y} - na - bn\bar{x} = 0$, or $a = \bar{y} - b\bar{x}$. The second equation implies $\sum_{i=1}^{n}(y_i - a - bx_i)x_i = 0$, and plugging in $a = \bar{y} - b\bar{x}$ yields

$$\sum_{i=1}^{n}(y_i - \bar{y} - b(x_i - \bar{x}))x_i = 0$$

or, equivalently,

$$\sum_{i=1}^{n}(y_i - \bar{y} - b(x_i - \bar{x}))(x_i - \bar{x}) = 0.$$

Solving this last equation for b yields $b = \frac{\sum_{i=1}^{n}(y_i-\bar{y})(x_i-\bar{x})}{\sum_{i=1}^{n}(x_i-\bar{y})(x_i-\bar{x})} = \frac{s_{xy}}{s_x^2} = \hat{\beta}$. Finally, plugging $\hat{\beta}$ for b in $a = \bar{y} - b\bar{x}$ yields $a = \bar{y} - \hat{\beta}\bar{x} = \hat{\alpha}$.
6. Although not explicitly stated in proposition 17.9, the additional technical assumption that X and U have finite variances is required to apply CLT results and prove the result.
7. The interested reader can confirm that the nonrobust standard errors obtained using `lm` are $se(\hat{\alpha}) = 3.24$ and $se(\hat{\beta}) = 1.51$. To do so, store the results with the command `results <- lm(cigsales~cigtax, data=cigdata)` and view a summary of the results, including the standard errors, with `summary(results)`.

Exercises

1. Consider a sample $\{(x_i, y_i)\}_{i=1}^{n}$, where x and y are standardized variables with sample correlation $r_{xy} = 0.7$.
   a. For the SLR model $Y = \alpha + \beta X + U$, what are the least-squares estimates $\hat{\alpha}$ and $\hat{\beta}$?
   b. Under the exogeneity assumption $E(U|X) = 0$, what is the interpretation of the least-squares slope estimate $\hat{\beta}$?
2. A manufacturing company has 100 factories spread across the US. It has purchased a new technology that it can deploy at 30 of its factories. Assume that the company assigns the technology randomly to 30 factories. In the subsequent year, the company collects data on total production ($prod_i$, in thousands of units) at each factory, with $tech_i = 1$ if the factory has the new technology and $tech_i = 0$ otherwise. Let $PROD$ and $TECH$ be the associated random variables, and assume that the following SLR model holds:

$$PROD = 10 + 3TECH + U \text{ with } E(U|TECH) = 0,$$

with an additional assumption that $U|TECH \sim N(0,4)$.

a. Explain why the exogeneity assumption $E(U|TECH) = 0$ is likely to hold.

b. What is the conditional distribution of *PROD* given $TECH = 0$? What is the conditional distribution of *PROD* given $TECH = 1$?

c. Determine $P(PROD > 12|TECH = 1) - P(PROD > 12|TECH = 0)$.

d. If factory *A* has the new technology and factory *B* does not, what is the distribution of the difference between factory *A* production and factory *B* production?

3. Use the **hrs** data set for this question. The data consist of 6,052 nonmarried individuals who are 50 and older. Consider an SLR model, where the outcome variable is annual out-of-pocket medical costs (*outofpocket_costs*, in dollars) during 2000 and the explanatory variable is age (*age*, in years).

a. Use `lm` to estimate the SLR model, and store the results in `hrs_results`.

b. Interpret the slope estimate $\hat{\beta}$.

c. What is the estimated difference in expected out-of-pocket medical costs between a 70-year-old and a 60-year-old?

d. What percentage of the residuals are positive? Negative?

e. Which five observations are associated with the largest magnitudes of the estimated residuals?

f. What percentage of fitted values are negative? Is this problematic, since *outofpocket_costs* ≥ 0?

g. Draw a scatter plot of *outofpocket_costs* versus *age*.

h. Add the least-squares regression line to the plot in (g) with the command `abline(hrs_results, col="blue")`. Are the residuals left-skewed, right-skewed, or approximately symmetric?

i. Estimate the SLR model separately for men ($male = 1$) and women ($male = 0$). How do the slope estimates from the two regressions compare?

4. Use the **baseball** data set for this question. The data consist of 30 Major League Baseball (MLB) teams for the 2022 regular season. Consider an SLR model, where the outcome variable is a team's average attendance at its home games (*attend_home*) and the explanatory variable is the team's winning percentage for the season (*winpct_22*). A team winning half its games has $winpct\_22 = 0.5$, and a team winning 55% of its games has $winpct\_22 = 0.55$.

a. Use `lm` to estimate the SLR model, and store the results in `mlb_results`.

b. Draw a scatter plot of *attend_home* versus *winpct_22*. Add the least-squares regression line to the plot with the command `abline(mlb_results, col="blue")`.

c. What does the slope estimate $\hat{\beta}$ say about the difference between a team with a 55% winning percentage and a team with a 50% winning percentage?

d. How much of the variation in *attend_home* is left *unexplained* by the least-squares regression?

e. What is the estimated standard deviation of the SLR model residual?

f. Re-run the regression using the *previous* year's winning percentage (*winpct_21*) as the explanatory variable. Focusing on the slope estimate and the R-squared value, how do the results compare to the original regression?

g. Define a new outcome variable *pctattend* equal to *attend_home* divided by *capacity* (the size of the team's stadium). Estimate the SLR model with *pctattend* as the outcome variable and *winpct_22* as the explanatory variable. What does the slope estimate $\hat{\beta}$ say about the difference between a team with a 55% winning percentage and a team with a 50% winning percentage?

h. To determine the association between team payroll (*payroll*, in millions of dollars) and team performance (*winpct_22*), run a regression with *winpct_22* as the outcome variable and *payroll* as the explanatory variable. What is the estimated difference in expected winning percentage between a team with a payroll of $200 million and a team with a payroll of $150 million?

5. Use the **cps** data set for this question, and focus on the sample of 2,809 employed workers. You'll find it easiest to create a new data frame `cpsemployed` for the employed workers.

   a. Use `lm_robust` to estimate the SLR model from example 17.3 with *earnwk* as the outcome variable and *educ* as the explanatory variable.

   b. Interpret the slope estimate $\hat{\beta}$.

   c. Provide an asymptotic 95% confidence interval for the SLR slope parameter.

   d. Interpret the R-squared value.

   e. Use `plot` to plot the estimated residuals versus *educ*, and add a horizontal line at zero.

   f. From the plot in (e), do you think the residuals are homoskedastic or heteroskedastic? Explain.

   g. From the plot in (e), do you think the conditional expectation of the residuals is zero for all values of *educ*? Explain.

6. Use the **exams** data set for this question. You are interested in modeling *exam2* performance based on *exam1* performance.

   a. Use `lm_robust` to estimate the SLR model with *exam2* as the outcome variable and *exam1* as the explanatory variable.

   b. Interpret the slope estimate $\hat{\beta}$.

   c. Perform a *z*-test of $H_0 : \beta = 0$ at the 5% level. What do you conclude?

   d. Perform a *z*-test of $H_0 : \beta = 1$ at the 5% level. What do you conclude? What is the p-value for this test?

   e. Provide an estimate of the conditional expectation $E(exam2|exam1 = 80)$.

   f. Draw two histograms, one of the actual *exam2* values and one of the fitted values from the regression. Which histogram has lower dispersion, and why?

g. Standardize both variables (*exam*1 and *exam*2) and rerun the regression with `lm_robust`.

   i. Interpret the slope estimate of the new regression.

   ii. How does the R-squared value compare to the R-squared value of the original regression?

   iii. How does the z-statistic for testing $H_0 : \beta = 0$ compare to the original regression?

7. Consider a sample $\{(x_i, y_i)\}_{i=1}^n$, where $\bar{x} = 0$ and $\bar{y} = 0$. For the SLR model $Y = \alpha + \beta X + U$, show that the least-squares estimates are

$$\hat{\alpha} = 0 \text{ and } \hat{\beta} = \frac{\sum_{i=1}^n x_i y_i}{\sum_{i=1}^n x_i^2}.$$

8. A data set has IQ scores (*iq*) for 930 individuals between the ages of 20 and 24, along with the IQ scores of their mothers (*momiq*) and fathers (*dadiq*). The average parental IQ is $avgiq = \frac{momiq + dadiq}{2}$. With *iq* as the outcome variable, the following table summarizes the results from regressions for three SLR models: (i) *momiq* as the explanatory variable, (ii) *dadiq* as the explanatory variable, and (iii) *avgiq* as the explanatory variable.

	SLR: *iq* on *momiq*	SLR: *iq* on *dadiq*	SLR: *iq* on *avgiq*
$\hat{\alpha}$ (se)	69.585 (3.394)	77.156 (3.426)	56.487 (4.331)
$\hat{\beta}$ (se)	0.299 (0.034)	0.231 (0.035)	0.434 (0.044)
R^2	0.0925	0.0522	0.1161

Each of the IQ variables has a sample mean close to 100 and a sample standard deviation close to 15.

a. Which of the three explanatory variables explains the most variation in *iq*? What is the sample correlation between this variable and *iq*?

b. Interpret the slope estimate from the regression of *iq* on *momiq*.

c. Provide an asymptotic 95% confidence interval for the slope in the SLR model with *momiq* as the explanatory variable.

d. Provide an estimate of the expected difference in IQ between an individual whose mother's IQ is 110 and an individual whose mother's IQ is 105. What is the standard error of this estimate?

e. Thinking about an observation with $momiq = 120$, what is the fitted value from the regression of *iq* on *momiq*? Thinking about an observation with $dadiq = 120$, what is the fitted value from the regression of *iq* on *dadiq*? Thinking about an observation with $avgiq = 120$, what is the fitted value from the regression of *iq* on *avgiq*? Explain why one of these fitted values is markedly different from the others.

f. Focus on the SLR with *avgiq* as the explanatory variable. Copy and execute the *R* code here, which draws the solid estimated regression line, a dashed 45-degree line ($iq = avgiq$), and a dotted horizontal line at the outcome mean ($iq = 100$). The graph shows a phenomenon known as **regression to the mean**. For *avgiq* values between 110 and 130, how do the fitted values of *iq* compare to *avgiq* and the mean of *iq*?

For *avgiq* values between 70 and 90, how do the fitted values of *iq* compare to *avgiq* and the mean of *iq*?

```
# create a vector of IQ scores from 70 to 130
iqvec <- seq(70,130,1)

# plot the fitted regression line using avgiq
plot(iqvec, 56.487+0.434*iqvec, type="l", ylim=c(70,130), xlab="avgiq",
     ylab="iq")

# draw the iq=avgiq 45-degree line (dashed)
lines(iqvec, iqvec, lty=2)

# draw the line at iq=100 (dotted)
abline(h=100, lty=3)
```

g. Assume that the true SLR model is

$$IQ = 60 + 0.4AVGIQ + U,$$

where $U|AVGIQ \sim N(0, 13.75^2)$ is a homoskedastic normal random variable.

i. Provide a 95% probability interval for IQ, conditional on $AVGIQ = 110$.

ii. What is the probability that $IQ \geq 110$, conditional on $AVGIQ = 110$?

9. Suppose $Y = \alpha + \beta X + U$ is the SLR model. Reversing the roles of Y and X, we specify the **reverse regression model**

$$X = \gamma + \delta Y + V.$$

For a sample $\{(x_i, y_i)\}_{i=1}^n$, let $\hat{\alpha}$ and $\hat{\beta}$ denote the least-squares estimates of the SLR model and $\hat{\gamma}$ and $\hat{\delta}$ denote the least-squares estimates of the reverse regression model.

a. How do the R-squared values for the two regressions compare to each other?

b. Show that $\hat{\beta}\hat{\delta} = r_{xy}^2$.

10. For the **sp500** data set, example 17.12 reported $R^2 = 0.336$ for the least-squares regression of Home Depot (HD) returns on the market index (IDX) returns. The sample size is $n = 364$.

a. What is the sample correlation $r_{HD,IDX}$?

b. Provide an asymptotic 95% confidence interval for the population correlation $\rho_{HD,IDX}$.

c. $\rho_{HD,IDX}^2$ is the value to which R-squared converges as the sample size grows large. Use the end points of the interval in (b) to provide an asymptotic 95% confidence interval for $\rho_{HD,IDX}^2$.

d. *Use the delta method (section 14.5) to determine the asymptotic standard error of R^2, as an estimator of $\rho_{HD,IDX}^2$, and construct an asymptotic 95% confidence interval. How does it compare to the interval found in (c)?

11. For each of the following examples, do you think the least-squares slope estimate should be considered an estimate of the causal effect (β) of the explanatory variable on the outcome variable? If not, use proposition 17.13 to determine whether the least-squares regression is overestimaing or underestimating β.

a. A new medical treatment, targeted at 50-year-olds with a certain disease, is being tested. There is a sample of patients with the disease, but the medical treatment is provided only to patients who can afford it (i.e., patients over a certain income threshold). Data are collected after all the patients in the sample die. The outcome variable is the life span of the patient, and the explanatory variable is an indicator of whether they received treatment.

b. An economics department is interested in improving the postgraduation outcomes of its graduate students. The department chair decides to provide a fellowship, with extra monetary support and reduced teaching obligations, to the department's best students (specifically the top 20%) during the year prior to their graduation. The outcome variable is the income level of each graduate student in their initial job, and the explanatory variable is an indicator of whether they received the fellowship.

c. To assess the causal effect of an informational campaign about a new tax-credit policy, the Internal Revenue Service mails a postcard explaining the policy to a random subset of taxpayers. The new policy increases the tax refund for an individual who claims it. The outcome variable is the total refund amount for an individual, and the explanatory variable is an indicator of whether they received the postcard.

d. To assess the causal effect of policing on crime, a researcher collects data for hundreds of cities with the outcome variable being crimes per capita and the explanatory variable being police officers per capita.

e. To assess the causal effect of weather on crime, a researcher collects data for hundreds of days for a single city, with the outcome variable being daily crimes and the explanatory variable being daily rainfall total.

12. *For this question, you will conduct Monte Carlo simulations to illustrate the performance of the slope z-test in an SLR model $Y = \alpha + \beta X + U$. For all simulations, the marginal distribution of Y is $N(0, 1)$ and the marginal distribution of X is $N(0, 1)$.

    a. Consider the case where Y and X are independent, such that $\alpha = \beta = 0$. Conduct 10,000 Monte Carlo simulations, where for each simulation, you (i) create an i.i.d. sample with $n = 100$, $\{(y_i, x_i)\}_{i=1}^{100}$, with y_i drawn from $N(0, 1)$ and x_i drawn from $N(0, 1)$; (ii) estimate the least-squares regression; and (iii) test $H_0 : \beta = 0$ at the 1% level, the 5% level, and the 10% level (and keep track of the results). Over the 10,000 simulations, what is the percentage of times that the test is rejected at the 1% level, the 5% level, and the 10% level? Are your findings as expected?

    b. Consider the case where $Y = 0.2X + U$. Follow the process in (a), except with the following change in the sample creation step (i): Draw x_i from $N(0, 1)$ and u_i from $N(0, 1)$, and generate $y_i = 0.2x_i + \sqrt{0.96}u_i$. (This sampling ensures that y_i is drawn from a $N(0, 1)$ marginal distribution.) Over the 10,000 simulations, what is the percentage of times that the test is rejected at the 1% level, the 5% level, and the 10% level?

    c. Same as (b), except $Y = 0.5X + U$ and generate each y_i as $y_i = 0.5x_i + \sqrt{0.75}u_i$. Over the 10,000 simulations, what is the percentage of times that the test is rejected at the 1% level, the 5% level, and the 10% level?

d. Explain why the rejection rates change for (a)–(c).

e. Without actually doing the simulations, how do you think the results would change in (a)–(c) if you tested $H_0: \alpha = 0$ instead of $H_0: \beta = 0$?

13. Use the **sp500** data set for this question. We have considered the returns of two bank stocks, Bank of America (*BAC*) and Wells Fargo (*WFC*). Let's add a third bank stock, M&T Bank Corporation (*MTB*). You are interested in predicting *WFC* returns, but with only one explanatory variable, so the two possible SLR models are

$$WFC = \alpha_1 + \beta_1 BAC + U_1$$

and

$$WFC = \alpha_2 + \beta_2 MTB + U_2.$$

a. Use `lm_robust` to run the two least-squares regressions. How do $\hat{\beta}_1$ and $\hat{\beta}_2$ compare to each other? How do the two R-squared values compare?

b. *It's difficult to test $H_0: \beta_1 = \beta_2$ or form a confidence interval for $\beta_1 - \beta_2$ since (i) estimates $\hat{\beta}_1$ and $\hat{\beta}_2$ are not independent and (ii) the regression output from (a) doesn't provide the covariance of the two estimates. As an alternative, use the bootstrap (from chapter 15) to estimate the standard error of $\hat{\beta}_1 - \hat{\beta}_2$. Use the bootstrap with 5,000 iterations. During the bth iteration, for $b \in \{1, 2, \ldots, 5000\}$, run both regressions (on the same bootstrap sample) and calculate $\hat{\beta}_1^b - \hat{\beta}_2^b$. Then, calculate the bootstrap standard error. What is the normal-based bootstrap 95% confidence interval for $\beta_1 - \beta_2$? Do you reject $H_0: \beta_1 = \beta_2$ at a 5% level?

c. *As in (b), use the boottstrap to form a normal-based bootstrap 95% confidence interval for $\rho^2_{WFC,BAC} - \rho^2_{WFC,MTB}$, which is the limit to which the difference in R-squared values converges for a large sample. Rather than calculating $\hat{\beta}_1^b - \hat{\beta}_2^b$ in each iteration, calculate the difference in R-squared values between the first regression (*WFC* on *BAC*) and the second regression (*WFC* on *MTB*).

d. *As in (b), use the bootstrap to form a normal-based bootstrap 95% confidence interval for $\sigma_{U_1} - \sigma_{U_2}$, the difference between the standard deviations of the two regression residuals.

18

Multiple Linear Regression

Chapter 17 introduced the simple linear regression (SLR) model to model the relationship between outcome variable Y and a single explanatory variable X. The SLR model is often too simplistic since there may be multiple explanatory variables that should be included in a model describing the outcome variable. The regression approach is easily generalizable to allow more explanatory variables. This chapter introduces the multiple linear regression (MLR) model, a model that generalizes the SLR model from the previous chapter. While many of the ideas and results for the SLR model carry over to the MLR model, there are important new issues that arise with additional explanatory variables.

18.1 The Multiple Linear Regression Model

The **multiple linear regression (MLR) model** assumes a linear association between outcome variable Y and several explanatory variables. Let the explanatory variables be denoted as

$$X_1, X_2, \ldots, X_K,$$

where K is the number of explanatory variables in the MLR model. The notation $\mathbf{X}$ denotes the full collection of explanatory variables $X_1, X_2, \ldots, X_K$, which is particularly convenient when we need to condition on all the explanatory variables.

The population of interest is characterized by the joint distribution of $(Y, X_1, X_2, \ldots, X_K)$. The MLR model specifies that outcome variable Y is a linear function of the explanatory variables $X_1, X_2, \ldots, X_K$ plus a random noise term, represented by the random variable U:

$$Y = \alpha + \beta_1 X_1 + \beta_2 X_2 + \cdots + \beta_K X_K + U \text{ with } E(U|X_1, X_2, \ldots, X_K) = 0,$$

or using $\mathbf{X}$ as shorthand notation for the random variables $X_1, X_2, \ldots, X_K$:

$$Y = \alpha + \beta_1 X_1 + \beta_2 X_2 + \cdots + \beta_K X_K + U \text{ with } E(U|\mathbf{X}) = 0.$$

The MLR model has $K+1$ population parameters, the intercept parameter α and the K slope parameters $(\beta_1, \beta_2, \ldots, \beta_K)$. The units of the intercept parameter α are the same as the units of Y, whereas the units of each slope parameter β_k are the units of Y divided by the units of the corresponding explanatory variable X_k.

Like the SLR model, the MLR model consists of a linearity assumption and an exogeneity assumption, as follows:

- *Linearity assumption*: The association between Y and the explanatory variables is assumed to be linear, as captured by the

$$\alpha + \beta_1 X_1 + \beta_2 X_2 + \cdots + \beta_K X_K$$

part of the model. This assumption is not as restrictive as it may seem since, as discussed in section 18.6, the possibility of having multiple explanatory variables means that nonlinear functions of some variables, and even interactions between different variables, can be included in the MLR model.

- *Exogeneity assumption*: The exogeneity assumption $E(U|\mathbf{X}) = 0$ implies that there is no correlation between the unobservable U and *any* of the explanatory variables:

$$Cov(X_k, U) = 0 \text{ for any } k \in \{1, 2, \ldots, K\}.$$

Knowing the value of one or more of the explanatory variables tells us nothing about the expected value of U, as its conditional expectation is zero for all possible values. Importantly, the exogeneity assumption says nothing about the relationship *between* the explanatory variables. Any two explanatory variables X_k and X_ℓ may be correlated with each other and, in most cases, would be expected to be.

In one sense, the MLR exogeneity assumption seems stronger than the SLR exogeneity assumption since it requires that U be uncorrelated with a larger set of explanatory variables. In another sense, however, the inclusion of additional explanatory variables in the MLR model may make the exogeneity assumption more plausible than it was without those variables. For instance, in example 17.21, it was argued that the exogeneity assumption was not likely to hold for the SLR model of state-level cigarette sales and taxes, as the unobservable U is likely to be negatively related to cigarette taxes. If we also had a state-level variable that measured the pro-tax sentiment of residents (e.g., the percentage of residents who say yes to "Do you favor increased cigarette taxes?"), the inclusion of this additional explanatory variable as a second variable in an MLR model could make the $Cov(CIGTAX, U) = 0$ assumption more plausible.

For the MLR model with the exogeneity assumption, the conditional expectation of Y given the explanatory variables is

$$E(Y|\mathbf{X}) = \alpha + \beta_1 X_1 + \beta_2 X_2 + \cdots + \beta_K X_K$$

since $E(U|\mathbf{X}) = 0$. The interpretation of the parameters $(\alpha, \beta_1, \beta_2, \ldots, \beta_K)$ of the MLR model follows from this conditional expectation.

- *Meaning of the intercept* α: Note that

$$\alpha = E(Y|X_1 = X_2 = \cdots = X_K = 0),$$

which is the conditional expectation of Y when *all* the explanatory variables equal zero. Whether α has a practical interpretation depends on the specific MLR model being considered and, specifically, whether zero is a relevant value for each of the explanatory variables.

- *Meaning of the slope parameters*: To discuss the meaning of the slope parameters $(\beta_1, \beta_2, \ldots, \beta_K)$, consider some arbitrary *starting values* for the explanatory variables, such as $(X_1, X_2, \ldots, X_K) = (x_1^*, x_2^*, \ldots, x_K^*)$. Starting with the first explanatory variable X_1, what happens to the conditional expectation of Y if X_1 is increased by 1 unit *while the values of all the other explanatory variables are held fixed*? The change in the conditional expectation equals

$$\begin{aligned}&E(Y|X_1 = x_1^* + 1, X_2 = x_2^*, \ldots, X_K = x_K^*) - E(Y|X_1 = x_1^*, X_2 = x_2^*, \ldots, X_K = x_K^*)\\&= (\alpha + \beta_1(x_1^* + 1) + \beta_2 x_2^* + \cdots + \beta_K x_K^*) - (\alpha + \beta_1 x_1^* + \beta_2 x_2^* + \cdots + \beta_K x_K^*) = \beta_1.\end{aligned}$$

Therefore, slope parameter β_1 is the change in the expected outcome Y associated with a 1-unit change in X_1 while holding all the other explanatory variables fixed. This change is known as the **partial effect** of X_1 on $E(Y|\mathbf{X})$. Similarly, if X_1 is changed from x_1^* to x_1^{**} while the values of all the other explanatory variables are held fixed, the change in the expected outcome Y is $\beta_1(x_1^{**} - x_1^*)$. This same reasoning applies to the other slope parameters. So, in general for any $k \in \{1, 2, \ldots, K\}$, slope parameter β_k is the change in the expected outcome Y associated with a 1-unit change in X_k while holding all the other explanatory variables fixed. In some instances, it may be impossible to "hold the values of all the other explanatory variables fixed" since two or more of the explanatory variables in the MLR model may be direct functions of each other. Some examples, each of which are discussed later in this chapter, include MLR models that handle nonlinearity, MLR models that contain interaction variables, and MLR models with indicator variables representing a categorical variable (with more than two categories). For such cases, the MLR slope parameters still have meaning, but perhaps no longer individually. In fact, it is always possible to consider the more general case when the values of the explanatory variables change from $(X_1, X_2, \ldots, X_K) = (x_1^*, x_2^*, \ldots, x_K^*)$ to $(X_1, X_2, \ldots, X_K) = (x_1^{**}, x_2^{**}, \ldots, x_K^{**})$, in which case the change in the expected outcome Y is

$$\beta_1(x_1^{**} - x_1^*) + \beta_2(x_2^{**} - x_2^*) + \cdots + \beta_K(x_K^{**} - x_K^*).$$

This expression can be used to make statements about changes in the expected outcome Y when multiple explanatory variables change values, which includes cases where an explanatory variable may be a direct function of one or more other explanatory variables.

Example 18.1 (Monthly stock returns) *Example 17.5 considered an SLR model with an individual stock return as the outcome variable and a market-index return as the explanatory variable. This SLR model captures the association between the performance of the stock of an individual company and the overall performance of the stock market. Previously, in chapter 7 (see, e.g., example 7.13), we measured the correlation between the returns of several individual stocks. By using an MLR model, we can model an individual stock return with* both *a market-index return and the return of a related stock (or the returns of several related stocks). As an example, consider an MLR model with monthly returns of Home Depot (HD) as the outcome variable and monthly returns of the S&P 500 index (IDX) and monthly returns of Lowe's (LOW) as explanatory variables:*

$$HD = \alpha + \beta_1 IDX + \beta_2 LOW + U \text{ with } E(U|IDX, LOW) = 0$$

or, equivalently,

$$E(HD|IDX, LOW) = \alpha + \beta_1 IDX + \beta_2 LOW.$$

The intercept parameter $\alpha = E(HD|IDX = 0, LOW = 0)$ *is the expected monthly return for Home Depot when the monthly returns for both the S&P 500 index and Lowe's equal zero. Slope parameter* β_1 *measures how much the conditional expectation of Home Depot returns changes when the S&P 500 return changes, holding the Lowe's return fixed. For example, if the S&P return changes by* 0.01 *(1 percentage point) and the Lowe's return is unchanged, the conditional expectation of Home Depot returns changes by* $0.01\beta_1$. *This interpretation is different from the interpretation of the slope parameter in an SLR model with only the S&P 500 as an explanatory variable. Such a model would be* $HD = \alpha + \beta IDX + U$, *where* β *measures the association between Home Depot returns and S&P 500 returns but does not control for (or hold fixed) the returns for Lowe's. Similarly, slope parameter* β_2 *measures how much the conditional expectation of Home Depot returns changes when the Lowe's return changes, holding the S&P 500 return fixed. If the Lowe's return changes by* 0.01 *(1 percentage point) and the S&P return is unchanged, the conditional expectation of Home Depot returns changes by* $0.01\beta_2$. *The interpretation of this parameter is different from an SLR model with only the Lowe's return as an explanatory variable, where the slope parameter measures the association between Home Depot and Lowe's without controlling for the S&P return.*

18.2 The Least-Squares Estimator

For the population described by the joint distribution of $(Y, X_1, X_2, \ldots, X_K)$, assume that the observed sample consists of i.i.d. draws

$$(y_1, x_{11}, x_{12}, \ldots, x_{1K}), (y_2, x_{21}, x_{22}, \ldots, x_{2K}), \ldots, (y_n, x_{n1}, x_{n2}, \ldots, x_{nK}),$$

where n is the sample size. y_i is the realized outcome for the ith observation, and x_{ik} is the value of the kth explanatory variable for the ith observation. For the x_{ik} notation, the first subscript indicates the number of the observation, ranging from 1 to n, and the second subscript indicates the number of the explanatory variable, ranging from 1 to K. Each observation, indexed by i, therefore has $(K+1)$ values associated with it: the single outcome variable y_i and the K explanatory variables $x_{i1}, x_{i2}, \ldots, x_{iK}$. We assume that $n > K+1$, so there are enough observations to estimate the MLR model parameters, although we will not explicitly state this assumption in the results described next, in the interest of brevity.

If the MLR model

$$Y = \alpha + \beta_1 X_1 + \beta_2 X_2 + \cdots + \beta_K x_K + U$$

holds for $(Y, X_1, X_2, \ldots, X_K)$, then for each observation $i = 1, \ldots, n$,

$$y_i = \alpha + \beta_1 x_{i1} + \beta_2 x_{i2} + \cdots + \beta_K x_{iK} + u_i.$$

For the data-generating process, it is useful to think of the explanatory variable values $(x_{i1}, x_{i2}, \ldots, x_{iK})$ as draws from the joint distribution of $(X_1, X_2, \ldots, X_K)$. Then residual u_i is a draw from the conditional distribution of U given $X_1 = x_{i1}, X_2 = x_{i2}, \ldots, X_K = x_{iK}$, and the realized outcome y_i is generated by the previous equation. As in the SLR model, the residual can be written in terms of the outcome variable and explanatory variables:

$$u_i = y_i - \alpha - \beta_1 x_{i1} - \beta_2 x_{i2} - \cdots - \beta_K x_{iK}.$$

Residual u_i is positive when outcome y_i is greater than the conditional expectation

$$E(y_i|X_1 = x_{i1}, X_2 = x_{i2}, \ldots, X_K = x_{ik}) = \alpha + \beta_1 x_{i1} + \beta_2 x_{i2} + \cdots + \beta_{iK} x_{iK},$$

and residual u_i is negative when outcome y_i is less than this conditional expectation. For the SLR model, there was a nice graphical intuition for the sign and magnitude of the residuals, with data points above the SLR line ($E(Y|X) = \alpha + \beta X$) having positive residuals, data points below the SLR line having negative residuals, and the vertical distance of the data point to the SLR line indicating the magnitude of the residual. Things are more complicated for the MLR model since a graph of the $(y_i, x_{i1}, x_{i2}, \ldots, x_{iK})$ data points requires $(K+1)$ dimensions, so we focus more on analytical expressions than graphical intuition.[1]

Turning to least-squares estimation, the least-squares estimators of the MLR parameters are denoted as

$$(\hat{\alpha}_{\mathbf{X}Y}, \hat{\beta}_{1,\mathbf{X}Y}, \hat{\beta}_{2,\mathbf{X}Y}, \ldots, \hat{\beta}_{K,\mathbf{X}Y}).$$

These $(K+1)$ estimators are random variables, as they are random before the realized sample is observed. After the sample is observed, the realizations of the estimators are the least-squares estimates, denoted as

$$(\hat{\alpha}, \hat{\beta}_1, \hat{\beta}_2, \ldots, \hat{\beta}_K).$$

For the MLR model, the closed-form expressions for the least-squares estimators and estimates require the use of linear algebra. Therefore, instead of providing these expressions, we define least-squares estimators and estimates directly as the minimizers of a least-squares objective function that generalizes the SLR least-squares objective function from proposition 17.6. Specifically, the least-squares estimates $(\hat{\alpha}, \hat{\beta}_1, \hat{\beta}_2, \ldots, \hat{\beta}_K)$ are the values of $(a, b_1, b_2, \ldots, b_K)$ that minimize the function

$$S(a, b_1, b_2, \ldots, b_K) = \sum_{i=1}^{n} (y_i - a - b_1 x_{i1} - b_2 x_{i2} - \cdots - b_K x_{iK})^2.$$

To guarantee that $(\hat{\alpha}, \hat{\beta}_1, \hat{\beta}_2, \ldots, \hat{\beta}_K)$ is the *unique* solution to this minimization problem, the following two assumptions are required:

- **Assumption MLR-VarX**: Each explanatory variable x_k has a positive sample variance. That is, $s_{x_k}^2 > 0$ for each $k \in \{1, 2, \ldots, K\}$.
- **Assumption MLR-NPC**: For each explanatory variable x_k, it is not possible to write x_k as a linear combination of the other explanatory variables.

Assumption MLR-VarX is very weak, as it just requires that each explanatory variable is nonconstant. This assumption generalizes proposition 17.6 for the SLR model ($s_x^2 > 0$ for the single explanatory variable x). Assumption MLR-NPC requires that there is **no perfect colinearity** among the explanatory variables, which means that it is not possible to express one explanatory variable perfectly as a linear combination of one or more other explanatory variables. For instance, if one explanatory variable is *height* (in inches), having another explanatory variable *heightft* (height in feet), would violate the assumption of no perfect colinearity since $height = 12heightft$. As another example, if the indicator variable *union* is an explanatory variable equal to 1 for union members and 0 for nonmembers, having another explanatory variable *nonunion* (equal to 1 for nonmembers and 0 for union members) would violate the assumption of no perfect colinearity since $nonunion = 1 - union$. This assumption is also not very restrictive, as it is generally violated only when a practitioner makes a mistake in specifying the set of explanatory variables, as would be the case, for example, when including both a *union* indicator variable and a *nonunion* indicator variable in the MLR model. Proposition 18.1 states that these two assumptions are sufficient to ensure that the least-squares estimates uniquely maximize the $S(a, b_1, b_2, \ldots, b_K)$ function.[2]

Proposition 18.1 *If assumption MLR-VarX and assumption MLR-NPC hold, the minimization of*

$$S(a, b_1, b_2, \ldots, b_K) = \sum_{i=1}^{n} (y_i - a - b_1 x_{i1} - b_2 x_{i2} - \cdots - b_K x_{iK})^2$$

has a unique solution.

Example 18.2 (Monthly stock returns) *Using the* ***sp500*** *data set, the parameters $(\alpha, \beta_1, \beta_2)$ of the MLR model from example 18.1,*

$$HD = \alpha + \beta_1 IDX + \beta_2 LOW + U \text{ with } E(U|IDX, LOW) = 0,$$

can be estimated. The first column of the following table shows the least-squares estimates $(\hat{\alpha}, \hat{\beta}_1, \hat{\beta}_2)$, obtained by minimizing the function $S(a, b_1, b_2)$. For comparison, the second column shows the least-squares estimates for an SLR model having only IDX (S&P 500 monthly returns) as an explanatory variable, and the third column shows the least-squares estimates for an SLR model having only LOW (Lowe's monthly returns) as an explanatory variable.

	MLR estimates	*SLR estimates*	*SLR estimates*
α *(intercept)*	0.004	0.009	0.006
β_1 *(slope on IDX)*	0.595	1.020	
β_2 *(slope on LOW)*	0.384		0.522

The MLR estimates can be obtained from the `lm` *or* `lm_robust` *function in R. Since standard errors are not being calculated yet, the* `lm` *function is sufficient:*

```
lm(HD~IDX+LOW, data=sp500)

##
## Call:
## lm(formula = HD ~ IDX + LOW, data = sp500)
##
## Coefficients:
## (Intercept)          IDX          LOW
##     0.00425      0.59491      0.38425
```

The syntax to include multiple explanatory variables is to use a plus sign (+) between the variable names, so `IDX+LOW` *indicates two explanatory variables,* `IDX` *and* `LOW`*, are in the model.*

For the MLR estimates, the intercept estimate $\hat{\alpha} = 0.004$ *(or* 0.4%*) is an estimate of* $E(HD|IDX = LOW = 0)$*, the expected monthly return of Home Depot when both the S&P 500 and Lowe's have zero monthly returns. The slope estimate* $\hat{\beta}_1 = 0.595$ *estimates the expected change in Home Depot returns given a 1-unit change in S&P 500 returns, holding the returns of Lowe's fixed. Therefore, for a* 0.01 *increase in S&P 500 returns (IDX) with Lowe's returns (LOW) held fixed, we estimate that the Home Depot returns will, on average, increase by* 0.00595*. Similarly, for a* 0.01 *increase in Lowe's returns (LOW) with S&P 500 returns (IDX) held fixed, we estimate that the Home Depot returns will, on average, increase by* 0.00384*. The estimates can also be used to directly estimate the conditional expectation of HD for any specific values of IDX and LOW. For instance, in a month where the S&P 500 market return is 4% (IDX* = 0.04*) and Lowe's return is 2% (LOW* = 0.02*), the estimated conditional expectation of HD is*

$$\hat{E}(HD|IDX = 0.04, LOW = 0.02) = \hat{\alpha} + \hat{\beta}_1(0.04) + \hat{\beta}_2(0.02)$$
$$= 0.004 + (0.595)(0.04) + (0.384)(0.02) \approx 0.035,$$

or 3.5%*. Although we do not yet know whether there is a statistically significant difference between the two slope estimates* $\hat{\beta}_1$ *and* $\hat{\beta}_2$*, the estimates suggest that the S&P 500 returns have a stronger association (slope of* 0.595*) than Lowe's returns (slope of* 0.384*) have with expected Home Depot returns.*

Why is the estimated slope on IDX ($\hat{\beta} = 1.020$*) in the SLR model, in the second column of the table, so much larger than the MLR estimated slope of* 0.595*? First the underlying population parameters are measuring different things, with the SLR slope measuring the association of a 1-unit change in IDX with expected HD but completely ignoring LOW, and the MLR slope measuring the association of a 1-unit change in IDX with expected HD and with LOW held fixed. For the SLR model, the least-squares estimate of the IDX slope parameter is essentially forced to pick up any association with HD that may be coming from LOW since LOW is not in the model. It turns out that, perhaps unsurprisingly, there is a positive correlation between the S&P returns (IDX) and Lowe's returns (LOW), with* $r_{IDX,LOW} = 0.506$*. Therefore, when IDX increases by 1 unit, LOW also tends to increase, so the SLR slope parameter on IDX is expected to be larger than the MLR slope parameter on IDX, as the SLR slope parameter captures the direct association of IDX with expected HD, but also the indirect association of LOW with expected HD. Similar reasoning can be applied to explain why the SLR slope estimate for LOW, in the third column of the table, is larger than the MLR slope estimate for LOW.*

Example 18.3 (Cigarette sales and cigarette taxes) *In examples 17.4 and 17.7, we introduced an SLR model to explain state-level cigarette sales (CIGSALES) with state-level cigarette taxes (CIGTAX). We now add a variable to the model, yielding an MLR model with two explanatory variables. The new variable is the binary variable PRODUCER, which is equal to* 1 *for any state producing more than 20 million pounds of tobacco in 2019 and* 0 *otherwise.*

$$CIGSALES = \alpha + \beta_1 CIGTAX + \beta_2 PRODUCER + U$$
$$\text{with } E(U|CIGTAX, PRODUCER) = 0.$$

The PRODUCER variable is included to allow the possibility that cigarette sales, even after controlling for cigaratte taxes, may be higher in states that produce tobacco (e.g., due to greater acceptance of tobacco and smoking). In the **cigdata** *data set, there are seven states (Georgia, Kentucky, North Carolina, Pennsylvania, South Carolina, Tennessee, and Virginia) with producer* $=1$. *The following table reports the least-squares estimates* $(\hat{\alpha}, \hat{\beta}_1, \hat{\beta}_2)$ *of the MLR model parameters in the first column and, for comparison, the SLR estimates from example 17.7 in the second column:*

	MLR estimates	*SLR estimates*
α *(intercept)*	54.28	55.95
β_1 *(slope on CIGTAX)*	−8.97	−9.49
β_2 *(slope on PRODUCER)*	5.37	

The MLR estimates are obtained using the `lm` *function in R:*

```
lm(cigsales~cigtax+producer, data=cigdata)

##
## Call:
## lm(formula = cigsales ~ cigtax + producer, data = cigdata)
##
## Coefficients:
## (Intercept)       cigtax     producer
##       54.28        -8.97         5.37
```

There is not much change in the CIGTAX slope estimate in the MLR model (-8.97*) compared to the SLR model (*-9.49*). Again, the meaning of the two parameters and estimates is different, as the MLR estimate measures what happens when holding PRODUCER fixed and the SLR estimate does not. For the MLR model, the slope estimate* $\hat{\beta}_1 = -8.97$ *implies that a 1-unit ($1) change in state cigarette taxes, holding fixed whether the state is a tobacco producer, is associated with an estimated decrease of* 8.97 *packs per capita, on average. For the tobacco-producer variable, a 1-unit change in the variable involves going from a nonproducing state (PRODUCER* $=0$*) to a producing state (PRODUCER* $=1$*). The slope estimate* $\hat{\beta}_2 = 5.37$ *estimates the difference in expected per-capita packs sold between a tobacco-producing state and a nonproducing state, holding state cigarette taxes fixed. Whether this difference is statistically significant depends on the standard error of the estimate, an issue that we revisit later in this chapter.*

Example 18.4 (Weekly earnings) *In the previous chapter, SLR models were used to model weekly earnings in terms of union status (example 17.2) and educational attainment (example 17.3). With the MLR model, we can model weekly earnings in terms of several explanatory variables simultaneously. We use the* **cps** *data on* $n=2{,}809$ *employed individuals and the following four explanatory variables (*$K=4$*) to model the outcome earnwk (weekly earnings):*

$$
\begin{aligned}
&educ &&= \text{years of education}\\
&exper &&= \text{years of experience}\\
&union &&= 1 \text{ if individual is a union member, } 0 \text{ otherwise}\\
&female &&= 1 \text{ if individual is female, } 0 \text{ otherwise}
\end{aligned}
$$

For the experience (exper) variable, we adopt a standard definition used by labor economists, which defines[3]

$$exper = age - educ - 6, \text{ where age is age in years.}$$

With both exper and educ in the MLR model, it is not possible to also include age since, by the definition of exper, the age variable is a perfect linear combination of educ and exper, which would violate assumption MLR-NPC.

Here is the R code to construct the female and exper variables and calculate the MLR least-squares estimates:

```
# define the female indicator variable
cps$female <- 1*(cps$gender=="Female")

# define the experience variable
cps$exper <- cps$age-cps$educ-6

# least-squares estimation of the MLR model
lm(earnwk~educ+exper+union+female, data=cps)

##
## Call:
## lm(formula = earnwk ~ educ + exper + union + female, data = cps)
##
## Coefficients:
## (Intercept)         educ         exper         union        female
##      -431.0        110.9           4.9         142.7        -344.8
```

The interpretation of the slope estimates are as follows:

- $\hat{\beta}_1 = 110.9$*: An additional year of education is associated with an estimated increase of* \$110.90 *in weekly earnings, on average, holding all other variables (experience, union, female) fixed.*
- $\hat{\beta}_2 = 4.9$*: An additional year of experience is associated with an estimated increase of* \$4.90 *in weekly earnings, on average, holding all other variables (education, union, female) fixed.*
- $\hat{\beta}_3 = 142.7$*: A union member is estimated to earn* \$142.70 *more than a nonmember, on average, holding all other variables (education, experience, female) fixed.*
- $\hat{\beta}_4 = -344.8$*: A female worker is estimated to earn* \$344.80 *less than a non-female worker, on average, holding all other variables (education, experience, union) fixed.*

The estimates also allow us to make statements about changing two or more variables at once. For example, let's say that we are interested in the four possible combinations of the union and female variables, while holding educ and exper fixed. The estimated difference in average weekly earnings between a female union member ($female = union = 1$*) and a male nonmember (*$female = union = 0$*) equals* $\hat{\beta}_3 + \hat{\beta}_4 = 142.7 - 344.8 = -202.1$. *The estimated difference in average weekly earnings between a female union member and a female nonmember (*$female = 1$*,* $union = 0$*) equals* $\hat{\beta}_3 = 142.7$. *The estimated difference in average weekly earnings between a female union member and a male union member (*$female = 0$*,* *union* = 1*) equals* $\hat{\beta}_4 = -344.8$. *Estimated differences in average weekly earnings between other groups can be calculated similarly.*

The least-squares minimization problem in proposition 18.1 yields the least-squares estimates for the observed sample drawn from the population. The values of the least-squares estimates depend on the particular sample that happened to be drawn from the population. In thinking about the sampling distribution associated with least-squares estimation, the distribution of the least-squares estimators $(\hat{\alpha}_{\mathbf{X}Y}, \hat{\beta}_{1,\mathbf{X}Y}, \hat{\beta}_{2,\mathbf{X}Y}, \ldots, \hat{\beta}_{K,\mathbf{X}Y})$ is described by the distribution of all possible least-squares estimates that could arise from every possible n-observation i.i.d. sample drawn from the population. Like the least-squares estimators for the SLR model, the least-squares estimators for the MLR model have the desirable properties of consistency and asymptotic normality, with the latter allowing large-sample inference using the normal distribution. Proposition 18.2 states the consistency and asymptotic normality properties of the least-squares estimators of the MLR model:[4]

Proposition 18.2 *If the MLR model holds and assumptions MLR-VarX and MLR-NPC hold for any possible sample drawn from the population, least-squares estimation is consistent, with* $\hat{\alpha}_{\mathbf{X}Y}$ *being a consistent estimator of* α *and each* $\hat{\beta}_{k,\mathbf{X}Y}$ *being a consistent estimator of* β_k *for* $k \in \{1, 2, \ldots, K\}$. *Moreover, the least-squares estimators are asymptotically normal, with*

$$\sqrt{n}(\hat{\alpha}_{\mathbf{X}Y} - \alpha) \overset{a}{\sim} N(0, V_\alpha)$$

and

$$\sqrt{n}(\hat{\beta}_{k,\mathbf{X}Y} - \beta_k) \overset{a}{\sim} N\left(0, V_{\beta_k}\right) \text{ for each } k \in \{1, 2, \ldots, K\}.$$

The asymptotic variance of the intercept estimator $\hat{\alpha}_{\mathbf{X}Y}$ *is* $\frac{V_\alpha}{n}$, *and the asymptotic variance of each slope estimator* $\hat{\beta}_{k,\mathbf{X}Y}$ *is* $\frac{V_{\beta_k}}{n}$ *for* $k \in \{1, 2, \ldots, K\}$.

Proposition 18.2 says that the least-squares estimators get arbitrarily close to the underlying MLR parameters as the sample size increases. The asymptotic normality of the estimators allows construction of confidence intervals and hypothesis testing based on the normal distribution, as discussed in further detail in section 18.3.

18.2.1 Fitted Values, Estimated Residuals, and Regression fit

The concepts of fitted values and estimated residuals generalize to the MLR model. We start by describing fitted values in definition 18.1.

Definition 18.1 *The* ***fitted value*** *for observation i, denoted as* $\hat{y}_i$, *is*

$$\hat{y}_i = \hat{\alpha} + \hat{\beta}_1 x_{i1} + \hat{\beta}_2 x_{i2} + \cdots + \hat{\beta}_K x_{iK},$$

where $(\hat{\alpha}, \hat{\beta}_1, \hat{\beta}_2, \ldots, \hat{\beta}_K)$ *are the least-squares estimates of* $(\alpha, \beta_1, \beta_2, \ldots, \beta_K)$.

The fitted value y_i provides an in-sample estimate of the conditional expectation of the outcome variable, given the values of observation *i*'s explanatory variables. The true conditional expectation is

$$E(Y|X_1 = x_{i1}, X_2 = x_{i2}, \ldots, X_K = x_{ik}) = \alpha + \beta_1 x_{i1} + \beta_2 x_{i2} + \cdots + \beta_K x_{iK},$$

and the fitted value provides an estimate by plugging in the least-squares estimates of the MLR model parameters,

$$\hat{E}(Y|X_1 = x_{i1}, X_2 = x_{i2}, \ldots, X_K = x_{ik}) = \hat{y}_i = \hat{\alpha} + \hat{\beta}_1 x_{i1} + \hat{\beta}_2 x_{i2} + \cdots + \hat{\beta}_K x_{iK}.$$

We move next to estimated residuals, in definition 18.2.

Definition 18.2 *The **estimated residual** for observation i, denoted as $\hat{u}_i$, is*

$$\hat{u}_i = y_i - \hat{y}_i = y_i - \hat{\alpha} - \hat{\beta}_1 x_{i1} - \hat{\beta}_2 x_{i2} - \cdots - \hat{\beta}_K x_{iK},$$

where $(\hat{\alpha}, \hat{\beta}_1, \hat{\beta}_2, \ldots, \hat{\beta}_K)$ are the least-squares estimates of $(\alpha, \beta_1, \beta_2, \ldots, \beta_K)$.

The estimated residual is the difference between realized outcome y_i and its fitted value $\hat{y}_i$. When an estimated residual is small in magnitude, the realized outcome is close to the value predicted by the fitted value. Since the least-squares estimators are consistent, the fitted values and estimated residuals calculated from the least-squares estimates are also consistent. Proposition 18.3 states this result, generalizing the SLR result (proposition 17.5).

Proposition 18.3 (Consistency of fitted values and estimated residuals) *Assume that the MLR model and assumptions MLR-VarX and MLR-NPC hold for any possible sample drawn from the population. Then, for any $i \in \{1, 2, \ldots, n\}$, the fitted value $\hat{y}_i$ is a consistent estimate of*

$$E(Y|X_1 = x_{i1}, X_2 = x_{i2}, \ldots, X_K = x_{ik}) = \alpha + \beta_1 x_{i1} + \beta_2 x_{i2} + \cdots + \beta_K x_{iK},$$

and the estimated residual $\hat{u}_i$ is a consistent estimate of the population residual

$$u_i = y_i - \alpha - \beta_1 x_{i1} - \beta_2 x_{i2} - \cdots - \beta_K x_{iK}.$$

Example 18.5 (Weekly earnings) *Continuing example 18.4, the following table shows detailed information about the first 10 observations in the **cps** data set, including the values of the four explanatory variables (educ, exper, union, female), the observed outcome value (earnwk), and the fitted values and estimated residuals:*

i	$educ_i$	$exper_i$	$union_i$	$female_i$	y_i ($earnwk_i$)	$\hat{y}_i$	$\hat{u}_i$
1	14	30	0	0	577	1,269	−692
2	18	10	0	1	3,049	1,270	1,779
3	18	6	0	1	2,500	1,250	1,250
4	12	17	0	0	300	983	−683
5	12	38	0	0	1,000	1,086	−86
6	7.5	28.5	0	1	1,000	196	804
7	12	37	0	1	650	737	−87
8	13	39	0	1	1712	857	854
9	16	17	0	1	820	1,082	−262
10	12	41	1	0	1240	1,244	−4

We calculate the fitted values and estimated residuals in R in the same way as seen for the SLR model:

```
# store results of least-squares estimation
results <- lm(earnwk~educ+exper+union+female, data=cps)

# access estimated residuals and fitted values
uhat <- results$residuals
yhat <- results$fitted.values

# output first 10 estimated residuals and fitted values, rounded to nearest dollar
round(uhat[1:10])

## -692 1779 1250 -683  -86  804  -87  854 -262   -4

round(yhat[1:10])

## 1269 1270 1250  983 1086  196  737  857 1082 1244
```

The fitted values are based on the least-squares parameter estimates, using the formula in definition 18.1. Looking at the sixth observation ($i=6$), for instance, the fitted value is very low since the education level is so low; the fitted value indicates an estimate of \$196 for the estimated expected earnings of an individual with those characteristics (female, nonunion worker with 7.5 years of education and 28.5 years of experience). The fitted values for the first three observations happen to be very close to each other due to the values of their explanatory variables. Looking at the second and third observations, the only difference is a 4-year difference in experience ($exper = 10$ for $i = 2$ and $exper = 6$ for $i = 3$), leading to a difference in fitted values of $4\hat{\beta}_{exper} = (4)(4.9) \approx 20$. The estimated residuals $\hat{u}_i$ exhibit a lot of variation. Some observations have fitted values quite close to the actual outcomes and estimated residuals close to zero, most notably for $i \in \{5, 7, 10\}$, while other observations have fitted values very far from the actual outcomes and estimated residuals with large magnitudes, most notably for $i \in \{2, 3\}$.

Summary measures of the overall size of the model's residuals can be obtained by generalizing the residual variance and standard deviation estimates introduced for the SLR model. Here are their definitions for the MLR model.

Definition 18.3 *The **residual variance estimate**, denoted as $\hat{\sigma}_U^2$, is*

$$\hat{\sigma}_U^2 = \frac{1}{n-K-1} \sum_{i=1}^{n} \hat{u}_i^2.$$

Definition 18.4 *The **residual standard deviation estimate**, denoted as $\hat{\sigma}_U$, is*

$$\hat{\sigma}_U = \sqrt{\hat{\sigma}_U^2} = \sqrt{\frac{1}{n-K-1} \sum_{i=1}^{n} \hat{u}_i^2}.$$

Compared to the $\hat{\sigma}_U^2$ and $\hat{\sigma}_U$ formulas for least-squares estimation of the SLR model, the formulas for the MLR model involve a scaling of $\frac{1}{n-K-1}$ rather than $\frac{1}{n-2}$. This scaling accounts for the estimation of the $(K+1)$ parameters in the MLR model, with the denominator equal to $n-(K+1) = n-K-1$. The SLR formulas are a special case of the MLR formulas with $K=1$. While the $\frac{1}{n-K-1}$ scaling does not differ much numerically from either $\frac{1}{n-1}$ or $\frac{1}{n}$ scaling when n is large, this scaling does ensure that the residual variance estimator is unbiased.

Proposition 18.4 summarizes the properties of the estimated residuals from least-squares estimation of the MLR model. This proposition generalizes proposition 17.7, which considered the properties of the estimated residuals for the SLR model.

Proposition 18.4 (Properties of estimated residuals) *The estimated residuals*

$$\hat{u}_i = y_i - \hat{\alpha} - \hat{\beta}_1 x_{i1} - \hat{\beta}_2 x_{i2} - \cdots - \hat{\beta}_K x_{iK},$$

based on the least-squares estimates $(\hat{\alpha}, \hat{\beta}_1, \hat{\beta}_2, \ldots, \hat{\beta}_K)$ *have the following properties:*

i. *The sample average of the estimated residuals is zero:*

$$\frac{1}{n}\sum_{i=1}^{n} \hat{u}_i = 0.$$

ii. *The sample correlation between the values of any explanatory variable and the estimated residuals is zero:*

$$r_{x_k \hat{u}} = 0 \text{ for every } k \in \{1, 2, \ldots, K\}.$$

iii. *The sample correlation between the fitted values and the estimated residuals is zero:*

$$r_{\hat{y}\hat{u}} = 0.$$

iv. *The residual variance estimate* $\hat{\sigma}_U^2 = \frac{1}{n-K-1}\sum_{i=1}^{n} \hat{u}_i^2$ *is a consistent (and unbiased) estimate of* σ_U^2*, the unconditional variance of the random variable* U.

v. *The residual standard deviation estimate* $\hat{\sigma}_U = \sqrt{\frac{1}{n-K-1}\sum_{i=1}^{n} \hat{u}_i^2}$ *is a consistent estimate of* σ_U*, the unconditional standard deviation of the random variable* U.

Properties (i), (iii), (iv), and (v) are basically the same as those stated for the SLR model in proposition 17.7, with the different scaling $\frac{1}{n-K-1}$ used for properties (iv) and (v). Property (ii) states that the sample correlation between estimated residuals and *any* of the explanatory variables equals zero. That is, the explanatory variable x_k is uncorrelated with the estimated residual $\hat{u}$ for any $k \in \{1, 2, \ldots, K\}$. This property is a sample analog of the corresponding population property—that the population residual U is uncorrelated with each random variable X_k for $k \in \{1, 2, \ldots, K\}$, which is implied by the exogeneity assumption $E(U|\mathbf{X}) = 0$. Since property (ii) is a by-product of least-squares estimation, it is true *even if the exogeneity assumption doesn't actually hold.* As such, it is not fruitful to use a sample correlation between an explanatory variable x_k and the estimated residual $\hat{u}$ to test whether X_k is related to U, as the sample correlation is always zero.

With estimated residuals and fitted values defined for least-squares estimation of the MLR model, we can generalize the R-squared measure that was introduced in definition 17.5 for the SLR model.

Definition 18.5 *The* ***R-squared*** *value associated with least-squares estimation of the MLR model, denoted as* R^2*, is*

$$R^2 = \frac{s_{\hat{y}}^2}{s_y^2} = 1 - \frac{s_{\hat{u}}^2}{s_y^2}.$$

This definition is identical to the definition of R-squared for least-squares estimation of the SLR model. In terms of intepretation, the only difference for the MLR model is that there may be multiple explanatory variables explaining the variation in the outcome variable. For example, if $K=3$ and $R^2=0.36$, then the three explanatory variables x_1, x_2, and x_3 explain 36% of the variation in the y-variable.

Proposition 18.5 summarizes the important properties of the R-squared value for MLR models.

Proposition 18.5 (Properties of R-squared) *The R-squared value, based on the least-squares estimates* $(\hat{\alpha}, \hat{\beta}_1, \hat{\beta}_2, \ldots, \hat{\beta}_K)$, *has the following properties:*

i. $R^2 = r^2_{y\hat{y}}$.

ii. $0 \leq R^2 \leq 1$.

iii. *If an explanatory variable* x_{K+1} *is added to the MLR model, least-squares estimation of the new MLR model (with* $K+1$ *explanatory variables) has an R-squared value (call it* R^2_{K+1}*) that is at least as large as the original* R^2 *value (from the MLR model with K explanatory variables):* $R^2_{K+1} \geq R^2$.

Properties (i) and (ii) have been seen previously for SLR models. R^2 equals the square of the correlation between the values of outcome variable y and the fitted values $\hat{y}$. To the degree that the explanatory variables provide a better fit or prediction for the outcome variable, the correlation between y and $\hat{y}$ is larger in magnitude, and therefore, R^2 is higher. The extreme of $R^2=1$ corresponds to 100% of the variation in y being explained by the explanatory variables, which happens when y is a perfect linear function of the explanatory variables. The extreme of $R^2=0$ corresponds to none of the variation in y being explained by the explanatory variables, which happens only when all the slope estimates $\hat{\beta}_1, \hat{\beta}_2, \ldots, \hat{\beta}_K$ are exactly equal to zero. Property (ii) follows directly from property (i) since the correlation $r_{y\hat{y}}$ satisfies $-1 \leq r_{y\hat{y}} \leq 1$.

Property (iii) states that the R-squared value (weakly) increases when a variable is added to the MLR model and new least-squares estimates are calculated. The intuition here is that you can't do worse in explaining the variation in outcome variable y when you add an explanatory variable to the model. (Mathematically, if the estimated slope on the added variable is exactly equal to zero, the fitted values would be the same as the original least-squares estimation, so R^2 is unchanged. For any other nonzero estimate of the slope of the added variable, the R^2 value increases compared to the original R^2 value.) By continually applying property (iii), R^2 increases if we add more and more explanatory variables to the MLR model. This property should be considered a cautionary tale. Specifically, if a practitioner is too focused on the R^2 value, it may cause them to add too many explanatory variables to a model. In choosing which explanatory variables belong in the model, R^2 should not be the guiding force; instead, variables should be included based on prior knowledge that the researcher has about their practical or economic relevance to explaining the outcome variable.

Example 18.6 (Monthly stock returns) *Continuing example 18.2, the following table reports the R-squared values and residual standard deviation estimates for least-squares estimation of three models for the outcome HD: (1) MLR with IDX and LOW as explanatory variables, (2) SLR with IDX as the explanatory variable, and (3) SLR with LOW as the explanatory variable:*

	MLR estimates	*SLR estimates*	*SLR estimates*
α *(intercept)*	0.004	0.009	0.006
β_1 *(slope on IDX)*	0.595	1.020	
β_2 *(slope on LOW)*	0.384		0.522
R^2	0.505	0.336	0.421
$\hat{\sigma}_U$	0.052	0.060	0.056

Let's first compare the results from the two SLR models. There is a higher R-squared value for the SLR model with LOW than for the SLR model with IDX. Lowe's returns explain 42.1% *of the variation in Home Depot returns, while the S&P 500 returns explain* 33.6% *of the variation in Home Depot returns. From these values, we know that the correlation between HD and LOW* ($r_{HD,LOW} = \sqrt{0.421} \approx 0.65$) *is larger than the correlation between HD and IDX* ($r_{HD,IDX} = \sqrt{0.336} \approx 0.58$)*. Since LOW explains more of the variation of HD than IDX does, the residual standard deviation estimate* $\hat{\sigma}_U$ *is lower for the SLR model with LOW* ($\hat{\sigma}_U = 0.056$) *than for the SLR model with IDX* ($\hat{\sigma}_U = 0.060$)*. That is, there is less variation left unexplained by LOW compared to IDX. To put these* $\hat{\sigma}_U$ *estimates in perspective, the overall sample standard deviation of the outcome variable HD is* $s_{HD} = 0.074$.

Looking at the estimates for the MLR model that includes both IDX and LOW, the R-squared value is 0.501*, meaning S&P 500 returns and Lowe's returns together explain* 50.1% *of the variation in Home Depot returns. This R-squared value is larger than either of the R-squared values for the SLR models, which is expected from property (iii) of proposition 18.5 since the MLR model has an additional explanatory variable. For instance, moving from the SLR model with IDX to the MLR model with IDX and HD, the R-squared value increases from* 0.336 *to* 0.501*; that is, with IDX already in the model, the addition of LOW to the model explains an additional* 16.5% *of the variation in HD. The residual standard deviation estimate* $\hat{\sigma}_U$ *for the MLR model is* 0.052*, which is lower than each of the two corresponding estimates for the SLR models. The two explanatory variables explain more of the variation in HD than either explanatory variable individually, leaving less variation in the residual. But even with the two explanatory variables in the MLR model, a lot (*49.9%*) of the variation in Home Depot returns remains left unexplained.*

Here is the R code to calculate the R^2 *and* $\hat{\sigma}_U$ *values reported in the table:*

```
# store MLR estimation results (HD on IDX and LOW)
results_mlr <- lm(HD~IDX+LOW, data=sp500)

# store SLR estimation results (HD on IDX)
results_slr1 <- lm(HD~IDX, data=sp500)

# store SLR estimation results (HD on LOW)
results_slr2 <- lm(HD~LOW, data=sp500)

# output the R-squared values and estimated residual standard deviations
print(paste("MLR results (IDX, LOW)", "R-sq",
            round(summary(results_mlr)$r.squared,3),
            "sighat_u", round(summary(results_mlr)$sigma,4)))

## [1] "MLR results (IDX, LOW) R-sq 0.505 sighat_u 0.052"

print(paste("SLR results (IDX)", "R-sq", round(summary(results_slr1)$r.squared,3),
            "sighat_u", round(summary(results_slr1)$sigma,4)))
```

```
## [1] "SLR results (IDX) R-sq 0.336 sighat_u 0.0602"

print(paste("SLR results (LOW)", "R-sq", round(summary(results_slr2)$r.squared,3),
            "sighat_u", round(summary(results_slr2)$sigma,4)))

## [1] "SLR results (LOW) R-sq 0.421 sighat_u 0.0562"
```

Example 18.7 (Weekly earnings) *Revisiting example 18.4, the following table shows how R^2 changes when explanatory variables are added one by one to the model that explains weekly earnings (earnwk):*

Explanatory variables	R^2
educ	0.106
educ, exper	0.108
educ, exper, union	0.115
educ, exper, union, female	0.166

Property (iii) of proposition 18.5 states that R^2 increases as variables are added to the model. The first row corresponds to an SLR model with only educ as an explanatory variable. Education alone explains 10.6% *of the variation in weekly earnings. Adding experience (exper) to the model provides a small incremental increase in R^2 to* 0.108, *as does adding union, which increases R^2 to* 0.115. *Adding the female variable provides a larger increase, to an R^2 of* 0.166, *meaning that the four explanatory variables together explain* 16.6% *of the variation in weekly earnings.*

18.3 Standard Errors and Confidence Intervals

This section focuses on statistical inference based on least-squares estimation of the MLR model. Proposition 18.2 states that the least-squares estimators are consistent and asymptotically normal. As such, with a statistical package like *R* that calculates the standard errors of the least-squares estimates, we can use standard approaches based on the normal distribution to construct confidence intervals for the MLR population parameters and perform hypothesis tests involving these parameters.

As with SLR models in chapter 17, we rely on standard errors that are robust to heteroskedasticity of the MLR model residuals. However, we also discuss the case of homoskedasticity since this case provides useful insight into the features of the population and sample that affect the magnitude of the standard errors, and therefore the precision of the least-squares estimates. Unlike the SLR model with only a single explanatory variable, the possibility of multiple explanatory variables in the MLR model means that the relationships *among* the explanatory variables can affect the standard errors. Proposition 18.6 provides a framework for examining these relationships by decomposing each explanatory variable into two parts—one that is a linear function of the other explanatory variables and another that is a random variable uncorrelated with the other explanatory variables.

Proposition 18.6 *If assumptions MLR-VarX and MLR-NPC hold for any possible sample drawn from the population, the random variable X_1 can be decomposed into* *a linear function of the other explanatory variables and another random variable* $\tilde{X}_1$ *such that*

$$X_1 = \gamma + \delta_1 X_2 + \delta_2 X_3 + \cdots + \delta_{K-1} X_K + \tilde{X}_1,$$

with

$$Cov(\tilde{X}_1, X_2) = Cov(\tilde{X}_1, X_3) = \cdots = Cov(\tilde{X}_1, X_K) = 0 \text{ and } E(\tilde{X}_1) = 0.$$

Similarly, for any other $k \in \{2, 3, \ldots, K\}$*, the random variable* X_k *can be decomposed into a linear function of the explanatory variables and another random variable* $\tilde{X}_k$ *such that* $\tilde{X}_k$ *is uncorrelated with each of the other explanatory variables and has a population mean of zero.*

Let's focus on the first explanatory variable X_1 in the MLR model. From proposition 18.6, the random variable $\tilde{X}_1$ is the part of the explanatory variable X_1 that is not explained by (or correlated with) the other $(K-1)$ explanatory variables $X_2, X_3, \ldots, X_K$. Recall that slope parameter β_1 indicates how the expected outcome Y changes when X_1 changes and all the other explanatory variables are held fixed. As such, since the other explanatory variables are being held fixed, we care about the part of X_1 that is not related to those other explanatory variables when we are estimating β_1. This part of X_1 is the residual $\tilde{X}_1$ from the decomposition of X_1 given in the equation in proposition 18.6. When there is a lot of possible variation in the $\tilde{X}_1$ random variable, which happens when $Var(\tilde{X}_1) = \sigma^2_{\tilde{X}_1}$ is high, it should be possible to have a more precise estimate of β_1. The variation in $\tilde{X}_1$ is sometimes called the **independent variation** of X_1, since $\tilde{X}_1$ is the part of X_1 that is uncorrelated with the other explanatory variables.

This same reasoning applies to any other explanatory variable X_k in the MLR model. For any $k \in \{1, 2, \ldots, K\}$, the random variable $\tilde{X}_k$ is the part of X_k that is not related to the other explanatory variables in the model. To precisely estimate β_k, it is better to have a lot of independent variation in the X_k variable, which happens when the variance of $\tilde{X}_k$, $Var(\tilde{X}_k) = \sigma^2_{\tilde{X}_k}$, is high.

As we did for the SLR model (proposition 17.10), we consider the case of homoskedastic residuals for the MLR model as a pedagogical tool to see the factors that affect the standard errors of the slope estimates. To be clear, homoskedasticity is not required to perform least-squares estimation, as the standard errors of the least-squares estimates can and should be calculated under the more relaxed case of heteroskedasticity.

Homoskedasticity and heteroskedasticity are generalized to the MLR model as in definitions 18.6 and 18.7.

Definition 18.6 *The residuals of an MLR model are* ***homoskedastic*** *if the conditional variance*

$$Var(U|X_1 = x_1, X_2 = x_2, \ldots, X_K = x_K) = \sigma^2_U$$

is constant and does not depend on the values $x_1, x_2, \ldots, x_K$*. The residuals are said to exhibit* ***homoskedasticity****.*

Definition 18.7 *The residuals of an MLR model are* ***heteroskedastic*** *if the conditional variance*

$$Var(U|X_1 = x_1, X_2 = x_2, \ldots, X_K = x_K)$$

is nonconstant and depends on the values $x_1, x_2, \ldots, x_K$. The residuals are said to exhibit ***heteroskedasticity***.

In the case of homoskedastic residuals, proposition 18.7 provides the asymptotic variance formulas for the least-squares slope estimators.

Proposition 18.7 *If the MLR model holds, assumptions MLR-VarX and MLR-NPC hold for any possible sample drawn from the population, and the residuals are homoskedastic, then the asymptotic variance of $\hat{\beta}_{k,XY}$ (the least-squares slope estimator of β_k) is*

$$\frac{V_{\beta_k}}{n} = \frac{\sigma_U^2}{n\sigma_{\tilde{X}_k}^2} \text{ for each } k \in \{1, 2, \ldots, K\}.$$

The variance in the denominator of the asymptotic-variance expression is the independent variation in X_k, given by $\sigma_{\tilde{X}_k}^2$, *not* the overall variation in X_k, given by $\sigma_{X_k}^2$. To obtain the standard errors of the least-squares slope estimates, sample descriptive statistics can be plugged in for the population statistics, so

$$se(\hat{\beta}_k) = \sqrt{\frac{\hat{V}_{\beta_k}}{n}} = \sqrt{\frac{s_{\hat{u}}^2}{ns_{\tilde{x}_k}^2}} = \frac{s_{\hat{u}}}{\sqrt{n}\, s_{\tilde{x}_k}} \text{ for each } k \in \{1, 2, \ldots, K\},$$

where $s_{\hat{u}}$ is the sample standard deviation of the estimated residuals and $s_{\tilde{x}_k}$ is the sample standard deviation of the variable that measures the part of x_k that is not linearly related to the other explanatory variables.[5] Thus, there are three factors that affect the standard errors of the least-squares slope estimates, with the first two being the same as for the SLR model:

- *Sample size*: Larger n leads to a smaller standard error $se(\hat{\beta}_k)$, with the usual $\frac{1}{\sqrt{n}}$ scaling.
- *Residual noise*: A smaller residual variance σ_U^2, as estimated by $s_{\hat{u}}^2$, leads to a smaller standard error $se(\hat{\beta}_k)$.
- *Independent variation of the explanatory variable*: Whereas the overall variation of the explanatory variable affects the standard error for the SLR model, it is the *independent* variation of an explanatory variable x_k that affects $se(\hat{\beta}_k)$ for the MLR model. The independent variation of the random variable X_k is given by the variance of the random variable $\tilde{X}_k$ introduced here and measures how much variation is left in X_k after the linear relationship with other explanatory variables has been accounted for. When this independent variation, as estimated by $s_{\tilde{x}_k}^2$, is larger, the standard error $se(\hat{\beta}_k)$ is smaller.

For the third factor, while a lot of independent variation in x_k is good for the precision of $\hat{\beta}_k$ (i.e., a low $se(\hat{\beta}_k)$), it is also true that the precision of $\hat{\beta}_k$ is poor (i.e., a high $se(\hat{\beta}_k)$) when there is little independent variation in x_k. When variable x_k is highly correlated with and well explained by the other explanatory variables in the model, variance $\sigma_{\tilde{X}_k}^2$ can be very low, which can cause the asymptotic variance of slope estimator $\hat{\beta}_{XY,k}$ and the standard error of the estimate $\hat{\beta}_k$ to be very large since this variance appears in the denominator of the asymptotic-variance formula in proposition 18.7. This issue, when high correlation among two or more explanatory variables causes standard errors to be large, is known as

multicollinearity. While some textbooks offer so-called solutions to multicollinearity, there really is no solution if we wish to keep all the affected variables in the model. For instance, suppose $K=2$ and the variables x_1 and x_2 are highly correlated with each other. In such a case, it is possible that x_1 and x_2 together provide a good prediction of outcome variable y, as reflected by a high R^2 value, even if both $\hat{\beta}_1$ and $\hat{\beta}_2$ have very high standard errors due to multicollinearity. While the two variables are jointly important in explaining y, their high correlation means that it is difficult or even impossible to disentangle the separate effects of x_1 and x_2 on the expected outcome. β_1 measures how X_1 affects $E(Y|\mathbf{X})$ while holding X_2 fixed, but estimating β_1 is difficult because there is little variation in x_1 once x_2 is held fixed due to the high value of $r_{x_1x_2}$.

Example 18.8 (Cigarette sales and cigarette taxes) *The following table reproduces the least-squares estimates from example 18.3, now with heteroskedasticity-robust standard errors reported in parentheses:*

	MLR estimates	*SLR estimates*
α *(intercept)*	54.28 (3.51)	55.95 (2.92)
β_1 *(slope on CIGTAX)*	−8.97 (1.19)	−9.49 (1.06)
β_2 *(slope on PRODUCER)*	5.37 (5.08)	

The standard error for the cigarette-tax slope in the MLR model is $se(\hat{\beta}_1)=1.19$. *This standard error is only slightly higher than the standard error of* 1.06 *for the cigarette-tax slope in the SLR model, suggesting that even after controlling for producer (i.e., whether a state is a tobacco producer), there is still a lot of independent variation left in cigtax for precisely estimating* β_1 *in the MLR model.*

The MLR estimates in the table are obtained in R using the `lm_robust` *function as follows:*

```
results <- lm_robust(cigsales~cigtax+producer, data=cigdata)
results

##              Estimate Std. Error t value     Pr(>|t|) CI Lower CI Upper DF
## (Intercept)   54.2762     3.5128 15.4511 2.9333e-20   47.213   61.339 48
## cigtax        -8.9687     1.1935 -7.5146 1.2014e-09  -11.368   -6.569 48
## producer       5.3669     5.0785  1.0568 2.9589e-01   -4.844   15.578 48
```

Example 18.9 (Monthly stock returns) *The following table reproduces the least-squares estimates from example 18.2, now with heteroskedasticity-robust standard errors reported in parentheses:*

	MLR estimates	*SLR estimates*	*SLR estimates*
α *(intercept)*	0.004 (0.003)	0.009 (0.003)	0.006 (0.003)
β_1 *(slope on IDX)*	0.595 (0.091)	1.020 (0.082)	
β_2 *(slope on LOW)*	0.384 (0.047)		0.522 (0.040)

For the MLR model, the standard error of the slope on IDX (S&P 500 returns) is $se(\hat{\beta}_1)=0.091$. *This standard error is roughly* 11% *higher than the standard error of* 0.082 *obtained for the slope on IDX in the SLR model. The higher standard error for the MLR model* *is not too surprising since β_1 is estimated using only the independent variation of IDX after controlling for LOW. Since IDX and LOW are correlated, with $r_{IDX,LOW} = 0.5061$, the amount of independent variation in IDX used in the estimation of the MLR model is somewhat lower than the overall variation of IDX used in the estimation of the SLR model. Similarly, for the same reason, the standard error on LOW (Lowe's returns) is higher for the estimation of the MLR model, with $se(\hat{\beta}_2) = 0.047$, as compared to estimation of the SLR model with only LOW as an explanatory variable, where the standard error is 0.040.*

Here is the R code to produce the estimates and standard errors in the table:

```
# MLR model of HD on IDX and LOW
mlr_results <- lm_robust(HD~IDX+LOW, data=sp500)
mlr_results

##              Estimate Std. Error t value   Pr(>|t|)   CI Lower  CI Upper  DF
## (Intercept) 0.0042476  0.0027743  1.5311 1.2663e-01 -0.0012081 0.0097034 361
## IDX         0.5949112  0.0906584  6.5621 1.8444e-10  0.4166262 0.7731962 361
## LOW         0.3842477  0.0474368  8.1002 8.5485e-15  0.2909606 0.4775348 361

# SLR model of HD on IDX
slr_results1 <- lm_robust(HD~IDX, data=sp500)
slr_results1

##              Estimate Std. Error t value   Pr(>|t|)  CI Lower CI Upper  DF
## (Intercept) 0.0087316  0.0032119  2.7185 6.8739e-03 0.0024152 0.015048 362
## IDX         1.0204494  0.0816816 12.4930 5.0256e-30 0.8598195 1.181079 362

# SLR model of HD on LOW
slr_results2 <- lm_robust(HD~LOW, data=sp500)
slr_results2

##              Estimate Std. Error t value   Pr(>|t|)   CI Lower CI Upper  DF
## (Intercept) 0.0060877  0.0030005  2.0289 4.3199e-02 0.00018713 0.011988 362
## LOW         0.5218212  0.0404116 12.9127 1.2310e-31 0.44235020 0.601292 362
```

The asymptotic normality of the least-squares estimators (proposition 18.2) allows practitioners to use the normal distribution for calculating confidence intervals for the MLR parameters and conducting hypothesis tests involving the MLR parameters. The $1-\alpha$ confidence interval for the intercept parameter α is

$$(\hat{\alpha} - z_{\alpha/2} se(\hat{\alpha}), \hat{\alpha} + z_{\alpha/2} se(\hat{\alpha})),$$

and the $1-\alpha$ confidence interval for each slope parameter β_k, for $k \in \{1, 2, \ldots, K\}$, is

$$(\hat{\beta}_k - z_{\alpha/2} se(\hat{\beta}_k), \hat{\beta}_k + z_{\alpha/2} se(\hat{\beta}_k)).$$

The specific case of 95% confidence intervals for α and β_k are, respectively,

$$(\hat{\alpha} - 1.96 se(\hat{\alpha}), \hat{\alpha} + 1.96 se(\hat{\alpha}))$$

and

$$(\hat{\beta}_k - 1.96 se(\hat{\beta}_k), \hat{\beta}_k + 1.96 se(\hat{\beta}_k)).$$

Example 18.10 (Weekly earnings) *The following R code shows the least-squares estimates from example 18.4, with heteroskedasticity-robust standard errors and 95% confidence intervals also reported:*

```
results <- lm_robust(earnwk~educ+exper+union+female, data=cps)
results

##               Estimate Std. Error  t value   Pr(>|t|)  CI Lower  CI Upper   DF
## (Intercept) -431.031   100.1611   -4.3034 1.7398e-05 -627.4277 -234.6338 2804
## educ         110.940     7.2783   15.2425 1.8421e-50   96.6684  125.2113 2804
## exper          4.897     1.3157    3.7221 2.0146e-04    2.3173    7.4768 2804
## union        142.748    44.4518    3.2113 1.3363e-03   55.5862  229.9094 2804
## female      -344.791    25.8271  -13.3500 1.8495e-39 -395.4331 -294.1491 2804
```

The 95% *confidence interval for the education slope* β_1 *is*

$$(110.90-(1.96)(7.28), 110.90+(1.96)(7.28)) \approx (96.67, 125.21).$$

We are 95% *confident that the change in expected weekly earnings associated with a 1-year change in education, holding all other variables fixed, is between* \$96.67 *and* \$125.21. *For the union-membership slope* β_3, *the* 95% *confidence interval is*

$$(142.70-(1.96)(44.45), 142.70+(1.96)(44.45)) \approx (55.59, 229.91).$$

We are 95% *confident that the difference in expected weekly earnings between a union member and a nonmember, holding all other variables fixed, is between* \$55.59 *and* \$229.91. *While this interval for* β_3 *is quite wide (approximately* \$174 *wide), it does not include zero, supporting the idea that the union differential is statistically meaningful even if it is not very precisely estimated.*

We can obtain other confidence intervals for the parameters by changing the optional `alpha` *argument for the* `lm_robust` *function. Here is output with* 90% *confidence intervals:*

```
results <- lm_robust(earnwk~educ+exper+union+female, data=cps, alpha=0.10)
results

##               Estimate Std. Error  t value   Pr(>|t|)  CI Lower  CI Upper   DF
## (Intercept) -431.031   100.1611   -4.3034 1.7398e-05 -595.8356 -266.2259 2804
## educ         110.940     7.2783   15.2425 1.8421e-50   98.9641  122.9156 2804
## exper          4.897     1.3157    3.7221 2.0146e-04    2.7322    7.0618 2804
## union        142.748    44.4518    3.2113 1.3363e-03   69.6069  215.8887 2804
## female      -344.791    25.8271  -13.3500 1.8495e-39 -387.2869 -302.2953 2804
```

The 90% *confidence interval for the education slope* β_1 *is*

$$(110.90-(1.645)(7.28), 110.90+(1.645)(7.28)) \approx (98.96, 122.92),$$

meaning that we are 90% *confident that the change in expected weekly earnings associated with a 1-year change in education, holding all the other variables fixed, is between* \$98.96 *and* \$122.92.

How about a change in education that is not a 1-unit change? For instance, for a 2-year change in education, the associated change in $E(EARNWK|\mathbf{X})$ *is* $2\beta_1$. *The estimate is* $2\hat{\beta}_1$, *and its standard error is* $se(2\hat{\beta}_1) = 2se(\hat{\beta}_1)$, *which is 2 times the original standard error of* $\hat{\beta}_1$. *This standard error can be used to construct confidence intervals, so, for example, the* 90% *confidence interval for* $2\beta_1$ *is*

$$((2)(110.90)-(1.645)(2)(7.28), (2)(110.90)+(1.645)(2)(7.28)) \approx (197.9, 245.8),$$

which can also be obtained by multiplying the upper and lower limits of the 90% *confidence interval for* β_1 *by* 2.

18.4 Inference for Linear Combinations of Regression Parameters

The previous section considered asymptotic inference, including standard errors and confidence intervals, for individual regression model parameters. In this section, we consider asymptotic inference for linear combinations of regression model parameters. To motivate the importance of these linear combinations, we consider an MLR model with two explanatory variables under the exogeneity assumption:

$$E(Y|\mathbf{X}) = \alpha + \beta_1 X_1 + \beta_2 X_2.$$

Based on least-squares estimates $(\hat{\alpha}, \hat{\beta}_1, \hat{\beta}_2)$ and their standard errors, inference for the individual parameters α, β_1, and β_2 proceeds as in section 18.3. Here are some examples with more complicated inference since they involve linear combinations of more than one model parameter:

- *Combined effects of variables*: If both X_1 and X_2 increase by 1 unit, the relevant change in $E(Y|\mathbf{X})$ is the linear combination $\beta_1 + \beta_2$. The estimate of this change is straightforward and given by $\hat{\beta}_1 + \hat{\beta}_2$. The standard error is more complicated since, in general,

  $$se(\hat{\beta}_1 + \hat{\beta}_2) \neq \sqrt{se(\hat{\beta}_1)^2 + se(\hat{\beta}_2)^2}$$

  due to possible correlation between estimators $\hat{\beta}_{1,\mathbf{X}Y}$ and $\hat{\beta}_{2,\mathbf{X}Y}$. More generally, if X_1 increases by v_1 units and X_2 increases by v_2 units, the relevant change in $E(Y|\mathbf{X})$ is the linear combination $v_1\beta_1 + v_2\beta_2$, with estimate $v_1\hat{\beta}_1 + v_2\hat{\beta}_2$ and standard error $se(v_1\hat{\beta}_1 + v_2\hat{\beta}_2)$.
- *Difference in effects*: If X_1 and X_2 are comparable variables (e.g., they are measured in the same units), we might be interested in the difference between the two variables' effects on $E(Y|\mathbf{X})$. This difference is given by the linear combination $\beta_1 - \beta_2$. (This turns out to be a special instance of the previous case, thinking about increasing X_1 by 1 unit and decreasing X_2 by 1 unit.)
- *Estimating the conditional expectation*: The conditional expectation $E(Y|\mathbf{X})$ is itself a linear combination of the three model parameters. For $X_1 = x_1^*$ and $X_2 = x_2^*$, the conditional expectation is

  $$E(Y|X_1 = x_1^*, X_2 = x_2^*) = \alpha + \beta_1 x_1^* + \beta_2 x_2^*.$$

  This linear combination can be rewritten as $1 \cdot \alpha + x_1^* \cdot \beta_1 + x_2^* \cdot \beta_2$, making it clear that the expression is a linear combination of the three parameters, with coefficients 1, x_1^*, and x_2^*. The estimate of the conditional expectation is obtained by plugging in the least-squares estimates,

  $$\hat{E}(Y|X_1 = x_1^*, X_2 = x_2^*) = \hat{\alpha} + \hat{\beta}_1 x_1^* + \hat{\beta}_2 x_2^*,$$

  and has the standard error $se(\hat{\alpha} + \hat{\beta}_1 x_1^* + \hat{\beta}_2 x_2^*) = se(1 \cdot \hat{\alpha} + x_1^* \cdot \hat{\beta}_1 + x_2^* \cdot \hat{\beta}_2)$.

To accommodate general forms of linear combinations, we introduce a row vector that contains the appropriate coefficients for any given linear combination of the model parameters. In the example given here, there are three model parameters $(\alpha, \beta_1, \beta_2)$ and three estimates $(\hat{\alpha}, \hat{\beta}_1, \hat{\beta}_2)$, so the vector has three elements. The first element corresponds to the constant multiplying $\hat{\alpha}$, the second element corresponds to the constant multiplying $\hat{\beta}_1$, and the third element corresponds to the constant multiplying $\hat{\beta}_2$. For the three examples here, the vectors would be

$$\begin{bmatrix} 0 & 1 & 1 \end{bmatrix} \text{ for } \hat{\beta}_1 + \hat{\beta}_2,$$

$$\begin{bmatrix} 0 & 1 & -1 \end{bmatrix} \text{ for } \hat{\beta}_1 - \hat{\beta}_2,$$

and

$$\begin{bmatrix} 1 & x_1^* & x_2^* \end{bmatrix} \text{ for } \hat{\alpha} + \hat{\beta}_1 x_1^* + \hat{\beta}_2 x_2^*.$$

More generally, the MLR model has $K+1$ model parameters $(\alpha, \beta_1, \beta_2, \ldots, \beta_K)$ and $K+1$ estimates $(\hat{\alpha}, \hat{\beta}_1, \hat{\beta}_2, \ldots, \hat{\beta}_K)$, so the vector will have $K+1$ elements. For example, with $K=4$ explanatory variables, the linear combination $\beta_1 + 3\beta_3 - 2\beta_4$ is represented by the vector

$$\begin{bmatrix} 0 & 1 & 0 & 3 & -2 \end{bmatrix}$$

since $\beta_1 + 3\beta_3 - 2\beta_4 = 0\alpha + 1\beta_1 + 0\beta_2 + 3\beta_3 + (-2)\beta_4$.

To facilitate the calculation of a standard error for a linear combination of least-squares estimates, we define an *R* function `linear_combination` as follows:

- `linear_combination(regresults, linvec)`: Takes `regresults`, the results from a `lm_robust` regression, and `linvec`, a vector that specifies the linear combination of interest, as arguments, and returns the estimate (`estimate`) of the linear combination and the standard error (`se`) of the estimate. The vector `linvec` has the same length $(K+1)$ as the parameter vector $(\alpha, \beta_1, \beta_2, \ldots, \beta_K)$, with its elements $(a_1, a_2, a_3, \ldots, a_{K+1})$ corresponding to the coefficients of each parameter that give the linear combination of the estimates, $a_1\hat{\alpha} + a_2\hat{\beta}_1 + a_3\hat{\beta}_2 + \cdots + a_{K+1}\hat{\beta}_K$.

Example 18.11 (Monthly stock returns) *Parameter estimates and standard errors for the MLR model* $E(HD|IDX, LOW) = \alpha + \beta_1 IDX + \beta_2 LOW$ *were provided in example 18.9. We use the* `linear_combination` *function to provide estimates and standard errors for some linear combinations of the parameters, specifically the following: (1)* $\beta_1 + \beta_2$*, representing the combined effect of increasing both IDX and LOW; (2)* $\beta_1 - \beta_2$*, representing the difference in the effects of IDX and LOW; and (3)* $\alpha + 0.05IDX + 0.03LOW$*, representing the conditional expectation* $E(HD|IDX = 0.05, LOW = 0.03)$*. The following R code provides output for these three cases:*

```
results <- lm_robust(HD~IDX+LOW, data=sp500)

# estimate beta(IDX)+beta(LOW) and its standard error
linear_combination(results, c(0,1,1))

## $estimate
## [1] 0.97916
```

```
##
## $se
## [1] 0.068797

# estimate beta(IDX)-beta(LOW) and its standard error
linear_combination(results, c(0,1,-1))

## $estimate
## [1] 0.21066
##
## $se
## [1] 0.1273

# estimate E(HD|IDX=0.05,LOW=0.03) and its standard error
linear_combination(results, c(1,0.05,0.03))

## $estimate
## [1] 0.045521
##
## $se
## [1] 0.0042493
```

We have $\hat{\beta}_1 + \hat{\beta}_2 = 0.979$*, with a standard error of* 0.069. *If IDX and LOW both increase by* 0.01 *(1 percentage point), these estimates imply an associated increase in the conditional expectation of HD of* 0.00979*, with a standard error of* 0.00069. *For the difference in the slope parameters, we have* $\hat{\beta}_1 - \hat{\beta}_2 = 0.211$*, with a standard error of* 0.127*, which implies the* 95% *confidence interval for* $\beta_1 - \beta_2$ *is* $(-0.039, 0.460)$. *And the estimated conditional expectation is*

$$\hat{E}(HD|IDX = 0.05, LOW = 0.03) = \hat{\alpha} + \hat{\beta}_1(0.05) + \hat{\beta}_2(0.03) = 0.0455,$$

with a standard error of 0.0042*, which implies the* 95% *confidence interval for the true conditional expectation* $E(HD|IDX = 0.05, LOW = 0.03)$ *is* $(0.0372, 0.0538)$.

18.5 Hypothesis Testing

For a hypothesis test involving an individual MLR parameter, the z-test is applicable. For instance, a null hypothesis involving just the intercept parameter,

$$H_0 : \alpha = c,$$

is tested with the z-statistic

$$z\text{-statistic} = \frac{\hat{\alpha} - c}{se(\hat{\alpha})}$$

or its associated p-value. Similarly, a null hypothesis involving one of the slope parameters,

$$H_0 : \beta_k = c,$$

is tested with the z-statistic

$$z\text{-statistic} = \frac{\hat{\beta}_k - c}{se(\hat{\beta}_k)}$$

or its associated p-value. It is very common to test whether an individual slope parameter equals zero ($c = 0$), as the null hypothesis $H_0 : \beta_k = 0$ implies that variable X_k does not belong in the MLR model. In fact, *R* provides the z-statistic and p-value associated with

$H_0 : \beta_k = 0$, for each $k \in \{1, 2, \ldots, K\}$, as part of the least-squares estimation output. If we do not reject $H_0 : \beta_k = 0$, there is no statistical support for its inclusion in the model, and we might say that "the variable X_k is not statistically significant in the model." If there was uncertainty about including X_k in the model at the outset, the failure to reject $H_0 : \beta_k = 0$ might prompt a practitioner to drop X_k from the model. That said, variables should be dropped from the MLR model with caution. First, it's possible that we just don't have a large enough sample to estimate a statistically significant slope on X_k. Second, it can be useful to have a variable in the model even if its slope estimate is not statistically significant, as someone who is looking at the results might have expected the variable to matter in the model and would be interested to see that it does not.

Example 18.12 (Monthly stock returns) *The following R code shows the MLR results from example 18.9:*

```
mlr_results <- lm_robust(HD~IDX+LOW, data=sp500)
mlr_results

##              Estimate Std. Error t value    Pr(>|t|)   CI Lower  CI Upper  DF
## (Intercept) 0.0042476  0.0027743  1.5311 1.2663e-01 -0.0012081 0.0097034 361
## IDX         0.5949112  0.0906584  6.5621 1.8444e-10  0.4166262 0.7731962 361
## LOW         0.3842477  0.0474368  8.1002 8.5485e-15  0.2909606 0.4775348 361
```

The output contains z-statistics and p-values for testing each individual MLR parameter versus zero. Since $\alpha = E(HD|IDX = LOW = 0)$, the null hypothesis $H_0 : \alpha = 0$ corresponds to testing whether the expected Home Depot return (HD) equals zero when the S&P 500 return (IDX) and Lowe's return (LOW) both equal zero. From the table, the z-statistic for this null hypothesis is 1.53, so $H_0 : \alpha = 0$ would not be rejected at a 5% level ($1.53 < z_{0.025} = 1.96$) or a 10% level ($1.53 < z_{0.05} = 1.645$). The p-value of 0.1266 indicates that $H_0 : \alpha = 0$ would not be rejected at any level below 12.66%. For both slope parameters, the test of $H_0 : \beta_k = 0$ has a p-value equal to zero to many decimal places (e.g., zero to 9 decimal places for the IDX slope). Therefore, $H_0 : \beta_1 = 0$ and $H_0 : \beta_2 = 0$ are both rejected at any level, indicating that IDX and LOW are both statistically significant variables in the MLR model.

Example 18.13 (Cigarette sales and cigarette taxes) *The following R code shows the MLR results from example 18.8, with the z-statistics and p-values for testing each parameter versus zero:*

```
mlr_results <- lm_robust(cigsales~cigtax+producer, data=cigdata)
mlr_results

##             Estimate Std. Error t value    Pr(>|t|) CI Lower CI Upper DF
## (Intercept)  54.2762     3.5128 15.4511 2.9333e-20   47.213   61.339 48
## cigtax       -8.9687     1.1935 -7.5146 1.2014e-09  -11.368   -6.569 48
## producer      5.3669     5.0785  1.0568 2.9589e-01   -4.844   15.578 48
```

The state-level cigarette tax (CIGTAX) variable is statistically significant, as the p-value of 0.0000 indicates that $H_0 : \beta_1 = 0$ is rejected at any level. On the other hand, the tobacco-producing state indicator (PRODUCER) variable has a z-statistic of 1.06 and a p-value of 0.296 associated with testing $H_0 : \beta_2 = 0$. The null hypothesis $H_0 : \beta_2 = 0$ corresponds *to PRODUCER not being in the MLR model, and* $H_0 : \beta_2 = 0$ *would not be rejected at any level below* 29.6%. *Given these results, whether to leave PRODUCER in the model is a choice that the practitioner needs to make. If PRODUCER is dropped from the model, the result is the SLR model with only CIGTAX as an explanatory variable; from example 18.8, the cigarette-tax slope estimate for that SLR model is* −9.49, *with a standard error of* 1.06.

Other hypothesis tests may be of interest for an MLR model, including the following:

Testing a single linear restriction: The simple null hypotheses given here are examples of single linear restrictions on the MLR parameters. What if we want to test whether the slope on one variable, X_1, is equal to the slope on another variable, X_2? If the two variables are measured in the same units, a test of $H_0 : \beta_1 = \beta_2$ assesses whether the partial effect of X_1 on $E(Y|\mathbf{X})$, holding all other variables fixed, is the same as the partial effect of X_2 on $E(Y|\mathbf{X})$, holding all other variables fixed. The null hypothesis $H_0 : \beta_1 = \beta_2$ is equivalent to

$$H_0 : \beta_1 - \beta_2 = 0,$$

which can be tested with the z-statistic

$$z\text{-statistic} = \frac{\hat{\beta}_1 - \hat{\beta}_2}{se(\hat{\beta}_1 - \hat{\beta}_2)}.$$

For example, for the MLR model in example 18.12, the null hypothesis $H_0 : \beta_1 = \beta_2$ is true if the partial effects of IDX and LOW on $E(HD|IDX, LOW)$ are the same. The main issue with calculating the z-statistic is that the standard error $se(\hat{\beta}_1 - \hat{\beta}_2)$ is not provided in the least-squares estimation output, so we use the `linear_combination` function discussed in section 18.4 to calculate $se(\hat{\beta}_1 - \hat{\beta}_2)$. Using the estimates $\hat{\beta}_1 - \hat{\beta}_2$ and the standard error $se(\hat{\beta}_1 - \hat{\beta}_2)$ calculated in example 18.11, the associated z-statistic is

$$\frac{\hat{\beta}_1 - \hat{\beta}_2}{se(\hat{\beta}_1 - \hat{\beta}_2)} = \frac{0.2107}{0.1273} \approx 1.66,$$

which has a p-value of 0.098, meaning that $H_0 : \beta_1 = \beta_2$ is rejected at a 10%, but not a 5% level.

Of course, $H_0 : \beta_1 = \beta_2$ is not the only linear restriction that might be of interest. How about the null hypothesis $H_0 : \beta_1 + \beta_2 = c$? If X_1 and X_2 are measured in the same units, this null hypothesis assesses whether the combination of the partial effects of X_1 and X_2 on $E(Y|\mathbf{X})$ (i.e., the change in $E(Y|\mathbf{X})$ when *both* X_1 and X_2 are increased by 1 unit) is equal to a constant c. This null hypothesis can be tested with z-statistic $\frac{\hat{\beta}_1 + \hat{\beta}_2 - c}{se(\hat{\beta}_1 + \hat{\beta}_2)}$, which can be calculated after the standard error $se(\hat{\beta}_1 + \hat{\beta}_2)$ is calculated by `linear_combination`.

Testing multiple restrictions: In the same way that $H_0 : \beta_k = 0$ can be tested if we are considering dropping X_k from the MLR model, we can also test whether *several* slope parameters equal zero if we are considering dropping a *group* of explanatory variables. As an example, if we are considering dropping the three variables X_2, X_4, and X_5 from an MLR model, the null hypothesis of interest is

$$H_0 : \beta_2 = \beta_4 = \beta_5 = 0$$

or, equivalently written explicitly as three linear restrictions,

$$H_0 : \beta_2 = 0, \beta_4 = 0, \beta_5 = 0.$$

The Wald test from section 16.4 can be used to test such a null hypothesis. If we do not reject H_0, that would provide some justification for dropping the three variables from the model. One important reason to do a test of the **joint significance** of variables is that testing significance of variables one by one, via z-tests, can be problematic if there is multicollinearity between variables. As an example, let's say that X_1 and X_2 are *very* highly correlated. The slope estimates $\hat{\beta}_1$ and $\hat{\beta}_2$, from the estimation of an MLR model with X_1 and X_2 as explanatory variables, might be statistically insignificant simply because X_1 and X_2 both have very little independent variation. This statistical insignificance would be reflected in high standard errors and low magnitudes of the z-statistics used to test $H_0 : \beta_1 = 0$ and $H_0 : \beta_2 = 0$. Even though X_1 and X_2 are individually insignificant, it is possible that they are jointly significant in explaining the outcome variable, which could be verified if $H_0 : \beta_1 = \beta_2 = 0$ is rejected.

To facilitate the Wald test, we define an *R* function `test_linear_restrictions`, summarized as follows:

- `test_linear_restrictions(regresults, R, c)`: Takes `regresults` (results from a `lm_robust` regression), a matrix `R` that specifies one or more linear combinations of interest, as well as a vector `c` specifying the constants associated the linear combinations as arguments, and returns both the Wald statistic and the p-value for the test. Matrix `R` has Q rows, where Q is the number of restrictions being tested, and $(K+1)$ columns. For each row, the elements $(a_1, a_2, a_3, \ldots, a_{K+1})$ correspond to the coefficients of each parameter, leading to the linear combination $a_1\alpha + a_2\beta_1 + a_3\beta_2 + \cdots + a_{K+1}\beta_K$. The corresponding element of the vector `c` is the constant being tested against.

For example, for the null hypothesis $H_0 : \beta_2 = 0, \beta_4 = 0, \beta_5 = 0$ considered here in a model with six explanatory variables, matrix `R` has $Q = 3$ rows and $K + 1 = 7$ columns. In the first row, the third element (for β_2) would be `1` and the other elements `0`; in the second row, the fifth element (for β_4) would be `1` and the other elements `0`; and, in the third row, the sixth element (for β_5) would be `1` and the other elements `0`. The `c`-vector would consist of three `0` elements since each of the parameters is being tested against zero. Using the notation from the appendix to chapter 16, the number of restrictions is $Q = 3$ and the number of parameters is $L = K + 1 = 7$, and we have

$$\mathbf{R} = \begin{bmatrix} 0 & 0 & 1 & 0 & 0 & 0 & 0 \\ 0 & 0 & 0 & 0 & 1 & 0 & 0 \\ 0 & 0 & 0 & 0 & 0 & 1 & 0 \end{bmatrix} \text{ and } \mathbf{c} = \begin{bmatrix} 0 \\ 0 \\ 0 \end{bmatrix}.$$

Each row of matrix $\mathbf{R}$ is a specific linear combination of the $L = K + 1$ model parameters. For the specific $\mathbf{R}$ matrix shown here, the first row is β_2, the second row is β_4, and the third row is β_5. Each element of the $\mathbf{c}$ (column) vector provides the constant against which the corresponding row of $\mathbf{R}$ is being tested. For the example shown here, these elements are all zero.

Suppose that the MLR model for Home Depot (*HD*) monthly returns is augmented to include the monthly returns of Bank of America (*BAC*) and Wells Fargo (*WFC*) as explanatory variables, with the results provided here:

```
results <- lm_robust(HD~IDX+LOW+BAC+WFC, data=sp500)
results

##                 Estimate Std. Error  t value   Pr(>|t|)   CI Lower CI Upper  DF
## (Intercept)  0.0042975  0.0027741  1.54913 1.2223e-01 -0.0011581 0.009753 359
## IDX          0.6567405  0.1158056  5.67106 2.9223e-08  0.4289978 0.884483 359
## LOW          0.3876343  0.0478259  8.10512 8.3756e-15  0.2935802 0.481688 359
## BAC         -0.0393008  0.0382543 -1.02736 3.0494e-01 -0.1145315 0.035930 359
## WFC         -0.0067428  0.0503518 -0.13391 8.9355e-01 -0.1057644 0.092279 359
```

Based on the z-statistics and p-values, the *IDX* and *LOW* variables remain statistically significant, but the p-values on the two added variables indicate a lack of significance for each variable individually (p-value of 0.30 for *BAC* and p-value of 0.89 for *WFC*). These results suggest that the two added variables are not valuable in the model for explaining Home Depot's returns, and a Wald test of $H_0 : \beta_3 = \beta_4 = 0$ provides a p-value for their joint significance:

```
results <- lm_robust(HD~IDX+LOW+BAC+WFC, data=sp500)

# use rbind function to create a matrix consisting of the two restrictions:
#   first restriction is beta3=0, second restriction is beta4=0
R <- rbind(c(0,0,0,1,0), c(0,0,0,0,1))

# create a vector of zeros for the two restrictions
c <- c(0,0)

# conduct the Wald test
test_linear_restrictions(results, R, c)

## $W
## [1] 1.4126
##
## $p_value
## [1] 0.49346
```

The `rbind` function stacks the rows of linear restrictions to form matrix R. The resulting p-value of 0.49 implies that $H_0 : \beta_3 = \beta_4 = 0$, corresponding to *BAC* and *WFC* not being in the model, would not be rejected at any reasonable level. This test, therefore, provides support for dropping the two variables from the model.

18.6 Modeling Approaches and Explanatory Variables

The examples of MLR models considered thus far have been rather simplistic in their specification of the explanatory variables. First, while some models considered indicator variables, we have not included categorical variables with more than two categories. Such categorical variables are the focus of section 18.6.1. Second, each explanatory variable has been considered only in isolation, as none of the models has had an underlying variable showing up as part of more than one explanatory variable. For instance, if X is a variable, we can include both X and X^2 as explanatory variables to allow Y to be a nonlinear function of X. Such nonlinear specifications are considered in section 18.6.2. Also, if X_1 and X_2 are explanatory variables, we can include the interaction X_1X_2 in the model. Such interaction variables, considered in section 18.6.3, allow more flexibility in how the expected outcome is associated with the explanatory variables.

18.6.1 Categorical Variables

The binary variables considered in previous examples in this chapter are special cases of categorical variables. For example, in the weekly earnings examples, the *union* indicator variable is equal to 1 for the category of union workers and 0 for the category of nonunion workers. For a more general categorical variable with C categories, where the categories are mutually exclusive and exhaustive, the MLR model can contain $C-1$ indicator variables representing $C-1$ of the C categories. It is not possible to include C indicator variables, as that would result in perfect colinearity (violating assumption MLR-NPC). Example 18.14 shows how indicator variables are used to represent a categorical variable and how the slope estimates associated with these indicator variables should be interpreted.

Example 18.14 (Earnings and marital status) *The **cps** data set contains a categorical variable, marstatus, that measures marital status, with four possible values: "Married," "Divorced," "Widowed," and "Never married." Define the following four indicator variables corresponding to the four categories of the marstatus:*

$$\begin{aligned}
&\textit{married} &&= 1 \textit{ if marstatus} = \textit{"Married"}, \; 0 \textit{ otherwise}\\
&\textit{divorced} &&= 1 \textit{ if marstatus} = \textit{"Divorced"}, \; 0 \textit{ otherwise}\\
&\textit{widowed} &&= 1 \textit{ if marstatus} = \textit{"Widowed"}, \; 0 \textit{ otherwise}\\
&\textit{nevermarried} &&= 1 \textit{ if marstatus} = \textit{"Never married"}, \; 0 \textit{ otherwise}
\end{aligned}$$

These categories are mutually exclusive and exhaustive, meaning that one and only one of the four indicator variables is equal to 1, so

$$\textit{married} + \textit{divorced} + \textit{widowed} + \textit{nevermarried} = 1$$

is always true. Therefore, only three of the indicator variables can be included in the MLR model, as the fourth one is a perfect linear combination of the other three, which would be a violation of assumption MLR-NPC. As a practical matter, it does not matter which of the four indicator variables is omitted; it only affects how we use the estimates to interpret the model estimates. Let's choose "Never married" as the omitted category, with the three variables married, divorced, and widowed included in the MLR model. The following output shows least-squares estimates for two models involving these three indicator variables,[6] Model I, which has only the three indicator variables and no other explanatory variables, and model II, which has the three indicator variables in addition to the explanatory variables educ, exper, union, and female:

```
# create marital-status indicator variables
cps$married <- 1*(cps$marstatus=="Married")
cps$divorced <- 1*(cps$marstatus=="Divorced")
cps$widowed <- 1*(cps$marstatus=="Widowed")
cps$nevermarried <- 1*(cps$marstatus=="Never married")
```

```
# estimate Model I
model1 <- lm_robust(earnwk~married+divorced+widowed, data=cps)
model1

##               Estimate Std. Error t value     Pr(>|t|)  CI Lower CI Upper   DF
## (Intercept)    820.512     27.929 29.3785 1.2281e-165  765.7488  875.276 2805
## married        226.198     33.675  6.7170  2.2374e-11  160.1667  292.229 2805
## divorced        81.235     43.091  1.8852  5.9504e-02   -3.2573  165.728 2805
## widowed       -159.499     58.786 -2.7132  6.7040e-03 -274.7669  -44.231 2805

# estimate Model II
model2 <- lm_robust(earnwk~married+divorced+widowed+educ+exper+union+female,
data=cps)
model2

##               Estimate Std. Error   t value   Pr(>|t|)   CI Lower   CI Upper   DF
## (Intercept) -500.4410    98.1084  -5.10090 3.6061e-07 -692.81309 -308.0690 2801
## married      204.3868    33.2880   6.13994 9.4230e-10  139.11518  269.6583 2801
## divorced      86.3465    41.5833   2.07647 3.7941e-02    4.80958  167.8834 2801
## widowed      -50.4511    64.6772  -0.78005 4.3543e-01 -177.27084   76.3685 2801
## educ         108.4820     7.2819  14.89757 2.3140e-48   94.20361  122.7603 2801
## exper          3.3677     1.3992   2.40697 1.6150e-02    0.62425    6.1112 2801
## union        134.2413    44.7788   2.99787 2.7426e-03   46.43840  222.0441 2801
## female      -341.3955    25.6764 -13.29607 3.6589e-39 -391.74215 -291.0489 2801
```

In model I, with no other explanatory variables, the intercept estimate has a direct interpretation as the estimate of the expected weekly earnings for the omitted category since

$$\alpha = E(earnwk|married = divorced = widowed = 0).$$

The intercept estimate $\hat{\alpha} = 820.51$ *implies that the average weekly earnings for never-married individuals is* \$820.51. *Each of the three slope estimates is interpreted as the difference in expected weekly earnings for the corresponding category and the omitted category. For instance, the slope on the married variable is*

$$\beta_{married} = E(earnwk|married = 1) - E(earnwk|nevermarried = 1),$$

with the estimate $\hat{\beta}_{married} = 226.20$ *implying that the estimated difference in expected weekly earnings between married and never-married individuals is* \$226.20. *Similarly,* $\hat{\beta}_{divorced} = 81.24$ *implies that the estimated difference in expected weekly earnings between divorced and never-married individuals is* \$81.24, *and* $\hat{\beta}_{widowed} = -159.50$ *implies that the estimated difference in expected weekly earnings between widowed and never-married individuals is* $-\$159.50$. *For any of these differences with the omitted category, we can test the statistical significance of the difference with a z-test for the appropriate slope estimate. For example, the null hypothesis* $H_0 : \beta_{married} = 0$, *corresponding to married and never-married individuals having the same expected weekly earnings, has a z-statistic of*

$$\frac{\hat{\beta}_{married}}{se(\hat{\beta}_{married})} = \frac{226.20}{33.68} \approx 6.72$$

and p-value of 0.0000. *Thus,* $H_0 : \beta_{married} = 0$ *is rejected at any level, indicating a statistically significant difference between the expected weekly earnings of married and never-married individuals. The z-tests of* $H_0 : \beta_{divorced} = 0$ *and* $H_0 : \beta_{widowed} = 0$ *have z-statistics of* 1.89 *and* -2.71, *respectively, and p-values of* 0.060 *and* 0.007, *respectively. For a 5% level,* $H_0 : \beta_{divorced} = 0$ *is not rejected, while* $H_0 : \beta_{widowed} = 0$ *is rejected.*

How about differences for two of the categories included in the MLR model? Let's say we are interested in the difference in expected earnings between married and divorced individuals:

$$\beta_{married} - \beta_{divorced} = E(earnwk|married = 1) - E(earnwk|divorced = 1).$$

This difference can be estimated directly by

$$\hat{\beta}_{married} - \hat{\beta}_{divorced} = 226.20 - 81.24 = 144.96.$$

Testing whether this difference is statistically significant is more difficult since the z-statistic for testing $H_0 : \beta_{married} - \beta_{divorced} = 0$ requires the standard error $se(\hat{\beta}_{married} - \hat{\beta}_{divorced})$. There are two options for conducting this test. The first is to use the `linear_combination` *function to calculate $se(\hat{\beta}_{married} - \hat{\beta}_{divorced})$. The second is to reestimate the model and omit either married or divorced, rather than nevermarried, so a z-test of $H_0 : \beta_{divorced} = 0$ (if married is omitted) or a z-test of $H_0 : \beta_{married} = 0$ (if divorced is omitted) can be easily conducted.*

For model II, the important difference from model I is that interpretation of estimates now involves holding the other variables (educ, exper, union, female) fixed. The intercept estimate no longer has the simple interpretation that it had in model I since the intercept parameter is the expected weekly earnings when all the explanatory variables equal zero. The slope estimates do have a similar interpretation, but the other variables are held fixed. The married slope estimate $\hat{\beta}_{married} = 204.40$ indicates that the estimated difference in expected weekly earnings for married and never-married individuals, holding the other variables fixed, is equal to \$204.39. This estimate is slightly lower than the estimate from model I. The estimate is statistically significant with a z-statistic of

$$\frac{\hat{\beta}_{married}}{se(\hat{\beta}_{married})} = \frac{204.39}{33.29} \approx 6.14$$

and p-value of 0.0000. For the divorced variable, the z-statistic and p-value are 2.08 and 0.038, respectively, so the estimated difference in expected weekly earnings between divorced and never-married individuals, holding the other variables fixed, is statistically significant at a 5% level. For the widowed variable, the z-statistic and p-value are −0.78 and 0.435, so the estimated difference in expected weekly earnings between widowed and never-married individuals, holding the other variables fixed, is not statistically significant at any level less than 43.5%.

18.6.2 Nonlinear Functions of Explanatory Variables

In the MLR model

$$E(Y|\mathbf{X}) = \alpha + \beta_1 X_1 + \beta_2 X_2 + \cdots + \beta_K X_K,$$

when each of the K explanatory variables enters into only a single $\beta_k X_k$ term, there is an implicit restriction on how $E(Y|\mathbf{X})$ is related to changes in the explanatory variables. Specifically, a 1-unit change of any X_k variable, holding the other variables fixed, is associated with a change in $E(Y|\mathbf{X})$ equal to β_k, which is a *constant* that does not depend on the values of the explanatory variables at all. In this section, we relax this restriction by considering adding nonlinear functions of an explanatory variable to the model. We start by considering polynomial functions. Suppose we would like to allow for a quadratic function of X_1 rather than a linear function, so the MLR model takes the form

$$E(Y|\mathbf{X}) = \alpha + \beta_1 X_1 + \beta_2 X_1^2 + \beta_3 X_3 + \cdots + \beta_K X_K.$$

When X_1 changes, the variable $X_2 = X_1^2$ must also change. For a 1-unit change in X_1, from $X_1 = x_1^*$ to $X_1 = x_1^* + 1$, the associated change in $E(Y|\mathbf{X})$, holding the other explanatory variables fixed, is[7]

$$\begin{aligned}
&E(Y|X_1 = x_1^* + 1, X_3 = x_3^*, \ldots, X_K = x_K^*) - E(Y|X_1 = x_1^*, X_3 = x_3^*, \ldots, X_K = x_K^*) \\
&= \beta_1 + \beta_2 \left((x_1^* + 1)^2 - (x_1^*)^2\right) \\
&= \beta_1 + \beta_2 (2x_1^* + 1).
\end{aligned}$$

Unlike the model with only *educ*, where the partial effect of X_1 on $E(Y|\mathbf{X})$ is constant, the partial effect of X_1 on $E(Y|\mathbf{X})$ is a function of x_1^* itself. The sign of β_2 indicates how the partial effect of X_1 on $E(Y|\mathbf{X})$ changes with x_1^*. When $\beta_2 > 0$, the partial effect $\beta_1 + \beta_2(2x_1^* + 1)$ increases when x_1^* increases, and when $\beta_2 < 0$, the partial effect $\beta_1 + \beta_2(2x_1^* + 1)$ decreases when x_1^* increases.

The model can include even higher powers of the explanatory variable, although in practice, it's unusual to see terms beyond a quadratic or cubic term. To allow for a cubic polynomial of the X_1 variable, the variables $X_2 = X_1^2$ and $X_3 = X_1^3$ would be in the model. A 1-unit change in X_1, from $X_1 = x_1^*$ to $X_1 = x_1^* + 1$, the associated change in $E(Y|\mathbf{X})$, holding the other explanatory variables fixed, is[8]

$$\begin{aligned}
&E(Y|X_1 = x_1^* + 1, X_4 = x_4^*, \ldots, X_K = x_K^*) - E(Y|X_1 = x_1^*, X_4 = x_4^*, \ldots, X_K = x_K^*) \\
&= \beta_1 + \beta_2 \left((x_1^* + 1)^2 - (x_1^*)^2\right) + \beta_3 \left((x_1^* + 1)^3 - (x_1^*)^3\right) \\
&= \beta_1 + \beta_2 (2x_1^* + 1) + \beta_3 (3x_1^{*2} + 3x_1^* + 1).
\end{aligned}$$

For this cubic model, the partial effect of X_1 on $E(Y|\mathbf{X})$ is a quadratic function of x_1^*. This same idea can be generalized to higher-order polynomials.

Example 18.15 (Earnings with a quadratic in education) *Example 18.10 provided least-squares estimates and standard errors for an MLR model of weekly earnings (earnwk) with educ, exper, female, and union as explanatory variables. To allow the possibility of a nonlinear association between $E(earnwk|\mathbf{X})$ and educ, we add the quadratic variable $educ^2$ to the model. In R, there are two alternative methods for including the quadratic variable. The first is to define a new variable equal to $educ^2$ and add it to the list of variables in the* `lm_robust` *function. The second is to use the* `I()` *syntax in the* `lm_robust` *function, as illustrated here:*

```
results <- lm_robust(earnwk~educ+I(educ^2)+exper+union+female, data=cps)
results

##               Estimate Std. Error   t value   Pr(>|t|)  CI Lower  CI Upper   DF
## (Intercept)   883.6885   123.1009    7.1786 8.9823e-13  642.3109 1125.0661 2803
## educ         -122.7119    23.5915   -5.2015 2.1190e-07 -168.9703  -76.4534 2803
## I(educ^2)       9.8213     1.1688    8.4031 6.8198e-17    7.5296   12.1130 2803
## exper           5.2867     1.2943    4.0846 4.5384e-05    2.7488    7.8245 2803
## union         143.9565    43.9820    3.2731 1.0767e-03   57.7163  230.1968 2803
## female       -345.5150    25.2119  -13.7044 2.0057e-41 -394.9508 -296.0792 2803
```

```
summary(results)$r.squared

## [1] 0.19813
```

The second variable is specified as `I(educ^2)`, *and the* `I()` *syntax tells the* `lm_robust` *function to do the calculation within parentheses for each observation and include the resulting variable in the least-squares estimation. The following table reports these results, with the original model in the first column and the new model, with* $educ^2$ *included, in the second column. Standard errors are reported in parentheses:*

	MLR model with educ	*MLR model with educ,* $educ^2$
Intercept	−431.03 (100.08)	883.69 (123.10)
educ	110.94 (7.27)	−122.71 (23.59)
$educ^2$		9.821 (1.169)
exper	4.897 (1.315)	5.287 (1.294)
union	142.75 (44.39)	143.96 (43.98)
female	−344.79 (25.82)	−345.52 (25.21)
R^2	0.166	0.198
$\hat{\sigma}_U$	685.78	672.64

For a z-test of $H_0 : \beta_{educ^2} = 0$*, the z-statistic is* $\frac{9.821}{1.169} \approx 8.4$ *and p-value* 0.000*, indicating that the* $educ^2$ *variable belongs in the model. Including* $educ^2$ *in the model increases the R-squared value from* 16.6% *to* 19.8% *and decreases the residual standard deviation estimate from* \$685.78 *to* \$672.64*. On the other hand, adding* $educ^2$ *to the model has very little effect on the estimates of the slopes for exper, union, and female.*

Let's take a closer look at the partial effects of education implied by the two models. For the model with only educ, the estimated partial effect of a 1-year change in education on $E(earnwk|\mathbf{X})$ *is constant and equal to* $\hat{\beta}_{educ} = 110.94$*. For the model with* $educ^2$ *included, the estimated partial effect of a 1-year change in education, from* $educ^*$ *to* $educ^* + 1$*, on* $E(earnwk|\mathbf{X})$ *is*

$$\hat{\beta}_{educ} + \hat{\beta}_{educ^2}(2educ^* + 1) = -122.71 + 9.821(2educ^* + 1).$$

To calculate a standard error for this partial effect, we use the `linear_combination` *function to calculate* $se(\hat{\beta}_{educ} + \hat{\beta}_{educ^2}(2educ^* + 1))$ *for a given value of* $educ^*$*. The following table summarizes the partial effects estimated by the two models for four possible values of* $educ^*$*, with standard errors reported in parentheses:*

Change in $E(earnwk\|\mathbf{X})$ *when*	*MLR model with educ*	*MLR model with educ,* $educ^2$
educ changes from 10 *to* 11	110.94 (7.27)	83.54 (4.84)
educ changes from 12 *to* 13	110.94 (7.27)	122.82 (7.65)
educ changes from 14 *to* 15	110.94 (7.27)	162.11 (11.71)
educ changes from 16 *to* 17	110.94 (7.27)	201.39 (16.12)

*For the quadratic model, the estimated partial effect of education increases as the level of education increases. For example, the estimated partial effect at educ** $= 16$ *is equal to* \$201.39 *and is roughly* 64% *larger in magnitude than the estimated partial effect at educ** $=$ 12, *which is equal to* \$122.82.

Here is the R code to calculate standard errors for the estimated partial effects in the quadratic model:

```
# consider educ values of 10, 12, 14, and 16
educ_vec <- c(10,12,14,16)

# calculate the partial effect and standard error for each educ value
for (educ_star in educ_vec) {
    print(linear_combination(results, c(0,1,2*educ_star+1,0,0,0)))
}

## $estimate
## [1] 83.535
##
## $se
## [1] 4.8381
##
## $estimate
## [1] 122.82
##
## $se
## [1] 7.6458
##
## $estimate
## [1] 162.11
##
## $se
## [1] 11.714
##
## $estimate
## [1] 201.39
##
## $se
## [1] 16.115
```

18.6.3 Interactions of Explanatory Variables

In the previous section, the use of polynomial functions of an explanatory variable X_1 led to partial effects that were functions of the value of X_1 rather than a constant. In this section, this idea is generalized to allow partial effects of one explanatory variable to depend on *another* explanatory variable. For instance, to allow the partial effect of X_1 to depend on the value of X_2, an **interaction variable** X_1X_2 is added to the MLR model. Consider the following MLR model, which contains X_1, X_2, the interaction variable $X_3 = X_1X_2$, and additional explanatory variables:

$$E(Y|\mathbf{X}) = \alpha + \beta_1X_1 + \beta_2X_2 + \beta_3X_1X_2 + \beta_4X_4 + \cdots + \beta_KX_K.$$

The partial effect of increasing X_1 by 1 unit, from x_1^* to $x_1^* + 1$, on $E(Y|\mathbf{X})$ is equals

$$\begin{aligned}
&E(Y|X_1 = x_1^* + 1, X_2 = x_2^*, \ldots, X_K = x_K^*) - E(Y|X_1 = x_1^*, X_2 = x_2^*, \ldots, X_K = x_K^*) \\
&= \beta_1 + \beta_3\left((x_1^* + 1)x_2^* - x_1^*x_2^*\right) \\
&= \beta_1 + \beta_3x_2^*.
\end{aligned}$$

Therefore, the partial effect of X_1 on $E(Y|\mathbf{X})$ is a function of x_2^*, the value of X_2. When β_3 is positive, the partial effect of X_1 on $E(Y|\mathbf{X})$ is an increasing function of x_2^*, and when β_3 is negative, the partial effect of X_1 on $E(Y|\mathbf{X})$ is a decreasing function of x_2^*.

Similarly, we can determine the partial effect of increasing X_2 by 1 unit, from x_2^* to $x_2^* + 1$, on $E(Y|\mathbf{X})$:

$$\begin{aligned}
&E(Y|X_1 = x_1^*, X_2 = x_2^* + 1, \ldots, X_K = x_K^*) - E(Y|X_1 = x_1^*, X_2 = x_2^*, \ldots, X_K = x_K^*) \\
&= \beta_2 + \beta_3 \left(x_1^*(x_2^* + 1) - x_1^* x_2^*\right) \\
&= \beta_2 + \beta_3 x_1^*.
\end{aligned}$$

The partial effect of X_2 on $E(Y|\mathbf{X})$ depends on x_1^*, with the partial effect increasing in x_1^* if β_3 is positive and decreasing in x_1^* if β_3 is negative. Thus, a feature of a model with the interaction variable X_1X_2 is that, so long as $\beta_3 \neq 0$, it must be the case that *both* the partial effect of X_1 depends on X_2 and the partial effect of X_2 depends on X_1. When including X_1X_2 in a model, it is also good practice to *always* include both the original variables X_1 and X_2 in the model, even if it turns out that one or both appear to be insignificant based on z-tests.

To test whether an interaction variable is statistically significant, a z-test can be used for testing the null hypothesis $H_0 : \beta_3 = 0$. Rejection of $H_0 : \beta_3 = 0$ indicates statistical significance of the interaction variable, and a failure to reject $H_0 : \beta_3 = 0$ would support dropping the interaction variable, especially if the p-value is very high.

Example 18.16 (Weekly earnings) *Example 18.10 provided least-squares estimates and standard errors for an MLR model of weekly earnings (earnwk) with educ, exper, female, and union as explanatory variables. To allow the possibility that the partial effect of education on weekly earnings depends on a worker's experience level, the interaction variable educ · exper can be added to the MLR model. This inclusion also allows the partial effect of experience on weekly earnings to depend on a worker's education level. The* `I()` *syntax for* `lm_robust` *can be used to include the interaction variable in the model without creating a new variable. For the educ · exper interaction, the added variable is* `I(educ*exper)`*:*

```
results <- lm_robust(earnwk~educ+exper+I(educ*exper)+union+female, data=cps)
results

##                    Estimate Std. Error t value  Pr(>|t|)  CI Lower  CI Upper   DF
## (Intercept)       -1569.625   237.6139  -6.606 4.714e-11 -2035.541 -1103.709 2803
## educ                198.198    18.6682  10.617 7.650e-26   161.593   234.803 2803
## exper                45.915     6.5496   7.010 2.966e-12    33.073    58.758 2803
## I(educ * exper)      -3.194     0.5293  -6.034 1.814e-09    -4.232    -2.156 2803
## union               145.437    44.1914   3.291 1.010e-03    58.786   232.088 2803
## female             -350.552    25.4446 -13.777 7.837e-42  -400.444  -300.660 2803

summary(results)$r.squared

## [1] 0.1785
```

The following table shows the least-squares estimates for the original model without the interaction variable and the new model with the added interaction variable. Standard errors are reported in parentheses:

	MLR model without interaction	*MLR model with educ · exper*
Intercept	−431.03 (100.08)	−1,569.63 (237.61)
educ	110.94 (7.27)	198.20 (18.67)
exper	4.897 (1.315)	45.915 (6.550)
educ · exper		−3.194 (0.529)
union	142.75 (44.39)	145.44 (44.19)
female	−344.79 (25.82)	−350.55 (25.44)
R^2	0.166	0.179
$\hat{\sigma}_U$	685.78	680.84

The interaction variable seems to belong in the model, as the z-statistic for testing $H_0: \beta_{educ \cdot exper} = 0$ *is* $\frac{-3.194}{0.529} \approx -6.04$. *The negative* $\hat{\beta}_{educ \cdot exper}$ *implies that (1) the estimated partial effect of education on weekly earnings* ($\hat{\beta}_{educ} + \hat{\beta}_{educ \cdot exper} exper^*$) *is a decreasing function of experience; and (2) the estimated partial effect of experience on weekly earnings* ($\hat{\beta}_{exper} + \hat{\beta}_{educ \cdot exper} educ^*$) *is a decreasing function of education. We calculate standard errors for these partial effects with the* `linear_combination` *function. The following table compares some estimated partial effects based on the two models, with standard errors reported in parentheses:*

Change in $E(earnwk\|\mathbf{X})$ *when*	*MLR model without interaction*	*MLR model with educ · exper*
educ changes by 1 unit, exper = 10	110.94 (7.27)	166.26 (13.79)
educ changes by 1 unit, exper = 20	110.94 (7.27)	134.32 (9.38)
exper changes by 1 unit, educ = 12	4.897 (1.315)	7.591 (1.265)
exper changes by 1 unit, educ = 14	4.897 (1.315)	1.204 (1.601)

For the model without the interaction variable educ · exper, the estimated partial effects of the two variables are constant, with the estimated partial effect of education given by $\hat{\beta}_{educ} = 110.94$ *and the estimated partial effect of experience given by* $\hat{\beta}_{exper} = 4.897$. *For the model with the interaction variable, the estimated partial effect of educ on* $E(earnwk|\mathbf{X})$ *declines from* \$166.26 *at 10 years of experience* ($exper = 10$) *to* \$134.32 *at 20 years of experience* ($exper = 20$). *Since* $\hat{\beta}_{educ \cdot exper}$ *is negative, this estimated partial effect would continue to decrease at even higher values of exper. The estimated partial effect of exper on* $E(earnwk|\mathbf{X})$ *declines from* \$7.59 *at 12 years of education* ($educ = 12$) *to* \$1.20 *at 14 years of education. In fact, the estimated partial effect of exper at* $educ = 14$ *is not statistically significant, as the z-statistic for testing the partial effect against zero is* $\frac{1.204}{1.601} \approx 0.75$, *which is associated with a p-value of* 0.45.

Here is the R code to calculate standard errors for the estimated partial effects in the interaction model:

```
# consider exper values of 10 and 20
exper_vec <- c(10,20)

# calculate the partial effect and standard error for each exper value
```

```
for (exper_star in exper_vec) {
    print(linear_combination(results, c(0,1,0,exper_star,0,0)))
}

## $estimate
## [1] 166.26
##
## $se
## [1] 13.792
##
## $estimate
## [1] 134.32
##
## $se
## [1] 9.3784

# consider educ values of 12 and 14
educ_vec <- c(12,14)

# calculate the partial effect and standard error for each educ value
for (educ_star in educ_vec) {
    print(linear_combination(results, c(0,0,1,educ_star,0,0)))
}

## $estimate
## [1] 7.5911
##
## $se
## [1] 1.2648
##
## $estimate
## [1] 1.2038
##
## $se
## [1] 1.6005
```

When one of the variables in an interaction variable is an indicator variable, we can determine whether, and by how much, the partial effect of the other variable depends on the two possible values of the indicator variable. For the MLR model

$$E(Y|\mathbf{X}) = \alpha + \beta_1 X_1 + \beta_2 X_2 + \beta_3 X_1 X_2 + \beta_4 X_4 + \cdots + \beta_K X_K,$$

consider the case where X_1 is an indicator variable, with $X_1 \in \{0, 1\}$. The only relevant 1-unit change for X_1 is going from 0 to 1, and the partial effect of X_1 on $E(Y|\mathbf{X})$ equals

$$\beta_1 + \beta_3 x_2^*$$

when $X_2 = x_2^*$. The partial effect of X_2 on $E(Y|\mathbf{X})$ equals

$$\beta_2 + \beta_3 x_1^* = \begin{cases} \beta_2 & \text{if } x_1^* = 0 \\ \beta_2 + \beta_3 & \text{if } x_1^* = 1, \end{cases}$$

meaning that β_3 measures the difference between the partial effect of X_2 at $X_1 = 1$ and the partial effect of X_2 at $X_1 = 0$. For instance, for the weekly earnings model considered in example 18.16, we could allow the partial effect of education on weekly earnings to depend on union membership by adding an interaction between *educ* and *union*; alternatively, we could allow the partial effect of experience on weekly earnings to depend on union membership by adding an interaction between *exper* and *union*.

Example 18.17 (Monthly stock returns and sample-splitting) *The* ***sp500*** *data set has* 364 *monthly observations, spanning* 30 *years between* 1991 *and* 2021. *A possible concern with using a regression model for such a long time horizon is that the "true model" might have changed over time. For instance, what if the relationship between an individual stock's return and the market index return has not remained the same over the 30 years observed in the data? To address that concern, one approach is to split the sample by creating an indicator variable* $post2005$ *that indicates which observations are after* 2005*:*

$$post2005 = \begin{cases} 1 & \text{if observation is after 2005 (2006–2021)} \\ 0 & \text{if observation is 2005 or earlier (1991–2005).} \end{cases}$$

The MLR model

$$E(HD|IDX, post2005) = \alpha + \beta_1 post2005 + \beta_2 IDX + \beta_3 post2005 \cdot IDX$$

includes the interaction variable $post2005 \cdot IDX$*, allowing the partial effect of the market index monthly return (IDX) to depend on whether* $post2005 = 0$ *or* $post2005 = 1$. *Specifically, if IDX increases by one unit, the expected HD return changes by* β_2 *if* $post2005 = 0$ *and by* $\beta_2 + \beta_3$ *if* $post2005 = 1$. *The parameter* β_3 *is, therefore, the difference in the partial effects of the market index return on the Home Depot return between the* 2006–2021 *period and the* 1991–2005 *period. There's nothing special about Home Depot here, so we can replace the outcome variable by the monthly return of any individual stock.*

The following table shows the least-squares estimates of this model for four stocks: Home Depot (HD), Lowe's (LOW), Bank of America (BAC), and ConocoPhillips (COP). Standard errors are reported in parentheses, and a row with the p-value for the test of the statistical significance of the interaction variable ($H_0 : \beta_3 = 0$) *is included:*

	HD	*LOW*	*BAC*	*COP*
Intercept	0.009 (0.005)	0.018 (0.007)	0.008 (0.005)	0.008 (0.005)
$post2005$	0.000 (0.006)	−0.012 (0.008)	−0.012 (0.009)	−0.010 (0.007)
IDX	1.150 (0.134)	1.070 (0.174)	0.998 (0.138)	0.697 (0.118)
$post2005 \cdot IDX$	−0.239 (0.164)	0.068 (0.214)	0.904 (0.272)	0.635 (0.202)
p-value for $H_0 : \beta_3 = 0$	0.147	0.752	0.001	0.002

Starting with the Home Depot (HD) model, the estimated slope on the interaction variable, $\hat{\beta}_3 = -0.239$*, indicates that the association between expected HD returns and IDX returns is lower in the post* 2005 *period. Specifically, when IDX goes up by* 0.01*, expected HD is estimated to increase by* $0.01\hat{\beta}_2 = 0.01150$ *in the pre-2005 period and* $0.01(\hat{\beta}_2 + \hat{\beta}_3) = 0.00911$ *in the post* 2005 *period. But, in looking at the p-value of* 0.147 *for the test of* $H_0 : \beta_3 = 0$*, we do not reject that the interaction variable has a population slope of zero at the* 10% *level. There is limited statistical evidence that there is meaningful difference in the partial effects of IDX on HD for the pre* 2005 *and post* 2005 *periods. For the Lowe's (LOW) model, the p-value of* 0.784 *for testing* $H_0 : \beta_3 = 0$ *is very high, meaning that we do not reject* $H_0 : \beta_3 = 0$ *at any reasonable level. The picture is quite different for the other two stocks, Bank of America (BAC) and ConocoPhillips (COP), where the p-values for* $H_0 : \beta_3 = 0$ *equal* 0.001 *and* 0.002*, respectively. These low p-values indicate that the interaction variable is* *statistically significant in both the BAC and the COP models. For the BAC model, when IDX goes up by* 0.01, *expected BAC is estimated to go up by* 0.00998 *(with standard error* 0.00138*) in the pre-*2005 *period and by* 0.01902 *(with standard error* 0.00235*) in the post* 2005 *period.*

Here is the R code to calculate the least-squares estimates given here, along with the standard error for the post 2005 *period IDX partial effect for the BAC model:*

```
# create the post-2005 indicator variable
sp500$post2005 <- c(rep(0,180),rep(1,184))

# least-squares estimation for models with post-2005 interaction variable
results_HD <- lm_robust(HD~post2005+IDX+I(post2005*IDX), data=sp500)
results_HD

##                     Estimate Std. Error  t value  Pr(>|t|)  CI Lower CI Upper DF
## (Intercept)        0.0086138   0.005211  1.65305 9.919e-02 -0.001634 0.01886 360
## post2005           0.0000169   0.006427  0.00263 9.979e-01 -0.012622 0.01266 360
## IDX                1.1500041   0.133917  8.58747 2.722e-16  0.886647 1.41336 360
## I(post2005 * IDX) -0.2392878   0.164496 -1.45467 1.466e-01 -0.562782 0.08421 360

results_LOW <- lm_robust(LOW~post2005+IDX+I(post2005*IDX), data=sp500)
results_LOW

##                   Estimate Std. Error t value  Pr(>|t|)  CI Lower CI Upper  DF
## (Intercept)        0.01754   0.006829  2.5688 1.061e-02  0.004113 0.030972 360
## post2005          -0.01154   0.008369 -1.3788 1.688e-01 -0.027999 0.004919 360
## IDX                1.06958   0.174243  6.1384 2.202e-09  0.726914 1.412240 360
## I(post2005 * IDX)  0.06781   0.214007  0.3169 7.515e-01 -0.353050 0.488672 360

results_BAC <- lm_robust(BAC~post2005+IDX+I(post2005*IDX), data=sp500)
results_BAC

##                    Estimate Std. Error t value  Pr(>|t|)  CI Lower CI Upper  DF
## (Intercept)        0.007928   0.004899   1.618 1.065e-01 -0.001706 0.017562 360
## post2005          -0.012168   0.008931  -1.362 1.739e-01 -0.029732 0.005397 360
## IDX                0.998438   0.137935   7.238 2.764e-12  0.727179 1.269697 360
## I(post2005 * IDX)  0.903626   0.272383   3.317 1.001e-03  0.367963 1.439288 360

results_COP <- lm_robust(COP~post2005+IDX+I(post2005*IDX), data=sp500)
results_COP

##                   Estimate Std. Error t value  Pr(>|t|)   CI Lower CI Upper  DF
## (Intercept)        0.00835   0.004624   1.806 7.182e-02 -0.0007446 0.017444 360
## post2005          -0.01036   0.007065  -1.466 1.435e-01 -0.0242519 0.003537 360
## IDX                0.69732   0.117650   5.927 7.223e-09  0.4659514 0.928687 360
## I(post2005 * IDX)  0.63502   0.201638   3.149 1.773e-03  0.2384856 1.031556 360

# partial effect (with standard error) of IDX for post-2005 period in BAC model
linear_combination(results_BAC,c(0,0,1,1))

## $estimate
## [1] 1.902
##
## $se
## [1] 0.2349
```

18.7 Log-Transformed Outcome Variable

In many economics and business applications, regression models use a log-transformed outcome variable rather than the outcome variable itself. There are several reasons for considering a log-transformed outcome variable:

1. For variables with right-skewed distributions (e.g., wages, sales, profits, and website visits), the log transformation can make the unconditional distribution of the outcome variable symmetric. In example 11.7, for instance, the distribution of log-transformed weekly earnings was quite symmetric and bell-shaped, even though the distribution of weekly earnings itself was extremely right-skewed.
2. For variables with right-skewed distributions, the log transformation decreases the magnitude of extreme outliers, meaning that the influence of outliers on regression estimates may be reduced with a log-transformed outcome variable.
3. A log-transformed outcome variable leads to an interpretation in terms of percentage changes, rather than absolute changes, in the outcome variable.

For an outcome variable Y with only positive values ($Y > 0$), the MLR model based on the log-transformed outcome is

$$\ln(Y) = \alpha + \beta_1 X_1 + \beta_2 X_2 + \cdots + \beta_K X_K + U \text{ with } E(U|\mathbf{X}) = 0$$

or, equivalently,

$$E(\ln(Y)|\mathbf{X}) = \alpha + \beta_1 X_1 + \beta_2 X_2 + \cdots + \beta_K X_K.$$

Estimation of this MLR model involves using $\ln(Y)$ as the outcome variable rather than Y. The interpretation of the model parameters, however, is more complicated. Thinking about the partial effect of the X_1 variable, for instance, a 1-unit change from $X_1 = x_1^*$ to $X_1 = x_1^* + 1$, holding all other variables fixed, leads to the conditional expectation $E(\ln(Y)|\cdot)$ changing by β_1:

$$E(\ln(Y)|X_1 = x_1^* + 1, X_2 = x_2^*, \ldots, X_K = x_K^*)$$
$$- E(\ln(Y)|X_1 = x_1^*, X_2 = x_2^*, \ldots, X_K = x_K^*) = \beta_1.$$

Unfortunately, this partial effect is difficult to describe since it is in terms of the log-outcome rather than the outcome itself. For instance, if Y is weekly earnings, the partial effect of a 1-unit increase in X_1 is a change of β_1 in the conditional expectation of *log-earnings*. To get an interpretation in terms of Y itself, we use the following calculus-based approximation:[9]

$$\Delta \ln(Y) \approx \frac{\Delta Y}{Y} \text{ when } \Delta Y \text{ is small.}$$

Since $\frac{\Delta Y}{Y}$ is the percentage change in variable Y, the partial-effect formula here says that a 1-unit change in X_1, holding all other variables fixed, is associated with an expected percentage change in Y that is approximately equal to β_1.

Example 18.18 (Weekly earnings) *Using the* ***cps*** *data set, here are the results from least-squares estimation of an MLR model using log-earnings as the outcome variable and educ, exper, union, and female as the explanatory variables:*

```
logwage_results <-lm_robust(I(log(earnwk))~educ+exper+union+female,data=cps)
logwage_results

##              Estimate Std. Error t value  Pr(>|t|)  CI Lower  CI Upper   DF
## (Intercept)  5.385820   0.087332  61.671 0.000e+00  5.214579  5.557061 2804
## educ         0.100279   0.005494  18.252 2.039e-70  0.089506  0.111052 2804
```

```
## exper         0.005667   0.001361    4.163 3.241e-05  0.002997  0.008336 2804
## union         0.198673   0.034718    5.722 1.161e-08  0.130597  0.266748 2804
## female       -0.400754   0.023896 -16.771 3.127e-60 -0.447610 -0.353898 2804

logwage_results$r.squared

## [1] 0.1943
```

The use of `I(log(earnwk))` *specifies the natural logarithm of weekly earnings as the outcome variable. Using the approximation result, a 1-year change in education, holding the other variables fixed, is associated with an expected change of* 0.100 *or* 10.0% *in weekly earnings. Similarly, a 1-year change in experience, holding the other variables fixed, is associated with an expected change of* 0.0057, *or* 0.57%, *in weekly earnings. For the union indicator variable, where a 1-unit change means going from a nonunion worker to a union worker, union workers are expected to earn approximately* 0.199, *or* 19.9%, *more than nonunion workers, holding all else fixed. And, female workers are expected to earn approximately* 0.401, *or* 40.1%, *less than male workers, holding all else fixed.*

The R-squared value is 0.194, *or* 19.4%. *Importantly, this value is* not *comparable to values for regressions with weekly earnings as the outcome variable. Since the outcome variable here is log-earnings, the R-squared value says that the explanatory variables explain* 19.4% *of the variation in* log-earnings, *not* 19.4% *of the variation in earnings.*

In some cases, a regression model may have an explanatory variable that is also log-transformed. Without loss of generality, let's say that the first explanatory variable is positive-valued ($X_1 > 0$) and log-transformed,

$$\ln(Y) = \alpha + \beta_1 \ln(X_1) + \beta_2 X_2 + \cdots + \beta_K X_K + U \text{ with } E(U|\mathbf{X}) = 0$$

or, equivalently,

$$E(\ln(Y)|\mathbf{X}) = \alpha + \beta_1 \ln(X_1) + \beta_2 X_2 + \cdots + \beta_K X_K + U.$$

The approximation $\Delta \ln(X_1) \approx \frac{\Delta X_1}{X_1}$ (when ΔX_1 is small) also holds here. To have the approximation perform well, it is common practice to use $\frac{\Delta X_1}{X_1} = 0.01$, which is equivalent to a 1% change in X_1. For $\frac{\Delta X_1}{X_1} = 0.01$, holding all other variables fixed, the expected percentage change in Y is then approximately equal to $0.01\beta_1$.

Example 18.19 (Weekly earnings) *We modify the MLR model from example 18.18 to use a log-transformation of the experience (exper) variable. Here are the results from least-squares estimation of the model:*

```
logwage_results <-lm_robust(I(log(earnwk))~educ+I(log(exper))+union+female,data=cps)
logwage_results

##                  Estimate Std. Error t value     Pr(>|t|) CI Lower CI Upper   DF
## (Intercept)        5.1063    0.13613  37.511 6.411e-250  4.83934   5.3732 2804
## educ               0.1006    0.00553  18.186  5.959e-70  0.08972   0.1114 2804
## I(log(exper))      0.1329    0.03061   4.342  1.465e-05  0.07287   0.1929 2804
## union              0.1974    0.03476   5.679  1.490e-08  0.12924   0.2655 2804
## female            -0.4003    0.02390 -16.747  4.488e-60 -0.44721  -0.3535 2804
```

```
logwage_results$r.squared

## [1] 0.1946
```

The slope estimates for educ, union, and female are very similar to the estimates from example 18.18, and their interpretations are similar as well. The R-squared is also virtually identical to that found in example 18.18, suggesting that the overall fit from this model specification is similar to the specification that used exper without a log transformation. For the log-transformed exper variable, the slope estimate is 0.133. *Therefore, a 1-percent change in experience, holding all other variables fixed, is estimated to be associated with an expected percentage change in weekly earnings of* $(0.01)(0.133) = (0.00133)$, *or* 0.133%.

Example 18.20 (Cigarette sales and cigarette taxes) *To get a partial effect of taxes in terms of percentages, the SLR model from example 17.4 can be changed to have log transformations of both the outcome variable (CIGSALES) and the explanatory variable (CIGTAX). Here are the results from least-squares estimation of the model with both variables log-transformed:*

```
lm_robust(I(log(cigsales))~I(log(cigtax)), data=cigdata)

##                    Estimate Std. Error t value Pr(>|t|) CI Lower CI Upper DF
## (Intercept)          3.7086    0.04616  80.341 1.105e-53   3.6158   3.8013 49
## I(log(cigtax))      -0.4277    0.06354  -6.732 1.726e-08  -0.5554  -0.3001 49
```

A 1-percent change in state-level cigarette taxes is estimated to be associated with an expected percentage change in cigarette sales of $(0.01)(-0.428) = -0.00428$, *or* -0.428%. *This estimate quantifies the* price elasticity of demand *for cigarettes.*

18.8 Asymptotic Predictive Intervals

This section considers the following prediction problem: Given known values of the explanatory variables, can we provide a predictive interval for a new realization of the outcome variable (call it Y^*)? Outcome Y^* has not yet been observed, but the values $X_1 = x_1^*, X_2 = x_2^*, \ldots, X_K = x_K^*$ are known. For example, in the context of a model for weekly earnings, we might want to provide a predictive interval for the weekly earnings of an individual, newly drawn from the population, who has $educ = 12$, $exper = 10$, $female = 1$, and $union = 0$.

From the MLR model, the outcome variable is given by

$$Y = \alpha + \beta_1 X_1 + \beta_2 X_2 + \cdots + \beta_K X_K + U \text{ with } E(U|\mathbf{X}) = 0.$$

While the exogeneity assumption $E(U|X) = 0$ is assumed to simplify exposition, it is not necessary for the purposes of predicting the value of the outcome variable.[10] The approaches described next can be applied even if there is doubt about the exogeneity assumption holding. That is, even in a model where it may not be possible to establish causality due to failure of the exogeneity assumption, we can still use least-squares estimation for predictive purposes.

If the values of the explanatory variables are $X_1 = x_1^*, X_2 = x_2^*, \ldots, X_K = x_K^*$, the new outcome Y^* can be written as

$$Y^* = \alpha + \beta_1 x_1^* + \beta_2 x_2^* + \cdots + \beta_K x_K^* + U^*,$$

where U^* is the population residual associated with Y^*. There are two parts of outcome Y^*: the part linearly related to the explanatory variables, which is the conditional expectation

$$E(Y^*|X_1 = x_1^*, X_2 = x_2^*, \ldots, X_K = x_K^*) = \alpha + \beta_1 x_1^* + \beta_2 x_2^* + \cdots + \beta_K x_K^*,$$

and the part unrelated to the explanatory variables, which is the population residual U^*. The least-squares estimates can be used to estimate the conditional-expectation part, with

$$\hat{E}(Y^*|X_1 = x_1^*, X_2 = x_2^*, \ldots, X_K = x_K^*) = \hat{\alpha} + \hat{\beta}_1 x_1^* + \hat{\beta}_2 x_2^* + \cdots + \hat{\beta}_K x_K^*.$$

Due to the consistency of least-squares estimation, this estimate of the conditional-expectation part gets arbitrarily close to the true conditional expectation of Y^* as the sample size grows. In this section, we assume that the sample size n is large enough that we can ignore any estimation imprecision for the conditional-expectation part of Y^*, meaning the uncertainty associated with the *asymptotic predictive interval* for Y^* comes only from the uncertainty associated with the population residual U^*.[11]

Since $Y^* = E(Y^*|\mathbf{X}) + U^*$, the distribution of Y^* conditional on $\mathbf{X}$ is the same as the distribution of U^* conditional on $\mathbf{X}$ but shifted by an amount equal to $E(Y^*|\mathbf{X})$. The shape of the conditional distribution of Y^* given $\mathbf{X}$ is identical to the shape of the conditional distribution of U^* given $\mathbf{X}$. Therefore, determining a predictive interval for Y^* given $\mathbf{X}$ simplifies to determining a predictive interval for U^* given $\mathbf{X}$ and then adding $\hat{E}(Y^*|\mathbf{X})$, which is a consistent estimate of $E(Y^*|\mathbf{X})$. While the appealing properties of the least-squares estimators, including consistency and asymptotic normality, do not require any distributional assumptions on the population residuals, imposing additional assumptions can lead to simplified predictive intervals. To illustrate, we focus on a specific distribution assumption namely, the assumption of normally distributed residuals.

The remainder of this section considers how a predictive interval can be constructed for Y^* given $\mathbf{X}$ in four cases: (1) U^* is normally distributed and homoskedastic, (2) U^* is normally distributed and heteroskedastic, (3) U^* has an unspecified distribution that does not depend on $\mathbf{X}$, and (4) U^* has an unspecified distribution that does depend on $\mathbf{X}$.

Case (1): U^* is normally distributed and homoskedastic. In this scenario, the conditional distribution of U^* does not depend on the explanatory variables and is always $N(0, \sigma_U^2)$, where σ_U^2 is the unconditional variance of U. The $1-\alpha$ probability interval for U^* is

$$(-z_{\alpha/2}\sigma_U, z_{\alpha/2}\sigma_U).$$

Based on the least-squares estimates, this $1-\alpha$ probability interval can be consistently estimated by

$$(-z_{\alpha/2}\hat{\sigma}_U, z_{\alpha/2}\hat{\sigma}_U),$$

where $\hat{\sigma}_U = \sqrt{\frac{1}{n-K-1}\sum_{i=1}^n \hat{u}_i^2}$ is the residual standard deviation estimate. Then, the $1-\alpha$ asymptotic predictive interval for Y^*, given the values $(x_1^*, x_2^*, \ldots, x_K^*)$ for the explanatory variables, is

$$(\hat{\alpha}+\hat{\beta}_1x_1^*+\hat{\beta}_2x_2^*+\cdots+\hat{\beta}_Kx_K^*-z_{\alpha/2}\hat{\sigma}_U, \hat{\alpha}+\hat{\beta}_1x_1^*+\hat{\beta}_2x_2^*+\cdots+\hat{\beta}_Kx_K^*+z_{\alpha/2}\hat{\sigma}_U).$$

Example 18.21 (Monthly stock returns) *Suppose that we assume homoskedastic and normally distributed residuals in the MLR model, with Home Depot returns (HD) as the outcome variable and S&P 500 index returns (IDX) and Lowe's returns (LOW) as the explanatory variables. From example 18.6, the residual standard deviation estimate is* $\hat{\sigma}_U=0.052$. *A 95% predictive interval for HD given* $IDX=0$ *and* $LOW=0$ *is*

$$(\hat{\alpha}-z_{0.025}\hat{\sigma}_U, \hat{\alpha}+z_{0.025}\hat{\sigma}_U)=(0.004-(1.96)(0.052), 0.004+(1.96)(0.052))$$
$$\approx(-0.098, 0.106),$$

meaning that there is a 95% probability that the Home Depot monthly return is between −9.8% and 10.6% when the S&P 500 and Lowe's returns equal zero. Asymptotic predictive intervals can be formed for other choices of the values of IDX and LOW. For example, if IDX and LOW both equal 0.05, the 95% predictive interval for HD is

$$0.004+(0.595)(0.05)+(0.384)(0.05)\pm(1.96)(0.052)\approx(-0.049, 0.155),$$

and if IDX and LOW both equal −0.05, the 95% predictive interval for HD is

$$0.004+(0.595)(-0.05)+(0.384)(-0.05)\pm(1.96)(0.052)\approx(-0.147, 0.057).$$

Case (2): U^* **is normally distributed and heteroskedastic.** In this scenario, the conditional distribution of U^* depends on the explanatory variables. While this conditional distribution is normal, by assumption, the conditional variance of U^* depends on the values of the explanatory variables. To provide a predictive interval for U^*, then, we need to explicitly model the conditional variance of U^* as a function of the explanatory variables. A particularly simple approach is to adopt a model with a linear specification like the original MLR model, with

$$Var(U|\mathbf{X})=\gamma+\delta_1X_1+\delta_2X_2+\cdots+\delta_KX_K.$$

The drawback to this model is that, depending on the parameter values, the conditional variance is not guaranteed to be positive for all the values of the explanatory variables. Note that $Var(U|\mathbf{X})=E((U-E(U))^2|\mathbf{X})=E(U^2|\mathbf{X})$ since $E(U|\mathbf{X})=0$, so the conditional-variance model becomes

$$E(U^2|\mathbf{X})=\gamma+\delta_1X_1+\delta_2X_2+\cdots+\delta_KX_K.$$

To estimate parameters $(\gamma, \delta_1, \delta_2, \ldots, \delta_K)$, the least-squares estimator can be applied to a model with $\hat{u}_i^2$ as the outcome variable and $(x_{i1}, x_{i2}, \ldots, x_{iK})$ as the explanatory variables.[12] Then the conditional variance of U given $\mathbf{X}$ is consistently estimated by

$$\widehat{Var}(U|\mathbf{X})=\hat{\gamma}+\hat{\delta}_1X_1+\hat{\delta}_2X_2+\cdots+\hat{\delta}_KX_K,$$

and the conditional standard deviation of U given $\mathbf{X}$ is consistently estimated by

$$\widehat{sd}(U|\mathbf{X})=\sqrt{\widehat{Var}(U|\mathbf{X})}=\sqrt{\hat{\gamma}+\hat{\delta}_1X_1+\hat{\delta}_2X_2+\cdots+\hat{\delta}_KX_K} \text{ if } \widehat{Var}(U|\mathbf{X})>0.$$

Then the estimated $1-\alpha$ probability interval of U^*, given values $(x_1^*, x_2^*, \ldots, x_K^*)$ for the explanatory variables, is

$$\left(-z_{\alpha/2}\sqrt{\hat{\gamma}+\hat{\delta}_1 x_1^*+\cdots+\hat{\delta}_K x_K^*}, z_{\alpha/2}\sqrt{\hat{\gamma}+\hat{\delta}_1 x_1^*+\cdots+\hat{\delta}_K x_K^*}\right).$$

The $1-\alpha$ asymptotic predictive interval of Y^*, given values $(x_1^*, x_2^*, \ldots, x_K^*)$ for the explanatory variables, is

$$\left(\hat{\alpha}+\hat{\beta}_1 x_1^*+\cdots+\hat{\beta}_K x_K^* - z_{\alpha/2}\sqrt{\hat{\gamma}+\hat{\delta}_1 x_1^*+\cdots+\hat{\delta}_K x_K^*},\right.$$
$$\left.\hat{\alpha}+\hat{\beta}_1 x_1^*+\cdots+\hat{\beta}_K x_K^* + z_{\alpha/2}\sqrt{\hat{\gamma}+\hat{\delta}_1 x_1^*+\cdots+\hat{\delta}_K x_K^*}\right).$$

Example 18.22 (Birthweight data) *The **births** data set contains information on 50,249 births in the US during December 2021. Birth outcomes are of interest to health economists since adverse birth outcomes, like low birth weight, are often associated with high health-care costs. The MLR model has the outcome variable bweight, the baby's birth weight measured in grams, modeled as a function of the following explanatory variables:*

age = *mother's age (in years)*
hsgrad = 1 *if mother is high-school grad and not beyond,* 0 *otherwise*
somecoll = 1 *if mother has some college but not college grad,* 0 *otherwise*
collgrad = 1 *if mother is a 4-year college grad,* 0 *otherwise*
married = 1 *if mother is married,* 0 *otherwise*
smoke = 1 *if mother smoked during pregnancy,* 0 *otherwise*
male = 1 *if baby is male,* 0 *otherwise*

This model specification has three indicator variables associated with educational-attainment categories, with the omitted category being nonhsgrad (nonhigh-school graduates). Their estimates, therefore, should be interpreted as differences from nonhigh-school graduates.

For the conditional-variance model, we use a specification that, like the MLR model, is a linear function of the explanatory variables:

$$Var(U|\mathbf{X}) = E(U^2|\mathbf{X}) = \gamma + \delta_1 age + \delta_2 hsgrad + \delta_3 somecoll + \delta_4 collgrad$$
$$+ \delta_5 married + \delta_6 smoke + \delta_7 male.$$

To estimate the parameters of the conditional-variance model, the least-squares estimates $(\hat{\alpha}, \hat{\beta}_1, \ldots, \hat{\beta}_7)$ *of the MLR model are used to construct the estimated residuals* $\hat{u}_i$ *for* $i=1,\ldots,n$. *Then, the* $\hat{u}_i^2$ *values for* $i=1,\ldots,n$ *are obtained by squaring each* $\hat{u}_i$. *The squared estimated residuals* $\hat{u}_i^2$ *are used as the outcome variable, and least-squares estimation yields the estimates* $(\hat{\gamma}, \hat{\delta}_1, \ldots, \hat{\delta}_7)$ *of the parameters for the conditional-variance model.*

The following table shows the least-squares estimates of the MLR model alongside the estimates of the conditional-variance model. Standard errors are reported in parentheses, and the additional columns provide the p-values for testing the MLR slopes (β_k) *against zero and the conditional-variance slopes* (δ_k) *against zero:*

	MLR model	*p-value for* $H_0: \beta_k = 0$
α *(intercept)*	3,262.93 (28.13)	
β_1 *(age)*	−5.377 (0.828)	0.000
β_2 *(hsgrad)*	32.69 (16.57)	0.048
β_3 *(somecoll)*	66.59 (16.06)	0.000
β_4 *(collgrad)*	92.56 (15.77)	0.000
β_5 *(married)*	44.73 (5.15)	0.000
β_6 *(smoke)*	−168.68 (19.94)	0.000
β_7 *(male)*	99.39 (4.68)	0.000

	Var$(U\|\mathbf{X})$ *model*	*p-value for* $H_0: \delta_k = 0$
γ *(intercept)*	194,758.2 (28,661.5)	
δ_1 *(age)*	3,889.3 (850.5)	0.000
δ_2 *(hsgrad)*	−13,472.8 (17,101.7)	0.431
δ_3 *(somecoll)*	−16,022.7 (16,554.2)	0.333
δ_4 *(collgrad)*	−59,650.4 (16,193.9)	0.000
δ_5 *(married)*	−14,443.6 (5,270.0)	0.006
δ_6 *(smoke)*	42,771.7 (19,780.7)	0.031
δ_7 *(male)*	29,592.6 (4,774.4)	0.000

Looking at the conditional-variance model estimates, the p-values indicate that age, collgrad, married, smoke, and male all have statistically significant associations, at a 5% level, with the conditional variance of the MLR residual. These p-values provide strong evidence that the residual variance depends on the explanatory variables, and therefore the MLR model residuals are heteroskedastic. For the age variable, the estimate $\hat{\delta}_1 = 3{,}889.3$ *means that a 1-year increase in age is associated with an estimated increase in the residual variance of* 3,889.3, *holding all other variables fixed. The residual variance is in the units of grams squared, so this* 3,889.3 *estimate is in the units of grams squared. For the smoke variable, the estimate* $\hat{\delta}_6 = 42{,}771.7$ *means that a mother who smokes during pregnancy has a residual with a variance estimated to be* 42,771.7 *larger than the residual variance for a nonsmoking mother, holding all other variables fixed.*

The MLR model and Var$(U|\mathbf{X})$ *model estimates can be used together to provide a predictive interval for birth weight based on any specific values of the explanatory variables. For instance, consider a 30-year-old mother (age* $= 30$*) who is a college graduate (collgrad* $= 1$*), is married (married* $= 1$*), doesn't smoke during pregnancy (smoke* $= 0$*), and has a male child (male* $= 1$*). The estimated conditional expectation of birth weight, based on the MLR model, is*

$$3262.93 + (-5.377)(30) + (92.56)(1) + (44.73)(1) + (99.39)(1) \approx 3338.3.$$

The estimated conditional variance of the MLR residual, based on the Var$(U|\mathbf{X})$ *model, is*

$$194758.2 + (3889.3)(30) + (-59650.4)(1) + (-14443.6)(1) + (29592.6)(1) \approx 266934.9.$$

Therefore, the estimated conditional standard deviation of the MLR residual is

$$\sqrt{266934.9} \approx 516.7,$$

and a 95% predictive interval for birth weight, given these values of the explanatory variables, is

$$(3338.3 - (1.96)(516.7), 3338.3 + (1.96)(516.7)) \approx (2325.6, 4351.0).$$

The following R code calculates the estimates for the MLR model and Var$(U|\mathbf{X})$ *model reported here:*

```
# create the squared residuals
lmresults <- lm(bweight~age+hsgrad+somecoll+collgrad+married+smoke+male, data=births)
births$uhatsq <- (lmresults$residuals)^2
```

```
# MLR model with birthweight as the outcome
lm_robust(bweight~age+hsgrad+somecoll+collgrad+married+smoke+male, data=births)

##              Estimate Std. Error  t value   Pr(>|t|)   CI Lower  CI Upper    DF
## (Intercept) 3262.9269   28.13280 115.9830 0.0000e+00 3207.78625 3318.0675 50241
## age           -5.3767    0.82824  -6.4918 8.5630e-11   -7.00007   -3.7534 50241
## hsgrad        32.6901   16.56687   1.9732 4.8476e-02    0.21888   65.1614 50241
## somecoll      66.5883   16.05817   4.1467 3.3786e-05   35.11414   98.0625 50241
## collgrad      92.5560   15.76927   5.8694 4.4013e-09   61.64810  123.4640 50241
## married       44.7337    5.15184   8.6830 3.9667e-18   34.63600   54.8313 50241
## smoke       -168.6788   19.93603  -8.4610 2.7213e-17 -207.75361 -129.6039 50241
## male          99.3946    4.68139  21.2318 1.3318e-99   90.21898  108.5701 50241

# Var(U|X) model with squared residuals as the outcome
lm_robust(uhatsq~age+hsgrad+somecoll+collgrad+married+smoke+male, data=births)

##             Estimate Std. Error  t value   Pr(>|t|) CI Lower CI Upper    DF
## (Intercept) 194758.2   28661.53  6.79511 1.0944e-11 138581.3 250935.1 50241
## age           3889.3     850.48  4.57306 4.8180e-06   2222.3   5556.2 50241
## hsgrad      -13472.8   17101.69 -0.78780 4.3082e-01 -46992.3  20046.7 50241
## somecoll    -16022.7   16554.24 -0.96789 3.3310e-01 -48469.2  16423.8 50241
## collgrad    -59650.4   16193.90 -3.68351 2.3029e-04 -91390.6 -27910.2 50241
## married     -14443.6    5269.99 -2.74073 6.1325e-03 -24772.8  -4114.4 50241
## smoke        42771.7   19780.65  2.16230 3.0600e-02   4001.4  81542.0 50241
## male         29592.6    4774.35  6.19824 5.7539e-10  20234.8  38950.4 50241
```

Case (3): U* has an unspecified distribution that does not depend on X. In this scenario, the conditional distribution of U^* given **X** is the same as the unconditional distribution of U^* since it is assumed that the distribution of U^* does not depend on **X**. Without the assumption of normality, the quantiles of the U^* distribution cannot be determined using the $z_{\alpha/2}$ critical values for the normal distribution. Instead, we directly estimate the desired quantiles of the U^* distribution using the corresponding sample quantiles of the estimated residuals $\hat{u}_i$. Let $\hat{v}_q$ denote the sample qth quantile of the distribution of the estimated residuals $\{\hat{u}_1, \hat{u}_2, \ldots, \hat{u}_n\}$. Then the estimated $1-\alpha$ probability interval for U^* is

$$(\hat{v}_{\alpha/2}, \hat{v}_{1-\alpha/2}),$$

and the estimated $1-\alpha$ asymptotic predictive interval for Y^*, given the values $(x_1^*, x_2^*, \ldots, x_K^*)$ for the explanatory variables, is

$$(\hat{\alpha} + \hat{\beta}_1 x_1^* + \hat{\beta}_2 x_2^* + \cdots + \hat{\beta}_K x_K^* + \hat{v}_{\alpha/2}, \hat{\alpha} + \hat{\beta}_1 x_1^* + \hat{\beta}_2 x_2^* + \cdots + \hat{\beta}_K x_K^* + \hat{v}_{1-\alpha/2}).$$

Example 18.23 (Monthly stock returns) *Revisiting example 18.21, we consider predictive intervals for Home Depot returns (HD), based on S&P 500 index returns and Lowe's returns, but without assuming the normality of the residuals. The estimated residuals* $\{\hat{u}_1, \hat{u}_2, \ldots, \hat{u}_n\}$ *have the sample* 2.5% *and* 97.5% *quantiles*

$$\hat{v}_{0.025} = -0.113 \text{ and } \hat{v}_{0.975} = 0.104, \textit{respectively},$$

which are used to construct 95% *predictive intervals for HD. When* $IDX = LOW = 0$, *the* 95% *predictive interval for HD is*

$$(0.004 - 0.113, 0.004 + 0.104) \approx (-0.109, 0.106).$$

When $IDX = LOW = 0.05$, *the* 95% *predictive interval for HD is*

$$\begin{matrix}(0.004+(0.595)(0.05)+(0.384)(0.05)-0.113, \\ 0.004+(0.595)(0.05)+(0.384)(0.05)+0.104)\end{matrix} \approx (-0.060, 0.157).$$

When IDX = LOW = −0.05, the 95% predictive interval for HD is

$$\begin{matrix}(0.004+(0.595)(-0.05)+(0.384)(-0.05)-0.113, \\ 0.004+(0.595)(-0.05)+(0.384)(-0.05)+0.104)\end{matrix} \approx (-0.158, 0.059).$$

These predictive intervals are similar to those in example 18.21 under the assumption of normality. To construct predictive intervals for other confidence levels, we use different sample quantiles of the estimated residuals. For example, for 90% predictive intervals, the appropriate sample quantiles are $\hat{v}_{0.05}$ and $\hat{v}_{0.95}$, which are −0.087 and 0.083, respectively.

Case (4): U* has an unspecified distribution that depends on X. This scenario is the most general, as it involves no assumptions on the conditional distribution of U^* given **X** and therefore is always applicable. Unfortunately, it is also the most difficult to handle and requires methods beyond the scope of this book. While we used unconditional quantile estimates of the residuals for case (3), the basic idea is that *conditional* quantile estimates of the residuals should be used in this general case. That is, what is the qth quantile of U^* given the values $(x_1^*, x_2^*, \ldots, x_K^*)$ for the explanatory variables? This qth conditional quantile can be modeled as[13]

$$v_q(U^*|\mathbf{X}) = \gamma^q + \delta_1^q X_1 + \delta_2^q X_2 + \cdots + \delta_K^q X_K,$$

where $v_q(U^*|\mathbf{X})$ represents the qth quantile of U^* given **X**. Estimation of this model, for any $q \in (0, 1)$, requires a method called **quantile regression**. If the parameters are estimated consistently for $q = \alpha/2$ and $q = 1 - \alpha/2$, the estimated $1 - \alpha$ predictive interval of Y^*, given the values $(x_1^*, x_2^*, \ldots, x_K^*)$ for the explanatory variables, is

$$\Big(\hat{\alpha} + \hat{\beta}_1 x_1^* + \cdots + \hat{\beta}_K x_K^* + \hat{\gamma}^{\alpha/2} + \hat{\delta}_1^{\alpha/2} x_1^* + \cdots + \hat{\delta}_K^{\alpha/2} x_K^*,$$
$$\hat{\alpha} + \hat{\beta}_1 x_1^* + \cdots + \hat{\beta}_K x_K^* + \hat{\gamma}^{1-\alpha/2} + \hat{\delta}_1^{1-\alpha/2} x_1^* + \cdots + \hat{\delta}_K^{1-\alpha/2} x_K^*\Big).$$

To learn more about quantile regression, see Koenker (2005) and the *R* package `quantreg`.

18.9 Linear Probability Model

The SLR and MLR models considered thus far have involved a continuous outcome variable Y. What if we want to model a noncontinuous or discrete outcome variable? This section focuses on a special case that is useful in many empirical settings, which is a binary outcome variable $Y \in \{0, 1\}$.

In the absence of explanatory variables, the unconditional model for Y is the Bernoulli model (section 9.1). If π is the probability that Y is equal to 1, $\pi = P(Y = 1)$, recall that

$$E(Y) = 1 \cdot \pi + 0 \cdot (1 - \pi) = \pi,$$

so the expected value of Y is equal to the probability that Y equals 1. This model can be extended to include explanatory variables by having the probability $P(Y = 1)$ depend on the explanatory variables **X**. The most general way to formulate such a model is to write the conditional probability as

$$P(Y=1|\mathbf{X})=\pi(\mathbf{X})$$

for a function $\pi(\cdot)$. For this *conditional Bernoulli model*, similar to the unconditional model,

$$E(Y|\mathbf{X})=1\cdot\pi(\mathbf{X})+0\cdot(1-\pi(\mathbf{X}))=\pi(\mathbf{X}),$$

so the conditional expectation of $E(Y|\mathbf{X})$ is equal to the conditional probability $P(Y=1|\mathbf{X})$.

A simple version of the conditional Bernoulli model is the **linear probability model (LPM)**, where the conditional probability of $Y=1$ given $\mathbf{X}$ is modeled as a *linear* function of the explanatory variables, similar to the MLR model:

$$P(Y=1|\mathbf{X})=\alpha+\beta_1X_1+\beta_2X_2+\cdots+\beta_KX_K.$$

The interpretation of the LPM parameters is most easily understood in terms of conditional probabilities:

- *Meaning of the intercept* α: $\alpha=P(Y=1|X_1=X_2=\cdots=X_K=0)$, which is the conditional probability of $Y=1$ when all explanatory variables equal zero. Whether α has a practical interpretation depends on whether zero is a relevant value for each of the explanatory variables.
- *Meaning of the slope parameters*: Consider the simplest case where each explanatory variable X_k only enters into the β_kX_k term, which rules out polynomials and interaction variables. Similar to the approach for MLR slope parameters, consider specific values $(x_1^*,x_2^*,\ldots,x_K^*)$ for the explanatory variables. Then the partial effect of increasing X_k by 1 unit, from x_k^* to x_k^*+1, is to change the conditional probability $P(Y=1|\mathbf{X})$ by β_k. For instance, if $\beta_2=0.03$, the partial effect of increasing X_2 by 1 unit is to change $P(Y=1|\mathbf{X})$ by 0.03, or 3 percentage points. Interpretation of partial effects in the presence of other types of variables, like categorical variables, polynomial variables, and interaction variables, is similar to the interpretation for MLR parameters (section 18.6), except that the partial effects are interpreted in terms of $P(Y=1|\mathbf{X})$.

Since $E(Y|\mathbf{X})=P(Y=1|\mathbf{X})$, the LPM can be rewritten as

$$E(Y|\mathbf{X})=\alpha+\beta_1X_1+\beta_2X_2+\cdots+\beta_KX_K.$$

This equation has exactly the same form as the MLR model, and therefore, least-squares estimation can be used to estimate the LPM parameters. Let $(\hat{\alpha},\hat{\beta}_1,\hat{\beta}_2,\ldots,\hat{\beta}_K)$ denote the least-squares estimates of the LPM parameters, and let $se(\hat{\alpha})$ and $se(\hat{\beta}_k)$, for $k\in\{1,2,\ldots,K\}$, denote the corresponding heteroskedasticity-robust standard errors. In addition to providing estimates of the partial effects discussed here, the least-squares estimates also provide estimates of the in-sample predicted probabilities. Specifically, the fitted values from least-squares estimation are consistent estimates of the conditional probability of $Y=1$, given the observed values of the explanatory variables:

$$\hat{y}_i=\hat{\alpha}+\hat{\beta}_1x_{i1}+\hat{\beta}_2x_{i2}+\cdots+\hat{\beta}_Kx_{iK}=\hat{P}(Y=1|X_1=x_{i1},X_2=x_{i2},\ldots,X_K=x_{iK}).$$

A drawback of the LPM is that it can imply conditional probabilities that are less than 0 and/or greater than 1 when there are continuous explanatory variables. As a simple example, consider the case of a single explanatory variable X_1, so $P(Y|X_1)=\alpha+\beta_1X_1$. If $\beta_1\neq0$ and

X_1 is a continuous variable whose values may extend forever in both directions, it must be the case that $P(Y|X_1) < 0$ for a range of X_1 and $P(Y|X_1) > 1$ for a range of X_1. To deal with this issue formally, practitioners sometimes use a nonlinear model for $P(Y|\mathbf{X})$ that restricts the probability to be strictly between 0 and 1. The most popular versions of the nonlinear model are the *probit model* and the *logit model*. That said, the LPM is frequently used since it leads to easily interpretable estimates of partial effects and, even for many empirical applications with continuous explanatory variables, most or all of the estimated in-sample probabilities fall between 0 and 1.[14]

Example 18.24 (Widget website) *We consider the use of the LPM for A/B testing when the outcome variable is binary. Specifically, let Y be the binary outcome indicating whether a purchase is made ($Y = 1$) or not ($Y = 0$). The explanatory variables are emailA, which is a binary variable indicating whether a user receives e-mail A ($emailA = 1$) or not ($emailA = 0$), and emailB, which is a binary variable indicating whether a user receives e-mail B ($emailB = 1$) or not ($emailB = 0$). The omitted category, corresponding to $emailA = emailB = 0$, is for users who do not receive an e-mail, and therefore are in the control group. As described in example 2.1, there are a total of $n = 3{,}000$ observations, of which 300 have $emailA = 1$,300 have $emailB = 1$, and 2,400 have $emailA = emailB = 0$. Of the $emailA = 1$ observations, 20% (or 60 out of 300) have $Y = 1$ and 80% have $Y = 0$. Of the $emailB = 1$ observations, 22% (or 66 out of 300) have $Y = 1$ and 78% have $Y = 0$. And, of the $emailA = emailB = 0$ observations, 15% (or 360 out of 2,400) have $Y = 1$ and 85% have $Y = 0$. The* ***widgets*** *data set contains the emailA, emailB, and purchase (Y) variables for the 3,000 observations.*

The LPM is $P(Y = 1|emailA, emailB) = \alpha + \beta_1 emailA + \beta_2 emailB$. The least-squares estimates of the LPM parameters are provided in the following table, with standard errors and z-statistics and p-values for the z-test of each parameter being equal to zero:

	Estimate	*Standard error*	*z-statistic*	*p-value*
α	0.1500	0.0073	20.58	0.000
β_1	0.0500	0.0243	2.06	0.039
β_2	0.0700	0.0250	2.80	0.005

The parameter estimates are as expected, with

$$\hat{P}(Y = 1|emailA = emailB = 0) = \hat{\alpha} = 0.15,$$

$$\hat{P}(Y = 1|emailA = 1) = \hat{\alpha} + \hat{\beta}_1 = 0.15 + 0.05 = 0.20,$$

and

$$\hat{P}(Y = 1|emailB = 1) = \hat{\alpha} + \hat{\beta}_2 = 0.15 + 0.07 = 0.22.$$

The estimates $\hat{\beta}_1 = 0.05$ and $\hat{\beta}_2 = 0.07$ indicate that e-mail A recipients and e-mail B recipients are 5 percentage points and 7 percentage points more likely, respectively, than non-recipients to make a purchase. The p-value of 0.039 for $H_0 : \beta_1 = 0$ indicates that, at a 5% level, there is a statistically significant difference between the purchase probability for e-mail A recipients and the control group. Similarly, the p-value of 0.005 for $H_0 : \beta_2 = 0$ indicates *that, for any level above 0.5%, there is a statistically significant difference between the purchase probability for e-mail B recipients and the control group. A z-test of $H_0 : \beta_1 = \beta_2$, which can be conducted by either making e-mail A recipients the omitted category or by directly calculating $se(\hat{\beta}_1 - \hat{\beta}_2)$, has a p-value of 0.548. Therefore, there is no statistically meaningful difference between the purchase probabilities of e-mail A recipients and e-mail B recipients since $H_0 : \beta_1 = \beta_2$ cannot be rejected at any reasonable level.*

The following R code calculate the LPM estimates here, with the p-value for the $H_0 : \beta_1 = \beta_2$ test:

```
# estimate the LPM model
lpm_results <- lm_robust(purchase~emailA+emailB, data=widgets)
lpm_results

##              Estimate Std. Error t value    Pr(>|t|)  CI Lower CI Upper   DF
## (Intercept)      0.15  0.0072902 20.5755 4.2270e-88 0.1357057 0.164294 2997
## emailA           0.05  0.0242542  2.0615 3.9341e-02 0.0024435 0.097556 2997
## emailB           0.07  0.0250412  2.7954 5.2166e-03 0.0209004 0.119100 2997

# test the equality of the emailA and emailB slopes
abdiff <- linear_combination(lpm_results, c(0,1,-1))
zstat_abdiff <- abdiff$estimate/abdiff$se
2*(1-pnorm(abs(zstat_abdiff)))

## [1] 0.54813
```

Example 18.25 (Union membership) *The previous examples using the* **cps** *data set considered weekly earnings as the outcome variable. In this example, we instead consider union membership, measured by the indicator variable union, as the outcome variable. The least-squares estimates of the LPM describe how other variables are associated with, and can be used to predict, union membership for the sample of 2,809 employed individuals. The following R code provides the least-squares estimates for one such LPM, with explanatory variables educ, exper, $exper^2$, and female:*

```
union_lpm <- lm_robust(union~educ+exper+I(exper^2)+female, data=cps)
union_lpm

##                 Estimate Std. Error t value  Pr(>|t|)   CI Lower   CI Upper   DF
## (Intercept) -0.1343665  0.0495898  -2.710 6.778e-03 -0.2316027 -3.713e-02 2804
## educ         0.0098938  0.0022852   4.329 1.547e-05  0.0054129  1.437e-02 2804
## exper        0.0084330  0.0030392   2.775 5.562e-03  0.0024736  1.439e-02 2804
## I(exper^2)  -0.0001111  0.0000588  -1.890 5.893e-02 -0.0002264  4.193e-06 2804
## female      -0.0609701  0.0110421  -5.522 3.667e-08 -0.0826215 -3.932e-02 2804
```

The LPM slope estimates are interpreted as follows:

- $\hat{\beta}_{educ} = 0.0099$*: An additional year of education, holding exper and female fixed, is associated with an increase of 0.99 percentage points in the probability of union membership.*
- $\hat{\beta}_{female} = -0.0610$*: Female workers are 6.10 percentage points less likely to be union members than male workers, holding educ and exper fixed.*
- $\hat{\beta}_{exper} = 0.00843$, $\hat{\beta}_{exper^2} = -0.00011$*: Experience is included in the model as a quadratic form, with exper and $exper^2$ as explanatory variables. Therefore, the partial*

effect of exper on the conditional probability of union membership depends on the value of exper. For example, with exper = 15, *the estimated partial effect is*

$$\hat{\beta}_{exper}(16-15)+\hat{\beta}_{exper^2}(16^2-15^2)=0.00843+(-0.00011)(31)\approx 0.0050,$$

which means that an additional year of experience for a worker with 15 *years of experience, holding educ and female fixed, is associated with an increase of* 0.50 *percentage points in the probability of union membership. The standard error,* $se(\hat{\beta}_{exper}+31\hat{\beta}_{exper^2})$, *is calculated to be* 0.0013 *(see the R output that follows). If the starting value is exper* = 20, *the estimated partial effect is lower and equal to*

$$\hat{\beta}_{exper}(21-20)+\hat{\beta}_{exper^2}(21^2-20^2)=0.00843+(-0.00011)(41)\approx 0.0039,$$

with a standard error, $se(\hat{\beta}_{exper}+41\hat{\beta}_{exper^2})$, *of* 0.0008 *(see the R output that follows).*

Here is the R code to calculate the standard errors associated with the estimated partial effects of exper:

```
# standard error for exper partial effect at exper=15
linear_combination(union_lpm, c(0,0,1,31,0))

## $estimate
## [1] 0.0049889
##
## $se
## [1] 0.0013044

# standard error for exper partial effect at exper=20
linear_combination(union_lpm, c(0,0,1,41,0))

## $estimate
## [1] 0.0038779
##
## $se
## [1] 0.00082957
```

We can also use the LPM estimates to directly predict the probability of union membership for any specific values of the explanatory variables. For example, for a female worker (female = 1*) with* 12 *years of education (educ* = 12*) and* 15 *years of experience (exper* = 15, $exper^2=225$*), the predicted probability of union membership is*

$$-0.1344+(0.0099)(12)+(0.00843)(15)+(-0.00011)(225)+(-0.0610)(1)$$
$$\approx 0.0249 \text{ or } 2.49\%,$$

with a standard error of 0.0097 *(or* 0.97%*).*

```
# estimated union conditional probability and its standard error
linear_combination(union_lpm, c(1,12,15,225,1))

## $estimate
## [1] 0.024886
##
## $se
## [1] 0.0096934
```

Notes

1. For two explanatory variables ($K = 2$), the graph is three-dimensional. The MLR conditional expectation, $E(Y|\mathbf{X}) = \alpha + \beta_1 X_1 + \beta_2 X_2$, is a two-dimensional plane that extends forever in a three-dimensional space. (Think of a sheet of paper that may be tilted and extends forever.) If y is measured vertically, positive residuals are associated with data points that are above the MLR plane, negative residuals are associated with data points that are below the MLR plane, and the magnitude of a residual is the vertical distance from the data point to the plane.
2. If either assumption MLR-VarX or assumption MLR-NPC is violated, the minimization problem has an infinite number of possible solutions. For instance, if x_1 has zero variance, there are an infinite number of combinations of a and b_1 that could be chosen to minimize $S(a, b_1, b_2, \ldots, b_K)$.
3. Six years of age is roughly when children in the US start their education, so the experience variable approximates the number of post education years. The variable is not a perfect measure of experience since it does not take into account any periods of unemployment.
4. Although not explicitly stated in proposition 18.2, additional technical assumptions are required to prove the result. Specifically, we require that U has a finite variance, each component of $\mathbf{X}$ has a finite variance, and the covariance between any two components of $\mathbf{X}$ is finite.
5. $s_{\tilde{x}_k}$ can be calculated as the sample standard deviation of the estimated residuals from least-squares estimation of a model that has x_k as the outcome variable and all other x variables as the explanatory variables, which corresponds to the decomposition described in proposition 18.6.
6. Alternatively, `lm_robust(earnwk~marstatus, data=cps)` automatically creates three indicator variables from *marstatus*.
7. An alternative approach is to use derivatives to approximate the change in $E(Y|\mathbf{X})$. A small change dx_1 in X_1, from x_1^* to $x_1^* + dx_1$, is associated with a change of $E(Y|\mathbf{X})$ equal to $(\beta_1 + 2\beta_2 x_1^*)dx_1$ since $\frac{\partial E(Y|\mathbf{X})}{\partial x_1} = \beta_1 + 2\beta_2 x_1^*$.
8. For the derivative approach, a small change dx_1 in X_1, from x_1^* to $x_1^* + dx_1$, is associated with a change of $E(Y|\mathbf{X})$ equal to $(\beta_1 + 2\beta_2 x_1^* + 3\beta_3 x_1^{*2})dx_1$ since $\frac{\partial E(Y|\mathbf{X})}{\partial x_1} = \beta_1 + 2\beta_2 x_1^* + 3\beta_3 x_1^{*2}$.
9. This approximation is based on the derivative of the natural logarithm, $\frac{dln(Y)}{dY} = \frac{1}{Y}$.
10. Proposition 17.11 can be generalized to the case of multiple explanatory variables, with the decomposition of Y into a linear function of the explanatory variables and a random variable uncorrelated with the explanatory variables: $Y = \alpha^* + \beta_1^* X_1 + \beta_2^* X_2 + \cdots + \beta_K^* + V$, with $Cov(X_k, V) = 0$ for $k \in \{1, 2, \ldots, K\}$ and $E(V) = 0$. For this decomposition, the least-squares estimates $(\hat{\alpha}, \hat{\beta}_1, \hat{\beta}_2, \ldots, \hat{\beta}_K)$ consistently estimate $(\alpha^*, \beta_1^*, \beta_2^*, \ldots, \beta_K^*)$ even if the exogeneity assumption of the MLR model doesn't hold.
11. For small sample sizes, where there may be imprecision in the estimate of $E(Y^*|\mathbf{X})$, the resulting asymptotic predictive interval is not wide enough. One approach for gauging whether imprecision in the estimate of $E(Y^*|\mathbf{X})$ affects the predictive interval is to use the bootstrap (chapter 15). Specifically, a predictive interval can be constructed for each bootstrap sample to see how much the predictive interval varies over bootstrap samples. If the estimates of $E(Y^*|\mathbf{X})$ are very precise, there should be little difference in the predictive intervals over bootstrap samples.
12. There are alternative conditional-variance models that guarantee positive estimated conditional variances. An oft-used model is the nonlinear (exponential) model,

$$E(U^2|\mathbf{X}) = e^{\gamma + \delta_1 X_1 + \delta_2 X_2 + \cdots + \delta_K X_K},$$

for which parameters $(\gamma, \delta_1, \delta_2, \ldots, \delta_K)$ can be estimated by many statistical packages. With consistent estimates $(\hat{\gamma}, \hat{\delta}_1, \hat{\delta}_2, \ldots, \hat{\delta}_K)$, the estimated conditional standard deviation of U^* given $(x_1^*, x_2^*, \ldots, x_K^*)$ is $\sqrt{e^{\hat{\gamma} + \hat{\delta}_1 x_1^* + \cdots + \hat{\delta}_K x_K^*}}$.
13. An alternative approach is to directly model the conditional quantiles $v_q(Y|\mathbf{X})$ rather than using the $E(Y|\mathbf{X})$ model.
14. With only discrete explanatory variables, $P(Y = 1|\mathbf{X})$ may be guaranteed to be between 0 and 1. A simple example is an LPM with a single binary variable X_1, for which $P(Y = 1|X_1 = 0) = \alpha$ and $P(Y = 1|X_1 = 1) = \alpha + \beta$.

Exercises

1. Use the **widgets** data set for this question. These data are for 3,000 users, 300 of whom receive e-mail A ($emailA = 1$), 300 of whom receive e-mail B ($emailB = 1$), and 2,400 of whom receive neither ($emailA = emailB = 0$). The outcome variable of interest is *amount*, which is the total amount purchased (in dollars) by the user.
   a. How many users have $amount = 0$?
   b. What are the sample averages of *amount* for the three subsamples corresponding to e-mail A recipients, e-mail B recipients, and nonrecipients?
   c. Use `lm_robust` to estimate the multiple regression of *amount* on *emailA* and *emailB*. Interpret the intercept estimate and the two slope estimates. How do these estimates relate to the sample averages in (b)?
   d. What is the p-value for the z-test of $H_0 : \beta_{emailA} = 0$? What do you conclude from this p-value?
   e. Create a binary variable *nonrecipient* equal to 1 for nonrecipients and 0 for e-mail A and e-mail B recipients. Rerun the regression using *emailA* and *nonrecipient* as the explanatory variables. Test whether there is a significant difference, at the 5% level, between average purchases for e-mail A users and e-mail B users.
2. Use the **metricsgrades** data set for this question. These data are from a graduate econometrics course with 68 students, containing the following variables:

   total = overall composite course grade (out of 100 points)
   gre_quant = score on Graduate Record Examinations (GRE) quantitative test (out of 170 points)
   gre_verbal = score on GRE (English) verbal test (out of 170 points)
   domestic = 1 if domestic (US) student, 0 otherwise

   a. Provide the sample correlation matrix for the four variables. Which variable has the largest correlation (in magnitude) with *total*?
   b. Use `lm_robust` to estimate the multiple regression, with *total* as the outcome variable and the other three variables as explanatory variables.
   c. Interpret the estimate of β_{gre_quant}.
   d. What is the estimated conditional expectation of *total* for a nondomestic student with GRE quantitative score of 160 and GRE verbal score of 150?
   e. What is the estimated standard deviation of the regression model's residual?
   f. Test $H_0 : \beta_{domestic} = 0$ at a 10% level. What do you conclude?
   g. Drop *domestic* from the regression and rerun it. How do the results compare to the original regression?
   h. Now put *domestic* back in the regression and drop *gre_verbal*. Rerun the regression. What happens to the statistical significance of *domestic*, and why?
3. Use the **cigdata** data set for this question. Example 18.8 provided the results from a regression of *cigsales* on *cigtax* and *producer*.

a. Add the variable *price_pack* (equal to the total price per pack) to the model and rerun the regression using `lm_robust`.

b. How does the R-squared value of this regression compare to the R-squared value of the regression without the *price_pack* variable?

c. What happens to the statistical significance of the slope on the state tax (*cigtax*)?

d. Considering the correlation between *cigtax* and *price_pack*, explain the result in (c).

e. Now drop the state-tax (*cigtax*) variable, and rerun the regression with *price_pack* and *producer* as the explanatory variables. How do the results compare to the regression in example 18.8?

f. Do you prefer the MLR model with *cigtax* and *producer*, or the MLR model with *price_pack* and *producer*?

4. Use the **mutualfunds** data set for this question. The sample, consisting of 206 mutual funds categorized as "Large Blend Equity" by Morningstar, includes the following variables:

return_10yr	= ten-year annualized return
expense_ratio	= annual fee (e.g., 0.005 is an annual fee of 0.5%)
manager_tenure	= tenure of current fund manager (in years)
fund_age	= age of fund (in years)
load	= "Y" if fund has a sales charge, "N" otherwise

a. Add the binary variable *hasload*, equal to 1 if the fund has a sales charge and 0 otherwise, to the data frame.

b. Use `lm_robust` to estimate the multiple regression with *return_10yr* as the outcome variable and the four explanatory variables *expense_ratio*, *manager_tenure*, *fund_age*, and *hasload*.

c. Interpret the R-squared value.

d. If the *expense_ratio* value increases by 0.001 (0.1%), what does the estimate of $\beta_{expense\_ratio}$ imply about the conditional expectation of *return_10yr*?

e. Do any of the variables appear to be statistically significant at a 5% level? If so, which ones?

f. Provide an asymptotic 90% confidence interval for $\beta_{hasload}$.

g. *You are considering dropping the *fund_age* and *manager_tenure* variables from the regression. You are worried about multicollinearity, so you want to test them jointly. Use the `test_linear_restrictions` function to determine the p-value for the test of $H_0 : \beta_{fund\_age} = \beta_{manager\_tenure} = 0$.

5. Use the **congress** data set for this question. The data consist of congressional election outcomes in the US between 1948 and 1990. For this question, you will focus on the subsample of 476 congressional district elections that occurred in 1990. Each election in these data is between a Democrat and a Republican, where *demvoteshare* (between 0 and 1) gives the fraction of votes received by the Democrat, meaning that the Democrat won the election if *demvoteshare* > 0.5. The explanatory variables of interest are:

medianincome = median income within the district
pcturban = fraction (between 0 and 1) of district residents who live in an urban area
pctblack = fraction (between 0 and 1) of district residents who are black
pcthighschl = fraction (between 0 and 1) of district residents who have a high-school degree

a. Plot the histogram of *demvoteshare*. Is the distribution unimodal or bimodal?

b. The variable *democrat* is equal to 1 if the Democrat won and 0 if the Republican won. Of the 476 elections in 1990, what fraction were won by Democrats?

c. Provide the sample averages of the explanatory variables separately for elections won by Democrats and elections won by Republicans.

d. Run the necessary simple linear regressions (with *democrat* as the explanatory variable) to test, at a 5% level, whether each of the explanatory variables has a different population mean in the two subsamples.

e. Use `lm_robust` to estimate the multiple regression with *demvoteshare* as the outcome variable and the four explanatory variables given in this question.

f. Interpret the slope estimate for *medianincome*, thinking about a $1,000 change.

g. Interpret the slope estimate for *pcturban*, thinking about a change of 10 percentage points.

h. Add *lagdemocrat*, which is equal to 1 if the Democrat won the previous election and 0 otherwise, and rerun the regression. How do the results change?

i. Interpret the slope estimate for *lagdemocrat* in the regression in (h), and provide an asymptotic 95% confidence interval for the $\beta_{lagdemocrat}$.

j. Without actually doing it, imagine rerunning the regression in (e) using $repvoteshare = 1 - demvoteshare$ as the outcome variable and the same explanatory variables. Describe how the slope estimates, their z-statistics, and their p-values would change.

6. Use the **hrs** data set for this question. The data consist of 6,052 nonmarried individuals who are 50 and older. For this question, focus on the subsample of 3,983 individuals with positive (nonzero) out-of-pocket medical costs (*outofpocket_costs*, in dollars) during 2000. The explanatory variables of interest are:

age = individual's age (in years)
ins_none = 1 if individual has no health insurance, 0 otherwise
ins_medicare = 1 if individual has Medicare insurance, 0 otherwise
male = 1 if individual is male, 0 otherwise

a. Use `lm_robust` to estimate the multiple regression with *outofpocket_costs* as the outcome variable and the four explanatory variables given in this question. Interpret the slope estimates for *age* and *ins_none*.

b. Draw a histogram of *outofpocket_costs*.

c. Create a new variable, *ln_oopc*, equal to the natural logarithm of *outofpocket_costs*.

d. Draw a histogram of *ln_oopc*.

e. What are the sample average and sample standard deviation of *ln_oopc*?

f. Use `lm_robust` to estimate the multiple regression with *ln_oopc* as the outcome variable and the four explanatory variables in this question. Interpret the slope estimates for *age* and *ins_none*.

g. What is the R-squared value for the regression in (f)?

h. What is the estimated residual standard deviation for the regression in (f)?

7. Use the **cps** data set for this question.

a. Create a table like the one in example 18.15, except use *hrslastwk* (hours worked last week) as the outcome variable rather than *earnwk*. The sample is still the 2,809 employed workers.

b. Comparing your table to the one in example 18.15, do the explanatory variables explain more of the variation in weekly earnings or hours worked?

c. What is the p-value associated with the test of the null hypothesis that the slope on $educ^2$ is zero? From this p-value, does the quadratic variable belong in the MLR model?

d. For the two MLR models, what is the estimated partial effect of *educ* on the conditional expectation of *hrslastwk*, evaluated at $educ = 12$? Evaluated at $educ = 16$?

8. Use the **auctions** data set for this question. The data consist of 684 eBay auctions for Apple iPod Mini devices in June and July 2006. The outcome variable of interest is *finalprice* (the sales price, in dollars). The binary variables *new*, *used*, and *refurb* indicate the condition of the device (e.g., a new device has $new = 1$ and $used = refurb = 0$), and *bidders* indicates the number of bidders in the auction.

a. Use `lm_robust` to estimate the multiple regression with *finalprice* as the outcome variable and *bidders* and $bidders^2$ as the explanatory variables. Is the quadratic variable statistically significant at a 5% level?

b. Plot the fitted values from the regression in (a) versus *bidders*.

c. Add two indicator variables, *used* and *refurb*, to the model and rerun the regression. Is the statistical significance of the two *bidders* variables affected at all? How does the R-squared value of this regression compare to the one in (a)?

9. Use the **inflation_expectations** data set for this question. The data consist of survey results for 290 individuals in the US in 2007 and 2008, where each individual was asked to predict inflation in the next year. The outcome variable is *inflation_pred*, which is an individual's inflation prediction measured in percentage points (e.g., a value of 5 corresponds to a predicted 5% inflation). The explanatory variables of interest are:

age = individual's age (in years)
finlit_score = financial literacy score (out of 12 points)
collgrad = 1 if individual is a college graduate, 0 otherwise
famincome_hi = 1 if family income > $75,000, 0 otherwise

a. Use `lm_robust` to estimate the multiple regression with *inflation_pred* as the outcome variable and the four explanatory variables in this question. Which slope estimates are statistically significant at a 5% level?

b. Interpret the slope estimate for *finlit_score*.

c. Interpret the slope estimate for *collgrad*.

d. Draw a histogram of the fitted values.

e. Add the interaction variable *finlit_score* × *collgrad* and rerun the regression.

   i. What is the p-value for the test of the null hypothesis that the slope on the interaction variable is zero?

   ii. What does the sign of the interaction-variable slope say about the partial effect of *collgrad* as a function of *finlit_score*?

   iii. What is the estimated partial effect of *finlit_score* (on the conditional expectation of *inflation_pred*) for a college graduate? What is the estimated partial effect of *finlit_score* for a noncollege graduate?

   iv. *Use the `linear_combination` function to determine the standard errors for the two partial effects in (e)(iii).

f. The actual inflation, for a benchmark, was 3.2% in 2006, 2.9% in 2007, and 3.8% in 2008. Therefore, a sensible forecast for inflation should probably fall in the 2% to 5% range. Create a binary variable *accurate* equal to 1 if $2 \leq \textit{inflation\_pred} \leq 5$ and 0 otherwise. Run an LPM regression with *accurate* as the outcome variable, using the same explanatory variables as in (a). Interpret the slope estimates for *finlit_score* and *collgrad*.

10. You have a sample of 1,000 graduating seniors from a certain university. The outcome variable y equals 1 if the student has a job offer and 0 otherwise. The explanatory variable x equals 1 if the student is an economics major and 0 otherwise. The joint sample counts are given by the following table:

		econ (x)	
		0	1
offer (y)	0	270	20
	1	630	80

a. For the LPM model $P(Y=1|X)=\alpha+\beta X$, what are the least-squares estimates $\hat{\alpha}$ and $\hat{\beta}$? Use only this table to answer this part of the question.

b. Interpret the slope estimate $\hat{\beta}$.

c. In *R*, create a data frame with 1,000 rows and two columns that corresponds to the table of joint sample counts. Use `lm_robust` to confirm the answer to (a). What is the p-value for testing $H_0 : \beta = 0$?

11. Use the **births** data set considered in example 18.22. Since the health-care costs associated with births are mostly concentrated on babies with low birth weight, public health researchers and economists use a specific definition of "low birth weight" for babies having birth weights less than 2,500 grams.

a. Create a new variable *lowbwt* equal to 1 if *bweight* is less than 2,500 and 0 otherwise.

b. Run an LPM regression with *lowbwt* as the outcome, using the explanatory variables from example 18.22.

c. What are the highest and lowest LPM fitted values (predicted probabilities)?

d. Plot the LPM fitted values (predicted probabilities) versus *age*.

e. Interpret the LPM slope estimate for *smoke*.

f. What do the LPM results say about the difference in low-birth-weight probabilities for college-graduate mothers ($collgrad = 1$) versus high-school-graduate mothers ($hsgrad = 1$), holding all other variables fixed?

g. What is the p-value associated with testing that there is no difference in between the probabilities (f)? For this part, either test the linear combination directly or rerun the LPM with a different omitted education category.

h. Provide an asymptotic 95% confidence interval for the difference in low-birth-weight probabilities between a 30-year-old mother and a 25-year-old mother, holding all other variables fixed. Is the difference statistically significant at a 5% level?

i. Add a quadratic age variable (*age* squared) to the LPM and rerun the regression. Provide an asymptotic 95% confidence interval for the difference in low-birth-weight probabilities between a 30-year old mother and a 25-year-old mother, holding all other variables fixed. Is the difference statistically significant at a 5% level?

j. For the LPM regression in (i), you should see very high p-values for the two age-variable slopes. Does this result suggest that the two age variables should be dropped from the model? Explain why or why not.

k. *For the LPM regression in (i), use the `test_linear_restrictions` function to determine the p-value for testing that the slopes for the two age variables both equal zero.

12. *Use the **brands** data set for this question. Refer to exercise 16.12. Run an LPM regression with *purchase* as the outcome variable and four indicator variables (for four of the five possible values of *last_brand*) as explanatory variables. Use the regression results and the `test_linear_restrictions` function as necessary to answer parts (b) and (e) of exercise 16.12.

References

Babu, G. Jogesh, and C. Radhakrishna Rao. 1988. "Joint Asymptotic Distribution of Marginal Quantiles and Quantile Functions in Samples from a Multivariate Population." *Journal of Multivariate Analysis* 27: 15–23.

Bowley, A. L. 1928. "The Standard Deviation of the Correlation Coefficient." *Journal of the American Statistical Association* 23: 31–34.

Cramér, Harald. 1946. *Mathematical Methods of Statistics*. Princeton University Press.

Davison, A. C., and D. V. Hinkley. 1997. *Bootstrap Methods and their Application*. Cambridge University Press.

Efron, Bradley, and Robert J. Tibshirani. 1993. *An Introduction to the Bootstrap*. Chapman & Hall.

Fisher, R. A. 1915. "Frequency Distribution of the Values of the Correlation Coefficient in Samples from an Indefinitely Large Population." *Biometrika* 10: 507–521.

Hogg, Robert V., Joseph W. McKean, and Allen T. Craig. 2019. *Introduction to Mathematical Statistics*. 8th ed. Pearson.

Koenker, Roger. 2005. *Quantile Regression*. Cambridge University Press.

Stock, James H., and Mark W. Watson. 2018. *Introduction to Econometrics*. 4th ed. Pearson.

Wald, Abraham. 1943. "Tests of Statistical Hypotheses Concerning Several Parameters When the Number of Observations Is Large." *Transactions of the American Mathematical Society* 54: 426–482.

Wooldridge, Jeffrey M. 2025. *Introductory Econometrics: A Modern Approach*. 8th ed. Cengage Learning.

Index

Pages numbers followed by an *f* denote a figure.